Polymetallic Metallogenic System

Polymetallic Metallogenic System

Commemorative Issue in Honor of Academician Yu-Sheng Zhai on the Occasion of His 90th Birthday

Special Issue Editor

Liqiang Yang

MDPI • Basel • Beijing • Wuhan • Barcelona • Belgrade

Special Issue Editor
Liqiang Yang
China University of Geosciences
China

Editorial Office
MDPI
St. Alban-Anlage 66
4052 Basel, Switzerland

This is a reprint of articles from the Special Issue published online in the open access journal *Minerals* (ISSN 2075-163X) from 2018 to 2019 (available at: https://www.mdpi.com/journal/minerals/special_issues/metallogenic_system).

For citation purposes, cite each article independently as indicated on the article page online and as indicated below:

LastName, A.A.; LastName, B.B.; LastName, C.C. Article Title. *Journal Name* **Year**, *Article Number*, Page Range.

ISBN 978-3-03921-293-4 (Pbk)
ISBN 978-3-03921-294-1 (PDF)

Contents

About the Special Issue Editor

Liqiang Yang, a professor and doctoral supervisor in China University of Geosciences, Beijing. His research interests in recent years are the tectonic-structural controls and fluid evolution of orogenic gold deposits and polymetallic metallogenic system.

Preface to "Polymetallic Metallogenic System"

Yu-Sheng Zhai is an academician of the Chinese Academy of Sciences and a Professor and Doctoral Supervisor of China University of Geosciences, Beijing. He was born in Wen'an County, Heibei Province in February, 1930. He received his bachelor degree in Geology from Peking University in 1952, and received masters degree from Changchun College of Geology in 1957. He has been engaged in geology for 70 years since he entered the university.

Yu-Sheng Zhai, as a famous economic geologist and geological educator, has made outstanding contributions in the fields of mineral deposits, ore-field structure, metallogenic systems, and regional metallogeny. His representative treatises include "Mineral Deposit Geology", "Fe-Cu-Au Metallogenic Regularity of the Middle-Lower Yangtze River Metallogenic Belt", "The Ore-Field Structure", "Metallogenic System in the Paleo-Continental Margin", "Research Methods of Regional Metallogeny", and "The Metallogenic System". His great achievements covered many aspects of mineral deposits. Here, we emphasize his pioneering research and contributions to the construction of "metallogenic system" theory and its applications.

In the 1990s, following the rise of earth system science, Yu-Sheng Zhai firstly proposed the "Metallogenic System and Its Evolution", and published a series of papers and treatises. The paper entitled "Outline of the Mineral Resources of China and their Tectonic Setting" published in "Australian Journal of Earth Sciences" constructed the tempo-spatial framework of metallogenic systems in different tectonic units and different eras of China, which had attracted the attention of international economic geologists.

The published paper entitled "On the Metallogenic System", published in 1999, proposed the definition for the term "metallogenic system". The metallogenic system is a natural system with an ore-forming function. It includes geological factors controlling ore-formation and preservation, and ore-forming processes and their products—ore deposit series and anomaly series. This paper also proposed the metallogenic mechanisms as the main classification standards of metallogenic systems.

The treatise "Regional Metallogeny", published in 1999, mainly focused on the regional metallogenic system, and comprehensively discussed the research objectives, contents, methods, and thoughts. Under the guidance of systematic theory and the historical view, this treatise systematically presented the global tectonic settings, controlling factors, mechanisms, and metallogenic products for regional metallogeny, as well as the conditions and mechanisms of post-mineralization transformation and preservation. The metallogenic characteristics of six metallogenic domains and their metallogenic belts in China were also presented in this book.

The treatise "Metallogenic System in the Paleo-Continental Margin", published in 2002, proposed that the paleo-continental margin was one of the regions with the most active crust–mantle interaction, the most complex tectonic activities, and the most prominent metallogenic processes. Based on the investigations on the formation and evolution of polymetallic mineralization in the paleo-continental margin of North China Craton, this treatise constructed different types of metallogenic systems in the paleo-continental margin, summarized the tectonic evolution of continental margin and its tempo-spatial linkage with different metallogenic systems, and proposed geodynamic models for the evolution of the metallogenic system under distinctive tectonic settings. Ten elements of the metallogenic system in the paleo-continental margin were then summarized, and the theoretical framework and method to study metallogeny in paleo-continental margin were developed. Thereby, the regional metallogenic theory was deepened and applied regarding exploration practice.

"Research Methods of Regional Metallogeny", published in 2004, focused on the metallogenic system and its evolution, emphasizing the important implications of the integrated formation–transformation–preservation process of mineral deposits. By combining theoretical developments and practical applications, this book comprehensively presented the research contents and methods of regional metallogeny; stressed the compilation of various types of geological information; and proposed a method that investigated the regional mineralization network and distribution characteristics of mineral deposits from the material structure, spatial structure, and temporal structure. This book presented the regional metallogenic characteristics of China, and provided potential research methods for studying the regional metallogenic regularity.

In 2010, Yu-Sheng Zhai continued to write and complete the "metallogenic system theory" at the age of 80. The theory and method of metallogenic system were comprehensively summarized by applying systematic scientific thinking. This book finally clarified the core of the metallogenic system, which was "source-transportation-trap-transformation preservation". It proposed five basic types of metallogenic systems; emphasized the combination of the earth system, metallogenic system, and exploration system; and proposed several new ideas, including the metallogenic mechanism of "multifactor coupling and critical transition" and the dual effects of resources and the environment in the metallogenic system.

The representative treatises mentioned above reflected the brewing, formation, and development of the metallogenic system. Research and practice over the years have proved that the metallogenic system has not only affected studies of mineral deposits, it has also effectively guided prospecting by enhancing traditional studies of mineral deposits into the level of earth system science. Studies on the metallogenic system have accommodated the developing trend of earth system science, which has become a framework for realizing the breakthrough of metallogenic theory. Based on these achievements regarding the metallogenic system, Yu-Sheng Zhai and his group received second prize in the National Science and Technology Process, entitled "The Construction of Metallogenic System and Its Applications to the Prospecting Practices in the North China Paleo-Continent".

In his nineties, Yu-Sheng Zhai is still working on the front line of teaching and research. He has devoted himself to teaching and educating, and has been teaching courses on regional metallogeny to undergraduate and graduate students for a long time. He had guided some excellent post-doc research fellows, doctors, and masters, and successfully cultivated a large number of excellent exploration geologists for the country. As an exploration geologist, he enjoys working in the mines and at the front line of exploration. His footprints have not only covered the mainland of China, but have also involved some overseas mines. A series of breakthroughs in mineral exploration, such as the discovery of the large Gaoerni Pb-Zn deposit in Inner Mongolia, were motivated by the theory of the metallogenic system.

Under the guidance of the metallogenic system defined by Yu-Sheng Zhai, his student Jun Deng firstly proposed the term of "composite metallogenic system" in 2016. This term accounts for the fact that two metallogenic processes occurred in different episodes of the superimposed orogeny, and that these occurred simultaneously in one tectonic belt. The former is called multiperiod type and the latter is called synchronous type. Composite metallogenesis is characterized by the inheritance and overlapping of ore metals in different episodes, and by a combination of various synchronous metallogenesis.

On the occasion of the 90th birthday of academician Yu-Sheng Zhai and his 70 years worthy of hard work throughout his geological career, the research group, impressed by his outstanding achievements in mineral deposits, has proposed to put together a Special Issue to celebrate and express their respect and gratitude to him. This Special Issue focuses on the polymetallic metallogenic system, including Au-Cu-Mo-Sn-REE polymetallic mineralization, and shows how the definition and theory of the metallogenic system influences the current scientific research in mineral deposits.

This Special Issue "Polymetallic Metallogenic System" is dedicated to Yu-Sheng Zhai to express our best wishes and admiration.

Liqiang Yang
Special Issue Editor

Editorial

Editorial for Special Issue "Polymetallic Metallogenic System"

Liqiang Yang

State Key Laboratory of Geological Processes and Mineral Resources, China University of Geosciences (Beijing), Beijing 100083, China; lqyang@cugb.edu.cn

Received: 5 July 2019; Accepted: 10 July 2019; Published: 15 July 2019

In the last century, following the development of Earth System Science, the metallogenic system has become an important topic in the study of mineral deposits. The term "metallogenic system" firstly presented in 1970s, connotes a natural system with ore-forming functions [1]. The metallogenic system includes geological factors controlling ore-formation and preservation, and ore-forming processes and their products—ore deposit series and anomaly series [1]. It emphasizes the integration of tectonic settings, controlling factors, metallogenic mechanisms, and conditions and mechanisms of post-mineralization transformation and preservation (i.e., the source, transport, trap, transformation, and preservation). On the basis of the theory of the metallogenic system, Deng et al. [2–4] defined the composite metallogenic system and summarized the distribution and characteristics of composite metallogenic systems in China.

While economic geology typically focuses on the differences between individual deposits, the metallogenic system looks for broad similarities in each system [5]. The metallogenic system model allows for multiple mineralization styles to be identified within a single system [5]. For example, a porphyry copper system may develop associated high and low sulfidation epithermal gold mineralization, and skarn-type or hydrothermal vein-type Cu-Pb-Zn-Ag mineralization [6–8]. By integrating all geological ingredients and processes that are necessary for the formation of mineral deposits, the investigation of the metallogenic system is more useful for revealing the geodynamic evolution of a region, and more effective for regional prospecting exploration [9–11]

From the perspective of regional metallogeny, the study by Niu et al. [12] relates deep dynamic processes in the Earth to gold deposits at the crustal surface (core → mantle → crust) in Jiaodong Province. Zircon Hf and whole-rock Nd isotopic mapping presented in Zhang et al. [13] precisely illustrate the controls of lithospheric architecture on the distribution of regional polymetallic mineralization. Liu et al. [14] and Wei et al. [15] propose the main gold deposition mechanisms by studying the fluid inclusions and corresponding H–O–S isotopic compositions of different ore-forming stages in the Sanshandao and Sizhuang gold deposits, respectively. Zhao et al. [16] classifies the Koka gold deposit in NE Africa as an orogenic gold deposit based on the similar features of geology and ore-forming fluids.

The study by Chen et al. [17] describes the occurrence of texturally heterogeneous gold-hosting quartz types and variations in trace element geochemistry, offering a more multi-generational perspective on precipitation mechanisms that fluid inclusion studies may not be able to offer. Guo et al. [18] investigates the rare earth element geochemistry and C–O isotope characteristics of hydrothermal calcites to provide new insights into the fluid–rock interactions and ore-forming processes. Li N. et al. [19] present a textural and trace-element analysis of sulfides, and investigate the controls on the distribution of invisible and visible gold in the ores to further improve our genetic understanding of the deposits in the Yangshan gold belt. The arsenopyrite and chlorite mineral thermometers are applied in Sun et al. [20] to constrain the process and mechanism of gold mineralization in the Zhengchong gold deposit–Jiangnan orogenic belt.

For the strategies of mineral exploration, Li et al. [21] recognize the negligible wall–rock alteration of high-grade auriferous quartz vein-type ores and increase the target mineralized zone and the volume of potential ore for gold deposits that are hosted in Archean metamorphic rocks. A geostatistical analysis of data is presented in Wang et al. [22] to determine the geometry of mineralized zones and the structural control of ore shoot plunge in the Sizhuang Au ore deposit, exemplifying an effective exploration strategy for studying the structural control of the geometry, orientation, and grade distribution of orebodies via the integration of geostatistical tools and structural analysis.

The studies by Yu et al. [23], Wang et al. [24], Yang Q. et al. [25], and Yang F. et al. [26] provide geological, geochronological, and geochemical insights into the formation of polymetallic deposits, including the timing of mineralization and ore genetic types. In addition, garnet Sm–Nd dating is conducted on the Hongshan skarn deposits to further constrain the timing of skarn formation in Zu et al. [27]. The emplacement ages and geochemical and isotopic signatures of ore-related intrusions are also proved to be genetically linked with the formation of mineral deposits [28–34].

In order to reveal the importance of water content and the oxidation state for the formation of porphyry Au mineralization, Bao et al. [35] analyze the compositions of amphiboles and zircons to evaluate the physicochemical conditions (e.g., pressure, temperature, fO_2, and water content) of the ore-fertile porphyries and ore-barren porphyries identified in the Beiya Au deposit. To evaluate the relative contribution of ore-forming elements from granites and from the surrounding strata, Jiang et al. [36] compare the whole-rock geochemistry of ore-related skarns to the composition of associated granites and strata and suggest that the strata played an important role in the formation of W-Sn polymetallic deposits in South China.

Overall, we hope this Special Issue will contribute further to the development of metallogenic system theory, enhance scientific thinking, and suggest ways forward to investigate polymetallic mineral deposits.

Funding: This research is funded by the National Key Research Program of China (Grant No. 2016YFC0600107-4), the National Natural Science Foundation of China (Grant No. 41572069), the National Science Foundation of China (Grant No. 2015CB2605 and 2009CB421008), and the 111 Project under the Ministry of Education and the State Administration of Foreign Experts Affairs, China (Grant No. B07011).

Acknowledgments: L.Y. is a professor at the China University of Geosciences, Beijing (CUGB). I would like to thank my research group at CUGB, especially for Yu-Sheng Zhai and Jun Deng.

Conflicts of Interest: The author declares no conflict of interest.

References

1. Zhai, Y.S. On the metallogenic system. *Earth Sci. Front.* **1999**, *6*, 13–27.
2. Deng, J.; Wang, Q.F.; Li, G.J. Composite orogenesis and composite metallogenic system: Case study from the Sanjiang Tethyan belt, SW China. *Acta Petrol. Sin.* **2016**, *32*, 2225–2247. (In Chinese with English abstract)
3. Deng, J.; Wang, Q.F.; Li, G.J. Tectonic evolution, superimposed orogeny, and composite metallogenic system in China. *Gondwana Res.* **2017**, *50*, 216–226. [CrossRef]
4. Deng, J.; Zhang, J.; Wang, Q.F. Research advances of composite metallogenic system and deep driving mechanism in the Tethys, SW China. *Acta Petrol. Sin.* **2018**, *34*, 1229–1238. (In Chinese with English abstract)
5. Ford, A.; Peters, K.J.; Partington, G.A.; Blevin, P.L.; Downes, P.M.; Fitzherbert, J.A.; Greenfield, J.E. Translating expressions of intrusion-related mineral systems into mappable spatial proxies for mineral potential mapping: Case studies from the Southern New England Orogen, Australia. *Ore Geol. Rev.* **2019**. [CrossRef]
6. Sillitoe, R.H. Porphyry copper systems. *Econ. Geol.* **2010**, *105*, 3–41. [CrossRef]
7. Chen, L.; Qin, K.; Li, G.; Li, J.; Xiao, B.; Zhao, J.; Fan, X. Sm–Nd and Ar–Ar isotopic dating of the Nuri Cu–W–Mo deposit in the southern Gangdese, Tibet: Implications for the porphyry–Skarn metallogenic system and metallogenetic epochs of the Eastern Gangdese. *Resour. Geol.* **2016**, *66*, 259–273. [CrossRef]
8. Yang, L.Q.; He, W.Y.; Gao, X.; Xie, S.X.; Yang, Z. Mesozoic multiple magmatism and porphyry–skarn Cu-polymetallic systems of the Yidun Terrane, Eastern Tethys: Implications for subduction- and transtension-related metallogeny. *Gondwana Res.* **2018**, *62*, 144–162. [CrossRef]
9. Gardiner, N.J.; Robb, L.J.; Morley, C.K.; Searle, M.P.; Cawood, P.A.; Whitehouse, M.J.; Kirkland, C.L.; Roberts, N.L.M.; Myint, T.A. The tectonic and metallogenic framework of Myanmar: A Tethyan mineral system. *Ore Geol. Rev.* **2016**, *79*, 26–45. [CrossRef]

10. Reimann, C.; Ladenberger, A.; Birke, M.; de Caritat, P. Low density geochemical mapping and mineral exploration: Application of the mineral system concept. *Geochem. Explor. Environ. Anal.* **2016**, *16*, 48–61. [CrossRef]

11. McCuaig, T.C.; Beresford, S.; Hronsky, J. Translating the mineral systems approach into an effective exploration targeting system. *Ore Geol. Rev.* **2010**, *38*, 128–138. [CrossRef]

12. Niu, S.; Chen, C.; Zhang, J.; Zhang, F.; Wang, F.; Sun, A. The Thermal and Dynamic Process of Core → Mantle → Crust and the Metallogenesis of Guojiadian Mantle Branch in Northwestern Jiaodong. *Minerals* **2019**, *9*, 249. [CrossRef]

13. Zhang, Z.; Wang, Y.; Li, D.; Lai, C. Lithospheric Architecture and Metallogenesis in Liaodong Peninsula, North China Craton: Insights from Zircon Hf-Nd Isotope Mapping. *Minerals* **2019**, *9*, 179. [CrossRef]

14. Liu, Y.; Yang, L.; Wang, S.; Liu, X.; Wang, H.; Li, D.; Wei, P.; Cheng, W.; Chen, B. Origin and Evolution of Ore-Forming Fluid and Gold-Deposition Processes at the Sanshandao Gold Deposit, Jiaodong Peninsula, Eastern China. *Minerals* **2019**, *9*, 189. [CrossRef]

15. Wei, Y.-J.; Yang, L.-Q.; Feng, J.-Q.; Wang, H.; Lv, G.-Y.; Li, W.-C.; Liu, S.-G. Ore-Fluid Evolution of the Sizhuang Orogenic Gold Deposit, Jiaodong Peninsula, China. *Minerals* **2019**, *9*, 190. [CrossRef]

16. Zhao, K.; Yao, H.; Wang, J.; Ghebretnsae, G.F.; Xiang, W.; Xiong, Y.-Q. Genesis of the Koka Gold Deposit in Northwest Eritrea, NE Africa: Constraints from Fluid Inclusions and C–H–O–S Isotopes. *Minerals* **2019**, *9*, 201. [CrossRef]

17. Chen, B.; Deng, J.; Wei, H.; Ji, X. Trace Element Geochemistry in Quartz in the Jinqingding Gold Deposit, Jiaodong Peninsula, China: Implications for the Gold Precipitation Mechanism. *Minerals* **2019**, *9*, 326. [CrossRef]

18. Guo, L.; Hou, L.; Liu, S.; Nie, F. Rare Earth Elements Geochemistry and C–O Isotope Characteristics of Hydrothermal Calcites: Implications for Fluid-Rock Reaction and Ore-Forming Processes in the Phapon Gold Deposit, NW Laos. *Minerals* **2018**, *8*, 438. [CrossRef]

19. Li, N.; Deng, J.; Groves, D.I.; Han, R. Controls on the Distribution of Invisible and Visible Gold in the Orogenic Gold Deposits of the Yangshan Gold Belt, West Qinling Orogen, China. *Minerals* **2019**, *9*, 92. [CrossRef]

20. Sun, S.-C.; Zhang, L.; Li, R.-H.; Wen, T.; Xu, H.; Wang, J.-Y.; Li, Z.-Q.; Zhang, F.; Zhang, X.-J.; Guo, H. Process and Mechanism of Gold Mineralization at the Zhengchong Gold Deposit, Jiangnan Orogenic Belt: Evidence from the Arsenopyrite and Chlorite Mineral Thermometers. *Minerals* **2019**, *9*, 133. [CrossRef]

21. Li, R.; Albert, N.N.; Yun, M.; Meng, Y.; Du, H. Geological and Geochemical Characteristics of the Archean Basement-Hosted Gold Deposit in Pinglidian, Jiaodong Peninsula, Eastern China: Constraints on Auriferous Quartz-Vein Exploration. *Minerals* **2019**, *9*, 62. [CrossRef]

22. Wang, S.-R.; Yang, L.-Q.; Wang, J.-G.; Wang, E.-J.; Xu, Y.-L. Geostatistical Determination of Ore Shoot Plunge and Structural Control of the Sizhuang World-Class Epizonal Orogenic Gold Deposit, Jiaodong Peninsula, China. *Minerals* **2019**, *9*, 214. [CrossRef]

23. Yu, H.; Guo, C.; Qiu, K.; McIntire, D.; Jiang, G.; Gou, Z.; Geng, J.; Pang, Y.; Zhu, R.; Li, N. Geochronological and Geochemical Constraints on the Formation of the Giant Zaozigou Au-Sb Deposit, West Qinling, China. *Minerals* **2019**, *9*, 37. [CrossRef]

24. Wang, F.; Li, Q.; Liu, Y.; Jiang, S.; Chen, C. Geochronology of Magmatism and Mineralization in the Dongbulage Mo-Polymetallic Deposit, Northeast China: Implications for the Timing of Mineralization and Ore Genesis. *Minerals* **2019**, *9*, 255. [CrossRef]

25. Yang, Q.; Ren, Y.-S.; Chen, S.-B.; Zhang, G.-L.; Zeng, Q.-H.; Hao, Y.-J.; Li, J.-M.; Yang, Z.-J.; Sun, X.-H.; Sun, Z.-M. Geological, Geochronological, and Geochemical Insights into the Formation of the Giant Pulang Porphyry Cu (–Mo–Au) Deposit in Northwestern Yunnan Province, SW China. *Minerals* **2019**, *9*, 191. [CrossRef]

26. Yang, F.; Sun, J.; Wang, Y.; Fu, J.; Na, F.; Fan, Z.; Hu, Z. Geology, Geochronology and Geochemistry of Weilasituo Sn-Polymetallic Deposit in Inner Mongolia, China. *Minerals* **2019**, *9*, 104. [CrossRef]

27. Zu, B.; Xue, C.; Dong, C.; Zhao, Y. Mineralogy and Garnet Sm–Nd Dating for the Hongshan Skarn Deposit in the Zhongdian Area, SW China. *Minerals* **2019**, *9*, 243. [CrossRef]

28. Mao, C.; Lü, X.; Chen, C. Geochemical Characteristics of A-Type Granite near the Hongyan Cu-Polymetallic Deposit in the Eastern Hegenshan-Heihe Suture Zone, NE China: Implications for Petrogenesis, Mineralization and Tectonic Setting. *Minerals* **2019**, *9*, 309. [CrossRef]

29. Wei, P.; Yu, X.; Li, D.; Liu, Q.; Yu, L.; Li, Z.; Geng, K.; Zhang, Y.; Sun, Y.; Chi, N. Geochemistry, Zircon U–Pb Geochronology, and Lu–Hf Isotopes of the Chishan Alkaline Complex, Western Shandong, China. *Minerals* **2019**, *9*, 293. [CrossRef]

30. Jia, R.-Y.; Wang, G.-C.; Geng, L.; Pang, Z.-S.; Jia, H.-X.; Zhang, Z.-H.; Chen, H.; Liu, Z. Petrogenesis of the Early Cretaceous Tiantangshan A-Type Granite, Cathaysia Block, SE China: Implication for the Tin Mineralization. *Minerals* **2019**, *9*, 257. [CrossRef]

31. Xiao, C.; Shen, Y.; Wei, C. Petrogenesis of Low Sr and High Yb A-Type Granitoids in the Xianghualing Sn Polymetallic Deposit, South China: Constrains from Geochronology and Sr–Nd–Pb–Hf Isotopes. *Minerals* **2019**, *9*, 182. [CrossRef]

32. Ju, N.; Ren, Y.-S.; Zhang, S.; Bi, Z.-W.; Shi, L.; Zhang, D.; Shang, Q.-Q.; Yang, Q.; Wang, Z.-G.; Gu, Y.-C.; et al. Metallogenic Epoch and Tectonic Setting of Saima Niobium Deposit in Fengcheng, Liaoning Province, NE China. *Minerals* **2019**, *9*, 80. [CrossRef]

33. Gou, Z.; Yu, H.; Qiu, K.; Geng, J.; Wu, M.; Wang, Y.; Yu, M.; Li, J. Petrogenesis of Ore-Hosting Diorite in the Zaorendao Gold Deposit at the Tongren-Xiahe-Hezuo Polymetallic District, West Qinling, China. *Minerals* **2019**, *9*, 76. [CrossRef]

34. Wang, J.; Wang, X.; Liu, J.; Liu, Z.; Zhai, D.; Wang, Y. Geology, Geochemistry, and Geochronology of Gabbro from the Haoyaoerhudong Gold Deposit, Northern Margin of the North China Craton. *Minerals* **2019**, *9*, 63. [CrossRef]

35. Bao, X.; Yang, L.; He, W.; Gao, X. Importance of Magmatic Water Content and Oxidation State for Porphyry-Style Au Mineralization: An Example from the Giant Beiya Au Deposit, SW China. *Minerals* **2018**, *10*, 441. [CrossRef]

36. Jiang, W.; Li, H.; Evans, N.J.; Wu, J.; Cao, J. Metal Sources of World-Class Polymetallic W–Sn Skarns in the Nanling Range, South China: Granites versus Sedimentary Rocks? *Minerals* **2018**, *8*, 265. [CrossRef]

Article

The Thermal and Dynamic Process of Core → Mantle → Crust and the Metallogenesis of Guojiadian Mantle Branch in Northwestern Jiaodong

Shuyin Niu, Chao Chen *, Jianzhen Zhang, Fuxiang Zhang, Fengxiang Wang and Aiqun Sun

College of Resources, Key Laboratory of Regional Geology and Mineralization, Hebei GEO University, Shijiazhuang 050031, China; niusy@hgu.edu.cn (S.N.); zjzhen@hgu.edu.cn (J.Z.); sdzfx@hgu.edu.cn (F.Z.); fengxiang@hgu.edu.cn (F.W.); sunaq@hgu.edu.cn (A.S.)
* Correspondence: chenchao@hgu.edu.cn; Tel.: +86-0311-8720-8285

Received: 12 March 2019; Accepted: 18 April 2019; Published: 24 April 2019

Abstract: The Jiaodong gold mineral province, with an overall endowment estimated as >3000 t, located at the eastern segment of the North China Craton (NCC), ranks as the greatest source of Au in China. The structural evolution, magmatic activity and metallogenesis during the Mesozoic played important roles in the large scale regional gold, silver and polymetallic mineralization in this area; among them, the intensive activation of fault structures is the most important factor for metallogenesis. This study takes the regional deep faults as main thread to discuss the controlling role of faults in large scale metallogenesis. The Jiaojia fault and Sanshandao faults in the northwest margin of the Guojiadian mantle branch not only are dominant migration channels for hydrothermal fluid but are very important favorable spaces for ore-forming and ore-hosting during the formation of world-class super large gold deposits in this area. The deep metallogenic process can be summarized as involving intensive Earth's core, mantle and crust activity → magmatism → uplifting of metamorphic complex → detachment of cover rocks → formation of mantle branch → penetration of hydrothermal fluid along deep faults → concentration of metallogenic materials → formation of super large deposits.

Keywords: mantle branch; metallogenesis; ore-controlling structures; metallogenic rule; ore prospecting target; Jiaodong area

1. Introduction

The Early Cretaceous' extensive magmatism and gold mineralization of the Jiaodong Peninsula, on the eastern part of the North China Craton (NCC) defining China's largest gold province, has attracted considerable attention over the last decades [1–15]. For example, Song et al. [13] summarized the deep ore prospecting results from the gold deposits in northwestern Jiaodong and discovered middle-shallow gold deposits conjunct as a whole in the deep ore. Especially for Jiaojia and Sanshandao metallogenic fracture zones, proven gold reserves have reached 1200 t and 1000 t respectively, which form two significant metallogenic belts of Jiaodong gold province.

Large-scale mineralization and its origin continue to be problematic and controversial. Based on their structural setting, mineralization style, and ore fluid geochemistry, they have been classified as (1) the orogenic gold deposit [16–23], (2) considered as world-class gold deposits as a class of unique "Jiaodong-type" gold deposits; [2,9,24], (3) the "decratonic gold deposits" [25], and the anorogenic gold deposit (e.g., intrusion-related gold deposit) [26]. Although the genesis of ore deposits is still under debate, most scholars hold that the gold deposits in these areas were caught up within the zone of a series of tectonic disturbances and over a highly perturbed mantle, according to mantle-crust movements [27–34]. Furthermore, more attention should be paid to following issues for a new prospective:

(1)   Distribution and migration of Earth's core, mantle and crust materials.

(2)   The formation and evolution of mantle plume.

(3)   The anti-gravity migration of metallogenic elements.

(4)   The concentration and positioning of metallogenic materials.

Above all, this contribution focuses on the relation between Cretaceous regional crust-mantle-core movement in Jiaodong and dynamic mechanism of large-scale gold mineralization based on successional researches and practices. Furthermore, this study mainly discusses the constraints of formation and the multiple evolutions of the mantle plume and its effects on regional gold metallogenesis.

2. Regional Geological and Metallogenetic Setting

Intensive crustal movement occurred in north China during the middle and late Mesozoic. Tancheng–Lujiang deep fault, which is called the Yisu fault in this area, also showed intensive activation. It is characterized by great horizontal and vertical extension and made great impact on multiple periods of structural movement, intensive magmatism and large scale metallogenesis [2,16,19,30,32,35–43].

The regional distribution and classification of rocks or strata often reflect the regional tectonic setting; it can be used to deduce the evolution process of regional structures. This region is located within the southeastern margin of the NCC, which is a block dominated by 3.8–2.5 Ga Archean rock units, that were significantly reworked during orogenies. The region witnessed migration and collision between the NCC and Yangtze Craton, and were temporally related to Early Cretaceous major mantle-plume activity.

Based on the summary report of the regional geological survey for Shandong province (2000), intensive structural movements and metallogenesis took place during the Cretaceous. The strata of the Cretaceous can be divided into four groups from bottom to top, which include the Laiyang group, the Qingshan group, the Dasheng group and the Wangshi group.

The Laiyang group in the lower part is composed of continental facies clastic rocks with some fold deformation, and the depression center is in the Laiyang and Zhucheng areas. The Qingshan group is a set of volcanic rocks and pyroclastic sedimentary rocks, which can be divided into three sections based on its lithology. The first and third sections are mainly composed of widely spread intermediate and acid intrusive rocks; the second section is composed of pyroclastic rocks that are obviously controlled in fracturing zones.

The Wangshi group is mainly distributed in the eastern Shandong stratigraphic division and is composed of red terrestrial clastic rocks. It generally is 2000–3000 m in thickness, the maximum is up to 3500–4000 m. In the Qingdao-Laiyang area, it interbeds with many basalt, andesite and andesitic turf, covered by thin layers of marl. Its thickness varies between 496 m and 1600 m. Both the Qingshan group and the Dasheng group belong to the middle Cretaceous, the former refers to the Qingshan group volcanic rock or intrusive rock, the later specially refers to the clastic rocks in northwestern Jiaodong. The thickness of Cretaceous volcanic-sedimentary formation in Shandong is close to, or even exceeds the thickness of sedimentary-volcanic formation (over 100 Ma) of the middle-late Proterozoic in the Jixian group.

Generally, lower and upper systems of Cretaceous rock are made up of clastic rocks, and the middle system is made up of volcanic rocks in this area. The sedimentary basin is associated with small extension and deep depression, and the dominant rocks are volcanic rocks and intrusive rocks, which demonstrates the intensive magmatic activity in that period. This is caused by the upwelling of the mantle plume and the rapid splitting and depression of the upper crust (Figure 1). At the same time, large scale gold mineralization took place, which is called "explosive metallogenesis" by some geologists [35].

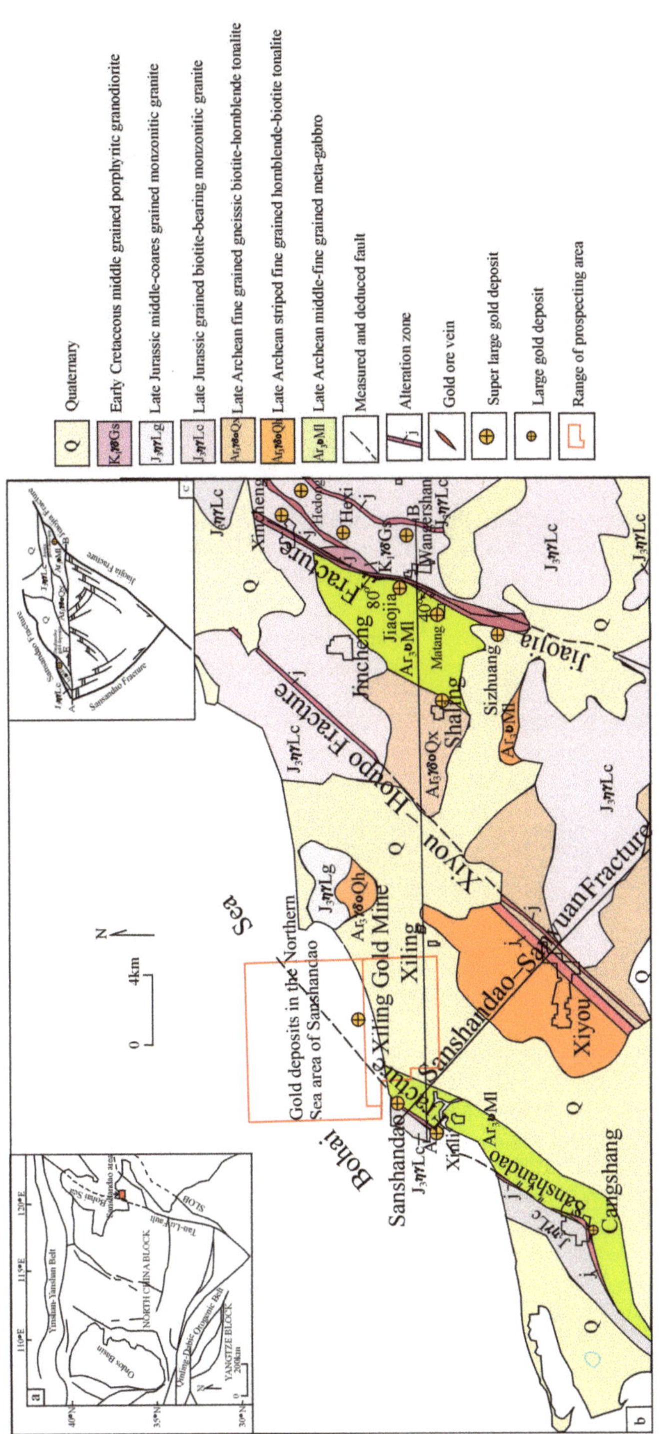

Figure 1. Geological map of the Sanshandao area. (**a**) Regional location map, from [1]; (**b**) geological map, modified after [44]; (**c**) geological structure sketch of A–B area).

3. Ore-Controlling Factors and Enlightenment of Current Prospecting Works

More than 100 large gold deposits (>20 t) and hundreds of middle-sized (5–20 t) and smaller (<5 t) gold deposits have been discovered. In northwestern Jiaodong in Shandong, which makes this area the third largest gold metallogenic area in the world. Sanshandao and Jiaojia gold deposits have become world class giant gold deposits (Table 1). Since 2017, proven gold reserves in the mining segment of north Sanshandao sea area have reached 470 t, among them, the Shaling mining segment has submitted 389 t of gold reserves [14].

The Jiaojia and Sanshandao giant gold deposits are distributed along the Jiaojia and Sanshandao fault belts, respectively. Two fault belts are nearly parallel on the surface and vertically decline in opposite direction with the stepped fault plane. Ore bodies are thickened in gently declined sections, are enriched in turning points and intersected parts of the diamond lattice faults [14,15]. Ore bodies are mainly hosted in Pre-Cambrian metamorphic rocks and Jurassic-Cretaceous terrestrial clastic rocks that have no metallogenic specificity.

3.1. The Spatial Distribution of Main Gold Ore Bodies

The ore bodies show regular distributions in Sanshandao and Jiaojia fault belts. Two gently declined ore-hosting fault benches have been exposed in both deposits. Based on the study of the spatial pattern of regional ore-controlling faults, the Jiaojia fault is the main detachment zone in the northwest margin of the Guojiadian mantle branch. The Sanshandao fault is an opposite listric fault in the hanging wall of the Jiaojia fault [45]. The relationship between the two faults, based on dynamic analysis, is that the Jiaojia fault is a backbone fault and the Sanshandao fault is a derived fault, which should be terminated on the Jiaojia fault and shows a "y-letter" spatial pattern.

The dip of main ore bodies in the Jiaojia giant deposit are same as those of the Jiaojia backbone fault. Ore bodies dip toward the northwest, and pitch toward the southwest at a 25°–30° pitch angle. The dip direction of main ore bodies in the Sanshandao giant deposit is the same as that of the Sanshandao fault. Ore bodies in the Sanshandao giant deposit dip toward the southeast, and pitch toward the northeast, the minimum pitch angle is about 35°. Both faults were sinistral strike-slip faults during the main metallogenic period; the pitch direction of the ore bodies is controlled by the strike-slip direction of the faults.

3.2. Rock-Controlling and Ore-Controlling Character of Fault Structures

There are continuously distributed 5–40 cm thick fault gouges on the main fault planes of the Jiaojia and Sanshandao faults. Most ore bodies are in the fracturing zone below the fault gouge. It is indicated that the fault gouge plays the role of a barrier layer, which makes the major ore bodies constrained below the fault gouge. Ore hosting rocks show zoning of fracturing and alteration, and they usually develop a beresitization cataclasite zone, beresitization granitic cataclasite zone, beresitization granite zone and a potassic-altered granite zone. The mineralization intensity is varied across different zones, and the major ore bodies are hosted in the altered zones [14,15].

Gold ore bodies in northwestern Jiaodong are obviously controlled by fault structures. Horizontally, the Jiaojia giant gold deposit formed in the turning part of the Jiaojia fault, which is the turning point from the South-North (SN) direction (Xuchunyuan-Sizhuang section) to the North 22.5°East-North 45°East (NNE-NE) direction (Sizhuang-Gaojiazhuangzi section). Vertically, the fault plane shows wave-shaped extension along the declined direction. The gently declined sections are usually taken as expanding spaces for the concentration of metallogenic materials. Moreover, many branch faults are developed in the foot wall of the backbone fault. The mineralization enrichment zone of the Sanshandao giant gold deposit is also in the turning part of the Sanshandao fault, which is the turn point from 80° strike (south Xinli section) to 40° strike (Sanshandao-Xinli section).

Table 1. Geological characteristics and metallogenic ages of typical Au deposits in northwestern Jiaodong.

Ore Deposit	Reserve and Grade	Wall Rocks and Intrusions and Age	Orebody and Main Metallic Minerals	Metallogenic Age	Reference
Sanshandao-Cangshang NE Ductile-Brittle Fracture Zone					
Northern sea area of Sanshandao	Au: 470 t, 4.23 ppm	Linglong gneissic granite, Guojiadian granite (zircon U-Pb 140–160 Ma); Guojialing granodiorite (zircon U-Pb 125–130 Ma)	Large veins and large lenses locally; Disseminated, einlet-reticulated pyrite-sericite cataclastic rock type ore; The main ore mineral in the ore is pyrite, followed by galena, Sphalerite, chalcopyrite, arsenopyrite, pyrrhotite, limonite, magnetite, etc.	-	[46,47]
Sanshandao	Au: >20 t, 3.40 ppm	Jurassic granite (zircon U-Pb 127 ± 2 Ma)	Flat or lenticular; Pyrite sericite type and gold-bearing pyrite sericite granitic cataclastic rock type; It is mainly pyrite, followed by arsenopyrite, chalcopyrite, galena and sphalerite, etc.	Altered Sericite Rb-Sr isochron age 117.6 ± 3.0 Ma	[48]
Xiling	Au: 383 t, 4.63 ppm	Linglong gneissic granite, Guojiadian granite (zircon U-Pb 140–160 Ma); Guojialing granodiorite (zircon U-Pb 125–130 Ma)	Large lenticular; Pyrite sericitized granitic cataclastic rocks type; pyrite, arsenopyrite, chalcopyrite, galena and sphalerite, etc.	-	[49]
Xinli	Au: 112 t, 2.78 ppm	Linglong gneissic granite, Guojiadian granite (zircon U-Pb 140–160 Ma); Guojialing granodiorite (zircon U-Pb 125–130 Ma)	Large vein-type orebodylenticular; Veinlet-disseminated pyrite-sericite cataclastic rocks type; pyrite, arsenopyrite, chalcopyrite, galena and sphalerite, etc.	-	
Cangshang	Au: 2 t, 4.93 ppm	Guojialing granodiorite (zircon U-Pb 127 ± 2 Ma)	lamellar; Pyrite sericitized cataclastic rocks type; pyrite, arsenopyrite, chalcopyrite, galena and sphalerite, etc.	Altered Sericite ^{40}Ar–^{39}Ar isochron age 121.3 ± 0.2 Ma	[50]
Longkou-Laizhou NE Ductile-Brittle Fracture Zone					
Jiaojia (deep)	Au: 73 t, 4.91 ppm	Linglong gneissic granite, Guojiadian granite (zircon U-Pb 153–160 Ma); Guojialing granodiorite (zircon U-Pb 126–130 Ma)	Stratiform and vein; Pyrite sericitized cataclastic rocks type; Pyrite, chalcopyrite, galena and sphalerite, etc.	Altered Sericite ^{40}Ar–^{39}Ar isochron age 120.5 ± 0.6 Ma; 120.1 ± 0.2 Ma; 120.2 ± 0.2Ma	[19,51,52]
Xincheng	Au: 12 t, 3.32 ppm	Guojialing granodiorite (zircon U-Pb 127 ± 2 Ma)	Lamellar, branched, compound and dilated with shrinkage; Veinlet disseminated and reticulated ores; Pyrite, chalcopyrite and galena, etc.	Altered Sericite ^{40}Ar–^{39}Ar isochron age 120.2–120.9 Ma; Beresite Rb-Sr isochron age 116. 6 ± 5.3 Ma	[51,53,54]
Hexi	Au: 13 t, 6.64 ppm	Linglong gneissic granite, Guojiadian granite (zircon U-Pb 153–160 Ma); Guojialing granodiorite (zircon U-Pb 126–130 Ma)	Lenticular, bifurcated along strike; Reticulated pyrite sericitized granitic cataclastic rock type type; Metal minerals are mainly pyrite, natural gold and bismuth tellurite, followed by chalcopyrite, galena, sphalerite, silver-gold ore, gold-silver ore, pyrrhotite, porphyry and malachite, etc.	Pyrite in ore Rb-Sr isochron age 112 Ma	[55,56]
Shaling	Au: 389 t, 2.92 ppm	Linglong gneissic granite, Guojiadian granite (zircon U-Pb 153–160 Ma); Guojialing granodiorite (zircon U-Pb 126–130 Ma)	Lamellar, and veined, branched, compound, inflated and pinched Pyrite sericitized cataclastic rocks type; pyrite, arsenopyrite, chalcopyrite, galena and sphalerite, etc.	-	[15,46]

3.3. The Constraint of the Backbone Faults on Ore Formation and Ore Controlling

Diamond lattice faults are recognized based on their different orientations and overprinting relationships, and Au mineralization occurs mainly as vein and breccia in the faults, where the regional deep fault—Jiaojia fault intersects with secondary faults (e.g., the Wangershan, the Hexi, the Houjia and the Qiujia fault). Additionally, The Sanshandao fault is a transitional part of NNE (NE) and NEE trending regional deep faults, and it controls the spatial distribution of major ore-bodies in the Changshang, Xinli and Sanshandao deposits (Figure 1).

It is obvious that ore-bodies are mainly controlled by a few ore-controlling structures. More than 180 mineralization zones have been delineated in the Shaling gold deposit with approximately 77 being economic with 389 t Au. However, the #1 and #2 are the largest ore-bodies with a 300 t gold reserve, those ore-bodies are up to >2000 m long and 1.06–125.64 m thickness.

Horizontally, the thick part of the ore body is at the southeast part of the ore body (Figure 2a). The thickness of the ore vein shows an alternative thick and thin pattern. The ore veins are branched as finger-like veins, which might be caused by the explosive hydraulic cracking of hydrothermal fluid. The grade from single engineering ore body varies between 1.00–11.37 ppm. The Au grade of bonanzas (>5 ppm) are located in the southeast part of the ore-bodies, and they also show a finger-like pattern, they are likely due to a combination of cryptoexplosion, where can be a diffusion center of ore-forming fluid. Hence, it is worthwhile to notice that some large thick ore-bodies and high grade ore-bodies in the main vein group might be caused by the pulsation cryptoexplosive metallogenesis of hydrothermal fluid.

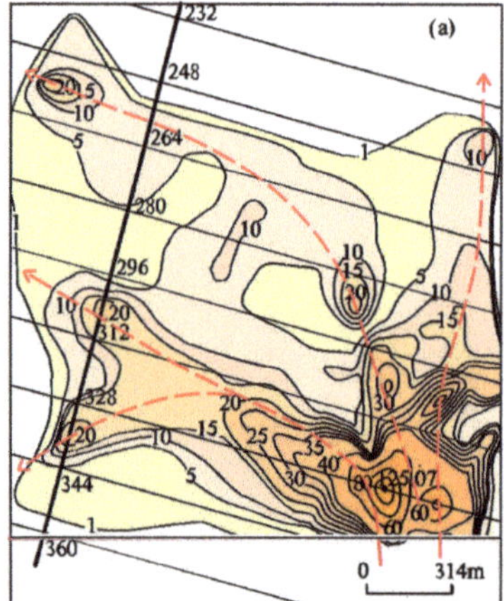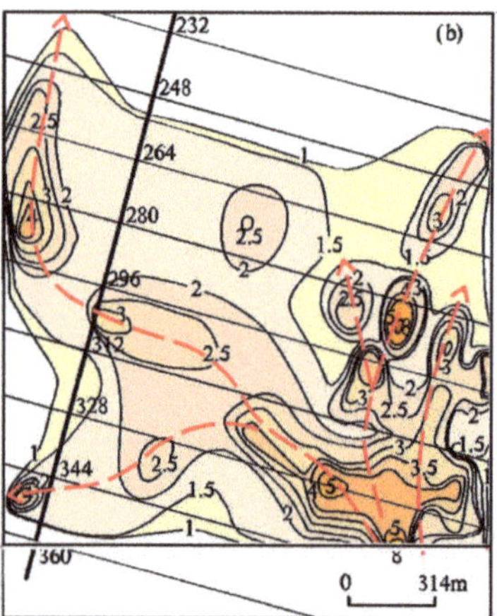

Figure 2. (a) Thickness isoline (unit: m) and (b) grade isoline (unit: ppm) for ore body I-2 in the Shaling deposit (modified after [15]). Red lines indicate the branches and extension direction of isoline.

Based on the analysis of metallogenesis of hydrothermal gold deposits, only few are mineralized in the mineralization zone, although most faults can be developed. The variation of physical and chemical conditions in deep zones can drive the ore-bearing hydrothermal fluid to migrate from deep earth to shallow crust, and some strata or rocks may stop or obstruct the migration of the hydrothermal fluid. Once the hydraulic cracking penetrates into a regional fracturing zone, the hydrothermal fluid would rapidly penetrate into the main structural zone. Because of the rapid decline of hydrodynamic pressure, the penetration of hydrothermal fluid into other secondary fractures is relatively weak, and the metallogenesis also is weak in these fractures.

Generally, if an ore concentration area develops wide migration channels for fluid, the ore-bodies will be thicker and of high grade. When hydrothermal fluid rapidly penetrates into some main fractures, penetration of hydrothermal fluid into other faults would greatly decrease because of the variation of the physicochemical conditions of the fluid. For this reason, the reserve of main ore bodies in a deposit may occupy more than 50% of the deposit reserve, even more than 90% of the deposit reserve. In the Jiaojia, Lingnan and Taishang gold deposits, gold grade isoline is positively correlated with the thickness of the ore vein [57,58]. The closer to the fluid cryptoexplosive point, the higher the gold grade (Figure 3). This ore-forming and ore-controlling character of structure also exists in the Xiaoqinling, North Hebei and East Hebei gold deposit concentration districts.

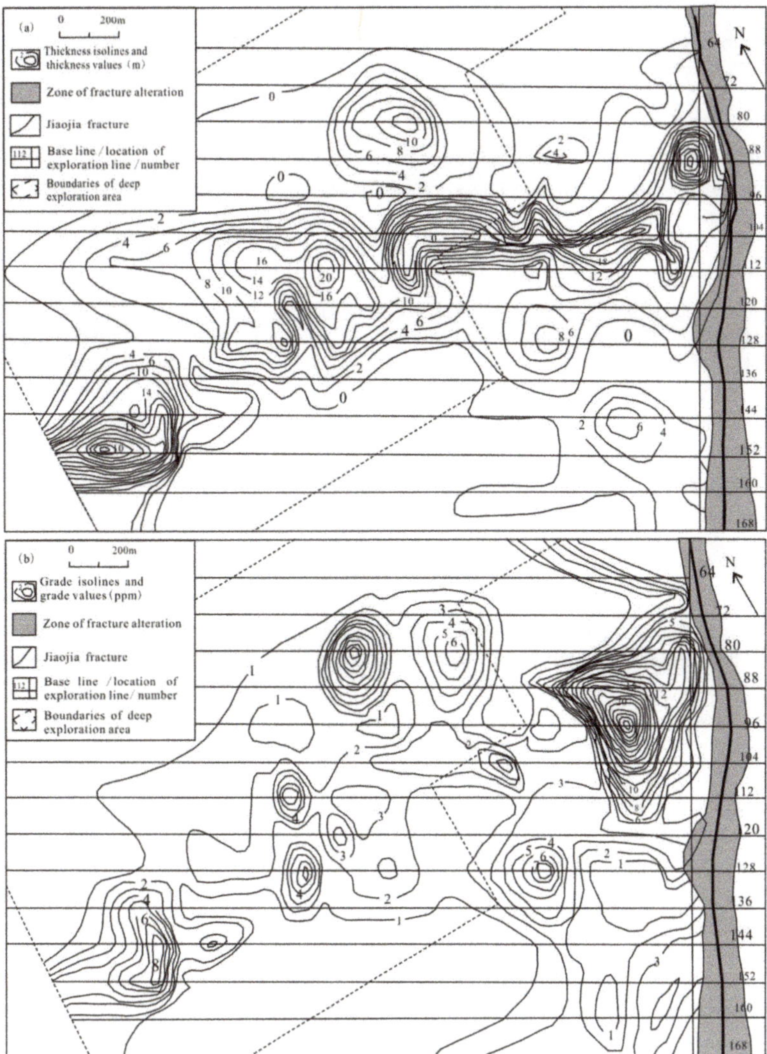

Figure 3. (a) Thickness isoline and (b) grade isoline of No.1 orebody in the Jiaojia gold deposit (from [57]).

3.4. Physicochemical Conditions of Metallogenesis

Homogenization temperature of fluid inclusions from Jiaodong gold deposits is in the range of 120–420 °C [9], most of them are concentrated in 230–320 °C [59], which is close to the metallogenic temperature of epithermal gold deposits (generally <300 °C) in the Pacific rim metallogenic belt. So, some scholars believe that Jiaodong gold deposits belong to epithermal gold deposits [60].

The ore bodies No. I-1 and No. I-2 in the Shaling gold deposit have great vertical extension. In adjacent ore districts, ore bodies also vertically extend over 2 km, which greatly exceeds the metallogenic depth of epithermal gold deposits (<1 km). It is indicated that the metallogenic condition of Jiaodong gold deposits are obviously different from that of epithermal deposits. In recent years, many researchers also believed that the Jiaodong gold deposit is different from other known gold deposits [12,15].

Compared with previous low-temperature hydrothermal gold deposits, the spatial distribution, magnitude and physicochemical conditions (such as Temperature (T), Pressure (P), Potential of Hydrogen (Ph), Oxidation-Reduction Potential (Eh)) of Jiaodong gold deposits obviously exceed the scope of known epithermal deposits, especially for the Jiaodong middle to deep gold deposits represented by the Shaling gold deposit prospected in the past two years.

In previous studies on metallogenesis, geologists believed that regional large faults play the role of conducting ore-forming materials, and secondary faults are ore-hosting structures. More and more studies indicate that main faults usually control main ore bodies in many deposits [34]. Main faults can both conduct ore-forming materials and host ore bodies. Whether or not faults are hosting ore bodies does not depend on the scale of fault but depends on physicochemical conditions. When ore-bearing hydrothermal fluid passes through a main fault, if the physicochemical condition can not satisfy the crystallization of ore-forming materials, hydrothermal fluid would continue to migrate. In that case, main faults only play the role of ore-conducting structures; if the main faults were elevated to shallow crust with a lower P-T condition, hydrothermal fluid would have crystallized and stopped migrating before it reached shallow crust. In that case, main faults are neither ore-conducting structures nor ore-hosting structures. So, the metallogenesis of main faults does not depend on its magnitude but depends on geologic condition.

4. Mantle-Crust Evolution and the Structural Movement of Metallogenic Districts

Many geologic factors have an impact on gold mineralization [12,61–63]. Magmatic activity is the important expression of structure movement. During a period of intensive crust movement, large scale of volcanic activity or/and igneous intrusions are developed, which are accompanied with different types of metallogenesis.

Since Wilson [64] and Morgan [65] put forward the theory of hot spots and mantle plumes successively, the study of the vertical motion of the earth has been gradually strengthened [66–73] and many scholars believe that the formation of polymetallic deposits is related to mantle plumes [74–79]. But it is obvious that the scale of a mantle plume is too large for explaining the formation process of a specific deposit or ore field. The mantle branch structure which was first put forward by Professor Niu in 1996 [80–85], is the third grade of structural units during multiple evolutions of mantle plumes and is mainly used to research ore-forming and the ore-controlling role of Au-Ag polymetallic ore fields. So, in this paper, the metallogenesis of gold deposits in the Northwest of Jiaodong are discussed from the perspective of a mantle branch structure.

4.1. Mantle Uplifting and the Formation of the Laiyang Sub-Mantle Plume

Magmatic activity can be divided into large structural cycles and small pulse periods. Large structural cycles are usually related with crust movement. For instance, the structural movement and magmatism during the Yanshanian movement are divided into four cycles, and then each cycle is divided into several periods. Dominated by the intensity of structural movement, magmatic activity can express as volcanic eruption or large scale of intrusion. Based on the time and intensity of a volcanic eruption and the time and scale of intrusion, it can be divided into several periods.

Magmatism often begins with basic-intermediate volcanic eruptions and ends with intermediate-acid eruption or intrusion. When the temperature and pressure of deep magma exceed the bearing capability of overlapped rocks, the magma would break through the rocks and bring out volcanic eruption. With the decrease of pressure, volcanic eruption would transform to intrusion [86].

During middle-late period of the Yanshanian movement, large scale magmatism occurred and a rift basin was developed in the Jiaodong area. Since the Cretaceous period, mantle magma under the Jiaolai basin showed significant upwelling, and the upper crust was rapidly extended to form the basin, which resulted in the formation of a typical sub-mantle plume. The main structural or sedimentary formation includes several geologic units: The Lower Cretaceous Laiyang group is composed of thick conglomerate, sandstone and argillaceous rocks (fluvio-lacustrine facies); The Middle-upper Cretaceous Qingshan group is composed of thick pyroclastic rocks interbedded with some terrigenous clastic rocks; the time of the Dasheng group is same as the Qingshan group, but it particularly refers to the fluvio-lacustrine sedimentary rocks in northwestern Jiaodong; The Late Cretaceous Wangshi group is made up of thick conglomerate and sandstone (fluvio-lacustrine facies). The basin takes Laiyang as its center, and it is characterized by rapid subsiding and rapid sediment accumulation. The palaeogeographic lithofacies of the whole Cretaceous can be divided into nine phases, and eight of them are in great depression. Total depression amplitude is up-to or even over 10,000 m (Table 2). Intermediate-basic dykes and intermediate-acid dykes are widely distributed in the sedimentary strata. The Guojiadian mantle branch, the Qipeng mantle branch and the Muru mantle branch are developed around the Laiyang sub-mantle plume. Generally speaking, the Qingshan period is the most intensive period of deep magmatism, and it is characterized by a short time, significant depression, intensive structure movement, thick volcanic material accumulation, a complex intrusive phase and varieties of dykes. Simultaneously, it is the most important period for metallogenesis.

Table 2. Division and correlation of Cretaceous strata in Jiaodong

Era	Period	Series	Group	Formation	Member	Grade 2	Grade 3	System	Biota / Radioactive Dating (Ma)	Sedimentary Environment
Mesozoic	Cretaceous	Upper aeries	Wangshi geoup	Jingangkou		Ms$_4$	Ksq$_7$	NLST	*Skacrium-Phyrsa* Mollusk fauna	Restrictedbasin
				Hongtuya			Ksq$_7$	NLST	*Protosaunus* Dinosaur fauna	River
				Hongtuya					68–70	Layeredvolcano
		88 Ma					XI		*Hadrosauridae* Dinosaur fauna	River
		Middle series		Xingezhuang	Shijiatun basalt		Ksq$_6$	LEST	*Viuikaruc-Skhacrium* Mollusk fauna / *Yanjiestheria* Estheria fauna	Shallow lake
				Linjiazhuang			X	LLST	*Psittacasaunue* Dinosaur fauna	River
		112 Ma	Pingshan group / Dasheng group	Fanggezhuan (Mengtuan, Siqiangcun)		Ms$_3$	IX	LEST / LLST	90–100	Volcanic basin; Lake / River
				Bamudi Shiqianzhuang (Tianjialou, Malanggou)			Ksq$_4$	LEST	*Yanjiestherla* Estheria fauna 100–110	Lake
							Ksq$_4$	LLST	110–120	River
				Datuling			VIII	LEST	110–120	Lake
				Xiaodian			VIII	LLST		Rier
				Houkuang			VII / VI	-	120–126	-
		Lower series	Laiyang group	Malianpo — Fajiaying			Ksq$_3$	-	Marine algae Bryozoon Foraminiferan	Restrictedgnlf / Restrictedlake
				Chengshanhou — Dncun — Qugezhuang -		Ms$_2$	V	NLST	*Nakamuranaia* Bivalvian fauna	Volcanicbasin / -
				Yangjiazhaung — Longwangzhuang			Ksq$_2$	LCST	-	River / Shallow lake
				Shuinan			Ksq$_2$	LCS	Laiyang biota	Deep lake
				Zhifengzhuang			IV	LEST	-	-
				Linsishan			IV	LLET	-	
				Wawukuang			Ksq$_1$	LEST		Deep lake
		137 Ma					Ksq$_1$	LLST		-

Note: Ms—Marine flooding surface; NLST—Non-Lacustrine system tract; LEST—Lacustrine extension system tract; LLST—Lacustrine lowstand system tract; LCST—Lacustrine contraction system tract.

4.2. Pulse Upwelling of Magma and the Formation of Mantle Branches

The density (specific gravity) of mantle magma is higher than that of crust rocks. Liguliform mantle magma can emplace into low-velocity and high-conductivity middle crust and make the sandwich structure of the crust intensively deformed. Decoupled lower crust would fall down in the form of residual structural blocks. Middle and upper crust would appear as a negative gravity anomaly caused by the emplacement of intermediate-acid magma, which results in the uplifting of igneous-metamorphic complexes and the formation of a mantle branch.

The igneous rocks in a mantle branch usually evolve from intermediate-basic to intermediate-acid. When liguliform mantle magma emplaces into middle crust, high temperature mantle magma (1300–1000 °C for ultra-basic and basic magma, [87]) would melt some middle-upper crust rocks and exchange solid or liquid materials with crust rocks, which forms mantle-crust magma with a percentage of gas-liquid solution. When the energy in magma chamber accumulates enough to break through overlapped rocks, the mantle-crust magma would explosively emplace toward upper crust. When the energy is released, magma emplacement stops and the magma chamber would begin next cycle to accumulate energy until the next magma emplacement takes place. This much likes the geothermal fountain in Yellow Stone of the USA in where energy accumulation and hot water eruptions are periodically repeated. Obviously, the periodical emplacement-heat melting produces igneous complex with multiple periods of pulsation intrusion. In the middle-late period of magmatism, intermediate-acid magma occupies the majority volume of the emplaced magma. Significant negative gravity anomalies drives the metamorphosed country rocks to uplift together with igneous rocks, accompanying the detachment of cover rocks toward the surrounding area. So far, the typical mantle branch structures are formed, it has an igneous-metamorphic complex core surrounded by ring-like cover rocks and a series of normal detachment faults. The combination of the Laiyang sub-mantle plume with the Guojiadain mantle branch in the west, the Qipeng mantle branch in the north and the Muru mantle branch in the east, perfectly reflects a kind of structural combination relationship.

4.3. The Spatial Pattern of the Mantle Branch

Mantle branch theory can be used to explain the multiple pulsation intrusion process and the spatial distribution of intrusive rocks during Yanshanian magmatism. The heat circulating driven by periodical magmatic activity brought out large scale gold mineralization.

There is the Guojiadain mantle branch, the Qipeng mantle branch and the Muru mantle branch around the Laiyang sun-mantle plume. Take the Guojiadian mantle branch as an example. It developed in the west of the Laiyang sub-mantle plume. Taking the "0-value" line of gravity anomaly as a division, the outer-ring is a positive gravity anomaly area in which the anomaly value increases away from the division line. The inner-ring is a negative gravity anomaly area. Gold deposits controlled by the mantle branch are distributed along the "0-value" line. It is indicated that gold deposits were mainly formed in major and secondary detachment zones around the mantle branch.

5. Multiple Evolutions of the Mantle Plume and the Metallogenesis of Mantle Branches

In the process of the Earth's evolution, heavy elements were mainly concentrated in the core. Small amount of heavy elements (such as gold) can migrate to shallow crust through mantle plume → sub-mantle plume → mantle branch in the form of gas, gas-liquid mixture or ore-bearing hydrothermal fluid [88]. Under the impact of a regional tectonic stress field, they can penetrate into fracture zones and accumulate to form ore bodies with a variation of physicochemical conditions. The ore-forming materials mainly come from deep Earth.

Based on the non-exclusive character and the atomic structure model of gold, it is speculated that 99% gold is concentrated in the core. Gold is mixed with iron and nickel in a purple gas state within the core. In the process of intensive outer core convection and differential rotation of the core-mantle, the majority of gold gas accumulates in the core–mantle boundary (D″ layer) that is the boundary between the core and mantle. Once the dynamic balance of the core-mantle boundary was damaged by some astronomical or intraterrestrial agents, the liquid materials in the outer-core would pass through the core-mantle boundary and migrate up in the form of a mantle plume. As a component of mantle plumes, gold gas moves up in a state of anti-gravity migration. The initial mantle plume has a small diameter and it gradually increases during its upwelling process. When it comes to the boundary between lower mantle and upper mantle, the top head of the mantle plume would be enlarged to form a mushroom-like crown because of the restriction and obstruction of the boundary. When accumulated energy in the head of a mantle plume exceeds the bearing capability of the boundary, the upwelling mantle materials would penetrate the boundary to form several sub-mantle plumes. Similarly, at the top of upper mantle, sub-mantle plume experiences same process as the mantle plume, a series of mantle branches are formed, which constitutes the typical multiple evolution sequence of mantle plumes.

Accompanied with the multiple evolution of mantle plumes, parts of gaseous gold could transform into liquid to form gas-liquid mixture when it passes through the asthenosphere. Together with the mantle methane material (CH_4), the gas-liquid mixed gold continues to migrate up with the evolution of the mantle plume [88]. In the process of upwelling of magma with a gas-liquid mixture along a deep fracture, two-thirds of the gaseous gold in mixture would enter into softened plastic country rocks, and one-third of the gaseous gold in gas-liquid mixture would experience differentiation with magmatic evolution. During the differentiation, liquid gold can directly crystallize into solid gold in near-surface freshwater environments. In surface salt water (sea water), gold migrates in a complex form, and it can aggregate into solid gold under the action of freshwater or bacteria. Table 3 shows the differential distribution of gold and silver in different spheres of the Earth.

Table 3. Distribution of gold and silver in the earth's layers [89].

The Earth's Layers	Mass Percentage of Crust-Mantle-Core (%)	Distribution of Gold			Distribution of Silver		
		Abundance ppb	Weight Ratio ppb	Volume Ratio (%)	Abundance ppb	Weight Ratio ppb	Volume Ratio (%)
Crust	0.4	3.0	1.2	0.00423	80.0	32.0	0.01005
Mantle	68.1	1.0	68.1	0.23962	55.0	3745.5	1.17495
Core	31.5	900.0	28,350.0	99.75615	10,000.0	315,000.0	98.81500
Earth	100.0	284.0	28,419.3	100.00000	3188.0	318,777.5	100.00000

Jiaodong gold deposits are characterized by their various deposit types, enormous gold reserves and rapid metallogenic process. It is necessary to study metallogensis and the metallogenic rule, conclude the metallogenic model and summarize metallogenic laws in order to guide a new round of gold prospecting works.

5.1. Heat Migration Ore-Conducting and Ore-Forming of Sub-Mantle Plumes

Traditional gold geological survey works pay much attention to looking for metamorphic rocks and granitic rocks with a high gold background value. Their theoretical idea for gold mineralization is that gold comes from the high background value country rocks and the gold is extracted from country rocks and accumulated into favorable structures as ore bodies under the impact of magmatism and structure movement. Based on the above idea, gold migration is from low P-T environments to high P-T environments and from low grade rocks to high grade rocks, this is difficult to realize in normal geological conditions [90].

From the viewpoint of the multiple evolutions of mantle plumes, heavy elements such as gold and silver mainly come from the boundary between the core and mantle. Through multiple evolutions of the mantle plume, they penetrate the major structural expanding zones of mantle branches in the shallow crust. In that case, ore prospecting work should follow the system of sub-mantle plume–mantle branch–structural expanding zone–ore-hosting fractures (Figure 4).

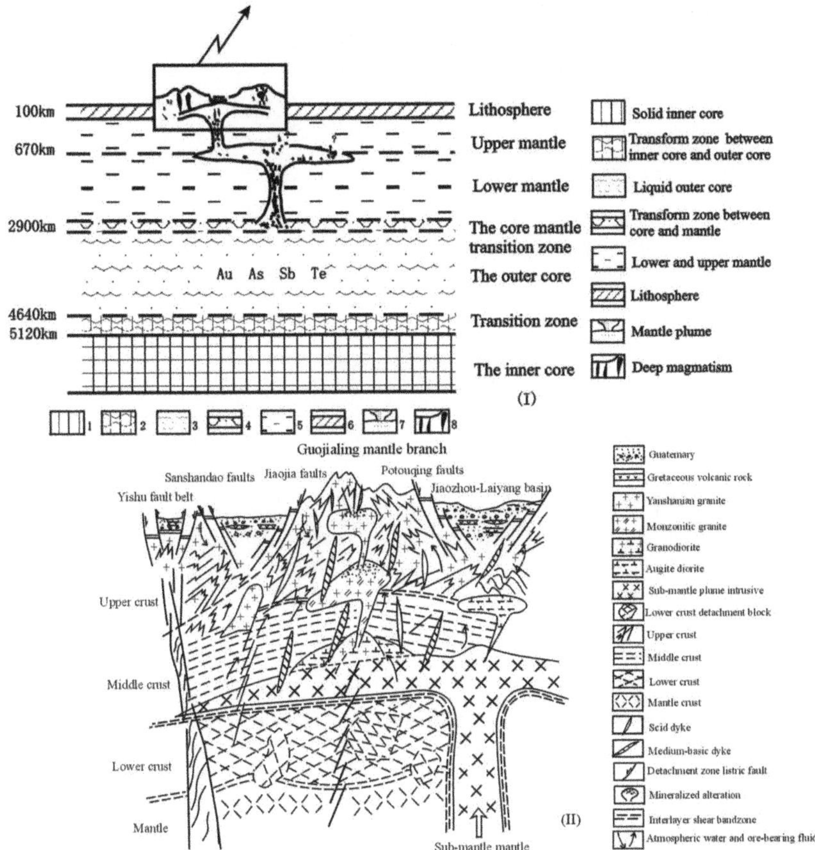

Figure 4. Sub-mantle plume-mantle branch evolution and its metallogenesis: Mantle plume-sub-mantle plume mode (**I**) and Sub-mantle plume-mantle branch metallogenic mode (**II**).

The sub-mantle plume is the secondary structure unit of a mantle plume, and it plays the role of connecting the lower mantle and the upper crust. It also is the pathway for magma to migrate from deep earth to shallow crust, and the main place for energy to accumulate and release. The pulsation of magmatic activity, energy release and accumulation, is much like the geothermal springs of Yellowstone Park in the USA. Intermittent igneous intrusion or eruptions produce multiple-phases of complexes. Early eruption and intrusion are dominated by intermediate-basic magmatic activity; late magmatic activity which evolves to intermediate-acid magma.

With the rapid uplifting of a sub-mantle plume, a large scale faulted subsidence basin rapidly forms above the top of the sub-mantle plume. The basin has a mirror symmetry relationship with the sub-mantle plume. Intense activity and large uplifting amplitudes of deep mantle magma are usually accompanied with a rapid fault basin development, large influence dimension and significant depression in the shallow crust. This shows a gravity equilibrium compensation in relation with deep mantle uplifting.

The Laiyang basin is a fault basin formed rapidly on the top of a sub-mantle plume. It developed in the Cretaceous and was widely spread in the Laiyang stratigraphic district which includes the Jiaolai basin and its surrounding elements. Sedimentary formation is composed of inland basin fluvio-lacustrine clastic rocks interbedded with terrestrial pyroclastic rocks that are overlapped non-conformingly on an ancient metamorphic basement. It is divided into the Laiyang group, the

Qingshan group and the Wangshi group, from bottom to top [58]. The Laiyang group exhibits obvious horizontal variation and a differential diagenetic environment, and it is made up of terrestrial clastic rocks interbedded with some volcanic rocks. The Qingshan group is widely spread in the Jiaolai basin and its surrounding elements. It is a set of terrestrial volcanic rocks and pyroclastic rocks formed in a volcanic basin. The rock formation includes dactitic pyroclastic rocks → andisitic-basaltic andesitic lava and pyroclastic rocks → rheolitic pyroclastic rocks and lava → trachytitic andesitic lava interbedded with pyroclastic rocks, from bottom to top. Subvolcanic intrusions were developed in each period of volcanic activity. The Wangshi group is distributed in the Jiaolai basin and its western elements. It is mainly distributed in Laixi, Laiyang and Jiaozhou in the Jiaodong area, especially in Laiyang. It is much same as molasses formations made up by a set of red terrestrial clastic rocks interbedded with some volcanic rocks.

Overall, above mentioned sedimentary formation in the Laiyang basin indicates that intensive structural movement during the Cretaceous had wide distributions and deep extensions. At the same time, large scale of metallogenesis took place in that time.

5.2. The Ore-Forming and Ore-Controlling of Mantle Branches

Geodynamics indicate that upwelling mantle magma from sub-mantle plumes comes to the surface or shallow crust in the form of volcanic eruptions or intrusions. Constrained by regional structural stress fields, magmatic activity shows alternative variations of intense and weak, large volume and small volume, eruptions and intrusions. The uplifting and evolution of magmatic-metamorphic complexes results in a series of mantle branches.

The crust has an obvious sandwich structure, and different layers show different characters. The upper crust is shallow and cold; the lower crust is deep and rigid; the middle crust is heat softened layer. The regional ductile shearing zones in the crust constrain the structural movement and magmatism. The density (specify) of mantle magma is much higher than that of the middle crust. The existence of detachment zones in the crust makes it easy for the mantle magma intrude into detachment zone in a large tongue shape. Because the tongue-shaped mantle intrusion could detach the sandwich structure of the middle-lower crust, parts of residual crust blocks would drop into the mantle magma. Meanwhile, the upwelling of the mantle magma results in faulting and subsiding of the upper crust, forming a fault basin controlled by listric faults around the basin margin.

With the increase of depth, the temperature and pressure of crust increases gradually, which causes the deep crust rock to be softened. Intruded high temperature mantle magma (1300–1000 °C for ultra-basic and basic magma) would heat and melt some crust rocks to form magma chambers that are filled with mantle-crust magma with a percentage of gas-liquid solution. The periodical energy accumulation and release in the magma chamber causes pulse eruptions or intrusions. The country rocks of magma chamber are progressively heated, so the temperature for melting country rocks decreases and the crust material in magma gradually increases. So, igneous rocks usually show the evolution sequence of basic → intermediate-basic → intermediate-acid → acid.

When the intermediate-acid igneous rocks reach a certain volume in the magmatic complex, a negative gravity anomaly would drive metamorphic country rocks to uplift together with the magmatic rocks, forming the magmatic-metamorphic complex in the core of the mantle branch. The uplifting of magmatic-metamorphic complexes either provides a connection for a sub-mantle plume and mantle branch or creates a migration channel for deep ore-bearing fluid and favorable structural expanding spaces for metallogenesis.

5.3. Mantle Branch Metallogenesis and Its Main Types of Mineralization

As mentioned above, once the mantle plume evolution links the crust to deep source, especially to the core-mantle boundary, the source material in the D″ layer would migrate up through the mantle plume → sub-mantle plume → mantle branch and concentrate to form a deposit in a favorable structural expanding zone of the mantle branch. These favorable structural expanding spaces include the brittle-ductile detachment zone in the core of the mantle branch, the contact zone between intrusion and country rock, the contact zone between metamorphic complex and cover rocks, faults and joints formed in the process of mantle branch uplifting. Different structural spaces result in different structural metallogenic types.

The core of the Guojiadian mantle branch (Linglong-type gold deposit): It mainly refers to the vein ore bodies in the core uplifting zone between the Jiaojia fault and the Potouqing fault. Ore bodies are hosted in Linglong granite and some metamorphic rocks. Fractures are well developed within ore-bearing geological bodies between the Jiaojia fault and the Linglong fault. The deposit type belongs to structural fracture veins. Ore veins are densely developed, and their gold grade is high. It is a major ore-forming and ore-controlling structure type in Jiaodong [14].

Detachment zone in the northwest margin of a mantle branch (Jiaojia-type gold deposit): The Jiaojia fault is a dominant ore-forming and ore-controlling structure in the northwest margin of the Guojiadian magmatic-metamorphic core complex. The fault has both great horizontal extension and vertical extension. It not only is a regional ore-conducting structure, but an important ore-forming and ore-controlling structure. Deposits controlled by this fault are the most famous fracture alteration gold deposits in Jiaodong. Recent ore prospecting works indicate that the vertical extension of ore bodies is usually larger than their horizontal extension. Previous proved gold deposits such as the Shizhuang deposit, the Matang deposit, and the Jiaojia deposit connect as a whole deeper down, which forms an ultra-large gold deposit. Deep prospecting works indicate that ore bodies show stable extension along the fault, which provides great prospecting potential for this gold deposit [14].

Sanshandao reverse listric fault belt (Sanshandao-type gold deposit): Recently, prospecting works for the Sanshandao metallogenic zone have made much progress. Viewing from metallogenesis, it is an ultra-large fracture alteration gold deposit that is strictly dominated by a reverse listric fault [13,44]. The giant gold mineralization zones, named the Sanshandao giant gold deposit, have been delineated in Sanshandao with several larger gold ore-bodies, which are thought to be an independent deposit previously [15].

Peripheral faults of a mantle branch (Pengjiakuang-type gold deposit): Due to the rapid splitting of the Laiyang fault basin, the structural contact belts around the Laiyang fault basin are developed between the Guojiadian mantle branch and the Laiyang fault basin. A certain thickness of ore-bearing conglomerate has developed in the structural contact belts. Studies indicate that conglomerate distribution belts are usually weak structure belts for ore-forming and ore-controlling, and they easily form typical conglomerate-type gold deposits.

Mantle branches mainly form under a tectonic setting transformed from regional compression to regional extension. A core complex, marginal main detachment zones, hanging wall cover rocks and related fault structures are the components of the mantle branch. The detachment zones and related faults are usually taken as main structural expanding zones for metallogenesis. Studying the fault distribution and ore-bearing potential, metallogenesis, metallogenic modes and metallogenic rule of mantle branches is important for guiding the next round of ore prospecting.

6. Conclusions

The Jiaodong mineralization concentration area is characterized by its large metallogenic scale, deep ore body extension, high ore grade, variety of deposits and great resource potential.

(1) The multiple evolutions of mantle plume break the restriction of earth spheres (core, mantle, crust) structure, and provide an important channel for deep ore-forming material to migrate up. The study on the multiple evolutions of mantle plume elaborates the migration mechanism of heavy elements migrating from the core to shallow crust.

(2) In the process of mantle plume multiple evolutions initiated from core-mantle boundary (D″ layer), heavy elements (such as gold and silver) which sank into the core by gravity differentiation migrate up in the form of a gas-liquid state. They are concentrated to form ore bodies in the

favorable structural expanding zones in mantle branch and different fractures. Ultra-large gold deposits with huge reserves can be formed.

(3) Metallogenic material came from deep earth and accumulated as ore-bodies in favorable structural expanding zones. Generally speaking, a favorable fault is not necessary to form deposits, but ore-bearing fault structure is linked to the deep earth. Favorable fault structures can be either migration channels for metallogenic material or hosting places for ore bodies.

(4) Structure movement, magmatism and metallogenesis are critical factors in the mineralization of large metal deposits. The structure movement is the most important factor in igneous activity related to gold metallogenic events. In the Jiaodong area, the main faults play both roles of fluid migration channels and ore-hosting places, forming large to ultra-large gold deposits.

Author Contributions: Conceptualization, S.N. and C.C.; Methodology, S.N.; Formal analysis, A.S.; Investigation, S.N., C.C., J.Z., F.Z., F.W. and A.S; Data curation, F.Z.; Writing—original draft preparation, S.N. and J.Z.; Writing—review and editing, C.C. and F.W.; Project administration, S.N.; Funding acquisition, S.N., C.C. and F.W.

Funding: This research was funded by the "Key laboratory of Ministry of Land and Resources for gold metallogenic process and resource utilization" and the "Key laboratory of Shandong for metallogenic process and resource utilization" (fund No. 2013001); A special fund project for the Public welfare industry, "Study on the scientific base of typical metal deposits in China" (fund No. 200911007); Graduate demonstration course of Hebei province "Plate tectonics and mantle branch structures"(fund No. KCJSX2018089), and Doctoral Scientific Research Foundation of Hebei GEO University (fund No. BQ2018032).

Acknowledgments: The latest research results from several scholars were consulted during the compilation of this paper. Special thanks are paid to leaders, experts and relevant technicians from the Shandong Institute of Geological Science, the Shandong Department of Land and Resources, the Shandong Bureau of Geological Mineral Resource Exploration and Development and Shandong Gold Group Co., Ltd. for their enthusiastic guidance and helps.

Conflicts of Interest: The authors declare no conflict of interest.

References

1. Deng, J.; Yang, L.Q.; Li, R.H.; Groves, D.I.; Santosh, M.; Wang, Z.L.; Sai, S.X.; Wang, S.R. Regional structural control on the distribution of world-class gold deposits: An overview from the Giant Jiaodong Gold Province, China. *Geol. J.* **2019**, *54*, 378–391. [CrossRef]

2. Deng, J.; Wang, C.M.; Bagas, L.; Carranza, E.; Lu, Y.J. Cretaceous–Cenozoic tectonic history of the Jiaojia Fault and gold mineralization in the Jiaodong Peninsula, China: Constraints from zircon U–Pb, illite K–Ar, and apatite fission track thermochronometry. *Miner. Depos.* **2015**, *50*, 987–1006. [CrossRef]

3. Deng, J.; Wang, Q.F. Gold mineralization in China: Metallogenic provinces, deposit types and tectonic framework. *Gondwana Res.* **2016**, *36*, 219–274. [CrossRef]

4. Li, S.R.; Santosh, M. Geodynamics of heterogeneous gold miner-alization in the North China Craton and its relationship to lithospheric destruction. *Gondwana Res.* **2017**, *50*, 267–292. [CrossRef]

5. Mills, S.E.; Tomkins, A.G.; Weinberg, R.F.; Fan, H.R. Implications of pyrite geochemistry for gold mineralisation and remobilisation in the Jiaodong gold district, northeast China. *Ore Geol. Rev.* **2015**, *71*, 150–168. [CrossRef]

6. Groves, D.I.; Santosh, M. The giant Jiaodong gold province: The key to a unified model for orogenic gold deposits? *Geosci. Front.* **2016**, *7*, 409–418. [CrossRef]

7. Goldfarb, R.J.; Santosh, M. The dilemma of the Jiaodong gold deposits: Are they unique? *Geosci. Front.* **2014**, *5*, 139–153. [CrossRef]

8. Yang, L.Q.; Deng, J.; Wang, Z.L.; Zhang, L.; Goldfarb, R.J.; Yuan, W.M.; Weinberg, R.F.; Zhang, R.Z. Thermochronologic constraints on evolution of the Linglong Metamorphic Core Complex and implications for gold mineralization: A case study from the Xiadian gold deposit, Jiaodong Peninsula, eastern China. *Ore Geol. Rev.* **2016**, *72*, 165–178. [CrossRef]

9. Yang, L.Q.; Deng, J.; Wang, Z.L.; Zhang, L.; Guo, L.N.; Song, M.C.; Zheng, X.L. Mesozoic gold metallogenic system of the Jiaodong gold province, eastern China. *Acta Geosci. Sin.* **2014**, *30*, 2447–2467. (In Chinese with English Abstract)

10. Zhang, L.; Li, G.W.; Zheng, X.L.; An, P.; Chen, B.Y. ^{40}Ar/^{39}Ar and fission-track dating constraints on the tectonothermal history of the world-class Sanshandao gold deposit, Jiaodong Peninsula, eastern China. *Acta Petrol. Sin.* **2016**, *32*, 2465–2476, (In Chinese with English Abstract).

11. Song, M.C.; Song, Y.X.; Cui, S.X.; Jiang, H.L.; Yuan, W.H.; Wang, H.J. Characteristic comparison between shallow and deep-seated gold ore bodies in Jiaojia superlarge gold deposit, northwestern Shandong peninsula. *Miner. Depos.* **2011**, *30*, 923–932, (In Chinese with English Abstract).

12. Song, M.C.; Zhang, J.J.; Zhang, P.J.; Yang, L.Q.; Liu, D.H.; Ding, Z.J.; Song, Y.X. Discovery and tectonic-magmatic background of superlarge gold deposit in offshore of northern Sanshandao, Shandong Peninsula, China. *Acta Geolo. Sin.* **2015**, *89*, 365–383, (In Chinese with English Abstract).

13. Song, M.C. The relevant issues of deep gold deposit exploration in China. *Gold Sci. Technol.* **2017**, *25*, 1–2. (In Chinese)

14. Song, Y.X.; Song, M.C.; Ding, Z.J.; Wei, X.F.; Xu, S.H.; Li, J.; Tan, X.F.; Li, S.Y.; Zhang, Z.L.; Jiao, X.M.; et al. Important progress of deep ore prospecting and metallogenic characteristics in Jiaodong gold deposit concentration area. *Gold Sci. Technol.* **2017**, *25*, 4–18, (In Chinese with English Abstract).

15. Song, G.Z.; Yan, C.M.; Cao, J.; Guo, Z.F.; Bao, Z.Y.; Liu, G.D.; Li, S.; Fan, J.M.; Liu, C.J. Breakthrough and significance of exploration at depth more than 1000 m in Jiaojia metallogenic belt, Jiaodong: A case of Shaling mining area. *Gold Sci. Technol.* **2017**, *25*, 19–27, (In Chinese with English Abstract).

16. Goldfarb, R.J.; Groves, D.I.; Gardoll, S. Orogenic gold and geologic time: A global synthesis. *Ore Geol. Rev.* **2001**, *18*, 1–75. [CrossRef]

17. Goldfarb, R.J.; Baker, T.; Dube, B.; Groves, D.I.; Hart, C.J.R.; Gosselin, P. Distribution, character, and genesis of gold deposits in metamorphic terranes. In *Economic Geology 100th Anniversary Volume*; Hedenquist, J.W., Thompson, J.F.H., Goldfarb, R.J., Richards, J.P., Eds.; Society of Economic Geologists: Littleton, CO, USA, 2005; pp. 407–450.

18. Goldfarb, R.J.; Hart, C.; Davis, G. East Asian gold: deciphering the anomaly of Phanerozoic gold in Precambrian cratons. *Econ. Geol.* **2007**, *102*, 341–345. [CrossRef]

19. Qiu, Y.M.; Groves, D.I.; Mcnaughton, N.J.; Wang, L.G.; Zhou, T.H. Nature, age, and tectonic setting of granitoid-hosted, orogenic gold deposits of the Jiaodong Peninsula, eastern North China craton, China. *Miner. Depos.* **2002**, *37*, 283–305. [CrossRef]

20. Mao, J.W.; Wang, Y.T.; Zhang, Z.H.; Yu, J.J.; Niu, B.G. Geodynamic settings of Mesozoic large-scale mineralization in North China and adjacent areas. *Sci. China (Ser. D Earth Sci.)* **2003**, *46*, 838–851. [CrossRef]

21. Zhou, T.H.; Lv, G.X. Tectonics, granitoids and Mesozoic gold deposits in East Shandong, China. *Ore Geol. Rev.* **2000**, *16*, 71–90. [CrossRef]

22. Ding, Z.J.; Sun, F.Y.; Liu, F.L.; Liu, J.H.; Peng, Q.M.; Ji, P.; Li, B.L.; Zhang, P.J. Mesozoic geodynamic evolution and metallogenic series of major metal deposits in Jiaodong Peninsula, China. *Acta Petrol. Sin.* **2015**, *31*, 3045–3080, (In Chinese with English Abstract).

23. Jiang, S.Y.; Dai, B.Z.; Jiang, Y.H.; Zhao, H.X.; Hou, M.L. Jiaodong and Xiaoqinling:two orogenic gold province formed in different tectonic settings. *Acta Petrol. Sin.* **2009**, *25*, 2727–2738, (In Chinese with English Abstract).

24. Li, L.; Santosh, M.; Li, S.R. The "Jiaodong-type gold" deposits—Characteristics, origin and prospecting. *Ore Geol. Rev.* **2015**, *65*, 589–611. [CrossRef]

25. Zhu, R.X.; Fan, H.R.; Li, J.W.; Meng, Q.R.; Li, S.R. Decratonic gold deposits. *Chin. Sci. Geosci.* **2015**, *58*, 1523–1537. [CrossRef]

26. Zhai, M.G.; Fan, H.R.; Yang, J.H.; Miao, L.C. Large-scale cluster in east Shandong: Anorogenic metallogenesis. *Earth Sci. Front.* **2004**, *11*, 85–98, (In Chinese with English Abstract).

27. Mao, J.W.; Wang, Y.T.; Li, H.M.; Pirajno, F.; Zhang, C.Q.; Wang, R.T. The relationship of mantle-derived fluids to gold metallogenesis in the Jiaodong Peninsula: Evidence from D–O–C–S isotope systematics. *Ore Geol. Rev.* **2008**, *33*, 361–381. [CrossRef]

28. Sun, F.Y.; Shi, Z.L.; Feng, B.Z. *Gold Deposit Geology and Differential Diagenesis and Mineralisation of Mantle-derived C–H–O Fluids in Jiaodong Peninsula*; Jilin People's Publishing House: Changchun, China, 1995; p. 170. (In Chinese with English Abstract)

29. Mao, J.W.; Wang, Y.T.; Ding, T.P.; Chen, Y.C.; Wei, J.X.; Yin, J.Z. Dashuigou tellurium deposit in Sichuan Province, China—An example of mantle fluid evolving in mineralized process: Evidences from C, O, H, S isotopes. *Resour. Geol.* **2002**, *52*, 15–23. [CrossRef]

30. Zhou, X.H.; Yang, J.H.; Zhang, L.C. Metallogenesis of super-large gold deposits in Jiaodong region and deep processes of subcontinental lithosphere beneath North China Craton in Mesozoic. *Sci. China (Ser. D Earth Sci.)* **2003**, *46*, 14–25.

31. Liu, J.M.; Ye, J.; Xu, J.H.; Sun, J.G.; Shen, K. C–O and Sr–Nd isotopic geochemistry of carbonate minerals from gold deposits in East Shandong, China. *Acta Petrol. Sin.* **2003**, *19*, 775–784, (In Chinese with English Abstract).

32. Fan, H.R.; Zhai, M.G.; Xie, Y.H.; Yang, J.H. Ore-forming fluids associated with granite-hosted gold mineralisation at the San shandao deposit, Jiaodong gold province, China. *Miner. Depos.* **2003**, *38*, 739–750. [CrossRef]

33. Wang, B.D.; Niu, S.Y.; Sun, A.Q.; Zhang, J.Z. *Deep Sources of Ore-forming Materials and Mantle Branch Structure in Association with Metallogenesis*; Geological Publishing House: Beijing, China, 2010; pp. 187–212. (In Chinese)

34. Niu, S.Y.; Hu, H.B.; Sun, A.Q. *Shandong Mantle Branch and Its Ore-Forming, Ore-Controlling, Ore-Prospecting*; Geological Publishing House: Beijing, China, 2017; pp. 1–256. (In Chinese)

35. Hua, R.M.; Mao, J.W. A preliminary discussion on the Mesozoic metallogenic explosion in east China. *Miner. Depos.* **1999**, *18*, 300–307, (In Chinese with English Abstract).

36. Lv, G.X.; Guo, T.; Shu, B.; Shen, Y.K.; Liu, D.J.; Zhou, G.F.; Ding, Y.X.; Wu, J.C.; Zhao, K.G.; Sun, Z.F.; et al. Study on the multi-level controlling rule for thctonic system in Jiaodong gold-centralized area. *Geotecton. Metallog.* **2007**, *2*, 193–204, (In Chinese with English Abstract).

37. Sun, A.Q.; Niu, S.Y.; Zhang, J.Z.; Chen, C.; Ma, B.J.; Wang, B.D.; Zhang, X.F.; Zhang, F.X.; Liu, C. Analysis on the ore-controlling of Guojialing mantle branch structure in northwest Shangdong. *Gold Sci. Technol.* **2012**, *20*, 1–7, (In Chinese with English Abstract).

38. Leech, M.L.; Webb, L.E. Is the HP–UHP Hong'an–Dabie–Sulu orogen a piercing point for offset on the Tan–Lu fault? *J. Asian Earth Sci.* **2013**, *63*, 112–129. [CrossRef]

39. Deng, J.; Yang, L.Q.; Sun, Z.S.; Wang, J.P.; Wang, Q.F.; Xin, H.B.; Li, X.J. Ametallogenic model of gold deposits of the Jiaodong granite–green-stone belt. *Acta Geol. Sin.* **2003**, *77*, 537–546.

40. Deng, J.; Yang, L.Q.; Ge, L.S.; Wang, Q.F.; Zhang, J.; Gao, B.F.; Zhou, Y.H.; Jiang, S.Q. Research advances in the Mesozoic tectonic regimes during the formation of Jiaodong ore cluster area. *Prog. Nat. Sci.* **2006**, *16*, 777–784.

41. Ling, Y.Y.; Zhang, J.J.; Liu, K.; Ge, M.H.; Wang, M.; Wang, J.M. Geochemistry, geochronology, and tectonic setting of Early Cretaceous volcanic rocks in the northern segment of the Tan–Lu Fault region, northeast China. *J. Asian Earth Sci.* **2017**, *144*, 303–322. [CrossRef]

42. Yan, G.H.; Cai, J.H.; Ren, K.X.; Mu, B.L.; Li, F.T.; Chu, Z.Y. Nd, Sr and Pb isotopic geochemistry of late-Mesozoic alkaline-rich intrusions from the Tan-lu Fault zone: Evidence of the magma source. *Acta Petrol. Sin.* **2008**, *24*, 1223–1236.

43. Zhu, G.; Liu, C.; Gu, C.C.; Zhang, S.; Li, Y.J.; Su, N.; Xiao, S.Y. Oceanic plate subduction history in the western Pacific Ocean: Constraint from late Mesozoic evolution of the Tan-Lu Fault Zone. *Sci. China (Earth Sci.)* **2018**, *6*, 386–405. [CrossRef]

44. Zhang, J.J.; Ding, Z.J.; Liu, D.H.; Zhang, P.J.; Zou, J.; Ma, B.; Luan, G.D. Exploration practice and prospecting results of super-large gold mine of Sanshandao norther sea area in Laizhou city, Shandong province. *Gold Sci. Technol.* **2016**, *24*, 1–10, (In Chinese with English Abstract).

45. Niu, S.Y.; Sun, A.Q.; Zhang, J.Z.; Ma, B.J.; Wang, B.D.; Chen, C.; Zhang, F.X.; Liu, C.; Zhang, X.F. Discussion on the deep ore-controlling structure in gold deposit concentrated area of Northwestern Jiaodong. *Acta Geol. Sin.* **2011**, *85*, 1094–1107, (In Chinese with English Abstract).

46. Song, M.C.; Song, Y.X.; Ding, Z.J.; Wei, X.F.; Sun, Z.L.; Song, G.Z.; Zhang, J.J.; Zhang, P.J.; Wang, Y.G. The discovery of the Jiaojia and the Sanshandao giant gold geposits in Jiaodong peninsula and Discussion on relevant issues. *Geotecton. Metallog.* **2019**, *43*, 92–110, (In Chinese with English Abstract).

47. Song, M.C. The main achievements and key theory and methods of deep-seated prospecting in the Jiaodong gold concentration, Shandong province. *Geol. Bull. China* **2015**, *34*, 1758–1771. (In Chinese with English Abstract)

48. Liu, L.L. Characteristics of Ore-Controlling Structures and Distribution of Mineralization Intensity in the Sanshandao Gold Deposit, Jiaodong Peninsula, China. Master's Thesis, China University of Geosciences (Beijing), Beijing, China, 2017. (In Chinese with English Abstract)

49. Liu, X.P.; Feng, T.; Deng, Q.H.; Lei, Y.X.; Wang, X. Geological characteristics and prospecting direction of Xiling gold deposit in northwest Jiaodong. *Gold* **2017**, *38*, 15–18, (In Chinese with English Abstract).

50. Zhang, X.O.; Cawood, P.A.; Wilde, S.A.; Liu, R.Q.; Song, H.L.; Li, W.; Snee, L.W. Geology and timing of mineralization at the Cangshang gold deposit, north-western Jiaodong Peninsula, China. *Miner. Depos.* **2003**, *38*, 141–153. [CrossRef]

51. Li, J.W. Mesozoic large scale mineralization in Jiaodong gold deposits: Chronology and geodynamics setting. In Proceedings of the Petrology and Geodynamics Seminar of China, Haikou, China, 19–24 November 2004; pp. 97–100. (In Chinese with English Abstract)

52. Wang, L.G.; Qiu, Y.M.; McNaughton, N.J.; Groves, D.I.; Luo, Z.K.; Huang, J.Z.; Miao, L.C.; Liu, Y.K. Constraints on crustal evolution and gold metallogeny in the Northwestern Jiaodong Peninsula, China, from SHRIMP U–Pb zircon studies of granitoids. *Ore Geol. Rev.* **1998**, *13*, 275–291. [CrossRef]

53. Li, R.H. Ore-controlling model of the Jiaojia gold belt, Shandong province. Ph.D. Thesis, China University of Geosciences (Beijing), Beijing, China, 2017. (In Chinese with English Abstract).

54. Yang, J.H.; Zhou, X.H.; Chen, L.H. Dating of gold mineralization for super-large tectonite-type gold deposits in northwestern Jiaodong peninsula and its implications for gold metallogeny. *Acta Petrol. Sin.* **2000**, *16*, 454–458. (In Chinese with English Abstract)

55. Zhang, A.N. Study on ore-controlling regularity of Hexi fault and metallogenic regularities of gold deposits in Jiaodong. Master's Thesis, China University of Geosciences (Beijing), Beijing, China, 2017. (In Chinese with English Abstract)

56. Hou, L.M.; Jiang, S.Y.; Jiang, Y.H.; Ling, H.F. S-Pb Isotope geochemistry and Rb-Sr geochronology of the Penglai gold field in the eastern Shandong province. *Acta Petrol. Sin.* **2006**, *22*, 2525–2533. (In Chinese with English Abstract)

57. Song, M.C.; Cui, S.X.; Yi, P.H.; Xu, J.X.; Yuan, W.H.; Jiang, H.L. *Prospecting and Metallogenetic Model of Deep Large-Superlarge Gold Deposit in the Gold Concentration Area of Northwest Jiaodong*; Geological Publishing House: Beijing, China, 2010; pp. 1–339. (In Chinese)

58. Li, S.X.; Liu, C.C.; An, Y.H.; Wang, W.C.; Huang, T.L.; Yang, C.H. *Jiaodong Gold Deposit Geology*; Geological Publishing House: Beijing, China, 2007; pp. 1–423. (In Chinese)

59. Sha, D.M.; Yuan, L.H. Thecharacteristics, distribution and prospect of epithermal gold deposits. *Geol. Resour.* **2003**, *12*, 115–124. (In Chinese with English Abstract)

60. Yang, Z.F.; Zhao, L.S.; Zhou, Q.M.; Xu, J.K. Physicochemical constrains on mineralization of epithermal gold deposits in Muping-Rushan mineralization zone, eastern Shandong Province, China. *Acta Mineral. Sin.* **1994**, *14*, 270–278. (In Chinese with English Abstract)

61. Ye, T.Z.; Xue, J.L. Geological research in the deep ore prospecting of metal deposit. *Geol. China.* **2007**, *34*, 855–869.

62. Teng, J.W. Material-energy exchange within the Earth and resources and disasters. *Earth Sci. Front.* **2001**, *8*, 1–8.

63. Niu, S.Y.; Sun, A.Q.; Liu, X.H.; Zhang, J.Z.; Zhang, F.X.; Chen, C.; Ma, B.J.; Hou, J.L. Structural characteristics and ore-controlling of Beibo gold deposit in Jiaodong. *Gold Sci. Technol.* **2016**, *24*, 1–7. (In Chinese with English Abstract)

64. Wilson, J.T. A possible origin of the Hawaiian Islands. *Can. J. Phys.* **1963**, *41*, 863–866. [CrossRef]

65. Morgan, W.J. Deep mantle convection plumes and plate motions. *Bull. Am. Assoc. Petrol. Geol.* **1972**, *56*, 203–213.

66. Zhang, Z.C.; Wang, F.S.; Hao, Y.L. Picrites from the Emeishan Large Igneous Province: Evidence for the Mantle Plume Activity. *Acta Metall. Sin.* **2005**, *24*, 17–22.

67. Niu, Y.L. On the great plume debate. *Chin. Sci. Bull.* **2005**, *50*, 1537–1540. [CrossRef]

68. Davies, G.F. Mantle plumes, mantle stirring and hotspot chemistry. *Earth Planet. Sci. Lett.* **1990**, *99*, 94–109. [CrossRef]

69. Hou, Z.Q.; Lu, J.R.; Lin, S.Z. The Axial Zone Consisting of Pyrolite and Eclogite in the Emei Mantle Plume: Major, Trace Element and Sr-Nd-Pb Isotope Evidence. *Acta Geol. Sin.* **2005**, *79*, 200–219. (In Chinese with English Abstract)

70. Boss, A.P.; Sack, I.S. Formation and growth of deep mantle plume. *Geophys. J. R. Astron. Soc.* **1985**, *80*, 241–255. [CrossRef]

71. Anderson, D.L. The thermal state of the upper mantle; no role for mantle plumes. *Geophys. Res. Lett.* **2000**, *27*, 3623–3626. [CrossRef]

72. French, S.W.; Romanowicz, B. Broad plumes rooted at the base of the Earth's mantle beneath major hotspots. *Nature* **2015**, *525*, 95–99. [CrossRef] [PubMed]

73. Gerya, T.V.; Stern, R.J.; Baes, M.; Sobolev, S.V.; Whattam, S.A. Plate tectonics on the Earth triggered by plume-induced subduction initiation. *Nature* **2015**, *527*, 221–225. [CrossRef]

74. Caruso, S.; Fiorentini, M.L.; Hollis, S.P.; Laflamme, C.; Baumgartner, R.J.; Steadman, J.A.; Savard, D. The fluid evolution of the Nimbus Ag-Zn-(Au) deposit: An interplay between mantle plume and microbial activity. *Precambrian Res.* **2018**, *317*, 211–229. [CrossRef]

75. Dobretsov, N.L.; Borisenko, A.S.; Izokn, A.E.; Zhmodik, S.M. A thermochemical model of Eurasian Permo-Triassic mantle plumes as a basis for prediction and exploration for Cu-Ni-PGE and rare-metal ore deposits. *Russ. Geol. Geophys.* **2010**, *51*, 903–924. [CrossRef]

76. Kuzmin, M.I.; Yarmolyuk, V.V. Mantle plumes of Central Asia (Northeast Asia) and their role in forming endogenous deposits. *Russ. Geol. Geophys.* **2015**, *55*, 120–143. [CrossRef]

77. Berzina, A.P.; Berzina, A.N.; Gimon, V.O. The Sora porphyry Cu-Mo deposit (Kuznetsk Alatau): Magmatism and effect of mantle plume on the development of ore-magmatic system. *Russ. Geol. Geophys.* **2011**, *52*, 1553–1562. [CrossRef]

78. Wang, D.H. Basic concept, classification, evolution of mantle plume and large scale mineralization- probe into Southwestern China. *Earth Sci. Front. (China Univ. Geosci. Beijing)* **2001**, *8*, 67–72. (In Chinese with English Abstract)

79. Xu, YG.; Wang, Y.; Wei, X.; He, B. Mantle plumerelated mineralization and their principal controlling factors. *Acta Petrol. Sin.* **2013**, *29*, 3307–3322. (In Chinese with English Abstract)

80. Niu, S.Y.; Cheng, G.S.; Zhang, J.Z.; Sun, A.Q.; Ma, B.J.; Zhang, F.X.; Wang, B.D.; Xu, M.; Wu, J.C.; Zhao, R.X.; et al. Study on the Metallogenetism of Sub-mantle Plume and Mantle Branches in the Gold Mineralization Concentration Area of Northwest Jiaodong Peninsula. *Acta Geol. Sin. (Engl. Ed.)* **2014**, *88*, 1409–1420. [CrossRef]

81. Niu, S.Y.; Hou, Q.L.; Hou, Z.Q.; Sun, A.Q.; Wang, B.D.; Li, H.Y.; Xu, C.S. Cascaded Evolution of Mantle Plumes and Metallogenesis of Core- and Mantle-derived Elements. *Acta Geol. Sin. (Engl. Ed.)* **2003**, *77*, 522–536.

82. Niu, S.Y.; Guo, L.J.; Liu, J.M.; Shao, J.A.; Wang, B.D.; Hu, H.B.; Sun, A.Q.; Wang, S. Tectonic ore-controlling in the middle-southern segment of Da Hinggan Ling, Northeast China. *Chin. J. Geochem.* **2006**, *25*, 62–70.

83. Wang, B.D.; Niu, S.Y.; Sun, A.Q.; Zhang, J.Z.; Wang, X.; Wang, C.E. Isotope tracers for deep-seated fluids and noble gases. *Chin. J. Geochem.* **2013**, *32*, 195–202. [CrossRef]

84. Wang, B.D.; Niu, S.Y.; Sun, A.Q.; Li, H.Y. The isotopic composition of noble gases in gold deposits and the source of ore-forming materials in the region of North Hebei, China. *Chin. J. Geochem.* **2003**, *22*, 313–319.

85. Chen, C.; Niu, S.Y.; Wang, Z.L.; Guo, Z.; Sun, A.Q.; Wang, B.D.; Gao, Y.C. Comparison between the Xishimen gold deposit the Shihu gold deposit in western Hebei analysis of the potential for ore exploration. *Chin. J. Geochem.* **2011**, *30*, 281–289. [CrossRef]

86. Zhu, G.; Liu, G.S.; Niu, M.L.; Song, C.Z.; Wang, D.X. Transcurrent movement and genesis of the Tan-Lu fault zone. *Geol. Bull. China* **2003**, *22*, 200–207.

87. Ye, J.L.; Huang, D.H.; Zhang, J.X. *Introduction to Geology*; Geological Publishing House: Beijing, China, 1996. (In Chinese)

88. Niu, S.Y.; Sun, A.Q.; Shao, A.G.; Wang, B.D.; Zhao, M.H.; Wang, L.F.; Jiang, W.; Xu, C.S. *The Multiple Evolution of Mantle Plume and Its Mineralization*; Seismological Press: Beijing, China, 2001; pp. 41–175. (In Chinese)

89. Huo, M.Y. A tomic structure model of gold and its mineralization model. In *Gold Economics*; Tu, G.C., Huo, M.Y., Eds.; Science Press: Beijing, China, 1991; pp. 1–7. (In Chinese)

90. Niu, S.Y.; Sun, A.Q.; Wang, B.D. *Mantle Plume and Natural Resources Environment*; Geological Publishing House: Beijing, China, 2007; pp. 1–183. (In Chinese with English abstract)

Article

Lithospheric Architecture and Metallogenesis in Liaodong Peninsula, North China Craton: Insights from Zircon Hf-Nd Isotope Mapping

Zhichao Zhang [1], Yuwang Wang [1,*], Dedong Li [1] and Chunkit Lai [2]

[1] Beijing Institute of Geology for Mineral Resources, Beijing 100012, China; ZZC_CUGB@126.com (Z.Z.); lidedong2005@126.com (D.L.)

[2] Faculty of Science, Universiti Brunei Darussalam, Gadong BE1410, Brunei; chunkit.lai@ubd.edu.bn

* Correspondence: wyw@cnncm.com

Received: 17 January 2019; Accepted: 8 March 2019; Published: 14 March 2019

Abstract: The Liaodong Peninsula is an important mineral province in northern China. Elucidating its lithospheric architecture and structural evolution is important for gold metallogenic research and exploration in the region. In this study, Hf-Nd isotope maps from magmatic rocks are constructed and compared to geological maps to correlate isotopic signatures with geological features. It is found that gold deposits of different age periods in Liaodong are located in areas with specific εHf(t) and εNd ranges (Triassic: from −8 to −4 and from −12 to −8, Jurassic: from −22 to −8 and from −14 to −8, Cretaceous: from −12 to −10 and from −22 to −20), respectively. This may reflect that when the Paleo-Pacific plate was subducted beneath the North China Craton, the magma was derived from the juvenile lower crust and the ancient lower crust, and formed the low-to-moderate hydrothermal Au-(Ag) and Pb-Zn deposits in the Triassic. In the Jurassic, continued subduction may have led to lithospheric thickening. Subsequently, the magma from the ancient lower crust upwelled and formed low-to-moderate hydrothermal Au deposits and porphyry Mo deposits. In the Cretaceous, crustal delamination may have taken place. The magma from the ancient lower crust upwelled and formed various low-to-moderate hydrothermal Au deposits.

Keywords: lithospheric architecture; metallogenesis; Hf-Nd isotopic mapping; Liaodong Peninsula; North China Craton

---

1. Introduction

The North China Craton (NCC), containing the Liaodong and Jiaodong gold provinces, is the top gold producer in northeast Asia [1–6]. The Liaodong Peninsula is located between the Yalujiang and TanLu fault zones (Figure 1) [7–9] and represents an important mineral province in the NCC. The peninsula has undergone complex magmato-tectonic modifications, during which many important polymetallic (Pb-Zn, Au, Ag, and Mo) deposits have been formed (Figures 1 and 2) [10,11]. Most of these deposits are interpreted to be genetically linked with granitoid in the peninsula [11]. Granitoid in the Liaodong peninsula include the diorite and the granite that formed in the Paleoproterozoic, Permian, Jurassic, and Cretaceous [7,12,13]. These many phases of magmatism provide a window into the study of the lithospheric architecture and its control on metallogenesis.

Tectonic evolution and the characteristics of gold deposits in Liaodong and Jiaodong are similar [14–18]; however, whether or not the lithospheric architecture played a role in controlling the tectonic evolution and gold ore formation remains poorly understood.

The Hf and Nd isotopes are powerful tools to trace the nature of basement rocks and the age of the continental crust [19–21], and Hf-Nd isotope mapping has been used to reveal the lithospheric architecture and evolution, and their control on the distribution of mineral deposits [22–29].

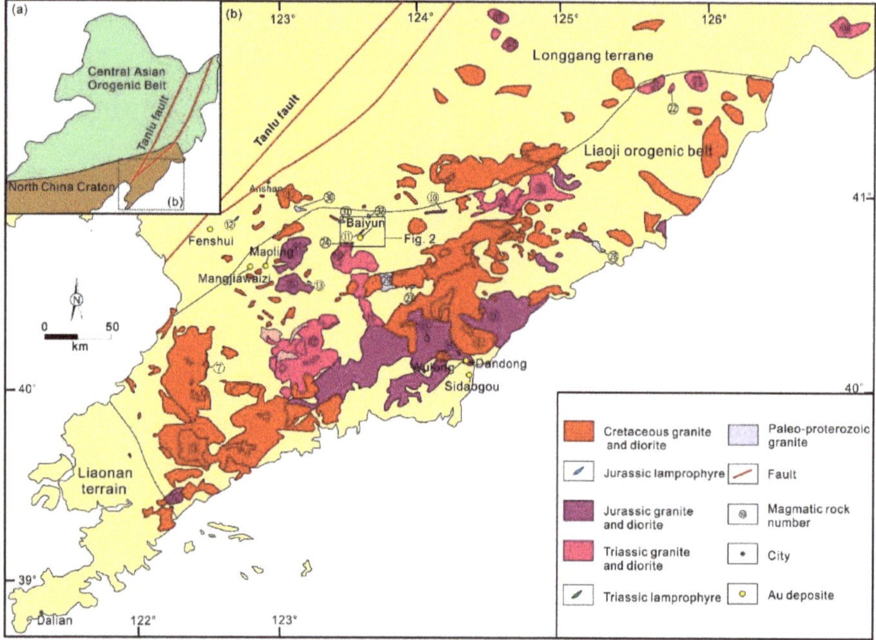

Figure 1. (**a**) Simplified tectonic map of the Liaodong Peninsula showing the major suture zones and blocks. (**b**) Geological map of the Liaodong Peninsula showing the distribution of magmatic rocks, and the locations of major mineral deposits [30].

In this study, we summarize the spatial distribution, age, and geochemical and isotopic data of the Paleoproterozoic to Cretaceous magmatic rocks in the Liaodong Peninsula, and we use Hf-Nd isotope mapping to reveal the crustal architecture and its controls on the regional mineralization.

2. Geological Setting

2.1. Regional Tectonics

The Liaodong Peninsula is located in the eastern margin of the NCC (Figure 1). It is bounded by the Yalujiang fault in the east and by the Tanlu fault in the north [31,32]. The Liaodong Peninsula can also be subdivided into the Longgang terrane in the north, the Liaoji orogenic belt in the middle, and the Langlin terrane in the south. This study only focuses on the Longgang terrane and the Liaoji orogenic belt. The Longgang terrane is composed of Archean to Paleoproterozoic basement rocks, and unmetamorphosed Mesoproterozoic to Cenozoic sedimentary and volcanic rocks [7]. The Liaoji orogenic belt consists mainly of Paleoproterozoic to Cretaceous magmatic rocks. In the Longgang terrane, the Paleoproterozoic sequences are missing, and the magmatic rocks are largely Triassic (Figure 3) [33].

2.2. Magmatism

The Liaodong Peninsula consists of Paleoproterozoic granite, Triassic granite and diorite, Jurassic granite and diorite, and Cretaceous granite and diorite (Table 1 and Table S1) (Figures 4 and 5) [34–39]. During the Paleoproterozoic, the voluminous granitoid and the mafic intrusions in the peninsula were emplaced (Figure 5) and then metamorphosed at 1.93 Ga [40], marking the cratonization of the NCC eastern block. The Triassic magmatism is characterized by metaluminous mafic and felsic magmatic rocks (Figure 5), which are also identified in the southern Liaodong Peninsula [7,30]. Late Mesozoic intrusive rocks include Jurassic (180–153 Ma) ductile-deformed, peraluminous/metaluminous granite (Figure 5), and undeformed to slightly deformed early Cretaceous (131–120 Ma) metaluminous granite and diorite (Figure 5) [7,41,42].

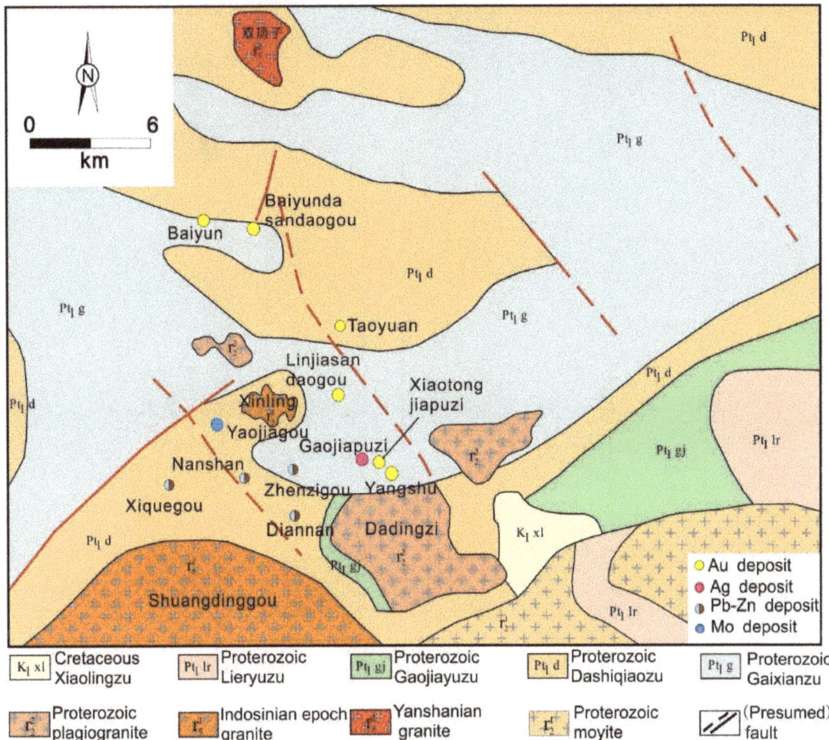

Figure 2. Simplified geologic map of the Qingchengzi orefield showing the distribution of deposits [11].

2.3. Mineralization

The Liaodong Peninsula contains Pb-Zn, Au, Ag, and Mo polymetallic deposits, which are mainly distributed in the Qingchengzi, Wulong, and Maoling orefields (Figure 1) (Table 2) [11,43,44]. The Qingchengzi orefield is in the northern part of the Liaodong Peninsula, which hosts a number of magmatic-hydrothermal (low-to-moderate hydrothermal) Au-(Ag) and Pb-Zn deposits and porphyry Mo deposits (Figure 2) [45,46]. The magmatic-hydrothermal Au-(Ag) deposits were mainly formed in the Triassic (225–240 Ma), as exemplified by the Baiyun and Yangshu deposits (Table 2). The mineralization of these deposits has been correlated to the granite and the diorite, which are the result of lithospheric thinning associated with the Paleo-Pacific plate subduction [30,35]. The magmatic-hydrothermal Pb-Zn deposits (e.g., Xiquegou and Zhenzigou) were also formed in the Triassic (221–232 Ma), whilst the Yaojiagou porphyry Mo deposit was formed in the Jurassic (168 Ma). The mineralization of these Pb-Zn deposits has been correlated to the granite and the diorite, and that of the Mo deposit has been correlated to the granite. The Pb-Zn and Mo deposits have been correlated to large-scale delamination [7,35].

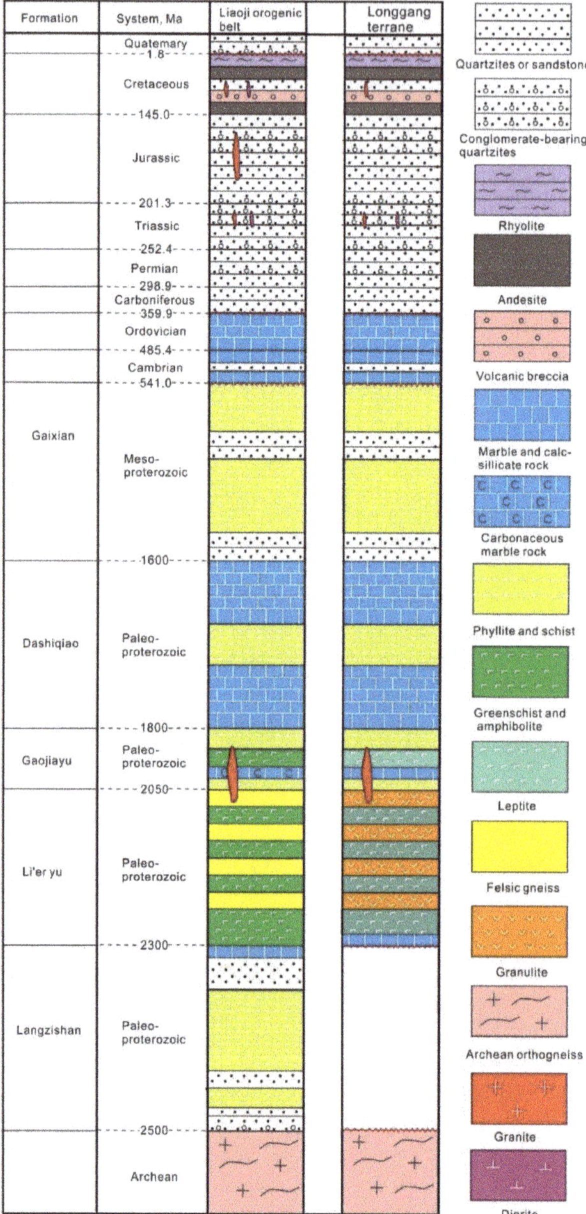

Figure 3. Stratigraphic columns showing the basement rocks, sedimentary cover, and magmatic history of the Longgang Terrane and the Liaoji orogenic belt.

Table 1. Zircon U-Pb ages for the magmatic rocks from the Liaodong Peninsula.

Period	No.	Pluton	Phase	Age/Ma	Sample	Method	References
Cretaceous	1	Wulongbei	Quartz diorite	126–127	Zircon	SHRIMP U-Pb	[7]
	2	Sanguliu	Porphyritic granite	125 ± 3	Zircon	SHRIMP U-Pb	[7]
	3	Yinmawanshan	Gneissic granodiorite	122 ± 2	Zircon	LA-ICP-MS U-Pb	[7]
		Yinmawanshan	Monzogranite (dike)	124 ± 5	Zircon	LA-ICP-MS U-Pb	[7]
		Yinmawanshan	Monzogranite	122 ± 6	Zircon	LA-ICP-MS U-Pb	[7]
	4	Qianshan	Granite	126 ± 2	Zircon	LA-ICP-MS U-Pb	[32]
Jurassic	5	Xiaoheshan	Granodiorite	$173–174 \pm 4$	Zircon	LA-ICP-MS U-Pb	[7]
	6	Hanjialing	Granodiorite	179 ± 3	Zircon	LA-ICP-MS U-Pb	[7]
		Hanjialing	Monzogranite	164 ± 4	Zircon	LA-ICP-MS U-Pb	[7]
	7	Yutun	Mylonitic granite	157 ± 3	Zircon	LA-ICP-MS U-Pb	[7]
	8	Heigou	Monzogranite	$161 \pm 6, 163 \pm 7$	Zircon	LA-ICP-MS U-Pb	[7]
	9	Jiuliancheng	Monzogranite	156 ± 3	Zircon	LA-ICP-MS U-Pb	[7]
	10	Gaoliduntai	Plagiogranite	156 ± 5	Zircon	LA-ICP-MS U-Pb	[7]
	11	Baiyun gold mine	Porphyritic dyke	168 ± 3	Zircon	LA-ICP-MS U-Pb	[7]
	12	Huaziyu	Lamprophyres	155 ± 4	Zircon	LA-ICP-MS U-Pb	[47]
	13	Waling	Monzonitic granite	162.4 ± 1.9	Zircon	SHRIMP U-Pb	[36]
	14	Dandong	Granite	157–167	Zircon	LA-ICP-MS U-Pb	[14]
Triassic	15	Shuangdinggou	biotite monzogranite	224.2 ± 1.2	Zircon	LA-ICP-MS U-Pb	[9]
	16	Xinling	Granites	225.3 ± 1.8	Zircon	SHRIMP U-Pb	[8]
	17	Xiuyan	Monzogranite	210 ± 1	Zircon	LA-ICP-MS U-Pb	[30]
	18	Nankouqian	Monzogranite	224 ± 2	Zircon	SIMS	[12]
		Nankouqian	Monzogranite	221 ± 2	Zircon	LA-ICP-MS U-Pb	[12]
		Nankouqian	Granite	224 ± 1	Zircon	LA-ICP-MS U-Pb	[48]

Table 1. *Cont.*

Period	No.	Pluton	Phase	Age/Ma	Sample	Method	References
Triassic	19	Mayihe	Pyroxene diorite	222 ± 2	Zircon	SIMS	[12]
			Pyroxene syenodiorite	223 ± 2	Zircon	SIMS	[12]
			Fine-grained diorite	222 ± 2	Zircon	SIMS	[12]
			Biotite monzogranite	$220 \pm 2, 223 \pm 3, 221 \pm 2$	Zircon	LA-ICP-MS U-Pb	[12]
	20	Xidadingzi	Monzogranite	$220 \pm 2, 221 \pm 2$	Zircon	LA-ICP-MS U-Pb	[12]
	21	Chaxinzi	Monzogranite	$222 \pm 2, 219 \pm 2$	Zircon	LA-ICP-MS U-Pb	[12]
			Diorite	219 ± 4	Zircon	LA-ICP-MS U-Pb	[12]
			Monzodiorite	222 ± 2	Zircon	LA-ICP-MS U-Pb	[12]
			Diorite	221 ± 2	Zircon	SIMS	[12]
			Granodiorite	222 ± 1	Zircon	SIMS	[12]
	22	Xiaoweishahe	Granodiorite	218 ± 2	Zircon	LA-ICP-MS U-Pb	[12]
			Quartz diorite	$220 \pm 2, 219 \pm 4$	Zircon	LA-ICP-MS U-Pb	[12]
	23	Longtou	Granodiorite	224 ± 2	Zircon	SIMS	[12]
			Granodiorite	220 ± 2	Zircon	LA-ICP-MS U-Pb	[12]
			Fine-grained granite	221 ± 2	Zircon	LA-ICP-MS U-Pb	[12]
	24	Qingchengzi	Lamprophyres	224–230	Zircon	LA-ICP-MS U-Pb	[9]
	25	Saima	Syenite	222 ± 3.4	Zircon	LA-ICP-MS U-Pb	[49]
			Syenite	221 ± 2.3	Zircon	LA-ICP-MS U-Pb	[49]
	26	Bailinchuan	Syenite	221 ± 2.3	Zircon	LA-ICP-MS U-Pb	[49]
Paleo-proterozoic	27	Jiguanshan	Granite	2175 ± 13	Zircon	SHRIMP U-Pb	[38,50]
	28	Laoheishan	Granite	2166 ± 14	Zircon	SHRIMP U-Pb	[38,50]
	29	Dadingzi	Granite	1869 ± 16	Zircon	SHRIMP U-Pb	[39]
	30	Wuleishan	Granite	1830.5 ± 5.9	Zircon	SHRIMP U-Pb	[39]
	31	Simenzi	Granite	2157 ± 14	Zircon	SHRIMP U-Pb	[39]
	32	Gujiapu	Granite	2169 ± 11	Zircon	SHRIMP U-Pb	[39]

Table 2. Summary of the geological characteristics of major ore deposits in the Liaodong Peninsula.

Number	Deposits	Orefield	Type	Metallic Comm.	Tonnage (t)	Grade	Host Rock	Age (Ma)	Data Source
1	Zhenzigou	Qingchengzi	Magmatic hydrothermal	Pb-Zn		0.37, 450	Marble, Amphibolite, Schist	221	[8,51]
2	Nanshan	Qingchengzi	Magmatic hydrothermal	Pb-Zn		0.5, 153		227	[9,11]
3	Diannan	Qingchengzi	Magmatic hydrothermal	Pb-Zn		0.08, 650		232	[11]
4	Xiquegou	Qingchengzi	Magmatic hydrothermal	Pb-Zn		0.28, 250		225	[8,11]
5	Baiyun	Qingchengzi	Magmatic hydrothermal	Au	31.7	2.85 g/t	Metamorphic rock and quartz veins	225	[10,11]
6	Xiaotongjiapuzi	Qingchengzi	Magmatic hydrothermal	Au-Ag	20-50	0.07~2.92, 0.14~6.12	Marble	239	[11,43,52,53]
7	Gaojiapuzi	Qingchengzi	Magmatic hydrothermal	Ag		312	Marble	240	[11,45,52,54]
8	Yangshu	Qingchengzi	Magmatic hydrothermal	Au-Ag	3.72	1.61, 3.72	Metamorphic rock and marble		[11,55]
9	Taoyuan	Qingchengzi	Magmatic hydrothermal	Au-Ag		0.005~0.06, 0.0025~0.1	Metamorphic rock		[11,56]
10	Baiyundasandaogou	Qingchengzi	Magmatic hydrothermal	Au-Ag		7.28, 1.28			[11,54]
11	Linjiasandaogou	Qingchengzi	Magmatic hydrothermal	Au		0.031, 0.034	Metamorphic rock		[11,56]
12	Yaojiagou	Qingchengzi	porphyry	Mo		0.34	Metamorphic rock and skarn	168	[57,58]
13	Sidaogou	Wulong	Magmatic hydrothermal	Au	20-50		Metamorphic rock		[43]
14	Wulong	Wulong	Magmatic hydrothermal	Au	>40		Metamorphic rock and quartz veins	122	[43,59]
15	Wangjiawaizi	Maoling	Magmatic hydrothermal	Au-Ag	>5	8.9, 16.9	Metamorphic rock, quartz veins and Breccia		[60]
16	Maoling	Maoling	Magmatic hydrothermal	Au	25	3.2 g/t	Metamorphic rock and quartz veins	196	[44]
17	Fenshui	Maoling	Magmatic hydrothermal	Au	1.8	3~5	Quartz veins	186	[61]

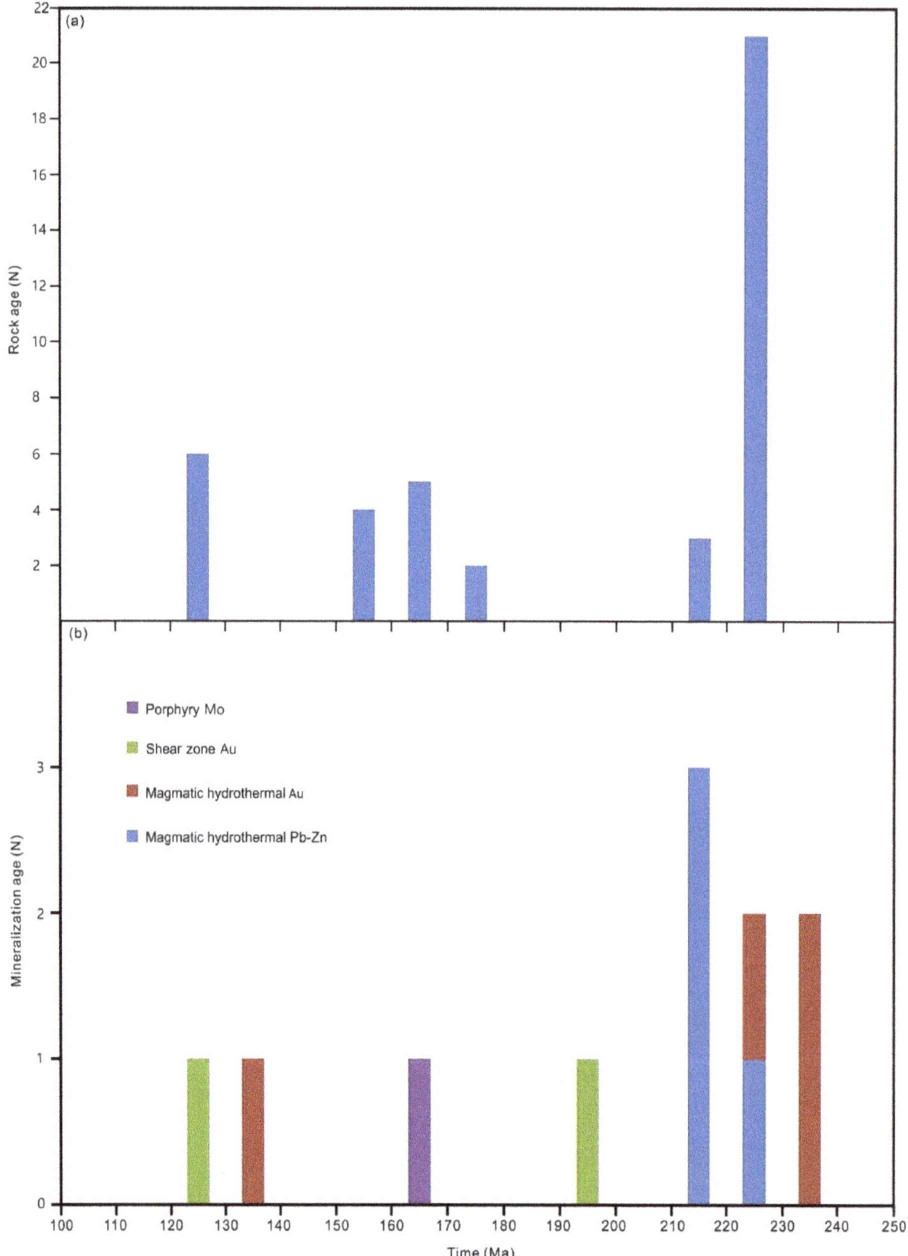

Figure 4. (**a**) Histogram of geochronological dating of zircon U-Pb ages of magmatic rocks. (**b**) Histogram of geochronological dating of mineralization.

The Wulong orefield contains the Wulong and Sidaogou magmatic-hydrothermal (low-to-moderate hydrothermal) Au deposits. The largest Wulong deposit was formed at 122 Ma [62], whilst the Sidaogou deposit in southern Liaodong has no reliable mineralization age data.

The Maoling orefield contains the Jurassic, Maoling, Fenshui, and Wangjiawaizi magmatic-hydrothermal (low-to-moderate hydrothermal) Au deposits (186–189 Ma) [44,60,61].

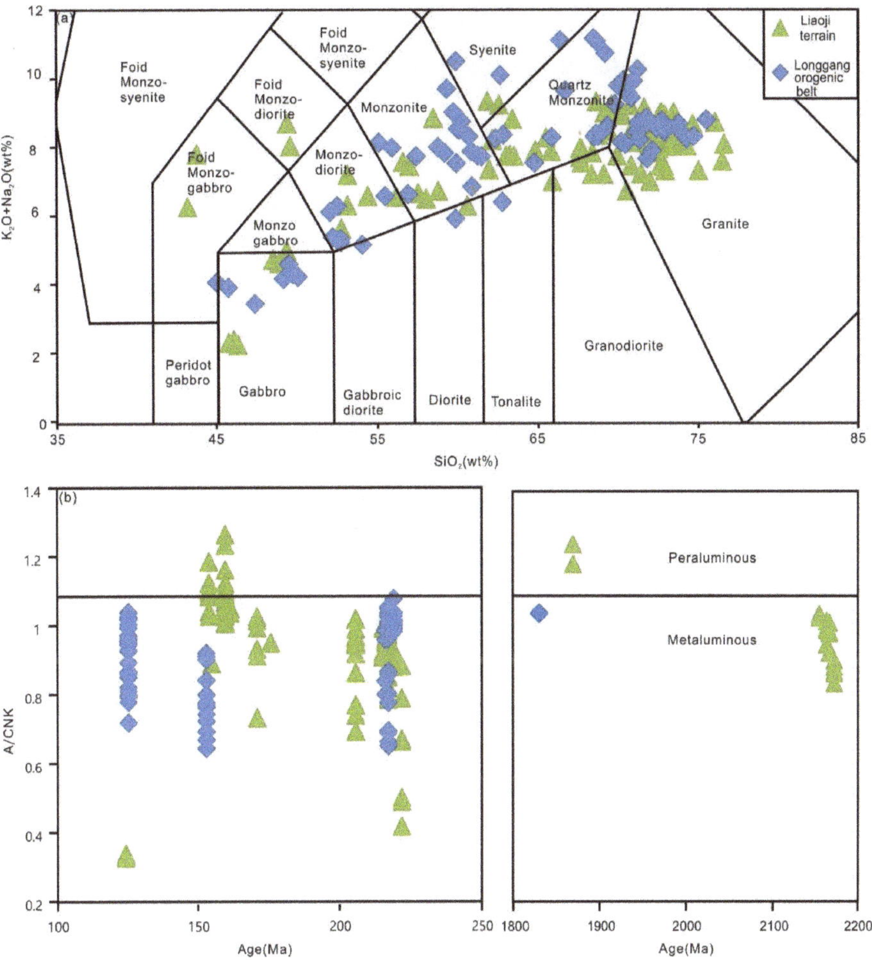

Figure 5. (**a**) Total alkali (Na$_2$O + K$_2$O) versus SiO$_2$ diagram. (**b**) A/CNK (molecular Al$_2$O$_3$/(CaO + Na$_2$O + K$_2$O)) versus zircon U-Pb age diagram. Dates are in Table 1 and the Supplementary Material Table S1.

3. Methods

Published zircon U-Pb age and Hf isotope data (35 samples) and whole-rock Sr-Nd isotope data (35 samples) from the Liaodong Peninsula have been compiled. Data compiled were from Paleoproterozoic to Cretaceous rocks, including (porphyritic) granite, monzogranite, lamprophyre, plagiogranite, (gneissic) granodiorite, (quartz) diorite, and syenodiorite. New data were also added in this study via collecting and analyzing (for zircon Lu-Hf isotopes) samples from the Qingchengzi orefield. The Zircon U-Pb age of the Miaonangou gabbro near the Baiyun gold deposit was in 1252 Ma [63], and the porphyrie (diorite, monzogranite) in the Baiyun deposit were emplaced in 229–222 Ma [63,64]. The Gujiapuzi granite porphyry in the Qingchengzi orefield was in 219 Ma [63].

Zircon Hf isotopes were analyzed using a 193-nm laser ablation (LA) system attached to a Neptune multi-collector (MC)-ICP-MS (Laboratory of Isotope Geology, Tianjin Institute of Geology and Mineral

Resources, China). A laser pulse (100 mJ energy, 10 Hz frequency, 50 μm beam size) was used for the laser ablation [65]. Isobaric interference of ^{176}Lu on ^{176}Hf was corrected on the basis of the measured ^{175}Lu value and the recommended ^{176}Lu/^{175}Lu ratio of 0.02655. Similarly, the ^{176}Yb/^{172}Yb value of 0.5887 and mean β Yb value obtained during Hf analysis on the same spot were used for interference correction of ^{176}Yb on ^{176}Hf [66,67]. A ^{176}Lu decay constant of 1.865 × 10^{-11}-year^{-1} [68] and the chondritic ratios of ^{176}Hf/^{177}Hf = 0.282785 and ^{176}Lu/^{177}Hf = 0.0336 [69] were used to calculate the εHf(t) values [70–72].

The Hf-Nd contour maps were produced using the inverse distance weighted interpolation method in the MapGIS 6.7 (Manufacturer is Zondy Cyber, Wuhan, China) program to contour the Hf and Nd dataset, which accounts for the distance between sample points in the most representative manner [73–75]. In order to produce the most robust spatial representation of the isotopic dataset, this method used 12 nearest neighbors at a power of 2 following [24]. All isotope data were grouped by the geometric interval method designated for class breaks. This ensured that each class range had approximately the same number of values, and that the change between intervals was fairly consistent. All point data shown in the contour maps represent the median for a range of Hf-Nd isotope values from an individual sample, which helped to minimize data anomalies [23,28,76].

4. Results

4.1. Zircon Hf Isotope Features

Zircon εHf(t) values of the Longgang terrane vary from −18.9 to 5.8 (average −3.3), and the old crustal Hf model ages (T$_{DM}^{C}$) range from 994 to 2058 Ma (average 1349 Ma). Zircon εHf(t) values of the Liaoji orogenic belt vary from −33 to 11.7 (average −11.4), and the T$_{DM}^{C}$ range from 763 to 2785 Ma (average 1449 Ma).

For the Paleoproterozoic rocks, the εHf(t) and T$_{DM}^{C}$ ranged from −17.4 to 7.9 (average −0.8) and from 2036 to 3874 Ma (average 2948 Ma), respectively. For the Triassic rocks, the zircon εHf(t) and T$_{DM}^{C}$ ranged from −18.9 to 5.2 (average −11.5) and from 763 to 2613 Ma (average 1422 Ma), respectively. For the Jurassic rocks, the zircon εHf(t) and T$_{DM}^{C}$ range from −28.9 to −1.1 (average −16.3) and from 1505 to 2785 Ma (average 2041 Ma), respectively (Table S2).

Contour maps of the zircon εHf(t) values for the Paleoproterozoic-Cretaceous Liaodong magmatic rocks show four high εHf(t) domains and two low εHf(t) domains, among which two high εHf(t) domains are in the Longgang terrane, and the other two are in the Liaoji orogenic belt (Figure 6). There are two low εHf(t) domains in the Longgang terrane and the Liaoji orogenic belt, respectively (Figure 6).

4.2. Whole-Rock Sr-Nd Isotope Features

The (^{87}Sr/^{86}Sr)$_i$ ratios of the Liaoji orogenic belt and the Longgang terrane range from 0.7044 to 0.7215 (average 0.7098) and from 0.7037 to 0.7306 (average 0.7085), respectively. The εNd values of the Liaoji belt and the Longgang terrane range from −24.9 to −0.9 (average −14.1) and from −18.9 to 3.82 (average −5.1), respectively (Table S3).

For the Cretaceous rocks, the εNd and T$_{DM}$ values range from −19.3 to −11.9 (average −15.1) and from 1388 to 2191 Ma (average 1831 Ma). For the Jurassic rocks, the (^{87}Sr/^{86}Sr)$_i$, εNd, and T$_{DM}$ values range from 0.7044 to 0.7215 (average 0.7104), −24.9 to −9.6 (average −11.2), and 1110 to 2826 Ma (average 1888 Ma), respectively. For the Triassic rocks, the (^{87}Sr/^{86}Sr)$_i$, εNd, and T$_{DM}$ values range from 0.7037 to 0.7306 (average 0.7082), −18.9 to 3.82 (average −9.5), and 726 to 2290 Ma (average 1541 Ma), respectively. For the Paleoproterozoic rocks, the εNd and T$_{DM}$ values range from −16.2 to −0.9 (average −5.8) and from 2480 to 2813 Ma (average 2279 Ma), respectively (Table S3).

The Contour maps of whole-rock Nd isotopes for the Paleoproterozoic-Cretaceous magmatic rocks show three high εNd domains and four low εNd domains in the region. One high εNd and one low εNd domain are in the Longgang terrane, and the other two high εNd domains and three low εNd domains are in the Liaoji belt (Figure 7).

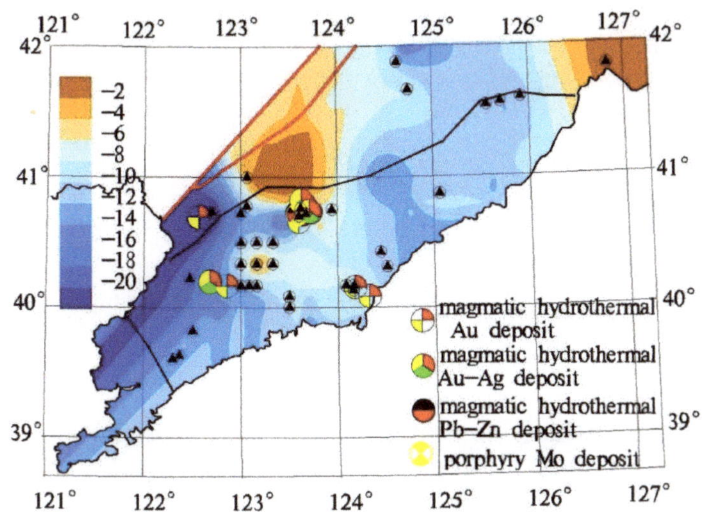

Figure 6. Contour maps of the Hf isotope for the magmatic rocks in the Liaodong Peninsula. Data are from Supplementary Material Table S2.

5. Discussion

5.1. Lithospheric Architecture of the Liaodong Peninsula

In the Longgang terrane, there are two domains characterized by high εHf values (Figure 6), that are present in the area of the Triassic and Paleoproterozoic granite and diorite (Figure 1) and indicate that the granite and the diorite of this area are derived from the mantle or juvenile lower crust. There are also two domains characterized by low-εHf in the Longgang terrane (Figure 6), that indicate the magmatic rocks of this area are derived from the lower crust. In the Liaoji orogenic belt, there are two domains characterized by high-εHf (Figure 6), which are present in the area of Triassic and Paleoproterozoic granite and diorite (Figure 1), and indicate that the rocks are derived from the mantle or juvenile lower crust. There are also two low-εHf domains in the Liaoji orogenic belt (Figure 6) that indicate the crustal origin of the magmatic rocks in this area.

There is one high εNd domain in the Longgang terrane, and there are two high εNd domains in the Liaoji orogenic belt (Figure 7). However, the εNd values of the two domains in the Liaoji orogenic belt are still below zero. Therefore, the magmatic rocks in the Liaoji orogenic belt are derived from the lower crust. The high εNd domain in the Longgang terrane is present in the area of the Triassic granite and the diorite (Figure 7). This also can indicate that the granite and the diorite are mostly derived from the mantle or juvenile lower crust.

In the Longgang terrane, the area of the Triassic granite and the diorite with high-εHf and high-εNd shows that the magmatic rocks are derived from the juvenile lower crust (Figure 8). The high $(^{87}Sr/^{86}Sr)_i$ ratios also support this evidence (Figure 9). The area of Paleoproterozoic granite and diorite with high-εHf in the Longgang terrane shows that the magmatic rocks are from juvenile lower crust (Figure 8). In the Liaoji orogenic belt, the area of Triassic granite and diorite with high-εHf shows that the magmatic rocks are from the juvenile lower crust (Figure 8). The area of Paleoproterozoic granite and diorite with high-εHf shows that the magmatic rocks are from the depleted mantle (Figure 8).

In the Paleoproterozoic, the Tanlu fault may have experienced dextral shear movement, and the intense regional extension creating the Liaodong rift valley [77], although the actual timing and number of stages (argued variably from four to six) of the rifting process remain controversial. The timing of rifting is also variably attributed from 2.3 to 1.7 Ga or from 2.3 to 1.8 Ga [9,77]. The magmatic rocks in the Longgang terrane are from the juvenile lower crust and the ancient lower crust, but the magmatic rocks in the Liaoji orogenic belt are from the depleted mantle and the ancient lower crust (Figures 6–8).

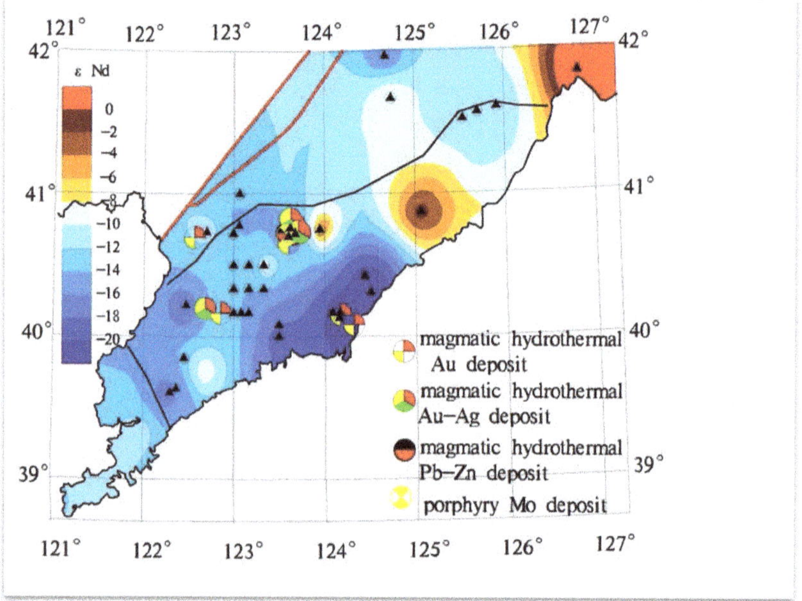

Figure 7. Contour maps of the Nd isotope for the magmatic rocks in the Liaodong Peninsula. Data are from Supplementary Material Table S3.

In the Mesozoic, the Liaodong Peninsula was likely in a post-collisional extensional setting [9,78,79], and the Paleo-Pacific plate may have subducted beneath the NCC [6,9]. Zircon and monazite SHRIMP U-Pb dating suggested that the continental collision took place in 220–240 Ma [80–82]. In the Triassic, the collision between North China and Paleo-Pacific plate likely caused the lithospheric thickening in the Liaoji rift [30,35,83]. The Triassic magmatic rocks are derived from the juvenile lower crust and the ancient lower crust (Figures 6–9). The rocks have positive whole rock εNd(t) and zircon εHf(t) values, indicating a juvenile lower crustal source. In addition, the Triassic magmatic rocks with high SiO_2 contents and low MgO concentrations have strong negative and variable whole rock εNd(t) and zircon εHf(t) values, indicating that they were derived from partial melting of the ancient lower crustal materials with involvement of mantle components [12]. In the Jurassic, the lithospheric thickening continued [1,84–88]. The sources of the magmatic rocks are from the ancient lower crust (Figures 6–9). The rocks with strong negative and variable whole rock εNd(t) and zircon εHf(t) values indicate that they were derived from partial melting of the Precambrian basement [7]. In the Cretaceous, large-scale delamination may have taken place [35,89]. The magmatic rocks have the same characteristics of whole rock εNd(t) and zircon εHf(t) values as those of Cretaceous magmatic rocks. Therefore, the sources of the magmatic rocks are also derived from the ancient lower crust (Figures 6, 7 and 9).

5.2. Regional Tectonic Evolution and Relation to Mineralization

The Triassic is the principal metallogenic epoch in Liaodong. Deposits formed in the Triassic include the Zhenzigou, Nanshan, Diannan, and Xiquegou low-to-moderate hydrothermal Pb-Zn deposits, the Baiyun and Xiaotongjiapuzi low-to-moderate hydrothermal Au-Ag deposits, and the Gaojiapuzi low-to-moderate hydrothermal Ag deposits, that are all located in the Qingchengzi orefield (Table 2). The mineralization ages (221–240 Ma) are consistent with the magmatic ages (210–230 Ma), suggestive of a magmatic-hydrothermal genesis for these deposits [8]. The Sr and Pb isotope characteristics of the deposits in the Qingchengzi orefield show that the ore-forming materials were derived from the magma and metamorphosed sequences [8,64]. The deposits of the Qingchengzi orefield are clustered in regions with high-εHf (Figures 6, 8 and 9). This infers that the deposits are correlated to the magma, and that the magma is derived from the juvenile lower crust and the ancient lower crust (Figure 10a).

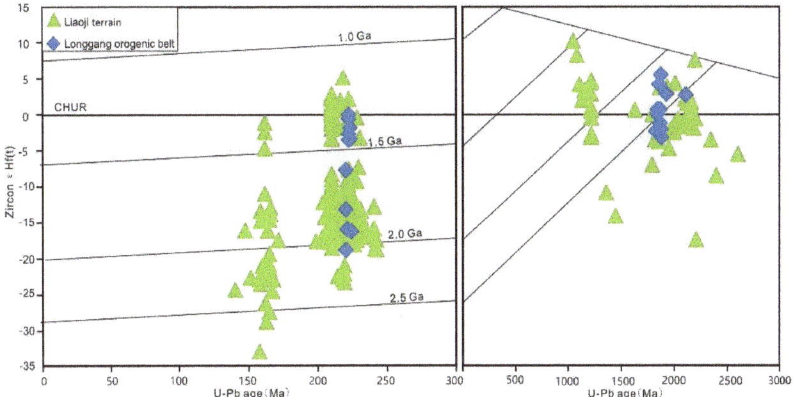

Figure 8. Plot of zircon U-Pb age versus εHf(t) values for the magmatic rocks from the Liaodong Peninsula. Data are from Supplementary Material Table S2.

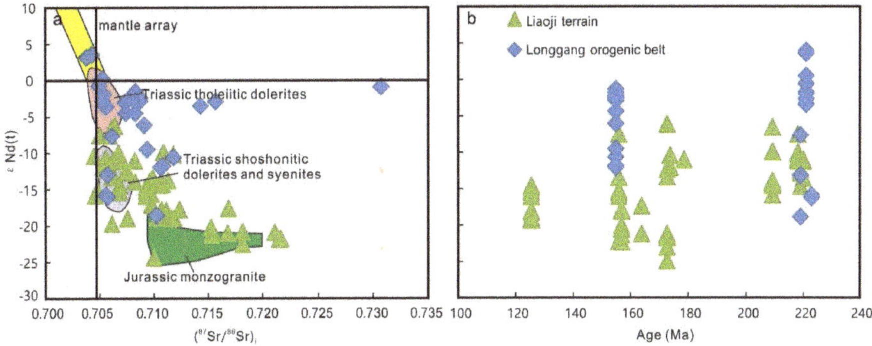

Figure 9. (**a**) Plot of whole-rock (^{87}Sr/^{86}Sr)$_i$ versus εNd(t) values for the magmatic rocks from the Liaodong Peninsula. (**b**) Plot of age versus whole-rock εNd(t) values for the magmatic rocks from the Liaodong Peninsula. Data are from Supplementary Material Table S3.

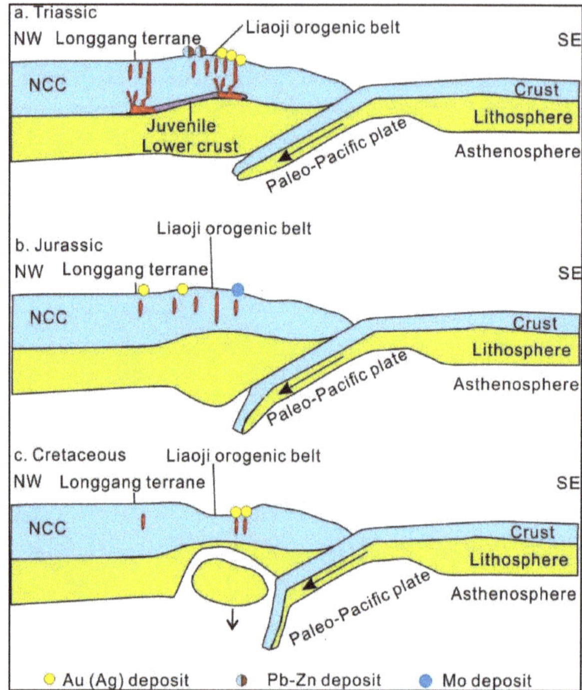

Figure 10. (**a**) Lithospheric architecture of the Liaodong Peninsula in the Triassic. (**b**) Lithospheric architecture of the Liaodong Peninsula in the Jurassic. (**c**) Lithospheric architecture of the Liaodong Peninsula in the Cretaceous.

In the Jurassic, the Maoling and Fenshui low-to-moderate hydrothermal Au deposits were formed in the Maoling orefield, and the Yaojiagou porphyry Mo deposit was formed in the Qingchengzi orefield (Table 2). Sulfur isotopes from the Miaoling deposit show typical magmatic sulfur characteristics [90]. The Pb isotope characteristics of the deposit show that the ore-forming materials came from the magma and the metamorphic sequences [90]. The deposits are clustered in regions with low εHf and εNd values (Figures 6–9). This infers that the deposits are correlated to the magma, which is derived from the ancient lower crust (Figure 10b).

The Cretaceous is another important metallogenic epoch. The Wulong orefield contains Wulong and Sidaogou low-to-moderate hydrothermal Au deposits. The characteristics of Sr and Pb isotopices suggest that the rock- and ore-forming and diagenetic materials of the Sanguliu granite near the Wulong orefield were derived from the magmatic rocks [62]. The H-O isotopes characteristics demonstrate that the ore-forming fluid came from magmatic fluid [62]. The deposits are clustered in regions with low εHf and εNd values (Figures 6–9). This infers that the deposits are correlated to the magma, which is derived from the ancient lower crust (Figure 10c).

6. Conclusions

In the Triassic, the Paleo-Pacific plate subducted beneath the NCC and caused the lithospheric thickening. The Triassic ore deposits are characterized by high εHf(t) values, and are correlated to the magma, which is derived from the juvenile lower crust and the ancient lower crust. In the Jurassic, the lithospheric thickening continued. The Jurassic ore deposits are characterized by low εHf(t) and εNd values and are correlated to the magma derived from the ancient lower crust. In the Cretaceous, large-scale delamination may have taken place in this period. The Cretaceous ore deposits are characterized by low εHf(t) and εNd values, and are correlated to the magma, which is derived from the ancient lower crust.

Supplementary Materials: The following are available online at http://www.mdpi.com/2075-163X/9/3/179/s1, Table S1: Major elements datas, Table S2: Hf isotope, Table S3: Sr-Nd isotope.

Author Contributions: Z.Z. and Y.W. conceived and designed the ideas; Z.Z. and D.L. collected and analyzed the data; Z.Z. wrote the manuscript; Y.W., D.L. and C.L. revised and edited the manuscript.

Funding: This study was financially supported by the National Key R&D Program of China (Grant No. 2018YFC0603804).

Acknowledgments: We thank the geologists from Team 103 of the Bureau of Non-ferrous Metal Geology (Dandong, Liaoning Province) for their field assistance.

Conflicts of Interest: The authors declare no conflict of interest.

References

1. Deng, J.; Wang, Q.F. Gold mineralization in China: Metallogenic provinces, deposit types and tectonic framework. *Gondwana Res.* **2016**, *36*, 219–274. [CrossRef]

2. Deng, J.; Liu, X.F.; Wang, Q.F.; Dilek, K.; Yang, L.Q. Isotopic characterization and petrogenetic modeling of early cretaceous mafic diking-lithospheric extension in the North China Craton, Eastern Asia. *GSA Bull.* **2017**, *129*, 1379–1407. [CrossRef]

3. Yang, L.Q.; Deng, J.; Guo, L.N.; Wang, Z.L.; Li, X.Z.; Li, J.L. Origin and evolution of ore fluid, and gold deposition processes at the giant Taishang gold deposit, Jiaodong Peninsula, Eastern China. *Ore Geol. Rev.* **2016**, *72*, 585–602. [CrossRef]

4. Yang, L.Q.; Deng, J.; Wang, Z.L.; Zhang, L.; Aufarb, R.J.; Yuan, W.M.; Weinberg, R.F.; Zhang, R.Z. Thermochronologic constraints on evolution of the Linglong Metamorphic Core Complex and implications for Au mineralization: A case study from the Xiadian Au deposit, Jiaodong Peninsula, eastern China. *Ore Geol. Rev.* **2016**, *72*, 165–178. [CrossRef]

5. Yang, L.Q.; Deng, J.; Wang, Z.L.; Guo, L.N.; Li, R.H.; Groves, D.I.; Danyushevsky, L.; Zhang, C.; Zheng, X.L.; Zhao, H. Relationships between gold and pyrite at the Xincheng gold deposit, Jiaodong Peninsula, China: Implications for gold source and deposition in a brittle epizonal environment. *Econ. Geol.* **2016**, *111*, 105–126. [CrossRef]

6. Yang, L.Q.; He, W.Y.; Gao, X.; Xie, S.X.; Yang, Z. Mesozoic multiple magmatism and porphyry-skarn Cu-polymetallic systems of the Yidun Terrane, Eastern Tethys: Implications for subduction- and transtension-related metallogeny. *Gondwana Res.* **2018**, *62*, 144–162. [CrossRef]

7. Wu, F.Y.; Yang, J.H.; Wilde, S.A.; Zhang, X.O. Geochronology, petrogenesis and tectonic implications of Jurassic granites in the Liaodong Peninsula, NE China. *Chem. Geol.* **2005**, *221*, 127–156. [CrossRef]

8. Yu, G.; Chen, J.F.; Xue, C.J.; Chen, Y.C.; Chen, F.K.; Du, X.Y. Geochronological framework and Pb, Sr isotope geochemistry of the Qingchengzi Pb-Zn-Ag-Au orefield, Northeastern China. *Ore Geol. Rev.* **2009**, *35*, 367–382. [CrossRef]

9. Duan, X.; Zeng, Q.; Yang, J.; Liu, J.; Wang, Y.; Zhou, L. Geochronology, geochemistry and Hf isotope of Late Triassic magmatic rocks of Qingchengzi district in Liaodong Peninsula, Northeast China. *J. Asian Earth Sci.* **2014**, *91*, 107–124. [CrossRef]

10. Zhang, P.; Li, B.; Li, J.; Chai, P.; Wang, X.; Sha, D.; Shi, J.M. Re-Os isotopic dating and its geological implication of gold bearing pyrite from the Baiyun gold deposit in Liaodong rift. *Geotecton. Metallog.* **2016**, *40*, 731–738. (In Chinese with English Abstract)

11. Wang, Y.W.; Xie, H.J.; Li, D.D.; Shi, Y.; Liu, F.X.; Sun, G.Q.; Sun, Q.M.; Zhou, G.C. Prospecting prediction of ore concentration area exemplified by Qingchengzi Pb-Zn-Au-Ag ore concentration area, Eastern Liaoning Province. *Miner. Depos.* **2017**, *36*, 1–24. (In Chinese with English Abstract)

12. Yang, J.H.; Sun, J.F.; Zhang, J.H.; Wilde, S.A. Petrogenesis of Late Triassic intrusive rocks in the Northern Liaodong Peninsula related to decratonization of the North China Craton: Zircon U-Pb age and Hf-O isotope evidence. *Lithos* **2012**, *153*, 108–128. [CrossRef]

13. Xiao, S.Y.; Zhu, G.; Zhang, S.; Liu, C.; Su, N.; Yi, H.; Wu, X.D.; Li, Y.J. Structural processes and dike emplacement mechanism in the Wulong gold field, eastern Liaoning. *Chin. Sci. Bull.* **2018**, *63*, 3022–3036. (In Chinese with English Abstract) [CrossRef]

14. Li, S.Z.; Liu, J.Z.; Zhao, G.C.; Wu, F.Y.; Han, Z.Z.; Yang, Z.Z. Key geochronology of Mesozoic deformation in the eastern block of the North China Craton and its constraints on regional tectonics: A case of Jiaodong and Liaodong Peninsula. *Acta Petrol. Sin.* **2004**, *20*, 633–646. (In Chinese with English Abstract)

15. Deng, J.; Wang, Q.F.; Li, G.J. Tectonic evolution, superimposed orogeny, and composite metallogenic systemin China. *Gondwana Res.* **2017**, *50*, 216–266. [CrossRef]

16. Deng, J.; Yang, L.Q.; Li, R.H.; Groves, D.I.; Santosh, M.; Wang, Z.L.; Sai, S.X.; Wang, S.R. Regional structural control on the distribution of world-class gold deposits: An overview from the giant Jiaodong Gold Province. *China Geol. J.* **2017**. [CrossRef]

17. Yang, L.Q.; Deng, J.; Qiu, K.F.; Ji, X.Z.; Santosh, M.; Song, K.R.; Song, Y.H.; Geng, J.Z.; Zhang, C.; Hua, B. Magma mixing and crust-mantle interaction in the Triassic monzogranites of Bikou Terrane, central China: Constraints from petrology, geochemistry, and zircon U-Pb-Hf isotopic systematics. *J. Asian Earth Sci.* **2015**, *98*, 320–341. [CrossRef]

18. Yang, L.Q.; Guo, L.N.; Wang, Z.L.; Zhao, R.X.; Song, M.C.; Zheng, X.L. Timing and mechanism of gold mineralization at the Wang'ershan gold deposit, Jiaodong Peninsula, Eastern China. *Ore Geol. Rev.* **2017**, *88*, 491–510. [CrossRef]

19. Griffin, W.L.; Wang, X.; Jackson, S.E.; Pearson, N.J.; O'Reilly, S.Y.; Xu, X.S.; Zhou, X.M. Zircon chemistry and magma mixing, SE China: In-situ analysis of Hf isotopes, Tonglu and Pingtan igneous complexes. *Lithos* **2002**, *61*, 237–269. [CrossRef]

20. Griffin, W.L.; Belousova, E.A.; Shee, S.R.; Pearson, N.J.; O'reilly, S.Y. Archean crustal evolution in the northern Yilgarn Craton: U-Pb and Hf-isotope evidence from detrital zircons. *Precambrian Res.* **2004**, *131*, 231–282. [CrossRef]

21. Kemp, A.I.S.; Hawkesworth, C.J.; Paterson, B.A.; Kinny, P.D. Episodic growth of the Gondwana supercontinent from hafnium and oxygen isotopes in zircon. *Nature* **2006**, *439*, 580–583. [CrossRef] [PubMed]

22. McCuaig, T.C.; Beresford, S.; Hronsky, J. Translating the mineral systems approach into an effective exploration targeting system. *Ore Geol. Rev.* **2010**, *38*, 128–138. [CrossRef]

23. Mole, D.R.; Fiorentini, M.L.; Thebaud, N.; McCuaig, T.C.; Cassidy, K.F.; Kirkland, C.L.; Wingate, M.T.D.; Romano, S.S.; Doublier, M.P.; Belousova, E.A. Spatiotemporal constraints on lithospheric development in the Southwest-Central Yilgarn Craton, Western Australia. Australian. *J. Earth Sci.* **2012**, *59*, 625–656.

24. Mole, D.R.; Fiorentini, M.L.; Thebaud, N.; Cassidy, K.F.; McCuaig, T.C.; Kirkland, C.L.; Romano, S.S.; Doublier, M.P.; Belousova, E.A.; Barnes, S.J.; et al. Archean komatiite volcanism controlled by the evolution of early continents. *Proc. Natl. Acad. Sci. USA* **2014**, *111*, 10083–10088. [CrossRef] [PubMed]

25. Mole, D.R.; Fiorentini, M.L.; Cassidy, K.F.; Kirkland, C.L.; Romano, N.; Maas, R.; Belousova, E.A.; Barnes, S.J.; Mill, J. Crustal Evolution, Intra-cratonic Architecture and the Metallogeny of an Archaean Craton. *Geol. Soc. Lond. Spec. Pub.* **2015**, *393*, 23–80. [CrossRef]

26. Belousova, E.A.; Kostitsyn, Y.A.; Griffin, W.L.; Begg, G.C.; OReilly, S.Y.; Pearson, N.J. The growth of the continental crust: Constraints from zircon Hf-isotope data. *Lithos* **2010**, *119*, 457–466. [CrossRef]

27. Hou, Z.Q.; Duan, L.F.; Lu, Y.J.; Zheng, Y.C.; Zhu, D.C.; Yang, Z.M.; Yang, Z.S.; Wang, B.D.; Pei, Y.R.; Zhao, Z.D.; et al. Lithospheric architecture of the Lhasa Terran and its control on ore deposits in the Himalayan-Tibetan Orogen. *Econ. Geol.* **2015**, *110*, 1541–1575. [CrossRef]

28. Wang, C.M.; Bagas, L.; Lu, Y.J.; Santosh, M.; Du, B.; McCuaig, T.C. Terrane boundary and spatio-temporal distribution of ore deposits in the Sanjiang Tethyan Orogen: Insights from zircon Hf-isotopic mapping. *Earth-Sci. Rev.* **2016**, *156*, 39–65. [CrossRef]

29. Wang, C.M.; Deng, J.; Bagas, L.; Wang, Q.F. Zircon Hf-isotopic mapping for understanding crustal architecture and metallogenesis in the Eastern Qinling Orogen. *Gondwana Res.* **2017**, *50*, 293–310. [CrossRef]

30. Yang, J.H.; Wu, F.Y.; Wilde, S.A.; Liu, X.M. Petrogenesis of Late Triassic granitoids and their enclaves with implications for post-collisional lithospheric thinning of the Liaodong Peninsula, North China Craton. *Chem. Geol.* **2007**, *242*, 155–175. [CrossRef]

31. Suo, A.; Zhao, D.Z.; Zhang, F.S.; Wang, H.R.; Liu, F.Q. Driving forces and management strategies for estuaries in Northern China. *Front. Earth Sci. China* **2010**, *4*, 51–58. [CrossRef]

32. Yang, J.H.; Wu, F.Y.; Chung, S.L.; Wilde, S.A.; Chu, M.F. A hybrid origin for the Qianshan A-type granite, Northeast China: Geochemical and Sr-Nd-Hf isotopic evidence. *Lithos* **2006**, *89*, 89–106. [CrossRef]
33. Li, S.Z.; Zhao, G.C.; Santosh, M.; Liu, X.; Dai, L.M. Palaeoproterozoic tectonothermal evolution and deep crustal processes in the Jiao-Liao-Ji belt, North China Craton: A review. *Geol. J.* **2011**, *46*, 525–543. [CrossRef]
34. Wang, Y.; Wang, F.; Wu, L.; Shi, W.; Yang, L. (U-Th)/He thermochronology of metallic ore deposits in the Liaodong Peninsula: Implications for orefield evolution in Northeast China. *Ore Geol. Rev.* **2018**, *92*, 348–365. [CrossRef]
35. Jiang, Y.H.; Jiang, S.Y.; Ling, H.F.; Ni, P. Petrogenesis and tectonic implications of Late Jurassic shoshonitic lamprophyre dikes from the Liaodong Peninsula, NE China. *Miner. Petrol.* **2010**, *100*, 127–151. [CrossRef]
36. Yang, F.C.; Song, Y.H.; Hao, L.B.; Chai, P. Late Jurassic shrimp U-Pb age and Hf isotopic characteristics of granite from the Sanjiazi area in Liaodong and their geological significance. *Acta Geol. Sin.* **2015**, *89*, 1773–1782. (In Chinese with English Abstract)
37. Pei, F.P. Zircon U-Pb Chronology and Geochemistry of Mesozoic Intrusive Rocks in Southern Liaoning and Jilin Provinces: Constraints on the Spatial-Temporal Extent of the North China Craton Destruction. Ph.D. Thesis, Jilin University, Changchun, China, 2008. (In Chinese with English Abstract).
38. Li, C.; Chen, B.; Li, Z.; Yang, C. Petrologic and geochemical characteristics of Paleoproterozoic monzogranitic gneisses from Xiuyan-Kuandian area in Liaodong Peninsula and their tectonic implications. *Acta Petrol. Sin.* **2017**, *33*, 963–977. (In Chinese with English Abstract)
39. Song, Y.; Yang, F.; Yan, G.; Wei, M.; Shi, S. Shrimp U-Pb ages and Hf isotopic compositions of Paleoproterozoic granites from the Eastern part of Liaoning Province and their tectonic significance. *Acta Geol. Sin.* **2016**, *90*, 2620–2636. (In Chinese with English Abstract)
40. Luo, Y.; Sun, M.; Zhao, G.C.; Li, S.Z.; Xu, P.; Ye, K.; Xia, X.P. LAICP-MS U-Pb zircon ages of the Liaohe Group in the Eastern Block of the North China Craton, constraints on the evolution of the Jiao-Liao-Ji Belt. *Precambrian Res.* **2004**, *134*, 349–371. [CrossRef]
41. Liu, J.L.; Ji, M.; Shen, L.; Guan, H.M.; Davis, G.A. Early Cretaceous extensional structures in the Liaodong peninsula: Structural associations, geochronological constraints and regional tectonic implications. *Sci. China* **2011**, *54*, 823–842. [CrossRef]
42. Wu, F.Y.; Lin, J.Q.; Wilde, S.A.; Zhang, X.O.; Yang, J.H. Nature and significance of the Early Cretaceous giant igneous event in Eastern China. *Earth Planet. Sci. Lett.* **2005**, *233*, 103–119. [CrossRef]
43. Yang, J.H.; Wu, F.Y.; Wilde, S.A. A review of the geodynamic setting of large-scale late mesozoic gold mineralization in the North China Craton: An association with lithospheric thinning. *Ore Geol. Rev.* **2003**, *23*, 125–152. [CrossRef]
44. Yu, G.; Yang, G.; Chen, J.; Qu, W.; Du, A.; He, W. Re-Os dating of gold-bearing arsenopyrite of the Maoling gold deposit, Liaoning province, Northeast China and its geological significance. *Chin. Sci. Bull.* **2005**, *50*, 1509–1514. [CrossRef]
45. Dai, J.Z. Characteristics of ore-forming fluids and discussion on the genesis of Au, Ag deposits in Qingchengzi region, Liaoning province. Master Dissertation, Jilin University, Changchun, China, 2005. (In Chinese with English abstract).
46. Li, D.D.; Wang, Y.W.; Zhou, G.C.; Shi, Y.; Xie, H.J. Preliminary analysis of the relationship between dikes and gold mineralization in Baiyun gold deposit, Liaoning. *Miner. Explor.* **2016**, *7*, 113–119, (In Chinese with English abstract).
47. Jing, Y.; Jiang, S.; Zhao, K.; Ni, P.; Ling, H.; Liu, D. Shrimp U-Pb zircon dating for lamprophyre from Liaodong Peninsula: Constraints on the initial time of Mesozoic lithosphere thinning beneath Eastern China. *Sci. Bull.* **2005**, *50*, 2612–2620. [CrossRef]
48. Liu, J.L.; Sun, F.Y.; Zhang, Y.J.; Ma, F.; Liu, F.X.; Zeng, L. Zircon U-Pb Geochronology, Geochemistry and Hf Isotopes of Nankouqian Granitic Intrusion in Qingyuan Region, Liaoning Province. *Earth Sci.* **2016**, *41*, 55–66. (In Chinese with English Abstract)
49. Song, J.Q. Geochemistry and Chronology of Alkaline Rocks in Fengcheng, Liaodong and Their Geological Implications. Master's Thesis, China University of Geosciences, Beijing, China, 2017. (In Chinese with English Abstract).
50. Li, S.Z.; Zhao, G.C. SHRIMP U-Pb zircon geochronology of the Liaoji granitoids: Constraints on the evolution of the Paleoproterozoic Jiao-Liao-Ji belt in the Eastern Block of the North China Craton. *Precambrian Res.* **2007**, *158*, 1–16. [CrossRef]

51. Ma, Y.B.; Bagas, L.; Xing, S.W.; Zhang, S.T.; Wang, R.J.; Li, N.; Zhang, Z.J.; Zou, Y.F.; Yang, X.Q.; Wang, Y.; et al. Genesis of the stratiform Zhenzigou Pb Zn deposit in the North China Craton: Rb Sr and C O S Pb isotope constraints. *Ore Geol. Rev.* **2016**, *79*, 88–104. [CrossRef]

52. Liu, G.P.; Ai, Y.F. Metamorphic rock-hosted disseminated gold deposits—A case study of the xiaotongjiapuzi gold deposit of eastern liaoning. *Acta Geol. Sin.-Engl.* **1999**, *73*, 429–437.

53. Xu, S.; Wang, M.; Liu, C.C.; Li, S.Y. Evaluation of gold geochemical anomalies in the liaodong paleorift. *Appl. Mech. Mater.* **2014**, *484–485*, 620–627. [CrossRef]

54. Chen, J.F.; Yu, G.; Xue, C.J.; Qian, H.; He, J.F.; Xing, Z.; Zhang, X. Pb isotope geochemistry of lead, zinc, gold and silver deposit clustered region, Liaodong rift zone, Northeastern China. *Sci. China* **2005**, *48*, 467–476. [CrossRef]

55. Zhang, W.Q.; Zheng, T.; Zhang, X.; Xu, D.N. Geological characteristics and metallogenic mechanism of Yangshu large-scale Au-Ag deposits in Qingchengzi orefield primary study. *Non-Ferr. Min. Metall.* **2007**, *23*, 4–6. (In Chinese with English Abstract)

56. Sun, G.Q.; Sun, Q.M.; Zheng, T. Geological characteristics and metallogenic mechanism of the Taoyuan gold deposit in Qingchengzi ore field, Liaoning. *Gansu Metall.* **2008**, *30*, 49–51. (In Chinese with English Abstract)

57. Fang, J.Q.; Nie, F.J.; Zhang, K.; Liu, Y.; Xu, B. Re-Os isotopic dating on molybdenite separates and its geological significance from the Yaojiagou molybdenum deposit, Liaoning Province. *Acta Petrol. Sin.* **2012**, *28*, 372–378. (In Chinese with English Abstract)

58. He, C.Z. The Characteristics of Metallogenic Fluid and Depths of Yaojiagou Porphyry-Mo Deposit in Qingchengzi Town, Liaoning Province. Master's Thesis, China University of Geosciences, Beijing, China, 2015. (In Chinese with English Abstract).

59. Yu, B.; Zeng, Q.D.; Frimmel, H.E.; Wang, Y.B.; Guo, W.K.; Sun, G.T.; Zhou, T.C.; Li, J.P. Genesis of the Wulong gold deposit, northeastern North China Craton: Constraints from fluid inclusions, H-O-S-Pb isotopes, and pyrite trace element concentrations. *Ore Geol. Rev.* **2018**, *102*, 313–337. [CrossRef]

60. Hao, R.X.; Peng, S.L. Geochemical study of Wangjiawaizi gold deposit in Liaoning. *Gold Geol.* **1999**, *5*, 47–51. (In Chinese with English Abstract)

61. Qu, F.F. Geology and genesis of Fenshui gold deposit. *J. Precious Met. Geol.* **2003**, *12*, 85–91. (In Chinese with English Abstract)

62. Wei, J.H.; Liu, C.Q.; Tang, H.F. Rb-Sr and U-Pb isotopic systematics of pyrite and granite in Liaodong gold province, North, China: Implication for the age and genesis of a gold deposit. *Geochem. J.* **2003**, *37*, 567–577. [CrossRef]

63. Zhou, G.C. Study on the Genesis of the Baiyun Gold Deposit in Liaodong, Liaoning Province. Master's Thesis, Kunming University of Science and Technology, Kunming, China, 2017. (In Chinese with English Abstract).

64. Liu, J.; Liu, F.X.; Li, S.H.; Lai, C.K. Formation of the Baiyun gold deposit, Liaodong gold province, NE China: Constraints from zircon U–Pb age, fluid inclusion, and C-H-O-Pb-He isotopes. *Ore Geol. Rev.* **2019**, *104*, 628–655. [CrossRef]

65. Qiu, K.F.; Taylor, R.D.; Song, Y.H.; Yu, H.C.; Song, K.R.; Li, N. Geologic and geochemical insights into the formation of the taiyangshan porphyry copper-molybdenum deposit, western qinling orogenic belt, China. *Gondwana Res.* **2016**, *35*, 40–58. [CrossRef]

66. Geng, J.Z.; Qiu, K.F.; Gou, Z.Y.; Yu, H.C. Tectonic regime switchover of Triassic Western Qinling orogen: Constraints from LA-ICP-MS zircon U–Pb geochronology and Lu–Hf isotope of Dangchuan intrusive complex in Gansu, China. *Chem. Erde Geochem.* **2017**, *77*, 637–651. [CrossRef]

67. Qiu, K.F.; Yu, H.C.; Gou, Z.Y.; Liang, Z.L.; Zhang, J.L.; Zhu, R. Nature and origin of triassic igneous activity in the western qinling orogen: The wenquan composite pluton example. *Int. Geol. Rev.* **2018**, *60*, 242–266. [CrossRef]

68. Scherer, E.; Münker, C.; Mezger, K. Calibration of the lutetium-hafnium clock. *Science* **2001**, *293*, 683–687. [CrossRef] [PubMed]

69. Blichert-Toft, J.; Albarède, F. The Lu–Hf isotope geochemistry of chondrites and the evolution of themantle–crust system. *Earth Planet. Sci. Lett.* **1997**, *148*, 243–258. [CrossRef]

70. Bouvier, A.; Vervoort, J.D.; Patchett, P.J. The Lu–Hf and Sm–Nd isotopic composition of CHUR: Constraints from unequilibrated chondrites and implications for the bulk composition of terrestrial planets. *Earth Planet. Sci. Lett.* **2008**, *273*, 48–57. [CrossRef]

71. Qiu, K.F.; Deng, J.; Taylor, R.D.; Song, K.R.; Song, Y.H.; Li, Q.Z.; Goldfarb, R.J. Paleozoic magmatism and porphyry cu-mineralization in an evolving tectonic setting in the north qilian orogenic belt, nw china. *J. Asian Earth Sci.* **2016**, *122*, 20–40. [CrossRef]

72. Qiu, K.F.; Deng, J. Petrogenesis of granitoids in the dewulu skarn copper deposit: Implications for the evolution of the paleotethys ocean and mineralization in western qinling, china. *Ore Geol. Rev.* **2017**, *99*, 1078–1098. [CrossRef]

73. Erdogan, S. A comparision of interpolation methods for producing digital elevation models at the field scale. *Earth Surf. Proc. Land.* **2010**, *34*, 366–376. [CrossRef]

74. Deng, J.; Wang, C.M.; Bagas, L.; Santosh, M.; Yao, E. Crustal architecture and metallogenesis in the south-eastern North China Craton. *Earth-Sci. Rev.* **2018**, *182*, 251–272. [CrossRef]

75. Gao, X.; Yang, L.Q.; Orovan, E.A. The lithospheric architecture of two subterranes in the eastern Yidun Terrane, East Tethys: Insights from Hf-Nd isotopic mapping. *Gondwana Res.* **2018**. [CrossRef]

76. Wang, C.M.; Bagas, L.; Deng, J.; Dong, M. Crustal architecture and its controls on mineralization in the North China Craton. *Ore Geol. Rev.* **2018**. [CrossRef]

77. Li, H.Y.; Xu, Y.G.; Liu, Y.M.; Huang, X.L.; He, B. Detrital zircons reveal no Jurassic plateau in the Eastern North China Craton. *Gondwana Res.* **2013**, *24*, 622–634. [CrossRef]

78. Hall, H.C.; Fahrig, W.F. Mafic dyke swarms. *Geol. Assoc. Can. Spec.* **1987**, *34*, 1–503.

79. Zhao, J.X.; McCulloch, M.T. Melting of a subduction-modified continental lithospheric mantle, evidence from late Proterozoic mafic dike swarms in central Australia. *Geology* **1993**, *21*, 463–466. [CrossRef]

80. Rowley, D.B.; Xue, F.; Tucker, R.D.; Peng, Z.X.; Baker, J.; Davis, A. Ages of ultrahigh pressure metamorphism and protolith orthogneisses from the Eastern Dabie Shan: U/Pb zircon geochronology. *Earth Planet. Sci. Lett.* **1997**, *151*, 191–203. [CrossRef]

81. Li, S.G.; Jagoutz, E.; Chen, Y. Sm-Nd and Rb-Sr isotopic chronology and cooling history of ultrahigh pressure metamorphic rocks and their country rocks at Shuanghe in the Dabie Mountains, central China. *Geochim. Cosmochim. Acta* **2000**, *64*, 1077–1093. [CrossRef]

82. Wan, Y.S.; Li, R.W.; Wilde, S.A.; Liu, D.Y.; Chen, Z.Y.; Yan, L.; Song, T.R.; Yin, X.Y. UHP metamorphism and exhumation of the Dabie Orogen, China, evidence from SHRIMP dating of zircon and monazite from UHP granitic gneiss cobble from the Hefei Basin. *Geochim. Cosmochim. Acta* **2005**, *69*, 4333–4348. [CrossRef]

83. Yang, J.H.; O'reilly, S.; Walker, R.J.; Griffin, W.; Wu, F.Y.; Zhang, M.; Pearson, N. Diachronous decratonization of the Sino-Korean Craton, geochemistry of mantle xenoliths from North Korea. *Geology* **2010**, *38*, 799–802. [CrossRef]

84. Gilder, S.A.; Gill, J.; Coe, R.S.; Zhao, X.X.; Liu, Z.; Wang, G. Isotopic and paleomagnetic constraints on the Mesozoic tectonic evolution of south China. *Geophys. Res.* **1996**, *101*, 16137–16154. [CrossRef]

85. Zhou, X.M.; Li, W.X. Origin of Late Mesozoic igneous rocks in Southeastern China: Implications for lithosphere subduction and underplating of mafic magmas. *Tectonophysics* **2000**, *326*, 269–287. [CrossRef]

86. Wang, R.; Tafti, R.; Hou, Z.Q.; Shen, Z.C.; Guo, N.; Evans, N.J.; Jeon, H.; Li, Q.Y.; Li, W.K. Across-arc geochemical variation in the Jurassic magmatic zone, Southern Tibet: Implication for continental arc-related porphyry Cu-Au mineralization. *Chem. Geol.* **2017**, *451*, 116–134. [CrossRef]

87. Li, X.H.; Chen, Z.G.; Liu, D.Y.; Li, W.X. Jurassic gabbro-granite-syenite suites from southern Jiangxi Province, SE China: Age, origin, and tectonic significance. *Int. Geol. Rev.* **2003**, *45*, 898–921. [CrossRef]

88. Zhuang, L.; Pei, F.P.; Meng, E. Zircon U-Pb age, geochemical and Nd isotopic data of Middle Jurassic high-mg dioritic dike in Liaodong peninsula, NE China. *Glob. Geol.* **2014**, *17*, 143–154.

89. Wilde, S.A.; Zhou, X.H.; Nemchin, A.A.; Sun, M. Mesozoic crust-mantle interaction beneath the North China craton: A consequence of the dispersal of Gondwanaland and accretion of Asia. *Geology* **2003**, *31*, 817–820. [CrossRef]

90. Zhang, P.; Zhao, Y.; Kou, L.L.; Bi, Z.W.; Yang, H.Z. Sulfur-lead Isotopic Composition of the Maoling Gold Deposit in Liaodong, and its Geological Implications. *Geol. Rev.* **2017**, *63*, 175–176. (In Chinese with English Abstract)

 *minerals*

Article

Origin and Evolution of Ore-Forming Fluid and Gold-Deposition Processes at the Sanshandao Gold Deposit, Jiaodong Peninsula, Eastern China

Yazhou Liu [1], Liqiang Yang [1,*], Sirui Wang [1], Xiangdong Liu [1], Hao Wang [1], Dapeng Li [2], Pengfei Wei [2], Wei Cheng [2] and Bingyu Chen [3]

[1] State Key Laboratory of Geological Processes and Mineral Resources, China University of Geosciences, Beijing 100083, China; 3001160020@cugb.edu.cn (Y.L.); creedwang@163.com (S.W.); lywhlxd@foxmail.com (X.L.); 2001180061@cugb.edu.cn (H.W.)

[2] Shandong Institute of Geological Sciences, Shandong Key Laboratory of Geological Processes and Resource Utilization in Metallic Minerals, Key Laboratory of Gold Mineralization Processes and Resources Utilization Subordinated to the Ministry of Land and Resources, Jinan 250013, China; dpengli@126.com (D.L.); pfeiwei@126.com (P.W.); weiwei0208@sohu.com (W.C.)

[3] Sanshandao Gold Company, Shandong Gold Mining Stock Co., Ltd., Laizhou 261442, China; cbysxf@sohu.com

* Correspondence: lqyang@cugb.edu.cn

Received: 14 January 2019; Accepted: 7 March 2019; Published: 19 March 2019

Abstract: The Early Cretaceous Sanshandao gold deposit, the largest deposit in the Sanshandao-Cangshang goldfield, is located in the northwestern part of the Jiaodong peninsula. It is host to Mesozoic granitoids and is controlled by the north by northeast (NNE) to northeast (NE)-trending Sanshandao-Cangshang fault. Two gold mineralizations were identified in the deposit's disseminated and stockwork veinlets and quartz–sulfide veins, which are typically enveloped by broad alteration selvages. Based on the cross-cutting relationships and mineralogical and textural characteristics, four stages have been identified for both styles of mineralization: Pyrite–quartz (stage 1), quartz–pyrite (stage 2), quartz–pyrite–base metal–sulfide (stage 3), and quartz–carbonate (stage 4), with gold mainly occurring in stages 2 and 3. Three types of fluid inclusion have been distinguished on the basis of fluid-inclusion assemblages in quartz and calcite from the four stages: Pure CO_2 gas (type I), CO_2–H_2O inclusions (type II), and aqueous inclusions (type III). Early-stage (stage 1) quartz primary inclusions are only type II inclusions, with trapping at 280–400 °C and salinity at 0.35 wt %–10.4 wt % NaCl equivalent. The main mineralizing stages (stages 2 and 3) typically contain primary fluid-inclusion assemblages of all three types, which show similar phase transition temperatures and are trapped between 210 and 320 °C. The late stage (stage 4) quartz and calcite contain only type III aqueous inclusions with trapping temperatures of 150–230 °C. The $\delta^{34}S$ values of the hydrothermal sulfides from the main stage range from 7.7‰ to 12.6‰ with an average of 10.15‰. The $\delta^{18}O$ values of hydrothermal quartz mainly occur between 9.7‰ and 15.1‰ (mainly 10.7‰–12.5‰, average 12.4‰); calculated fluid $\delta^{18}O$ values are from 0.97‰ to 10.79‰ with a median value of 5.5‰. The δD_{water} values calculated from hydrothermal sericite range from −67‰ to −48‰. Considering the fluid-inclusion compositions, $\delta^{18}O$ and δD compositions of ore-forming fluids, and regional geological events, the most likely ultimate potential fluid and metal would have originated from dehydration and desulfidation of the subducting paleo-Pacific slab and the subsequent devolatilization of the enriched mantle wedge. Fluid immiscibility occurred during the main ore-forming stage due to pressure decrease from the early stage (165–200 MPa) to the main stage (90–175 MPa). Followed by the changing physical and chemical conditions, the metallic elements (including Au) in the fluid could no longer exist in the form of complexes and precipitated from the fluid. Water–rock sulfidation and pressure fluctuations, with associated fluid unmixing and other chemical changes, were the two main mechanisms of gold deposition.

Keywords: fluid inclusion; stable isotopes; gold deposition; Sanshandao gold deposit; Jiaodong

---

1. Introduction

The Jiaodong gold province, being the largest gold producer in China, is located in the southeastern part of the North China Craton (NCC) (Figure 1) [1–4], with gold reserves above 3000 t [5]. The gold deposits in Jiaodong Province are hosted predominantly by Late Mesozoic granitoids [6–10] and are controlled by northeast to north by northeast (NE–NNE) trending faults, making Jiaodong one of the biggest known mesothermal lode-gold provinces hosted in granitoids in the world [8]. Two styles of gold mineralization have been identified in this area: The Linglong-type and the Jiaojia-type [11]. The Linglong-type gold mineralization is characterized by single or multiple quartz veins, hosted in second- or third-order faults cutting the Mesozoic granitoids (Figure 1). The Jiaojia-type gold deposits are dominated by disseminated and stockwork-style altered ores that generally occur along major regional faults (Figure 1) [12], which are commonly surrounded by widespread hydrothermal alteration.

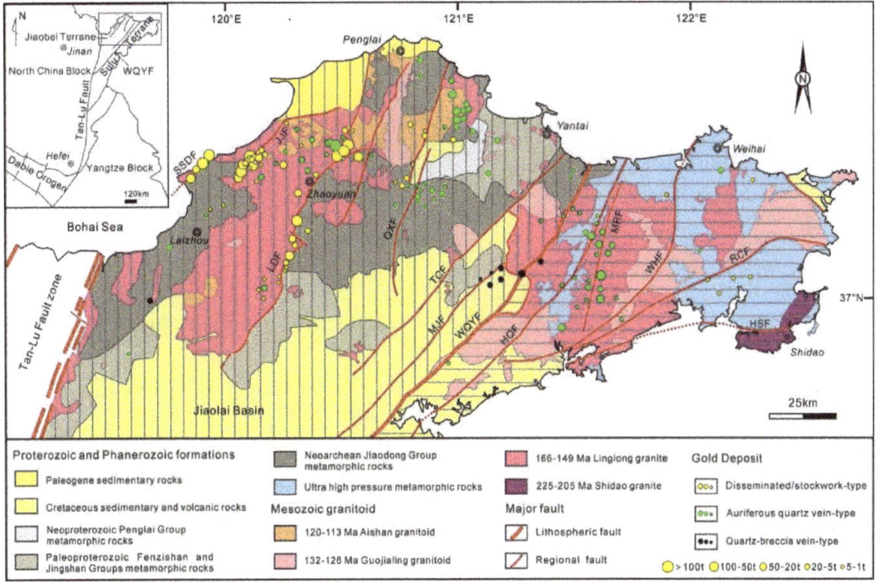

Figure 1. Simplified geological map of the Jiaodong gold province [13]. SSDF—Sanshandao fault; JJF—Jiaojia fault; LDF—Zhaoping fault; QXF—Qixia fault; TCF—Taocun fault; MJF—Muping-Jimo fault; WQYF—Wulian-Qingdao-Yantai fault; HQF—Haiyang-Qingdao fault; MRF—Muping-Rushan fault; WHF—Weihai fault; RCF—Rongcheng fault; HSF—Haiyang-Shidao fault. Reproduced with permission from [13]; published by Elsevier B.V., 2017.

Many studies have focused on geological features, ore-forming age, ore fluid characteristics, origin of the ore-forming fluid and metals, and geodynamic setting [7,12–21], all of which indicate that gold deposits throughout the whole Jiaodong Peninsula formed in broadly uniform hydrothermal fluid and physicochemical conditions [12]. Despite these intensive studies, a universal answer is not yet available, and the question of source of the ore fluids and metals is still open, as well as the genesis of gold deposit [22–30]. Fan et al. [17] concluded that the ore-forming fluids derived from mantle magma degassing in the shallow crust and are a combination of crustal and mantle components. Wang et al. [31] revealed the ore-fluid P–T–X (pressure–temperature–composition) conditions of most deposits are consistent with those typical of a metamorphic source. However, Yang et al. [32] and Deng et al. [33] proposed that the origin of the ore-forming fluid and materials may be associated with the Paleo-Pacific oceanic subducting slab.

As the Jiaojia-type deposit is now considered to be the most important and typical gold resource in this area, the distinctive geological background, structural setting, and metallogenic characteristics in the Sanshandao gold deposit provide an excellent opportunity to evaluate the gold depositional processes for the disseminated style mineralization in detail. Although certain studies have been completed on the ore fluid of the Sanshandao deposit, the P–T conditions of gold deposition have not been well defined [17]. If fluid immiscibility did occur during gold precipitation at the deposits [34], then the homogenization temperature and pressure would correspond to the trapping P–T conditions [35,36] and a pressure correction is unnecessary [37,38].

Fluid-inclusion data, ore petrography, and structural mapping are keys to determining ore genesis [39]. Fluid inclusions (FIs) trapped in hydrothermal minerals provide the best available method for defining the physical and chemical conditions of ore-forming fluids in fossilized hydrothermal systems [40]. Detailed FI study is essential to determine the origin and evolution of the ore-forming fluids. In this paper, based on detailed field and petrographic observations, FI analyses and isotope (H, O, S) compositions of the ores, the ore fluid, and metal origin and evolution together with the genesis of gold deposits were evaluated in detail.

2. Regional Geology

The Jiaodong Peninsula is located in the southeastern part of the NNC, bounded by the north to northeast (N–NE)-trending Tan-Lu fault to the west and by the Su-Lu Ultra-High Pressure (UHP) metamorphic belt to the south (Figure 1). The Peninsula comprises the Jiaobei terrane in the northwest and Sulu terrane in the southeast. The two terranes were sutured together along the Wulian–Qingdao–Yantai fault during continental collision in the Triassic to Early Jurassic [26,41] (Figure 1). Almost all the largest gold deposits and more than 90% gold resources are preserved in the Jiaobei terrane.

The Jiaobei terrane is composed of the Jiaobei uplift in the north and the Jiaolai basin in the south. The former is principally defined by Precambrian basement and supracrustal rocks cut by Mesozoic intrusions. The Precambrian basement consists of the Neoarchean Jiaodong Group (amphibolite and mafic granulite sequence), Paleoproterozoic Fenzishan and Jingshan Group (schist, paragneiss, calc–silicate rocks, marble, and minor mafic granulite and amphibolite), and Neoproterozoic Penglai Group, which is dominated by marble, slate, and quartzite [1,21,42–44]. The Jiaodong Group is unconformable, overlain by the Fenzishan and Jingshan Group. The Jiaolai Basin mainly consists of Mesozoic volcanic rocks, which refer to the Qingshan Formation; these Mesozoic volcanic rocks formed at ca. 108–110 Ma [11]. The Qingshan Formation comprises two units, with the upper assemblage composed of rhyolite flows and pyroclastic rocks, underlain by the lower assemblage characterized by trachybasalt, latite, and trachyte [45].

The widespread Mesozoic granitoids in the Jiaodong province are traditionally divided into the Linglong, Guojialing, and Aishan Suites [46,47], with the gold mineralization being mainly hosted in the former two suites (Figure 1). The Linglong Suite is distributed between the Jiaojia and Zhaoping fault zones, composed of garnet granite, biotite granite, and amphibole-bearing biotite granite [48,49], and considered to be generated by partial melting of remnant thickened Archaean lower crust during 165–150 Ma. The 132–123 Ma Guojialing Suite [13,50], which intruded into both the Late Jurassic granitoids and the Precambrian basement, mainly consists of porphyritic quartz monzonite, granodiorite, and monzogranite [46,51], and was likely derived from intense crust-mantle interaction [41,49]. The Guojialing Suite was cut by the later Early Cretaceous Aishan granitoids at ca. 118 to 110 Ma [52]; the Aishan granitoids mainly include monzogranites and syenogranites of I-type affinity with local alkali-feldspar granite of A-type affinity [13]. In addition to the granitoids introduced above, many mafic to felsic dikes were emplaced at ca. 122–114 Ma in the gold district. They are represented by dolerite, lamprophyre, granodiorite, and syenite. The genesis of these rocks was probably caused by magma activity associated with the Cretaceous lithospheric thinning and asthenospheric upwelling [19,53].

Regional structures mainly include the EW, NNE–NE, and NW trending fault zones. The EW tectonic belt is a basement structure, maintain long-term activity since the Yanshan movement, mainly characterized by folds and ductile shear zones. The NW trending fault zones formed in the Archean era, followed by inherited activities during the Late Proterozoic era and large-scale intense activities accompanied by magmatism and volcanism in the Mesozoic era [25]. The NE–NNE trending faults are the most important ore controlling structures in Jiaodong district. There are four fault systems from west to east: SSDF (Sanshandao fault), JJF (Jiaojia fault), LDF (Zhaoping fault), and QXF (Qixia fault) (Figure 1), corresponding to four gold belts. These faults are argued to be subsidiary structures to the regional Tan-Lu fault system. Two main stages of deformation during the late Mesozoic era have been identified. The first stage is characterized by brittle-ductile shear zones with sinistral oblique reverse movements, caused by northwest–southeast oblique compression [5,13,54]. This was followed by development of brittle normal faults [3,55], accompanied by hydrothermal alteration and gold mineralization [47].

The Sanshandao fault zone is located at the western part of the Jiaodong gold district; the majority of the fault zone is covered by Quaternary sedimentary rock, which is locally exposed to the surface; the fault is 12 km long with a width of 50–200 m. The distribution of the fault zone on the plane is S-type, and generally strikes 40° NE and dips 45°–75° SE. The fault zone mainly develops along the contact between the Linglong suite and Malianzhuang suite, with a continuous and stable main fault plane, characterized by mylonite and cataclastic rocks. More than 13% of the gold reserves (including the giant Sanshandao gold, Xinli, and Cangshang deposits) are hosted in the Sanshandao-Cangshang fault zone (Figures 1 and 2).

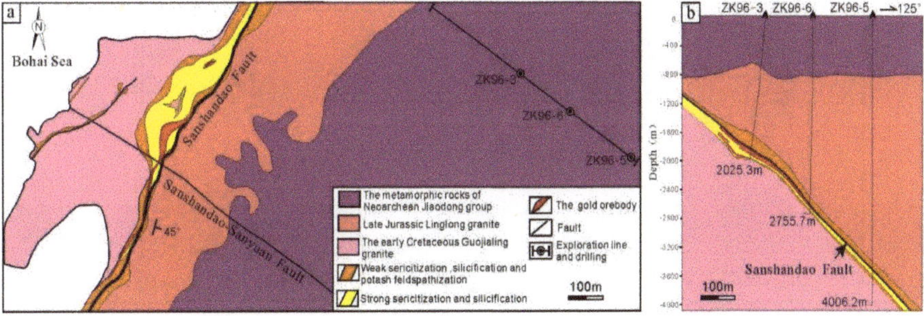

Figure 2. Simplified geological map (**a**) and geological cross section (**b**) of the Sanshandao gold deposit [21].

3. Deposit Geology

The Sanshandao gold deposit is located in the northwestern part of the Jiaobei uplift, about 25 km north of Laizhou City (Figure 1), with a proven gold reserve of >200 t. A total of 80 t gold has already been produced with an estimated annual gold production of 8 t (Data from Shandong Gold Group Co., Ltd.). In the deposit, the Sanshandao fault follows the contact between the Linglong granite and Jiaodong group metamorphic rocks in the hanging wall, and the Guojialing Granodiorite accompanied by minor Linglong granite in the footwall (Figure 2). Linglong granite is intruded by Guojialing Granodiorite, showing a medium-grained equigranular texture, consisting mainly of Biotite, K-feldspar, plagioclase, quartz, and a small amount of magnetite and apatite. Guojialing granodiorite is characterized by a porphyritic-like texture, and the mineral assemblages are analogous to that of the Linglong granite, but with more hornblende. Dynamic recrystallized quartz occurs commonly in the Linglong and Guojialing granitoids, indicating the two suites underwent subsolid ductile deformation before mineralization.

The Sanshandao gold deposit is controlled by the Sanshandao fault. This fault generally trends northeast, dips 50° southeast, and extends offshore into the Bohai Sea. The fault zone is characterized by fault gouge, mylonite, hydrothermal breccias, and cataclastic rocks, which comprise zones up to tens of meters thick of K-feldspar, quartz, and sericite, with both disseminated pyrite and pyrite–quartz stockworks [11], reflecting early shearing and late brittle fracturing of the structures along the fault zone, dominated by brittle failure. Two mineralization styles were identified at the gold deposit and all hosted in the footwall of the fault zone: Disseminated- and stockwork-style (Figure 3). The former contains disseminated gold within the granodiorite with strong hydrothermal alteration, including sericitization, silicification, sulfidation, and K-feldspar wall-rock halos; the degree of alteration and fracturing decrease with distance from the fault. This is the most common style mineralization in this deposit. The less common style is the gold–quartz vein/veinlet stockworks mainly controlled by the subsidiary fault.

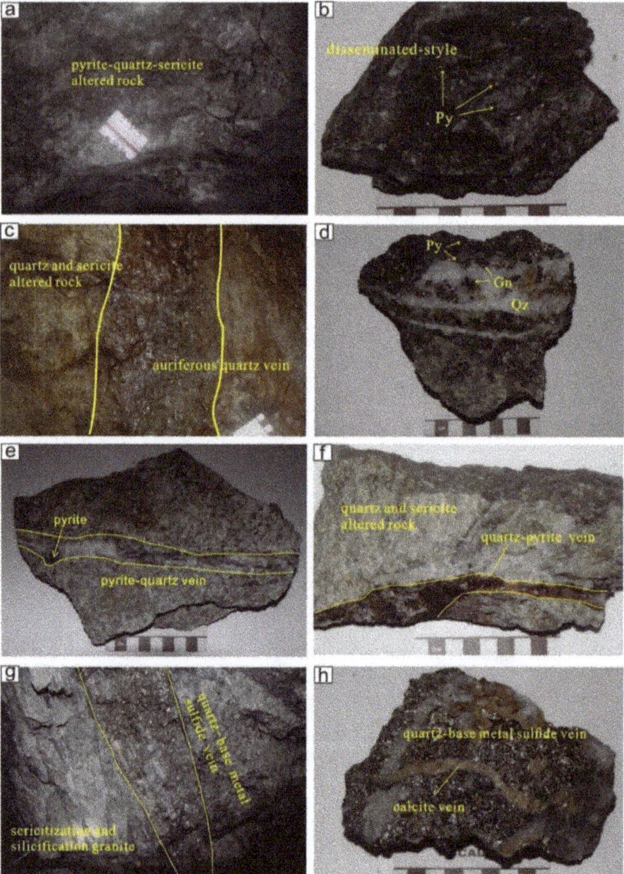

Figure 3. Field photographs demonstrating major geologic bodies with cross-cutting relationships and different mineral assemblages. (**a**,**b**) Disseminated-style ore of stage 1 (early stage); (**c**) auriferous quartz vein of stage 2 (main stage) in sericitized and silicified granite; (**d**) auriferous quartz vein-style ore of stage 3 (main stage); (**e**) the early-stage milky white quartz vein with disseminated pyrite; (**f**) main stage quartz–pyrite veinlet cut across the early-stage sericite–quartz altered rock; (**g**) quartz–base–metal–sulfide vein; (**h**) late stage (stage 4) quartz–calcite veins cutting the quartz–base–metal–sulfide vein. Qz—quartz; Gn—galena; Py—pyrite.

The combined disseminated and veinlet mineralization form tabular orebodies broadly parallel to the Sanshandao fault zone. Six orebodies have been defined in the Sanshandao gold deposit, all of which occur in the footwall of the fault zone. The orebodies generally strike NE, dip SE, with the No. 1 orebody being the largest and the No. 2 being second. The No. 1 orebody accounts for 70.35% of the known gold resources and occurs within the middle to upper part of the pyrite–sericite–quartz alteration zone. This orebody strikes 35° NE, dips 34°–44° SE, extends 1020 m along the strike and 700–1000 m along the dipping direction. It ranges in thickness from 0.95 to 12.08 m (average 6.65 m) and extends downdip from the −10 m level to below the −1050 m level, with gold grade ranging from 1.74 to 15.4 g/t (average 3.25 g/t). The No. 2 orebody underlays the No. 1 orebody, appearing as regular pulse, stratiform-like lenses, with a strike of 3°–40° and a dip of 32°–44°. It extends 335–380 m in length with an average thickness of 5.74 m. The gold grade varies from 1.55 to 24.8 g/t (average of 3.13 g/t).

The paragenetic assemblages of the disseminated- and stockwork–veinlet-style ores are shown in Figures 3 and 4. The ore minerals mainly include pyrite, chalcopyrite, galena, sphalerite, and arsenopyrite, with minor electrum. The gangue minerals are dominated by quartz, sericite, and calcite. Combining mineral assemblages and textural characteristics with crosscutting relationships (Figure 3), four mineralization stages have been identified in this deposit: Pyrite–quartz–sericite (stage 1/early stage), quartz–pyrite–arsenopyrite (stage 2/main stage), quartz–base metal–sulfide (stage 3/main stage), and quartz–carbonate (stage 4/late stage). Stage 1 is characterized by sericite–quartz-altered rocks or locally pyrite–quartz veins (Figure 3a,b,e). The mineral assemblages mainly consist of milky white subhedral quartz, sericite, and minor coarse-grained, irregular, or euhedral pyrite (Figure 4a–c). Little gold precipitated in this stage. Stage 2 is characterized by quartz–sulfide veins and stockworks and disseminated sulfides occurring in the sericite–quartz-altered rocks (Figure 4d), or as auriferous quartz–pyrite veins controlled by secondary structures (Figure 3c). The minerals are dominated by pyrite and white-gray quartz with a small amount of chalcopyrite, native gold, and pyrrhotite (Figure 4h,i). The gold minerals primarily occur as native gold wrapped in pyrite (Figure 4i) or fill within in the gaps and/or cracks of pyrite. Stage 3 is characterized by the assemblage of gold, quartz, base metal sulfide (Figure 3d,g). Large amounts of sulfide minerals precipitated during this stage, including pyrite, galena, sphalerite, and chalcopyrite (Figure 4e,g). These base metal sulfides minerals are fine-grained, occur as veinlets or disseminated aggregates, cutting the early-stage quartz–pyrite veins and veinlets. Gold minerals are mainly native gold, filling the open fractures in pyrite and/or galena. Stage 4 is characterized by quartz–calcite veinlets, commonly marking the termination of gold mineralization. Quartz and carbonates form veins and veinlets cutting the early quartz–pyrite veinlets (Figures 3h and 4f) or quartz–base–metal sulfide veins in the disseminated ores or subsidiary faults.

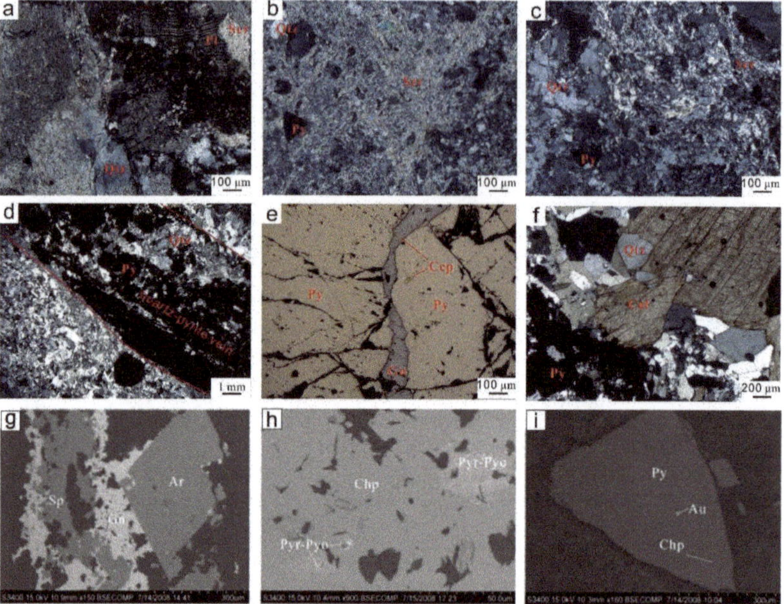

Figure 4. Photomicrographs under (**a–d,f**) transmitted light, (**e**) reflected light and (**g–i**) backscattered electron images, showing important ore minerals and alteration assemblages at the Sanshandao gold deposit. (a–c) Typical sericitization alteration consisting of sericite and quartz, with minor pyrite (stage 1); (d) quartz–pyrite vein (stage 2) occurring in sericite–quartz altered rock; (e) galena cemented with chalcopyrite replacing the pyrite grain (stage 3); (f) calcite (stage 4) cutting the quartz–pyrite vein (stage 2); (g,h) typical quartz–pyrite–base–metal–sulfide ore minerals consisting of pyrite, chalcopyrite, sphalerite, and galena; (i) native gold inclusion in pyrite. Cc—calcite; Ccp—chalcopyrite; Gn—galena; Py—pyrite; Qtz—quartz; Ser—sericite; Sp—sphalerite.

4. Sampling and Analytical Methods

Samples for fluid-inclusion analysis were selected from −150, −195, −255, −285, −330, −375, −390, −495, −510, and −600 m levels in the Sanshandao deposit, areas represented by quartz, quartz–pyrite, quartz–base–metal–sulfides, quartz–carbonate veins, and quartz–sericite–pyrite altered rocks. More than 100 doubly polished sections (about 0.2 mm thick) were petrographically examined in detail. Finally, 12 samples (Table 1) were selected for microthermometric and laser Raman spectroscopic investigations.

Table 1. Samples for microthermometry in the Sanshandao gold deposit.

Sample ID	Depth (m)	Mineral	Sample	Depth (m)	Mineral
SSD600C	−600	Quartz	SSD28505	−285	Quartz
SSD51001	−510	Quartz	SSD25508	−255	Quartz
SSD49501	−495	Quartz	SSD25505	−255	Quartz
SSD39006	−390	Quartz	SSD19504	−195	Quartz
SSD37501	−375	Quartz	SSD15003	−150	Quartz
SSD33004	−330	Quartz	SSD-02	0	Quartz

4.1. Fluid-Inclusion Analysis

Reconnaissance fluid-inclusion petrography was conducted on quartz veins from all paragenetic stages, recognized by observing intersection and overprinting relationships between different veins (veinlets) and alteration types (halos), and temporal–spatial relationships of gangue and ore minerals using a microscope. Subsequently, optical microscopy and detailed fluid-inclusion petrography were performed on selected quartz vein samples with a microscope using a series of doubly polished sections. Accuracy of the microthermometric data on selected fluid inclusions was ensured by calibration against the triple-point of pure CO_2 (–56.6 °C), the freezing point of water (0.0 °C) and the critical point of water (374.1 °C) using synthetic fluid inclusions supplied by FLUID INC (an incorporation in USA).

The microthermometric analysis of fluid inclusions was performed using a LinkamTHMSG 600 (Linkam Scientific Instruments Company, Surrey, UK) heating–cooling stage (from -198 to 600 °C) on a Leitz microscope (Wetzlar, Germany) at the Fluid Inclusion Laboratories, State Key Laboratory of Geological Processes and Mineral Resources, China University of Geosciences, Beijing, China. Synthetic fluid inclusions were used to calibrate the stage to ensure the precision of analysis. Measurements of the melting temperatures of the carbonic phase (Tm_{CO2}), ice final melting temperatures (Tm_{ice}) and clathrate melting temperatures of CO_2 (Tm_{clath}) were recorded at the heating rate of 0.1–0.2 °C/min. For the total homogenization and the CO_2 homogenization temperatures (Th_{CO2}), the heating rate was 0.2–0.4 °C/min. Salinities were then calculated using Tm_{ice} [35,40] and Tm_{clath} [56]. The CO_2 densities can be well defined through Th_{CO2}. Bulk composition and density were determined using the program of Bakker et al. [57] and software of Brown et al. [58].

4.2. Laser Raman Spectroscopy

Laser Raman spectroscopic analysis was carried out at the Institute of Geology and Geophysics of the Chinese Academy of Sciences (IGGCAS) by using Raman 2000 laser Raman spectrometer (British Ranishaw, London, UK). The wavelength was 514.5 nm for the Ar ion laser. The spectrum ranged from 50 to 9000 cm^{-1}, with a reproducibility of ± 0.2 cm^{-1} and a resolution of 2 cm^{-1}. The spatial resolution was better than 4 μm longitudinally and lower than 1 μm in the transverse direction.

4.3. Isotope Analysis

Isotopes of hydrogen, oxygen, and sulfur were analyzed at the Stable Isotope Laboratory, National Key Laboratory for Tectonic Evolution of Lithosphere, Institute of Geology and Geophysics, Chinese Academy of Sciences (IGGCAS, Beijing, China). Samples from different ore-forming stages were crushed to 40–60 mesh to separate the mineral grains. The quartz, calcite, and pyrite grains for isotopic analysis were handpicked under a binocular microscope to achieve 99% purity.

For hydrogen isotopic measurement, the water from secondary fluid inclusions were gradually removed in a vacuum system after drying the selected single mineral below 105 °C, then heating to 600 °C to cause the remaining inclusions to burst. The released water was disposed by collection, condensation, and purification, and replaced by zinc to obtain the hydrogen for mass spectrometry analysis [59]. Oxygen isotope analyses of quartz were conducted on the same samples for hydrogen isotopes using the following procedure. First, quartz grains were reacted with BrF_5 for 15 hours at 500–550 °C to produce oxygen. Then, the generating O_2 was purified using liquid nitrogen and converted to CO_2 on a platinum-coated carbon rod at 700 °C for mass spectrometry analysis. Both oxygen and hydrogen determinations were made on a MAT-252 (Finnigan company, San Francisco, CA, USA) mass spectrometer. The H–O isotope analysis results adopted the Vienna Standard Mean Ocean Water (V-SMOW), with a precision of $\pm 1‰$ for δD, and $\pm 0.2‰$ for $\delta^{18}O$ [59].

For sulfur isotope analyses, 36 pyrite samples from ores of the Sanshandao deposit were used. SO_2 was generated through the reaction of pyrite and cuprous oxide under a vacuum pressure of 2×10^{-2} Pa at 980 °C and was measured using a Delta-S (Finnigan company, San Francisco, CA, USA) mass spectrometer for sulfur isotope analysis. All the sulfur isotopic compositions are reported relative to CDT (Cañon Diablo Troilite). S isotope analysis results adopted Vienna Cañon Diablo Troilite (V-CDT) standards, with a precision of $\pm 0.2‰$ [59].

5. Results

5.1. Fluid-Inclusion Types

Three types of fluid inclusion were identified in this study based on microthermometry and Raman spectroscopy: Pure CO_2 gas (type I), CO_2–H_2O inclusions (type II), and aqueous inclusions (type III). Type I inclusions are completely filled with gas, darkened or even black, and are only light at the center of the bubble. Their size ranges from 5 to 10 μm. We observed that they coexist with the type II inclusions in the early stage and with type II and type III inclusions in the main stage (Figure 5g). These inclusions appear alone or scattered and are interpreted to be primary inclusions.

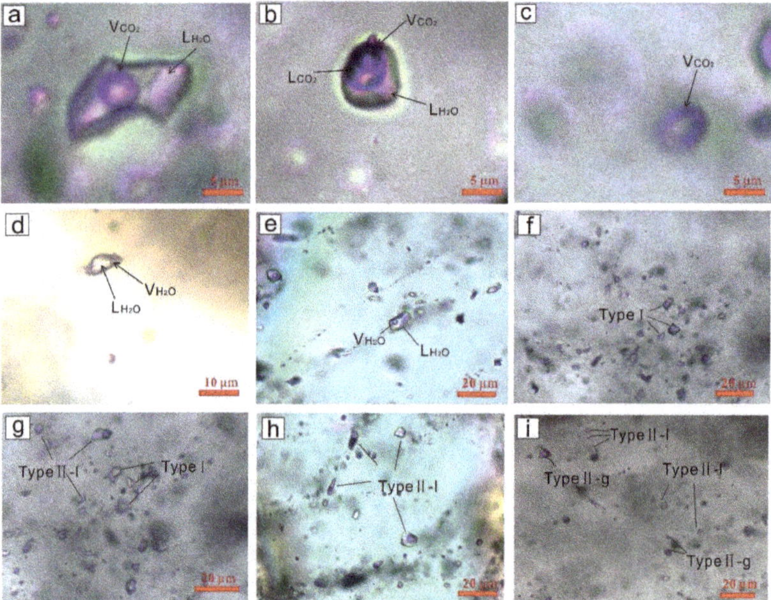

Figure 5. Photomicrographs of typical fluid inclusions in the Sanshandao gold deposit. (**a**) Type II inclusion consisting of two (H_2O liquid + CO_2 vapor/liquid) phases; (**b**) type II inclusion consisting of three (H_2O liquid + CO_2 liquid + CO_2 vapor) phases; (**c**) type I inclusion consisting of pure gas (CO_2 vapor) phase; (**d**) type III inclusion consisting of two (H_2O liquid + H_2O vapor) phases; (**e**) fluid-inclusion assemblage (FIA) consists of secondary type III inclusions; (**f**) FIA consists of type 1 inclusions; (**g**) FIA consists of type I and type II-l inclusions occurring together in stage 2 quartz; (**h**) FIA consists of type II-l inclusions; (**i**) FIA consists of type II-l and type II-g inclusions occurring together in stage 3 quartz.

Type II inclusions can be further divided into two subtypes, containing two-phase (V_{CO2} + L_{H2O}) or three-phase (V_{CO2} + L_{CO2} + L_{H2O}) inclusions at room temperature with sizes from 5 to 15 μm (Figure 5a). The two-phase inclusions (liquid phase and gas phase) have a boundary between the two phases denoted by a black circle. They can be subdivided into gas inclusions (II-g type) with gas filling degree greater than 50% (V/(V + L) > 50%) and aqueous inclusions (II-l type) with liquid filling degree lower than 50% (L/(V + L) < 50%). In some samples, we found the coexistence of type II-g and II-l inclusions, or type II and type I inclusions (Figure 5g,i). This reflects the fluid evolution. Components of the inclusions can be H_2O, CO_2, or saline solution. The three-phase inclusions usually occur with V_{CO2} + L_{CO2} accounts for 20%–40% of the total volume (Figure 5b).

Type III inclusions consist of two phases (H_2O liquid and H_2O vapor) and are mostly colorless (Figure 5d,e); they commonly occur between 5 and 10 μm, with the vapor occupying 5%–30% of the total cavity volume. These fluid inclusions have irregular or regular shapes and develop especially in late-stage quartz. The secondary type III inclusions are generally cut across the crystal boundaries of quartz with the primary ones occur as isolated singles or group.

The primary fluid inclusions are identified according to (1) the shape of the fluid inclusions. The shape of primary inclusions is usually regular and scattered, the secondary inclusions are usually irregular in shape, distributed along healing fissures, and sometimes healing fissures pass through different minerals; (2) the density, homogenization temperature, salinity, and composition of inclusions of the same origin are the same or similar, so the inclusions can be compared and classified with known primary and secondary inclusions. (3) In gold-bearing quartz veins, gold coexists with pyrite, whereas quartz in pyrite coexists with natural gold and pyrite, so fluid inclusions in quartz can represent the ore-forming fluids.

5.2. Microthermometry

The early-stage quartz grains mainly include the type II inclusions. The carbonic phase melting temperatures (Tm_{CO2}) range from –56.9 to –56.4 °C (Table 2, Figure 6), close to the triple point of pure CO_2 (–56.6 °C), indicating that the gas phase mainly comprises CO_2 with few other volatiles. The CO_2 clathrate melting temperature is between 7.4 °C and 7.7 °C, corresponding to salinities of 0.35 wt %–10.4 wt % NaCl equivalent. The carbonic phases (Th_{CO2}) partially homogenize to liquid at temperatures from 27.4 to 31.2 °C. Total homogenization (Th_{TOT}) into liquid of the carbonic and aqueous occurs between 280 and 400 °C, corresponding to densities of 0.61–1.07 g/cm^3, (Table 1, Figure 6).

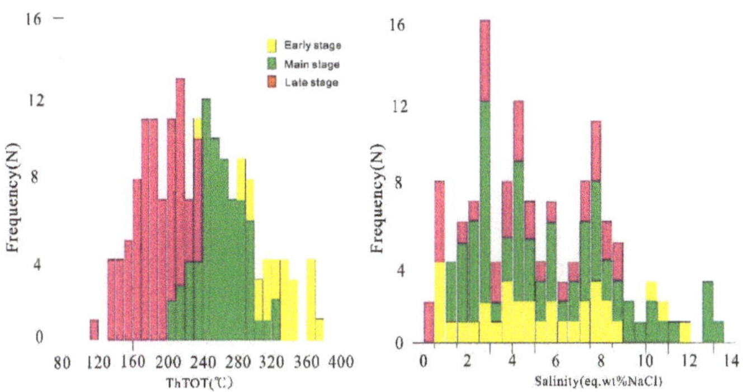

Figure 6. Histograms of microthermometric data showing temperatures of (**a**) Th_{TOT} and (**b**) salinity.

Quartz grains of the main stage include all three type FIs and are dominated by type II and type III. Melting temperatures of the carbonic phase (Tm_{CO2}) of type II inclusions occur between -59.3 and -57.6 °C, commonly below the pure CO_2 melting point (-56.6 °C), demonstrating that the existence of other gas components, such as CH_4 and N_2 [60], Melting temperatures of the CO_2 clathrate (Tm_{clath}) occur between 7.3 °C and 8.0 °C, corresponding to the fluid salinities from 2.2 wt % to 13.33 wt % NaCl equivalent. The total (Th_{TOT}) and carbonic phases (Th_{CO2}) homogenization into liquid range between 220 and 320 °C and 22 and 31.2 °C, respectively. The final ice melting temperature (Tm_{ice}) of type III is between –6.0 and –4.0 °C.

In the late stage, type I FIs were generally observed in the quartz–calcite veins. The Tm_{CO2} occurs between –56.8 and –56.6 °C, showing the gas components are almost pure CO_2. The Tm_{clath} ranges from 6.4 to 8.2 °C and Tm_{ice} is between –3.5 and –0.0 °C, yielding salinities of 0.1 wt %–12.5 wt % NaCl equivalent. Final homogenization into liquid of these inclusions is between 150 and 230 °C (Table 1, Figure 6), yielding densities of 0.82–1 g/cm^3.

Table 2. Summary of the microthermometric data for fluid inclusions trapped in quartz and calcite from the Sanshandao gold deposit.

Stage	Sample	Type	Tm_{CO2} (°C)	Tm_{clath} (°C)	Tm_{ice} (°C)	Th_{CO2} (°C)	Th_{tot} (°C)	Salinity (wt % NaCl eqv.)	Bulk Density (g/cm^3)	CO_2 Density (g/cm^3)
early	SSD28505	II								
	SSD-02	II	from −56.9 to −56.4 ($n = 19$)	7.4–7.7 ($n = 23$)		27.4–31.2 ($n = 9$)	280–400 ($n = 22$)	0.35–10.4 ($n = 23$)	0.61–1.07 ($n = 12$)	0.52–0.93 ($n = 6$)
	SSD33004	II								
	SSD19504	II								
	SSD51001	III			form −7.5 to −3.5 ($n = 13$)		210–300 ($n = 28$)	5.7–11.1 ($n = 13$)	0.77–0.94 ($n = 22$)	
main	SSD49501	II	from −59.3 to −56.6 ($n = 25$)	7.3–8.0 ($n = 31$)		22–31.2 ($n = 11$)	220–320 ($n = 32$)	2.2–13.33 ($n = 31$)	0.7–0.98 ($n = 25$)	0.64–0.90 ($n = 6$)
	SSD39006	II								
	SSD25505	I	from −57.6 to −56.4 ($n = 9$)			24.3–28.4 ($n = 8$)				0.52–0.84 ($n = 4$)
late	SSD15003	III								
	SSD25508	III			from −3.5 to 0.0 ($n = 33$)		100–230 ($n = 68$)	0.1–8.5 ($n = 26$)	0.82–1 ($n = 27$)	
	SSD600C	III								

5.3. Fluid-Inclusion Composition

Laser Raman spectroscopy indicated that the volatiles of the early mineralization stage are characterized by CO_2 and H_2O; no CH_4 or N_2 were detected (Figure 7b,d). However, in the main stage, abundant CH_4 was detected in many of the type II fluid inclusions (Figure 7a,c), but no H_2S or N_2 were detected. This is consistent with the microthermometric results that showed that melting temperatures of the carbonic phase are near $-56.6\ ^\circ C$ in early-stage mineralization and below $-56.6\ ^\circ C$ in main stage mineralization. For the type III FIs in the middle- and late-stage quartz, the component is composed of 100% liquid H_2O.

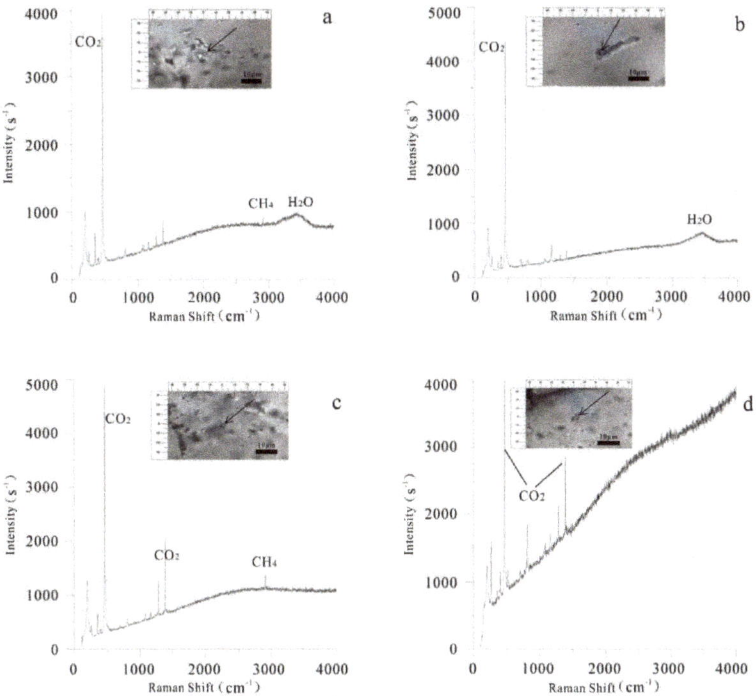

Figure 7. Representative Raman spectra of fluid inclusions in quartz. (**a**) Two-phase type II fluid inclusion, showing a small amount of CH_4; (**b**) three-phase type II fluid inclusion, containing water only; (**c**) two-phase type II fluid inclusion, showing much more CH_4; and (**d**) one-phase type I fluid inclusion.

5.4. Isotope Composition

The hydrogen and oxygen isotopic data obtained from this and previous studies in the Sanshandao deposit are listed in Table 3. The $\delta D_{V\text{-SMOW}}$ values for fluid inclusions in quartz range from $-62‰$ to $-100‰$, with a median value of $-86‰$. The $\delta D_{V\text{-SMOW}}$ values of fluid inclusions in the early stage are between $-86‰$ and $-100‰$, main stage from $-62‰$ to $-95‰$ (median $-84‰$), and late stage from $-70‰$ to $-99‰$. The $\delta D_{V\text{-SMOW}}$ for the sericite ranges from -48 to $-67‰$ in the main stage, with a median value of $-53‰$. The $\delta^{18}O$ values of hydrothermal quartz ($\delta^{18}O_{SMOW}$, ‰; quartz) occur between 9.7‰ and 15.1‰ (mainly 10.7‰–12.5‰) with a median of 12.4‰ (Table 3). Oxygen isotopic compositions of hydrothermal water in equilibrium with quartz were calculated based on an extrapolation of the fractionation formula: $1000\ln\alpha = \delta^{18}O_{quartz} - \delta^{18}O_{H2O} = 3.38 \times 10^6/T^2 - 3.40$ [61]. The fractionation factors were calculated using the mean value of the homogenization temperatures of fluid inclusions from the same ore-forming stage quartz samples. The calculated $\delta^{18}O_{water}$ values range from 0.97‰ to

10.79‰ with a median value of 5.5‰. The calculated $\delta^{18}O_{water}$ values of the early stage are 4.9‰–9.4‰ (median 7.0‰); main stage, 4.1‰–8.9‰ (median 6.6‰); and late stage, 1.2‰–2.6‰.

Table 3. Oxygen and hydrogen isotope analyses in the Sanshandao gold deposit.

Stage	Sample	Mineral	$\delta D_{V\text{-}SMOW}$ (‰)	δD_{water}	$\delta^{18}O_{mineral}$ (‰) SMOW	$\delta^{18}O_{water}$ (‰) SMOW	T (°C)	Reference
early	SSD28505	quartz	−100		13.3	4.9	315	This study
	SSD02	quartz	−86		13.5	9.4	350	This study
		quartz			14.0	7.1	320	
		quartz			15.0	7.0	290	[62]
		quartz			18.8	7.5	220	
		sericite	−52	−36	10.7	7.6	250	
		sericite	−48	−32	11.7	8.6	250	
		sericite	−52	−36	12.0	8.9	250	[35]
		sericite	−53	−37	9.7	6.6	250	
		sericite	−50	−33	10.9	7.8	250	
		sericite	−67	−51	11.0	7.9	250	
main		quartz	−92		12.5	4.1	280	[62]
		quartz	−72		12.8	4.8	290	
		quartz	−62		14.3	6.9	300	[63]
		quartz	−71		14.7	5.3	250	
	SSD600H	quartz	−65		13.2	4.7	284	
	SSD51001	quartz	−95		13.6	5.5	272	
	SSD49501	quartz	−86		13.9	5.9	258	This study
	SSD39006	quartz	−85		13.9	5.2	255	
	SSD39007	quartz	−84		12.6	4.2	250	
	SSD25505	quartz	−94		12.8	4.4	247	
late		quartz	−92		13.4	2.5	220	[63]
		quartz	−70		11.1	1.4	250	[62]
	SSD15003	quartz	−76		13.5	2.6	200	This study
	SSD25508	quartz	−99		13.5	1.2	177	

The $\delta^{34}S_{V\text{-}CDT}$ values of pyrite, sphalerite, and galena were all obtained from the main stage in the Sanshandao gold deposit, listed in Table 4 and graphically shown in Figure 8. The $\delta^{34}S_{CDT}$ values range from 7.8‰ to 12.6‰, with a median value of 11.4‰.

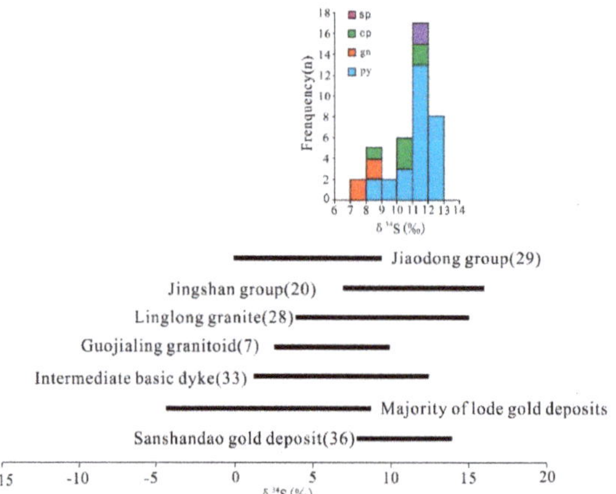

Figure 8. (a) Histograms of $\delta^{34}S$ values of sulfide minerals from the Sanshandao gold deposit; (b) comparison of sulfur isotopic compositions of main mineralization stages at Sanshandao and related major geologic bodies. The ranges of major geologic bodies were obtained from Yang et al. [12].

Table 4. Sulfur isotope analyses in the Sanshandao gold deposit.

Mineral	δ^{34}S (‰)	Reference	Mineral	δ^{34}S (‰)	Reference
Py	11.1		Gn	7.9	
Py	9.7		Sp	11.4	
Py	11.5		Sp	11.1	
Gn	7.8	This	Py	11.8	[64]
Py	11.8	study	Py	11.6	
Py	11.0		Py	11.1	
Py	11.0		Py	10.5	
Py	10.7		Py	10.4	
Py	11.8		Cp	8.3	[65]
Py	12.6		Py	9.5	
Py	12.6		Py	12.0	
Py	12.6	[63]	Py	12.0	
Cp	10.1		Py	11.7	[35]
Cp	10.7		Py	11.9	
Cp	10.5		Py	11.5	
Py	11.3		Py	12.6	
Cp	11.5	[64]	Py	12.5	[65]
Cp	11.4		Py	12.1	

6. Discussion

6.1. Fluid Immiscibility in the Main Stage

As described above, the type I, II, and III FIs coexisted during the main stage at the Sanshandao gold deposit (Figure 5). These FIs are spatially associated with each other, indicating that the ore-forming fluids were in a heterogeneous thermal condition when they were trapped [31]. This phenomenon can result from one of the three possible mechanisms: (1) Fluid immiscibility by unmixing from the homogeneous fluid [32], (2) mixing of different fluids [66], and (3) necking down as the fluids were not necessarily heterogeneous during trapping [57].

Mixing of two different fluids would cause the inclusions to show a wide range of compositions and degrees of filling [66], which was not observed here. The inclusions in the main-stage quartz grains of vein- and veinlet-style ores have two end-members (>30%V_{CO2} type II-l inclusions and >50% V_{CO2} type II-g inclusions), rather than a continuum of compositions, suggesting that mixing of two separate fluids was not the primary cause for the coexisting of types 1 and 2 inclusions. All the fluid inclusions selected for microthermometry and laser Raman studies were primary in origin. They commonly developed in undeformed quartz grains, with regular shape and constant degrees of gas fill (Figure 5), without any evidence of necking down [57], indicating that the fluid-inclusion assemblage in the main stage quartz of vein- and veinlet-style ores were unaffected (or minimally affected) by post-entrapment deformation.

The initial homogeneous ore-forming fluids underwent fluid immiscibility during the main stage in the Sanshandao gold deposit, as evidenced by: (1) All the three types fluid inclusions appear in the same growth phase of the quartz (Figure 5g–i), which shows that these FIs formed contemporaneously; (2) the total homogenization temperatures of type II inclusions are higher than those of aqueous inclusions (type III), and this also agrees with fluid immiscibility [64]; (3) the fluid salinity in the Sanshandao deposit shows a rising tend from the early stage to the main stage (Figure 9), consistent with fluid immiscibility (or phase separation) leading to higher salinity during evolution of the ore-forming liquid; (4) the type II-l inclusions and type II-g coexist as an isolated group, the type II-l inclusions homogenized to the liquid, and some of type II-g inclusions with high CO_2 phase(s) volumes homogenized to the vapor; and (5) two type II fluid inclusions with abnormal high density occur at about 300 °C in the Sanshandao gold deposit (Figure 10). The density of these two fluid inclusions is obviously higher than that of the other inclusions. Together with their high homogenization pressures, this shows that CO_2-rich overpressure fluid exists in the Sanshandao deposit, and the existence of such overpressure fluids is also likely to cause decompression boiling, which will also lead to fluid immiscibility. Finally, we think that fluid immiscibility was an important gold precipitation mechanism in the Sanshandao deposit.

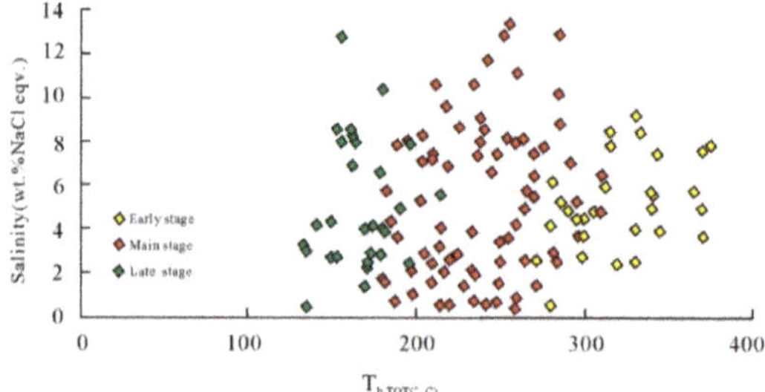

Figure 9. Evolution map of total homogenization temperatures (°C) vs. salinity of the ore-forming fluid in the Sanshandao gold deposit.

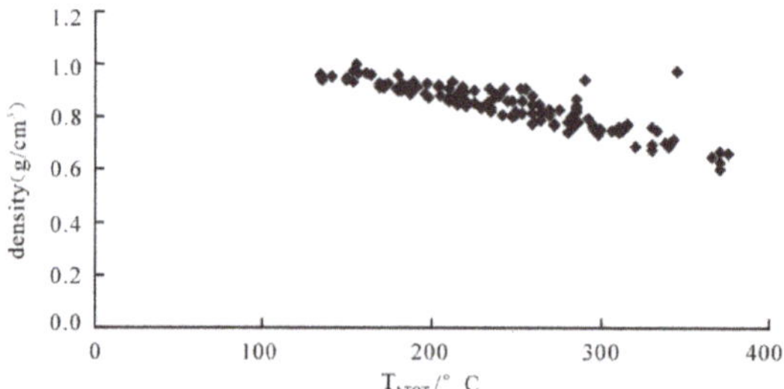

Figure 10. Evolution map of total homogenization temperatures (°C) vs. density of the ore-forming fluid in the Sanshandao gold deposit.

6.2. Source and Evolution of Ore-Forming Fluids

6.2.1. Source of Ore-Forming Fluids

The $\delta^{18}O$ and δD compositions of ore-forming fluids in the Sanshandao gold deposit are shown in Table 2 and Figure 11. The calculated $\delta^{18}O_{water}$ values of ore-forming fluids vary from 0.97‰ to 10.79‰, with the median values decreasing from 7.45‰, to 5.5‰, to 1.36‰ from the early stage to late stage, respectively, indicating these calculations are not credible because the calculated $\delta^{18}O_{water}$ values decreased from early stage to late stage mainly due to the decrease in equilibrium temperature, but the final homogenization temperature of fluid inclusions used for calculation is probably not the mineral–fluid equilibrium temperature [29]. Considering the similar median values of $\delta^{18}O_{SMOW}$ (quartz) for each hydrothermal stage (Table 3), the quartz in the four stages is probably deposited from the same fluid with a very constant $\delta^{18}O$ composition. The δD values of the fluid inclusions in the quartz range from −100‰ to −48‰, with nearly constant median values from early stage to late stage. The calculated δD_{water} values of hydrothermal waters in equilibrium with sericite are ~30‰ less than that of the fluid inclusions in the quartz; thus, the reliability of these two results must be questioned. The single primary fluid inclusions being selected for microthermometry, different generations of fluid inclusions being trapped in quartz grains, and the large amount of secondary fluid inclusions probably caused the δD values of the hydrothermal fluids in each stage to be incorrect. By contrast, the sericite in the main stage was defined by detailed petrographic observation. Therefore, the calculated δD_{water} values are more reliable. The most likely true $\delta^{18}O$ and δD compositions of the primary ore-forming fluid in the Sanshandao gold deposit were calculated from quartz and sericite, respectively (Figure 11).

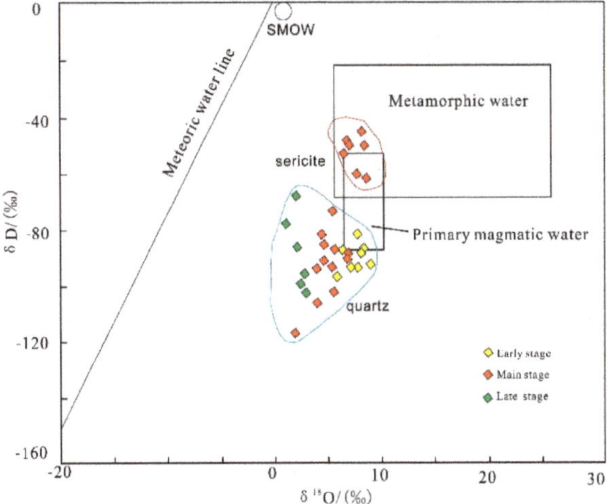

Figure 11. Isotopic compositions of oxygen and hydrogen in the Sanshandao gold deposit using data from Table 2. Blue circle and brown circles represent the data obtained from quartz and sericite, respectively.

Most of the data from sericite were recorded from the metamorphic water field, with some data from the field between metamorphic water and primary magmatic water. The ore-forming fluid of the Sanshandao gold deposit could have been derived from metamorphic fluid or primary magmatic water. The salinity of the ore-forming fluid in the Sanshandao deposit is generally lower (<10%) than that of the magmatic–hydrothermal fluids. The low salinity (<10 eqv. wt % NaCl) is different from the normally high salinity magmatic–hydrothermal fluids [29,67]. All the FI homogenization temperatures of the ore-forming fluid are less than 500 °C in this study, which is also inconsistent with the property of magmatic–hydrothermal fluids [68]. Deng et al. [25] studied the stable isotopes (H–O–C–S–Sr) in the Xinli gold deposit and revealed that δD_{SMOW} and $\delta^{18}O_{H2O}$ from the main stage are −69.6‰ to −88.3‰ and from 2.82‰ to 5.34‰, respectively; $\delta^{13}C_{PDB}$ and $\delta^{18}O_{SMOW}$ of late stage are from −6.4‰ to −2.4‰ and from 9.8‰ to 11.6‰, respectively; and $\delta^{34}S$ and $^{87}Sr/^{86}Sr$ ratios are from 9.42‰ to 11.62‰ and from 0.710657 to 0.711542, respectively. They suggested that the ore-forming fluid genetically originated from the subducting slab and metasomatized mantle, akin to the magmatic water. Yang et al. [29] reported consistent stable isotopic (H–O–S) compositions and discussed their result for the Taishang gold deposit. Wang [28] studied the REE (rare earth element) composition of fluids in different mineralization stages in the Jiaojia gold field and found that the total REE content of metallogenic pyrite demonstrated a trend of first increasing and then decreasing from the early to late stage. This indicates that the metamorphism in the early stage is weak and strengthened in the main stage, then decreased again in the late stage. They combined evolution of the REE composition and H–O isotope analysis also suggested that the ore-forming fluid originated from metamorphic water. As an exceptionally large fluid flux with a mixed H_2O–CO_2–CH_4 is required, the ore-forming fluid in the Sanshandao gold deposit probably had a metamorphic source.

The source of ore-forming fluids cannot be the Late Archean Jiaodong Group metamorphic rocks because the gold event occurred about 2 billion years later than the regional high-grade metamorphism. These components would have been lost from the rocks. As an exceptionally large fluid flux with a mixed H_2O–CO_2–CH_4 is required, the potential fluid reservoirs could be associated with the dehydration and decarbonization of the subducting Paleo-Pacific plate.

The $\delta^{34}S$ values of hydrothermal pyrite in the main stage have a narrow range, from 7.9‰ to 12.6‰ (Figure 8) with average of 10.25‰. These values are near or outside the upper ranges of the Linglong granite (6.1‰–10.1‰) [35], Guojialing granite (2.7‰–10.0‰) [29], Mesozoic intermediate-basic dykes

(5.3‰–10.8‰) [12], and the Jingshan Group [68] (Figure 8), but higher than that of the metamorphic rocks in the Jiaodong Group [35] (7.2‰–7.6‰) and the majority of lode gold deposits (Figure 8). The pyrite $\delta^{34}S$ from different sampling locations in the Sanshandao deposit are nearly invariable and consistent with other typical Jiaojia-type gold deposits. Similarly, in the Xinli gold deposit, the pyrites $\delta^{34}S$ display little variation [25]. The sulfur was suggested to be directly sourced from the Mesozoic granites, which probably scavenged the sulfur from the Jiaodong and Jingshan Groups [18,69,70]. Goldfarb and Santosh [6] ruled out the above argument on the basis of the evidence of the geological relationship and radiometric ages and proposed that the heavy sulfur could have been derived from the subducting marine sediments.

The enrichment of heavy sulfur isotopes is diagnostic of submarine exhalative deposits, the sulfur in which is mainly derived from non-bacterial reduction of isotopically heavy marine sulfate [71,72]. Zhu et al. [73] suggested that the heavy S isotope signature of both deposit types in Jiaodong indicates that the sulfur was derived from the reduction of marine sulfate. The heavy S isotope compositions originating from the reduction of seawater sulfate have been found in most sediment-hosted orogenic Au deposits, VHMS, and SEDEX Cu–Zn–Pb deposits [73–76]. However, the wall rocks of the Jiaodong gold deposits, dominated by Mesozoic granite and the Neoarchaean basement rocks, could not supply seawater sulfate, and the possibility of the heavy S isotope being derived from recycled seawater in ore-forming fluids was also ruled out by the H–O isotope compositions [17]. Considering the existence of many mafic dikes, which are associated with subduction of the Palaeo-Pacific plate nearly coeval with gold mineralization, sedimentary materials derived from the slab are a possible sulfur source. The possibility that the heavy sulfur could have been derived directly from evolved seawater is also worth considering.

Based on the H–O–S isotope analysis from this and previous studies [77–82], and given that an exceptionally large fluid flux with a mixed H_2O–CO_2–CH_4 is required, we think that the ore-forming fluids may be mainly associated with the dehydration and decarbonization of the subducting Paleo-Pacific plate, although devolatilization of a hydrated mantle wedge or subcontinental lithospheric mantle [6] cannot be excluded. Ore-forming materials bear both the mantle and the crustal fingerprints. This agrees with the subducted-related model proposed by Goldfarb and Santosh [6,83], and the sulfur was derived from subducted marine sediments.

6.2.2. Evolution of Ore-Forming Fluids

The evolutionary trend of ore fluid can be determined using the plot of homogenization temperature versus salinity [84]. The correlation between homogenization temperature and salinity of fluid inclusions trapped in fluid mixing is positive, and negative for fluid inclusions trapped in the process of fluid boiling or phase separation (Figure 9) [62]. According to fluid inclusion microthermometry and laser Raman spectroscopy, the dominant type II inclusions and subordinate type III inclusions developed in the early-stage quartz. Ore-forming fluids in this stage belong to H_2O–CO_2–NaCl system, which has moderate-high temperatures, high CO_2, and medium-to-low salinity (Figure 9). The water–rock reaction between metal-rich fluids and metamorphic rocks is a common process in each gold deposit (Water/Rock ratio of 0.1–0.5) [85], and the interaction was extensive at temperatures of 280–380 °C [85]. However, these temperatures coincide with the homogenization temperatures of early-stage FIs, in addition to widespread hydrothermal alteration in the early stage. The $\delta^{18}O$ values of hydrothermal quartz are clearly different among the ores, altered rocks, and wall rocks, indicating that water–rock interactions were common during the early stages, significantly affecting the ore-forming fluid of the deposit. In the main stage, all three types of inclusions were identified, but still dominated by type II inclusions. The CH_4 content in the carbonic phase increased dramatically, which led to the ore-forming system evolving from H_2O–CO_2–NaCl to H_2O–CO_2–NaCl ± CH_4. As mineralization progressed, the ore-forming fluids moved upward through the faults and underwent water–rock reactions and fluid immiscibility, resulting in a large number of volatiles escaping, accompanied by increasing salinity and density of the ore fluids (Figures 9 and 10).

Finally, the fluid evolved into an aqueous H_2O–NaCl system with low temperature, low salinity, high density, and no CO_2 in the late stage.

6.3. Mechanisms of Gold Deposition

In hydrothermal solutions, gold is mainly dissolved and transported in the form of gold bisulfide complexes ($Au(HS)_2^-$, $Au(HS)^0$) and gold chloride complexes ($AuCl_2^-$) [86–90]. Given the close relationship between gold and sulfides (especially pyrite), both in the alteration zones and in the stockwork quartz veins. The ore fluids were near-neutral to weakly acidic [29,91], mostly characterized by low salinity ($\leq$10 wt % NaCl eqv.) and moderate temperature (250–350 °C) at the Sanshandao gold deposit, we infer that $Au(HS)_2^-$ was most probably the gold transporting complex at Sanshandao.

The FIs homogenization temperatures of the Sanshandao gold deposit range from 220 to 320 °C, following a normal distribution (Figure 6). This indicates that mineralization occurred under medium-temperature conditions. In terms of gold mineralization at Sanshandao, the disseminated- and stockwork-style ores are within the hydrothermal breccias and cataclastic zones. The echelon tensile auriferous veins developed in subsidiary faults, suggesting the continuous cycle of fluid pressure fluctuation associated with incremental rupture during fault zone movement [92]. Sibson et al. described the mechanism of fluid pressure cycling that probably occurred at Sanshandao, and it was significant due to the high angle of the fault veins (Figure 3). The cycle of fluid pressure build-up, failure, fluid pressure drop, and re-sealing by hydrothermal deposition [93] is apparent during vein formation at the Sanshandao gold deposit. The estimated pressures of the type II FIs are shown in Figure 12. The trapping pressures of the Sanshandao gold deposit from the early stage to the main stage are 165–200 MPa and 90–175 MPa, respectively, showing a gradual decrease during the evolution of the ore-forming fluids (Figure 12). Subsequently, large-scale fluid immiscibility occurred during the main stage of gold mineralization due to pressure fluctuation and temperature decrease. Along with fluid immiscibility, H_2S was fractionated strongly into the vapor phase and the activity of HS^- lowered [94,95]. This process might have led to the decomposition of $Au(HS)_2^-$ and the concomitant deposition of gold [96]. Fluid immiscibility can be caused via fluid–pressure cycling due to fault failure during development of the vein system [97]. Fluid-inclusion results indicate that, similar to other gold deposits in Jiaodong [98,99], fluid immiscibility caused by seismic movement along fault zones, lowered the gold-complex solubility are interpreted to be the major precipitation mechanism of gold deposition [29], especially for the vein- and veinlet-style ores [100,101].

Fluid–rock interaction was intensive in the Sanshandao gold deposit, which resulted in widespread hydrothermal alteration, especially sericitization, silicification, and pyritization. Iron entered the ore-forming fluid continuously during the reaction between ore-bearing fluids and iron-bearing wall-rock, resulting in the formation of pyrite, which would decrease the H_2S content of the fluid, then decrease the solubility of $Au(HS)_2^-$, thereby causing gold deposition [96,102–107]. Previous work indicated that minor gold was deposited during the early stage when no fluid immiscibility occurred, and gold has a close relationship with the widespread sulfidation, especially pyritization. All this suggests that the wall-rock sulfidation also played a role in gold deposition [29]. Abundant CH_4 occurred in the main stage at the Sanshandao gold deposit (Figure 7) relative to other gold deposits in Jiaodong, and CH_4 can expand the temperature and pressure range of fluid immiscibility and facilitate the occurrence of fluid immiscibility [91]. This can also contribute to the precipitation of gold. The appearance of CH_4 indicates that the ore-forming fluid transformed into a more reducing fluid, which would cause CO_2 to be reduced to CH_4 and also cause gold precipitation. This is not a strong effect and is probably a second-order process for gold deposition.

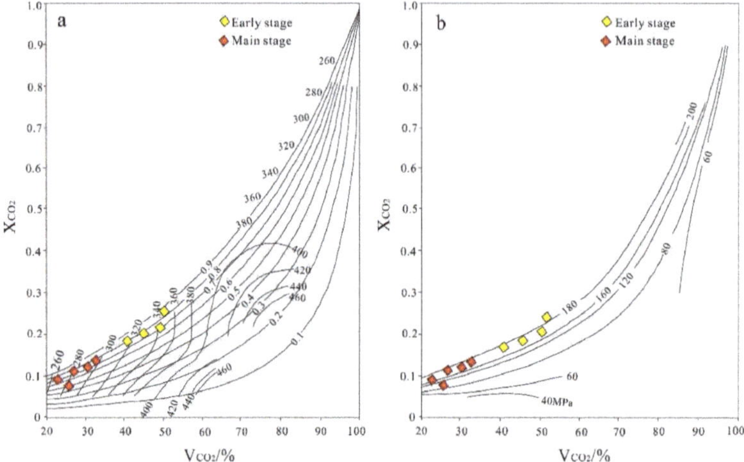

Figure 12. (**a**) Relationship of X_{CO2}, V_{CO2}, and CO_2 density, to Th_{TOT} of the CO_2–H_2O–NaCl fluid inclusions with salinity of 6.0 wt % NaCl eqv. [108]. (**b**) Relationship of X_{CO2}, V_{CO2} to pressure of the CO_2–H_2O–NaCl fluid inclusions with salinity of 6.0 wt % NaCl eqv. [108]. X_{CO2}—Mole fraction of CO_2; V_{CO2}—Volume of CO_2.

7. Conclusions

Three types of fluid inclusion were identified in the Sanshandao gold deposit: Type I pure CO_2, type II aqueous-carbonic, and type III aqueous fluid inclusions. The type II fluid inclusions can be subdivided into gas-rich inclusions (II-g type) and aqueous-rich inclusions (II-l type). The fluid-inclusion assemblage in the early stage mainly contains type II inclusions, whereas all three types of fluid inclusion coexist in the main stage quartz. In the late stage, only type III fluid inclusions were observed.

The ore-forming fluid was genetically derived from the subducting slab and metasomatized mantle. Ore-forming materials (gold and sulfur) were probably scavenged from the metasomatized mantle enriched with subduction-related components, predominantly originating from the Paleo-Pacific oceanic slab and its overlying sediments, which were thrust below the high-grade metamorphic rocks of the Jiaodong Peninsula. Both mantle and recycled crust components contributed ore-forming materials.

Gold was transported as the $Au(HS)_2^-$ complex in hydrothermal solution at the Sanshandao gold deposit. Fluid immiscibility caused by episodic pressure drops due to fault movements, together with sulfidation and other physicochemical changes resulting from fluid–rock interactions, were the two main gold deposition mechanisms.

Author Contributions: L.Y. conceived and designed the ideas; Y.L., S.W., X.L., H.W. and B.C. participated in field investigation and photo processing; Y.L. performed the data curation, and writing—original draft preparation; L.Y. and Y.L. analyzed the data; Y.L. and L.Y. reviewed and edited the draft. D.L., P.W. and W.C provided help in the observation of fluid inclusions. All the data were obtained from previous work of the project team.

Funding: This Study was financially supported by the National Key Research Program of China (Grant No. 2016YFC0600107), the National Natural Science Foundation of China (Grant No. 41572069), the MOST Special Fund from the State Key Laboratory of Geological Processes and Mineral Resources, China University of Geosciences (Grant No. MSFGPMR201804), Key Laboratory of Gold Mineralization Processes and Resource Utilization Subordinated to the Ministry of Natural Resources and Key Laboratory of Metallogenic Geological Process and Resources Utilization in Shandong Province, Shandong Institute of Geological Sciences (Grant No. KFKT201801, KFKT201802), and the 111 Project under the Ministry of Education and the State Administration of Foreign Experts Affairs, China (Grant No. B07011).

Acknowledgments: We thank to Kun Shen, who provided constructive comments on the fluid-inclusion experiment, as well as comments from Wei Shan and Min Li. We also thank Guangke Zhao and other staff from Sanshandao Gold Mine Corporation, as well as Jun Deng, Groves, Zhongliang Wang, Nan Li, Kunfeng Qiu and Liang Zhang for their help to improve the manuscript.

Conflicts of Interest: The authors declare no conflict of interest.

References

1. Deng, J.; Yang, L.Q.; Sun, Z.S.; Wang, J.P.; Wang, Q.F.; Xin, H.B.; Li, X.J. A metallogenic model of gold deposits of the Jiaodong granite green stone belt. *Acta Geol. Sin.* **2003**, *77*, 537–546.

2. Deng, J.; Yang, L.Q.; Li, R.H.; Groves, D.I.; Santosh, M.; Wang, Z.L.; Sai, S.X.; Wang, S.R. Regional structural control on the distribution of world-class gold deposits: An overview from the Giant Jiaodong Gold Province, China. *Geol. J.* **2018**. [CrossRef]

3. Yang, L.Q.; Deng, J.; Wang, Q.F.; Zhou, Y.H. Coupling effects on gold mineralization of deep and shallow structures in the northwestern Jiaodong Peninsula, eastern China. *Acta Geol. Sin.* **2006**, *80*, 400–411.

4. Yang, L.Q.; Dilek, Y.; Wang, Z.L.; Weinberg, R.F.; Liu, Y. Late Jurassic, High Ba-Sr Linglong granites in the Jiaodong Peninsula, East China: Lower crustal melting products in the Eastern North China Craton. *Geol. Mag.* **2018**, *155*, 1040–1062. [CrossRef]

5. Groves, D.I.; Santosh, M. The giant Jiaodong gold province: The key to a unified model for orogenic gold deposits? *Geosci. Front.* **2016**, *7*, 409–417. [CrossRef]

6. Goldfarb, R.J.; Santosh, M. The dilemma of the Jiaodong gold deposits: Are they unique? *Geosci. Front.* **2014**, *5*, 139–153. [CrossRef]

7. Goldfarb, R.J.; Hart, C.J.R.; Davis, G.; Groves, D.I. East Asian gold–deciphering the anomaly of Phanerozoic gold in Precambrian cratons. *Econ. Geol.* **2007**, *102*, 341–346. [CrossRef]

8. Yang, L.Q.; Deng, J.; Guo, R.P.; Guo, L.N.; Wang, Z.L.; Chen, B.H.; Wang, X.D. World-class Xincheng gold deposit: An example from the giant Jiaodong Gold Province. *Geosci. Front.* **2016**, *7*, 419–430. [CrossRef]

9. Wang, Z.L.; Yang, L.Q.; Deng, J.; Santosh, M.; Zhang, H.F.; Liu, Y.; Li, R.H.; Huang, T.; Zheng, X.L.; Zhao, H. Gold-hosting high Ba-Sr granitoids in the Xincheng gold deposit, Jiaodong Peninsula, East China: Petrogenesis and tectonic setting. *J. Asian Earth Sci.* **2014**, *95*, 274–299. [CrossRef]

10. Deng, J.; Liu, X.F.; Wang, Q.F.; Pan, R.G. Origin of the Jiaodong-type Xinli gold deposit, Jiaodong Peninsula, China: Constraints from fluid inclusion and C–D–O–S–Sr isotope compositions. *Ore Geol. Rev.* **2014**, *65*, 674–686. [CrossRef]

11. Qiu, Y.M.; Groves, D.I.; McNaughton, N.J.; Wang, L.Z.; Zhou, T.H. Nature, age and tectonic setting of granitoid-hosted orogenic gold deposits of the Jiaodong Peninsula, eastern North China craton, China. *Miner. Depos.* **2002**, *37*, 283–305. [CrossRef]

12. Yang, L.Q.; Deng, J.; Wang, Z.L.; Zhang, L.; Guo, L.N.; Song, M.C.; Zheng, X.L. Mesozoic gold metallogenic system of the Jiaodong gold province, eastern China. *Acta Petrol. Sin.* **2014**, *30*, 2447–2467. (In Chinese with English Abstract).

13. Yang, L.Q.; Guo, L.N.; Wang, Z.L.; Zhao, R.X.; Song, M.C.; Zheng, X.L. Timing and mechanism of gold mineralization at the Wang'ershan gold deposit, Jiaodong Peninsula, eastern China. *Ore Geol. Rev.* **2017**, *88*, 491–510. [CrossRef]

14. Deng, J.; Wang, Q.F.; Wan, L.; Yang, L.Q.; Gong, Q.J.; Zhao, J.; Liu, H. Self-similar fractal analysis of gold mineralization of Dayingezhuang disseminated-veinlet deposit in Jiaodong gold province, China. *J. Geochem. Explor.* **2009**, *102*, 95–102. [CrossRef]

15. Deng, J.; Yang, L.Q.; Gao, B.F.; Sun, Z.S.; Guo, C.Y.; Wang, Q.F.; Wang, J.P. Fluid evolution and metallogenic dynamics during tectonic regime transition: Example from the Jiapigou gold belt in Northeast China. *Resour. Geol.* **2009**, *59*, 140–152. [CrossRef]

16. Fan, H.R.; Zhai, M.G.; Xie, Y.H.; Yang, J.H. Ore-forming fluids associated with granite hosted gold mineralization at the Sanshandao deposit, Jiaodong gold province, China. *Miner. Depos.* **2003**, *38*, 739–750. [CrossRef]

17. Li, S.R.; Santosh, M. Metallogeny and craton destruction: Records from the North China Craton. *Ore Geol. Rev.* **2014**, *56*, 376–414. [CrossRef]

18. Tan, J.; Wei, J.H.; Audétat, A.; Pettke, T. Source of metals in the Guocheng gold deposit, Jiaodong Peninsula, North China Craton: Link to early Cretaceous mafic magmatism originating from Paleoproterozoic metasomatized lithospheric mantle. *Ore Geol. Rev.* **2012**, *48*, 70–87. [CrossRef]

19. Wang, L.; Sun, F.Y.; Wang, J.L. Geochemical features of ore-forming fluids of the Jinling gold deposit, Shandong Province. *Acta Petrol. Sin.* **2010**, *26*, 3735–3744. (In Chinese with English Abstract).

20. Yang, L.Q.; Deng, J.; Goldfarb, R.J.; Zhang, J.; Gao, B.F.; Wang, Z.L. 40Ar/39Ar geochronological constraints on the formation of the Dayingezhuang gold deposit: New implications for timing and duration of hydrothermal activity in the Jiaodong gold province, China. *Gonwana Res.* **2014**, *25*, 1469–1483. [CrossRef]

21. Zhang, L.; Li, G.W.; Zheng, X.L.; An, P.; Chen, B.Y. Tectonic-thermal history of Sanshandao gold deposit in eastern Jiaodong: Constraints of 40Ar/39Ar and fission track chronology. *Acta Petrol. Sin.* **2016**, *32*, 2465–2476. (In Chinese with English Abstract).

22. Zhai, M.G.; Yang, J.H.; Liu, W.J. Large clusters of gold deposits and large-scale metallogenesis in the Jiaodong Peninsula of Eastern China. *Sci. China Ser. D* **2001**, *44*, 758–768. [CrossRef]

23. Groves, D.I.; Goldfarb, R.J.; Robert, F.; Hart, C.J.R. Gold deposits in metamorphic belts: Overview of current understanding, outstanding problems, future research, and exploration significance. *Econ. Geol.* **2003**, *98*, 1–29.

24. Mao, J.W.; Wang, Y.T.; Li, H.M.; Pirajno, F.; Zhang, C.Q.; Wang, R.T. The relationship of mantle-derived fluids to gold metallogenesis in the Jiaodong Peninsula: Evidence from D–O–C–S isotope systematics. *Ore Geol. Rev.* **2008**, *33*, 361–381. [CrossRef]

25. Deng, J.; Wang, C.M.; Bagas, L.; Carranza, E.J.M.; Lu, Y.J. Cretaceous-Cenozoic tectonic history of the Jiaojia Fault and gold mineralization in the Jiaodong Peninsula, China: Constraints from zircon U-Pb, illite K-Ar and apatite fission track thermochronometry. *Miner. Depos.* **2015**, *50*, 987–1006. [CrossRef]

26. Chen, G.Y.; Shao, W.; Sun, D.S. *Genetic Mineralogy of Gold Deposits in Jiaodong Region with Emphasis on Gold Prospecting*; Chongqing Publishing House: Chongqing, China, 1989; pp. 181–224. (In Chinese with English Abstract).

27. Zhang, L.; Yang, L.Q.; Wang, Y.; Weinberg, R.F.; An, P.; Chen, B.Y. Thermochronologic constrains on the processes of formation and exhumation of the Xinli orogenic gold deposit, Jiaodong Peninsula, eastern China. *Ore Geol. Rev.* **2017**, *81*, 140–153. [CrossRef]

28. Goldfarb, R.J.; Groves, D.I.; Gardoll, S. Orogenic gold and geological time: A synthesis. *Ore Geol. Rev.* **2001**, *18*, 1–75. [CrossRef]

29. Deng, J.; Yuan, W.; Carranza, E.J.; Yang, L.; Wang, C.; Yang, L.; Hao, N. Geochronology and thermochronometry of the Jiapigou gold belt, northeastern China: New evidence for multiple episodes of mineralization. *J. Asian Earth Sci.* **2014**, *89*, 10–27. [CrossRef]

30. Zhang, X.O.; Cawood, P.A.; Wilde, S.A.; Liu, R.Q.; Song, H.L.; Li, W.; Snee, L.W. Geology and timing of mineralization at the Cangshang gold deposit, north-western Jiaodong Peninsula, China. *Miner. Depos.* **2003**, *38*, 141–153. [CrossRef]

31. Wang, Z.L.; Yang, L.Q.; Guo, L.N.; Marsh, E.; Wang, J.P.; Liu, Y.; Zhang, C.; Li, R.H.; Zhang, L.; Zheng, X.L.; et al. Fluid immiscibility and gold deposition in the Xincheng deposit, Jiaodong Peninsula, China: A fluid inclusion study. *Ore Geol. Rev.* **2015**, *65*, 701–717. [CrossRef]

32. Yang, L.Q.; Deng, J.; Guo, L.N.; Li, J.L. Origin and evolution of ore fluid, and gold deposition processes at the giant Taishang gold deposit, Jiaodong Peninsula, eastern China. *Ore Geol. Rev.* **2016**, *72*, 585–602. [CrossRef]

33. Deng, J.; Wang, Q.F.; Wan, L.; Liu, H.; Yang, L.Q.; Zhang, J. A mutifractal analysis of mineralization characteristics of the Dayinggezhuang disseminated-veinlet gold deposit in the Jiaodong gold province of China. *Ore Geol. Rev.* **2011**, *40*, 54–64. [CrossRef]

34. Mao, J.W.; Li, H.M.; Wang, Y.T.; Zhang, C.Q.; Wang, R.T. The relationship between mantle-derived fluid and gold ore-formation in the Eastern Shandong Peninsula: Evidences from D–O–C–S isotopes. *Acta Geol. Sin.* **2005**, *79*, 839–857. (In Chinese with English Abstract).

35. Brown, P.E. Fluid inclusion modeling for hydrothermal systems. *Rev. Econ. Geol.* **1998**, *10*, 151–171.

36. Uemoto, T.; Ridely, J.; Mikucki, E.; Groves, D.I. Fluid chemical evolution as a factor in controlling the distribution of gold at the Archean Golden Crown lode gold deposit, Murchison Province, Western Australia. *Econ. Geol.* **2002**, *97*, 1227–1248. [CrossRef]

37. Dugdale, A.L.; Hagemann, S.G. The Bronzewing lode gold deposit, eastern Australia: P–T–X evidence for fluid immiscibility caused by cyclic decompression in goldbearing quartz-veins. *Chem. Geol.* **2001**, *173*, 59–90. [CrossRef]

38. Hagemann, S.G.; Luders, V. P–T–X conditions of hydrothermal fluids and precipitation mechanism of stibnite–gold mineralization at the Wiluna lode-gold deposits, Western Australia: Conventional and infrared microthermometric constraints. *Mineral. Deposita* **2003**, *38*, 936–952. [CrossRef]

39. Hu, S.X. *Petrology of Metasomatic Rocks and Implications for Ore Exploration*; Science Press: Beijing, China, 2002; pp. 1–264. (In Chinese)

40. Bodnar, R.J.; Vityk, M.O. Interpretation of microthermometric data for H_2O-NaCl fluid inclusions. In *Fluid Inclusions in Minerals: Methods and Applications*; de Vivo, B., Frezzotti, M.L., Eds.; Verginia Tech: Blacksburg, VA, USA, 1994; pp. 117–130.

41. Deng, J.; Liu, X.F.; Wang, Q.F.; Dilek, Y.; Liang, Y.Y. Isotopic Characterization and Petrogenetic Modeling of the Early Cretaceous Mafic Diking-Lithospheric Extension in the North China craton, eastern Asia. *GSA Bull.* **2017**, *129*, 1379–1407. [CrossRef]

42. Faure, M.; Lin, W.; Monie, P.; Breton, N.L.; Poussineau, S.; Panis, D.; Deloule, E. Exhumation tectonics of the ultrahigh-pressure metamorphic rocks in the Qinling orogen in east China: New petrological–structural-radiometric insights from the Shandong Peninsula. *Tectonics* **2003**, *22*, 1018–1040. [CrossRef]

43. Tang, J.; Zheng, Y.F.; Wu, Y.B.; Gong, B.; Zha, X.P.; Liu, X.M. Zircon u–pb age and geochemical constraints on the tectonic affinity of the Jiaodong terrane in the Sulu orogen, China. *Precambrian Res.* **2008**, *161*, 389–418. [CrossRef]

44. Zhai, M.G.; Bian, A.G.; Zhao, T.P. The amalgamation of the supercontinent of north China craton at the end of Neo-archaean and its breakup during late Palaeoproterozoic and Mesoproterozoic. *Sci. China Earth Sci.* **2000**, *43*, 219–232. [CrossRef]

45. Li, Q.L.; Chen, F.; Wang, X.L.; Li, X.H.; Li, C.F. Ultra-low procedural blank and the single-grain mica Rb–Sr isochron dating. *Chin. Sci. Bull.* **2005**, *50*, 2861–2865.

46. Zhang, J.; Zhao, Z.F.; Zheng, Y.F.; Dai, M. Postcollisional magmatism: Geochemical constraints on the petrogenesis of Mesozoic granitoids in the Sulu orogen, China. *Lithos* **2010**, *119*, 512–536. [CrossRef]

47. Yang, L.Q.; Deng, J.; Wang, Z.L.; Guo, L.N.; Li, R.H.; Groves, D.I.; Danyushevskiy, L.; Zhang, C.; Zheng, X.L.; Zhao, H. Relationships between gold and pyrite at the Xincheng gold deposit, Jiaodong Peninsula, China: Implications for gold source and deposition in a brittle epizonal environment. *Econ. Geol.* **2016**, *111*, 105–126. [CrossRef]

48. Jiang, N.; Chen, J.; Guo, J.; Chang, G. In situ zircon U-Pb, oxygen and hafnium isotopic compositions of Jurassic granites from the North China craton: Evidence for Triassic subduction of continental crust and subsequent metamorphism-related 18O depletion. *Lithos* **2012**, *142*, 84–94. [CrossRef]

49. Yang, L.Q.; Deng, J.; Wang, Z.L.; Zhang, L.; Goldfarb, R.J.; Yuan, W.M.; Weinberg, R.F.; Zhang, R.Z. Thermochronologic constraints on evolution of the Linglong Metamorphic Core Complex and implications for gold mineralization: A case study from the Xiadian gold deposit, Jiaodong Peninsula, eastern China. *Ore Geol. Rev.* **2016**, *72*, 165–178. [CrossRef]

50. Liu, Y.; Deng, J.; Wang, Z.L.; Zhang, L.; Zhang, C.; Liu, X.D.; Zheng, X.L.; Wang, X.D. Zircon U-Pb age, Lu-Hf isotopes and petrogeochemistry of the monzogranites from Xincheng gold deposit, northwestern Jiaodong Peninsula, China. *Acta Petrol. Sin.* **2014**, *30*, 2559–2573. (In Chinese with English Abstract).

51. Hou, M.L.; Jiang, Y.H.; Jiang, S.Y.; Ling, H.F.; Zhao, K.D. Contrasting origins of late Mesozoic adakitic granitoids from the northwestern Jiaodong Peninsula, East China: Implications for crustal thickening to delamination. *Geol. Mag.* **2007**, *144*, 619–631. [CrossRef]

52. Goss, S.C.; Wilde, S.A.; Wu, F.; Yang, J. The age, isotopic signature and significance of the youngest Mesozoic granitoids in the Jiaodong terrane, Shandong province, North China craton. *Lithos* **2010**, *3*, 309–326. [CrossRef]

53. Yang, L.Q.; Deng, J.; Wang, Z.L. Ore-controlling structural pattern of Jiaodong gold deposits: Geological-geophysical integration constraints. In *The Deep-Seated Structures of Earth in China*; Chen, Y., Jin, Z., Shi, Y., Yang, W., Zhu, R., Eds.; Sciences Press: Beijing, China, 2014; pp. 1006–1030.

54. Faure, M.; Lin, W.; Chen, Y. Is the Jurassic (Yanshanian) intraplate tectonics of North China due to westward indentation of the North China block? *Terra Nova* **2012**, *24*, 456–466. [CrossRef]

55. Zhu, G.; Niu, M.L.; Xie, C.L.; Wang, Y.S. Sinistral to normal faulting along the Tan-Lu fault zone: Evidence for geodynamic switching of the East China continental margin. *J. Geol.* **2010**, *118*, 277–293. [CrossRef]

56. Collins, P.L.F. Gas hydrates in CO_2-bearing fluid inclusions and use freezing data for estimation of salinity. *Econ. Geol.* **1979**, *74*, 1435–1444. [CrossRef]

57. Bakker, R.J.; Jansen, J.B.H. A mechanism for preferential H$_2$O leakage from fluid inclusions in quartz, based on TEM observations. *Contrib. Mineral. Petrol.* **1994**, *116*, 7–20. [CrossRef]

58. Brown, P.E.; Hagemann, S.G. MacFlincor and its application to fluids in Archaean lode-gold deposits. Geochim. *Cosmochim. Acta* **1995**, *59*, 3943–3952. [CrossRef]

59. Liu, H.B.; Jin, G.S.; Li, J.J.; Han, J.; Zhang, J.F.; Zhang, J.; Zhong, F.W.; Guo, D.Q. Determination of stable isotope composition in uranium geological samples. *World Nucl. Geosci.* **2013**, *3*, 174–179.

60. Roedder, E. Fluid inclusions. *Rev. Mineral.* **1984**, *12*, 1–664.

61. Clayton, R.N.; O'Neil, J.R.; Mayeda, T.K. Oxygen isotope exchange between quartz and water. *J. Geophys. Res.* **1972**, *77*, 3057–3067. [CrossRef]

62. Li, X.M. Studies and implications of the oxygen, hydrogen and carbon stable isotopes of the Sanshandao gold deposit, Shandong province. *Contrib. Geol. Miner. Resour. Res.* **1988**, *3*, 62–71. (In Chinese)

63. Zhang, L.G.; Chen, Z.S.; Liu, J.X.; Yu, G.X.; Wang, B.C.; Xu, J.F.; Zheng, W.S. Water-Rock Exchange in Jiaojia Type Gold Deposit: Hydrogen and Oxygen Isotope Composition of Ore-forming Fluids. *Miner. Depos.* **1994**, *13*, 193–200. (In Chinese)

64. Wang, Y.W.; Zhu, F.S.; Gong, R.T. Tectonic isotope geochemistry-Restudy of sulfur isotope in Jiaodong gold province. *Gold* **2002**, *23*, 1–16. (In Chinese)

65. Huang, D.Y. Sulfur Isotope Study of Metallogenic Series of Jiaodong Gold Deposit. *Miner. Depos.* **1994**, *13*, 75–87. (In Chinese)

66. Xavier, R.P.; Foster, R.P. Fluid evolution and chemical controls in the FazendaMaria Preta (FMP) gold deposit, Rio Itapicuru greenstone belt, Bahia, Brazil. *Chem. Geol.* **1999**, *154*, 133–154. [CrossRef]

67. Rusk, B.G.; Reed, M.H.; Dilles, J.H. Fluid inclusion evidence for magmatic- hydrothermal fluid evolution in the porphyry copper–molybdenum deposit at butte, Montana. *Econ. Geol.* **2008**, *103*, 307–334. [CrossRef]

68. Goldfarb, R.J.; Groves, D.I. Orogenic gold: Common or evolving fluid and metal sources through time. *Lithos* **2015**, *233*, 2–26. [CrossRef]

69. Li, Y.P. The genesis of the Rushan gold deposits in east Shandong. *Miner. Depos.* **1992**, *11*, 165–172. (In Chinese with English Abstract).

70. Luo, Z.K.; Miao, L.C. *Granites and Gold Deposits in Zhaoyuan–Laizhou Area, Eastern Shandong Province*; Metallurgical Industry Press: Beijing, China, 2002; pp. 84–117. (In Chinese with English Abstract).

71. Ohmoto, H.; Rye, R. Isotopes of sulfur and carbon. In *Geochemistry of Hydrothermal Ore Deposits*, 2nd ed.; Bernes, H.L., Ed.; John Wiley and Sons: New York, NY, USA, 1979; pp. 509–567.

72. Zhong, R.C.; Li, W.B. The multistage genesis of the giant Dongshengmiao Zn-Pb-Cu deposit in western Inner Mongolia, China: Syngenetic stratabound mineralization and metamorphic remobilization. *Geosci. Front.* **2016**, *7*, 529–542. [CrossRef]

73. Zhu, Z.Y.; Jiang, S.Y.; Ryan, M.; Cook, N.J.; Yang, T.; Wang, M.L.; Ma Ciobanu, C.L. Iron isotope behavior during fluid/rock interaction in K-feldspar alteration zone—A model for pyrite in gold deposits from the Jiaodong Peninsula, East China. *Geochim. Cosmochim. Acta* **2018**, *222*, 94–116. [CrossRef]

74. Sangster, D.F. Relative sulphur isotope abundances of ancient seas and strata-bound sulphide deposits. *Proc. Geol. Assoc. Can.* **1968**, *17*, 79–91.

75. Goodfellow, W.D.; Lydon, J.W.; Turner, R. Geology and genesis of stratiform sediment-hosted (SEDEX) zinc-lead silver sulphide deposits. *Geol. Assoc. Can. Spec. Pap.* **1993**, *40*, 1–25.

76. Huston, D.L. Stable isotopes and their significance for understanding the genesis of volcanic-hosted massive sulfide deposits: A review. *Rev. Econ. Geol.* **1999**, *8*, 157–179.

77. Roedder, E. Fluid Inclusions as Samples of Ore Fluids. In *Geochemistry of Hydrothermal Ore Deposits*, 2nd ed.; Wiley: New York, NY, USA, 1979; pp. 684–737.

78. Gui, F.; Wang, L.; Ma, F. Study on Fluid Inclusions in Sanshandao Gold Deposit. *Gold* **2014**, *35*, 27–32. (In Chinese with English Abstract).

79. Shepherd, T.J.; Rankin, A.H.; Alderton, D.H.M. *A Practical Guide to Fluid Inclusion Studies*; Blackie & Son Limited: Glasgow, UK; London, UK, 1985; pp. 1–154.

80. Anderson, M.R.; Rankin, A.H.; Spiro, B. Fluid mixing in the generation of isothermal gold mineralization in the Transvaal Sequence, Transvaal, South Africa. *Eur. J. Mineral.* **1992**, *4*, 933–948. [CrossRef]

81. Yang, M.Z. *The Geochemistry of Wallrock Alteration Zone of Gold Deposits-As Exemplified by Jiaodong Gold Deposits*; Geological Publishing House: Beijing, China, 1998; pp. 1–120. (In Chinese)

82. Li, H.Q.; Liu, J.Q.; Wei, L. *Chronological Study of Fluid Inclusion for Hydrothermal Deposit and Its Geological Applications*; Geological Publishing House: Beijing, China, 1993; pp. 1–126. (In Chinese)

83. Deng, J.; Wang, Q.F. Gold mineralization in China: Metallogenic provinces, deposit types and tectonic framework. *Gondwana Res.* **2016**, *36*, 219–274. [CrossRef]

84. Zhai, D.G.; Liu, J.J.; Wang, J.P.; Yang, Y.Q.; Liu, X.W.; Wang, G.W.; Liu, Z.J.; Wang, X.L.; Zhang, Q.B. Characteristics of melt–fluid inclusions and sulfur isotopic compositions of the Hashitu molybdenum deposit, Inner Mongolia. *Earth Sci. J. China Univ. Geosci.* **2012**, *37*, 1279–1290. (In Chinese with English Abstract).

85. Yang, L.Y.; Yang, L.Q.; Yuan, W.M.; Zhang, C.; Zhao, K.; Yu, H.J. Origin and evolution of ore fluid for orogenic gold traced by D-O isotopes: A case from the Jiapigou gold belt, China. *Acta Petrol. Sin.* **2013**, *29*, 4025–4035. (In Chinese with English Abstract).

86. Seward, T.M. Thio complexes of gold and the transport of gold in hydrothermal ore solutions. *Geochim. Cosmochim. Acta* **1973**, *37*, 379–399. [CrossRef]

87. Seward, T.M. *The Hydrothermal Geochemistry of Gold, Gold Metallogeny and Exploration*; Springer: New York, NY, USA, 1990; pp. 37–62.

88. Hayashi, K.I.; Ohmoto, H. Solubility of gold in NaCl- and H2S-bearing aqueous olutions at 250–350 °C. *Geochim. Cosmochim. Acta* **1991**, *55*, 2111–2126. [CrossRef]

89. Gammons, C.H.; Williams-Jones, A.E.; Yu, Y. New data on the stability of gold I) chloride complexes at 300 °C. *Mineral. Mag. A* **1994**, *58*, 309–310. [CrossRef]

90. Benning, L.G.; Seward, T.M. Hydrosulphide complexing of gold (I) in hydrothermal solutions from 150 to 500 °C and 500 to 1500 bars. *Geochim. Cosmochim. Acta* **1996**, *60*, 849–1871. [CrossRef]

91. Yang, L.Q.; Badal, J. Mirror symmetry of the crust in the oil/gas region of Shengli, China. *J. Asian Earth Sci.* **2013**, *78*, 327–344. [CrossRef]

92. Cox, S.F. Faulting processes at high uid pressures: An example of fault valve behaviour from the Wattle Gully Fault, Victoria, Australia. *J. Geophys. Res.* **1995**, *100*, 12841–12859. [CrossRef]

93. Sibson, R.H.; Robert, F.; Poulsen, K.H. High angle reverse faults, uid-pressure cycling, and mesothermal gold–quartz deposits. *Geology* **1988**, *16*, 551–555. [CrossRef]

94. Naden, J.; Shepherd, T.J. Role of methane and carbon dioxide in gold depositions. *Nature* **1989**, *342*, 793–795. [CrossRef]

95. Cox, S.F.; Sun, S.S.; Etheridge, M.A.; Wall, V.J.; Potter, T.F. Structural and geochemical controls on the development of turbidite-hosted gold quartz vein deposits, Wattle Gully mine, central Victoria, Australia. *Econ. Geol.* **1995**, *90*, 1722–1746. [CrossRef]

96. Mikucki, E.J. Hydrothermal transport and depositional processes in Archean lode gold systems: A review. *Ore Geol. Rev.* **1998**, *13*, 307–321. [CrossRef]

97. Wilkinson, J.J.; Johnston, J.D. Pressure fluctuations, phase separation, and gold precipitation during seismic fracture propagation. *Geology* **1996**, *24*, 395–398. [CrossRef]

98. Yang, L.Q.; Deng, J.; Guo, C.Y.; Zhang, J.; Jiang, S.Q.; Gao, B.F.; Gong, Q.J.; Wang, Q.F. Ore-forming fluid characteristics of the Dayingezhuang gold deposit, Jiaodong gold province, China. *Resour. Geol.* **2009**, *59*, 182–195. [CrossRef]

99. Qiu, K.F.; Marsh, E.; Yu, H.C.; Pfaff, K.; Gulbransen, C.; Gou, Z.Y.; Li, N. Fluid and metal sources of the Wenquan porphyry molybdenum deposit, Western Qinling, NW China. *Ore Geol. Rev.* **2017**, *86*, 459–473. [CrossRef]

100. Qiu, K.F.; Deng, J.; Taylor, R.D.; Song, K.R.; Song, Y.H.; Li, Q.Z.; Goldfarb, R.J. Paleozoic magmatism and porphyry Cu-mineralization in an evolving tectonic setting in the North Qilian Orogenic Belt, NW China. *J. Asian Earth Sci.* **2016**, *122*, 20–40. [CrossRef]

101. Yang, Q.Y.; Santosh, M.; Shen, J.F.; Li, S.R. Juvenile vs. recycled crust in NE China: Zircon U–Pb geochronology, Hf isotope and an integrated model for Mesozoic gold mineralization in the Jiaodong Peninsula. *Gondwana Res.* **2014**, *25*, 1445–1468. [CrossRef]

102. Yang, L.Q.; Deng, J.; Qiu, K.F.; Ji, X.Z.; Santosh, M.; Song, K.R.; Song, Y.H.; Geng, J.Z.; Zhang, C.; Hua, B. Magma mixing and crust mantle interaction in the Triassic monzogranites of Bikou Terrane, central China: Constraints from petrology, geochemistry, and zircon U–Pb–Hf isotopic systematic. *J. Asian Earth Sci.* **2015**, *98*, 320–341. [CrossRef]

103. Qiu, K.F.; Taylor, R.D.; Song, Y.H.; Yu, H.C.; Song, K.R.; Li, N. Geologic and geochemical insights into the formation of the Taiyangshan porphyry copper–molybdenum deposit, Western Qinling Orogenic Belt, China. *Gondwana Res.* **2016**, *35*, 40–58. [CrossRef]

104. Yang, L.Q.; Deng, J.; Dilek, Y.; Qiu, K.F.; Ji, X.Z.; Li, N.; Taylor, R.D.; Yu, J.Y. Structure, Geochronology, and Petrogenesis of the Late Triassic Puziba Granitoid Dikes in the Mianlue Suture Zone, Qinling Orogen, China. *GSA Bulletin* **2015**, *127*, 1831–1854. [CrossRef]

105. Qiu, K.F.; Deng, J. Petrogenesis of granitoids in the Dewulu skarn copper deposit: implications for the evolution of the Paleotethys ocean and mineralization in Western Qinling, China. *Ore Geol. Rev.* **2017**, *90*, 1078–1098. [CrossRef]

106. Groves, D.I.; Phillips, G.; Ho, S.E.; Houstoun, S.M. The nature, genesis and regional controls of gold mineralization in Archaean greenstone belts of the Western Australian Shield; a brief review. *S. Afr. J. Geol.* **1985**, *88*, 135–148.

107. Qiu, K.F.; Song, K.R.; Song, Y.H. Magmatic-hydrothermal fluid evolution of the Wenquan porphyry molybdenum deposit in the north margin of the Western Qinling, China. *Acta Petrol. Sin.* **2015**, *31*, 3391–3404. (In Chinese with English Abstract).

108. Schwartz, M.O. Determining phase volumes of mixed CO_2-H_2O inclusions using microthermetric measurements. *Miner. Depos.* **1989**, *24*, 43–47. [CrossRef]

Article

Ore-Fluid Evolution of the Sizhuang Orogenic Gold Deposit, Jiaodong Peninsula, China

Yu-Ji Wei [1], Li-Qiang Yang [1,*], Jian-Qiu Feng [1], Hao Wang [1], Guang-Yao Lv [2], Wen-Chao Li [2] and Sheng-Guang Liu [2]

[1] State Key Laboratory of Geological Processes and Mineral Resources, China University of Geosciences, Beijing 100083, China; 2001170098@cugb.edu.cn (Y.-J.W.); fjq8710736@163.com (J.-Q.F.); 2001180061@cugb.edu.cn (H.W.)

[2] Sizhuang Gold Deposit, Shandong Gold Mining (Laizhou) Co., Ltd., Laizhou 261400, China; lvguangyao@sd-gold.com (G.-Y.L.); woliwenchao01@163.com (W.-C.L.); 13188781193@126.com (S.-G.L.)

* Correspondence: lqyang@cugb.edu.cn; Tel.: +86-010-82321937

Received: 14 January 2019; Accepted: 16 March 2019; Published: 21 March 2019

Abstract: The Sizhuang gold deposit with a proven gold resource of >120 t, located in northwest Jiaodong Peninsula in China, lies in the southern part of the Jiaojia gold belt. Gold mineralization can be divided into altered rock type, auriferous quartz vein type, and sulfide-quartz veinlet in K-feldspar altered granite. According to mineral paragenesis and mineral crosscutting relationships, three stages of metal mineralization can be identified: early stage, main stage, and late stage. Gold mainly occurs in the main stage. The petrography and microthermometry of fluid inclusion shows three types of inclusions (type 1 H_2O–CO_2 inclusions, type 2 aqueous inclusions, and type 3 CO_2 inclusions). Early stage quartz-hosted inclusions have a trapped temperatures range 303–390 °C. The gold-rich main stage contains a fluid-inclusion cluster with both type 1 and 2 inclusions (trapped between 279 and 298 °C), and a wide range of homogenization temperatures of CO_2 occurs to the vapor phase (17.6 to 30.5 °C). The late stage calcite only contains type 1 inclusions with homogenization temperatures between 195 and 289 °C. With evidences from the H–O isotope data and the study of water–rock interaction, the metamorphic water of the Jiaodong Group is considered to be the dominating source for the ore-forming fluid. The ore-fluid belonged to a CO_2–H_2O–NaCl system with medium-low temperature (160–360 °C), medium-low salinity (3.00–11.83 wt% NaCl eq.), and low density (1.51–1.02 g/cm^3). Fluid immiscibility caused by pressure fluctuation is the key mechanism in inducing gold mineralization in the Sizhuang gold deposit.

Keywords: fluid inclusions; H–O isotope; immiscibility; water–rock interaction; Sizhuang gold deposit; Jiaodong Peninsula; China

---

1. Introduction

The giant Jiaodong gold deposit, with more than 4500 t proven gold reserves [1], is the most important gold producer of China [2,3]. More than 150 deposits in Jiaodong area are generally divided into two styles, Jiaojia-style and Linglong-style [4]. The mineralization of the former is the disseminated-stockwork type which occurs in the pyrite-sericite-quartz alteration belt and the latter's is the auriferous quartz vein type which is mostly controlled by subsidiary faults [5]. However, both styles invariably occur in every single deposit in Jiaodong [6,7], with similar mineral paragenesis, alteration assemblages, and gold deposition ages [1,8].

Many previous inclusion studies show that three types of primary inclusions (H_2O–CO_2, aqueous, and CO_2 inclusions) have been founded in both the mineralization styles [6,9,10]. They belong to the CO_2–H_2O–NaCl $\pm$ CH_4 fluid system with a medium-high temperature (158–393 °C) and a medium-low salinity ($\leq$11.2 wt % NaCl eq.) [6–10]. The fluid immiscible system was discussed and considered as a significant role in gold mineralization [6,10]. The source of ore-forming fluids still remains a controversy: Magmatic water, mixing with meteoric water before and/or at mineralization, is considered the source based on a H–O isotope composition in many studies [11,12]. Deng et al. (2015) [13] and Yang et al. (2017) [14] considered that the source of Jiaodong gold deposit may be from both the crust and mantle by carbon and sulfur isotope study; The δD values of hydrothermal sericite together with the O–S isotope indicate that the source of ore-forming metal is derived from the paleo-Pacific oceanic slab and its overlying sediments [6].

For the Sizhuang gold deposit, there are few studies including the geology, structures, and hydrothermal alteration [15], and only one published paper includes a fluid inclusion study. Wei et al (2015) [16] identified H_2O–CO_2 and aqueous inclusions in quartz (hosted in auriferous quartz vein in early stage and main stage) and indicated a medium-low temperature (133–310 °C) and medium-low salinity (≤ 12 wt % NaCl eq.) H_2O–CO_2–NaCl fluid system. Just comparing the H–O isotope composition in the Sizhuang deposit with natural water, Wei et al (2015) [16] concluded that magmatic water was the dominating source of ore-fluid.

The origin of the Jiaodong gold province is still controversial. Only a simple comparison of isotopes cannot be convincing. In this study, three stages of quartz have been divided. Three mineralization styles have been identified and compared in the Sizhuang gold deposit which is well-known as a Jiaojia-style deposit. Three types of primary fluid inclusions were recognized by petrography and microthermometry. According to the H–O isotope, more comparisons with possibilities and the calculation of fluid compositions through a water–rock interaction will be done, the origin and evolution of ore-fluid in the Sizhuang gold deposit will be discussed.

2. Geological Background

2.1. Regional Geology

The Jiaodong Peninsula consists of two tectonic units, the Jiaobei Terrane of the North China Block and the Sulu Terrane of the Yangtze Block, which are bounded by Wulian-Qingdao-Yantai Fault [17] (Figure 1). The Jiaobei Terrane contains the Jiaobei Uplift in the north and Jiaolai Basin in the south. The Jiaobei Uplift includes more than 90% gold reserves and hosts the largest gold deposits of Jiaodong gold province [10,17] (Figure 1).

The ca. 2.9 to 2.5 Ga tonalite-trondhjemite-granodiorite (TTG) gneisses near Zhaoyuan city, which form the basement of the Jiaodong Group, are the oldest rocks in this area [18,19]. The ca. 2.5 Ga folding of TTG and Jiaodong Group [20] and the ca. 2.2–2.0 Ga intrusion of amphibolite-facies [21,22] happened one after another. These events were followed with the forming of schist, paragneiss, calc-silicate, marble, and minor mafic granulite and amphibolite [23,24]. After the ca. 1.9 Ga continental collision of the rocks of the Jingshan and Fenzishan Group, the Neoproterozoic sedimentary cover (mainly includes marble, slate, and quartzite) [25] deposited (Figure 1).

The Mesozoic granites, commonly existing in the Jiaobei Uplift, is divided into three groups (Figure 1): The Late Jurassic (165–150 Ma) Linglong granite, the middle Early Cretaceous (132–123 Ma) Guojialing granite, and the late Early Cretaceous (120–110 Ma) Aishan granite [26–28]. The Linglong granite, which is the product of widespread Mesozoic magmatism, hosts the Sizhuang gold deposit [16,27,29] (Figure 2). The Linglong granites are composed of garnet granite, biotite granite, amphibole-bearing biotite granite, and muscovite granite [29,30].

The major gold deposits in the Jiaobei Uplift are controlled by the NE- to NNE-trending fault system which cut through Linglong granite [4]. The main faults from the west to east are Sanshadao, Jiaojia, Zhaoping, and Qixia Fault. All these faults are the subsidiary faults of the Tan-Lu Fault Zone (Figure 1). The Jiaojia Fault, which is the ore-controlling fault at the Sizhuang gold deposit (Figure 2), is one of the most significant faults in Jiaobei Uplift.

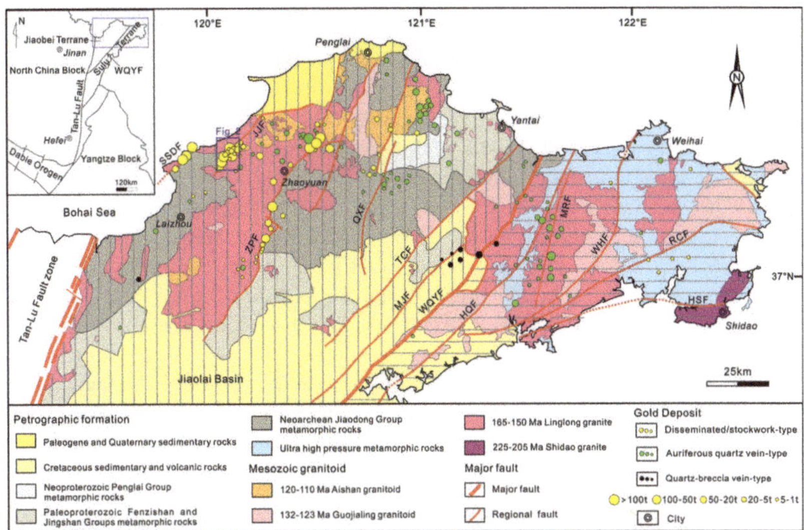

Figure 1. A simplified geological map of the Jiaodong gold province showing the distribution of major fault zones, Precambrian metamorphic rocks, Mesozoic granitoid intrusions, sedimentary rocks, and gold deposits: Modified from Yang et al. (2016) [17]. The double concentric circles represent cities. The areas covered with vertical grey lines and horizontal grey lines respectively represent Jiaobei Terrane and Sulu Terrane in Jiaodong. Abbreviations: SSDF, Sanshandao Fault; JJF, Jiaojia Fault; ZPF, Zhaoping Fault; QXF, Qixia Fault; TCF, Taocun Fault; MJF, Muping-Jimo Fault; WQYF, Wulian-Qingdao-Yantai Fault; HQF, Haiyang-Qingdao Fault; MRF, Muping-Rushan Fault; WHF, Weihai Fault; RCF, Rongcheng Fault; HSF, Haiyang-Shidao Fault.

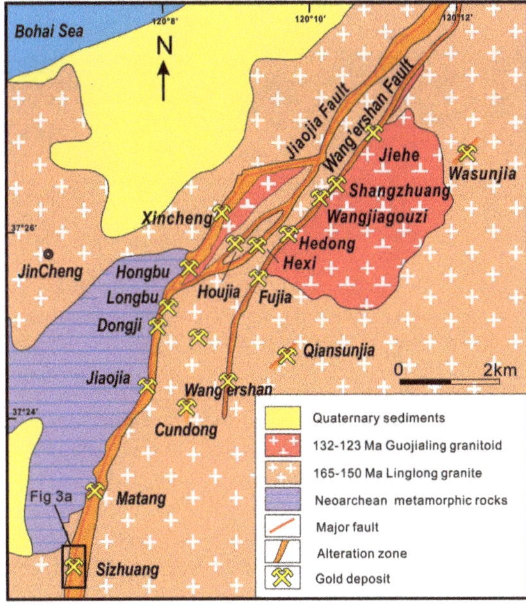

Figure 2. A simplified geological map of the Jiaojia gold belt [17] showing the distribution of major faults and gold deposits: The Sizhuang gold deposit is located at the south segment of the Jiaojia Fault.

2.2. Deposit Geology

The Sizhuang gold deposit is situated in the footwall of the southern section of the Jiaojia Fault, about 28 km northeast to Laizhou City (Figure 1). It is a world-class gold deposit with a proven reserve of >120 t Au. Xenoliths of Jiaodong Group metamorphic rocks, mostly biotite amphibolite, only occurs in the hanging wall of Jiaojia Fault. The Linglong biotite granite is widely exposed in the mine and mainly in the footwall of the Jiaojia Fault (Figure 3).

The hydrothermal alteration zones, which develops along the Jiaojia Fault, can be divided into the pyrite-sericite-quartz alteration zone, sericite-quartz alteration zone, and K-feldspar alteration zone. The extent of sericite-quartz alteration is weaker and the volume of metallic sulfide (mostly pyrite) is fewer with the increasing distance to the Jiaojia Fault in the footwall. The content of K-feldspar in the alteration zone, especially in the K-feldspar alteration zone, is macroscopically higher than in Linglong biotite granite.

All gold mineralization occurs in the footwall of Jiaojia fault. The orebody is mainly controlled by the NE- to NNE-trending and SE-dipping Jiaojia fault system (Figure 2). The secondary faults, which are often filled with quartz veins and are surrounded by hydrothermal zones with a width from 0.5 m to more than ten meters, can be several hundreds to one thousand meters long and 0.1–1.5 m wide (Figure 3).

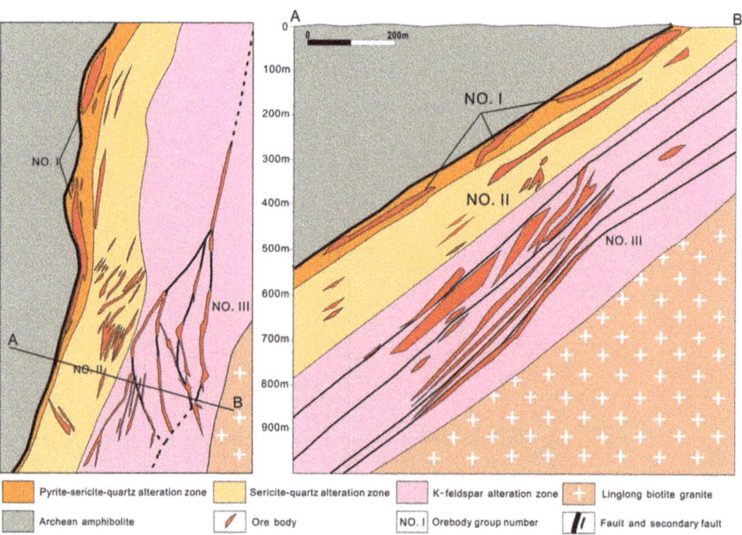

Figure 3. A sketch map of the Sizhuang gold deposit showing the orebodies and hydrothermal alteration: (**a**) The plan view of the Sizhuang deposit and (**b**) the geological cross sections along lines AB. The orebody was defined by the gold cut-off grade (1 g/t).

Although the Sizhuang gold deposit is known as Jiaojia-style mineralization, three mineralization styles can be identified in the Sizhuang gold deposit. The first one is disseminated-stockwork type (Jiaojia-style) mineralization which is hosted in the first-order Jiaojia Fault (Figure 4a,d). There is less Jiaojia-style mineralization surrounding the subsidiary second- or third-order faults. The second style is the auriferous quartz vein style (Linglong-style) which almost occurs along the subsidiary second- or third-order faults (Figure 4b,e). The third type is sulfide-quartz veinlet in the K-feldspar altered granite (Figure 4c,f). These three mineralization styles have almost the same mineral assemblages and mineral paragenesis (Figure 4g–i).

The No. I and No. II orebody, characterized by disseminated- and stockwork-type ores, are respectively hosted in the pyrite-sericite-quartz alteration zone and sericite-quartz alteration zone which are located in the footwall of the Jiaojia Fault. The biggest No. III orebody (reserves 57.91% Au) occurs in the K-feldspar altered alteration zone. It is characterized by both Jiaojia-style and Linglong-style mineralization. The auriferous quartz veins are surrounded by Sericite-quartz alterations.

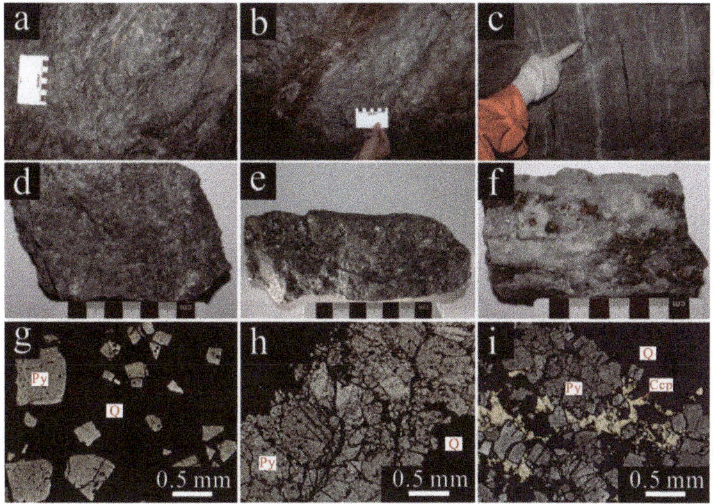

Figure 4. Photography of the three styles of ores: (**a,d**) Disseminated-stockwork style; (**b,e**) Auriferous quartz vein style; (**c,f**) sulfide-quartz veinlet in K-feldspar alteration granite; and (**g–i**) photomicrographs in the reflected light of the disseminated-stockwork type, auriferous quartz vein style and sulfide-quartz veinlet in K-feldspar alteration granite, respectively.

The relationships between different veins in the field clearly shows their relative formation timing (Figure 5): (1) Biotite granite; (2) earlier lamprophyre vein; (3) K-feldspar granite and granite pegmatite; (4) quartz sulfide vein; (5) later lamprophyre vein; and (6) calcite vein.

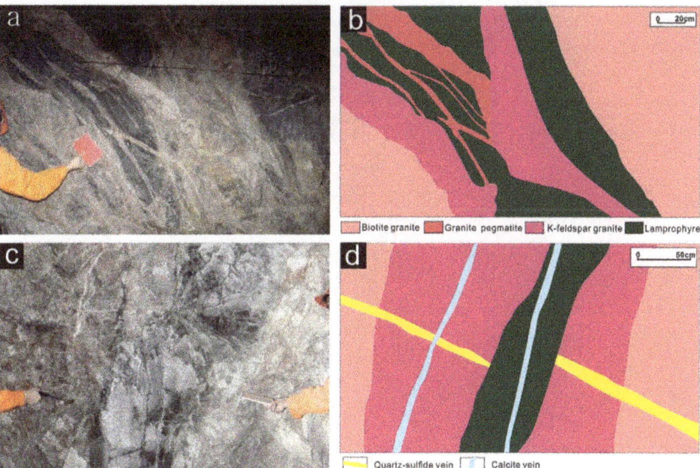

Figure 5. Field photographs showing the crosscutting relationships of the major geologic bodies from the Sizhuang gold deposit: (**a,b**) The K-feldspar altered granite crosscuts the biotite granite and (**c,d**) the quartz-sulfide vein crosscuts the K-feldspar altered granite, and the calcite vein crosscuts the quartz-sulfide vein.

As is pointed out above, there is no difference between different mineralization styles in mineral paragenesis in the Sizhuang gold deposit. The primary metallic mineral in ores is pyrite with lesser chalcopyrite, galena, sphalerite, pyrrhotite, and electrum. The gangue minerals mainly contain sericite, quartz, and calcite (Figure 6). With the study of cross-cutting relationships, textures, and mineral paragenesis, three mineralization stages have been divided which include the early stage, main stage, and late stage (Figure 7).

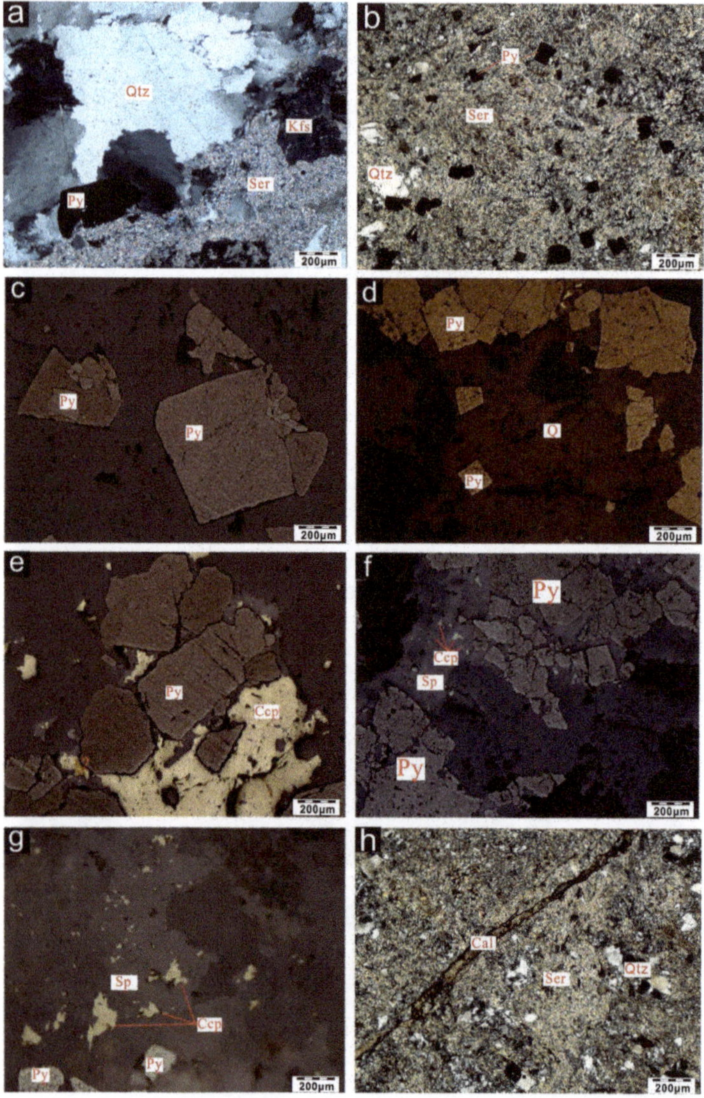

Figure 6. Photomicrographs under transmitted light (**a**,**b**,**h**) and reflected light (**c**–**g**), showing important mineral assemblages in the Sizhuang gold deposit: (**a**) Sericite-quartz alteration consists of sericite and quartz, with few pyrite; (**b**) disseminated pyrite occurs in the matrix of sericite and quartz; (**c**–**g**) the primary sulfides in the main stage of the Sizhaung deposit; and (**h**) the calcite vein crosscuts the sericite and quartz. Abbreviations: Kfs = K-feldspar, Ccp = chalcopyrite, Qtz = quartz, Ser = sericite, Py = pyrite, Sp = sphalerite, Cc = calcite.

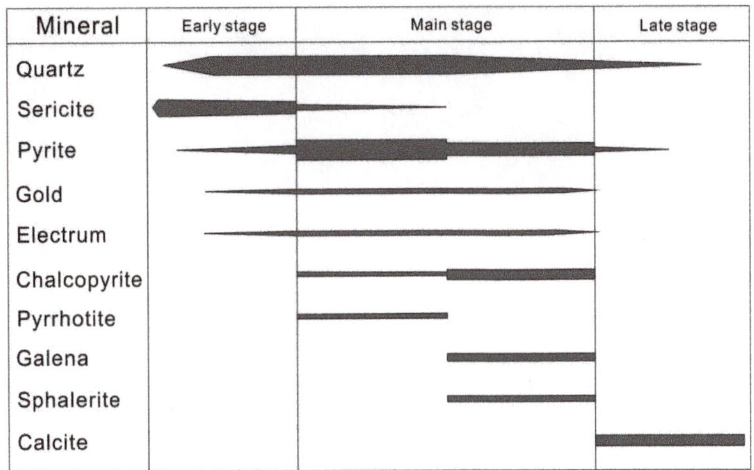

Figure 7. The paragenetic sequence of the hydrothermal alteration and mineralization at the Sizhuang gold deposit interpreted by crosscutting relationships and assemblages of minerals: The thickness means the abundance of minerals in the paragenetic sequence.

The early stage is characterized by sericite-quartz-altered rocks with magmatic K-feldspar and white-milky quartz veins. There are less pyrite and few gold in the early stage. Few gold is deposited in this stage.

The main stage is characterized by the formation of quartz-pyrite-base-metal-sulfide stockworks and veins/veinlets, with a mass of gold and grey quartz in it. Gold occurs along the crack of pyrite. According to mineral assemblage, the main stage can be divided into two parts. In the earlier part, pyrite is the dominating sulfide. The later part is featured with the sulfide assemblage of pyrite, chalcopyrite, galena, and sphalerite.

The late stage is characterized by calcite veins/veinlets crosscutting the veins of main stage, with less pyrite and no visible gold in it.

3. Sampling and Analytical Procedures

Samples with different stages and mineralization styles were chosen for the fluid inclusion analyses on the basis of detailed fieldwork (Table 1). Twenty 0.2 mm-thick polished sections were used in the petrographic study. Quartz is the host mineral of fluid inclusion in early stage and main stage (Figure 8a,c). Calcite is the host mineral in late stage (Figure 8b,d). At length, eight sections were selected for microthermometric analysis, and 168 sets of data were obtained.

Table 1. The basic information of the samples in this study.

Sample	Stage	Ore Style
SZ13D008B3	early	Sulfide-quartz veinlet in K-feldspar granite
SZ13D008B2	main	Sulfide-quartz veinlet in K-feldspar granite
SZ13D020B2	main	Disseminated-stockwork style
SZ14D019B2	main	Disseminated-stockwork style
SZ13D005B1	main	Auriferous quartz vein style
SZ13D008B3	main	Sulfide-quartz veinlet in K-feldspar granite
SZ14D016B1	main	Auriferous quartz vein style
SZ14D020B4	main	Disseminated-stockwork style
SZ14D017B2	late	Calcite vein

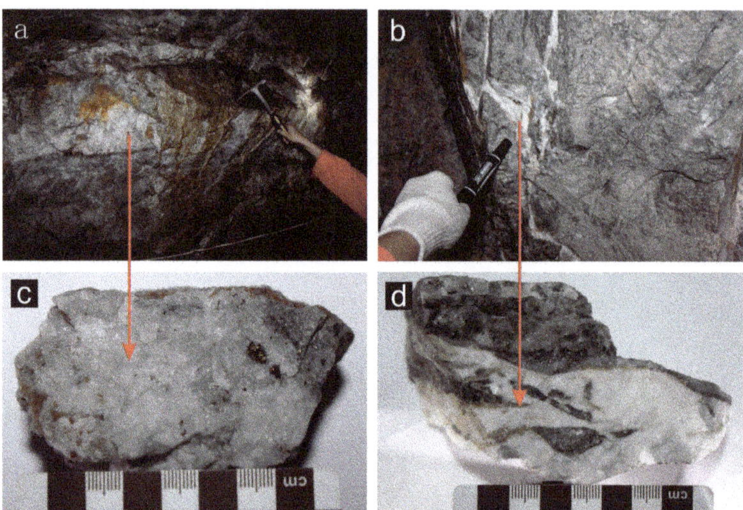

Figure 8. Photographs of the field view and ores shows the inclusion-hosted quartz (**a,c**) and calcite (**b,d**).

Fluid-inclusion microthermometry was made on the Linkam THMSG 600 heating-cooling stage ($-198\ ^\circ$C to 600 $^\circ$C) on a Leitz microscope at the Fluid Inclusion Laboratories, China University of Geosciences, Beijing. Synthetic fluid inclusions were used to calibrate the stage to make sure for an accuracy of which the error is $\pm 0.1\ ^\circ$C at temperatures <30 $^\circ$C and $\pm 1\ ^\circ$C at temperatures >30 $^\circ$C, the heating rat of carbonic phase (T_{mCO2}) and clathrate-melting temperatures (T_{mclath}) are 0.2 $^\circ$C/min, and it is 1.0 $^\circ$C/min for the carbonic phase (T_{hCO2}).

4. Results

4.1. Fluid Inclusion Petrography

Quartz grains border upon or included by pyrite were chosen, and 27 inclusion clusters were identified, which are mostly interpreted to be primary in origin. The inclusions generally growing along trails, which may be secondary or pseudosecondary, are not used in following analysis.

By detailed petrography, following microthermometry and a comparison with a previous study in the Sizhuang gold deposit [16], three types of primary fluid inclusions were identified (Figure 9): H_2O–CO_2–NaCl inclusions (type 1), aqueous inclusions (type 2) and CO_2 inclusions (type 3). Most type 1 inclusions contain two or three phases (liquid H_2O + CO_2 rich vapor or liquid H_2O-liquid CO_2– CO_2 rich vapor, respectively) at room temperature (Figure 9c,d). These inclusions (3–11 µm in diameter), the carbonic phases of which mainly account for 10–40%, commonly appear as groups or along growth zones and are widespread in any stages. Type 2 aqueous inclusions, which are commonly 2–13 µm in diameter, consist of H_2O liquid and H_2O vapor (Figure 9a,b). The degree of fill of H_2O vapor mostly range 10–40%. Like most of the gold deposit in Jiaodong [6,9–11], type 3 primary CO_2 inclusions are small (2–4 µm in diameter) and are infrequent in this study. They include CO_2 liquid and CO_2 vapor (Figure 9e). There is a phenomenon in main stage that both type 1 and type 2 fluid inclusions with similar total homogenization temperatures occur in the same inclusion cluster (Figure 9f), which suggests immiscibility during main stage.

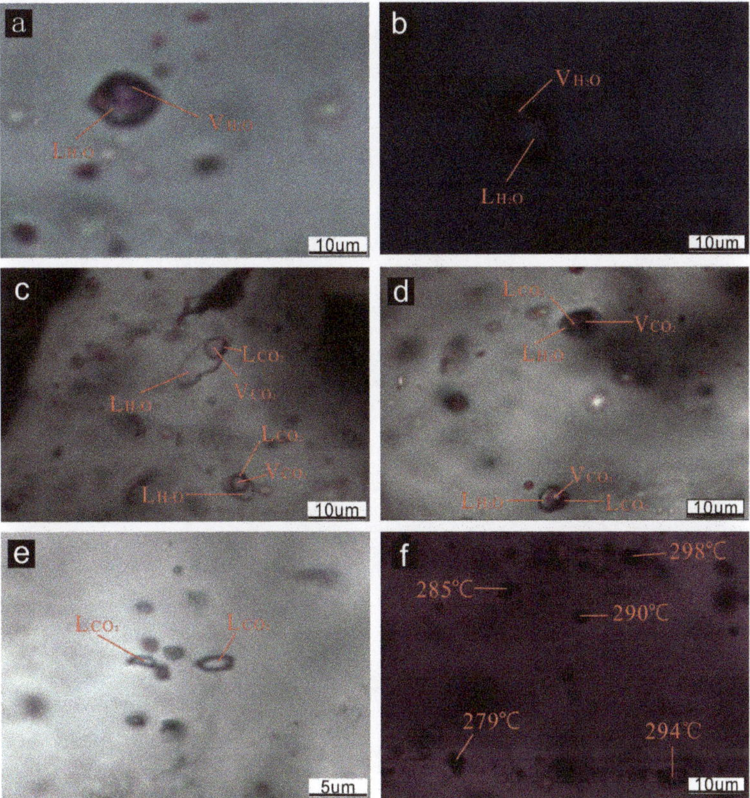

Figure 9. Photomicrographs of the fluid inclusions in the Sizhuang gold deposits: (**a,b**) Type 2 aqueous inclusions contain two phases (liquid H_2O and vapor H_2O); (**c,d**) Type 1 $H_2O–CO_2$ inclusions generally consist of three phases (liquid H_2O, liquid CO_2 and CO_2 rich vapor); (**e**) Type 3 CO_2 inclusions are small and infrequent in this study; and (**f**) the inclusions with a total homogenization temperature of 298 °C is a type 2 aqueous inclusion. The other inclusions are Type 1 $H_2O–CO_2$ inclusions.

4.2. Microthermometry of Fluid Inclusions

4.2.1. Early Stage

Type 1 $H_2O–CO_2$ inclusions and type 2 aqueous inclusions were identified in this stage. For type 1 inclusions, the carbonic phase melting temperatures (T_{mCO2}) of $H_2O–CO_2$ inclusions ranged between −57.5 and −56.6 °C (mean −56.9 °C, n = 6) (Figure 10; Table 2). The integral clathrate-melting temperature (T_{mclath}) was 5.2–7.9 °C (mean 7.3 °C, n = 8) (Figure 10; Table 2). The homogenization of CO_2 ($ThCO_2$) occurred in the vapor phase range from 26.9 to 29.5 °C (mean 28.0 °C, n = 5) (Figure 10; Table 2). The total homogenization temperatures (T_{hTOT}) occurred between 303 and 390 °C (mean 336 °C, n = 8) (Figure 10; Table 2). For type 2 inclusions, the final ice melting temperatures (Tmice) and total homogenization temperature (T_{hTOT}) of type 2 inclusions were −2.7 °C (n = 1) and 350 °C (n = 1) (Figure 10; Table 2).

4.2.2. Main Stage

In main stage, the three inclusion types mentioned above were all observed.

For type 1 inclusions (Figure 10; Table 2), the carbonic phase melting temperatures (T_{mCO2}) of $H_2O–CO_2$ inclusions ranged from −60.5 to −56.5 °C (mean −57.6 °C, n = 55).

The integral clathrate-melting temperature (T_{mclath}) was between 5.2–7.9 °C (mean 6.9 °C, n = 92). The homogenization of CO_2 (T_{hCO2}) occurred in the vapor phase range 17.6 to 30.5 °C (mean 25.7 °C, n = 24) which was much wider than the range of the type 1's T_{hCO2} in early stage, which suggested the possibility of a bigger density change in main stage. The total homogenization (T_{hTOT}) into liquid occurred between 165 and 342 °C (mean 272 °C, n = 100). Some of the type 1 inclusions contained second phases at room temperature, which made them look like aqueous inclusions. However, when the temperature approximately cools down to 16 °C, they became 3-phases inclusions.

For type 2 fluid inclusions (Figure 10; Table 2), the final ice melting temperatures (T_{mice}) ranged between −8.7 and −2.5 (mean −4.7 n = 12). The total homogenization (T_{hTOT}) into liquid occurred between 183 and 392 °C (mean 256, n = 13).

For CO_2 inclusions (Figure 10; Table 2), the carbonic phase melting temperatures (T_{mCO2}) of CO_2 inclusions were −56.6 °C (n = 2). The homogenization of CO_2 (T_{hCO2}) occurred in the vapor phase at 27.1 °C, nearly the same as T_{hCO2} of the mean of type 1.

4.2.3. Late Stage

Only type 1 H_2O–CO_2 inclusions were identified in this stage. The carbonic phase melting temperatures (T_{mCO2}) ranged between −58.3 and −56.6 °C (mean −57.5 °C, n = 12). The integral clathrate-melting temperature (T_{mclath}) was 5.0–9.6 °C (mean 7.2 °C, n = 12). The homogenization temperatures of CO_2 (T_{hCO2}) in the vapor phase ranged from 29.4 to 31.0 °C (mean 30.2 °C, n = 2), which was much wider than the range of T_{hCO2} in the early stage, suggesting that a larger density change occurred in this stage. The total homogenization temperatures (T_{hTOT}) occurred between 195 and 289 °C (mean 242 °C, n = 16) (Figure 10; Table 2).

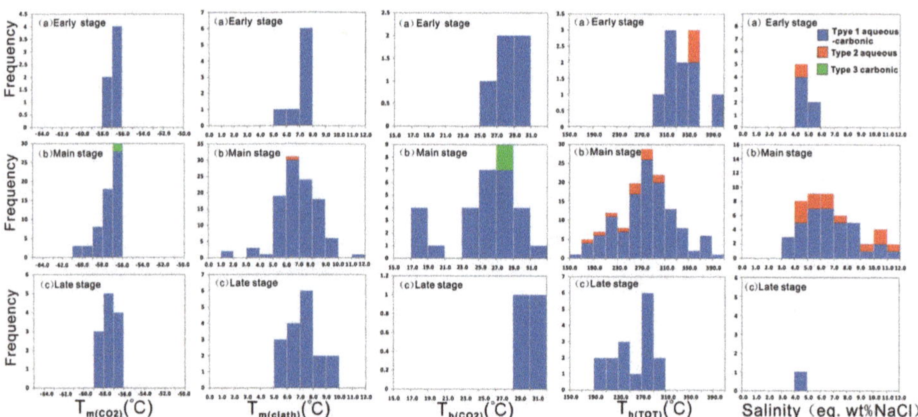

Figure 10. Histograms of the microthermometric data showing the temperatures of T_{mCO2}, T_{mclath}, T_{hCO2} and T_{hTOT} and the salinity.

Table 2. The results of fluid inclusions microthermometry.

Sample	Types	V_{CO2} (%)	T_{mCO2} (°C)	T_{mice} (°C)	T_{mclath} (°C)	T_{hCO2} (°C)	T_{hTOT} (°C)	Salinity (wt %)
Early Stage SZ13D008B3	type 1	20–50 (37, n = 8)	−57.5−−56.6 (−56.8, n = 5)		5.2–7.9 (7.3, n = 7)	26.9–29.5 (28.0, n = 5)	303–368 (331, n = 6)	4.14–8.77 (5.27, n = 6)
	type 2	20		−2.7(n = 1)			350(n = 1)	4.48
Main Stage SZ13D008B2	type 1	20–40 (33, n = 14)	−57−−56.6 (−56.9, n = 6)		5.3–7.8 (6.6, n = 9)	27.2–29.0 (26.9, n = 2)	285–321 (299, n = 10)	4.32–8.61 (6.38, n = 8)
SZ13D020B2	type 1	15–40 (28, n = 20)	−58.5−−57.2 (−57.6, n = 6)		5.1–8.9 (7.5, n = 15)	27.7	231–309 (272, n = 17)	4.22–8.93 (5.64, n = 14)
	type 2	20–40 (43, n = 4)		−7−−6.9 (−7, n = 2)	6.6 (n = 1)		285–298 (292, n = 2)	10.37–10.49 (10.43, n = 2)
SZ14D019B2	type 1	15–35 (28, n = 10)	−60.5−−56.9 (−58.6, n = 3)		5.1–6.5 (6.0, n = 3)	24–25.8 (25.2, n = 3)	296–332 (314, n = 7)	6.63–8.93 (7.52, n = 3)
	type 2	20–30 (23, n = 5)		−4.3−−3.7 (4.1, n=3)			251–294 (275, n = 4)	5.99–6.87 (6.58, n = 3)
	type 3	-	−56.6(n = 2)			27.1(n = 2)		
SZ13D005B1	type 1	20–45 (30, n = 18)	−59.8−−57 (−58.3, n = 8)		3.4–11.2 (7.7, n = 14)	17.8–28 (24.1, n = 3)	264–290 (276, n = 7)	4.32–11.43 (0.62, n = 7)
SZ13D008B3	type 1	5–35 (19, n = 7)	−56.9−−56.8 (−56.9, n = 2)		6.1–8.4 (7.5, n = 4)	-	224–275 (243, n = 6)	4.14–7.31 (5.38, n = 4)
	type 2	20–30 (27, n = 4)		−8.1−−4.5 (−6.3, n = 3)			270–271 (271, n = 2)	7.15–11.83 (9.57, n = 3)
SZ14D016B1	type 1	20–40 (30, n = 17)	−59.5−−56.7 (57.5, n = 10)		3.7–9.6 (6.8, n = 14)	18.3–28.3 (25.4, n = 6)	210–262 (232, n = 11)	4.32–11.01 (7.75, n = 11)
SZ14D020B4	type 1	10–40 (36, n = 19)	−57.8−−56.5 (−57.1, n = 11)		5.2–8.7 (6.7, n = 13)	17.6–29.7 (25.0, n = 4)	253–288 (266, n = 8)	4.14–8.77 (7.18, n = 10)
	type 2	10–40 (28, n = 5)		−3.2−−2.5 (−2.9, n = 4)			210–233 (219, n = 3)	4.17–5.25 (4.79, n = 3)
Late Stage SZ14D017B2	type 1	10–30 (33, n = 19)	−58.3−−56.9 (57.6, n = 10)		5–9.6 (7.1, n = 15)	29.4–31 (30.2, n = 2)	195–289 (240, n = 15)	4.69–9.08 (7.18, n = 10)

Note: V_{CO2}(%), degree of vapour filling; T_{mCO2}, melting temperature of CO_2 phase; T_{mice}, ice melting temperature; T_{mclath}, clathrate melting temperature; T_{hCO2}, homogenization temperature of CO_2; T_{hTOT}, total homogenization temperature; "n", number of measured inclusions; the number in brackets, average values.

4.2.4. Salinity and Density

The salinity (wt % NaCl eq.) and density of fluid inclusions were calculated by the software MacFlincor [31]. These calculations were applied to the H_2O–CO_2 inclusions (Type 1) and aqueous inclusions (Type 2). The measured salinities of type 1 H_2O–CO_2 inclusions were between 4.14 and 8.93 wt % NaCl eq. (meanly 5.99–8.93 wt % NaCl eq.). Those of type 2 aqueous inclusions were 4.17–11.83 wt % NaCl eq. (meanly 4.17–9.73 wt % NaCl eq.). On the whole, the range of fluid salinity in the Sizhuang gold deposit was 4.14–11.83 wt % equiv. NaCl (Figure 10; Table 2).

The range of densities of type 1 H_2O–CO_2 inclusions was 0.516–1.017 g/cm^3 (mean 0.886 g/cm^3). The range of type 2 aqueous inclusions was 0.506–0.937 g/cm^3. All in all, the density of fluid Sizhuang gold deposit ranged between 0.506 to 1.017 g/cm^3.

5. Discussion

5.1. Fluid Immiscibility and Pressure Fluctuation

As is mentioned above, the H_2O–CO_2 and aqueous inclusions hosted in pyrite-related quartz in main stage can occur in a same inclusion cluster (Figure 9f), which can be considered as evidence of a fluid immiscibility. However, fluid mixing and necking down can result in this phenomenon as well [32,33]. First and foremost, it should be eliminated that the fluid mixing and/or necking down may be the reasons for co-occurrences of the H_2O–CO_2 inclusions and aqueous inclusions.

In a mixing fluid, the homogenization temperatures and salinities of H_2O–CO_2 inclusions are different from those of aqueous inclusions [34]. However, in this study, the ranges of salinities of H_2O–CO_2 inclusions and aqueous inclusions are similar and the homogenization temperatures are discrepant, which indicates the fluid mixing is not the reason for type 1 and type 2 being in the same fluid cluster. Although necking down could generate this assemblage, the relatively constant degree of filling of vapor in each inclusion clusters and the low deformation level of quartz can even suggest that it should be eliminated.

Moreover, more evidences of immiscibility are shown in this study: (1) Gold mineralization related H_2O–CO_2 inclusions and aqueous inclusions occupy in same assemblages (Figure 9f); (2) compared with H_2O–CO_2 inclusions, aqueous inclusions have slightly lower total homogenization temperatures and higher salinities (Figure 11), which correspond with fluid immiscibility [35,36]; and (3) the similar homogenization temperatures of type 1 and type 3 suggest the enrichment of CO_2 may be caused by fluid immiscibility [37,38] (Figure 10; Table 2). Fluid immiscibility is vital in the Sizhuang deposit's gold mineralization.

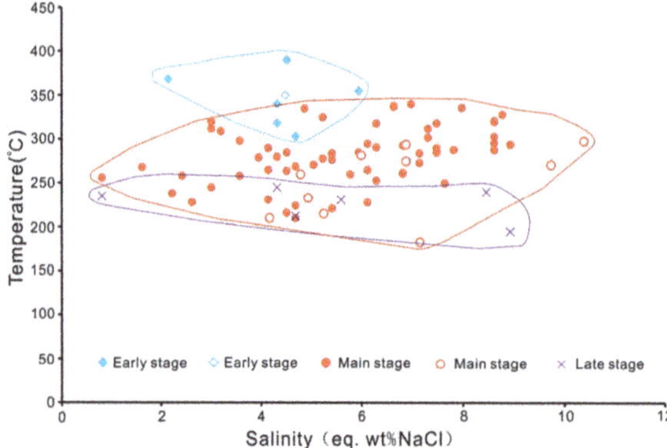

Figure 11. The temperatures (°C) of the total homogenization temperatures (T_{hTOT}) versus salinity (wt% NaCl eq.): The solid and hollow signs represent the H_2O–CO_2 inclusions and aqueous inclusions, respectively.

During fault movement, the fluid pressure decline can lead to the immiscibility of H_2O–NaCl–CO_2 crustal fluid [39]. Also, there is another argument that retrograde boiling can cause fluid immiscibility without a sharp decline in fluid pressure [40]. In this study, there is a phenomenon in main stage that the T_{hTOT} of type 2 aqueous inclusions is about 40 °C lower than that of the coexisting type 1 H_2O–CO_2 inclusions (Figure 11). It can be caused by the partial CO_2 boiling in saturated aqueous fluid [41]. Considering the Sizhuang gold deposit is controlled by a fault, CO_2 boiling may cause the fluid pressure fluctuation in the structural controlled Sizhuang gold deposit during faulting [41,42].

Fluid pressure can be reflected by the density of the fluid inclusions [43,44]. In the Sizhuang gold deposit, the density ranges of the fluid inclusions in different stages and mineralization styles are wide (Figure 12), which show the evidence of a fluid pressure fluctuation during gold mineralization.

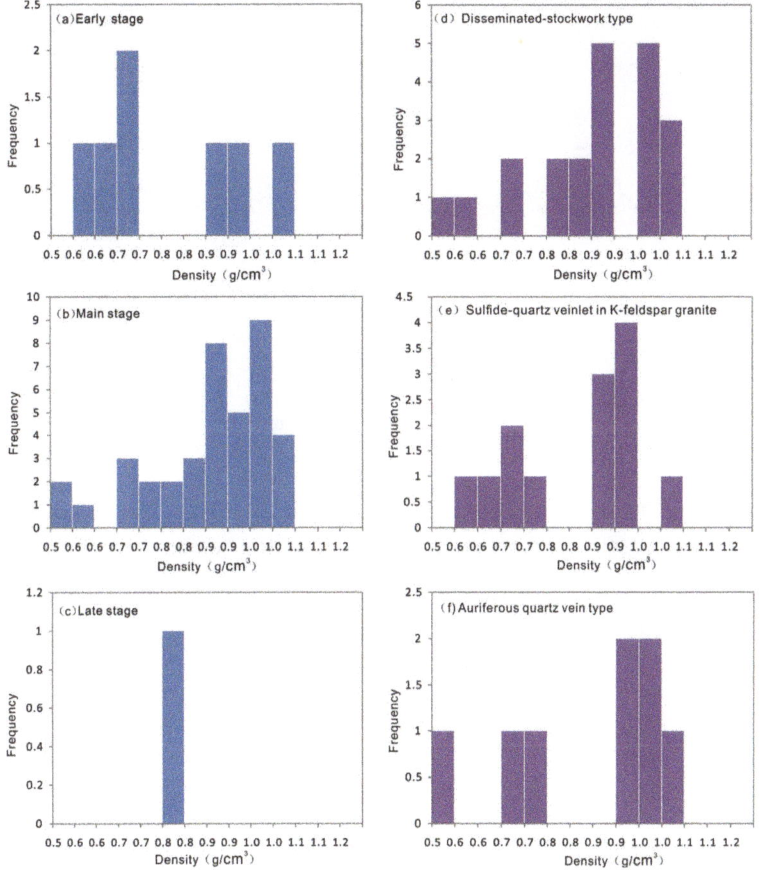

Figure 12. Histograms of the density of type 1 fluid inclusions: The range of inclusion density is wide in different stages or in different mineralization styles.

5.2. Source and Evolution of Ore-Fluid

5.2.1. Source of Ore-Fluid

H–O isotopes are commonly used for studying the source of fluid. Many previous studies compare the H–O isotopes compositions with those of magmatic water, metamorphic water and meteoric water in order to speculate the source of fluid. On the one hand, the range of natural H–O isotopes are wide, which makes the result inaccurate. On the other hand, ore-fluid may have a complicated water–rock interaction during transportation and contacting with rocks [45]. Therefore, more natural waters which may have a possibility of mineralization in the Sizhuang deposit (metamorphic water of the Jiaodong Group, magmatic water of the Linglong granite, magmatic water of the Guojialing granite) are compared with the H–O isotope composition in this study.

The H–O isotope composition of the Sizhuang deposit, which were collected from previous studies (Sample 10SZ06, Z4974-5, 10SZ01 and 10SZ31 are from Wei et al., 2015; sample 10SZ28 and 10SZ32 are from Jiang et al., 2011) [16,46], shows that $\delta^{18}O_{quaitz}$ values range from 9.7‰ to 14.2‰. The oxygen isotope composition of hydrothermal waters ($\delta^{18}O_{H2O}$) and the δD_{H2O} values of inclusions are calculated by a fractionation formula system of stable isotope from Clayton et al. (1972) [47] and Suzuoki and Epstein (1976) [48]. The $\delta^{18}O_{H2O}$ values range from −0.5‰ to 6.1‰, and δD_{H2O} values range from −77.7‰ to −55‰ (Table 3).

Table 3. The H–O isotope composition of the Sizhuang gold deposit.

Sample	Stage	$\delta^{18}O_{quaitz}$ (‰)	Equilibrium Temperature	$\delta^{18}O_{H2O}$ (‰)	δD_{H2O} (‰)
10SZ06	Early stage	10.3	320	3.6	−64.3
Z4974-5	Early stage	13.9	290	6.1	−55
10SZ01	Main stage	13.4	230	2.9	−74.7
10SZ31	Main stage	14.2	225	3.5	−73.8
10SZ28	Main stage	12.5	245	2.8	−77.7
10SZ32	Main stage	9.7	235	−0.5	−74.6

Note: δD_{H2O} is calculated by the equation: $\delta D_{H2O} = \delta^{18}O_{quaitz} + 22.1 \times 10^{6}T^{-2} − 19.1$ (Suzuoki and Epstein, 1976) [48]. δD_{H2O} is calculated by the equation: $\delta D_{H2O} = \delta^{18}O_{quaitz} − 33.8 \times 10^{6}T^{-2} + 3.40$ (Clayton et al., 1972) [47].

In general, similar to the majority of the gold deposits in Jiaodong, the spots of $\delta^{18}O$ and δD are located between metamorphic water/magmatic water and meteoric water (Figure 13). In view of the H–O isotope geochemical background in the Jiaojia gold belt (Table 4) and the absence of Guojialing granite in the Sizhuang god deposit, all spots located in region A and B suggest that the metamorphic water of the Jiaodong Group and the magmatic water of Linglong granite may take part in the mineralization (Figure 13). There is a trend of spots approach to Mesozoic meteoric water, which indicates meteoric water may be a part of mineralization. Therefore, the mixing fluid of magmatic water of Linglong granite, metamorphic water of the Jiaodong Group and meteoric water is the source of ore-fluid.

Table 4. The H–O isotope geochemical background in the Jiaojia gold belt.

Rocks/Fluid	$\delta^{18}O$ (‰)	δD (‰)	Data Resources
Jiaodong Group metamorphic rocks	8.2 ± 3.1	−88.5 ± 7.5	Chen, 1995 [49]
Linglong granite	7	−72 ± 11	Lin and Yin, 1998 [50] and Mao et al., 2005 [51]
Guojialing granodiorite	10.1 ± 0.4	−102 ± 15	Zhang et al., 1994 [52] and Mao et al., 2005 [51]
Metamorphic water of the Jiaodong Group	9.9 ± 0.7	−62 ± 21	Zhang et al., 1994 [52]
Magmatic water of Linglong granite	7.7 ± 1.0	−47 ± 11	Zhang et al., 1994 [52]
Magmatic water of Guojialing granite	9.3 ± 0.8	−77 ± 15	Mao et al., 2005 [51]
meteoric water	−15.6 ± 0.6	−115 ± 5	Zhang et al., 1994 [52]

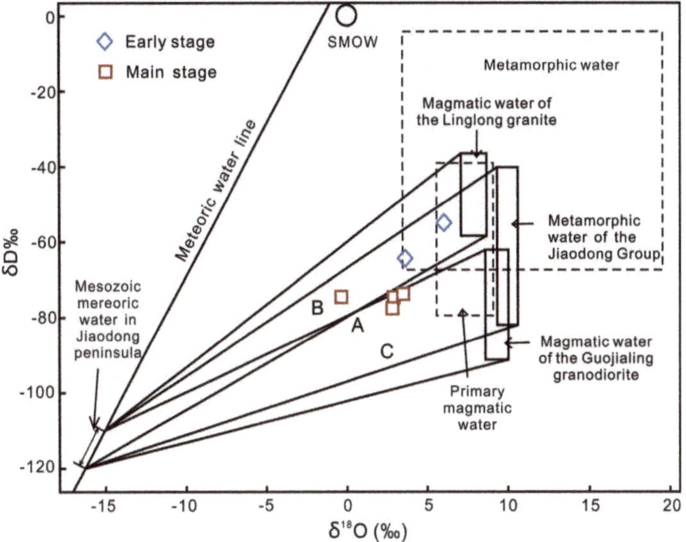

Figure 13. The H–O composition of ore-fluids in early and main stages: Area A means metamorphic water of the Jiaodong Group mixed with meteoric water; Area B means magmatic water of the Linglong granite mixed with meteoric water; and Area C means magmatic water of the Guojialing granite mixed with meteoric water. The base map is modified after Guo et al. (2014) [53].

5.2.2. Fluid Evolution and Water–Rock Interaction

As is noted above, ore-fluid may experience the water–rock interaction before mineralization. The water–rock interaction of ore-fluid and wall rocks follows the equation of meteoric-hydrothermal system [54]:

$$W \cdot \delta^i_{water} + R \cdot \delta^i_{Rock} = W \cdot \delta^f_{water} + R \cdot \delta^f_{Rock}$$

where δ^i_{water} and δ^i_{Rock} are the values of isotope compositions before the water–rock interaction; δ^f_{water} and δ^f_{Rock} are the final values after exchange; W and R are the numbers of atoms (oxygen or hydrogen in this study) in fluid and wall rock, respectively. When the water–rock exchange system is at equilibrium $(\Delta_{R-W} = \delta^f_{Rock} - \delta^f_{water})$, then [54]

$$\delta^f_W = \frac{\delta^f_{Rock} - \delta^i_{Rock}}{\delta^i_{Water} - \left(\delta^f_{Rock} - \Delta_{RW}\right)} = \frac{\delta^i_R (W/R)\delta^i_W - \Delta_{RW}}{1 + (W/R)}$$

The H–O isotope composition of granite and metamorphic rocks in the giant Jiaodong gold deposit can be replaced by $\delta D_{biotite}$ and $\delta^{18}O_{plagioclas}$ (An$_{30}$) [50]. Δ_{RW} of δO^{18} and δD were calculated and demonstrated in this system as follows [48,55]:

$$\Delta_{RW}{}^{18}O = 10^3 \ln \alpha_{plagioclas-water} = 2.68 \times 10^6 T^{-2} - 3.53$$

$$\Delta_{R-W}D = 10^3 \ln \alpha_{biotite-water} = -21.3 \times 10^6 T^{-2} - 2.8$$

The calculation of water–rock interaction uses the rate of quality of δD and $\delta^{18}O$ in this study. In intermediate-acidic granite, the quality rates of oxygen and hydrogen have a stable relationship with the rates of atoms' numbers as follows [56]:

$$O(W/R)_{atoms} = 2 \cdot O(W/R)_{quality}$$

$$D(W/R)_{atoms} = 100 \cdot D(W/R)_{quality}$$

Therefore, the final values of δO^{18} and δD can be worked out with detailed calculation (Table S1).

Figure 14a shows the H–O isotope composition and H–O isotope evolution after an exchange in the water–rock system in early stage (mean homogenization temperature 305 °C). Line 1 represents the evolution of exchange between the magmatic water of Linglong granite and Jiaodong Group metamorphic rocks. Line 2 represents the evolution of exchange between he metamorphic water of the Jiaodong Group and Linglong Granite. Line 3 represents the evolution of exchange between the meteoric water and Jiaodong Group metamorphic rocks. Line 4 represents the evolution of exchange between the meteoric water and Linglong granite.

Figure 14b shows the H–O isotope composition and H–O isotope evolution after exchange in the water–rock system in main stage (mean homogenization temperature 235 °C). Line 5 represents the evolution of exchange between the magmatic water of Linglong granite and Jiaodong Group metamorphic rocks. Line 6 represents the evolution of exchange between the metamorphic water of the Jiaodong Group and Linglong Granite. Lines 7–9 represents the evolution of exchange between the mixing fluid (metamorphic water of the Jiaodong Group and meteoric water mix with the rate 8:1, 4:1 and 2:1, respectively) and Linglong granite. Line 10 represents the evolution of exchange between the meteoric water and Linglong Granite.

In Figure 14a, the spots are under Line 1 distinctly, which suggests the magmatic water of Linglong granite may be not involved in fluid evolution in the early stage of the Sizhuang gold deposit. These spots are around Line 2, and one of them is on Line 3, indicating that the metamorphic water of the Jiaodong Group plays an important role in fluid evolution of the main stage.

In Figure 14b, the spots of the main stage mostly fall between Line 8 and Line 9, which suggest that the metamorphic water of the Jiaodong Group is the uppermost source of ore-fluid of the Sizhuang gold deposit. When the values of W/R are big (10–0.1), the mixing fluid of metamorphic water of the Jiaodong Group and meteoric water (with the mixing rate between 4:1 and 2:1) is propitious to gold mineralization.

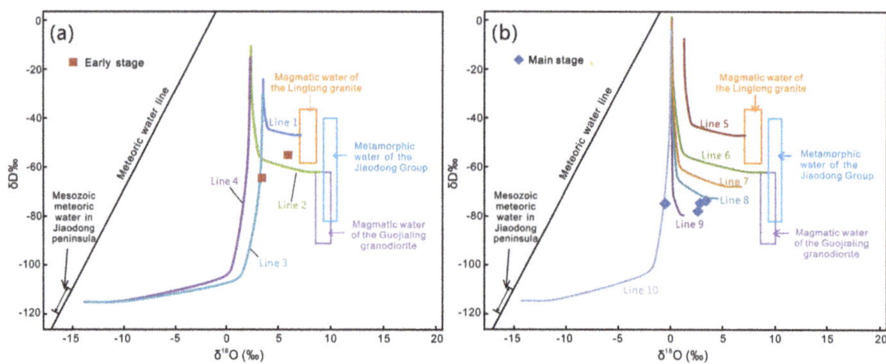

Figure 14. The evolution of the H–O composition during the water–rock interaction.

In conclusion, the source of the ore-fluid is the metamorphic water of the Jiaodong Group in early stage. In main stage, the metamorphic water of the Jiaodong Group and meteoric water mix with the rate between 4:1 and 2:1. The mixing fluid reacts with Linglong granite and extracts mineralization-related elements for gold deposition.

6. Conclusions

1. Microthermometry showed less obvious differences between different mineralization styles, which suggested a similar mechanism of gold deposition.

2. The ore-forming fluid was in a medium-high temperature (210–368 °C), medium-low salinity (4.14–11.83 wt% NaCl eq.) and low density (0.516–1.02 g/cm^3) CO_2–H_2O–NaCl system. The peak of T_{hTOT} during mineralization appeared in 270–300 °C.

3. The fluid immiscibility caused by pressure fluctuation at the main stage was the key mechanism of gold mineralization in the Sizhuang gold deposit.

4. The source of ore-forming fluid in the metallogenic early stage was mainly the original metamorphic water of Jiaodong group. In main stage, the mixing fluid of the original metamorphic water of Jiaodong group and meteoric water in mixing ranged between 4:1 to 2:1 was most conducive to mineralization.

Supplementary Materials: The following are available online at http://www.mdpi.com/2075-163X/9/3/190/s1, Table S1: The calculations of the water–rock interaction.

Author Contributions: Y.-J.W., L.-Q.Y., and J.-Q.F. conceived and designed the ideas; Y.-J.W. and J.-Q.F. performed the experiments and analyzed the data; Y.-J.W. prepared the original draft; Y.-J.W., L.-Q.Y, and H.W. reviewed and edited the draft; G.-Y.L., W.-C.L., and S.-G.L. provided the resources.

Funding: This study was financially supported by the National Key Research Program of China (Grant No. 2016YFC0600107); the National Natural Science Foundation of China (41572069); the MOST Special Fund from the State Key Laboratory of Geological Processes and Mineral Resources, China University of Geosciences (Grant No. MSFGPMR201804); the Key Laboratory of Gold Mineralization Processes and Resource Utilization Subordinated to the Ministry of Natural Resources and Key Laboratory of Metallogenic Geological Process and Resources Utilization in Shandong Province, Shandong Institute of Geological Sciences (Grant No. KFKT201801); the Key Laboratory of Gold Mineralization Processes and Resource Utilization Subordinated to the Ministry of Natural Resources and Key Laboratory of Metallogenic Geological Process and Resources Utilization in Shandong Province, Shandong Institute of Geological Sciences (Grant No. KFKT201802); and the 111 Project under the Ministry of Education and the State Administration of Foreign Experts Affairs, China (Grant No. B07011).

Acknowledgments: We would like to thank Rongxin Zhao at the Shandong Gold Mining (Laizhou) Corporation for his help during the field work, as well as Kunfeng Qiu and Liang Zhang for their suggestions and comments for this paper. We are grateful for constructive comments by the anonymous reviewers.

Conflicts of Interest: The authors declare no conflict of interest.

References

1. Yang, L.Q.; Deng, J.; Wang, Z.L.; Zhang, L.; Guo, L.N.; Song, M.C.; Zheng, X.L. Mesozoic gold metallogenic system of the Jiaodong gold province, eastern China. *Acta Petrol. Sin.* **2014**, *30*, 2447–2467. (In Chinese with English abstract)

2. Deng, J.; Yang, L.Q.; Li, R.H.; Groves, D.I.; Santosh, M.; Wang, Z.L.; Sai, S.X.; Wang, S.R. Regional structural control on the distribution of world-class gold deposits: An overview from the giant Jiaodong Gold Province, China. *Geol. J.* **2018**, *54*, 378–391. [CrossRef]

3. Deng, J.; Liu, X.F.; Wang, Q.F.; Dilek, Y.; Liang, Y.Y. Isotopic characterization and petrogenetic modeling of Early Cretaceous mafic diking–Lithospheric extension in the North China craton, eastern Asia. *Geol. Soc. Am. Bull.* **2017**, *129*, 1379–1407. [CrossRef]

4. Zhang, L.; Yang, L.Q.; Weinberg, R.F.; Groves, D.I.; Wang, Z.L.; Li, G.W.; Liu, Y.; Zhang, C.; Wang, Z.K. Anatomy of a world-class epizonal orogenic-gold system: A holistic thermochronological analysis of the Xincheng gold deposit, Jiaodong Peninsula, eastern China. *Gondwana Res.* **2019**, in press. [CrossRef]

5. Deng, J.; Wang, C.M.; Bagas, L.; Carranza, E.J.M.; Lu, Y.J. Cretaceous–Cenozoic tectonic history of the Jiaojia Fault and gold mineralization in the Jiaodong Peninsula, China: Constraints from zircon U–Pb, illite K–Ar, and apatite fission track thermochronometry. *Mineral. Depos.* **2015**, *50*, 987–1006. [CrossRef]

6. Yang, L.Q.; Deng, J.; Guo, L.N.; Wang, Z.L.; Li, X.Z.; Li, J.L. Origin and evolution of ore fluid, and gold deposition processes at the giant Taishang gold deposit, Jiaodong Peninsula, eastern China. *Ore Geol. Rev.* **2016**, *72*, 585–602. [CrossRef]

7. Goldfarb, R.J.; Santosh, M. The dilemma of the Jiaodong gold deposits: Are they unique? *Geosci. Front.* **2014**, *5*, 139–153. [CrossRef]

8. Deng, J.; Wang, Q.F. Gold mineralization in China: Metallogenic provinces, deposit types and tectonic framework. *Gondwana Res.* **2016**, *36*, 219–274. [CrossRef]

9. Wang, Z.L.; Yang, L.Q.; Guo, L.N.; Marsh, E.; Wang, J.P.; Liu, Y.; Zhang, C.; Li, R.H.; Zhang, L.; Zheng, X.L.; Zhao, R.X. Fluid immiscibility and gold deposition in the Xincheng deposit, Jiaodong Peninsula, China: A fluid inclusion study. *Ore Geol. Rev.* **2015**, *65*, 701–717. [CrossRef]

10. Guo, L.N.; Goldfarb, R.J.; Wang, Z.L.; Li, R.H.; Chen, B.H.; Li, J.L. A comparison of Jiaojia- and Linglong-type gold deposit ore-forming fluids: Do they differ? *Ore Geol. Rev.* **2017**, *88*, 511–533. [CrossRef]

11. Yang, L.Q.; Deng, J.; Zhang, J.; Wang, Q.F.; Gao, B.F.; Zhou, Y.H.; Guo, C.Y.; Jiang, S.Q. Preliminary studies of fluid inclusions in Damoqnjia gold deposit along Zhaoping fault zone, Shandong province, China. *Acta Petrol. Sin.* **2007**, *23*, 153–160.

12. Yang, L.Q.; Deng, J.; Zhang, J.; Guo, C.Y.; Gao, B.F.; Gong, Q.J.; Wang, Q.F.; Jiang, S.Q.; Yu, H.J. Decrepitation thermometry and compositions of fluid inclusions of the Damoqujia Gold Deposit, Jiaodong Gold Province, China: Implications for metallogeny and exploration. *J. China Univ. Geosci.* **2008**, *19*, 378–390.

13. Deng, J.; Liu, X.F.; Wang, Q.F.; Pan, R.G. Origin of the Jiaodong-type Xinli gold deposit, Jiaodong Peninsula, China: Constraints from fluid inclusion and C–D–O–S–Sr isotope compositions. *Ore Geol. Rev.* **2015**, *65*, 674–686. [CrossRef]

14. Yang, L.Q.; Dilek, Y.; Wang, Z.L.; Weinberg, R.F.; Liu, Y. Late Jurassic, high Ba-Sr Linglong granites in the Jiaodong Peninsula, East China: Lower crustal melting products in the Eastern North China Craton. *Geol. Mag.* **2018**, *155*, 1040–1062. [CrossRef]

15. Wang, Z.L. Metallogenic System of Jiaojia Gold Orefield, Shandong Province, China. Ph.D. Thesis, China University of Geosciences (Beijing), Beijing, China, 2012.

16. Wei, Q.; Fan, H.R.; Lan, T.G.; Liu, X.; Jiang, X.H.; Wen, B.J. Genesis of Sizhuang gold deposit, Jiaodong Peninsula: Evidences from fluid inclusion and quartz solubility modeling. *Acta Geol. Sin.* **2015**, *31*, 1049–1062. (In Chinese with English abstract)

17. Yang, L.Q.; Deng, J.; Wang, Z.L.; Guo, L.N.; Li, R.H.; Groves, D.I.; Danyushevskiy, L.; Zhang, C.; Zheng, X.L.; Zhao, H. Relationships between gold and pyrite at the Xincheng gold deposit, Jiaodong Peninsula, China: Implications for gold source and deposition in a brittle epizonal environment. *Econ. Geol.* **2016**, *111*, 105–126. [CrossRef]

18. Jahn, B.M.; Liu, D.Y.; Wan, Y.S.; Song, B.; Wu, J.S. Archean crustal evolution of the Jiaodong Peninsula, China, as revealed by zircon shrimp Geochronology lemental and Nd-isotope Geochemistry. *Am. J. Sci.* **2008**, *308*, 232–269. [CrossRef]

19. Deng, J.; Wang, Q.F.; Wan, L.; Liu, H.; Yang, L.Q.; Zhang, J. A multifractal analysis of mineralization characteristics of the Dayingezhuang disseminated-veinlet gold deposit in the Jiaodong gold province of China. *Ore Geol. Rev.* **2011**, *40*, 54–64. [CrossRef]

20. Geng, Y.S.; Du, L.L.; Ren, L.G. Growth and reworking of the early Precambrian continental crust in the North China Craton: Constraints from zircon Hf isotopes. *Gondwana Res.* **2012**, *21*, 517–529. [CrossRef]

21. Zhang, L.; Liu, Y.; Li, R.H.; Huang, T.; Zhang, R.Z.; Chen, B.H.; Li, J.K. Lead isotope geochemistry of Dayingezhuang gold deposit, Jiaodong Peninsula, China. *Acta Petrol. Sin.* **2014**, *30*, 2468–2480. (In Chinese with English abstract)

22. Zhang, L.; Li, G.W.; Zheng, X.L.; An, P.; Chen, B.Y. 40Ar/39Ar and fission-track dating constraints on the tectonothermal history of the world-class Sanshandao gold deposit, Jiaodong Peninsula, Eastern China. *Acta Petrol. Sin.* **2016**, *32*, 2465–2476. (In Chinese with English abstract)

23. Dong, C.Y.; Wang, S.J.; Liu, D.Y.; Wang, J.G.; Xie, H.Q.; Wang, W.; Song, Z.Y.; Wan, Y.S. Late palaeoproterozoic crustal evolution of the North China Craton and formation time of the Jingshan Group: Constraints from SHRIMP U-Pb zircon dating of meta-intermediate-basic intrusive rocks in eastern Shandong Province. *Acta Petrol. Sin.* **2010**, *27*, 1699–1706. (In Chinese with English abstract)

24. Liu, P.H.; Liu, F.L.; Wang, F.; Liu, J.H.; Cai, J. Petrological and geochronological preliminary study of the Xiliu ~2.1 Ga meta-gabbro from the Jiaobei terrane, the southern segment of the Jiao-Liao-Ji Belt in the North China Craton. *Acta Petrol. Sin.* **2013**, *29*, 2371–2390. (In Chinese with English abstract)

25. Faure, M.; Lin, W.; Monie, P.; Bruguier, O. Paleoproterozoic arc magmatism and collision in Liaodong Peninsula (north-east China). *Terra Nova* **2004**, *16*, 75–80. [CrossRef]

26. Zhang, L.; Yang, L.Q.; Wang, Y.; Weinberg, R.F.; An, P.; Chen, B.Y. Thermochronologic constrains on the processes of formation and exhumation of the Xinli orogenic gold deposit, Jiaodong Peninsula, eastern China. *Ore Geol. Rev.* **2017**, *81*, 140–153. [CrossRef]

27. Zhang, J.; Zhao, Z.F.; Zheng, Y.F.; Dai, M. Postcollisional magmatism: Geochemical constraints on the petrogenesis of Mesozoic granitoids in the Sulu orogen, China. *Lithos* **2010**, *119*, 512–536. [CrossRef]

28. Wang, Z.L.; Yang, L.Q.; Deng, J.; Santosh, M.; Zhang, H.F.; Liu, Y.; Li, R.H.; Huang, T.; Zheng, X.L.; Zhao, H. Gold-hosting high Ba-Sr granitoids in the Xincheng gold deposit, Jiaodong Peninsula, East China: Petrogenesis and tectonic setting. *J. Asian Earth Sci.* **2014**, *95*, 274–299. [CrossRef]

29. Deng, J.; Wang, C.M.; Bagas, L.; Santosh, M.; Yao, E. Crustal architecture and metallogenesis in the south-eastern North China Craton. *Earth-Sci. Rev.* **2018**, *182*, 251–272. [CrossRef]

30. Jiang, N.; Chen, J.; Guo, J.; Chang, G. In situ zircon U–Pb, oxygen and hafnium isotopic compositions of Jurassic granites from the North China craton: Evidence for Triassic subduction of continental crust and subsequent metamorphism-related 18O depletion. *Lithos* **2012**, *142*, 84–94. [CrossRef]

31. Brown, P.E.; Hagemann, S.G. MacFlincor and its application to fluids in Archaean lode-gold deposits. *Geochim. Cosmochim. Acta* **1995**, *59*, 3943–3952. [CrossRef]

32. Anderson, M.R.; Rankin, A.H.; Spiro, B. Fluid mixing in the generation of isothermal gold mineralization in the Transvaal Sequence, Transvaal, South Africa. *Eur. J. Mineral.* **1992**, *4*, 933–948. [CrossRef]

33. Bakker, R.J.; Jansen, J.B.H. A mechanism for preferential H_2O leakage from fluid inclusions in quartz, based on TEM observations. *Contrib. Mineral. Petrol.* **1994**, *116*, 7–20. [CrossRef]

34. Xavier, R.P.; Foster, R.P. Fluid evolution and chemical controls in the Fazenda Maria Preta (FMP) gold deposit, Rio Itapicuru greenstone belt, Bahia, Brazil. *Chem. Geol.* **1999**, *154*, 133–154. [CrossRef]

35. Roedder, E. Fluid inclusions. *Rev. Mineral.* **1984**, *12*, 1–664.

36. Jiang, N.; Xu, J.H.; Song, M.X. Fluid inclusion characteristics of mesothermal gold deposits in the Xiaoqinling District, Shaanxi and Henan Provinces, People's Republic of China. *Mineral. Depos.* **1999**, *34*, 150–162.

37. Oslen, S.N. High density CO_2 inclusions in the Colorado Front Range. *Contrib. Mineral. Petrol.* **1988**, *100*, 226–235.

38. Hollister, L.S. Enrichment of CO_2 in fluid inclusions in quartz by removal of H_2O during crystal–plastic deformation. *J. Struct. Geol.* **1990**, *12*, 895–901. [CrossRef]

39. Sibson, Y.C.; Robert, F.; Poulsen, K.H. High angle reverse faults, fluid-pressure cycling, and mesothermal gold quartz deposits. *Geology* **1988**, *16*, 551–555. [CrossRef]

40. Diamond, L.W. Fluid inclusion evidence for P–V–T–X evolution of hydrothermal solutions in late-Alpine gold–quartz veins at Brusson, Val d'Ayas, northwest Italian Alps. *Am. J. Sci.* **1990**, *290*, 912–958. [CrossRef]

41. Robert, F.; Kelly, W.C. Ore-forming fluids in Archean goldbearing quartz veins at the Sigma mine, Abitibi greenstone belt, Quebec, Canada. *Econ. Geol.* **1987**, *82*, 1464–1482. [CrossRef]

42. Dugdale, A.L.; Hagemann, S.G. The Bronzewing lode gold deposit, eastern Australia: P–T–X evidence for fluid immiscibility caused by cyclic decompression in gold-bearing quartz-veins. *Chem. Geol.* **2001**, *173*, 59–90. [CrossRef]

43. Robert, F.; Boullier, A.M.; Firdaous, K. Gold–quartz veins in metamorphic terranes and their bearing on the role of fluids in faulting. *J. Geophys. Res.* **1995**, *100*, 12861–12879. [CrossRef]

44. Qiu, K.F.; Marsh, E.; Yu, H.C.; Pfaff, K.; Gulbransen, C.; Gou, Z.Y.; Li, N. Fluid and metal sources of the Wenquan porphyry molybdenum deposit, Western Qinling, NW China. *Ore Geol. Rev.* **2017**, *86*, 459–473. [CrossRef]

45. Yang, Z.F.; Xu, J.K.; Zhao, L.S.; Wu, Y.B.; Shen, Y.L. Geochemical studies of hydrogen and oxygen isotopes and ore-forming fluid compositions of fluid inclusions in quartz from two types of gold deposits in Jiaodong. *Acta Mineral. Sin.* **1991**, *11*, 365–369. (In Chinese with English abstract)

46. Jiang, X.H. Fluid Evolution and Genesis of Altered Type Gold Deposits in Northwestern Jiaodong. Ph.D. Thesis, Institute of Geology and Geophysics, Chinese Academy of Sciences, Beijing, China, 2011.

47. Clayton, R.N.; O'Neil, J.R.; Mayeda, T.K. Oxygen isotope exchange between quartz and water. *J. Geophys. Res.* **1972**, *77*, 3057–3067. [CrossRef]

48. Suzuoki, T.; Epstein, S. Hydrogen isotope fractionation between OH-bearing minerals and water. *Geochim. Cosmochim. Acta* **1976**, *40*, 1229–1240. [CrossRef]

49. Chen, Z.S.; Zhang, L.G.; Liu, J.X.; Wang, B.C.; Xu, J.F.; Zheng, W.S. A preliminary study on hydrogen and oxygen isotope geochemical backgrounds of geological bodies in Jiaodong (eastern Shandong) gold metallogenic region. *Acta Petrol. Mineral.* **1995**, *14*, 211–218. (In Chinese with English abstract)

50. Lin, W.W.; Yin, X.L. Isotope geological characteristics of mineralizing fluids of gold deposits in Jiaodong area and a discussion on the application conditions of H.P. Taylor's equation. *Acta Petrol. Mineral.* **1998**, *17*, 248–259. (In Chinese with English abstract)

51. Mao, J.W.; Li, H.M.; Wang, Y.T.; Zhang, C.Q.; Wang, R.T. The relationship between mantle-derived fluid and gold ore-formation in the eastern Shandong peninsula: Evidences from D-O-C-S isotopes. *Acta Geol. Sin.* **2005**, *79*, 839–857. (In Chinese with English abstract)

52. Zhang, L.G.; Chen, Z.S.; Liu, J.X.; Yu, G.X.; Wang, B.C.; Xu, J.F.; Zheng, W.S. Water-rock exchange in the Jiaojia type gold deposit: A study of hydrogen and oxygen isotopic composition of ore-forming fluids. *Mineral. Depos.* **1994**, *13*, 193–200. (In Chinese with English abstract)

53. Guo, L.N.; Zhang, C.; Song, Y.Z.; Chen, B.H.; Zhou, Z.; Zhang, B.L.; Xu, X.L.; Wang, Y.W. Hydrogen and oxygen isotopes geochemistry of the Wang'ershan gold deposit, Jiaodong. *Acta Petrol. Sin.* **2014**, *30*, 2481–2494. (In Chinese with English abstract)

54. Taylor, H.P., Jr. The application of oxygen and hydrogen isotope studies to problems of hydrothermal alteration and ore deposition. *Econ. Geol.* **1974**, *69*, 843–883. [CrossRef]

55. O'Neil, J.R.; Taylor, H.P., Jr. The oxygen isotope and cation exchange chemistry of feldspars. *Am. Mineral.* **1967**, *52*, 1414–1437.

56. Zheng, Y.F.; Chen, J.F. *Stable Isotope Geochemistry*, 1st ed.; Science Press: Beijing, China, 2000; pp. 177–182.

Article

Genesis of the Koka Gold Deposit in Northwest Eritrea, NE Africa: Constraints from Fluid Inclusions and C–H–O–S Isotopes

Kai Zhao [1], Huazhou Yao [1], Jianxiong Wang [1], Ghebsha Fitwi Ghebretnsae [2], Wenshuai Xiang [1] and Yi-Qu Xiong [3,4,*]

[1] Wuhan Center, China Geological Survey (Central South China Innovation Center for Geosciences), Wuhan 430205, China; kabao_8725@163.com (K.Z.); yaohuazhou123@163.com (H.Y.); wangjianxiong00@163.com (J.W.); oldwenzi@163.com (W.X.)

[2] State Key Laboratory of Geological Processes and Mineral Resources, Faculty of Earth Sciences, China University of Geosciences, Wuhan 430074, China; ohghebsh@gmail.com

[3] State Key Laboratory of Geological Processes and Mineral Resources, Faculty of Earth Resources, Collaborative Innovation Center for Exploration of Strategic Mineral Resources, China University of Geosciences, Wuhan 430074, China

[4] Key Laboratory of Metallogenic Prediction of Nonferrous Metals and Geological Environment Monitoring, Ministry of Education, School of Geosciences and Info-Physics, Central South University, Changsha 410083, China

* Correspondence: xiongyiqu@126.com

Received: 14 January 2019; Accepted: 21 March 2019; Published: 27 March 2019

Abstract: The Koka gold deposit is located in the Elababu shear zone between the Nakfa terrane and the Adobha Abiy terrane, NW Eritrea. Based on a paragenetic study, two main stages of gold mineralization were identified in the Koka gold deposit: (1) an early stage of pyrite–chalcopyrite–sphalerite–galena–gold–quartz vein; and (2) a second stage of pyrite–quartz veins. NaCl-aqueous inclusions, CO_2-rich inclusions, and three-phase CO_2–H_2O inclusions occur in the quartz veins at Koka. The ore-bearing quartz veins formed at 268 °C from NaCl–CO_2–H_2O(–CH_4) fluids averaging 5 wt% NaCl eq. The ore-forming mechanisms include fluid immiscibility during stage I, and mixing with meteoric water during stage II. Oxygen, hydrogen, and carbon isotopes suggest that the ore-forming fluids originated as mixtures of metamorphic water and magmatic water, whereas the sulfur isotope suggests an igneous origin. The features of geology and ore-forming fluid at the Koka deposit are similar to those of orogenic gold deposits, suggesting that the Koka deposit might be an orogenic gold deposit related to granite.

Keywords: C–H–O isotopes; fluid inclusion; Koka deposit; orogenic gold deposit

---

1. Introduction

The Nubian Shield, located in northeastern Africa, is an important Gondwana metallogenic domain [1,2] that formed during the Neoproterozoic Pan-African orogenic cycle (ca. 900–550 Ma) [2,3]. Most of the Volcanic-associated Massive Sulfide (VMS) type and quartz vein-hosted gold ± sulfide deposits and occurrences in Eritrea are concentrated along NNW- and NNE-trending narrow zones in the south Nubian Shield [4]. Ghebreab et al. [5] named these zones, the Augaro-Adobha Belt (AAB) and the Asmara-Nakfa Belt (ANB). Both the world class Bisha VMS and Koka gold deposits are located in the AAB copper and gold metallogenic belt [6,7].

The Koka gold deposit is located in northwest Eritrea, which has a long mining history that extends back to the Egyptian Pharaohs. However, modern mining began in the early 20th century. During the Italian colonization, until it was terminated due to the war for independence. Following independence in 1991, several foreign mining companies from China, South Africa, India, Japan, the United Kingdom, Australia, and Canada have been involved in exploring the mineral potential of the country. As a result of several years of explorations, Koka gold still has promising prospects.

Previous studies have suggested that the Koka deposit is a vein-type gold deposit, which is controlled by the shear zone and Koka granite, and is considered to be an orogenic-like gold deposit [8]. However, orogenic gold deposits are commonly unrelated to the granite [9–11], making the genesis of Koka gold deposits enigmatic. The key question is the metal and fluid source, and the evolutionary history of ore-forming fluid.

Therefore, in order to understand the genesis of the Koka deposit, we conducted a detailed study of deposit geology, fluid inclusions (FIs), quartz C–H–O isotopes, and the sulfide S isotope of the Koka gold deposit to better constrain the fluid and metal source, ore-forming fluid evolutionary history, and genesis.

2. Regional Geology

Tectonically, the Koka gold deposit is located in the south Nubian Shield. The shield was formed by the collision between East and West Gondwana upon the closure of the Mozambique Ocean during the Neoproterozoic Pan-African orogenic cycle (ca. 900–550 Ma) (Figure 1a) [3]. The prolonged tectonomagmatic evolution of the Arabian-Nubian Shield involves the rifting and breakup of Rodinia, the formation and accretion of island-arcs, continental collision, extension, and orogenic collapse [2]. Voluminous magmatic activities have also been recorded that are associated with all phases of the tectonic evolution. High strain NNE to NNW trending brittle–ductile shear zones conformable with major fabrics of ANS terranes are dominant in the shield [12]. It is suggested that some of these shear zones that contain dismembered ophiolitic suites represent the major suture zones between terranes [13]. However, others are strike-slip faults and belts of shearing and folding that have modified older sutures [14,15]. Studies have shown that the later types of shear zones are known to host numerous VMS-type polymetallic and orogenic gold deposits and occurrences [15].

More than 60% of the territory of Eritrea is part of the southern Nubian Shield (Figure 1b). The geological setup of the country is made up of three major stratigraphic successions: the basement rocks constituting the Precambrian greenstone volcano-sedimentary assemblages, which are unconformably overlain by Paleozoic and Mesozoic sedimentary and volcanic rocks, which are in turn overlain by sedimentary and volcanic rocks of the Paleogene to Quaternary ages [7]. Based on lithological and structural characteristics, the Eritrean Neoproterozoic basement has been divided into five tectono-stratigraphic terranes: (1) the Barka terrane in the west, mainly composed of upper amphibolite to granulite metasedimentary and mafic gneiss complexes; (2) the Hagar terranes in the north are dominated by oceanic affinity supra-subduction mafic and felsic volcanic rocks; (3) the Adobha Abiy terrane in the central and western parts is principally composed of highly deformed ophiolites and post-accretionary basinal sediments, which are imbricated by the regional shear zones of the Elababu shear zone (ESZ) in the east and the Baden shear zone (BSZ)in the west; (4) the Nakfa terrane occupying more than half of the basement complex contains greenschist facies volcano-sedimentary and syn- to post-collision granitoid rocks; and (5) the easternmost Arig terrane is a narrow belt of high-grade gneiss and syn- to late-tectonic granitoid rocks along the Red Sea lowlands composition [16–18].

Regional structures including the NNW-striking brittle–ductile shear zones and strike-slip faults, low angle thrust faults, fold structures as well as local macroscopic en-echelon quartz veins and tension gashes with a general trend of NNE to NNW are commonly developed on the Precambrian granitoid-greenstone belt. Among these, the ductile strike-slip shear zones are the most prominent tectonic structures in the region and can be traced for several kilometers in length and several meters to several tens of meters in width. The dominantly sinistral AAB and dextral ANB are the two main transpressional strike-slip shear zones, along which the important mineral deposits occur [5]. Semi-brittle shear zones developed synchronously along axial planes of isoclinal folds are also common structures on the greenschist metamorphic rocks. Syn- to late-tectonic granitoid magmatic rocks intrude along the ductile shear zones as elliptical rigid bodies [5]. These magmatic rocks are dominated by granite, granodiorite, and diorite, and are accompanied by fine-grained rocks, dolerite, and quartz porphyry [7].

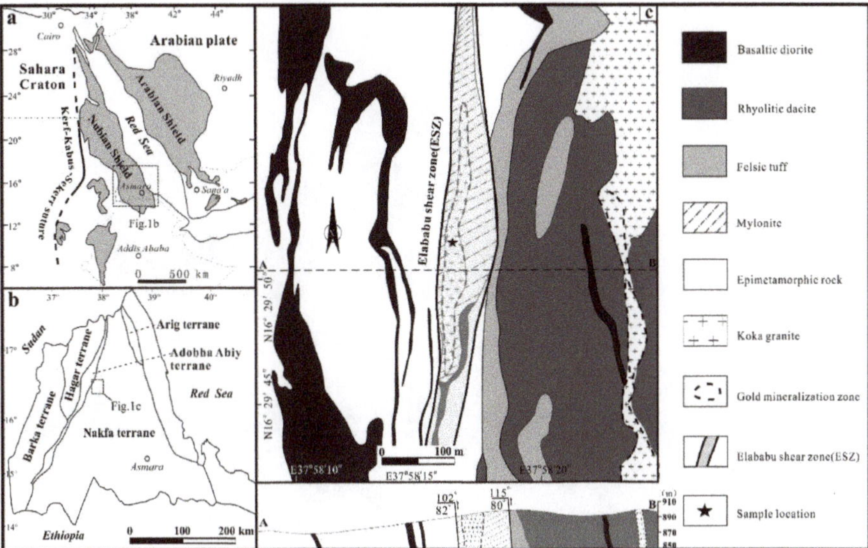

Figure 1. (**a**) Tectonic scheme of the Koka deposit (after Johnson et al. [15]). (**b**) Location of the Koka deposit (after Zhao et al. [19]). (**c**) Geological map of the Koka gold deposit (after Dean et al. [20]).

3. Geology of the Koka Deposit

The Koka gold mine is located in the Elababu shear zone (ESZ) that separates the Nakfa terrane in the east from the Adobha Abiy terrane in the west. The ESZ is the dominant ore- and granite-controlling structure in the area, where the ore bodies are strictly distributed in the SN-striking ESZ. The ESZ is characterized by the occurrence of thrust nappe faults (strike to 10~20°, tending to the southwest) and low grade metamorphic volcanic-sedimentary rocks upright folds (dip angle of limbs is 75°~90°) along the different lithologic interfaces. The formation of the ESZ is related to the SN-striking extrusion caused by the collision of the Neo-Proterozoic Gondwana continent. The gold-bearing quartz veins commonly occur as reticular veins, with branching and converging features, and have locally expanded to form lens-shaped bodies (Figure 2a). The granite occurring in the ESZ is characterized by strong mylonitization (Figure 2b,c). Preferred orientation of sericite defined the main foliation anastomozing around rotated K-feldspar porphyroclasts (Figure 2d).

The ores are hosted by a relatively fine to medium grained, gray colored, elongated nearly vertically dipping granitic body. The granitic host rock is dominated by primary minerals of plagioclase, quartz, and subordinate K-feldspar with alteration products dominated by sericite, micro-granular albite, and quartz. The host rock is bounded by greenschist facies metamorphic rocks, consisting of intermediate to felsic volcanic and pyroclastic rocks and post-tectonic granitoids in the east part of the Nakfa terrane and sequence of siliciclastic metasedimentary and metavolcanic rocks including tuffaceous greywacke, sandy mudstone, shales and mafic metabasaltic flows, and associated syn-tectonic granitoid rocks in the west part of the Adobha Abiy terrane (Figure 1c). The area is dominated by the NNE shear zones, which are comprised of a series of asymmetric overturned isocline folds and thrust faults, particularly prevailed on the fine grained volcanic and sedimentary rocks. The high-angle thrust faults are the main controlling structures for the ore-bearing hydrothermal activities in the mine area.

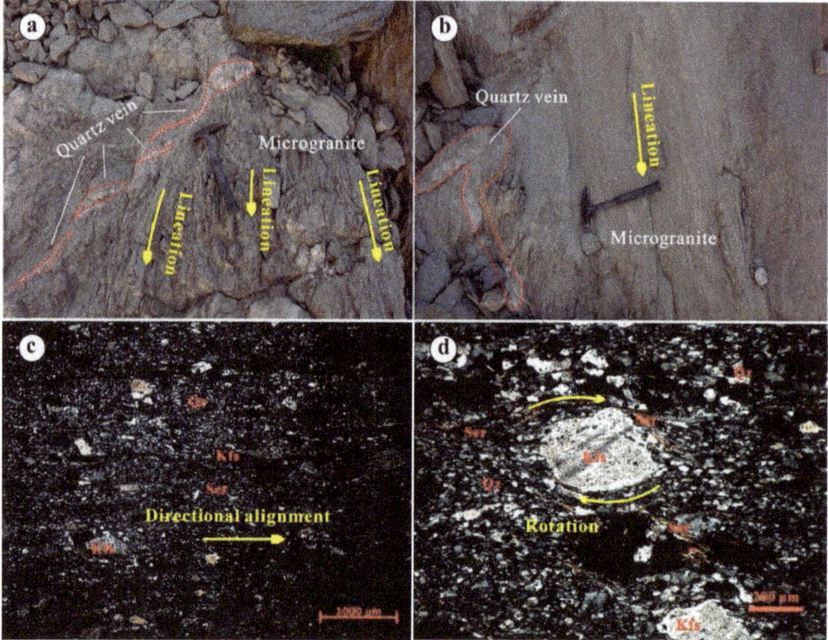

Figure 2. (**a**) Lens-shaped quartz veins occurring in the microgranite. (**b**) Microscopic aspects of foliated leucogranite. Foliation is defined by the preferred orientation of sericite. (**c**) Microscopic view of rotated K-feldspar porphyroclasts with asymmetric strain shadows. (**d**) Rotation of K-feldspar under microscope. Abbreviations: Kfs = K-feldspar; Qz = quartz; Ser = sericite.

In the Koka gold deposit, the mineralization developed within the relatively competent elongated NNE trending microganitic body. Post-magmatic deformations have fractured the Koka microgranite (851.2 ± 1.9 Ma, [19]). These tectonic-induced brittle fractures served as a pathway for the gold-bearing hydrothermal fluids and eventually became a mineralized stockworks of quartz veins. The main mineralization zone can be classified into two categories: the quartz vein type and wall-rock alteration type. The former is characterized by intense alteration and the main mineralization zone occurs from 50 to 80 m from the contact of the footwall within the microgranite. This zone is about 10 m wide and is characterized by stockworks of quartz veins with varying widths, generally no more than two meters. The second wall-rock alteration type mineralization mainly developed in the contact zone between the Koka microgranite and footwall metavolcano-sedimentary rocks. As of 2010, the diamond drilling had constrained the Koka gold orebody over a strike length of 650 m, with an average depth of more than 165 m below the surface, and an average grade of about 5~6 g/t, proving that the gold reserves are about 26.13 tons [20].

Gold is mainly formed in quartz veins in the form of native gold. Ore minerals are mainly native gold, pyrite, chalcopyrite, galena, and sphalerite. Gangue minerals are mainly quartz, calcite, and sericite. The types of alterations closely related to gold mineralization are silicification, sericitization, pyrite mineralization, and carbonation and these are superimposed on each other in the mining area and are generally zoning. In particular, silicification is consistent with the main mineralization. Sericitization and pyrite mineralization are beyond the distribution of the mineralization zone, and formed the sericite + pyrite altered halo boundary. The formation of the carbonation often occurred in the surrounding rock mainly as a carbonated vein, accompanied by other alterations or separately.

Based on microscopic observations, and a paragenetic study of the primary ore mineral assemblages, two main stages of mineralization are recognized in the Koka gold deposit: stage I is characterized by the development of quartz–sulfide–gold veins, of which sulfides are mainly pyrite, chalcopyrite, galena, and sphalerite (Figure 3a). The shape of natural gold is xenomorphic granular and is generally distributed between pyrite and quartz (Figure 3b,c). Pyrites mostly occur as clusters, which are hypautomorphic-cubic crystals with large particle size changes (Figure 3d). Galena is xenomorphic granular in shape and has a distinct triangular cleavage (Figure 3f). Chalcopyrite and sphalerite mostly show a solid solution structure, and aggregates can also be seen, occasionally as a single grain (Figure 3e,g). Stage II is characterized by the development of milky white quartz veins with poor sulfide (Figure 3h). The mineral combination is relatively simple, and only a small amount of its chalcopyrite and fine pyrite can be seen.

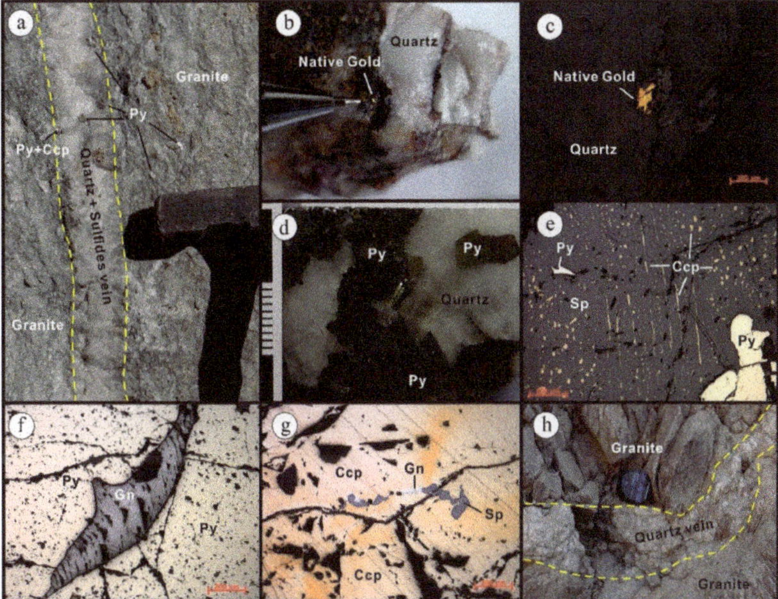

Figure 3. Microscopic photos of the orebodies, hand-specimens, and minerals of the Koka gold deposit. (**a**) Gold-bearing quartz veins of stage I occur in granite; (**b**) native gold occurs in quartz veins; (**c**) native gold occurs in the crack of quartz grains; (**d**) euhedral pyrite occurs in quartz veins; (**e**) pyrite coexists with chalcopyrite and sphalerite; (**f**) pyrite coexists with galena; (**g**) chalcopyrite coexists with galena and sphalerite; (**h**) quartz veins of stage II occur in granite. Abbreviations: Ccp = chalcopyrite; Gn = galena; Py = pyrite; Sp = sphalerite.

4. Sampling and Analytical Method

Five ore-bearing quartz samples were collected from the Koka gold deposit, namely KO-15, KO-16, KO-17, KO-3, KO-7, and KO-14. Sampling locations are shown in Figure 1c.

4.1. Fluid Inclusions (FIs)

Fluid inclusion assemblages were characterized prior to the selection of samples for microthermometry. Microthermometry analyses were completed at the Institute of Materials and Engineering, University of Science and Technology Beijing, by using the Linkam THMSG600 heating–freezing stage (−196 to 600 °C) (Linkam, Surrey, UK). The precision of each measurement was ±0.1 °C during the cooling cycles and ±1 °C during the heating cycles. The heating rate was held between 0.2 °C/min and 10 °C/min during these cycles. The temperatures of the phase transitions of the CO_2-bearing fluid inclusions and aqueous inclusions were determined at heating rates of 0.1 °C/min and 0.2 to 0.5 °C/min, respectively. The temperatures of the phase transitions were confirmed by the cycling technique to ensure the accuracy of the microthermometric data. For salinity determination of the CO_2–H_2O FIs, the equation was S = 15.52022 − 1.02342t − 0.05286t^2 (−9.6 ≤ t ≤ 10 °C), S = salinity wt% NaCl eqv, t represents the temperature of clathrate in CO_2 FIs [21]. For NaCl-aqueous FIs, the equation was S = 0.00 + 1.78t − 0.0442t^2 + 0.000557t^3, 0 < S < 23.3% wt% NaCl eqv, t represents the temperature of ice melting [22].

The chemical composition of the vapor phases in the fluid inclusions was determined by ion and gas chromatography and Raman spectroscopy at the Central South China of the Mineral Resources Supervision and Testing Center, Wuhan. The Raman spectroscopy instrument was a Renishaw inVia, UK. The spectrum ranged from 50 cm^{-1} to 4500 cm^{-1}. The operating conditions included a laser wavelength of 514.5 nm and laser power of 30 mW. The Raman shift was calibrated using a single crystal of silicon. The cation test instrument in the liquid phase composition of the group inclusions was a Hitachi Z-2300 (Hitachi, Tokyo, Japan), and a DIONEX ICS-3000 (Dionex, Sunnyvale, CA, USA) was used for ion chromatography. A GC-2014C was used for gas chromatography. The bursting temperature of the inclusions ranged from 100 to 550 °C, and the precision was 0.01 mg/L (0.01 μg/g).

4.2. Stable Isotope Analytical Methods

The H–O–C isotope analyses were accomplished with a MAT253 mass spectrometer (Finnigan, San Jose, CA, USA) at the Analytical Laboratory of the Beijing Research Institute of Uranium Geology. The accuracy of the O isotope analysis was better than ±0.2‰, and that of the H isotope analysis was better than ±2‰, and that of the C isotope analysis was typically better than ±0.1‰. The amount of O in quartz water (aquartz water) was calculated from the O isotope level of the analyzed quartz by using the fractionation equation $1000\ln\alpha_{quartz\ water} = (3.38 \times 10^6)T^{-2} - 3.40$, where T is the temperature in Kelvin [23], and the average fluid inclusion temperature of each stage was used to calculate the $\delta^{18}O_{water}$ value.

5. Fluid Inclusions

5.1. Fluid Inclusion Petrography

Fluid inclusions (FIs) of the Koka gold deposit are found in quartz. Quartz is mainly characterized by ductile deformation (quartz). A large number of fluid inclusions including primary, pseudosecondary, and secondary fluid inclusions [22,24] were identified in two mineralization stages of quartz using detailed petrographic observations. Primary and pseudosecondary inclusions occur in the growth zone of quartz (Figure 4), or are distributed as an isolated form. The microthermometric data in this study were basically from primary and pseudosecondary inclusions, and inclusions with only an aqueous liquid phase are the result of necking phenomena and cannot be used for microthermometry. In accordance with the classification principles of Roedder [24] and Lu et al. [22], vapor-to-liquid ratios at room temperature, heating–freezing behaviors, and results of laser Raman analysis were used to classify the fluid inclusions found in the Koka deposit into four types:

(1) Type I, liquid-rich aqueous fluid inclusions (L_{H2O} + V_{H2O}), are mainly found in stage II, and are rarely found in stage I. Type I FIs are dominantly liquid-rich inclusions with few pure liquid inclusions. The liquid-rich inclusions have ellipsoidal or irregular shapes with long axes of 8–16 μm and a 75–95 vol% liquid phase. These types of FIs are mostly primary and pseudosecondary inclusions that are distributed in isolation or along the crystal growth (Figure 4a,b).

(2) Type II, CO_2-rich fluid inclusions (L_{CO2} + V_{CO2}), are mainly found in the early stage quartz veins. Liquid-rich CO_2 inclusions have ellipsoidal or irregular shapes with long axes of 6–16 μm (Figure 4c–f) and a 40–95 vol% liquid phase. Type II FIs dominantly occurred as a liquid phase in room temperature (25 °C), whereas few occurred as two phases. Liquid-rich CO_2 inclusions will appear as vapor in the cooling process. Type II FIs commonly coexist with Type III FIs, and few coexist with Type I FIs (Figure 4d).

(3) Type III, three-phase CO_2-rich fluid inclusions (V_{CO2} + L_{CO2} + L_{H2O}) at room temperature, occurred in both the early and late stages, have ellipsoidal or irregular shapes with long axes of 6–24 µm (Figure 4a,g,h) and a 20–85 vol% liquid phase.

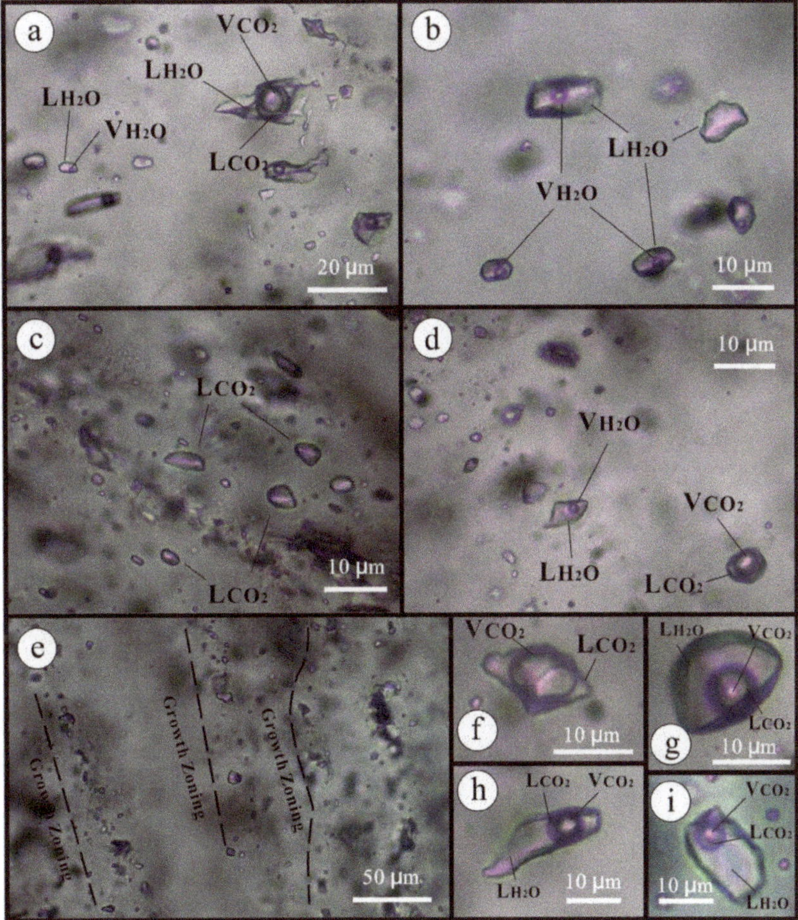

Figure 4. (**a**) Type I fluid inclusions (FIs) coexist with type III FIs in stage I veins; (**b**) Type I FIs occur in stage II veins; (**c**) Abundant Type II FIs occur in stage I veins; (**d**) Type I FIs coexist with type II FIs in stage I veins; (**e**) Primary FIs distributed along the crystal growth bands; (**f**) Type II of vapor-rich FI in stage I; (**g**) Type II of liquid-rich FI in stage I; (**h**) Type III FI in stage I; and (**i**) Type III FI in stage II.

5.2. Fluid Inclusion Microthermometry

Five ore-bearing quartz samples from the Koka gold deposit were selected for ice melting and homogenization temperatures analysis, namely KO-15, KO-16, KO-17, KO-3, KO-7, and KO-14. A total of 135 microthermometric data of all types of fluid inclusions are listed in Table 1 and Figure 5.

Table 1. Summary of microthermometric data of the fluid inclusions from the Koka gold deposit.

Sample No.	Stage	FI type	Size (μm)	Number	Tm, CO$_2$ (°C)	Tm, ice (°C)	Tm, Cl (°C)	Th, CO$_2$ (°C)	Th, (°C)	Salinity (wt% NaCl)	Density (g/cm^3)
KO-15		Type I, minor	6~12	4		−4.3~−2.4			254~341 (V)	4.0~6.9	0.757~0.822
		Type II, abundant	8~16	9	−56.9~−58.0			−2.3~17.2			0.801~0.904
		Type III, abundant	8~16	11	−56.7~−57.9		5.7~7.8	19.2~30.1	280~356 (V)	4.3~7.9	0.576~0.793
KO-16	Stage I	Type I, minor	6~12	5		−2.6~−1.9			243~288 (V)	3.2~4.3	0.711~0.831
		Type II, abundant	6~12	10	−56.7~−58.2			−4.9~18.9			0.757~0.898
		Type III, minor	6~20	14	−56.8~−58.1		5.8~7.9	19.9~29.6	235~295 (V)	4.1~7.7	0.608~0.783
KO-17		Type I, minor	6~12	6		−3.2~−1.8			223~275 (V)	3.1~5.3	0.783~0.858
		Type II, abundant	6~14	13	−56.8~−58.0			−4.7~16.4			0.812~0.901
		Type III, abundant	10~24	14	−56.6~−58.0		5.3~8.0	10.6~29.4	238~305 (V)	3.3~8.5	0.613~0.862
KO-3		Type I, abundant	8~16	6		−2.6~−1.5			221~288 (V)	2.6~4.3	0.765~0.872
		Type III, abundant	8~20	11	−57.0~−57.9		6.2~8.2	19.7~29.7	228~295 (V)	3.5~7.1	0.613~0.787
KO-7	Stage II	Type I, abundant	8~12	6		−2.8~−1.3			212~280 (V)	2.2~4.7	0.763~0.871
		Type III, abundant	8~24	11	−56.9~−57.4		6.6~8.5	18.5~29.7	232~318 (V)	3.0~7.5	0.597~0.793
KO-14		Type I, minor	8~14	5		−2.6~−0.9			235~267 (V)	2.2~4.3	0.786~0.842
		Type III, abundant	10~22	10	−56.7~−57.3		5.6~8.0	18.2~25.7	259~309 (V)	3.9~8.0	0.672~0.701

(1) Stage I

The fluid inclusions in stage I included types I, II, and III. Type I FIs homogenized to the liquid phase at temperatures ranging from 254 °C to 341 °C, predominantly at 260–280 °C. Their ice melting temperatures ranged from –4.3 °C to –1.8 °C, corresponding to salinities of 3.1–6.9 wt% NaCl eqv, with most around 3–4 wt% NaCl eqv. Type II fluid inclusions showed that the CO_2 phase melting temperatures ranged from –56.7 °C to –58.2 °C, with a peak at –57 °C, and the homogenization temperatures to the liquid phase were –4.9 °C to 18.9 °C. Type III (CO_2-rich) inclusions formed solid CO_2 upon cooling. The solid CO_2 melted at temperatures between –56.6 °C and –58.1 °C, lower than the triple point temperature of pure CO_2 (–56.6 °C). The CO_2-clathrate melting temperatures varied from 5.3 °C to 8.0 °C, corresponding to salinities of 3.3–8.5 wt% NaCl eqv, with most around 6–7 wt% NaCl eqv. Vapor- and liquid-CO_2 homogenized to the vapor phase at temperatures between 10.6 °C and 30.1 °C, higher than the type II FIs in stage I (Figure 6). The total homogenization temperatures of type III inclusions ranged from 235 °C to 356 °C, with most around 280–300 °C.

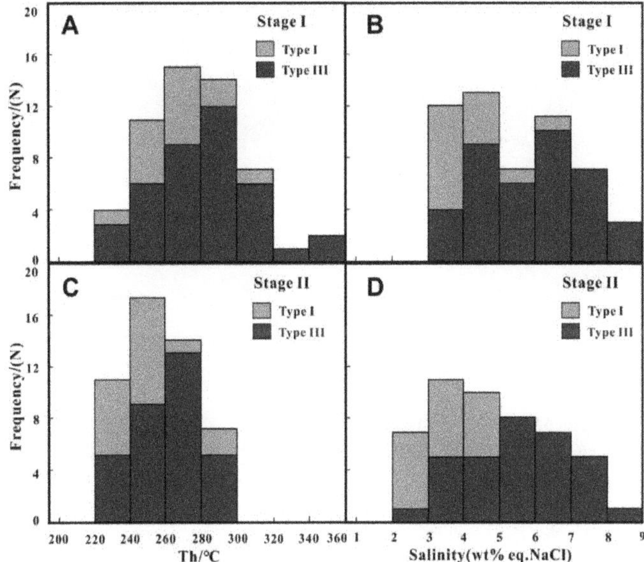

Figure 5. (**A,C**) Homogenization temperature of different stages of the Koka gold deposit; (**B,D**) Salinity of different stages of the Koka gold deposit.

(2) Stage II

The fluid inclusions in stage II included types I, III, and rare type II, and the inclusions of type III are commonly vapor-rich (V/(V + L) > 40%). The FIs of type I are dominantly liquid-rich, and homogenized to the liquid phase at temperatures varying from 212 °C to 288 °C, predominantly at 260–280 °C. Their ice melting temperatures ranged from −2.8 °C to −0.9 °C, corresponding to salinities of 2.2–4.7 wt% NaCl eqv, with most around 3–4 wt% NaCl eqv. Type II fluid inclusions showed the CO_2 phase melting temperatures ranged from −56.9 °C to −57.9 °C, and the homogenization temperatures to the liquid phase of 6.8 °C to 17.3 °C. The solid CO_2 of type III inclusions melted at temperatures between −56.9 °C and –57.9 °C, and the CO_2-clathrate melting temperatures varied from 6.2 °C to 8.5 °C, corresponding to salinities of 3.0–8.0 wt% NaCl eqv, with most around 5–7 wt% NaCl eqv. Vapor- and liquid-CO_2 homogenized to the liquid phase at temperatures between 18.5 °C and 29.7 °C, and the total homogenization temperatures of type III inclusions ranged from 232 °C to 318 °C, with most around 260–280 °C.

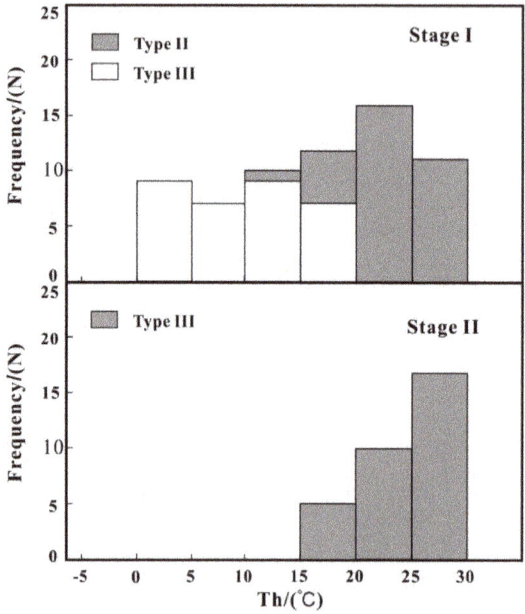

Figure 6. Homogenization temperature of the vapor phase CO_2 in different stages of the Koka gold deposit.

5.3. Laser Raman Spectroscopy

Selected samples based on the fluid inclusion petrology of the two stages were examined by laser Raman micro spectroscopy. The results showed that the composition of FIs in type I was relatively simple, dominated by H_2O (Figure 7d). The composition of type II FIs was basically pure CO_2 (Figure 7c). The inclusion gas of type III was dominated by CO_2 and H_2O, and the aqueous phase also contained some CO_2 (Figure 7a,b).

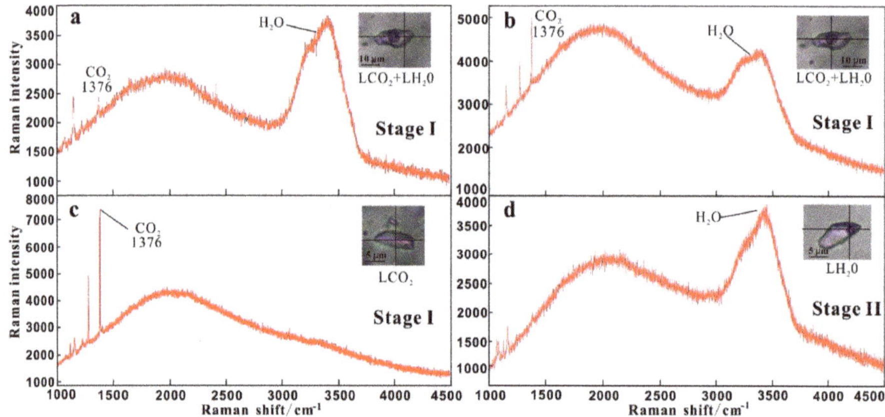

Figure 7. Raman spectra of the fluid inclusion of the Koka gold deposit. (**a,b**) Type III FI of stage I; (**c**) Type II FI of stage I; (**d**) Type I FI of stage II.

5.4. Ion and Gas Chromatography

The results of the composition of the fluid inclusions cluster in quartz based on ion and gas chromatography showed that CO_2 and H_2O were the dominant gaseous ingredients, and CH_4 was rare (Table 2). The average mole% of H_2O and CO_2 of stage I were 0.535 and 0.465, respectively, whereas the average mole% of H_2O increased to 0.753, and the average mole% of and CO_2 reduced to 0.245 in stage II. Only 0.11 mole% of CH_4 was detected in sample KO-16 of stage I. The mole% of CO and H_2 of all of the samples were below the detection limit.

The results of the ion chromatography showed that Na^+ was the dominant cation in both stages, with minor K^+ and Ca^+, and the ratio of Na^+/K^+ ranged from 20.13 to 26.33, and 12.75 to 17.15, respectively. The anion of the two stages was mainly Cl^- with minor SO_4^{2-}, and the ratio of Cl^-/SO_4^{2-} ranged from 11.52 to 17.42, and 19.24 to 31.26, respectively. Moreover, the average content of Na^+ and Cl^- in stage I was 5.40 mg/L and 6.96 mg/L, respectively. Compared with stage I, the average contents of Na^+ and Cl^- in stage II were 9.64 mg/L and 12.63 mg/L, respectively, higher than those in stage I. Li^+, F^-, and Br^- of all samples are below the detection limit.

Table 2. Results of the ion and gas chromatography of the fluid inclusions from the Koka gold deposit.

Sample No.	Stage	H_2O mol%	CO_2 mol%	CO mol%	CH_4 mol%	K^+ mg/L	Na^+ mg/L	Ca^{2+} mg/L	Mg^{2+} mg/L	Cl^- mg/L	SO_4^{2-} mg/L
KO-15		0.595	0.405	<0.01	<0.01	0.18	4.74	0.09	0.02	5.99	0.52
KO-16	Stage I	0.526	0.474	<0.01	0.11	0.31	6.24	0.07	<0.01	8.36	0.48
KO-17		0.484	0.516	<0.01	<0.01	<0.01	5.22	0.06	<0.01	6.52	0.56
KO-3	Stage II	0.741	0.259	<0.01	<0.01	0.47	8.06	0.08	<0.01	9.62	0.50
KO-7		0.765	0.235	<0.01	<0.01	0.88	11.22	0.10	0.02	15.63	0.50

6. C–H–O–S Isotopes

The carbon and oxygen isotopic data are listed in Table 3 and plotted in Figure 8. Three quartz samples from the early stage provided $\delta^{13}C$ value ranging from $-5.5‰$ to $-5.0‰$, and the $\delta^{18}O$ value ranged from $+9.0‰$ to $+9.7‰$. The $\delta^{13}C$ and $\delta^{18}O$ values of two quartz samples from the late stage fell within a range of $-4.4‰$ and $+10.1‰$ to $+10.8‰$, respectively.

Table 3. The oxygen, hydrogen, and carbon isotopic compositions of quartz from the Koka gold deposit.

Sample NO.	Stage	Mineral	δD_{smow} (‰)	$\delta^{18}O_{smow}$ (‰)	$\delta^{18}O_{H_2O}$ (‰)	$\delta^{13}C_{v\text{-}PDB}$ (‰)	Temperature (°C)
KO-15		Quartz	−52.5	9.7	2.4	−5.0	290
KO-16	Stage I	Quartz	−57.0	9.0	1.7	−5.4	290
KO-17		Quartz	−50.1	9.5	2.2	−5.5	290
KO-3	Stage II	Quartz	−53.1	10.8	2.7	−4.4	270
KO-7		Quartz	−54.1	10.1	2.0	−4.4	270

The oxygen and hydrogen isotopic data are listed in Table 3 and plotted in Figure 9. The measured $\delta^{18}O$ values of five quartz samples in the two stages ranged from +8.0 to +8.8‰. The $\delta^{18}O$ values of hydrothermal fluids were calculated using the equation of Clayton et al. [21], $1000\ln a_{quartz–water} = 3.38 \times 10^6 \times T^{-2} - 3.40$, together with the measured $\delta^{18}O_{quartz}$ values and the corresponding average homogenization temperatures of the FIs in the same stage of the same sample (Table 3). As a result, the $\delta^{18}O_{H_2O}$ values from the early stage and the late stage were $+1.7‰$ to $+2.4‰$, and $2.0‰$ to $+2.7‰$, respectively. All samples selected for $\delta^{18}O$ analysis were also analyzed for their hydrogen isotopic composition. The δD_{H_2O} values of the quartz samples in the early stage and the late stage were $-57.0‰$ to $-50.1‰$, and $-54.1‰$ to $-53.1‰$, respectively.

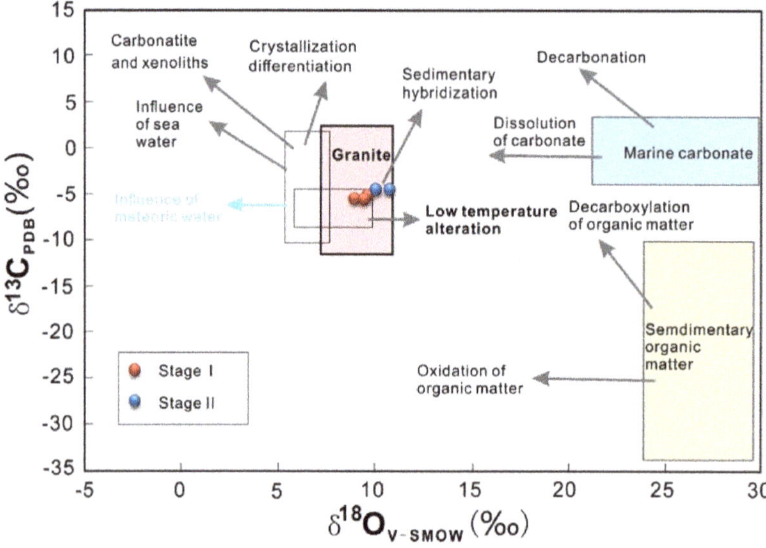

Figure 8. Calculated $\delta^{13}C$ and $\delta^{18}O$ values of fluids at Koka (after Chen et al., 2012 [25]).

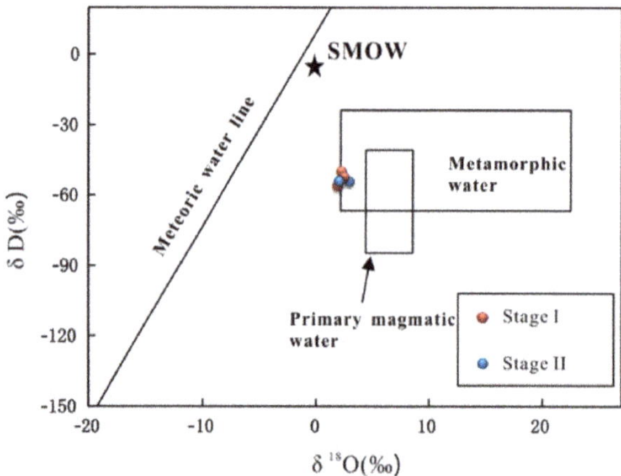

Figure 9. The $\delta^{18}O$ and δD values of the ore fluids at the Koka deposit. The metamorphic water field, primary magmatic water field, and meteoric water line are from Taylor (1974) [26].

The sulfur isotopic data are shown in Table 4 and plotted in Figure 10. The $\delta^{34}S$ values of ten pyrite samples from the Koka gold deposit ranged from −0.1‰ to +2.7‰, with an average of +1.6‰, and the $\delta^{34}S$ value of the samples of one chalcopyrite, one galena, and one sphalerite were +1.3‰, −1.3‰, and +1.2‰, respectively. Therefore, the sulfur isotopic compositions in the Koka gold deposit showed a relatively narrow range (−1.3 to +2.7‰, around zero), indicating a homogenous sulfur source.

Table 4. Sulfur isotopic compositions of sulfides from the Koka gold deposit.

Sample No.	Stage	Mineral	$\delta^{34}S_{CDT}$ (‰)
KO-11		Pyrite	1.7
KO-11		Pyrite	−0.1
KO-15		Pyrite	1.3
KO-16		Pyrite	1.4
KO-17	Stage I	Pyrite	1.6
KO-17		Pyrite	1.6
KO-17		Chalcopyrite	1.3
KO-17		Galena	−1.3
KO-17		Sphalerite	1.2
KO-3		Pyrite	1.9
KO-3	Stage II	Pyrite	1.9
KO-7		Pyrite	2.7
KO-14		Pyrite	1.8

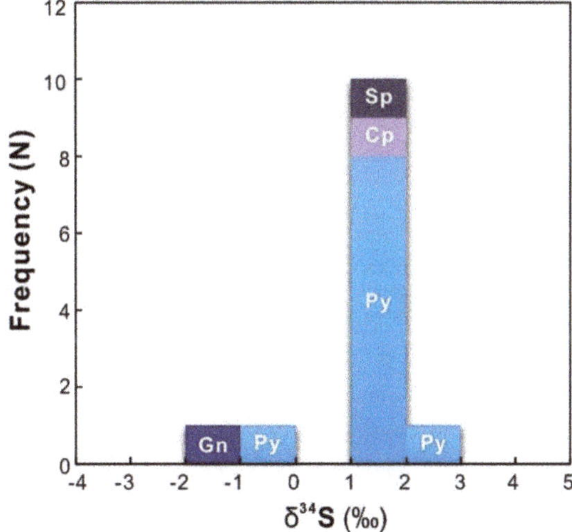

Figure 10. Histograms of the $\delta^{34}S$ values of sulfide for the Koka deposit.

7. Discussion

7.1. Nature and Evolution of Ore-Forming Fluid

CO_2-bearing FIs were the most abundant occurrence in the Koka gold deposit, with moderate liquid-rich aqueous FIs. Generally, the results of fluid inclusion petrography, microthermometry, and laser Raman micro-spectroscopy showed that the ore-forming fluid of the Koka gold deposit was a medium- to low-temperature and low-salinity CO_2–$NaCl$–H_2O system.

(1)   Stage I

The ore-forming fluid of this stage was a medium-temperature, low-salinity $NaCl$–H_2O–CO_2($-CH_4$) fluid (CH_4 was detected via gas chromatography). The majority of CO_2 and CH_4 are likely to be from the metamorphic strata [10,27–29]. Fluid immiscibility is one of the dominant ore-forming mechanisms in gold deposits [30–32]. The common coexistence of type I and II FIs during stage I suggests that they were entrapped simultaneously, and homogenized in different ways, which suggests that fluid immiscibility occurred prior to their entrapment [33]. The plot of salinity vs. the homogenization temperatures (Figure 11) indicates that the ore-forming fluid underwent a fluid immiscibility process during stage I at Xiangdong (three-phase CO_2-rich inclusions commonly coexist with two-phase aqueous inclusions in quartz, with similar homogenization temperatures [24]), which is likely to be due to the decrease in temperature and pressure as the ore-forming fluid ascended, leading to the escape of CO_2 from the fluid.

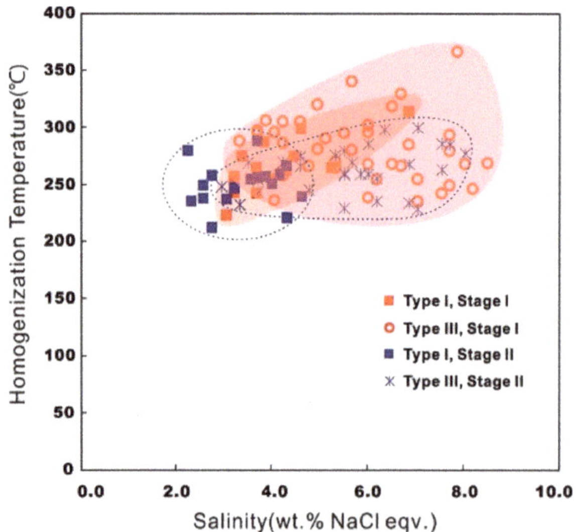

Figure 11. Plot of salinity vs. homogeneous temperature for fluid inclusions of different stages from the Koka deposit.

Moreover, varying degrees of CH_4 involvement in the fluid system were detected in the stage I FIs, and CH_4 may lead to fluid immiscibility of the $NaCl$–H_2O–CO_2 system at depth [34].

We calculated the densities of the type I (salinity <5 wt% NaCl eqv) and type II (CO_2-rich) inclusions in stage I via Flincor software [35] of 0.711–0.858 g/cm^3 and 0.757–0.904 g/cm^3, respectively, and then estimated the pressure in the range from 73 to 278 MPa, Figure 12 [11,36–38]). In addition, we can constrain the temperatures at ~290 °C from fluid inclusion microthermometry which is in agreement with the temperature obtained from sulfur isotope calculation of galena-sphalerite pair (see below). Then we obtained an estimated pressure between 106 and 168 MPa, which was similar to the estimated pressure between 105 and 250 MPa (Figure 13) using the method of Van den Kerhof and Thiéry [39]. Hence, we use 168 MPa to better constrain the estimated trapping pressure of stage I, corresponding to a maximum depth of 6.3 km (lithostatic pressure; rock density of 2.7 g/cm^3).

(2) Stage II

The occurrence and composition of different FIs types suggest that the ore-forming fluid from stage II is a medium-temperature, low-salinity $NaCl$–H_2O–CO_2 fluid. The temperature of the ore-forming fluid decreased from stage I to II, most likely due to the progressive cooling of the fluid system or/and mixing with relatively low-T fluid, whereas the meteoric water mixing is supported by the results of the hydrogen and oxygen isotopes. In addition, the water content increased in the fluid of stage II, suggesting a water mixing. Salinity from stage I to II changed slightly, most likely due to the precipitation of metal elements in stage II.

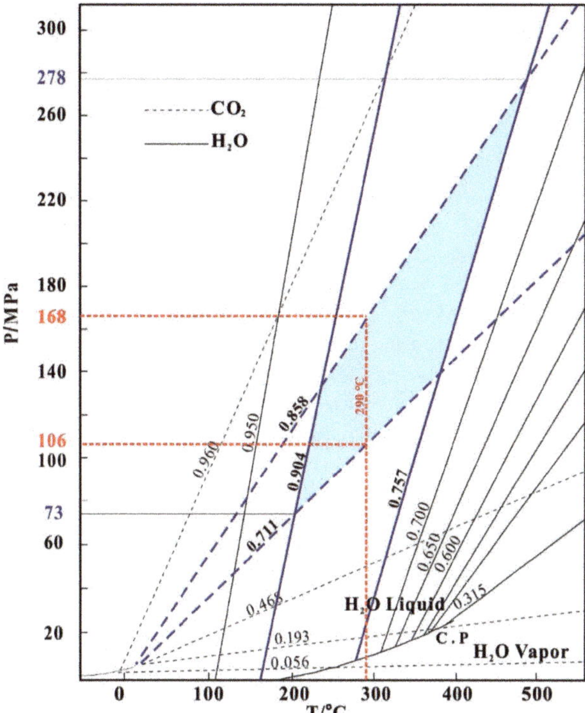

Figure 12. Pressure estimations for the primary fluid inclusions of the Koka deposit (after Roedder and Bodnar [36]).

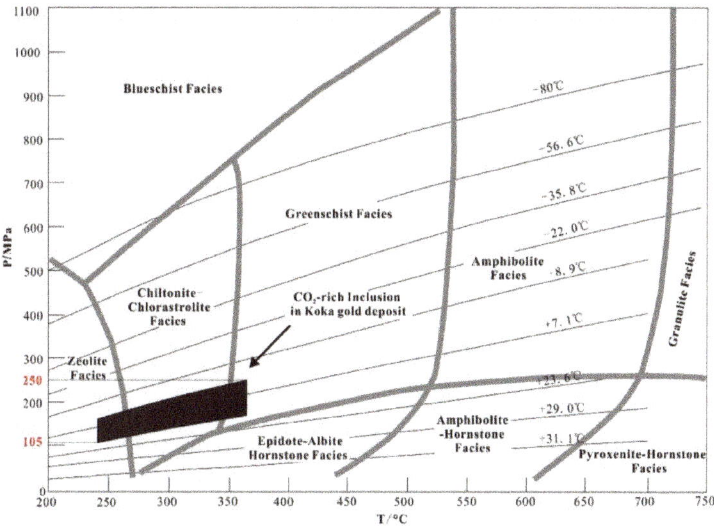

Figure 13. Pressure estimations for the primary fluid inclusions of the Koka deposit (after Van den Kerhof and Thiéry 2001 [39]).

7.2. Source of Ore-Forming Fluids and Metal

Most of the δD_{fluid} values (−57.0 to −50.1‰) and the $\delta^{18}O_{H2O}$ values (+1.7 to +2.7‰) of the ore quartz in the Koka gold deposit fell within the ranges of the isotopic compositions of metamorphic water [26,40], and near the primitive magmatic water field (Figure 9). The H–O data suggest that these ore-forming fluids were derived from metamorphic water, and that magmatic water could mix within the ore-forming fluids.

The $\delta^{13}C$ values of the quartz samples ranged from −5.5 to−4.4‰, which were higher than those of the organic matter (averaging −25‰ [41]), CO_2 dissolved in water (−9 to −20‰ [42]), atmospheric CO_2 (~−8‰ [43] or −7 to −11‰ [42]), and crust (−7‰ [44]), and lower than the marine carbonates (~0‰ [41]), but consistent with the C isotopic compositions of igneous/magma systems (−3 to −30‰ [42]; mantle (−5 to −7‰ [41]). In the $\delta^{13}C$ vs. $\delta^{18}O$ diagram, all of the quartz samples were plotted within the granite box field (Figure 8), indicating that carbon in the early ore-forming fluids was likely to be provided by magmatic water. Moreover, the late stage quartz samples showed an obvious trend of influence by low temperature alteration, which suggests a temperature cooling in the late ore-forming system. This interpretation is consistent with the results of the fluid inclusion microthermometry.

The $\delta^{34}S$ values of sulfides in the Koka gold deposit exhibited a narrow range of values that were close to 0‰ (−1.3 to +2.7‰, Figure 10), which suggests that the sulfides that precipitated from the fluid system originated from a single sulfur source that was primarily comprised of deep-seated magma ($\delta^{34}S = 0 \pm 3‰$ [45,46]). Moreover, the $\delta^{34}S$ values exhibited a trend of $\delta^{34}S_{pyrite} > \delta^{34}S_{sphalerite} > \delta^{34}S_{galena}$, which is consistent with the crystallization sequence of minerals in a hydrothermal system under the conditions of sulfur isotopic fractionation equilibrium [45,47]. These data suggest that the hydrothermal system reached a state of sulfur isotopic fractionation equilibrium before its ore minerals were precipitated [48–50], suggesting that these ore-forming materials were predominantly sourced from magma. Then we calculated the galena-sphalerite pair and obtained a temperature of 265±19 °C according to Ohmoto and Rye 1979 [45], which is in agreement with the temperatures obtained from oxygen isotopes.

7.3. Genetic Model for Ore Deposition

The continental collision of East and West Gondwana caused the reconstruction of the regional juvenile crust and lithosphere, and the formation of Eeat African Ogogenic Belt (EAOB) during the Neoproterozoic (~650 Ma). Most of the juvenile crust were conjoined along the arc–arc suture, and the subsequent orogeny lasted nearly 100 Ma [15]. Due to the rapid convergence between East Gondwana and West Gondwana, the Eritrea region in the southern part of the EAOB experienced regional metamorphism [51].

$^{40}Ar–^{39}Ar$ ages of sericite in the gold-bearing quartz veins of the Koka deposit were 600–580 Ma (unpub. data), suggesting that the Koka deposit most likely formed in the Neoproterozoic. Moreover, several gold deposits in the Nubian Shield were reported to have formed at this period such as Sukhaybarat [52], An Najadi [53], Ad Duwayhi [54], and Lega-Dembi [55]. Regionally, Post orogenic a-type granite was formed after 610 Ma [15], combined with the formation age of Kyanite schist (U–Pb age of monazite, 593 ± 5Ma [56]), suggesting that the time interval of 600–580 Ma was in the transition environment from crust compression to extension during the collision, accompanied by regional metamorphism.

The characteristics of geotectonic environment, occurrence of orebodies, mineral assemblages, and ore-forming fluid of the Koka gold deposit were similar to the features of orogenic gold deposits [11,30,57], whereas the carbon and sulfur isotopes suggested an igneous origin. Hence, the Koka gold deposit could be an orogenic gold deposit related to magmatism. The detailed ore-forming process is described as follows:

After the solidification of Koka granite, multi-fractures that existed in the granite became a relatively low-pressure zone in the district. Ore-bearing fluid, which was derived from the metamorphic strata and containing abundant CO_2 in high-pressure conditions, was driven by regional thermodynamic processes to flow into the fault system of Koka granite. The temperature and pressure decreased as the ore-forming fluid rose during stage I, which led to CO_2 separating from the fluid accompanied by fluid immiscibility, leading to the dissolution of the Au-bearing complexes and a pH change in the residual fluid. Then, ore-forming ions such as Au^+, Fe^{2+}, Cu^{2+}, Pb^{2+}, Zn^{2+}, and S^{2-} were precipitated at locations with advantageous structural conditions. During stage II, meteoric water was mixed in and cooled the fluid system, so that metal cations such as Fe^{2+} and S^{2-} remaining in the fluid were deposited in the structural fractures and micro-fractures of the existing veins.

8. Conclusions

(1) The ore-forming fluid of the Koka gold deposit is a medium- to low-temperature and low-salinity CO_2–$NaCl$–H_2O system, and ore-forming mechanisms include fluid immiscibility during an early stage and fluid mixing with meteoric water in subsequent stages at lower temperature. We used two methods to estimate a similar pressure at ~168 MPa, which corresponded to a depth of 6.3 km.

(2) The C–H–O isotopic compositions indicated that the ore-forming fluids of the Koka deposit could have originated from metamorphic strata and were likely to have made a considerable magmatic contribution. The S isotopic result suggest that the metals were derived from magma.

(3) Features of geology and ore-forming fluid at the Koka gold deposit were similar to those of orogenic gold deposits. Hence, the Koka deposit might be an orogenic gold deposit related to granite.

Author Contributions: K.Z., H.Y, J.W. and Y.-Q.X. conceived and designed the experiments; G.F.G., and W.X. performed the experiments; all authors wrote the paper.

Funding: This research was funded by (China Geological Survey, Mineral resources assessment of Egypt and adjacent areas) grant number (DD20160109), (the National Natural Science Foundation of China Project) grant number (41803044) and (the Construction Project of National Technical Standard System of Mineral Resources and Reserves) grant number (CB2017-4-10; 2017TP1029).

Acknowledgments: This work was supported by the China Geological Survey, Mineral resources assessment of Egypt and adjacent areas (DD20160109), Evaluation of large copper and gold resource bases in North Africa Project, and the National Natural Science Foundation of China Project (grant number 41803044).

Conflicts of Interest: The authors declare no conflict of interest.

References

1. Kröner, A.; Eyal, M.; Eyal, Y. Early Pan-African evolution of the basement around Elat, Israel and the Sinai Peninsula revealed by single-zircon evaporation dating and implication for crustal accretion rates. *Geology* **1990**, *18*, 545–548. [CrossRef]

2. Stern, R.J. Neoproterozoic (900–550 ma) arc assembly and continental collision in the east Africa orogen: Implications for the consolidation of Gondwanaland. *Annu. Rev. Earth & Planet. Sci.* **1994**, *22*, 319–351.

3. Stern, R.J. Neoproterozoic crustal growth: The solid Earth system during a critical episode of Earth history. *Gondwana Res.* **2008**, *14*, 33–50. [CrossRef]

4. Johnson, P.R.; Zoheir, B.A.; Ghebreab, W.; Stern, R.J.; Barrie, C.T.; Hamer, R.D. Gold-bearing volcanogenic massive sulfides and orogenic-gold deposits in the Nubian Shield. *S. Afr. J. Geol.* **2017**, *120*, 63–76. [CrossRef]

5. Ghebreab, W.; Greiling, R.O.; Solomon, S. Structural setting of Neoproterozoic mineralization, Asmara district, Eritrea. *J. Afr. Earth Sci.* **2009**, *55*, 219–235. [CrossRef]

6. Barrie, C.T.; Nielsen, F.W.; Aussant, C.H. The Bisha volcanic-associated massive sulfide deposit, western Nakfa terrane, Eritrea. *Econ. Geol.* **2007**, *102*, 717–738. [CrossRef]

7. Zhao, X.Z.; Duan, H.C.; Wang, F.X. General characteristics of geology and mineral resources in Eritrea and exploration progress. *Miner. Explor.* **2012**, *5*, 707–714. (In Chinese)

8. Xiang, P.; Wang, J.X. Ore geology character and type of Koka gold deposit, Eritrea. *Acta Mineral. Sin.* **2013**, *s2*, 1067–1068. (In Chinese)

9. Goldfarb, R.J.; Groves, D.I.; Gardoll, S. Orogenic gold and geologic time: A global synthesis. *Ore Geol. Rev.* **2001**, *18*, 12–75. [CrossRef]

10. Groves, D.I.; Goldfarb, R.J.; Robert, F. Gold deposits in metamorphic belts: Overview of current understanding, outstanding problems, future research, and exploration significance. *Econ. Geol.* **2003**, *98*, 1–29.

11. Goldfarb, R.J.; Groves, D.I. Orogenic gold: Common or evolving fluid and metal sources through time. *Lithos.* **2015**, *233*, 2–26. [CrossRef]

12. Abdelsalam, M.; Stern, R. Sutures and shear zones in the Arabian-Nubian Shield. *J. Afr. Earth Sci.* **1996**, *23*, 289–310. [CrossRef]

13. Stern, R.J.; Johnson, P.R.; Kröner, A.; Yibas, B. Neoproterozoic ophiolites of the Arabian-Nubian shield. *Dev. Precambrian Geol.* **2004**, *13*, 95–128.

14. Johnson, P.R.; Woldehaimanot, B. *Development of the Arabian–Nubian Shield: Perspectives on Accretion and Deformation in the Northern East African Orogen and the Assembly of Gondwana*; Special Publication: London, UK, 2003; pp. 290–325.

15. Johnson, P.R.; Andresen, A.; Collins, A.S.; Fowler, A.R.; Fritz, H.; Ghebreab, W.; Kusky, T.; Stern, R.J. Late Cryogenian–Ediacaran history of the Arabian–Nubian shield: A review of depositional, plutonic, structural, and tectonic events in the closing stages of the northern East African Orogen. *J. Afr. Earth Sci.* **2011**, *61*, 167–232. [CrossRef]

16. Drury, S.A.; Berhe, S.M. Accretion tectonics in Northern Eritrea revealed by remotely sensed imagery. *Geol. Mag.* **1993**, *130*, 177–190. [CrossRef]

17. Teklay, M. *Petrology, Geochemistry and Geochronology of Neoproterozoic Magmatic Arc Rocks from Eritrea: Implications for Crustal Evolution in the Southern Nubian Shield*; Department of Mines-Ministry of Energy Mines and Water Resources-State of Eritrea: Asmara, Eritrea, 1997; Volume 1, pp. 1–125.

18. Drury, S.A.; De Souza Filho, C.R. Neoproterozoic terrane assemblages in Eritrea: Review and prospects. *J. Afr. Earth Sci.* **1998**, *27*, 331–348. [CrossRef]

19. Zhao, K.; Yao, H.Z.; Wang, J.X.; Ghebsha, F.G.; Xiang, W.S.; Yang, Z. Zircon U-Pb geochronology and geochemistry of Koka granite and its geological significances, Eritrea. *Earth Sci.*. in press. (In Chinese)

20. Dean, C.; David, L.; David, G. *Technical Report on the Koka Gold Deposit, Eritea*; Chalice Gold Mine Limited: Asmara, Eritrea, 2010; pp. 1–111.

21. Bozzo, A.T.; Chen, H.S.; Kass, J.R.; Barduhn, A.J. The properties of the hydrates of chlorine and carbon dioxide. *Desalination* **1975**, *16*, 303–320. [CrossRef]

22. Lu, H.Z.; Fan, H.R.; Ni, P.; Ou, G.X.; Shen, K.; Zhang, W.H. *Fluid Inclusions*; Science Press: Beijing, China, 2004; pp. 406–419. (In Chinese)

23. Clayton, J.; Tretiak, D.N. Amine-citrate buffers for pH control in starch gel electrophoresis. *J. Fish. Board Can.* **1972**, *29*, 1169–1172. [CrossRef]

24. Roedder, E. *Fluid Inclusions. Review in Mineralogy*; Mineralogical Society of America: Chantilly, VA, USA, 1984; pp. 1–644.

25. Chen, H.Y.; Chen, Y.J.; Baker, M.J. Evolution of ore-forming fluids in the Sawayaerdun gold deposit in the Southwestern Chinese Tianshan metallogenic belt. *J. Asian Earth Sci.* **2012**, *49*, 131–144. [CrossRef]

26. Taylor, H.P. The application of oxygen and hydrogen isotope studies to problems of hydrothermal alteration and ore deposition. *Econ. Geol.* **1974**, *69*, 843–883. [CrossRef]

27. Yang, L.Q.; Deng, J.; Guo, L.N.; Wang, Z.L.; Li, X.Z.; Li, J.L. Origin and evolution of ore fluid, and gold-deposition processes at the giant Taishang gold deposit, Jiaodong Peninsula, eastern China. *Ore Geol. Rev.* **2016**, *72*, 585–602. [CrossRef]

28. Yang, L.Q.; Guo, L.N.; Wang, Z.L.; Zhao, R.X.; Song, M.C.; Zheng, X.L. Timing and mechanism of gold mineralization at the Wang'ershan gold deposit, Jiaodong Peninsula, eastern China. *Ore Geol. Rev.* **2017**, *88*, 491–510. [CrossRef]

29. Qiu, K.F.; Taylor, R.D.; Song, Y.H.; Yu, H.C.; Song, K.R.; Li, N. Geologic and geochemical insights into the formation of the Taiyangshan porphyry copper–molybdenum deposit, Western Qinling Orogenic Belt, China. *Gondwana Res.* **2016**, *35*, 40–58. [CrossRef]

30. Yang, L.Q.; Deng, J.; Wang, Z.L.; Guo, L.N.; Li, R.H.; Groves, D.I.; Danyushevsky, L.V.; Zhang, C.; Zheng, X.L.; Zhao, H. Relationships between gold and pyrite at the Xincheng gold deposit, Jiaodong Peninsula, China: Implications for gold source and deposition in a brittle epizonal environment. *Econ. Geol.* **2016**, *111*, 105–126. [CrossRef]

31. Yang, L.Q.; Deng, J.; Li, R.P.; Guo, L.N.; Wang, Z.L.; Chen, B.H.; Wang, X.D. World-class Xincheng gold deposit: An example from the giant Jiaodong gold province. *Geosci. Front.* **2016**, *7*, 419–430. [CrossRef]

32. Groves, D.I.; Goldfarb, R.J.; Gebre-Mariam, M.; Hagemann, S.G.; Robert, F. Orogenic gold deposits: A proposed classification in the context of their crustal distribution and relationship to other gold deposit types. *Ore Geol. Rev.* **1998**, *13*, 7–27. [CrossRef]

33. Shepherd, T.J.; Rankin, A.H.; Alderton, D.H.M. *A Practical Guide to Fluid Inclusion Studies*; Chapman & Hall: Abingdon, UK, 1985; pp. 1–239.

34. Naden, J.; Shepherd, T.J. Role of methane and carbon dioxide in gold deposition. *Nature* **1989**, *342*, 793–795. [CrossRef]

35. Brown, P.E. FLINCOR: A microcomputer program for the reduction and investigation of fluid-inclusion data. *Am. Mineral.* **1989**, *74*, 1390–1393.

36. Roedder, E.; Bodnar, R.J. Geologic pressure determinations from fluid inclusion studies. *Annu. Rev. Earth Planet. Sci.* **1980**, *8*, 263–301. [CrossRef]

37. Xiong, Y.Q.; Shao, Y.J.; Zhou, H.D.; Wu, Q.H.; Liu, J.P.; Wei, H.T.; Zhao, R.C.; Cao, J.Y. Ore-forming mechanism of quartz-vein-type W-Sn deposits of the Xitian district in SE China: Implications from the trace element analysis of wolframite and investigation of fluid inclusions. *Ore Geol. Rev.* **2017**, *83*, 152–173. [CrossRef]

38. Xiong, S.; He, M.; Yao, S.; Cui, Y.; Shi, G.; Ding, Z.; Hu, X. Fluid evolution of the Chalukou giant Mo deposit in the northern Great Xing'an Range, NE China. *Geol. J.* **2015**, *50*, 720–738. [CrossRef]

39. Van den Kerkhof, A.; Thiéry, R. Carbonic inclusions. *Lithos* **2001**, *55*, 49–68. [CrossRef]

40. Qiu, K.F.; Marsh, E.; Yu, H.C.; Pfaff, K.; Gulbransen, C.; Gou, Z.Y.; Li, N. Fluid and metal sources of the Wenquan porphyry molybdenum deposit, Western Qinling, NW China. *Ore Geol. Rev.* **2017**, *86*, 459–473. [CrossRef]

41. Hoefs, J. *Stable Isotope Geochemistry*, 6th ed.; Springer: Berlin/Heidelberg, Germany, 2009; pp. 130–135.

42. Hoefs, J. *Stable Isotope Geochemistry*, 3rd ed.; Springer: Berlin/Heidelberg, Germany, 1997; pp. 1–201.

43. Schidowski, M.; Hayes, J.M.; Kaplan, I.R. Isotopic inferences of ancient biochemistry: Carbon, sulfur, hydrogen and nitrogen. In *Earth's Earliest Biosphere*; Schopf, J.W., Ed.; Princeton University Press: Princeton, NJ, USA, 1983; pp. 149–186.

44. Faure, G. *Principles of Isotope Geology*, 2nd ed.; Wiley: New York, NY, USA, 1977; p. 589.

45. Ohmoto, H.; Rye, R.O. Isotopes of sulfur and carbon. In *Geochemistry of Hydrothermal Ore Deposits*, 2nd ed.; Barnes, H.L., Ed.; John Wiley and Sons: New York, NY, USA, 1979; pp. 509–567.

46. Chaussidon, M.; Lorand, J.P. Sulphur isotope composition of orogenic spinel lherzolite massifs from Ariege (North-Eastern Pyrenees, France): An ion microprobe study. *Geochim. Cosmochim. Acta.* **1990**, *54*, 2835–2846. [CrossRef]

47. Xiong, Y.Q.; Shao, Y.J.; Mao, J.W.; Wu, S.C.; Zhou, H.D.; Zheng, M.H. The polymetallic magmatic-hydrothermal Xiangdong and Dalong systems in the W–Sn–Cu–Pb–Zn–Ag Dengfuxian orefield, SE China: Constraints from geology, fluid inclusions, H–O–S–Pb isotopes, and sphalerite Rb–Sr geochronology. *Miner. Depos.* **2019**. [CrossRef]

48. Zheng, Y.F.; Xu, B.L.; Zhou, G.T. Geochemical studies of stable isotopes in minerals. *Earth Sci. Front.* **2000**, *7*, 299–320. (In Chinese)

49. Deng, J.; Liu, X.F.; Wang, Q.F.; Pan, R.G. Origin of the Jiaodong-type Xinli gold deposit, Jiaodong peninsula, China: Constraints from fluid inclusion and C–D–O–S–Sr isotope compositions. *Ore Geol. Rev.* **2015**, *65*, 674–686. [CrossRef]

50. Yang, L.Q.; Deng, J.; Li, N.; Zhang, C.; Yu, J.Y. Isotopic characteristics of gold deposits in the Yangshan Gold Belt, West Qinling, central China: Implications for fluid and metal sources and ore genesis. *J. Geochem. Explor.* **2016**, *168*, 103–118. [CrossRef]

51. Ghebreab, W. Tectono-metamorphic history of Neoproterozoic rocks in eastern Eritrea. *Precambrian Res.* **1999**, *98*, 83–105. [CrossRef]

52. Albino, G.V.; Jalal, S.; Christensen, K. Neoproterozoic mesothermal gold mineralization at Sukhaybarat East mine. *Trans. Inst. Min. Metall. (Sect. B Appl. Earth Sci.)* **1995**, *104*, 157–170.

53. Walker, B.M.; Lewis, R.S.; Al Otaibi, R.; Ben Talib, M.; Christian, R.; Gabriel, B.R. *An Najadi Gold Prospect, Kingdom of Saudi Arabia*; Geology and Gold-Resource Assessment; Saudi Arabian Deputy Ministry for Mineral Resources Technical Report, USGS-TR-94-5; Reston Publishing Service Center: Reston, VA, USA, 1994; pp. 1–89.

54. Doebrich, J.L.; Zahony, S.G.; Leavitt, J.D.; Portacio, J.S., Jr.; Siddiqui, A.A.; Wooden, J.L.; Fleck, R.J.; Stein, H.J. Ad Duwayhi, Saudi Arabia: Geology and geochronology of a Neoproterozoic intrusion-related gold system in the Arabian shield. *Econ. Geol.* **2004**, *99*, 713–741. [CrossRef]
55. Billay, A.Y.; Kisters, A.F.M.; Meyer, F.M.; Schneider, J. The geology of the Lega Dembi gold deposit, southern Ethiopia: Implications for Pan-African gold exploration. *Miner. Depos.* **1997**, *32*, 491–504. [CrossRef]
56. Andersson, U.B.; Ghebreab, W.; Teklay, M. Crustal evolution and metamorphism in east-central Eritrea, south-east Arabian-Nubian Shield. *J. Afr. Earth Sci.* **2006**, *44*, 45–65. [CrossRef]
57. Chen, Y.J.; Ni, P.; Fan, H.R.; Prajno, F.; Nai, Y.; Su, W.C.; Zhang, H. Diagnostic fluid inclusions of different types hydrothermal gold deposits. *Acta Petrol. Sin.* **2007**, *23*, 2085–2108. (In Chinese)

 *minerals* 

Article

Trace Element Geochemistry in Quartz in the Jinqingding Gold Deposit, Jiaodong Peninsula, China: Implications for the Gold Precipitation Mechanism

Binghan Chen [1,2], Jun Deng [2,*], Hantao Wei [1] and Xingzhong Ji [1]

[1] MNR Key Laboratory of Metallogeny and Mineral Assessment, Institute of Mineral Resources, CAGS, Beijing 100037, China; chenbinghan@cags.ac.cn (B.C.); csuwht@126.com (H.W.); jxz_cugb@126.com (X.J.)

[2] State Key Laboratory of Geological Processes and Mineral Resources, China University of Geosciences, Beijing 100083, China

* Correspondence: djun@cugb.edu.cn; Tel.: +86-10-82322301

Received: 16 March 2019; Accepted: 7 May 2019; Published: 27 May 2019

Abstract: Lots of studies on gold precipitation mechanisms have focused on fluid inclusions within quartz. However, the trace elements in quartz reflect the properties of the ore fluid, and a comparison of the trace element content in different types of quartz can reveal the precipitation mechanism. The Jinqingding gold deposit is the largest gold deposit in the Muping–Rushan gold belt and contains the largest single sulfide–quartz vein type orebody in the gold belt. This study distinguished four types of quartz in this orebody through field work and investigations of the mineralogy and cathodoluminescence (CL) of the quartz and crosscutting relationships as seen under a microscope. In situ studies via electron probe micro-analyzer (EPMA) and laser ablation-inductively coupled plasma-mass spectrometry (LA-ICP-MS) were used to determine the trace element content of the different quartz types. Type Qa displayed a comb structure in the field and zoning under the microscope and in CL. Milky white and smoke grey Qb was the most common quartz type and hosted the most sulfide and gold. Qc was Qa and Qb quartz that recrystallized around pyrite or overgrew and appeared different from Qa and Qb in CL images. Qd occurred within fractures in pyrite. Qa formed prior to the mineralization of gold, and Qd formed post-mineralization. Qb and Qc provided information regarding the ore fluid during mineralization. Sericites occurred with pyrite in fractures in the quartz, and some, along with free gold, filled in fractures in pyrite. Free gold occurred within Qa, Qb, Qc, and in brittle fractures in pyrite. Qc had the lowest Al content of all of the quartz types. As Al content is related to the acidity of the ore fluid in previous study, this indicated an acidity decrease during mineralization, which could be attributed to the sericitization. Sericitization could indicate a potential gold occurrence. The Ti content decreased from Qb to Qc, indicating a decrease in temperature during quartz overgrowth formation. Change in acidity and cooling can therefore be identified as possible causes of gold precipitation in the sulfide–quartz vein type in the Jinqingding gold deposit.

Keywords: quartz; trace element; precipitation mechanism; in situ study; sericite; acidity; formation temperature

---

1. Introduction

The Jiaodong Peninsula is the largest gold province in China [1–5]. Its gold deposits can be divided into fractured–altered rock type [6], mainly in the western Jiaodong, and sulfide–quartz vein type, mainly in the eastern Jiaodong [7–11] (Figure 1a). Quartz and pyrite are the main gold-bearing minerals [7]. Previous studies on the gold precipitation mechanism have focused on fluid inclusions and consideration of gold precipitation mechanisms such as phase separation, unmixing, and water–rock reactions [12–16]. However, hydrothermal activity can be multi-pulse, and evidence from fluid inclusions only represents a certain phase of activity. Gold-bearing quartz has the potential to record a more comprehensive history of the hydrothermal activity, but when dividing the mineralization stage, each stage is usually regarded as a whole and the relative timing of quartz and sulfide crystallization is ignored within each stage.

Quartz is the predominant mineral in the gold deposit and occurs in every stage of the mineralization [7]. As well as the fluid inclusion assemblage in the quartz, the trace elements in each stage of quartz can reflect the growth environment in the mineralization system. Furthermore, geochemical study of the patterns in trace elements such as Li, Sn, Sb, and Rb can be used to establish genetic relationships within magmatic and hydrothermally derived quartz groups [17]. The trace elements such as Al and Ti in quartz that formed during different stages can reflect changes in the properties of the ore fluid, such as acidity and formation temperature of quartz [18]. Research into trace element variations of this kind has previously been carried out for porphyry and skarn deposits through the use of in situ studies with laser ablation-inductively coupled plasma-mass spectrometry (LA-ICP-MS) and electron probe micro-analyzer (EPMA) on quartz [17,18].

The Muping–Rushan gold belt is one of the three largest gold belts in the Jiaodong Peninsula and lies in its eastern part. The ores are mainly sulfide–quartz vein type. Jinqingding is the largest gold deposit and has the largest single sulfide–quartz vein type orebody in the gold belt (No. II, >32 t). The quartz within it has the potential to preserve multi-pulse hydrothermal activity, which can be best studied in orebody No. II [19].

In this paper, we defined the different types of quartz present based on fieldwork, and petrographic and CL observations, and conducted in situ measurements to determine the trace element content. We then used the trace element contents of Al and Ti in different generations of quartz to infer the acidity changes in ore fluid and formation temperature of quartz, and talked about the possible gold precipitation mechanisms in the Jinqingding gold deposit. The results of this study have wider implications for gold exploration.

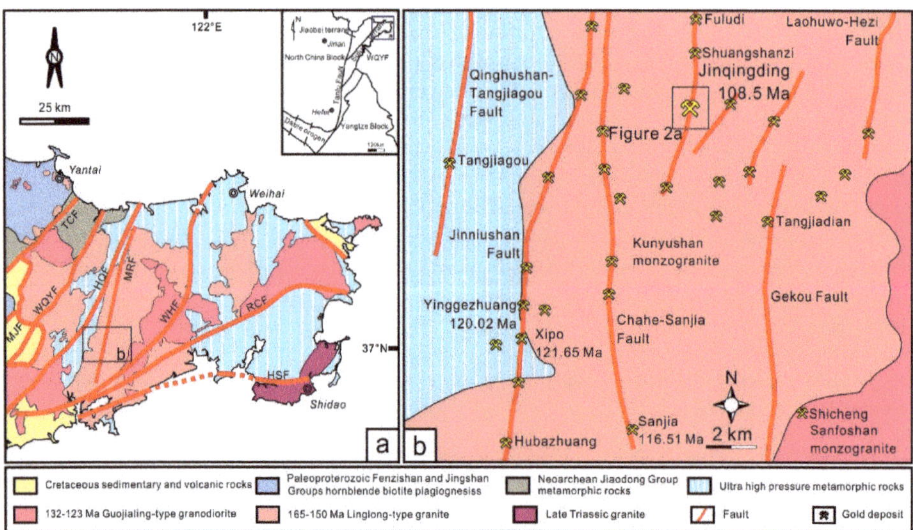

Figure 1. Regional geological map: (**a**) Geological map of the Eastern Jiaodong Peninsula; (**b**) geological map of the Muping–Rushan gold belt (modified from [15]; ages are from [20,21]). HQF—Haiyang–Qingdao fault; HSF—Haiyang–Shidao fault; MRF—Muping–Rushan fault; TCF—Taocun fault; WQYF—Wulian–Qingdao–Yantai fault; WHF—Weihai fault; RCF—Rongcheng fault.

2. Regional Geological Setting

The Jiaodong Peninsula is located on the southeast margin of the North China Craton; the Wulian–Qingdao–Yantai fault divides the peninsula into the Jiaobei and Sulu Terranes [22] (Figure 1a). Jiaodong has experienced both continental collision and subduction of oceanic lithosphere, with the Yangtze Craton subducting below the North China Craton during the Triassic and the Pacific plate subducting beneath the North China Craton in the Jurassic [21,23–25]. More than 150 gold deposits have been found, with more than 4000 tons of proven reserves [7,15]. Most of the gold deposits are located in the Mesozoic granites, and a minority are in Archaen metamorphic rocks [7,14]. Most of the gold deposits formed from 120 Ma onwards and are controlled by NNE-trending faults and secondary faults; a smaller number of gold deposits are controlled by the detachment fault in the Jiaolai Basin [7,14].

The regional basement is tonalite–trondhjemite–granodiorite (TTG) gneisses (2.9–2.5 Ga) [26]. The Neoarchean Jiaodong Group and Proterozoic Fenzishan, Jingshan and Penglai Groups overlie the basement. The Jiaodong Group is made up of metasedimentary and meta-igneous rocks [27,28], the Fenzishan and Jingshan Groups are composed of schist, calc-silicates, marble, and amphibolite [29–31], and the Neoproterozoic Penglai Group comprises marble, slate and quartzite [32]. Rocks of the Jiaodong, Fenzishan, Jingshan, and Penglai Groups occur as xenoliths in the wallrock of most of the gold deposits in Muping–Rushan gold belt.

Tectonic and magmatic events were frequent in the Mesozoic. The magmatic activity can be divided into four periods, namely late Triassic, Late Jurassic (165–150 Ma), Early Cretaceous (110–120 Ma), and Late Early Cretaceous [7]. In the Late Triassic event, mantle-derived syenite–granitic–miscellaneous rock intruded into Shidao after the North China Craton–Yangtze Craton collision [23].

In the Middle to Late Jurassic (165–150 Ma), biotite monzogranite, monzogranodiorite, quartz diorite and granodiorite intrusions derived from partially melted Neoarchaen lower crust intruded to form the Linglong, Kunyushan and Queshan granitoids [33,34]. The late Jurassic Kunyushan granite is the main wallrock in the gold belt (Figure 1b). This can be sub-divided into the Duogushan granite, Washan granite, and Wuzhuashan granite, which have sensitive high resolution ion microprobe (SHRIMP) zircon ages of 161 ± 1 Ma, 138–146 Ma and 160 ± 3 Ma [35] and ^{40}Ar/^{39}Ar biotite cooling ages of 135–147 Ma, 126–131 Ma and 120–123 Ma, respectively [36].

In the Early Cretaceous, granite–granodiorite–alkaline granite derived from partially melted lower crust, and Precambrian metamorphic basement intruded into Sanshandao, Shangzhuang, Sanfoshan and Guojialing and have been dated via SHRIMP U–Pb to 132–123 Ma [37,38]. Several gold deposits within the Muping–Rushan gold belt, such as the Shicheng and Tongling deposits, are hosted within the Sanfoshan monzogranite.

In the late Early Cretaceous, monzogranite intrusions with mixed crust–mantle and partly depleted mantle sources intruded Aishan [35] (Figure 1b).

Most of the gold mineralization is controlled by NE–NNE trending faults [7,39]. There are four main ore-controlling faults in the Muping–Rushan gold belt, namely the Qinghushan–Tangjiagou fault, Jinniushan fault, Jiangjunshi–Quhezhuang fault and Hezi–Gekou fault (Figure 1b). More than 20 gold deposits are located in the Muping–Rushan gold belt. The ores are dominantly of sulfide–quartz vein type.

3. Ore Deposit Geology

The Jinqingding gold deposit (37°06′40″–37°07′30″N, 121°38′06″–121°38′52″E) is the largest gold deposit in the Muping–Rushan gold belt (Figures 1b and 2a) with a proven reserve of >32 t and an average grade of 6.44 g/t [19,20]. It is located 25 km to the NE of Rushan city and was discovered by the Jinzhou Mining Company in the 1970s. ^{40}Ar/^{39}Ar dating of sericite in the quartz–sulfide veins has yielded an age of 107.7 ± 0.5–109.3 ± 0.3 Ma [21].

Sixteen orebodies have been discovered in the Jinqingding deposit; orebody No. II is the largest and contains 96% of the proven reserve [19,20]. Recent engineering activity has revealed that orebody No. II occurs at depths between +120 m and −1220 m, strikes NE, is 120–600 m long along strike, averaging 300 m, and is 100–610 m long along dip (Figure 2b). Auriferous quartz pyrite veins occur within secondary fractures (Figure 2b). The orebodies are located in fault bends where the dip angle changes from moderate to steep. The thickness of orebody No. II is 0.2–6.73 m, averaging 1.65 m, the grade is usually 1.50–30 g/t, averaging 10.40 g/t [19].

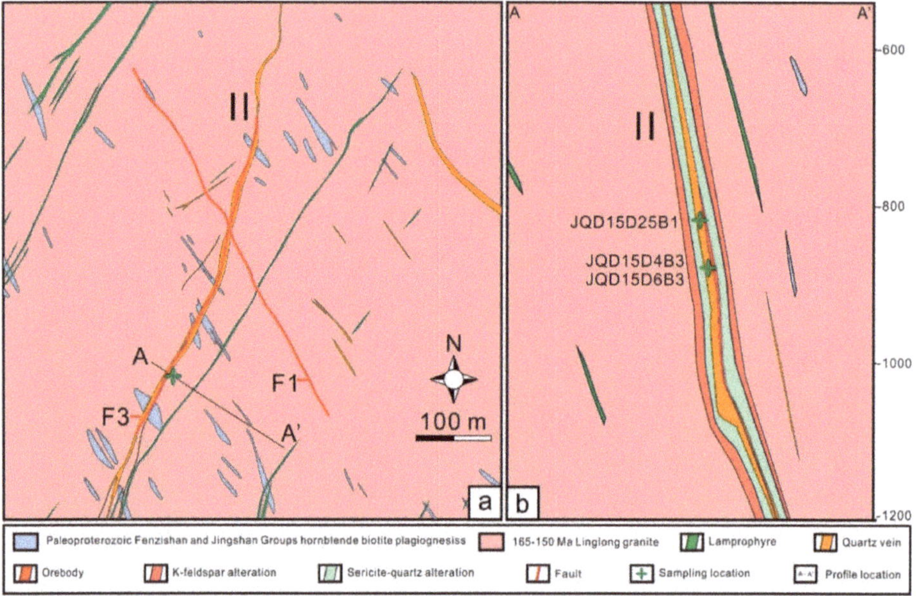

Figure 2. Geological maps of the Jinqingding gold deposit: (**a**) plan view; (**b**) section (modified from [20]).

The ore-controlling fault (F3) is part of the Jiangjunshi–Quhezhuang fault. Its general trend is 20° and it dips to the SE (Figure 2a,b). Based on the field work, F3 is a dextral normal fault with horizontal movement, the orebodies occur in extensional zones, while the other zones are dominated by transpression. F1 is the post-ore fault and no gold mineralization is found in it. Kunyushan monzogranite is the main wallrock (Figure 2a,b and Figure 3a), dikes are mainly lamprophyre, and Paleo–Proterozoic Jingshan Group amphibolite occurs as xenoliths in the deposit (Figure 2a,b). Sulfide–quartz veins have filled fractures in the monzogranites. There is usually a clear boundary between the vein and the wallrock (Figure 3a). Alteration can be seen in the wallrocks of the orebody, and the degree and scale of the alteration are controlled by the fault. The alteration types include K-feldspar alteration (Figure 3b), sericitization (Figure 3c), and silicification. There are gradual boundaries between the different types of alteration. Potassic alteration is the most common type; it can range in width up to 3–4 m and is most intense near the orebody. Pyrite–sericitization forms a narrow belt and usually coexists with a belt of sericitization; these occur in the deeper level of the gold deposit. Only a small amount of silicification is seen near quartz–sulfide veins in the field. The orebody contains some potassic altered wallrock breccias.

The structure of the ore rock varies between massive (Figure 3d), disseminated (Figure 3d), veinlet (Figure 3e), and brecciated, and it features textures such as fissure-intersertal texture and replacement texture. Pyrite and quartz are the main gold-bearing minerals. Gold mainly occurs in fractures in the pyrite and quartz and more rarely as inclusions within those minerals. Smaller amounts of galena, sphalerite, chalcopyrite, and pyrrhotite are also present. Sericites are common in the wallrock and ore. The early milky quartz has a comb structure (Figure 3f).

Pyrite is the most important metal mineral, and both massive pyrite and vein-type pyrite are seen in the field and in hand sample (Figure 3e,f). Pyrite occurrence can be found from the orebody to the wallrock, with massive and disseminated pyrites occurring in the middle part of quartz veins and pyrite veins occurring near the wallrock.

Figure 3. Alteration and mineralization types in the Jinqingding gold deposit: (**a**) A quartz–sulfide vein with unaltered wallrock; (**b**) K-feldspar alteration in the wallrock; (**c**) sericitization in the wallrock of a quartz–sulfide vein; (**d**) milky and smoke grey quartz (Qb + Qc + Qd) plus massive, disseminated pyrites in a hand sample of quartz–sulfide vein; (**e**) smoke grey quartz (Qb + Qc + Qd) in a hand sample of quartz–sulfide vein in which pyrites occur as veins (Py); (**f**) The earliest quartz (Qa) has a comb structure. Py—pyrite; Q—quartz.

In the sulfide–quartz veins, quartz appears milky and smoke grey (Figure 3d–f) and is medium- to coarse-grained (5 mm–1 cm). It can be divided into four main types, namely Qa, Qb, Qc and Qd (Figure 4). In the field and in hand sample, Qa is milky white (Figure 3f). It characteristically has a comb structure (Figures 3f and 4a,b), is euhedral, and has a grain size of several millimeters to several centimeters. No sulfide is found in Qa. Qb is very common and contains gold, many types of pyrites, and base metals (Figure 3d,e and Figure 4c–e). In hand sample, Qb looks milky white when there is no pyrite and smoke grey when pyrites are present (Figure 3d,e). Its grain size may vary under different conditions. Some recrystallized quartz and quartz overgrowths occur around the sulfide (Qc). Some quartz also occurs in fractures in the pyrites (Qd). These last two types are best identified under a microscope. Similar quartz types have been found in other gold deposits of Muping–Rushan gold belt.

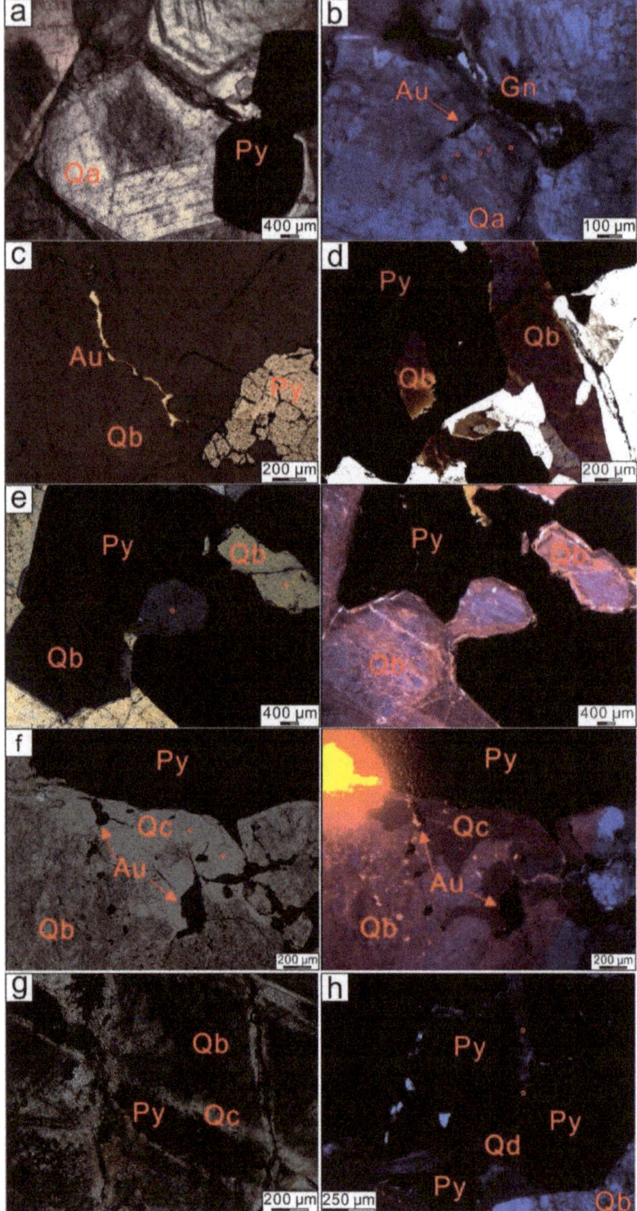

Figure 4. Features of different types of quartz: (**a**) Qa is zoned and is crosscut by pyrite; (**b**) cathodoluminescence (CL) shows zoning in Qa and gold grains within it; (**c**) gold occurs in Qb; (**d**) quartz outside and inside the pyrite goes extinct at the same angle; (**e**) quartz inside and outside pyrite has similar CL features; (**f**) recrystallized (Qc) quartz contains free gold and has darker CL features than other quartz types; (**g**) recrystallized quartz around a crosscutting pyrite vein; (**h**) quartz (Qd) in fractures in pyrite shows different CL features from Qb. Py—Pyrite; Q—Quartz; Au—Gold; Gn—Galena. See Section 4 for description of red circles.

4. Sampling and Analytical Techniques

Representative quartz–sulfide vein samples were collected from depths of −825 m and −865 m within orebody No. II (Figure 2b). Samples JQD15D4B3, JQD15D6B3 and JQD15D25B1 were chosen for further detailed research (Figure 2b). Quartz of each generation was selected with the understanding that multiple hydrothermal activities may be preserved within any one grain. In addition to observations on hand samples and under the microscope, several methods were applied to differentiate the quartz types present and detect their trace element content. Spot locations (red circle) are shown in Figure 4. Samples of each type of quartz and overgrowth quartz were selected for analysis.

Cathodoluminescence (CL) is an effective way to study growth zonation and interior structure in quartz [40], as the color and intensity are affected by structurally bound trace elements [40]. Quartz formation conditions and generations can be reflected in variations in CL [40–44]. Different types of quartz from the different samples were studied with CL and then subjected to further study. CL imaging was carried out with a HC5-LM hot-cathode CL microscope at the Colorado School of Mines, Golden, Colorado, USA. Its operating conditions were a voltage of 14 keV and a current density of approximately 10 $\mu A/mm^2$. A modified Olympus BXFM-S optical microscope was used to locate the quartz. To capture short-lived and long-lived CL signals, a high sensitivity, double-stage Peltier cooled Kappa DX40C CCD camera was used to capture images every seven seconds automatically following initial exposure. The CL images usually show colors changing from dark blue to light red during exposure.

The composition of the selected samples was analyzed by automated scanning electron microscopy (QEMSCAN) at the Colorado School of Mines, Golden, CO, USA. The samples were first loaded into the QEMSCAN instrument, and the analysis was initiated using the control program (iDiscover, FEI). Under the working conditions of a beam step interval of 5 μm or 10 μm, accelerating voltage of 25 keV and beam current of 5 nA, spectra were acquired from each particle by four energy dispersive X-ray (EDX) spectrometers. Monte Carlo simulation was used to model the interactions between the beam and the sample. The EDX spectra at each acquisition point were compared with spectra held in a look-up table based on an assignment that had been made as to composition. The QEMSCAN software calculates the area percentage of each composition in the look-up table and outputs the results as a spreadsheet.

Laser ablation-inductively coupled plasma-mass spectrometry (LA-ICP-MS) carried out at the United States Geological Survey (USGS), Denver, CO, USA was used to determine the trace element concentrations in the quartz. The LA-ICP-MS system is composed of a Horiba Xplora Raman Spectrometer (532 nm) and a Photon Machines Analyte G2 LA system (193 nm, 4 ns excimer). He gas (0.85 L/min) carries the ablated material to a modified glass mixing bulb, and the sample is mixed coaxially with Ar (0.6 L/min) at the ICP torch [45]. The thin section should be prepared carefully as the quartz from gold deposits is usually fragile and brittle and may break without the assistance of the glass. It is also likely to contain small fluid inclusions, which may cause it to explode when heated by the laser. The spots were chosen on the basis of the CL results and microscopic observation. Only two selected quartz grains (JQD15D25B1) were studied (Table S2). The detection limit for Ti and Al is 1 ppm and 2 ppm, respectively.

During electron probe micro-analyzer (EPMA) work on the quartz, mineral inclusions in the quartz were excluded so that the Al content would reflect the concentration of structurally bound Al (Table S1). The trace element contents of the quartz were studied quantitatively with a JEOL JXA 8800 electron microprobe analyzer at the USGS, Denver, CO, USA. The operating conditions were 20 kV, 100 nA, and a focused beam 2 μm in diameter. The detection limit of Al is 0.001 wt %, and the standard sample used was USGS-Menlo #5-168.

5. Mineralogy and Paragenesis

Previous studies have divided the mineralization process in the Jiaodong peninsula into four main stages, namely the gold–pyrite–quartz stage (I), gold–quartz–pyrite stage (II), gold–pyrite–base sulfide–quartz stage (III), and quartz–pyrite–calcite stage (IV). The second and third stages are regarded as the main stages of mineralization [6,12,15]. There has been little study of the sequence of mineral formation within each stage; the mineral assemblages are generally discussed as one. This paper presents a new paragenesis based on field work and the petrographic investigation.

The field, hand sample, and microscopic observations and short-lived CL results from this study indicated that there were differences between quartz grains including their crosscutting relationships, CL features, timing relative to pyrite, relationships with mineralization, and fluid inclusions. The detailed features of each type of quartz were therefore studied in more detail (Figure 4).

Under the microscope, Qa cores were turbid, and the grains had clear zoning. The zoned part was crosscut by pyrite; cross-cutting relationships indicated that Qa formed prior to all of the sulfides and mineralization (Figure 4a,b). CL revealed zonation in Qa grains and that the Au occurred in fractures perpendicular to the zoning (Figure 4b).

Au occurred in fractures in Qb (Figure 4c). Pyrites were present in broken quartz grains or at contacts between coarse and fine quartz. On some occasions, euhedral or subhedral quartz occurred inside or partly included in the sulfide (Figure 4d,e). Quartz outside and inside the pyrite goes extinct at the same angle (Figure 4d), which means that the quartz grains were structurally the same and crystallized simultaneously. Similar CL features were seen even where extinction did not occur at the same angle, indicating that the quartz was of the same origin (Figure 4e). Quartz overgrowths may have occurred during the period of Qb and Qc formation.

Qc forms by recrystallization and overgrowth of Qb. Recrystallized quartz occurred beside massive pyrite and pyrite veins; this quartz may have resulted from hydrothermal activities of sulphide formation (Figure 3e,f and Figure 4f,g) [20]. Recrystallization of quartz may have resulted from many causes, such as fault movement and hydrothermal activity. Qc was different from the recrystallization derived from fault movement. Qc was best distinguished under a microscope and via CL. Au was found in Qc, which made Qc mineralization-relevant gold-bearing quartz (Figure 4f). The CL color of Qc was darker than that of Qa and Qb (Figure 4f). Some quartz around pyrite veins was recrystallized (Figure 4g). Qb and Qc had the potential to preserve information regarding the ore fluid during mineralization. Overgrowths were rare in the sample, but two grains were selected for LA-ICP-MS study.

Qd occurred in fractures in the sulfide (Figure 4h) and obviously formed after gold mineralization. The grain size depended on the sizes of the pyrite fractures. No gold was found in Qd. It showed different CL features from the other types of quartz.

Sericitization is a common alteration type in the gold deposit, and sericite has several types of occurrence. Field and microscopic observations indicated that sericite may have derived from feldspar alteration (Figures 3c and 5a), which means the sericitization may have occurred during mineralization. According to mineral mapping by QEMSCAN, sericite and gold filled in fractures in pyrite (Figure 5b), indicating that the gold precipitation may be related to sericite formation. A high proportion of sericite was also seen in a pyrite vein filling in the fractures of quartz (Figure 5c), indicating that sericite was a major syn-mineralization mineral.

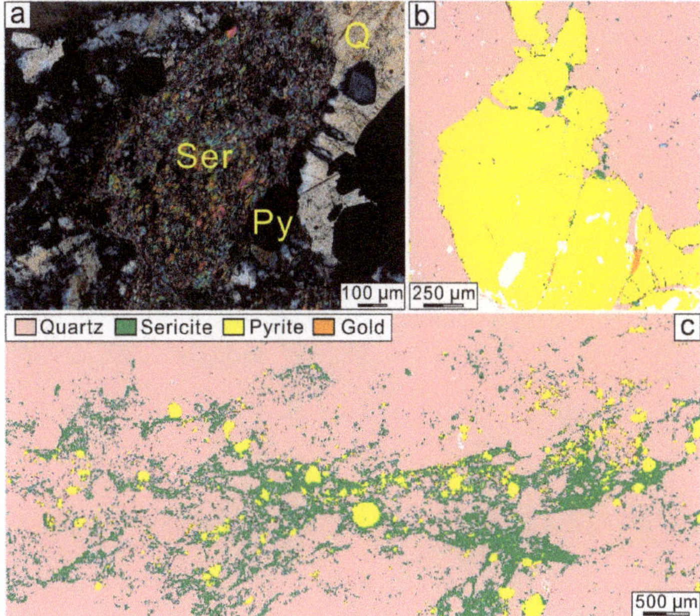

Figure 5. Sericite occurrence in the Jinqingding gold deposit: (**a**) Feldspar altered to sericite; (**b**) a QEMSCAN mineral map showing sericite coexisting with gold in pyrite fractures; (**c**) a QEMSCAN map showing the coexistence of sericite and pyrite in Qb. White color part is the glue on the thin section.

Visible gold occured in brittle fractures in the quartz and pyrite (Figure 4b,c,f and Figure 5b), meaning that this gold formed after or at the same time as these quartz and pyrite grains. Other sulfides such as galena and chalcopyrite sometimes filled in these fractures or were located around pyrites, indicating that they may have formed at the same time or after the pyrite. A new paragenesis has been concluded on the combined basis of all of the mineralogical investigations (Figure 6).

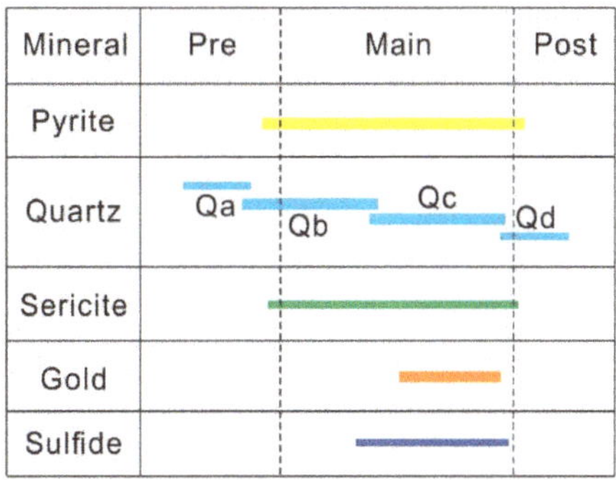

Figure 6. Paragenesis of quartz and other minerals. The thickness of the line indicates scale and intensity of mineral occurrence.

6. In Situ Study Results

The Al and Ti contents of the different types of quartz were analyzed by EPMA and LA-ICP-MS. The EPMA results are shown in Table 1 and Figure 7. The LA-ICP-MS results are shown in Table 2 and Figure 8.

Table 1. Electron probe micro-analyzer (EPMA)-derived Al content in different types of quartz in the Jinqingding gold deposit.

Al (wt %)	Qa ($n = 5$)	Qb ($n = 20$)	Qc ($n = 7$)	Qd ($n = 2$)
Max	0.086	0.284	0.069	0.167
Average	0.043	0.079	0.023	0.101
Min	0.008	0.004	0.001	0.036

The number in brackets is the number of spots on quartz.

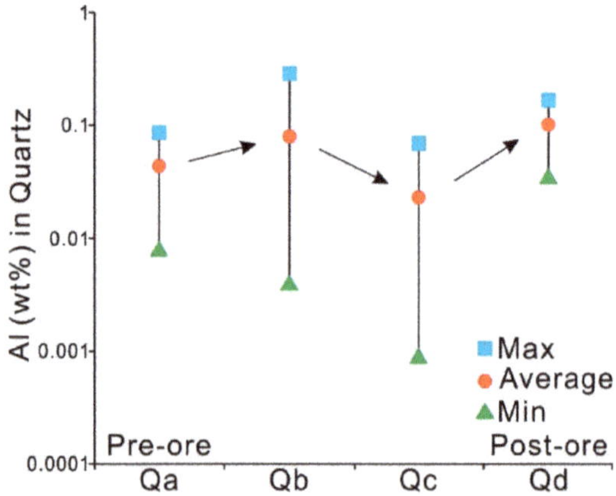

Figure 7. EPMA-derived Al content in different types of quartz.

Table 2. Laser ablation-inductively coupled plasma-mass spectrometry (LA-ICP-MS)-derived trace element content in quartz and estimated formation temperatures of quartz in the Jinqingding gold deposit.

Position	Element and Formation Temperature Under Different Pressures	Qb			Qc		
		Max	Average	Min	Max	Average	Min
Grain A	Al (ppm)	3310	2242.667	1548	393	192.667	54
	Ti (ppm)	12.1	9.433	7.9	5.7	5.2	4.6
	T (°C, 1 Kb)	217	209	204	194	192	188
	T (°C, 7.77 Kb)	310	300	294	282	279	275
Grain B	Al (ppm)	61	55	48	423	156.4	26
	Ti (ppm)	5.8	5.737	5.61	5.24	4.864	4.3
	T (°C, 1 Kb)	195	195	194	192	190	186
	T (°C, 7.77 Kb)	283	283	282	279	277	273

The EPMA results have a higher detection limit, so only Al could be detected. The Al content varied in the different types of quartz. The Al content varied between the five spots within Qa, ranging from 0.008 wt % to 0.086 wt % and averaging 0.043 wt % (Figure 7; Table 1). The Al content in Qb ranged between 0.284 wt % and 0.004 wt % among 15 spots, with an average of 0.073 wt % (Figure 7; Table 1). Qc was studied at seven spots that yield Al contents from 0.001 wt % to 0.689 wt %, averaging 0.023 wt % (Figure 7; Table 1). The two spots in Qd had Al contents of 0.036 wt % and 0.167 wt %, giving a mean value of 0.101 wt % (Figure 7; Table 1).

LA-ICP-MS has a lower detection limit and so could also detect Ti. CL showed clear cores and rims in the quartz grains analyzed (Figure 8a). Qb and Qc were distinct within each quartz grain, with Qb forming the core and Qc the rim; cavities within grains were filled with calcite. The results for spots on the rims and cores are shown in Figure 8 and listed in Table 2. In grain A, Qb had both higher Al and Ti content than Qc, with the Al content ranging as high as 3310 ppm and the Ti content reaching

12.1 ppm (Figure 8b,c; Table 2). In grain B, the range in Al content in Qb overlapped the range in Qc (Figure 8b). The Ti content in Qb was slightly higher than that in Qc (Figure 8c).

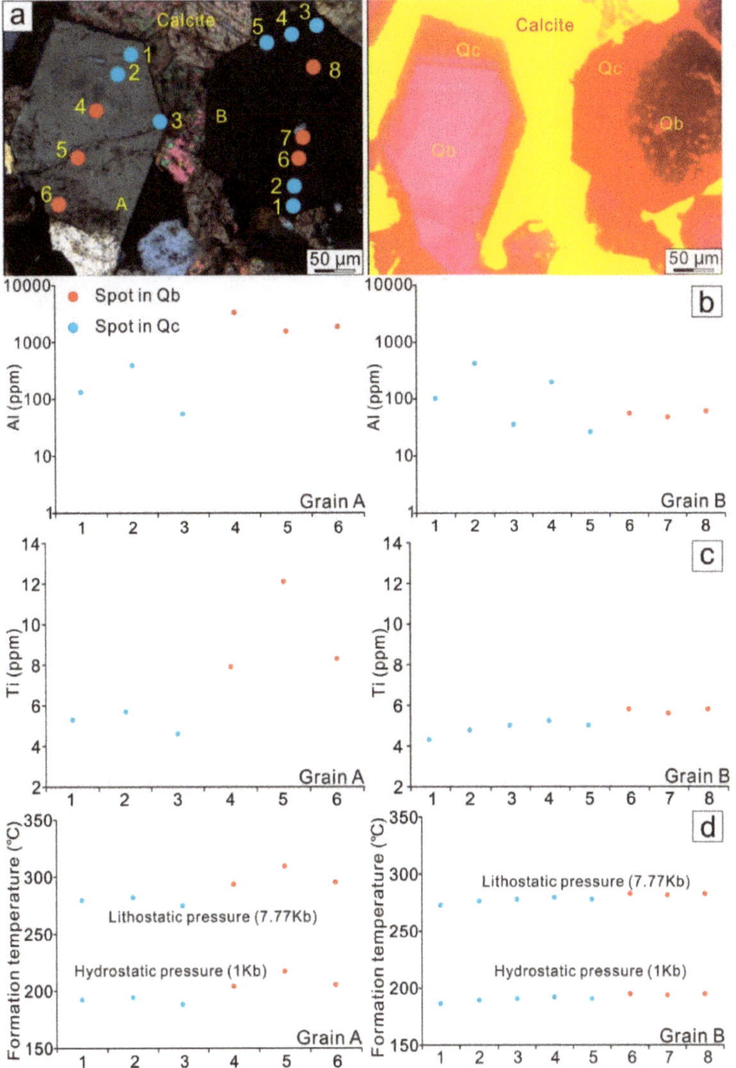

Figure 8. Trace element content in quartz as obtained via LA-ICP-MS. (**a**) CL shows differences between quartz cores and rims; (**b**) Al content in the quartz; (**c**) Ti content in the quartz; (**d**) estimated formation temperature of quartz under lithostatic and hydrostatic pressure.

Previous ore fluid studies have indicated that the fluid system was an H_2O–CO_2–NaCl system with a homogenization temperatures varying between 170 °C and 350 °C, a freezing temperature from −0.1 °C to −9.8 °C, a calculated salinity of from 0.2 wt % to 13.73 wt % NaCl eq., no salt daughter minerals, and a calculated hydrostatic pressure of 0.619 Kb to 1.532 Kb, averaging 1 Kb [46]. Combining the fully homogenization temperatures (268 °C), partly homogenization temperatures of

CO_2 fluid inclusions (23.6 °C), fluid inclusion explosion temperature (345 °C) and CO_2-saturated vapor pressure (8.5 Kb), Li [46] calculated the lithostatic pressure as 7.77 Kb.

Ti content is controlled by temperature and pressure [47,48]. According to the equation (RT $\ln X_{TiO2}{}^{quartz} = -60{,}952 + 1.520T(K) - 1741P(kbar) + RT \ln a_{TiO2}$) [47], if one is established, the other is directly proportional to the Ti content. When the system becomes saturated in Ti, the formation temperature of quartz can be estimated on the basis of the hydrostatic and lithostatic pressure [47,48]. Rutile was found in cavities within the pyrite and quartz, indicating Ti-saturation. Combining the hydrostatic pressure and lithostatic pressure, the formation temperature was calculated and is shown in Figure 8d and Table 2.

7. Discussion and Summary

Four main types of quartz are distinguished in the Jinqingding gold deposit. Quartz formation spans the pre-ore to post-ore periods. Qa has apparent zoning microscopically and in CL images, and formed prior to the mineralization. Qb is the most common quartz type and hosts the most sulfide and gold. Qc is recrystallized Qa and Qb around pyrites, which may form during mineralization. Qd occurs in fractures in pyrite and formed after the mineralization.

Two of the most efficient triggers of gold precipitation are acidity changes and cooling [49]. Alteration, which may result from water–rock reaction, is common in the gold deposit, and the acidity of ore fluid can change during alteration. Fluid inclusion studies, meanwhile, are not able to detect acidity changes whereas trace elements such as Al and Ti can. Additionally, changes in trace element content can reveal a decrease in the homogenization temperatures and can thus be used to infer the action of the cooling–related gold precipitation mechanism.

7.1. Acidity Changes

Precipitation mechanisms have been proposed for each gold deposit within the Muping–Rushan gold belt, including water–rock reactions and fluid immiscibility [50–52], fluid boiling [51,53,54], mixing and immiscibility [55], unmixing [56], and changes in oxygen fugacity [57]. These mechanisms were inferred based on the fluid inclusion assemblage in quartz belonging to different mineralization stages. Thus fluid mixing, phase separation and other hydrothermal activities may indeed have occurred, and the quartz has preserved evidence of this activity. If gold did precipitate due to phase separation, no gold is found in the fluid inclusions.

Quartz and pyrite are the main gold-bearing minerals. According to the mineralogical analysis, gold occurs in fractures in pyrite and in several kinds of quartz. Sericitization is a common alteration type in the gold deposit (Figures 2b and 3c), and the sericite coexists with pyrite and gold (Figure 5b,c). Some feldspar and albite will have been altered to sericite (Figure 5a), and sericite formation will have consumed H^+, reducing acidity.

The Al concentration in hydrothermal quartz does not reflect its formation temperature, however, previous study indicates the aqueous Al concentration shows a negative correlation with pH [2]. Of the four main kinds of quartz in the Jinqingding gold deposit, Qb and Qc may form during mineralization, while the other types apparently formed prior to or after the mineralization. A comparison of the Al contents of the different types of quartz shows an anomalously low Al concentration in Qc (Figure 7), which indicates a decrease in acidity. In Figure 8b, Qc overgrowing Qb grains is shown to have a lower Al content than that of Qb in grain A and a narrower range in Al content than Qb in grain B. The total Al content in quartz is very limited, so the Al content of Qc decreases or narrows within a small range.

Free gold precipitation may occur due to several factors such as a change in acidity, cooling, unmixing, or other mechanisms [49]. A change in pH is reflected by variation in the Al content of the quartz and sericites in fractures. Changes in acidity do not affect the solubility of gold in the ore fluid [58] but instead lead to the precipitation of sulfide. The reducing sulfur in the ore fluid is consumed when the wallrock reacts with the ore fluid, and chalcopyrite and other sulfides may precipitate [12]. Gold hydrosulfide then becomes unstable, and gold precipitates.

7.2. Cooling

In grain A (Figure 8), under lithostatic pressure (7.77 Kb), the estimated formation temperature of Qb is about 294–310 °C and that of Qc is about 275–282 °C; under hydrostatic pressure (1 Kb), these ranges change to 204–217 °C and 188–194 °C, respectively (Figure 8d). In grain B, under lithostatic pressure (7.77 Kb), the estimated formation temperature of Qb is about 282–283 °C and that of Qc is about 273–279 °C; under hydrostatic pressure (1 Kb), the ranges become about 194–195 °C and 186–192 °C (Figure 8d). Under hydrostatic pressure and lithostatic pressure, the estimated formation temperatures correspond to homogenization temperatures (268 °C) and fluid inclusion explosion temperature (345 °C) [46].

The temperature gap between the Qb and Qc indicates cooling during quartz overgrowth. These two grains formed after Qa but during the Qb and Qc stages, and this cooling was likely an important means of gold precipitation. The addition of rainwater or extension of ore-controlling faults could lead to cooling. When the temperature decreases, the solubility of gold in the ore fluid reduces [49], so gold will precipitate.

7.3. Water–Rock Reaction in Sulfide–Quartz Vein Type Jinqingding Gold Deposit

Variations in the trace element contents in four main types of quartz indicate that gold precipitated due to changes in acidity and cooling of the ore fluid. An Al content decrease corresponds to sericitization in the wallrock and orebody, which is consistent with an acidity change. A difference in the Ti values of Qb and Qc shows a reduction in the formation temperature of the quartz.

In the Jiaodong peninsula, water–rock reaction and phase separation are considered to be the main gold precipitation mechanism in the altered rock type and sulfide–quartz vein type gold deposit, respectively [13,16,59,60]. In the Jinqingding gold deposit, sericitization may have occurred during mineralization (Figures 3c and 5a), sericites appear with pyrite in fractures in the quartz, and some, along with free gold, fill in fractures in pyrite (Figure 5b,c). The Al content decrease corresponds to sericitization in the wallrock and orebody, which is consistent with the acidity change. Acidity changes and cooling may be two results of water–rock reaction. This research provides evidence of water–rock reaction in the sulfide–quartz vein type Jinqingding gold deposit, combining the sericitization, acidity changes and cooling, water–rock reaction is likely an efficient means of gold precipitation. Sericitization could be used as indicator of gold occurrence or mineralization in the Jinqingding gold deposit.

Supplementary Materials: The following are available online at http://www.mdpi.com/2075-163X/9/5/326/s1, Table S1: EPMA-derived Al content in different types of quartz in the Jinqingding gold deposit, Table S2: LA-ICP-MS-derived trace element content in quartz and estimated formation temperatures of quartz in the Jinqingding gold deposit.

Author Contributions: B.C. and J.D. conceived and designed the ideas; B.C. performed the experiments; B.C., H.W. and X.J. analyzed the data; B.C. prepared the original draft; H.W. and X.J. reviewed and edited the draft.

Funding: This research was funded by the National Key Research Program of China (Grant No. 2017YFC0601501, Grant No. 2016YFC0600107-4), the National Mineral Resources Assessment Project (41372007, DD20190193), the National Natural Science Foundation of China (Grant No. 41572069), funds from the State Key Laboratory of Geological Processes and Mineral Resources, the China University of Geosciences (Grant No. MSFGPMR201804), and Project 111 under the Ministry of Education and the State Administration of Foreign Experts Affairs, China (Grant No. B07011) and the China Scholarship Council (No. 201506400013).

Acknowledgments: Many thanks to Zhongliang Wang, Yue Liu, Shengxun Sai and Sirui Wang from the China University of Geosciences (Beijing) for help during fieldwork. Many thanks to the geologists at the Jinzhou Mining Company for their assistance in fieldwork at the Jinqingding gold deposit. Many thanks to Erin Marsh and David Adams from USGS in Denver, and Thomas Monecke from the Colorado School of Mines for the help during the in situ study. Many thanks to the reviewers.

Conflicts of Interest: The authors declare no conflict of interest.

References

1. Goldfarb, R.J.; Santosh, M.; Deng, J.; Yang, L.-Q. The giant Jiaodong gold deposits, China: Orogenic gold in a unique tectonic setting or a unique gold deposit type? In Proceedings of the 2013 GSA Annual Meeting in Denver: 125th Anniversary of GSA, Denver, CO, USA, 27–30 October 2013; The Geological Society of America: Boulder, CO, USA, 2013; p. 298.

2. Goldfarb, R.J.; Groves, D.I. Orogenic gold: Common or evolving fluid and metal sources through time. *Lithos* **2015**, *233*, 2–26. [CrossRef]

3. Deng, J.; Wang, C.; Bagas, L.; Santosh, M.; Yao, E. Crustal architecture and metallogenesis in the south–eastern north china craton. *Earth Sci. Rev.* **2018**, *182*, 251–272. [CrossRef]

4. Yang, L.-Q.; Deng, J.; Goldfarb, R.J.; Zhang, J.; Gao, B.-F.; Wang, Z.-L. ^{40}Ar/^{39}Ar geochronological constraints on the formation of the Dayingezhuang gold deposit: New implications for timing and duration of hydrothermal activity in the Jiaodong gold province, China. *Gondwana Res.* **2014**, *25*, 1469–1483. [CrossRef]

5. Yang, L.-Q.; Deng, J.; Wang, Z.-L.; Zhang, L.; Goldfarb, R.J.; Yuan, W.-M.; Weinberg, R.F.; Zhang, R.-Z. Thermochronologic constraints on evolution of the Linglong Metamorphic Core Complex and implications for gold mineralization: A case study from the Xiadian gold deposit, Jiaodong Peninsula, eastern China. *Ore Geol. Rev.* **2016**, *72*, 165–178. [CrossRef]

6. Chen, B.H.; Wang, Z.L.; Li, H.L.; Li, J.K.; Li, J.L.; Wang, G.Q. Evolution of ore fluid of the Taishang gold deposit, Jiaodong: Constraints on REE and trace element component of auriferous pyrite. *Acta Petrol. Sin.* **2014**, *30*, 2518–2532. (In Chinese with English abstract)

7. Yang, L.Q.; Deng, J.; Wang, Z.L.; Zhang, L.; Guo, L.N.; Song, M.C; Zheng, X.L. Mesozoic gold metallogenic system of the Jiaodong gold province, eastern China. *Acta Petrol. Sin.* **2014**, *30*, 2447–2467. (In Chinese with English abstract)

8. Mills, S.E.; Tomkins, A.G.; Weinberg, R.F.; Fan, H.-R. Implications of pyrite geochemistry for gold mineralisation and remobilisation in the Jiaodong gold district, northeast China. *Ore Geol. Rev.* **2015**, *71*, 150–168. [CrossRef]

9. Zhang, L.; Yang, L.-Q.; Wang, Y.; Weinberg, R.F.; An, P.; Chen, B.-Y. Thermochronologic constrains on the processes of formation and exhumation of the Xinli orogenic gold deposit, Jiaodong Peninsula, eastern China. *Ore Geol. Rev.* **2017**, *81*, 140–153. [CrossRef]

10. Zhang, L.; Yang, L.-Q.; Deng, J.; Weinberg, R.F.; Groves, D.I.; Wang, Z.-L.; Li, G.-W.; Liu, Y.; Zhang, C.; Wang, Z.-K. Anatomy of a world-class epizonal orogenic-gold system: A holistic thermochronological analysis of the Xincheng gold deposit, Jiaodong Peninsula, eastern China. *Gondwana Res.* **2019**, *70*, 50–70. [CrossRef]

11. Zhang, L.; Li, G.W.; Zheng, X.L.; An, P.; Chen, B.Y. ^{40}Ar/^{39}Ar and fission-track dating constraints on the tectonothermal history of the world-class Sanshandao gold deposit, Jiaodong Peninsula, eastern China. *Acta Petrol. Sin.* **2016**, *32*, 2465–2476. (In Chinese with English abstract)

12. Wang, Z.-L.; Yang, L.-Q.; Guo, L.-N.; Marsh, E.; Wang, J.-P.; Liu, Y.; Zhang, C.; Li, R.-H.; Zhang, L.; Zheng, X.-L.; et al. Fluid immiscibility and gold deposition in the Xincheng deposit, Jiaodong Peninsula, China: A fluid inclusion study. *Ore Geol. Rev.* **2015**, *65*, 701–717. [CrossRef]

13. Wen, B.-J.; Fan, H.-R.; Santosh, M.; Hu, F.-F.; Pirajno, F.; Yang, K.-F. Genesis of two different types of gold mineralization in the linglong gold field, china: Constrains from geology, fluid inclusions and stable isotope. *Ore Geol. Rev.* **2015**, *65*, 643–658. [CrossRef]

14. Deng, J.; Wang, Q.F. Gold mineralization in China: Metallogenic provinces, deposit types and tectonic framework. *Gondwana Res.* **2016**, *36*, 219–274. [CrossRef]

15. Yang, L.-Q.; Deng, J.; Wang, Z.-L.; Guo, L.-N.; Li, R.-H.; Groves, D.I.; Danyushevsky, L.V.; Zhang, C.; Zheng, X.-L.; Zhao, H. Relationships Between Gold and Pyrite at the Xincheng Gold Deposit, Jiaodong Peninsula, China: Implications for Gold Source and Deposition in a Brittle Epizonal Environment. *Econ. Geol.* **2016**, *111*, 105–126. [CrossRef]

16. Guo, L.-N.; Goldfarb, R.J.; Wang, Z.-L.; Li, R.-H.; Chen, B.-H.; Li, J.-L. A comparison of Jiaojia–and Linglong–type gold deposit ore–forming fluids: Do they differ? *Ore Geol. Rev.* **2017**, *88*, 511–533. [CrossRef]

17. Monnier, L.; Lach, P.; Salvia, S.; Melleton, J.; Bailly, L.; Béziata, D.; Monnier, Y.; Gouy, S. Quartz trace-element composition by LA-ICP-MS as proxy for granite differentiation, hydrothermal episodes, and related mineralization: The Beauvoir Granite (Echassières district), France. *Lithos* **2018**, *320*, 355–377. [CrossRef]

18. Rusk, B.G.; Lowers, H.A.; Reed, M.H. Trace elements in hydrothermal quartz: Relationships to cathodoluminescent textures and insights into vein formation. *Geology* **2008**, *36*, 547–550. [CrossRef]

19. Chen, H.Y. Genetic Mineralogy and Deep Prospects of Jinqingding Gold Deposit in Rushan, East Shandong Province. Master's Thesis, China University of Geosciences, Beijing, China, 2012. (In Chinese with English abstract)

20. Chen, B.H. Gold Mineralization Geochemistry in Muping–Rushan Gold Belt, Jiaodong Peninsula, China. Ph.D. Thesis, China University of Geosciences, Beijing, China, 2017. (In Chinese with English abstract)

21. Li, J.-W.; Vasconcelos, P.; Zhou, M.-F.; Zhao, X.-F.; Ma, C.-Q. Geochronology of the Pengjiakuang and Rushan gold deposits, eastern Jiaodong gold province, northeastern China: Implications for regional mineralization and geodynamic setting. *Econ. Geol.* **2006**, *101*, 1023–1038. [CrossRef]

22. Deng, J.; Yang, L.-Q.; Li, R.-H.; Groves, D.I.; Santosh, M.; Wang, Z.-L.; Sai, S.-X.; Wang, S.-R. Regional structural control on the distribution of world–class gold deposits: An overview from the giant Jiaodong Gold Province, China. *Geol. J.* **2018**, *54*, 378–391. [CrossRef]

23. Wang, Z.L. Metallogenic system of Jiaojia gold ore field, Shandong Province, China. Ph.D. Thesis, China University of Geosciences, Beijing, China, 2012. (In Chinese with English abstract)

24. Tan, J.; Wei, J.; Li, Y.; Tan, W.; Guo, D.; Yang, C. Geochemical characteristics of late Mesozoic dikes, Jiaodong Peninsula, North China Craton: Petrogenesis and geodynamic setting. *Int. Geo. Rev.* **2007**, *49*, 931–946. [CrossRef]

25. Qiu, Y.; Groves, D.I.; McNaughton, N.J.; Wang, L.-G.; Zhou, T. Nature, age, and tectonic setting of granitoid–hosted, orogenic gold deposits of the Jiaodong Peninsula, eastern North China craton, China. *Miner. Deposita* **2002**, *37*, 283–305. [CrossRef]

26. Jahn, B.-M.; Liu, D.; Wan, Y.; Song, B.; Wu, J. Archean crustal evolution of the Jiaodong Peninsula, China, as revealed by zircon shrimp Geochronology elemental and Nd–isotope Geochemistry. *Am. J. Sci.* **2008**, *308*, 232–269. [CrossRef]

27. Deng, J.; Wang, Q.; Wan, L.; Liu, H.; Yang, L.; Zhang, J. A multifractal analysis of mineralization characteristics of the Dayingezhuang disseminated–veinlet gold deposit in the Jiaodong gold province of China. *Ore Geol. Rev.* **2011**, *40*, 54–64. [CrossRef]

28. Wan, Y.S.; Dong, C.Y.; Xie, H.Q.; Wang, S.J.; Song, M.C.; Xu, Z.Y.; Wang, S.Y.; Zhou, H.Y.; Ma, M.Z.; Liu, D.Y. Formation Ages of Early Precambrian BIFs in the North China Craton: SHRIMP Zircon U–Pb Dating. *Acta Geol. Sin.* **2012**, *86*, 1447–1478. (In Chinese)

29. Dong, C.Y.; Wang, S.J.; Liu, D.Y.; Wang, J.G.; Xie, H.Q.; Wang, W.; Song, Z.Y.; Wan, Y.S. Late Paleoproterozoic crustal evolution of the North China Craton and formation time of the Jingshan Group: Constraints from SHRIMP U–Pb zircon dating of meta–intermediate–basic intrusive rocks in eastern Shandong Province. *Acta Petrol. Sin.* **2010**, *27*, 1699–1706. (In Chinese with English abstract)

30. Li, X.P.; Liu, Y.J.; Guo, J.H.; Li, H.K.; Zhao, G. Petrogeochemical characteristics of the Paleoproterozoic high–pressure mafic granulite and calc–silicate from the Nanshankou of the Jiaobei terrane. *Acta Petrol. Sin.* **2013**, *29*, 2340–2352. (In Chinese with English abstract)

31. Liu, P.H.; Liu, F.L.; Wang, F.; Liu, J.H.; Cai, J. Petrological and geochronological preliminary study of the Xiliu ~2.1Ga meta–gabbro from the Jiaobei terrane, the southern segment of the Jiao–Liao–Ji Belt in the North China Craton. *Acta Petrol. Sin.* **2013**, *29*, 2371–2390. (In Chinese with English abstract)

32. Faure, M.; Lin, W.; Monie, P.; Bruguier, O. Paleoproterozoic arc magmatism and collision in Liaodong Peninsula (north–east China). *Terra Nova* **2004**, *16*, 75–80. [CrossRef]

33. Jiang, N.; Chen, J.; Guo, J.; Chang, G. In situ zircon U–Pb, oxygen and hafnium isotopic compositions of Jurassic granites from the North China craton: Evidence for Triassic subduction of continental crust and subsequent metamorphism–related ^{18}O depletion. *Lithos* **2012**, *142*, 84–94. [CrossRef]

34. Yang, K.-F.; Fan, H.-R.; Santosh, M.; Hu, F.-F.; Wilde, S.A.; Lan, T.-G.; Lu, L.-N.; Liu, Y.S. Reactivation of the Archean lower crust: Implications for zircon geochronology, elemental and Sr–Nd–Hf isotopic geochemistry of late Mesozoic granitoids from northwestern Jiaodong Terrane, the North China Craton. *Lithos* **2012**, *146*, 112–127. [CrossRef]

35. Guo, J.H.; Chen, F.K.; Zhang, X.M.; Siebel, W.; Zhai, M.G. Evolution of syn–to post–collisional magmatism from north Sulu UHP belt, eastern China: Zircon U–Pb geochronology. *Acta Petrol. Sin.* **2005**, *21*, 1281–1301. (In Chinese with English abstract)

36. Zhang, D.; Xu, H.; Sun, G. Emplacement ages of the Denggezhuang gold deposit and the Kunyushan granite and their geological implications. *Geol. Rev.* **1995**, *41*, 415–425. (In Chinese)

37. Hu, S.; Wang, S.; Sang, H.; Qiu, J.; Zhang, R. Isotopic ages of Linglong and Guojialing batholiths in Shandong Province and their geological implication. *Acta Petrol. Sin.* **1987**, *3*, 83–89. (In Chinese with English abstract)

38. Guan, K.; Luo, Z.K.; Miao, L.C.; Huang, J.Z. SHRIMP in zircon chronology for Guojialing suite granite in Jiaodong Zhaoye district. *Sci. Geol. Sin.* **1998**, *33*, 318–328. (In Chinese with English abstract)

39. Deng, J.; Liu, X.; Wang, Q.; Pan, R. Origin of the Jiaodong–type Xinli gold deposit, Jiaodong Peninsula, China: Constraints from fluid inclusion and C–D–O–S–Sr isotope compositions. *Ore Geol. Rev.* **2015**, *65*, 674–686. [CrossRef]

40. Götze, J.; Plötze, M.; Habermann, D. Origin, spectral characteristics and practical applications of the cathodoluminescence (CL) of quartz—A review. *Mineral. Petrol.* **2001**, *71*, 225–250. [CrossRef]

41. Penniston-Dorland, S.C. Illumination of vein quartz textures in a porphyry copper ore deposit using scanned cathodoluminescence: Grasberg Igneous Complex, Irian Jaya, Indonesia. *Am. Mineral.* **2001**, *86*, 652–666. [CrossRef]

42. Müller, A.; René, M.; Behr, H.-J.; Kronz, A. Trace elements and cathodoluminescence of igneous quartz in topaz granites from the Hub Stock (Slavkovský Les Mts., Czech Republic). *Mineral. Petrol.* **2003**, *79*, 167–191. [CrossRef]

43. Landtwing, M.R.; Pettke, T. Relationships between SEM–cathodoluminescence response and trace–element composition of hydrothermal vein quartz. *Am. Mineral.* **2005**, *90*, 122–131. [CrossRef]

44. Müller, A.; Herrington, R.; Armstrong, R.; Seltmann, R.; Kirwin, D.J.; Stenina, N.G.; Kronz, A. Trace elements and cathodoluminescence of quartz in stockwork veins of Mongolian porphyry–style deposits. *Miner. Depos.* **2010**, *45*, 707–727. [CrossRef]

45. Longerich, H.P.; Jackson, S.E.; Günther, D. Laser Ablation Inductively Coupled Plasma Mass Spectrometric Transient Signal Data Acquisition and Analyte Concentration Calculation. *J. Anal. Atom. Spectrom.* **1996**, *11*, 899–904. [CrossRef]

46. Li, Y.P. The genesis of the Rushan gold deposits in east Shandong. *Miner. Depos.* **1992**, *11*, 165–172. (In Chinese)

47. Thomas, J.B.; Watson, E.B.; Spear, F.S.; Shemella, P.T.; Nayak, S.K.; Lanzirotti, A. TitaniQ under pressure: The effect of pressure and temperature on the solubility of Ti in quartz. *Contrib. Mineral. Petrol.* **2010**, *160*, 743–759. [CrossRef]

48. Thomas, J.B.; Watson, E.B. Application of the Ti–in–quartz thermobarometer to rutile–free systems. Reply to: A comment on: "TitaniQ under pressure: the effect of pressure and temperature on the solubility of Ti in quartz" by Thomas et al. *Contrib. Mineral. Petrol.* **2012**, *164*, 369–374. [CrossRef]

49. Pokrovski, G.S.; Akinfiev, N.N.; Borisova, A.Y.; Zotov, A.V; Kouzmanov, K. Gold speciation and transport in geological fluids: Insights from experiments and physical–chemical modelling. *Geol. Soc. Spec. Publ.* **2014**, *402*, 9–70. [CrossRef]

50. Hu, F.-F.; Fan, H.-R.; Zhai, M.-G.; Jin, C.-W. Fluid evolution in the Rushan lode gold deposit of Jiaodong Peninsula, eastern China. *J. Geochem. Explor.* **2006**, *89*, 161–164. [CrossRef]

51. Hu, F.F.; Fan, H.R.; Yang, K.F.; Shen, K.; Zhai, M.G.; Jin, C.W. Fluid inclusions in the Denggezhuang lode gold deposit at Muping, Jiaodong Peninsula. *Acta Petrol. Sin.* **2007**, *23*, 2155–2164. (In Chinese with English abstract)

52. Zhou, Q.F. Genetic Mineralogy and Deep Prospects of the Yinggezhuang Gold Deposit in Rushan County, Jiaodong. Master Thesis, China University of Geosciences, Beijing, China, 2010. (In Chinese with English abstract)

53. Zeng, Q.; Liu, J.; Liu, H.; Shen, P.; Zhang, L. The Ore–forming Fluid of the Gold Deposits of Muru Gold Belt in Eastern Shandong, China—A Case Study of Denggezhuang Gold Deposit. *Resour. Geol.* **2006**, *56*, 375–384. [CrossRef]

54. Hu, F.F.; Fan, H.R.; Yu, H.; Liu, Z.H.; Song, L.F.; Jin, C.W. Fluid inclusions in the Sanjia lode gold deposit, Jiaodong Peninsula of eastern China. *Acta Petrol. Sin.* **2008**, *24*, 2037–2044. (In Chinese with English abstract)

55. Shen, P.; Shen, Y.; Li, G.; Liu, T.; Zeng, Q.; Li, H. A study on structure–fluid–mineralization system in the Jinniushan Gold deposit, East Shandong. *Chin. J. Geol.* **2004**, *39*, 272–283. (In Chinese with English abstract)

56. Cai, Y.C.; Fan, H.R.; Yang, K.F.; Hu, F.F.; Yu, H.; Liu, Y.M.; Lan, T.G. Ore–forming fluids, stable isotope and mineralizing age of the Hubazhuang gold deposit, Jiaodong Peninsula of eastern China. *Acta Petrol. Sin.* **2011**, *27*, 1341–1351. (In Chinese with English abstract)

57. Lan, T.G.; Fan, H.R.; Hu, F.F.; Yang, K.F.; Liu, X.; Liu, Z.H.; Song, Y.B.; Yu, H. Characteristics of ore–forming fluids and ore genesis in the Shicheng gold deposit, Jiaodong Peninsula of eastern China. *Acta Petrol. Sin.* **2010**, *26*, 1512–1522. (In Chinese with English abstract)

58. Gibert, F.; Pascal, M.L.; Pichavant, M. Gold solubility and speciation in hydrothermal solutions: Experimental study of the stability of hydrosulphide complex of gold (AuHS) at 350 to 450 °C and 500 bars. *Geochem. Cosmochim. Acta* **1998**, *62*, 2931–2947. [CrossRef]

59. Fan, H.R.; Zhai, M.G.; Xie, Y.H.; Yang, J.H. Ore–forming fluids associated with granite–hosted gold mineralization at the Sanshandao deposit, Jiaodong gold province, China. *Miner. Depos.* **2003**, *38*, 739–750. [CrossRef]

60. Yang, L.-Q.; Deng, J.; Guo, L.-N.; Wang, Z.-L.; Li, X.-Z.; Li, J.-L. Origin and evolution of ore fluid, and gold–deposition processes at the giant Taishang gold deposit, Jiaodong Peninsula, eastern China. *Ore Geol. Rev.* **2016**, *72*, 585–602. [CrossRef]

 *minerals*

Article

Rare Earth Elements Geochemistry and C–O Isotope Characteristics of Hydrothermal Calcites: Implications for Fluid-Rock Reaction and Ore-Forming Processes in the Phapon Gold Deposit, NW Laos

Linnan Guo *, Lin Hou, Shusheng Liu and Fei Nie

Chengdu Center, China Geological Survey, Chengdu 610081, China; houlin_aaron@163.com (L.H.); lshusheng@cgs.cn (S.L.); niefei_cugb@163.com (F.N.)
* Correspondence: linnanguo@163.com; Tel.: +86-028-8232-1057

Received: 26 August 2018; Accepted: 6 October 2018; Published: 9 October 2018

Abstract: The Phapon gold deposit is located in the northern Laos and the northern segments of the Luang Prabang–Loei metallogenic belt. The lode-gold orebodies consist of auriferous calcite veins in the middle, and the surrounding siderite alteration and hematite alteration zones in red color. This deposit is hosted in Lower Permian limestone and controlled by a NE-trending ductile–brittle fault system, and it is characterized by the wallrock alteration of carbonatization and lack of quartz and metal sulfides. The hydrothermal calcite from auriferous calcite veins and red alteration zone, as well as the wall rocks of limestone and sandstone were selected for rare earth elements (REE) and C–O isotope analyses. The two types of calcite and limestone have generally consistent REE patterns and δEu and δCe values, which are completely different from those of sandstone. Calcites from the auriferous vein show slight light rare earth elements (LREE)-depleted patterns and higher Tb/La and Sm/Nd ratios than the ones from the red alteration zone with slight LREE-enriched patterns. These values indicate that the calcites from the auriferous veins and the red alteration zones are products of homologous fluids, but the former ones are generally likely to form later than the latter ones. The hydrothermal calcites have C–O isotope compositions within the range of marine carbonate, and markedly different from the magmatic or mantle reservoir values. Taking the Y/Ho–La/Ho and Tb/Ca–Tb/La variations into consideration, we believe the hydrothermal calcites could be formed from remobilization and recrystallization of the ore-hosted limestone, and the fluid-wallrock interaction played a major role in the gold mineralization in Phapon. In combination with the regional and local geology, the ore-forming process is suspected to be primarily associated with dehydration and decarbonisation of the Lower Permian limestone and Middle–Upper Triassic sandstones. The Phapon gold deposit could have been formed during the Late Triassic–Jurassic regional dynamic metamorphism driven by Indochina–Sibumasu post-collisional magmatism. A number of features in Phapon are similar to epizonal orogenic deposit, but it is still a unique calcite vein type gold deposit in the Luang Prabang-Loei metallogenic belt.

Keywords: Rare-Earth Elements; C–O isotopes; hydrothermal calcite; ore-forming processes; Phapon gold deposit

---

1. Introduction

The Phapon gold deposit is a calcite-vein type deposit located in the northern segments of the Luang Prabang–Loei metallogenic belt (Figure 1). This is characterized by multiple generations of arc-related magmatic events and associated copper, gold and other polymetallic deposits [1–5]. The Phapon gold deposit, hosted in the Lower Permian limestone and controlled by a NE-trending ductile–brittle fault system, has reached a large scale since it was found in northern Laos in the 1990s [6]. The deposit is characterized by lode-gold orebodies surrounded by carbonatization zones, and absence of quartz and metal sulfides. It is unique compared to the porphyry-related skarn and epithermal systems along the metallogenic belt. Previous studies on Phapon mainly focused on deposit geology [6,7], nature of ore-forming fluids [8], and hydrothermal alteration [9]. Some researchers, based on individual hydrogen and oxygen isotopic data, suggested that the ore-forming fluids derived from deep-seated magmatic water mixed with shallower meteoric water during mineralization [10,11]. On the other hand, microstructural studies by Li et al. [12] and Yang et al. [13] argued that the ore-forming fluids were sourced from pressure dissolution in shear dynamic metamorphism. Comprehensive studies on deposit geology, ore-controlling structures, and ore deposit geochemistry suggested the Phapon deposit is a shear zone-controlled low-temperature hydrothermal deposit [14,15]. However, there is little detailed study on the genetic relations between hydrothermal calcite and wall rocks, or the relationships between the origin of the Phapon gold deposit and the regional geotectonic evolution and magmatic events. The ore-forming processes, as a result, still remain controversial.

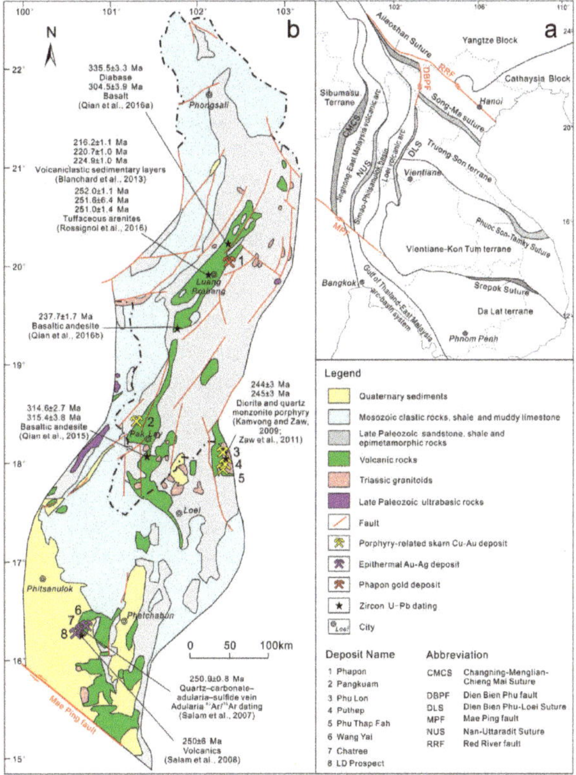

Figure 1. Regional geological map. (**a**) Simplified tectonic map showing structural units of Laos and adjacent regions [4,16]; (**b**) Simplified geological map of the Luang Prabang–Loei metallogenic belt (modified from [5]). RRF, Red River fault; DBPF, Dien Bien Phu fault; MPF, Mae Ping fault; CMCS, Changning-Menglian-Chieng Mai Suture; NUS, Nan-Uttaradit Suture; DLS, Dien Bien Phu-Loei Suture.

Rare earth elements (REE) generally migrate together during geological processes, but they have slightly different geochemical behaviors due to nuances of ionic radius and electronic configuration [17,18]; thus, REE usually contain important geochemical information [19] and have now become a powerful tool for geochemical research in hydrothermal deposits [20]. Previous studies indicated that REE^{3+} can go into the mineral lattice by replacement of Ca^{2+} [21,22], so calcium-bearing minerals usually have a strong holding capacity for REE [23]. The REE composition and distribution patterns of hydrothermal Ca-bearing minerals are, at present, widely used to trace ore-forming fluids and ore-forming process of hydrothermal deposits (e.g., Bau et al.; Zhang et al.; Wang et al.; Wu et al.; Cao et al. [24–28]). In this study, we carried out detailed field work, and REE and C–O isotope geochemistry analyses, to indicate genetic relationships of the hydrothermal calcites and limestone wallrock, to define the source of ore-forming fluids, and to discuss the ore-forming processes in the Phapon gold deposit.

2. Regional Geology

The Luang Prabang–Loei metallogenic belt, lies along the western periphery of the Indochina Terrane, is one of the important copper–gold–silver polymetallic metallogenic belts in East Tethys metallogenic region [29–31] (Figure 1a). The metallogenic belt is bounded to the west by the Nan-Uttaradit Suture and to the east by the Dien Bien Phu–Loei Suture, with a length of ~800 km from north to south and a width of ~200 km from east to west. The belt extends from northern Laos to central-northern Thailand (Figure 1b), which can be divided into the eastern Loei volcanic arc and Simao-Phitsanulok basin in the west (Figure 1a).

The Loei volcanic arc is distributed parallel with the Dien Bien Phu-Loei Suture on the east side (Figure 1a). It developed a Late Carboniferous–Middle Triassic island arc calc–alkaline series, with volcanic rocks ranging from basic to neutral to acidic [32]. It contains multiple generations of successive arc-related magmatic events, which separated by periods of exhumation and erosion [33]. The Carboniferous sedimentary sequence, consisting of clastic rocks, carbonate rocks and volcanic rocks, is overlain by thick middle Permian limestone marked by basal conglomerate, siltstone and shale [34,35].

The Simao-Phitsanulok basin is located to the east of the Nan-Uttaradit Suture (Figure 1a). It is an intracontinental basin formed during the basin-to-mountain transition from late Permian to early Triassic, after the Nan-Uttaradit back-arc ocean basin closed, and showed foreland basin in nature in middle–late Triassic. The basin is mainly composed of upper Triassic coal-bearing sandstone, shale, marl, conglomerate, and tuff sandstone, belonging to continental molasse deposits of the foreland basin, and Jurassic–Cretaceous continental clastic rock series, consist of siltstone, sandstone, shale, and marl [16] (Figure 1b).

Three main periods of volcanism have been identified based on zircon U–Pb ages, which can be divided into Carboniferous, late Permian to early Triassic, and middle–late Triassic (Figure 1b). Previous studies showed that the Carboniferous volcanic rocks of the Loei volcanic arc are widespread from Luang Prabang to Pak Lay in Laos [36,37]. The Late Permian to Early Triassic volcanic rocks are exposure around Luang Prabang [38], and to the west of Phetchabun [33]. The Middle–Late Triassic volcanic rocks are confined to the southwest of Luang Prabang [39], or around Luang Prabang [40]. The intrusive rocks, dominated by Triassic island arc I-type granitoids [32,41], are normally distributed to the northeast of Luang Prabang and around Pak Lay in Laos, and in Loei Province of Thailand.

The Luang Prabang–Loei metallogenic belt is dominated by porphyry-related skarn type copper-gold deposits, and epithermal gold-silver deposits (Table 1; Figure 1b). The former ones are associated with the Early–Middle Triassic subalkalic to calc-alkaline quartz monzonite and diorite–monzodiorite (granodiorite) porphyry [42,43], while the later ones are associated with Early Permian to Late Triassic andesitic to rhyolitic volcanic/volcaniclastic units [44,45]. The Phapon gold deposit, nevertheless, shows significant differences from the above skarn or epithermal-type gold deposits (e.g., Qiu et al. [46,47]), including ore-forming environment, wall rock alteration, mineralization style, and mineral assemblage (Table 1). Thus, it can be regarded as a unique type of gold deposit in the Luang Prabang-Loei metallogenic belt, named calcite vein type gold deposit.

Table 1. Basic information of major gold deposits in the Luang Prabang–Loei metallogenic belt (collated from Zaw et al., 2014; Zhao et al., 2017) [3,44].

No. in Map	Deposit	Deposit Type	Tonnage (Mt)	Grade	Host Rocks	Intrusions/Ages	Alteration Minerals
1	Phapon (Au)	Calcite vein-type	No data	6.28 g/t Au	Limestone	No known intrusion	Calcite, siderite, magnitite, realgar
2	Pangkuam (Cu–Au)	Porphyry-related skarn	No data	0.69% Cu, 2.41 g/t Au	Limestone, argillaceous siltstone	Intermediate-mafic intrusions	Quartz, sericite, epidote, chlorite, K-feldspar, garnet
3	Phu Lon (Cu–Au)	Porphyry-related skarn	5.4	2.4% Cu, 0.64 g/t Au	Limestone, volcaniclastics	Diorite and quartz monzonite porphyry/244 ± 3 Ma	Garnet, pyroxene, K-feldspar, tremolite, epidote, chlorite, calcite
4	Puthep (Cu–Au)	Porphyry-related skarn	164	0.53% Cu, 0.09 g/t Au	Sandstone, siltstone and sandstone	Diorite and monzodiorite porphyry/242.4 ± 1.3 Ma	K-feldspar, sericite, garnet, epidote, chlorite, calcite
5	Phu Thap Fah (Cu–Au–Ag)	Skarn	6.4	0.14% Cu, 2.19 g/t Au, 3.9 g/t Ag	Siliciclastics and limestone	Granodiorite/245 ± 3 Ma	garnet, pyroxene, quartz, epidote, calcite, chlorite
6	Wang Yai (Au–Ag)	Low-S epithermal	No data	No data	Volcaniclastics, rhyolite breccia	No known intrusion	Quartz, calcite, adularia, sericite, chlorite
7	Chatree (Au–Ag)	Low-S epithermal	81.7	1.18 g/t Au, 9 g/t Ag	Andesite breccias, volcanogenic sedimentary rocks	Diorite dyke and granodiorite-bearing basaltic dyke/244 ± 7 Ma	Quartz, calcite, adularia, sericite, chlorite, illite, smectite
8	LD Prospect (Au–Ag)	Low-S epithermal	2.9	1.1 g/t Au, 10 g/t Ag	Andesite	No known intrusion	Quartz, pyrite, calcite, adularia, sericite, chlorite

3. Local Geology

3.1. Ore Deposit Geology

The Phapon gold deposit is located about 30 km northeast to the Luang Prabang (Figure 1b). The exposed strata around the mining area are mainly Carboniferous, Lower–Middle Permian, Middle–Upper Triassic and Quaternary (Figure 2). Carboniferous sedimentary formation consists of mottled sandstone, siltstone, mudstone, gray-black claystone and limestone with banded chert. Strong ductile shearing occurred during the Variscan orogeny, showing obvious plastic flow in the carboniferous rock. The main lithology of the Lower Permian is thick layer microcrystalline limestone and bioclastic limestone with marlstone and siltstone at the bottom. The limestone masses occur as a lens which strikes 35–50° and dips 15–45° towards the NE, and show strong ductile–brittle shear deformation in the ore-bearing section. The Upper Permian is composed of gray-green andesite, andesitic tuff and basalt, and show unconformable or fault contact relationship with the underlying strata. The Middle–Upper Triassic extends northeast in the mining area and it consists mainly of purplish-red conglomerate, sandstone, siltstone and feldspar sandstone, representing dry and oxidizing deposition in the foreland basin and intermountain basin. The Quaternary is mainly composed of residual red clay, with discontinuous distribution. The area is lenticular in shape with locally distributed epigenetic lateritic gold deposits. There is no large intrusive rock mass in the mining area, except for effusive-facies volcanic rocks exposed in the northwestern part. Its lithology contains mainly andesite and andesitic tuff, with locally diabase and dioritic porphyrite. The andesite is cryptocrystalline with few phenocrysts, and is gray-green in color, showing weak propylitization.

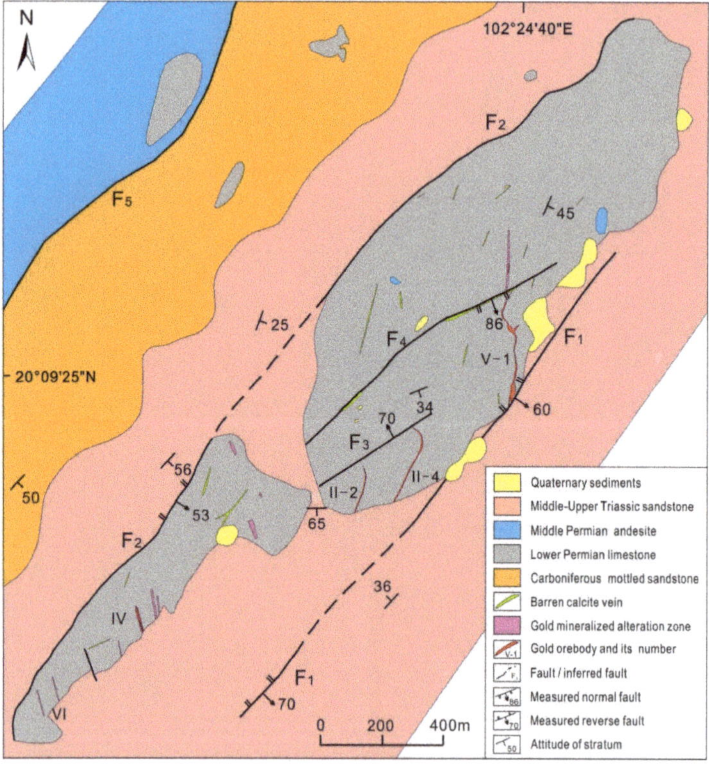

Figure 2. Sketch map of the Phapon gold deposit [13].

The structures in the mining area are mainly characterized by NE-trending ductile–brittle shear zones and brittle faults in different directions [13] (Figure 2). The NE-trending Luang Prabang ductile–brittle shear zone runs through the whole area and has a width of several hundred meters with multi-stage activities. The ductile–brittle shear zone developed during the early stage of Carboniferous and was dominated by ductile deformation. It developed strong plastic deformation including strong mylonitization, plasticity rheology, sheath fold, book slant tectonics and pyrite stress shadow during the Variscan orogeny. In the late stage, the reactivation Luang Prabang ductile–brittle shear zone in the Upper Permian limestone led to secondary faults with different orientations and controlled the NE-zonal distributed gold mineralization.

The main brittle faults in the mining area can be divided into NE-trending and NNW-trending groups, and the latter group is the main ore-controlling structure of the Phapon gold deposit. The NE-trending faults are well developed with multi-period activities, including F_1, F_2, F_3, F_4, and F_5 fault (Figure 2). The F_1 and F_2 are compressive-torsional reverse faults and along the contact zones of the Lower Permian limestone and the Middle–Upper Triassic sandstone on the southeast and northwest sides, respectively. It is marked by crushing limestone with carbonatization and calcite veins, showing a post-mineralization feature. The ore-hosting Lower Permian limestone overlay the Middle–Upper Triassic sandstone, and the ore-hosting limestone show differences in lithofacies with the local Lower Permian limestone. This infers a thrust nappe structure and the main thrust planes is the F_2 fault (Figure 2). The limestone along the thrust zone is generally broken. Compare to the more ductile argillaceous limestone, the sandy limestone appears relatively brittle, and normally show ductile deformation such as boudinage (Figure 3a). Schistosity zones and mylonitic lens also exist near the thrust plane (Figure 3b).

Figure 3. Typical outcrops of deformation, alteration and mineralization in Phapon gold deposit. (**a**) Boudinage deformation found in limestone; (**b**) schistosity zones and mylonitic lens near the thrust plane; (**c**) auriferous calcite vein and surrounding red alteration zone; (**d**) carbonation clumps locally present in the red alteration zone; (**e**) calcite veins branched and intercalated into the red alteration zone; (**f**) breccia of red altered rock.

There are five orebodies found in the Phapon gold deposit, which from east to west are No. V-1, II-4, II-2, IV, and VI, respectively (Figure 2). These vein-shaped orebodies are subparallel with a spacing of about ~300 m, and controlled from east to west by a series of NNW-trending faults. They generally vary from 300 to 800 m in length, 1 to 10 m in width, and strike 330–350° and dip 45–65° towards the SWW (Figure 2). The No. V-1 orebody, which comprises more than 90% of the proven reserves in the Phapon deposit, shows typical vein-shaped and locally cystic or lenticular-shaped orebodies. It is approximately 650 m long, generally strikes 330–355°, and dips 40–50° SW with large variation locally. It ranges in thickness from 0.3 to 10 m, averaging approximately 3.4 m, and extends down-dip approximately 350 m. The thickness of the orebody and the ore type change with its occurrence, and the average gold grade is 6.28 g/t.

3.2. Alteration and Mineralization

The hydrothermal alteration at Phapon developed in the lower Permian limestone and is controlled by the NNW-trending fault zones. It is characterized by medium–low temperature alteration including carbonation, siderite alteration, magnetite alteration, hematite alteration, and silicification, and shows macroscopic red-color alteration and light gray-color carbonation (Figure 3c–e). The red-color alteration is mainly caused by siderite and hematite alteration that infiltrate the surrounding limestone. It extends from centimeters to several meters along the ore-bearing faults (Figure 3c). It has a close relationship with gold mineralization and it represents a special alteration in the area. Siderite and hematite are normally distributed as disseminations, orlocally fill micro fractures in limestone (Figure 4d,e). Carbonation occurred mainly as planar alteration, which is characterized by the in situ dissolution and recrystallization of limestone and the formation of fine calcite grains, showing pale in color from the surrounding limestone (Figure 3d). Carbonation spread outside of the red-color alteration and is one of the main alteration of the deposit. The silicification is normally weak and intergrown with carbonation, and the anhedral quartz is distributed as disseminations in the wallrock. The alteration strength is weakened from the center and to the two sides of the fault zone. It mainly consists of a narrow proximal zone of red alteration, and a wide outer zone of carbonation (Figure 3d). In some cases, carbonation clumps locally existed in the red alteration zone (Figure 3d), and the calcite veins branched and intercalated into the red alteration zone (Figure 3e).

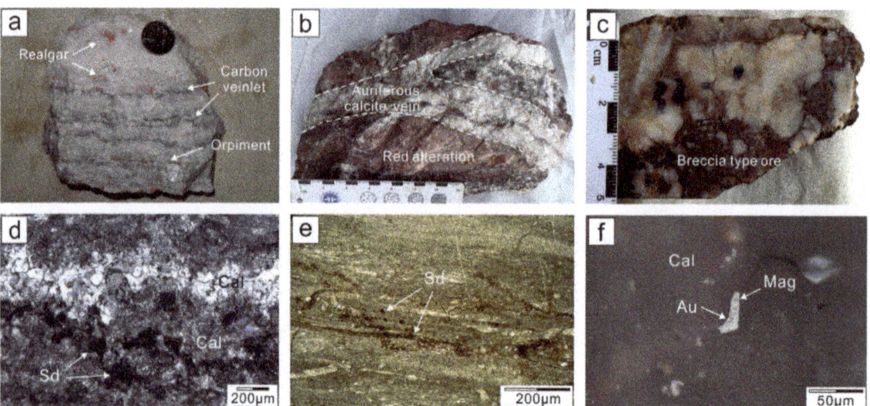

Figure 4. Hand samples and microscopic photographs of typical ores in Phapon gold deposit. (**a**) Auriferous calcite veins filled with dark-colored carbon layers and red alteration veinlets; (**b**) red alteration with calcite veinlets; (**c**) breccia type ore; (**d**) calcite veinlet in limestone; (**e**) visible siderite disseminated in limestone; (**f**) native gold together with magnetite appears as fine disseminations in calcite. Au, gold; Cal, calcite; Mag, magnetite; Sd, siderite.

There are three types of primary ore in the Phapon gold deposit: auriferous calcite vein type (Figure 4a), the red altered rock type (Figure 4b), and breccia type (Figures 3f and 4c). The auriferous calcite veins filled the NNW-trending tensile fractures and are dominated by grey-white calcite and in place dark-colored carbon veinlets (Figure 4a). The red altered ores occurred in red alteration zone (Figure 4c). The breccia type ore locally developed on one side of the calcite veins, and the breccia includes red altered limestone (Figure 3f) and auriferous calcite veins (Figure 4c), whereas the cements are carbonate minerals such as calcite and siderite. The breccia type gold ore has higher gold grade, with gold minerals in both breccia and cements. The ores from the red alteration zone generally have high gold grade, and their gold grade has a positive correlation with the strength of the alteration. The auriferous calcite vein type ore has a large variable range, but generally lower gold grade. The coarser the grains of calcite are, the lower the gold grade. The auriferous calcite vein type and red altered rock type ores are the two main ore types in Phapon.

Mineral paragenesis and mineralization stages are difficult to define and remain controversial. It is generally believed the auriferous calcite vein type and red altered rock-type ores formed in the same metallogenic event and through similar gold-deposition processes. The calcite veins normally show crosscutting, variation of calcite sizes (including miarolitic calcite, Figure 3e), or filled with dark-colored carbon veinlets (Figure 4a), indicating a characteristic of muti-stage mineralization. The ore minerals are mainly siderite, magnetite, and hematite, with minor native gold. The gangue minerals are mainly calcite with minor realgar, orpiment and quartz (Figure 4), and the realgar and orpiment are only found in the calcite veins. Calcite is the main gold-bearing mineral. Native gold appears as fine disseminations or in microfractures in calcite (Figure 4f).

4. Sampling and Analytical Methods

A total of seventeen typical samples were collected for this study. Calcite grains from auriferous calcite vein type ores (4 samples) and red altered rock type ores (4 samples), and the ore-hosting limestone (7 samples) and the surrounding sandstone (2 samples) were selected for trace element analyses. Except the sandstone, the other 15 samples were also used for carbon and oxygen isotope analyses. The ore hand samples were crushed to 40–60 mesh for mineral separation. Calcite grains were handpicked under a binocular microscope to achieve >99% purity. All the analyses were performed at the Analytical Laboratory of the Beijing Research Institute of Uranium Geology, Beijing, China.

4.1. Trace Element Analyses

The pure hydrothermal calcite grains and the limestone and sandstone samples were crushed to 200 meshes, and were dissolved in a Teflon bomb using a mixture of HF (1 mL) and HNO_3 (1 mL). Then the sealed Teflon bomb was heated at 190 °C for more than 12 h. The dried sample was refluxed with 30% HNO_3 (1 mL) and heated in the sealed Teflon bomb again at 190 °C for another 12 h with mixture of HNO_3 (1 mL), MQ water (1 mL) and internal standard In. The sample solution was then diluted to 100 g with 2% HNO_3 in a polyethylene bottle, and measured using an Inductively Coupled Plasma-Mass Spectrometry (ICP-MS) (Agilent 7700e). Detailed operating conditions of the ICP-MS instrument and data reduction processes have been described by Liu et al. [48].

4.2. C–O Isotope Analyses

Both calcite grains from ores and ore-hosting limestone samples were crushed to 200 meshes, and water was released from the carbonate sample by heating to 105 °C for 2 h in an oven. The dried samples then reacted with phosphoric acid to produce CO_2, and CO_2 was then measured for C and O isotopic composition by a MAT-253 mass spectrometer. The C and O isotope analysis results adopted the Pee Dee Belemnite (PDB) and Standard Mean Ocean Water (SMOW) standards, respectively, with a precision of $\pm 0.1‰$ for $\delta^{13}C$, and $\pm 0.2‰$ for $\delta^{18}O$ [49].

5. Analytical Results

5.1. Rare Earth Elements

The concentrations of REE for calcites, limestone and sandstone are presented in Table 2, and displayed in Figure 5. The calcites from the auriferous calcite vein type ores contain relatively low total REE abundances (ΣREE = 1.50–4.98 ppm, average of 2.67 ppm), and show heavy rare earth elements (HREE) flat patterns (Figure 5a). The LREE/HREE ratios vary from 0.88 to 1.89, with an average of 1.25. Eu is slightly depleted with δEu ranges between 0.69 and 0.81 (average of 0.77), and Ce is distinctively depleted with δCe ranges between 0.35 and 0.46 (average of 0.39).

All calcite samples contain relatively low Y content (2.14–7.55 ppm), with identical Y/Ho ratios (40.2–49.6). The calcites from red altered rock type ores have similar REE pattern with the vein-type ones (Figure 5a). They contain relatively higher total REE abundances (ΣREE = 3.69–7.57 ppm, average of 2.67 ppm), Y content (4.88–10 ppm) and Y/Ho ratios (52.3–61.0). They have constant LREE/HREE ratios varying from 1.68 to 1.80 with an average of 1.75. The Eu and Ce anomalies are somewhat stronger (δEu = 0.61–0.71 and δCe = 0.33–0.38).

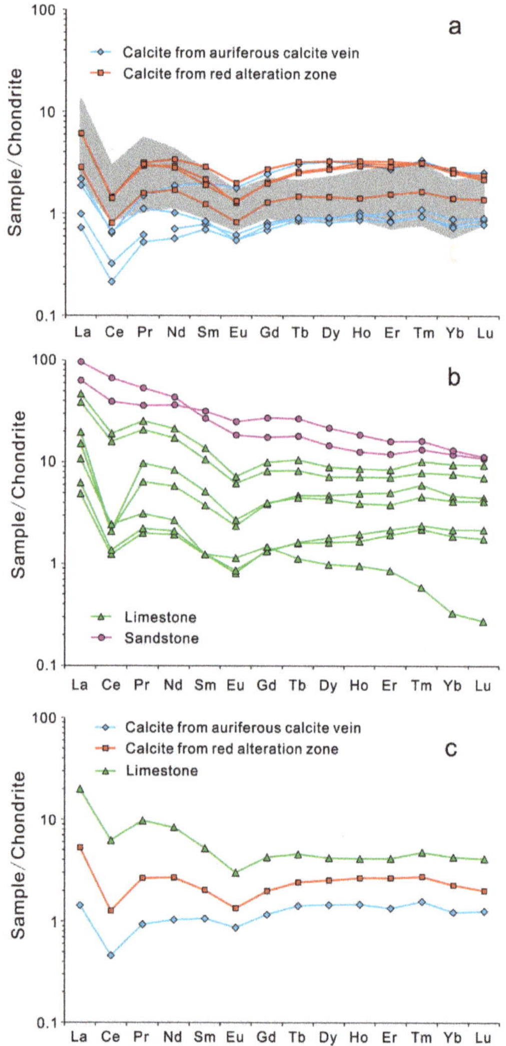

Figure 5. Rare Earth Elements (REE) distribution patterns of hydrothermal calcite and country rocks in Phapon. All samples are normalized to chondrite [50]. (**a**) Calcites respective from auriferous calcite vein and red alteration zone, shadow area referenced from [11]; (**b**) limestone and sandstone; (**c**) average value of two types of calcite and limestone.

Table 2. Trace element and REE concentrations ($\times 10^{-6}$) in calcite, limestone and sandstone from the Phapon deposit.

Sample	Calcite from Auriferous Calcite Vein				Calcite from Red Alteration Zone				Limestone							Sandstone	
	PB-PDB04-1	PB-PDB04-2	PB-PDB04-3	PB-ZKB04	PB-ZKB05-1	PB-ZKB05-2	PB-ZKB05-3	PB-ZKB05-4	PB-PDB03	PB-PDB05-1	PB-PDB05-2	PB-B01-1	PB-B01-2	PB-ZKB02-1	PB-ZKB02-2	PB-PDB06	PB-ZKB01
Li	<0.002	<0.002	<0.002	<0.002	<0.002	<0.002	<0.002	<0.002	2.02	2.24	1.89	5.28	5.76	1.82	2.18	23.6	38
Be	0.017	0.002	0.024	0.005	<0.002	0.025	<0.002	0.033	0.085	0.036	0.061	0.431	0.348	0.063	0.022	1.53	1.19
Sc	0.431	1.05	0.469	0.509	0.469	0.48	0.949	0.581	1.03	1.05	1.13	4.31	4.12	2.16	1.06	14.4	4.65
V	1.76	5.01	1.61	1.38	1.69	1.39	3.04	1.4	3.94	5.2	6.32	25.5	23.1	7.28	2.75	103	36.5
Cr	1.27	2.54	1.49	3.27	3.12	3.24	3.9	3.6	8.75	8.62	12.3	9.83	11.4	6.7	3.18	73	20.5
Co	1.13	1.14	1.31	1.2	1.18	1.21	1.28	1.26	1.52	1.5	1.57	3	3.41	1.75	1.66	14.2	4.44
Ni	11.8	13.7	13.5	14.3	15.3	15.1	16.6	16.3	12.2	12.6	13.2	7.62	10	13.7	13	17.9	15.4
Cu	0.057	0.099	0.061	0.149	0.026	0.196	0.104	0.024	0.585	0.429	0.474	16.5	18.2	1.61	0.207	9.08	11.6
Zn	2.53	2.1	2.1	2.22	6.05	3.34	2.9	4.11	4.22	7.68	7.35	25.3	33.1	18.6	4.98	68.1	48.3
Ga	0.019	0.01	<0.002	0.011	0.02	0.019	0.022	0.039	0.067	0.19	0.249	3.43	3.12	0.287	0.066	13.4	19.5
Rb	0.03	0.029	0.027	0.034	0.026	0.045	0.037	0.062	0.285	0.968	1.23	21.6	18.7	1.08	0.192	23.1	45.6
Sr	674	451	686	456	430	376	485	398	181	401	409	777	965	207	452	400	106
Y	2.33	7.55	2.14	2.68	4.88	10	9.67	9.47	12.1	6.07	6.42	13.9	16.3	8.72	4.62	20.1	41.7
Mo	<0.002	<0.002	<0.002	<0.002	<0.002	<0.002	<0.002	<0.002	<0.002	0.032	0.037	0.283	0.258	0.008	<0.002	0.284	0.193
Cd	0.01	0.009	0.036	0.175	0.013	0.056	0.076	0.025	0.573	0.418	0.383	0.2	0.247	1.28	0.088	0.22	0.617
In	<0.002	0.002	<0.002	0.002	<0.002	0.003	0.005	0.003	<0.002	0.004	0.005	0.017	0.018	<0.002	<0.002	0.058	0.016
Sb	0.028	0.012	0.038	0.055	0.04	0.04	0.051	0.053	0.369	0.206	0.301	0.622	0.737	0.569	0.128	0.65	0.613
Cs	0.011	0.011	0.012	0.018	0.012	0.013	0.012	0.025	0.022	0.058	0.057	1.06	0.98	0.194	0.027	1.03	10.1
Ba	4.25	2.59	3.75	3.83	3.07	7.67	3.79	7.7	6.98	5.78	5.87	458	401	10.2	3.37	153	185
La	0.235	0.515	0.173	0.444	0.671	1.45	1.44	1.45	3.54	1.15	1.47	9.03	10.9	4.58	2.52	22.2	14.7
Ce	0.199	0.393	0.131	0.408	0.495	0.885	0.875	0.864	1.28	0.757	0.835	9.54	11.4	1.28	1.49	40	23.6
Pr	0.059	0.142	0.05	0.106	0.15	0.277	0.298	0.286	0.6	0.19	0.211	1.93	2.35	0.911	0.294	4.95	3.35
Nd	0.332	0.873	0.267	0.475	0.794	1.36	1.58	1.31	2.68	0.905	0.983	7.91	9.74	3.84	1.25	19.9	16.7
Sm	0.12	0.3	0.107	0.128	0.189	0.331	0.437	0.291	0.571	0.189	0.19	1.6	2.06	0.781	0.189	4.02	4.77
Eu	0.036	0.103	0.032	0.032	0.048	0.074	0.115	0.077	0.137	0.047	0.05	0.355	0.413	0.156	0.066	1.05	1.42
Gd	0.167	0.501	0.142	0.156	0.264	0.415	0.558	0.404	0.792	0.279	0.272	1.66	2.02	0.814	0.304	3.53	5.46
Tb	0.034	0.114	0.032	0.034	0.055	0.096	0.12	0.093	0.175	0.06	0.061	0.304	0.388	0.166	0.042	0.662	0.976
Dy	0.225	0.822	0.207	0.232	0.37	0.706	0.825	0.686	1.19	0.413	0.458	1.8	2.25	1.09	0.25	3.63	5.39
Ho	0.058	0.173	0.049	0.054	0.08	0.176	0.185	0.165	0.277	0.095	0.111	0.401	0.481	0.221	0.054	0.703	1.03
Er	0.141	0.452	0.138	0.167	0.257	0.502	0.538	0.475	0.823	0.319	0.359	1.16	1.38	0.628	0.142	1.95	2.6
Tm	0.024	0.085	0.024	0.028	0.042	0.081	0.079	0.08	0.151	0.056	0.061	0.198	0.256	0.116	0.015	0.334	0.405
Yb	0.131	0.438	0.125	0.151	0.24	0.428	0.461	0.424	0.785	0.32	0.369	1.28	1.59	0.7	0.056	1.99	2.19
Lu	0.022	0.064	0.02	0.023	0.035	0.06	0.055	0.055	0.113	0.045	0.055	0.177	0.234	0.104	0.007	0.272	0.282
W	2.95	0.768	2.64	0.669	42.3	0.712	0.666	0.502	5.42	0.505	0.265	0.289	0.313	0.398	0.247	0.751	2.63
Re	<0.002	<0.002	<0.002	<0.002	<0.002	<0.002	<0.002	<0.002	<0.002	<0.002	<0.002	<0.002	0.002	<0.002	<0.002	<0.002	<0.002
Tl	0.01	0.007	0.008	0.05	0.186	0.016	0.1	0.003	0.11	0.03	0.044	0.143	0.107	0.099	0.01	0.13	0.409
Pb	0.176	0.038	0.122	0.128	0.358	0.577	0.284	0.242	0.92	1.1	5.87	2.65	2.68	1.64	0.84	9.26	2.62

Table 2. *Cont.*

Sample	Calcite from Auriferous Calcite Vein				Calcite from Red Alteration Zone								Limestone				Sandstone	
	PB-PDB04-1	PB-PDB04-2	PB-PDB04-3	PB-ZKB04	PB-ZKB05-1	PB-ZKB05-2	PB-ZKB05-3	PB-ZKB05-4	PB-PDB03	PB-PDB05-1	PB-PDB05-2	PB-B01-1	PB-B01-2	PB-ZKB02-1	PB-ZKB02-2	PB-PDB06	PB-ZKB01
Bi	<0.002	<0.002	<0.002	<0.002	<0.002	<0.002	<0.002	<0.002	0.012	0.06	0.036	0.051	0.071	0.018	0.014	0.207	0.011
Th	0.007	0.031	0.013	0.024	0.023	0.029	0.043	0.044	0.249	0.197	0.191	1.25	1.23	0.195	0.115	5.52	5.58
U	0.099	0.432	0.095	0.123	0.272	0.218	0.679	0.209	0.763	0.821	1.15	0.651	0.699	0.526	0.063	1.09	1.32
Nb	0.009	<0.002	0.004	0.005	<0.002	0.003	0.005	0.027	0.268	0.432	0.405	1.38	1.4	0.433	0.343	5.07	7.83
Ta	<0.002	<0.002	<0.002	0.002	0.002	<0.002	<0.002	0.004	0.104	0.321	0.214	0.115	0.108	0.114	0.147	0.418	0.756
Zr	0.104	0.094	0.125	0.075	0.057	0.062	0.035	0.067	1.35	1.96	2.47	17	18.3	3.46	0.89	40.4	68.9
Hf	<0.002	0.005	<0.002	<0.002	0.006	0.004	0.009	0.008	0.021	0.056	0.07	0.52	0.519	0.11	0.034	1.45	1.96
ΣREE	1.78	4.98	1.50	2.44	3.69	6.84	7.57	6.66	13.11	4.83	5.49	37.35	45.46	15.39	6.68	105.19	82.87
LREE	0.98	2.33	0.76	1.59	2.35	4.38	4.75	4.28	8.81	3.24	3.74	30.37	36.86	11.55	5.81	92.12	64.54
HREE	0.80	2.65	0.74	0.85	1.34	2.46	2.82	2.38	4.31	1.59	1.75	6.98	8.60	3.84	0.87	13.07	18.33
LREE/HREE	1.22	0.88	1.03	1.89	1.75	1.78	1.68	1.80	2.05	2.04	2.14	4.35	4.29	3.01	6.68	7.05	3.52
δEu	0.78	0.81	0.79	0.69	0.66	0.61	0.71	0.69	0.62	0.63	0.67	0.67	0.62	0.60	0.84	0.85	0.85
δCe	0.41	0.36	0.35	0.46	0.38	0.34	0.33	0.33	0.22	0.40	0.37	0.56	0.55	0.15	0.42	0.94	0.82
La/Ho	4.05	2.98	3.53	8.22	8.39	8.24	7.78	8.79	12.78	12.11	13.24	22.52	22.66	20.72	46.67	31.58	14.27
Y/Ho	40.2	43.6	43.7	49.6	61.0	56.8	52.3	57.4	43.7	63.9	57.8	34.7	33.9	39.5	85.6	28.6	40.5
Tb/La	0.14	0.22	0.18	0.08	0.08	0.07	0.08	0.06	0.05	0.05	0.04	0.03	0.04	0.04	0.02	0.03	0.07
Sm/Nd	0.36	0.34	0.40	0.27	0.24	0.24	0.28	0.22	0.21	0.21	0.19	0.20	0.21	0.20	0.15	0.20	0.29

Notes: δCe = Ce$_N$/$\sqrt{\text{La}_N \times \text{Pr}_N}$, δEu = Eu$_N$/$\sqrt{\text{Sm}_N \times \text{Gd}_N}$.

In contrast to hydrothermal calcites, the limestone have a relatively higher and wider range of the ΣREE contents of 4.83–45.46 ppm (average of 18.33 ppm), and show slight REE fractionation with LREE enrichments and HREE depletions (LREE/HREE = 2.04–6.68, average of 3.51; Figure 5b). They have relatively higher Y contents of 4.62–16.3 ppm, but similar Y/Ho ratios of 33.9–85.6. The negative anomalies of Eu and Ce are similar to the calcite from ores, with δEu = 0.60–0.84 and δCe = 0.15–0.56. The two sandstone samples have significantly higher ΣREE contents of 82.9–105.2 ppm, and show LREE enrichment pattern (Figure 5b), with nearly no Eu and Ce anomalies (δEu = 0.85 and δCe = 0.82–0.94). These REE characteristics of sandstone are obviously different from the limestone, as well as the hydrothermal calcites.

5.2. The Other Trace Elements

The contents of other trace elements (excluding REE) in calcites, limestone and sandstone are given in Table 2, and displayed in Figure 6. The high field-strength elements (HFSEs) such as Nb, Ta, Zr and Hf are extremely depleted in both types of hydrothermal calcites (Figure 6a), while the limestone has only a slight depletion of Nb, Zr and Hf (Figure 6b). The large ion lithophile elements (LILEs), such as Rb, U and Th are enriched in limestone, but Rb is depleted in the hydrothermal calcites. Even though the contents of multi-trace elements decrease from limestone, to calcite from red alteration zone, and to calcite from auriferous calcite vein, the characteristics of trace elements are similar in all these three kinds of samples (Figure 6c). The sandstone shows obviously higher trace element contents and different characteristics compared to the limestone and hydrothermal calcites (Figure 6b).

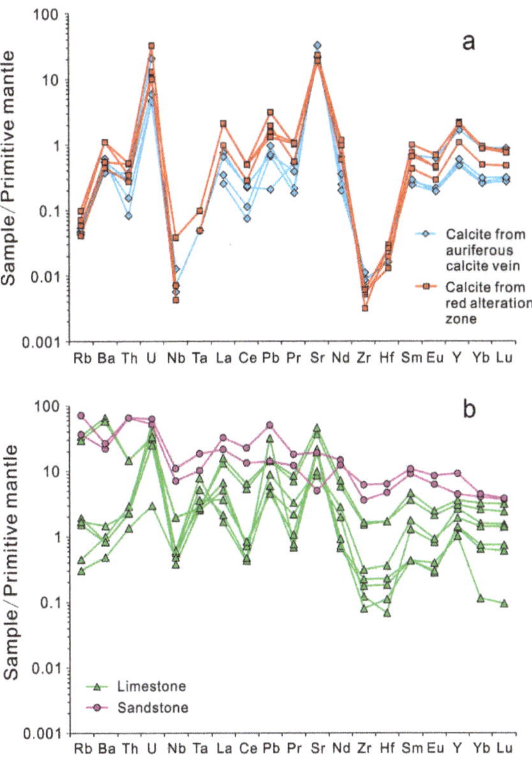

Figure 6. *Cont.*

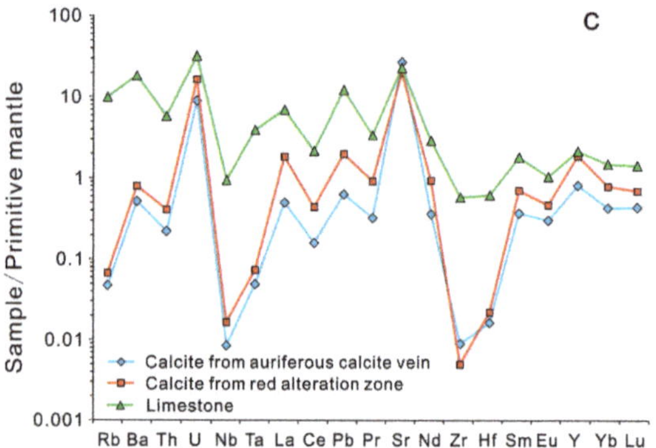

Figure 6. Trace elements distribution patterns of hydrothermal calcite and country rocks in Phapon. All samples are normalized to primitive mantle [50]. (**a**) Calcites from auriferous calcite vein and red alteration zone, respectively; (**b**) limestone and sandstone; (**c**) average value of two types of calcite and limestone.

5.3. C–O Isotopic Compositions

The carbon and oxygen isotopic composition of all the hydrothermal calcite and limestone are listed in Table 3.

Table 3. Carbon and oxygen isotope compositions of calcite and limestone from the Phapon deposit.

Ore Type	Sample No.	$\delta^{13}C_{V\text{-PDB}}$ (‰)	$\delta^{18}O_{V\text{-SMOW}}$ (‰)
Auriferous calcite vein	PB-PDB04-1	1.5	29.6
	PB-PDB04-2	1.6	28.5
	PB-PDB04-3	1.6	28.7
	PB-ZKB04	1.7	23.8
Red alteration zone	PB-ZKB05-1	1.2	27.5
	PB-ZKB05-2	1.2	28.2
	PB-ZKB05-3	0.7	28.7
	PB-ZKB05-4	1.5	28.8
Limestone	PB-PDB03	3	25.6
	PB-PDB05-1	3.3	25.5
	PB-PDB05-2	3.3	27.4
	PB-B01-1	2.7	18.8
	PB-B01-2	2.6	20.3
	PB-ZKB02-1	3.6	26.1
	PB-ZKB02-2	3.3	26.2

The hydrothermal calcites have a narrow $\delta^{13}C_{V\text{-PDB}}$ range of 0.7–1.7‰, with an average value of 1.4‰. The limestone differs from calcites by slightly increased $\delta^{13}C_{V\text{-PDB}}$ (2.6–3.6‰, average of 3.1‰). For two types of the calcite, the vein-type calcite (1.5‰ to 1.7‰) is slightly higher than the calcite from red alteration zone (0.7‰ to 1.5‰).

The hydrothermal calcites generally have a narrow $\delta^{18}O_{V\text{-SMOW}}$ values of 27.5–29.6‰, except for one lighter vein-type calcite (23.8‰). The limestone has a wider range of oxygen isotopic compositions, and $\delta^{18}O_{V\text{-SMOW}}$ varies between 18.8‰ and 27.4‰ (average 24.3‰).

6. Discussion

6.1. Genetic Relationships of the Two Types of Hydrothermal Calcite

The geochemistry of REE is useful in investigating hydrothermal mineralization, as well as understanding the genesis of calcite in different geological environments [51]. Two types of hydrothermal calcite, from the calcite vein and red alteration zones respectively, have similar REE distribution patterns (Figure 5a), and previous studies also show a consistent REE characteristic of the hydrothermal calcites [11]. The calcites, moreover, show constant Y/Ho values and are roughly horizontally distributed on the Y/Ho–La/Ho diagram (Figure 7a), indicating that both types of the calcite are probably contemporaneous crystallization and originating from the same fluid system [52,53]. The REE patterns of limestone are generally consistent with that of calcite, but the sandstones show a completely different REE pattern (Figure 5b), indicating that the calcite may have a genetic relationship with the limestone, rather than the sandstone.

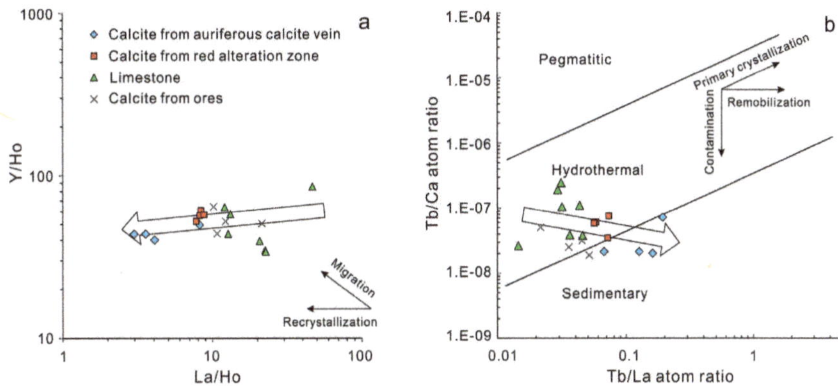

Figure 7. Plots of (**a**) Y/Ho versus La/Ho ratios and (**b**) Tb/Ca versus Tb/La ratios for the Phapon calcite and limestone. (Base maps respectively after [52,54], the grey × symbols show hydrothermal calcite from Phapon referenced from [11]).

The two types of hydrothermal calcite, in fact, have a slight difference on LREE/HREE fractionation, except for the Eu and Ce anomalies. The calcites from red altered rock type ores have constant LREE/HREE ratios varying from 1.68 to 1.80, showing a LREE-enriched pattern. And the calcites from the auriferous calcite vein type ores have a relatively variable LREE/HREE fractionation (with minimum LREE/HREE ratio of 0.88), showing a slight LREE-depleted pattern (Figure 5a). Morgan and Wandless [55] believed that the REE pattern in hydrothermal minerals is governed by the ionic radius of the major cation and that the LREE are more easily incorporated into the calcite crystal lattice than the HREE, because the differences in ionic radii between $LREE^{3+}$ and Ca^{2+} are smaller than these between $HREE^{3+}$ and Ca^{2+}. Therefore, the calcite should have an LREE-enriched pattern as usual, but evidently, this is not so. The LREE/HREE fractionation of hydrothermal calcite is primarily controlled by physico-chemical conditions governing the fluid during leaching of REE from source rocks, as well as fluid migration [56]. Bau and Möller [57] suggested that leaching of REE and fluid migration may occur under two different states: sorption and complexation, and specifically, complexation is more prevalent than sorption, along with an increasing pH and a decreasing temperature. The LREE-enriched patterns are produced in fluid under conditions favored by sorption (i.e., low pH and high T), alternatively, the LREE-depleted pattern should be controlled by complexation processes, due to the ligand-enriched fluid [58,59]. The complexation mechanism is considered to be responsible for the fractionation in Phapon, owing to REE complexation with ligands such as CO_3^{2-} and OH^- which form more stable complexes with HREE. Therefore, the early stage fluid should be LREE-enriched, and HREE preferentially precipitated in the late stage. Thus, the calcite from red altered rock type ores with a LREE-enriched pattern likely formed earlier than the auriferous calcite vein, which shows a slightly LREE-depleted pattern. During the fluids evolution and calcite precipitation, there are differences on not only LREE/HREE fractionation, but also Tb/La and Sm/Nd ratios. Constantopoulos [60] and Chesley [61] indicate that the early stage calcite has lower Tb/La and Sm/Nd ratios than the late stage ones. The Tb/La ratio of vein-type calcite ranges from 0.08 to 0.22, with an average value of 0.16, which is higher than the calcite from the red alteration zone (0.06–0.08). The vein-type calcites have also higher Sm/Nd ratios (0.27–0.40) than the calcite from red alteration zone (0.22–0.28), which also support the contention that the auriferous calcite vein may have formed later than the red alteration in Phapon.

Both types of calcite display a weak Eu negative anomaly and moderate Ce negative anomaly, suggesting low oxygen fugacities and low temperature during calcite deposition [51,62] The REE signatures of the two types of calcite are similar to the limestone, and the ΣREE of calcite from red alteration zone contains falls in between limestone and vein-type calcite. Given that the two types of the calcites have only a little difference on REE patterns, Tb/La and Sm/Nd ratios and δEu and δCe values, we suspect that the two types of calcite are products of different stages in the evolution of homologous fluids. The calcite from the red alteration zone formed earlier and was more affected from the limestone wallrock, whereas the auriferous calcite vein formed later and shows more characteristics from the ore fluids.

6.2. Nature and Sources of Ore-Forming Fluids

The Y and Ho generally show similar geochemical behavior and therefore the Y/Ho ratio is widely used for tracing fluid processes as a significant parameter (e.g., Xu et al.; Pei et al. [63,64]). For both types of the calcite, their Y/Ho ratios vary over narrow ranges in 40.2–49.6 and 52.3–61.0, respectively, indicating a hydrothermal origin (ca. 20–110, [52]). Considering the nearly constant Y/Ho ratios of hydrothermal calcite and the limestone wallrock, the La/Ho ratios decrease from limestone, to red alteration zone, and to auriferous calcite vein (Figure 7a), indicating the hydrothermal calcite successively formed from recrystallization of the limestone. On the other hand, the Tb/Ca–Tb/La variation established by Möller et al., [54], in the form of a discriminating diagram, represents different occurrences of fluorite, calcite and other Ca-bearing minerals, according to their sedimentary, hydrothermal, and pegmatitic affinities (Figure 7b). The Tb/Ca and Tb/La ratios are, respectively, criterions of the environment in which calcite forms and the crystallization proceeds degree of calcite differentiation [65]. The limestone falls across the hydrothermal and sedimentary field (Figure 7b), the calcite fall into the field of sedimentary, and no data plot in the pegmatitic field. Even though previous studies have indicated a deep-seated magmatic source based on H-O isotopic data, the data generally fall outside the magmatic area (δD range from $-92‰$ to $-75‰$ and δ^{18}O range from 9‰ to 14‰) [11]. Thus, the magmatic fluid can be excluded. As expected, the limestone and calcite distribute along remobilization trends in Figure 7b, which may imply a succession from the sedimentary limestone to hydrothermal calcite.

The limestone and hydrothermal calcite respectively have narrow $\delta^{13}C_{\text{V-PDB}}$ ranges of 2.6–3.6‰ (average 3.1‰) and 0.7–1.7‰ (average 1.4‰), which are higher than organic material in sedimentary rocks (-30 to $-10‰$; [66]), generally higher than mantle reservoir values [67], and basically within the range of marine carbonate [68] (Figure 8). The $\delta^{18}O_{\text{V-SMOW}}$ of limestone and hydrothermal calcite ranges from 18.8 to 29.6‰, which are significantly higher than igneous carbonate and mantle xenoliths, and can be preclude from the process of crystallization differentiation.

The two types of hydrothermal calcite and limestone have similar trace elements (exclude REE) characteristics (Figure 6), and the HFSEs such as Nb, Ta, Zr and Hf are extremely depleted, which may indicate the mixing of crustal-source fluid. Thus, the ore-forming fluids should be a single fluid that was not associated with magmatic or mantle fluid. As stated above, the contents of multi-trace elements decrease from limestone, to calcite from red alteration zone, and to calcite from auriferous calcite vein (Figure 6c), which is the same as those of the ΣREE signatures. The later vein-type calcite contains lower trace element content than earlier calcite from the red alteration zone, which could be explained as the lower total trace element concentrations in the fluids in late stage, which means the ore-forming fluids have been diluted by meteoric waters during late stage [69]. Combined with the Y/Ho–La/Ho and Tb/Ca–Tb/La diagram (Figure 7) that argue the hydrothermal calcite may formed from remobilization and recrystallization of the limestone, and the similar REE patterns between hydrothermal calcite and limestone, the ore-forming fluids are suspected to be primarily associated with dehydration and decarbonization of the Lower Permian limestone and Middle–Upper Triassic sandstones during a regional dynamic metamorphism, shown as a thrust nappe structure in the shallow ground.

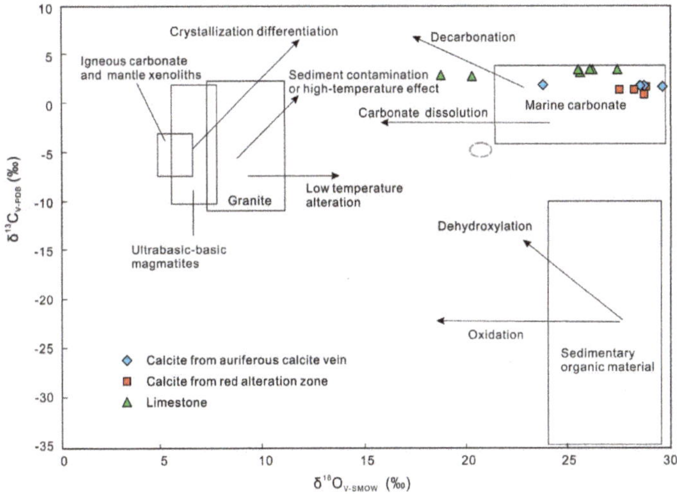

Figure 8. The $\delta^{13}C_{V-PDB}$ and $\delta^{18}O_{V-SMOW}$ values of calcite from the Phapon gold deposit. The area in grey dashed line shows $\delta^{13}C_{V-PDB}$–$\delta^{18}O_{V-SMOW}$ values of hydrothermal calcite from Phapon [15]. References for the fields of major carbon reservoirs and other deposits are from [68].

6.3. Implications for Ore-Forming Processes

During the migration of hydrothermal solutions up in the fractured zone, gold is normally dissolved and transported in the form of gold chloride complexes ($AuCl^{2-}$) and gold bisulfide complexes [$Au(HS)_2^-$] [70,71]. Gold chloride complexes ($AuCl^{2-}$) normally exist in near-neutral to weakly alkaline ore fluids at >400 °C [72], whereas the $Au(HS)_2^-$ are important in near-neutral to weakly acidic ore fluids with temperatures of <400 °C [73]. The calcite–quartz mineral assemblages in the Phapon gold deposit indicate that the pH values of ore fluids were near-neutral to weakly acidic. The ore-forming temperature should be lower than 400 °C, because of the negative anomaly of Eu and Ce of hydrothermal calcite, as well as fluid inclusion microthermometry [10]. Thus gold bisulfide complexes, mainly $Au(HS)_2^-$, were the gold-transporting species (e.g., Pokrovski et al. [74]), and gold-deposition may be the result of the breakdown of $Au(HS)_2^-$ [75,76].

Hydrothermal alteration, particularly the siderite alteration, is widely developed in the Phapon gold deposit (Figure 3b). Interaction between Fe-bearing orefluids and carbonate minerals in wall-rocks, to form siderite is in terms of the following equations [77]:

$$Fe^{2+} + (HCO_3)^- = FeCO_3 + H^+ \tag{1}$$

$$H_2S\ (g) = H^+ + HS^- \tag{2}$$

$$4Au(HS)_2^- + 2H_2O + 8H^+ \text{ (aq.)} = 4Au^0 \text{ (s)} + 8H_2S \text{ (aq.)} + O_2 \tag{3}$$

The processes would decrease the solubility of $Au(HS)_2^-$ and lead to gold deposition (e.g., Guo et al.; Yang et al. [78–80]), indicating that fluid-wallrock interaction played a major role in gold mineralization.

As discussed above, the potential fluid reservoirs could be associated with dehydration and decarbonization of Lower Permian–Middle Triassic sedimentary formations; the mineralization events cannot be earlier than Early to Middle Triassic. The Loei belt mineralization systems are mainly linked to one of the two subduction-related systems [36]: (1) Late Permian–Middle Triassic continental arcs, and (2) Late Triassic–Jurassic post-collisional magmatism. As there was no geochronological data reported in the Phapon deposit, the gold deposition in the Phapon deposit occurred, more likely, during the regional dynamic metamorphism, driven by Late Triassic–Jurassic post-collisional magmatism. Combined with the geological and geochemical studies, the process of geotectonic evolution and gold mineralization in Phapon is presumed as follows:

The basin of Luang Prabang in Laos subsided deeper in Early Permian, and sedimented thickly-bedded carbonate rocks. The Nan-Sra Kaeo back arc basins may have begun to form close to the Late Permian to Early Triassic, through east-vergent subduction of Sibumasu underneath the Indochina terrane, forming Nan-Uttaradit Suture [34], which created the Loei continental arc magmatism and its associated porphyry-related skarn and epithermal deposits (Figure 1), as well as continental clastic rocks and NE-trending brittle–ductile shear zones in the Phapon area (Figure 2). Continuous subduction along the Loei fold belts finally brought about the Indochina–Sibumasu collision during the Triassic–Jurassic, which represent a significant metallogenic epoch for many orogenic gold deposits along the Sukhothai- and East Malaya-terrane/fold belts [36] (Figure 1).

In Phapon area, the Late Triassic–Jurassic collision and post-collision caused the southeastern Permian limestone to be thrusted northward and superimposed on the Triassic sandstone. Strong tectonic activities led to ductile–brittle shearing along the lithological difference surface (Figure 3a,b), followed by dynamic metamorphism, and dehydration and decarbonation of Lower Permian limestone which created metamorphic fluids. At the same time, the NE-trending faults, formed during Late Permian–Early Triassic, reactivated with compression–torsion activities, forming deep faults and regional magmatism (locally diabase and dioritic porphyrite). A series of NNW-trending extension secondary faults, formed simultaneously, are identified as fluid channels and important ore-bearing spaces. The metamorphic fluid began to enter the limestone along the extension secondary faults. Even though the magmatic fluids may not have played a role in the gold mineralization, the regional magmatic events were likely to provide a heat source for driving fluid circulation. The fluids reacted with the surrounding sedimentary formation with relatively high gold concentration (4.56×10^{-9}, [13]), and mixed with near-surface meteoric waters. Gold deposited primary in the red alteration zones, as a result of the interaction between Fe-bearing orefluids and carbonate minerals in wall-rocks, and subsequently deposited in massive calcite veins.

As mentioned above, the gold mineralization in the Phapon gold deposit is completely different from the porphyry-related skarn and epithermal systems. It is also different from the sediment-hosted carlin-type gold that has submicron gold in pyrite [81], and the orogenic gold deposits in the East Malaya terrane that output ores with quartz–sulfide assemblage [82]. However, a number of features in Phapon are similar to orogenic gold systems: (1) strong structural control on the deposits by NNW-trending brittle faults (e.g., Deng et al. [83–85]); (2) lack of metal zonation; (3) occurrence of free gold in calcite; (4) extremely high Au/Ag ratio (much higher than 10, with average silver grade of 0.05 g/t) of the ores [14]; and (5) relative low fluid salinity (average 7.64%, [10]). Combined with the relatively shallower metallogenic depth (ca. 0.48–1.77 km, [8]), the appearance of low-temperature minerals such as realgar and orpiment, and the predominantly brittle deformation textures in ores, the Phapon gold deposit is compatible with an epizonal orogenic deposit (e.g., Gebre-Mariam et al.; Groves et al.; Yang et al.; [86–89]).

7. Conclusions

(1) The two types of hydrothermal calcite, respectively from auriferous calcite veins and red alteration zone, show a nearly consistent REE characteristic, and could be formed from remobilization and recrystallization of the ore-hosted limestone.

(2) The two types of calcite are products of homologous fluids. The calcite from red alteration zone formed earlier and was more affected by the limestone wallrock, whereas the auriferous calcite vein formed later and showed more characteristics from the ore fluids.

(3) The ore-forming fluids are suspected to be primarily associated with dehydration and decarbonization of the Lower Permian limestone and Middle–Upper Triassic sandstones during a regional dynamic metamorphism, and were probably mixed by meteoric waters during the formation of the massive calcite veins.

(4) Gold mineralization in Phapon probably occurred during the Late Triassic–Jurassic regional dynamic metamorphism driven by Indochina–Sibumasu post-collisional magmatism. A number of features in Phapon are compatible with an epizonal orogenic deposit, but it is still a unique calcite vein type gold deposit in the Luang Prabang-Loei metallogenic belt.

Author Contributions: Conceptualization, L.G. and L.H.; Data curation, L.G.; Formal analysis, L.G.; Funding acquisition, S.L.; Investigation, L.G. and L.H.; Project administration, S.L.; Writing—original draft, L.G.; Writing—review and editing, L.G., L.H. and F.N.

Funding: This research was funded by the China Geological Survey Project, grant No. 121201010000150013.

Acknowledgments: We thank the geologists from the Tianjin Huakan Mining Investment, Co, Ltd with their assistance in the field work for the Phapon gold deposit.

References

1. Goldfarb, R.J.; Taylor, R.D.; Collins, G.S.; Goryachev, N.A.; Orlandini, O.F. Phanerozoic continental growth and gold metallogeny of Asia. *Gondwana Res.* **2014**, *25*, 48–102. [CrossRef]

2. Deng, J.; Wang, Q.F. Gold mineralization in China: Metallogenic provinces, deposit types and tectonic framework. *Gondwana Res.* **2016**, *36*, 219–274. [CrossRef]

3. Zhao, Y.P.; Kang, T.S.; Ning, G.C.; Ge, H.; Pan, H. Geochemical characteristics of the volcanic intrusive complex in the Pangkuam copper-gold deposit of Laos and its geological significance. *Acta Petrol. Mineral.* **2017**, *36*, 281–294. (In Chinese with English Abstract).

4. Shi, M.F.; Lin, F.C.; Fan, W.Y.; Deng, Q.; Cong, F.; Tran, M.D.; Zhu, H.P.; Wang, H. Zircon U–Pb ages and geochemistry of granitoids in the Truong Son terrane, Vietnam: Tectonic and metallogenic implications. *J. Asian Earth Sci.* **2015**, *101*, 101–120. [CrossRef]

5. Lin, F.C.; Shi, M.F.; Li, X.Z. *Geological Background and Metallogenic Regularities of the Sanjiang—Mekong Metallogenic Belt*; Internal materials of Chengdu Center, China Geological Survey: Chengdu, China, 2010; p. 437. (In Chinese)

6. Shao, C.L. The geological characteristics and prospecting criteria of the Phabon gold deposit, Laos. *Geol. Surv. Res.* **2011**, *34*, 203–209. (In Chinese with English Abstract).

7. Liu, X.C.; Zhang, R.H.; Che, L.K. Geological characteristics of B. Phatem primary gold deposit, Laos and the ore-searching directions. *Contrib. Geol. Miner. Resour. Res.* **2010**, *25*, 171–176. (In Chinese with English Abstract).

8. Dai, F.Y.; Niu, Y.J. Characters of mineralogy and genesis of Phabon gold deposit in Luang Prabang Province, Laos. *Miner. Explor.* **2014**, *5*, 511–518. (In Chinese with English Abstract).

9. Shi, L.H.; Xue, L.H.; Sun, H.Y. Characteristic of wall rock alteration and its relation with gold mineralization of the Phapon gold deposit in Laos. *Geol. Surv. Res.* **2016**, *39*, 184–190. (In Chinese with English Abstract).

10. Lu, F.F.; Yang, H.L.; Tong, W.H.; Zhao, M.S. Inclusion characteristics of Phabon gold deposit, Laos. *Henan Sci.* **2015**, *33*, 1985–1989. (In Chinese with English Abstract).

11. Niu, Y.J.; Liu, W.; Gao, Y.L.; Liu, Z.Y.; Yang, H.L.; Yu, W.X. Geochemical characteristics of stable isotopes and REE at the PHAPUN gold deposit in Laos. *Geol. Surv. Res.* **2015**, *3*, 277–283. (In Chinese with English Abstract).
12. Li, H.K.; Zhang, X.J.; Wang, J. Genesis of Phapon Au deposit in Luangprabang, Laos. *Yunnan Geol.* **2011**, *30*, 280–284. (In Chinese with English Abstract).
13. Yang, C.Z.; Shen, L.X.; Zhou, L.; Feng, J.Z.; Liu, X.W. Characteristics of geology and structural geochemistry and metallogenic mechanism of Phapon gold deposit in Laos. *Miner. Resour. Geol.* **2017**, *31*, 11–22. (In Chinese with English Abstract).
14. Xue, L.H.; Shi, L.H. Mineralization and metallogenic evolution of the Phapon gold deposit, Laos. *Geol. Surv. Res.* **2016**, *39*, 191–203. (In Chinese with English Abstract).
15. Niu, Y.J.; Sun, H.Y.; Wang, J.S.; Chen, J.Y.; Liu, Z.Y.; Wang, K. Study on feature of ore-forming fluid and ore genesis of Phapon gold deposit, Luangprobang, Laos. *Contrib. Geol. Miner. Resour. Res.* **2017**, *32*, 317–323. (In Chinese with English Abstract).
16. Wang, H.; Lin, F.C.; Li, X.Z.; Shi, M.F. The division of tectonic units and tectonic evolution in Laos and its adjacent regions. *Geol. China* **2015**, *42*, 71–83. (In Chinese with English Abstract).
17. Hinton, R.W.; Upton, B.J.G. The chemistry of zircon: Variations within and between large crystals from syenite and alkali basalt xenoliths. *Geochim. Cosmochim. Acta* **1991**, *55*, 3287–3302. [CrossRef]
18. Orman, J.A.V.; Grove, T.L.; Shimizu, N.; Graham, D.L. Rare earth element diffusion in diopside: Influence of temperature, pressure and ionic radius and an elastic model for diffusion in silicates. *Contrib. Mineral. Petrol.* **2001**, *141*, 687–703. [CrossRef]
19. Brugger, J.; Etschmann, B.; Pownceby, M.; Liu, W.; Grundler, P.; Brewe, D. Oxidation state of europium in scheelite: Tracking fluidrock interaction in gold deposits. *Chem. Geol.* **2008**, *257*, 26–33. [CrossRef]
20. Peng, J.T.; Zhang, D.L.; Hu, R.Z.; Wu, M.J.; Liu, X.M.; Qi, L.; Yu, Y.G. Inhomogeneous distribution of rare earth elements (REEs) in scheelite from the Zhazixi W-Sb deposit, western Hunan and its geological implications. *Geol. Rev.* **2010**, *56*, 810–820. (In Chinese with English Abstract).
21. Schönenberger, J.; Köhler, J.; Markl, G. REE systematics of fluorides, calcite and siderite in peralkaline plutonic rocks from the Gardar Province, South Greenland. *Chem. Geol.* **2008**, *247*, 16–35. [CrossRef]
22. Dostal, J.; Kontak, D.; Chatterjee, A.K. Trace element geochemistry of scheelite and rutile from metaturbidite-hosted quartz vein gold deposits, Meguma Terrane, Nova Scotia, Canada: Genetic implications. *Mineral. Petrol.* **2009**, *97*, 95–109. [CrossRef]
23. Brugger, J.; Lahaye, Y.; Costa, S.; Lambert, D.; Bateman, R. Inhomogeneous distribution of REE in scheelite and dynamics of Archaean hydrothermal systems (Mt. Charlotte and Drysdale gold deposits, Western Australia). *Contrib. Mineral. Petrol.* **2000**, *139*, 251–264. [CrossRef]
24. Bau, M.; Romer, R.L.; Lüders, V.; Dulski, P. Tracing element sources of hydrothermal mineral deposits: REE and Y distribution and Sr-Nd-Pb isotopes in fluorite from MVT deposits in the Pennine orefield, England. *Miner. Depos.* **2003**, *38*, 992–1008. [CrossRef]
25. Zhang, D.L.; Peng, J.T.; Fu, Y.Z.; Peng, G.X. Rare-earth element geochemistry in Ca-bearing minerals from the Xianghuapu tungsten deposit, Hunan Province, China. *Acta Petrol. Sin.* **2012**, *28*, 65–74. (In Chinese with English Abstract).
26. Wang, J.S.; Wen, H.J.; Fan, H.F.; Zhu, J.J.; Zhang, J.R. Sm-Nd geochronology, REE geochemistry and C and O isotope characteristics of calcites and stibnites from the Banian antimony deposit, Guizhou Province, China. *Geochem. J.* **2013**, *46*, 393–407. (In Chinese with English Abstract). [CrossRef]
27. Wu, Y.; Zhang, C.Q.; Tian, G. REE geochemistry of fluorite from Paoma lead-zinc deposit in Sichuan Province, China and its geological implications. *Acta Mieral. Sin.* **2013**, *33*, 295–301. (In Chinese with English Abstract).
28. Cao, H.W.; Zhang, W.; Pei, Q.M.; Zhang, S.T.; Zheng, L. Trace element geochemistry of fluorite and calcite from the Xiaolonghe Tin deposits and Lailishan Tin deposits in Western Yunnan, China. *Bull. Mineral. Petrol. Geochem.* **2016**, *35*, 925–935. (In Chinese with English Abstract).
29. Deng, J.; Wang, Q.F.; Li, G.J.; Li, C.S.; Wang, C.M. Tethys tectonic evolution and its bearing on the distribution of important mineral deposits in the Sanjiang region, SW China. *Gondwana Res.* **2014**, *26*, 419–437. [CrossRef]
30. Deng, J.; Wang, Q.F.; Li, G.J.; Santosh, M. Cenozoic tectono-magmatic and metallogenic processes in the Sanjiang region, southwestern China. *Earth-Sci. Rev.* **2014**, *138*, 268–299. [CrossRef]

31. Yang, L.Q.; He, W.Y.; Gao, X.; Xie, S.X.; Yang, Z. Mesozoic multiple magmatism and porphyry-skarn Cu-polymetallic systems of the Yidun Terrane, Eastern Tethys: Implications for subduction- and transtension-related metallogeny. *Gondwana Res.* **2018**, *62*, 144–162. [CrossRef]
32. Phajuy, B.; Panjasawatwong, Y.; Osataporn, P. Preliminary geochemical study of volcanic rocks in the Pang Mayao area, Phrao, Chiang Mai, Northern Thailand: tectonic setting of formation. *J. Asian Earth Sci.* **2005**, *24*, 765–776. [CrossRef]
33. Zaw, K.; Meffre, S. *Metallogenic Relations and Depositscale Studies, Final Report: Geochronology, Metallogenesis and Deposit Styles of Loei Fold Belt in Thailand and Laos PDR*; ARC Linkage Project, CODES with Industry Partners; University of Tasmania: Hobart, Australia, 2007.
34. Salam, A. A Geological, Geochemical and Metallogenic Study of the Chatree Epithermal Deposit, Petchabun Province, Central Thailand. Ph.D. Thesis, ARC Centre of Excellence in Ore Deposits (CODES), University of Tasmania, Hobart, Australia, 2013; p. 250.
35. Salam, A.; Zaw, K.; Meffre, S.; Mcphie, J.; Lai, C.K. Geochemistry and geochronology of epithermal Au-hosted Chatree volcanic sequence: Implication for tectonic setting of the Loei Fold Belt in central Thailand. *Gondwana Res.* **2014**, *26*, 198–217. [CrossRef]
36. Qian, X.; Feng, Q.L.; Wang, Y.J.; Chonglakmani, C.; Monjai, D. Geochronological and geochemical constraints on the mafic rocks along the Luang Prabang zone: Carboniferous back-arc setting in northwest Laos. *Lithos* **2016**, *245*, 60–75. [CrossRef]
37. Qian, X.; Feng, Q.L.; Yang, W.Q.; Wang, Y.J.; Chonglakmani, C.; Monjai, D. Arc-like volcanic rocks in NW Laos: Geochronological and geochemical constraints and their tectonic implications. *J. Asian Earth Sci.* **2015**, *98*, 342–357. [CrossRef]
38. Rossignol, C.; Bourquin, S.; Poujol, M.; Hallot, E.; Dabard, M.P.; Nalpas, T. The volcaniclastic series from the Luang Prabang Basin, Laos: A witness of a triassic magmatic arc? *J. Asian Earth Sci.* **2016**, *120*, 159–183. [CrossRef]
39. Qian, X.; Feng, Q.L.; Wang, Y.J.; Yang, W.Q.; Chonglakmani, C.; Monjai, D. Petrochemistry and tectonic setting of the Middle Triassic arc-like volcanic rocks in the Sayabouli Area, NW Laos. *J. Earth Sci.* **2016**, *27*, 365–377. [CrossRef]
40. Blanchard, S.; Rossignol, C.; Bourquin, S.; Dabard, M.P.; Hallot, E.; Nalpas, T. Late Triassic volcanic activity in South-East Asia: New stratigraphical, geochronological and paleontological evidence from the Luang Prabang Basin (Laos). *J. Asian Earth Sci.* **2013**, *70–71*, 8–26. [CrossRef]
41. Kamvong, T.; Zaw, K.; Meffre, S.; Maas, R.; Stein, H.; Lai, C.K. Adakites in the Truong Son and Loei fold belts, Thailand and Laos: Genesis and implications for geodynamics and metallogeny. *Gondwana Res.* **2014**, *26*, 165–184. [CrossRef]
42. Kamvong, T.; Zaw, K. The origin and evolution of skarn-forming fluids from the Phu Lon deposit, northern Loei Fold Belt, Thailand: Evidence from fluid inclusion and sulfur isotope studies. *J. Asian Earth Sci.* **2009**, *34*, 624–633. [CrossRef]
43. Zaw, K.; Kamvong, T.; Khositanont, S.; Mernagh, T.P. Oxidized vs. reduced Cu–Au skarn formation and implication for exploration, Loei and Truong Son fold belts, SE Asia. In Proceedings of the International Conference on Geology, Geotechnology and Mineral Resources of Indochina (GEOINDO 2011), Khon Kean, Thailand, 1–3 December 2011; pp. 97–100.
44. Zaw, K.; Meffre, S.; Lai, C.K.; Burrett, C.; Santosh, M.; Graham, I.; Manaka, T.; Salam, A.; Kamvong, T.; Cromie, P. Tectonics and metallogeny of mainland Southeast Asia—A review and contribution. *Gondwana Res.* **2014**, *26*, 5–30. [CrossRef]
45. Tangwattananukul, L.; Ishiyama, D.; Matsubaya, O.; Mizuta, T.; Charusiri, P. Gold mineralization of Q prospect at Chatree deposit, central Thailand. *NMCC Annu. Rep.* **2009**, *16*, 70–75.
46. Qiu, K.F.; Taylor, R.D.; Song, Y.H.; Yu, H.C.; Song, K.R.; Li, N. Geologic and geochemical insights into the formation of the Taiyangshan porphyry copper–molybdenum deposit, Western Qinling Orogenic Belt, China. *Gondwana Res.* **2016**, *35*, 40–58. [CrossRef]
47. Qiu, K.F.; Marsh, E.; Yu, H.C.; Pfaff, K.; Gulbransen, C.; Gou, Z.Y.; Li, N. Fluid and metal sources of the Wenquan porphyry molybdenum deposit, Western Qinling, NW China. *Ore Geol. Rev.* **2017**, *86*, 459–473. [CrossRef]

48. Liu, Y.S.; Hu, Z.C.; Gao, S.; Günthe, D.; Xu, J.; Gao, C.G.; Chen, H.H. In situ analysis of major and trace elements of anhydrous minerals by LA-ICP-MS without applying an internal standard. *Chem. Geol.* **2008**, *257*, 34–43. [CrossRef]

49. Liu, H.B.; Jin, G.S.; Li, J.J.; Han, J.; Zhang, J.F.; Zhang, J.; Zhong, F.W.; Guo, D.Q. Determination of stable isotope composition in uranium geological samples. *World Nucl. Geosci.* **2013**, *3*, 174–179. (In Chinese with English Abstract).

50. Sun, S.S.; McDonough, W.F. Chemical and isotopic systematics of oceanic basalts: Implications for mantle composition and processes. *Geol. Soc. Lond. Spec. Publ.* **1989**, *42*, 313–345. [CrossRef]

51. Schwinn, G.; Markl, G. REE systematics in hydrothermal fluorite. *Chem. Geol.* **2005**, *216*, 225–248. [CrossRef]

52. Bau, M.; Dulski, P. Comparative study of yttrium and rare-earth element behavior in fluorine-rich hydrothermal fluids. *Contrib. Mineral. Petrol.* **1995**, *119*, 213–223. [CrossRef]

53. Shuang, Y.; Bi, X.W.; Hu, R.Z.; Peng, J.T.; Li, Z.L.; Li, X.M.; Yuan, S.D.; Qi, Y.Q. REE geochemistry of hydrothermal calcite from tin-polymetallic deposit and its indication of source of hydrothermal ore-forming fluid. *J. Mineral. Petrol.* **2006**, *26*, 57–65. (In Chinese with English Abstract).

54. Möller, P.; Parekh, P.P.; Schneider, H.J. The application of Tb/Ca–Tb/La abundance ratios to problems of fluorspar genesis. *Miner. Depos.* **1976**, *11*, 111–116. [CrossRef]

55. Morgan, J.W.; Wandless, G.A. Rare earth elements in some hydrothermal minerals: Evidence for crystallographic control. *Geochim. Cosmochim. Acta* **1980**, *44*, 973–980. [CrossRef]

56. Michard, A. Rare earth element systematics in hydrothermal fluids. *Geochim. Cosmochim. Acta* **1989**, *53*, 745–750. [CrossRef]

57. Bau, M.; Möller, P. Rare earth element fractionation in metamorphogenic hydrothermal calcite, magnesite and siderite. *Mineral. Petrol.* **1992**, *45*, 231–246. [CrossRef]

58. Bau, M. Rare-earth element mobility during hydrothermal and metamorphic fluid-rock interaction and the significance of the oxidation state of europium. *Chem. Geol.* **1991**, *93*, 219–230. [CrossRef]

59. Subías, I.; Fernández-Nieto, C. Hydrothermal events in the Valle de Tena (Spanish Western Pyrenees) as evidenced by fluid inclusions and trace-element distribution from fluorite deposits. *Chem. Geol.* **1995**, *124*, 267–282. [CrossRef]

60. Constantopoulos, J. Fluid inclusions and rare-earth element geochemistry of fluorite from south-central Idaho. *Econ. Geol.* **1988**, *88*, 626. [CrossRef]

61. Chesley, J.T.; Halliday, A.N.; Scrivener, R.C. Samarium-Neodymium Direct of Fluorite. *Science* **1991**, *252*, 949–951. [CrossRef] [PubMed]

62. Ghaderi, M.; Palin, J.M.; Campbell, I.H.; Sylvester, P.J. Rare earth element systematics in scheelite from hydrothermal gold deposits in the Kalgoorlie-Norseman region, Western Australia. *Econ. Geol.* **1999**, *94*, 423–437. [CrossRef]

63. Xu, C.; Taylor, R.N.; Li, W.; Kynicky, J.; Chakhmouradian, A.R.; Song, W. Comparison of fluorite geochemistry from REE deposits in the Panxi region and Bayan Obo, China. *J. Asian Earth Sci.* **2012**, *57*, 76–89. [CrossRef]

64. Pei, Q.M.; Zhang, S.T.; Santosh, M.; Cao, H.W.; Zhang, W.; Hu, X.K.; Wang, L. Geochronology, geochemistry, fluid inclusion and C, O and Hf isotope compositions of the Shuitou fluorite deposit, Inner Mongolia, China. *Ore Geol. Rev.* **2017**, *83*, 174–190. [CrossRef]

65. Möller, P.; Morteani, G. On the chemical fractionation of REE during the formation of Ca-minerals and its application to problems of the genesis of ore deposits. In *The Significance of Trace Elements in Solving Petrogenetic Problems*; Augustithis, S., Ed.; Theophrastus Publications: Athens, Greece, 1983; pp. 747–791.

66. Faure, G. *Principles of Isotope Geology*, 2nd ed.; Wiley: New York, NY, USA, 1986; pp. 1–589.

67. Ohmoto, H. Systematics of sulfur and carbon isotopes in hydrothermal ore deposits. *Econ. Geol.* **1972**, *67*, 551–578. [CrossRef]

68. Liu, J.M.; Liu, J.J.; Zheng, M.H.; Gu, X.X. Stable isotope compositions of micro-disseminated gold and genetic discussion. *Geochimica* **1988**, *27*, 585–591.

69. Assadzadeh, G.E.; Samson, I.M.; Gagnon, J.E. The trace element chemistry and cathodoluminescence characteristics of fluorite in the Mount Pleasant Sn–W–Mo deposits: Insights into fluid character and implications for exploration. *J. Geochem. Explor.* **2017**, *172*, 1–19. [CrossRef]

70. Seward, T.M. Thio complexes of gold and the transport of gold in hydrothermal ore solutions. *Geochim. Cosmochim. Acta* **1973**, *37*, 379–399. [CrossRef]

71. Hayashi, K.I.; Ohmoto, H. Solubility of gold in NaCl- and H2S-bearing aqueous solutions at 250–350 °C. *Geochim. Cosmochim. Acta* **1991**, *55*, 2111–2126. [CrossRef]

72. Gammons, C.H.; Williams-Jones, A.E.; Yu, Y. New data on the stability of gold (I) chloride complexes at 300 °C. *Mineral. Mag. A* **1994**, *58*, 309–310. [CrossRef]

73. Benning, L.G.; Seward, T.M. Hydrosulphide complexing of gold (I) in hydrothermal solutions from 150 to 500 °C and 500 to 1500 bars. *Geochim. Cosmochim. Acta* **1996**, *60*, 1849–1871. [CrossRef]

74. Pokrovski, G.S.; Tagirov, B.R.; Schott, J.; Hazemann, J.L.; Proux, O. A new view on gold speciation in sulfur-bearing hydrothermal fluids from in situ X-ray absorption spectroscopy and quantum-chemical modeling. *Geochim. Cosmochim. Acta* **2009**, *73*, 5406–5427. [CrossRef]

75. Cox, S.F.; Sun, S.S.; Etheridge, M.A.; Wall, V.J.; Potter, T.F. Structural and geochemical controls on the development of turbidite-hosted gold quartz vein deposits, Wattle Gully mine, central Victoria, Australia. *Econ. Geol.* **1995**, *90*, 1722–1746. [CrossRef]

76. Mikucki, E.J. Hydrothermal transport and depositional processes in Archean lode-gold systems: A review. *Ore Geol. Rev.* **1998**, *13*, 307–321. [CrossRef]

77. Williams-Jones, A.E.; Bowell, R.J.; Migdisov, A.A. Gold in solution. *Elements* **2009**, *5*, 281–287. [CrossRef]

78. Guo, L.N.; Zhang, C.; Song, Y.Z.; Chen, B.H.; Zhou, Z.; Zhang, B.L.; Xu, X.L.; Wang, Y.W. Hydrogen and oxygen isotopes geochemistry of the Wang'ershan gold deposit, Jiaodong. *Acta Petrol. Sin.* **2014**, *30*, 2481–2494. (In Chinese with English Abstract).

79. Guo, L.N.; Goldfarb, R.J.; Wang, Z.L.; Li, R.H.; Chen, B.H.; Li, J.L. A comparison of Jiaojia-and Linglong-type gold deposit ore-forming fluids: Do they differ? *Ore Geol. Rev.* **2017**, *88*, 511–533. [CrossRef]

80. Yang, L.Q.; Deng, J.; Guo, L.N.; Wang, Z.L.; Li, X.Z.; Li, J.L. Origin and evolution of ore fluid, and gold deposition processes at the giant Taishang gold deposit, Jiaodong Peninsula, eastern China. *Ore Geol. Rev.* **2016**, *72*, 585–602. [CrossRef]

81. Muntean, J.L.; Cline, J.S.; Simon, A.C.; Longo, A.A. Magmatic-hydrothermal origin of Nevada's Carlin-type gold deposits. *Nat. Geosci.* **2011**, *4*, 122–127. [CrossRef]

82. Makoundi, C.; Zaw, K.; Large, R.R.; Meffre, S.; Lai, C.K.; Hoe, T.G. Geology, geochemistry and metallogenesis of the Selinsing gold deposit, central Malaysia. *Gondwana Res.* **2014**, *26*, 241–261. [CrossRef]

83. Deng, J.; Wang, Q.F.; Li, G.J.; Zhao, Y. Structural control and genesis of the Oligocene Zhenyuan orogenic gold deposit, SW China. *Ore Geol. Rev.* **2015**, *65*, 42–54. [CrossRef]

84. Deng, J.; Yang, L.Q.; Li, R.H.; Groves, D.I.; Santosh, M.; Wang, Z.L.; Sai, S.X.; Wang, S.R. Regional structural control on the distribution of world-class gold deposits: An overview from the giant Jiaodong Gold Province, China. *Geol. J.* **2018**, 1–14. [CrossRef]

85. Deng, J.; Wang, C.M.; Bagas, L.; Santosh, M.; Yao, E. Crustal architecture and metallogenesis in the south-eastern North China Craton. *Earth-Sci. Rev.* **2018**, *182*, 251–272. [CrossRef]

86. Gebre-Mariam, M.; Groves, D.I.; Mcnaughton, N.J.; Mikucki, E.J.; Vearncombe, J.R. Archaean Au–Ag mineralisation at Racetrack, near Kalgoorlie, Western Australia: A high crustal-level expression of the Archaean composite lode-gold system. *Miner. Depos.* **1993**, *28*, 375–387. [CrossRef]

87. Groves, D.I.; Goldfarb, R.J.; Gebre-Mariam, M.; Hagemann, S.G.; Robert, F. Orogenic gold deposits: A proposed classification in the context of their crustal distribution and relationship to other gold deposit types. *Ore Geol. Rev.* **1998**, *13*, 7–27. [CrossRef]

88. Yang, L.Q.; Deng, J.; Wang, Z.L.; Guo, L.N.; Li, R.H.; Groves, D.I.; Danyushevskiy, L.; Zhang, C.; Zheng, X.L.; Zhao, H. Relationships between gold and pyrite at the Xincheng gold deposit, Jiaodong Peninsula, China: Implications for gold source and deposition in a brittle epizonal environment. *Econ. Geol.* **2016**, *111*, 105–126. [CrossRef]

89. Yang, L.Q.; Guo, L.N.; Wang, Z.L.; Zhao, R.X.; Song, M.C.; Zheng, X.L. Timing and mechanism of gold mineralization at the Wang'ershan gold deposit, Jiaodong Peninsula, eastern China. *Ore Geol. Rev.* **2017**, *88*, 491–510. [CrossRef]

 *minerals* 

Article

Controls on the Distribution of Invisible and Visible Gold in the Orogenic Gold Deposits of the Yangshan Gold Belt, West Qinling Orogen, China

Nan Li [1], Jun Deng [1,*], David I. Groves [1,2] and Ri Han [1,3,4,5]

[1]   State Key Laboratory of Geological Processes and Mineral Resources, China University of Geosciences, Beijing 100083, China; sutoby@126.com (N.L.); di_groves@hotmail.com (D.I.G.); hanri@mail.iggcas.ac.cn (R.H.)
[2]   Centre for Exploration Targeting, University of Western Australia, Crawley 6009 WA, Australia
[3]   Key Laboratory of Mineral Resources, Institute of Geology and Geophysics, Chinese Academy of Sciences, Beijing 100029, China
[4]   Institutions of Earth Science, Chinese Academy of Sciences, Beijing 100864, China
[5]   University of Chinese Academy of Sciences, Beijing 100049, China
*   Correspondence: djun@cugb.edu.cn; Tel.: +86-(010)-82322301

Received: 11 December 2018; Accepted: 1 February 2019; Published: 4 February 2019

Abstract: Six orogenic gold deposits constitute the Yangshan gold belt in the West Qinling Orogen. Gold is mostly invisible in solid solution or in the sulfide lattice, with minor visible gold associated with stibnite and in quartz-calcite veins. Detailed textural and trace-element analysis of sulfides in terms of a newly-erected paragenetic sequence for these deposits, together with previously published data, demonstrate that early magmatic-hydrothermal pyrite in granitic dike host-rocks has much higher Au contents than diagenetic pyrite in metasedimentary host rocks, but lower contents of As, Au, and Cu than ore-stage pyrite. Combined with sulfur isotope data, replacement textures in the gold ores indicate that the auriferous ore-fluids post-dated the granitic dikes and were not magmatic-hydrothermal in origin. There is a strong correlation between the relative activities of S and As and their total abundances in the ore fluid and the siting of gold in the Yangshan gold ores. Mass balance calculations indicate that there is no necessity to invoke remobilization processes to explain the occurrence of gold in the ores. The only exception is the Py_{1-2} replacement of Py_{1m}, where fluid-mediated coupled dissolution-reprecipitation reactions may have occurred to exchange Au between the two pyrite phases.

Keywords: invisible gold; visible gold; LA-ICP-MS; Yangshan gold belt; West Qinling

---

1. Introduction

Gold commonly has a complex distribution in gold deposits [1,2]. The gold that occurs in the structure of common sulfide minerals, or as discrete inclusions smaller than 100 nm, is referred to as 'invisible gold' [3]. Worldwide, pyrite and arsenopyrite are the most common hosts for invisible gold in gold deposits [4–6]. The other most common occurrence of gold is in the form of visible native gold, normally as distinct inclusions or filling fractures within these sulfide minerals [7–12], but also less commonly as nuggety gold in quartz veins. In orogenic gold systems, invisible gold within sulfides is normally followed by later visible gold [13–16]. Some authors have suggested that the visible native gold was remobilized from the invisible gold in sulfides [7,14,17–20], but it is unclear how gold could be first precipitated in pyrite and then become soluble in that pyrite, without transformation to pyrrhotite, and subsequently be re-precipitated from essentially the same ore fluid just a short distance away in the same ore zone [20]. In addition to this uncertainty, factors controlling the distribution of invisible and visible gold are still unclear, with metamorphic grade; arsenic concentrations; and fluid temperature, pH, and oxidation or sulfidation state all interpreted to play important roles [14,21,22].

LA-ICP-MS is now a preferred technique for trace-element analysis of sulfides and other ore minerals, offering high sensitivities and an excellent spatial resolution [23–25]. Detailed interpretation of pyrite textures is critical in terms of interpreting their trace-element compositions [26–32]. Moreover, the correlation between the distribution of invisible/visible gold and the textures of sulfides and related ore assemblages has implications for defining ore paragenesis and the mechanisms of incorporation and/or release of invisible gold from the sulfides [24].

The West Qinling Orogen is the third largest gold province in China, with a total of >1100 t gold resources and >50 gold deposits [33,34]. Yangshan is a world-class gold belt in the southern part of the West Qinling Orogen and comprises six gold deposits, including the Anba (280 t gold), Getiaowan, Guanyinba, Gaoloushan, Nishan, and Zhangjiashan deposits. These deposits share many common features. They are all structurally-controlled, orogenic gold deposits [35] in metasedimentary rocks and granitic dikes, and have similar alteration mineralogies of quartz, sericite, and calcite and ore mineralogies of auriferous arsenian pyrite and arsenopyrite, with stibnite and visible gold. The paragenesis and geochemistry of ore minerals in the gold deposits of the Yangshan gold belt have been studied previously [36,37]. However, backscattered images (BSE) show that these previous studies ignored: (1) partial dissolution textures in the early pyrite; (2) the euhedral pyrite rims associated with the replacement of early pyrite cores by main ore-stage pyrite; and (3) the complex textures of two different pyrite generations within the main ore-stage pyrite, which can constrain the mechanisms of Au incorporation in, or release of Au from, the pyrite [17,24,32]. This study presents detailed textural and trace-element analysis of sulfides in terms of a newly erected paragenetic sequence, designed together with previously published data, to investigate the factors controlling the distribution of invisible and visible gold in the Yangshan ores to further improve genetic understanding of the deposits in the gold belt.

2. Geological Background

2.1. Regional Geology

The West Qinling Orogen is the western part of the Qinling Orogen Belt, which is viewed as the product of long-lived accretion and collision between the North China Craton, North Qinling Block, South Qinling Block, and South China Craton. The Kuanping, Shangdan, and Mianlue suture zones, respectively, separate the above continental blocks [38–41] (Figure 1).

Proterozoic metavolcanic-sedimentary formations (846–776 Ma), the oldest strata in the region, are located in the southeast of the West Qinling Orogen [42]. Cambrian, Silurian, Devonian, Carboniferous, Permian, Triassic, and Cretaceous strata are widespread, with Devonian and Triassic strata being important host rocks for gold mineralization (Figure 1) [43–45]. Mesozoic granitoids are widespread in the West Qinling Orogen (Figure 1). They mainly occur between the Shangdan and Mianlue suture zones or within the Mianlue suture zone and have zircon U-Pb ages of 245–185 Ma [46].

The Yangshan gold belt, along with other gold districts, is located in the Mianlue suture zone within the West Qinling Orogen and formed during post-collisional tectonism between the North and South China Craton (Figure 1) [37,47]. Large igneous intrusions are absent in the gold belt. The ~215 Ma Puziba dikes, comprising granite, aplite, and porphyry dikes, were sheared into lenses along the E-W-trending Anchanghe-Guanyinba Fault, which is a secondary fault of the Mianlue suture zone [46].

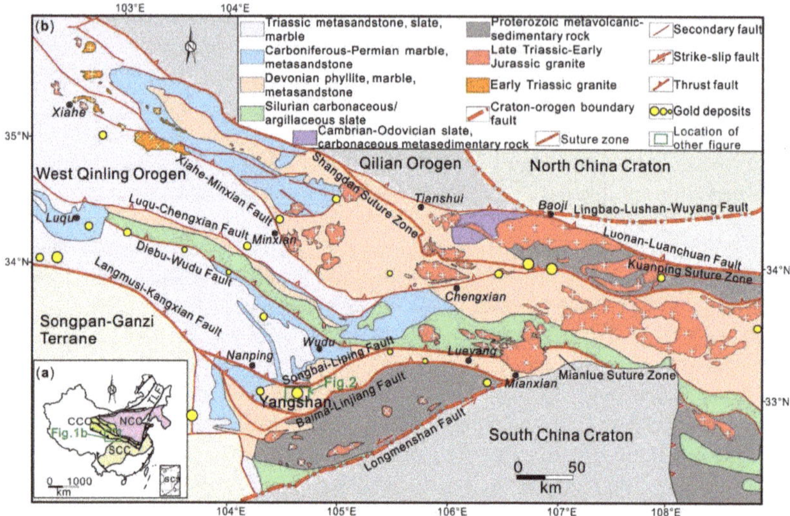

Figure 1. Simplified geological map of the West Qinling Orogen, showing the distribution of major crustal blocks, fault systems, and Mesozoic granitoids [47–49]. CCO, Central China Orogen; M, Mianlue; NCB, North China Block; SCB, South China Block; SCS, South China Sea; TLF, Tancheng–Lujiang Fault. (**a**): Tectonic overviews showing the position of West Qinling Orogen: (**b**): Schematic map showing the tectonic setting and distribution of gold deposits in West Qinling.

2.2. Deposit Geology

The deposits of the Yangshan gold belt are mainly hosted by Devonian phyllite and Triassic granitic dikes. Among the six gold deposits in the Yangshan gold belt, the Anba gold deposit hosts more than 90% of the gold resources in the district. However, all the deposits in the gold belt have similar geological and mineralogical characteristics [50,51].

Ore bodies in the gold belt are structurally controlled by the three splays (F1, F2, and F3) of the Anchanghe-Guanyinba Fault (Figure 2). Some of the ore bodies are also controlled by competency contrasts along contact zones between granitic dikes and phyllite (Figure 3a) [37].

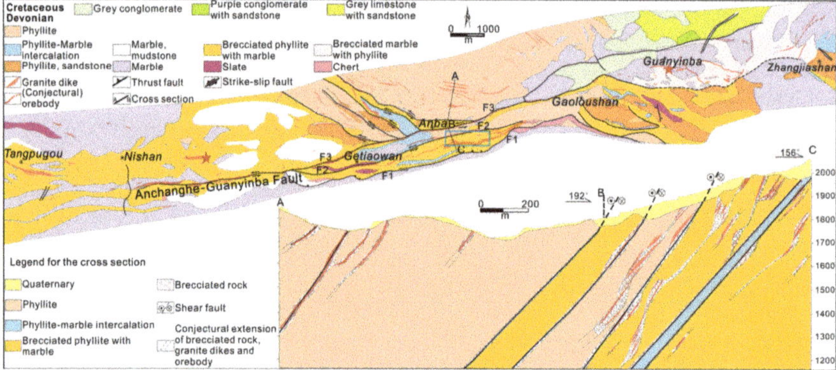

Figure 2. Geological map and cross-section of the Yangshan gold belt. Red stars and blue box are the sampling locations [37].

Figure 3. Ore styles and ore minerals from the Anba gold deposit. (**a**) The fault contact between granite porphyry dike and phyllite. (**b**) Pyrite is developed along the fault contact between granite porphyry and phyllilte. (**c**) Pyrite in phyllite and quartz vein. (**d**) Pyrite in granite porphyry. (**e**) Silicification, sericitization, and pyrite mineralization in granite porphyry. (**f**) Arsenopyrite coexists with the outer rim of pyrite. Apy, arsenopyrite; Cal, calcite; Py, pyrite; Qz, quartz; Ser, sericite.

Major ore minerals in the gold belt include pyrite, arsenopyrite, and stibnite. Disseminated ores in phyllite or granitic dikes are the dominant ore type, with minor pyrite-arsenopyrite quartz veins, stibnite-bearing quartz veinlets, and quartz-calcite veins (Figure 3b–f) [36,37,52]. The diagenetic pyrite (Py_0) has a framboidal or colloform texture and is disseminated in the metasedimentary host rocks. Hydrothermal mineralization is divided into an early ore-stage (Py_1–quartz), a main ore-stage (Py_2–Apy_2–quartz–sericite), and a late ore-stage (Py_3–Apy_3–stibnite–native gold–calcite–quartz). Pyrite is rare in the post-ore-stage quartz–calcite veins [37]. The alteration developed in the deposits is dominated by silicification, carbonation, sulfidation, and sericitization (Figure 3e). Gold is mostly invisible in solid solution or in the sulfide lattice, with minor visible gold associated with stibnite [36].

3. Analytical Techniques

Determination of mineral paragenesis was complemented by petrological work under transmitted and reflected light, and SEM. Based on careful examination of the mineral paragenesis, a suite of eight samples of pyrite from the granitic dikes (Table 1) was selected for LA-ICP-MS analyses at the U.S. Geological Survey, Denver, USA. Six samples are from the Anba gold deposit, and one sample each from the Guanyinba and Nishan gold deposits (Table 1), which are located to the east and west of Anba, respectively (Figure 2). These three deposits are considered representative of the Yangshan gold belt. Grey-green equigranular to porphyritic granitic dikes include granite, plagioclase granite,

granite porphyry, and plagioclase granite porphyry, which are common to all three deposits. The primary minerals in the dikes are plagioclase, quartz, and biotite, which are commonly altered to sericite, chlorite, epidote, and clays.

Table 1. Descriptions of the samples analysed in this study. Samples listed in a hierarchical manner under Deposit and then Sample Number.

Deposit	Sample Number	Lithology	Pyrite Stage
Anba	AB10PD4-102	Granite porphyry dike with quartz-calcite vein	Py_{1m}
Anba	AB10PB4-108_2	Granite porphyry dike with quartz vein	Py_{2a}, Py_{2b}
Anba	SM1-3	Granite	Py_{1m}, Py_{1-2}, Py_{2a}, Py_{2b}
Anba	YS-AB-10-PD1-01	Plagioclase granite dike with quartz-calcite veins	Py_{1m}, Py_{2b}
Anba	YS-AB-10-PD1-03	Granite dike in fault zone with quartz vein	Py_{1m}, Py_{1-2}, Py_{2a}, Py_{2b}
Anba	YS-AB-10-PD1-04	Plagioclase granite porphyry dike with quartz vein	Py_{1m}, Py_{2b}
Guanyinba	ZK054-4	Granite	Py_{2a}, Py_{2b}
Nishan	YS-NS-10-05	Plagioclase granite porphyry dike	Py_{2b}

The LA-ICP-MS analyses on pyrite were conducted on a Photon Machines Analyte G2 193 nm laser ablation system attached to a Perkin Elmer ELAN DRC-e ICP-MS. Laser ablation ICP-MS methods for pyrite are based on [26,53]. Both 15- and 30-micron diameter laser spots were used for the analyses. A laser fluence of 2 J/cm^2 and frequency of 5 Hz were used for the analyzed spots. Helium was used as a carrier gas. Calibration was conducted using the USGS MASS-1 sulfide reference material run five to ten times at the beginning of each session, following the procedures of [54] and using Fe as the internal standard element [26]. Concentration calculations were carried out using off-line data processing following the equations of [54]. The MASS-1 reference material was run periodically to monitor for drift. During these analytical sessions, drift was less than five percent for all elements. A stoichiometric value of 46% Fe was used for the LA-ICP-MS concentration calculations. Detection limits were calculated as three times the standard deviation of the blank [54] and are shown in Table S1. Data were examined for the presence of mineral inclusions or zoning recorded in the time resolved spectra as deviations from a stable signal (e.g., [26,53]). Because some data are below the detection limits of the analytical techniques, corrections are necessary for the incorporation into the statistical method. Data qualified with a "less than" value were replaced with 0.7 times the detection limit.

4. Textures of Pyrite and Mineral Paragenesis

Core-rim zonation is common in pyrite from deposits in the Yangshan gold belt, with the earlier-formed Py_0 and Py_1 as the core, and the main ore-stage Py_2 as the rim of the pyrite [36,37]. In this study, a more detailed analysis of pyrite textures improves paragenetic understanding of the deposits from the Yangshan gold belt.

4.1. Partial Dissolution Textures of Py_{1m}

To distinguish different Py_1 cores, the Py_1 hosted by the granitic dikes is named Py_{1m} and Py_1 from metasedimentary rocks is named Py_{1s}. Besides the normal core-rim texture formed by Py_{1m} and Py_2, the partial dissolution texture of Py_{1m} is recognized in this study. The inner Py_{1m} core has been partially replaced and overgrown by a Py_2 rim and formed a new pyrite generation (Py_{1-2}), which has a brighter response in BSE images than Py_{1m} and is therefore distinguishable from Py_{1m} and Py_2 in such images. Furthermore, Py_{1-2} retains the shape of Py_{1m} through pseudomorphic replacement (Figure 4a–h).

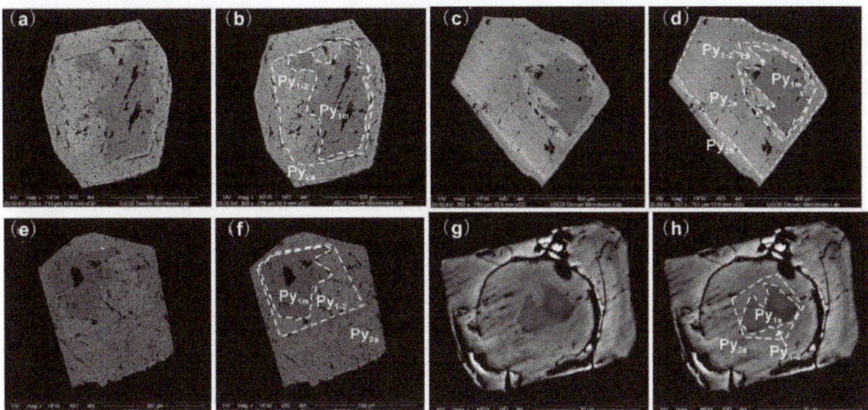

Figure 4. BSE images showing the pseudomorphous replacement textures of Py_1 and Py_2. (**a,b**): Sample number SM1-3-2; (**c,d**): Sample number SM1-3-3; (**e,f**): Sample number SM1-3-4; (**g,h**): Sample number SM4-2-25-2.

4.2. Complex Textures within Py2

Within the traditional main ore-stage Py_2, an irregular bright pyrite band (a few micrometres wide) with As enrichment is recognized in BSE images. The pyrite within the bright band is named Py_{2a} and the pyrite exterior to it is named Py_{2b} (Figure 4c,d; Figure 5a–h). Fine-grained arsenopyrite is located around the bright band, whereas relatively coarse-grained prismatic arsenopyrite is paragenetically associated with Py_{2b} (Figure 5a–h).

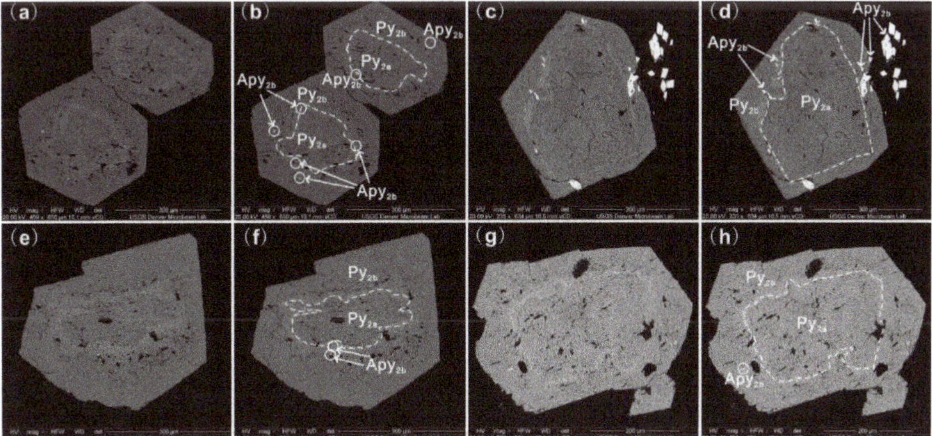

Figure 5. BSE images showing the bright band and replacement textures of Py_{2a} and Py_{2b}. (**a,b**): Sample number AB10PD4-108(2)-5; (**c,d**): Sample number YS-AB-10-PD1-03-4; (**e,f**): Sample number AB10PD4-108(2)-11; (**g,h**): Sample number AB10PD4-108(2)-12.

4.3. Mineral Paragenesis

Recognition of the pseudomorphic replacement texture between Py_1 and Py_2, and the bright band with arsenopyrite enrichment within Py_2, leads to the improvement of the paragenetic sequence diagram of the Yangshan gold belt (Figure 6). The multiple generations of pyrite from early diagenesis through the peak- to post- ore-stage, as well as trace-element evolution of the sulfides during the pre- to main- ore-stage described below, indicate more complicated ore-forming processes than were previously supposed to exist in the Yangshan gold belt.

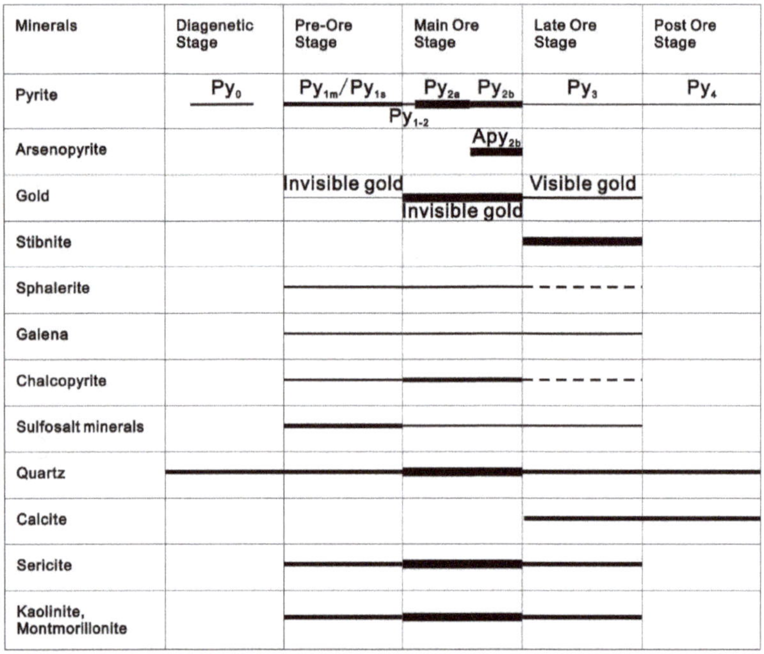

Minerals	Diagenetic Stage	Pre-Ore Stage	Main Ore Stage	Late Ore Stage	Post Ore Stage
Pyrite	Py_0	Py_{1m}/Py_{1a} Py_{1-2}	Py_{2a} Py_{2b}	Py_3	Py_4
Arsenopyrite			Apy_{2b}		
Gold		Invisible gold	Invisible gold	Visible gold	
Stibnite					
Sphalerite					
Galena					
Chalcopyrite					
Sulfosalt minerals					
Quartz					
Calcite					
Sericite					
Kaolinite, Montmorillonite					

Figure 6. Paragenetic sequence of gold mineralization and alteration of the gold deposits of the Yangshan gold belt. The bold lines indicate high abundance, the thin lines represent the minor amounts, and the discontinuous lines indicate uncertainty in the determination of the paragenetic sequence due to the lack of a clear textural relationship [35].

5. Trace Element Compositions of Pyrite and Arsenopyrite

As there is only minor pyrite in the diagenetic, late-, and post-ore stages, and their geochemistry is summarized by [36], the geochemistry of newly recognized Py_{1m}, Py_{1-2}, Py_{2a}, and Py_{2b} from the pre- and main-ore stage mineralization is emphasized in this paper (Table 1). A total of 119 LA-ICP-MS spot analyses were completed on various pyrite generations from Yangshan. A total of nine elements were analysed and interpreted in this study: Ag, As, Au, Bi, Co, Cu, Ni, Pb, and Sb. The pyrite results, together with Apy_{2b} and stibnite data from [37], are given in Table S1 and elemental variation is presented in Figure 7. Representative pyrite textures and geochemistry within single pyrite grains are shown in Figures 8 and 9.

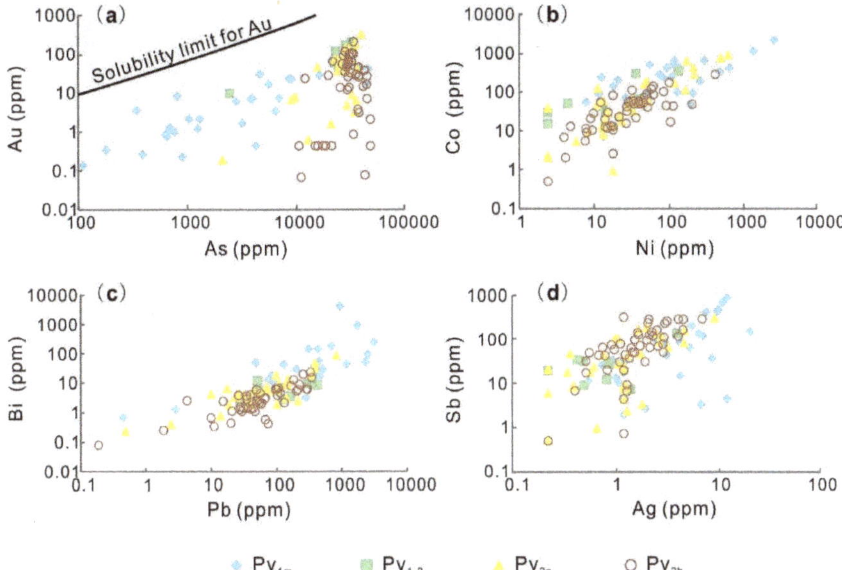

Figure 7. Trace-element binary plots of Au against (**a**) As, (**b**) Ni, (**c**) Pb, and (**d**) Ag for Yangshan pyrites. The line in (**a**) is the inferred solubility limit of gold in arsenian pyrite [21]. The trace-element concentrations are from Table S1, or calculated using 0.7 times mdl (minimum detection limit) values for all measurements <mdl.

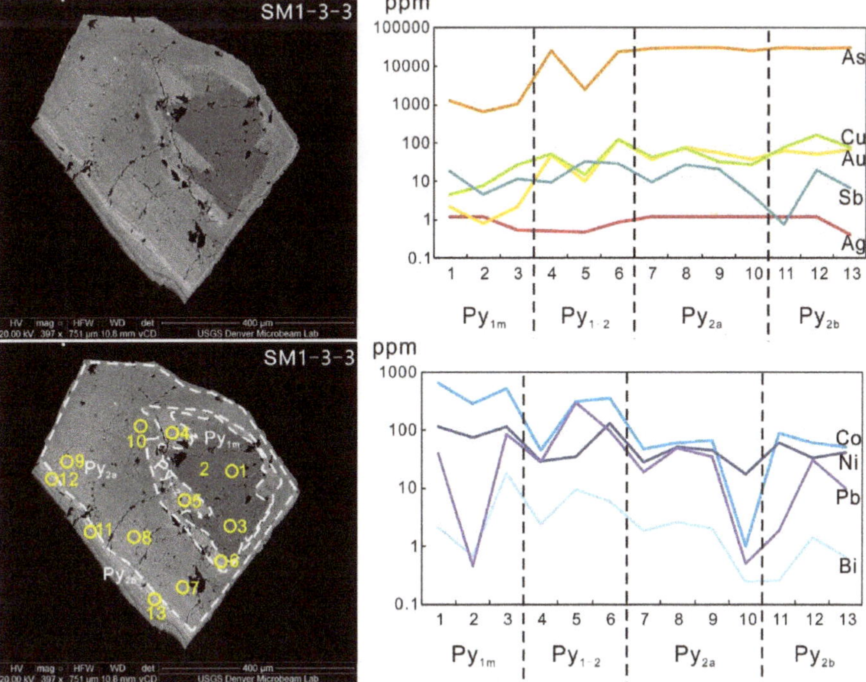

Figure 8. The core-rim texture of pyrite in sample SM1-3-3 and the trace-element contents analysed by LA-ICP-MS within each zone.

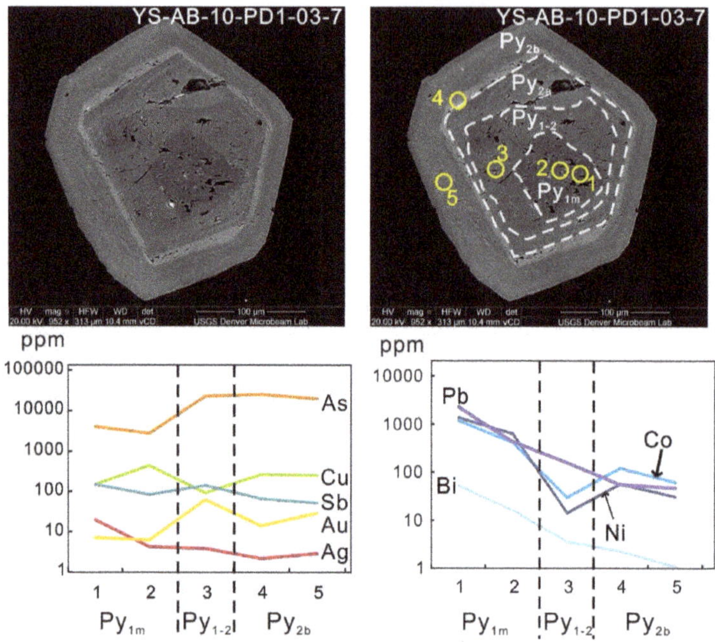

Figure 9. The core-rim texture of pyrite in sample YS-AB-10-PD1-03-7 and the trace-element contents analysed by LA-ICP-MS within each zone.

The median As contents in Py_{1m}, Py_{1-2}, Py_{2a}, and Py_{2b} are 3221, 28672, 28583, and 30,868 ppm, with ranges of 109-45803, 2454-32613, 2103-39768, and 10501-48107 ppm, respectively. The median Au contents in Py_{1m}, Py_{1-2}, Py_{2a}, and Py_{2b} are 3.7, 71, 56, and 28.2 ppm, with ranges of <0.39–54, 9.88–207, 0.2–337, and <0.64–123 ppm, respectively (Table S1). Although there is considerable variation, the As and Au contents in Py_{1-2}, Py_{2a}, and Py_{2b} are higher than Py_{1m} (Figures 7a and 10; Table S1), as also shown by the As and Au contents of single pyrite grains (Figures 8 and 9). Py_{1-2} has the highest Au content compared with Py_{1m}, Py_{2a}, and Py_{2b} (Figures 9 and 10). There is a decreasing trend of Au contents from Py_{1-2}, through Py_{2a}, to Py_{2b} (Figure 10; Table S1). Py_{2b} has the highest As, Cu (median 347 ppm), and Sb (median 67.5 ppm) contents compared to Py_{1m}, Py_{1-2}, and Py_{2a} (Figure 10; Table S1).

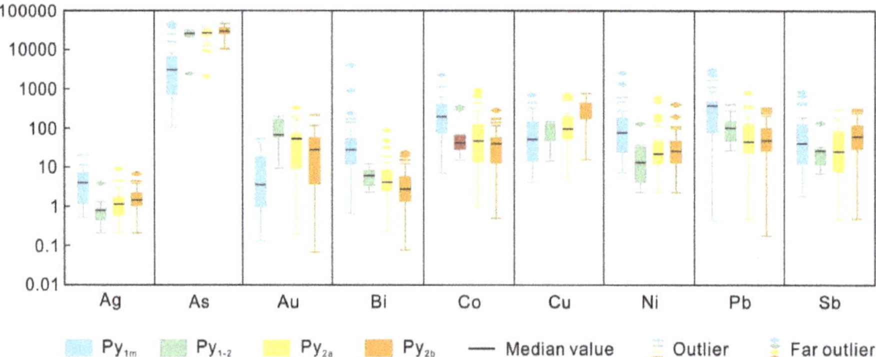

Figure 10. Box plot of trace-element concentrations of Yangshan pyrite. Boxes represent interquartile range (data between 25th and 75th percentiles), with top and bottom lines extending 1.5 times the interquartile range toward the maximum and minimum, respectively.

The median Cu contents in Py_{1m}, Py_{1-2}, Py_{2a}, and Py_{2b} are 55.2, 91.8, 103, and 347 ppm, with ranges of <13.9–708, 15.2–159, 5.43–730, and 16.5–826 ppm, respectively (Table S1). Generally, Cu contents increase from early Py_{1m}, through Py_{1-2} and Py_{2a}, to Py_{2b}.

The median Pb contents in Py_{1m}, Py_{1-2}, Py_{2a}, and Py_{2b} are 401, 108, 48.4, and 53.4 ppm, with ranges of 0.46–3089, 28–418, 0.5–825, <0.27–338 ppm, respectively (Table S1). The median Bi contents in Py_{1m}, Py_{1-2}, Py_{2a}, and Py_{2b} are 28.6, 6.3, 4.3, and 3 ppm, with ranges of <0.98–4052, 2.41–12.4, 0.24–89.1, and <0.11–23.4 ppm, respectively (Table S1). Lead and Bi contents decrease from Py_{1m}, through Py_{1-2} and Py_{2a}, to Py_{2b}.

The median Co contents in Py_{1m}, Py_{1-2}, Py_{2a}, and Py_{2b} are 209, 44.3, 50.2, and 44 ppm, with ranges of 7.32–2262, 15.8–354, 0.99–947, and 2.11–291 ppm, respectively. The median Ni contents in Py_{1m}, Py_{1-2}, Py_{2a}, and Py_{2b} are 82.1, 14, 23.1, and 27.7 ppm, with ranges of <18.8–2582, <3.41–133, <3.41–606, and <3.41–410 ppm, respectively (Table S1). Cobalt and Ni contents broadly parallel Pb and Bi, but there is lower Co and Ni in Py_{1-2} (Figures 7 and 10; Table S1).

The Au-As correlation becomes weaker from early Py_{1m}, through Py_{1-2} and Py_{2a}, to Py_{2b} (Table S2). For Py_{1-2}, there is a strong positive correlation between Au and Cu, which is unique among all pyrite stages (Table S2). Cobalt and Ni contents show a positive correlation among all pyrite generations (Figure 7b; Table S2). Lead and Bi in Py_{2a} and Py_{2b} have a strong positive correlation compared with that of the same elements in Py_1 and Py_{1-2} (Figure 7c; Table S2).

As summarized in [37], the Au content of Apy_{2b} ranges from <0.046 to 1205 ppm (median 15.4 ppm, average 143 ppm), whereas the Au content of late ore-stage stibnite ranges from 0.02 to 0.42 ppm (median 0.08 ppm, average 0.12 ppm).

6. Discussion

6.1. Invisible and Visible Gold in the Gold Ores

Py_{1m} within granitic dike host-rocks has a low Au content (<0.39~54 ppm, median 3.7 ppm), which is lower than the Au values in later generations of pyrite, but is higher than the gold values in diagenetic pyrite in phyllite (median 0.32 ppm; [37]). The $\delta^{34}S$ values of pyrite cores (Py_{1m} in this study) in granitic dikes from the Yangshan gold belt have a very limited range from 0 to 1.3‰, consistent with a magmatic–hydrothermal origin [55]. Similar magmatic-hydrothermal pyrite from the Wallaby gold deposit in Western Australia has median gold values of 0.35 ppm, with individual spot data as high as 78 ppm [56]. LA-ICP-MS analysis also shows that gold contents range from <0.03 to 2.1 ppm in the magmatic-hydrothermal cores of pyrites in syenite from the Chang'an gold deposit, China [57].

The magmatic-hydrothermal Py_{1m} has lower contents of As, Au, and Cu than ore-stage pyrite, but the values are higher than the As, Au, and Cu contents in other magmatic-hydrothermal pyrites described by [56,57]. The contrast in element patterns between Py_{1m} and ore-stage pyrites, combined with their erratic, more negative $\delta^{34}S$ values [50,58], indicate that the auriferous ore-fluids were not magmatic-hydrothermal in origin. Considering the sharp contact between Py_{1m} and Py_{1-2}, and the low metamorphic grade in the region, the replacement textures clearly shown by Py_{1-2} raise the possibility that the enhanced Au values of Py_{1m} resulted from fluid-mediated coupled dissolution-reprecipitation reactions [17,20,24,32,59,60].

According to the flotation test report, the contents of pyrite, arsenopyrite, and stibnite in the primary ores are 2.02%, 0.93%, and 0.18%, respectively [61]. Based on observations under the microscope, the contents of Py_0, Py_{1s}, Py_{1m}, Py_{1-2}, Py_{2a}, Py_{2b}, Apy_{2b}, and stibnite in the ores are estimated to be 0.005%, 0.005%, 0.005%, 0.005%, 1.5%, 0.5%, 0.93%, and 0.18%, respectively. Using the average gold contents of pyrite, arsenopyrite, and stibnite from each ore stage (0.26 ppm, 0.82 ppm, 11.81 ppm, 105.98 ppm, 70.05 ppm, 36.91 ppm, 143 ppm, 0.12 ppm, for Py_0, Py_{1s}, Py_{1m}, Py_{1-2}, Py_{2a}, Py_{2b}, Apy_{2b}, and stibnite, respectively; Table S1) [37], the invisible gold contained in sulfide in one tonne of ore is calculated to be 1.3×10^{-5} g, 5.9×10^{-4} g, 4.1×10^{-5} g, 5.3×10^{-3} g, 1.05 g, 0.18 g,

1.33 g, and 2.2×10^{-4} g for Py_0, Py_{1s}, Py_{1m}, Py_{1-2}, Py_{2a}, Py_{2b}, Apy_{2b}, and stibnite, respectively. Clearly, Py_{2a}, Py_{2b}, and Apy_{2b} are the only significant contributors to the gold budget. The total invisible gold locked in the structures of pyrite, arsenopyrite, and stibnite in one tonne of ore is 2.57 g. As 83.73% of gold is located in sulfides with 16.27% as visible gold that is enclosed in other minerals [61], there is 0.50 g native gold in one tonne of ore. Therefore, there is a calculated 3.07 g gold in one tonne of ore. The calculated gold grade is lower than the reported average gold grade (4.76 g/t) [62], which may be caused by the limited number of arsenopyrite samples in this study. In previous studies of Yangshan gold deposits, arsenopyrite has been shown to have Au contents as high as 4719 ppm (average 2073 ppm) [63]. This average is about 15 times greater than that of this study. Using the average Au content in [63] would add 18 g Au to the total gold budget, which would clearly exceed the known gold grade. The true Au content of arsenopyrite must lie between that of [63] and this study, but closer to this study.

6.2. Controls on the Distribution of Invisible and Visible Gold

The replacement of Au-As-poor Py_1 by Au-As-rich Py_2 and the formation of As-Au-rich Py_{1-2} indicate that the pre-ore magmatic-hydrothermal Py_{1m} became unstable and was partially replaced during ingress of the auriferous ore fluid, with Py_{1-2} rimming the Py_{1m} cores and tending to form crystal shapes. No arsenopyrite was developed in this stage to compete with pyrite for Au distribution, leading to the high Au concentrations in Py_{1-2}, which was then overgrown by $Py_{2a,\,b}$.

The bright pyrite band between Py_{2a} and Py_{2b} in BSE images, which is the reaction front of replacement [17], has a high As content (Figure 8), indicating that the ore fluid evolved to be As-rich during the late main-ore stage. Py_{2b} has the highest As content, but the lowest Au content, of all main ore-stage pyrites, consistent with its very low Au-As correlation coefficient (Table S2). This is due to the fact that Au preferentially entered the lattice of arsenopyrite that was intimately intergrown with Py_{2b}, as also shown by [51]. Py_{2b} contains more Sb than Py_{1m}, Py_{1-2}, and Py_{2a}, which indicates Sb enrichment during the late main ore-stage. Once stibnite formed in the absence of pyrite or arsenopyrite, there was no suitable host for invisible gold and free gold was deposited in cracks or on grain surfaces of the earlier-formed sulfides, as also described by [36,37].

The physicochemical parameters of ore formation in the Yangshan gold belt are estimated as follows. Based on the micro-thermometry of primary fluid inclusion, the mineralization temperatures of the main ore-stage and late ore-stage were 288~300 °C, and 271~288 °C, respectively [51,52]. The sulfur fugacity during the main ore-stage at the Yangshan gold belt can be estimated to have been $10^{-10.3}$~$10^{-11.1}$, which is similar to the calculated sulfur fugacity ($10^{-10.4}$) based on the As contents in arsenopyrite [51]. The sulfur fugacity during the late ore-stage is estimated to have been $10^{-13.5}$~$10^{-11.1}$ (Figure 11). Temperature and sulfur fugacity are two important factors controlling the stability of stibnite [64,65]. During the main ore-stage, pyrite and arsenopyrite precipitated at temperatures between 288 and 300 °C, and the majority of gold occurred within the lattices of pyrite and arsenopyrite. The precipitation of pyrite and arsenopyrite reduced the concentration of available reduced sulfur in the ore fluid. During the late ore-stage, the temperature decreased to 271~288 °C and the sulfur fugacity to $10^{-13.5}$~$10^{-11.1}$, leading to the decrease of antimony solubility and precipitation of stibnite and native gold [64,66].

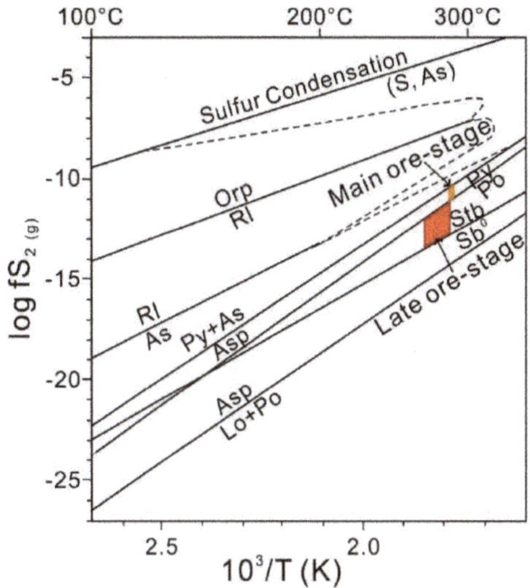

Figure 11. Temperature versus $\log f_{S_2 (g)}$ diagram showing stability fields of minerals recorded in main and late-ore stages at the Yangshan gold belt [67]. *Orange* represents the stability field of main-ore stage pyrite and arsenopyrite, and *red* represents the stability field of late ore-stage stibnite. As-arsenic, Asp-arsenopyrite, Lo-loellingite, Orp-orpiment, Po-pyrrhotite, Py-pyrite, Rl-realgar, S-sulfur, Sb^0-native antimony, Stb-stibnite.

7. Conclusions

The textural and geochemical characteristics of sulfides indicate multiple pulses of evolving hydrothermal fluids, with As becoming more abundant and S species less abundant with time until cessation of the main ore-stage. Arsenian pyrite was the dominant host for invisible gold until arsenopyrite became a common ore mineral and became the main host for invisible gold. Only once both pyrite and arsenopyrite ceased to be deposited did free gold become the dominant gold phase. There is thus a strong correlation between the relative activities of S and As and their total abundances in the ore fluid and the siting of gold in the Yangshan gold ores. Mass balance calculations indicate that there is no necessity to invoke remobilization processes to explain the occurrence of either invisible or visible gold in the ores. The only exception is for the Py_{1-2} replacement of Py_{1m}, where fluid-mediated coupled dissolution-reprecipitation reactions may have occurred to exchange Au between the two pyrite phases.

Supplementary Materials: The following are available online at http://www.mdpi.com/2075-163X/9/2/92/s1, Table S1: LA-ICP-MS elemental data for pyrite in typical gold deposits from Yangshan gold belt (in parts per million); Table S2: Correlation coefficient matrix diagram of trace-element contents of pyrite in this study.

Author Contributions: N.L. and J.D. conceived this contribution; N.L. and R.H. prepared the experiments; N.L. performed the experiments and wrote the original draft; N.L. and D.I.G. reviewed and edited the draft, with D.I.G. making several significant additions to interpretations.

Funding: This study was financially supported by the National Natural Science Foundation of China (Grant No. 41702072); the National Basic Research Program of China (Grant No. 2015CB452605); the Public Welfare Scientific Research Funding of China (Grant No. 201411048); the Geological investigation work project of the China Geological Survey (Grant No. 12120114038301); 111 Project of the Ministry of Education, China (Grant No. B07011); and the MOST Special Fund from the State Key Laboratory of Geological Processes and Mineral Resources, China University of Geosciences (Grant No. MSFGPMR201804).

Acknowledgments: We are grateful to Liqiang Yang, who made constructive comments on the manuscript. We would like to thank Xingzhong Ji, Zhichao Zhang, and Yuchen Zhao from China University of Geosciences, Beijing, for their assistance with field work. We are grateful for constructive comments by the reviewers.

Conflicts of Interest: The authors declare no conflict of interest.

References

1. Pitcairn, I.K.; Olivo, G.R.; Teagle, D.A.H.; Craw, D. Sulfide evolution during prograde metamorphism of the Otago and Alpine schists, New Zealand. *Can. Mineral.* **2010**, *48*, 1267–1295. [CrossRef]
2. Yang, L.Q.; Deng, J.; Wang, Z.L.; Guo, L.N.; Li, R.H.; Groves, D.I.; Danyushevsky, L.V.; Zhang, C.; Zheng, X.L.; Zhao, H. Relationships between gold and pyrite at the Xincheng gold deposit, Jiaodong Peninsula, China: Implications for gold source and deposition in a brittle epizonal environment. *Econ. Geol.* **2016**, *111*, 105–126. [CrossRef]
3. Cook, N.J.; Chryssoulis, S.L. Concentrations of invisible gold in the common sulfides. *Can. Mineral.* **1990**, *28*, 1–16.
4. Deng, J.; Wang, Q.F.; Li, G.J.; Zhao, Y. Structural control and genesis of the Oligocene Zhenyuan orogenic gold deposit, SW China. *Ore Geol. Rev.* **2015**, *65 Pt 1*, 42–54. [CrossRef]
5. Groves, D.I.; Goldfarb, R.J.; Robert, F.; Hart, C.J. Gold deposits in metamorphic belts: Overview of current understanding, outstanding problems, future research, and exploration significance. *Econ. Geol.* **2003**, *98*, 1–29.
6. Groves, D.I.; Santosh, M. The giant Jiaodong gold province: The key to a unified model for orogenic gold deposits? *Geosci. Front.* **2016**, *7*, 409–417. [CrossRef]
7. Velásquez, G.; Béziat, D.; Salvi, S.; Siebenaller, L.; Borisova, A.Y.; Pokrovski, G.S.; De Parseval, P. Formation and deformation of pyrite and implications for gold mineralization in the El Callao district, Venezuela. *Econ. Geol.* **2014**, *109*, 457–486. [CrossRef]
8. Yang, L.Q.; Deng, J.; Goldfarb, R.J.; Zhang, J.; Gao, B.F.; Wang, Z.L. 40Ar/39Ar geochronological constraints on the formation of the Dayingezhuang gold deposit: New implications for timing and duration of hydrothermal activity in the Jiaodong gold province, China. *Gondwana Res.* **2014**, *25*, 1469–1483. [CrossRef]
9. Deng, J.; Liu, X.F.; Wang, Q.F.; Pan, R.G. Origin of the Jiaodong-type Xinli gold deposit, Jiaodong Peninsula, China: Constraints from fluid inclusion and C–D–O–S–Sr isotope compositions. *Ore Geol. Rev.* **2015**, *65 Pt 3*, 674–686. [CrossRef]
10. Yang, L.Q.; Deng, J.; Guo, L.N.; Wang, Z.L.; Li, X.Z.; Li, J.L. Origin and evolution of ore fluid, and gold-deposition processes at the giant Taishang gold deposit, Jiaodong Peninsula, eastern China. *Ore Geol. Rev.* **2016**, *72*, 585–602. [CrossRef]
11. Yang, L.Q.; Guo, L.N.; Wang, Z.L.; Zhao, R.X.; Song, M.C.; Zheng, X.L. Timing and mechanism of gold mineralization at the Wang'ershan gold deposit, Jiaodong Peninsula, eastern China. *Ore Geol. Rev.* **2017**, *88*, 491–510. [CrossRef]
12. Deng, J.; Wang, C.M.; Bagas, L.; Carranza, E.J.M.; Lu, Y.J. Cretaceous–Cenozoic tectonic history of the Jiaojia Fault and gold mineralization in the Jiaodong Peninsula, China: Constraints from zircon U–Pb, illite K–Ar, and apatite fission track thermochronometry. *Miner. Depos.* **2015**, *50*, 987–1006. [CrossRef]
13. Mumin, A.H.; Fleet, M.E.; Chryssoulis, S.L. Gold mineralization in As-rich mesothermal gold ores of the Bogosu-Prestea mining district of the Ashanti Gold Belt, Ghana: Remobilization of "invisible" gold. *Miner. Depos.* **1994**, *29*, 445–460. [CrossRef]
14. Morey, A.A.; Tomkins, A.G.; Bierlein, F.P.; Weinberg, R.F.; Davidson, G.J. Bimodal Distribution of Gold in Pyrite and Arsenopyrite: Examples from the Archean Boorara and Bardoc Shear Systems, Yilgarn Craton, Western Australia. *Econ. Geol.* **2008**, *103*, 599–614. [CrossRef]
15. Lu, H.Z.; Zhu, X.Q.; Shan, Q.; Wang, Z.G. Hydrothermal evolution of gold-bearing pyrite and arsenopyrite from different types of gold deposits. *Miner. Depos.* **2013**, *32*, 823–842, (In Chinese with English Abstract).
16. Goldfarb, R.; Baker, T.; Dube, B.; Groves, D.I.; Hart, C.J.R.; Gosselin, P. Distribution, character and genesis of gold deposits in metamorphic terranes. In *Economic Geology 100th Anniversary*; Hedenquist, J., Thompson, J., Goldfarb, R.J., Eds.; Society of Economic Geologists: Littleton, CO, USA, 2005; pp. 407–450.
17. Sung, Y.-H.; Brugger, J.; Ciobanu, C.L.; Pring, A.; Skinner, W.; Nugus, M. Invisible gold in arsenian pyrite and arsenopyrite from a multistage Archaean gold deposit: Sunrise Dam, Eastern Goldfields Province, Western Australia. *Miner. Depos.* **2009**, *44*, 765–791. [CrossRef]
18. Deol, S.; Deb, M.; Large, R.R.; Gilbert, S. LA-ICPMS and EPMA studies of pyrite, arsenopyrite and loellingite from the Bhukia-Jagpura gold prospect, southern Rajasthan, India: Implications for ore genesis and gold remobilization. *Chem. Geol.* **2012**, *326–327*, 72–87. [CrossRef]

19. Li, W.; Cook, N.J.; Xie, G.-Q.; Mao, J.-W.; Ciobanu, C.L.; Li, J.-W.; Zhang, Z.-Y. Textures and trace element signatures of pyrite and arsenopyrite from the Gutaishan Au–Sb deposit, South China. *Miner. Depos.* **2018**. [CrossRef]

20. Fougerouse, D.; Micklethwaite, S.; Tomkins, A.G.; Mei, Y.; Kilburn, M.; Guagliardo, P.; Fisher, L.A.; Halfpenny, A.; Gee, M.; Paterson, D.; et al. Gold remobilisation and formation of high grade ore shoots driven by dissolution-reprecipitation replacement and Ni substitution into auriferous arsenopyrite. *Geochim. Cosmochim. Acta* **2016**, *178*, 143–159. [CrossRef]

21. Reich, M.; Kesler, S.E.; Utsunomiya, S.; Palenik, C.S.; Chryssoulis, S.L.; Ewing, R.C. Solubility of gold in arsenian pyrite. *Geochim. Cosmochim. Acta* **2005**, *69*, 2781–2796. [CrossRef]

22. Pitcairn, I.K.; Skelton, A.D.L.; Wohlgemuth-Ueberwasser, C.C. Mobility of gold during metamorphism of the Dalradian in Scotland. *Lithos* **2015**, *233*, 69–88. [CrossRef]

23. Large, R.R.; Danyushevsky, L.; Hollit, C.; Maslennikov, V.; Meffre, S.; Gilbert, S.; Bull, S.; Scott, R.; Emsbo, P.; Thomas, H.; et al. Gold and trace element zonation in pyrite using a laser imaging technique: Implications for the timing of gold in orogenic and Carlin-style sediment-hosted deposits. *Econ. Geol.* **2009**, *104*, 635–668. [CrossRef]

24. Cook, N.J.; Ciobanu, C.L.; Meria, D.; Silcock, D.; Wade, B. Arsenopyrite-pyrite association in an orogenic gold ore: Tracing mineralization history from textures and trace elements. *Econ. Geol.* **2013**, *108*, 1273–1283. [CrossRef]

25. Deng, J.; Wang, Q.; Li, G.; Hou, Z.; Jiang, C.; Danyushevsky, L. Geology and genesis of the giant Beiya porphyry–skarn gold deposit, northwestern Yangtze Block, China. *Ore Geol. Rev.* **2015**, *70*, 457–485. [CrossRef]

26. Large, R.R.; Maslennikov, V.V.; Robert, F.; Danyushevsky, L.V.; Chang, Z.S. Multistage sedimentary and metamorphic origin of pyrite and gold in the giant Sukhoi Log deposit, Lena gold province, Russia. *Econ. Geol.* **2007**, *102*, 1233–1267. [CrossRef]

27. Large, R.R.; Bull, S.W.; Maslennikov, V.V. A carbonaceous sedimentary source-rock model for Carlin-type and orogenic gold deposits. *Econ. Geol.* **2011**, *106*, 331–358. [CrossRef]

28. Thomas, H.V.; Large, R.R.; Bull, S.W.; Maslennikov, V.; Berry, R.F.; Fraser, R.; Froud, S.; Moye, R. Pyrite and pyrrhotite textures and composition in sediments, laminated quartz veins, and reefs at Bendigo gold mine, Australia: Insights for ore genesis. *Econ. Geol.* **2011**, *106*, 1–31. [CrossRef]

29. Steadman, J.A.; Large, R.R. Synsedimentary, diagenetic, and metamorphic pyrite, pyrrhotite, and marcasite at the Homestake BIF-hosted gold deposit, South Dakota, USA: Insights on Au-As ore genesis from textural and LA-ICP-MS trace element studies. *Econ. Geol.* **2016**, *111*, 1731–1752. [CrossRef]

30. Gao, S.; Huang, F.; Gu, X.; Chen, Z.; Xing, M.; Li, Y. Research on the growth orientation of pyrite grains in the colloform textures in Baiyunpu Pb–Zn polymetallic deposit, Hunan, China. *Mineral. Petrol.* **2017**, *111*, 69–79. [CrossRef]

31. Dubosq, R.; Lawley, C.J.M.; Rogowitz, A.; Schneider, D.A.; Jackson, S. Pyrite deformation and connections to gold mobility: Insight from micro-structural analysis and trace element mapping. *Lithos* **2018**, *310–311*, 86–104. [CrossRef]

32. Li, X.-H.; Fan, H.-R.; Yang, K.-F.; Hollings, P.; Liu, X.; Hu, F.-F.; Cai, Y.-C. Pyrite textures and compositions from the Zhuangzi Au deposit, southeastern North China Craton: Implication for ore-forming processes. *Contrib. Mineral. Petrol.* **2018**, *173*, 73. [CrossRef]

33. Goldfarb, R.J.; Taylor, R.D.; Collins, G.S.; Goryachev, N.A.; Orlandini, O.F. Phanerozoic continental growth and gold metallogeny of Asia. *Gondwana Res.* **2014**, *25*, 48–102. [CrossRef]

34. Deng, J.; Wang, Q.F. Gold mineralization in China: Metallogenic provinces, deposit types and tectonic framework. *Gondwana Res.* **2016**, *36*, 219–274. [CrossRef]

35. Groves, D.I.; Goldfarb, R.J.; Gebre-Mariam, M.; Hagemann, S.G.; Robert, F. Orogenic gold deposits: A proposed classification in the context of their crustal distribution and relationship to other gold deposit types. *Ore Geol. Rev.* **1998**, *13*, 7–27. [CrossRef]

36. Li, N.; Deng, J.; Yang, L.Q.; Goldfarb, R.J.; Zhang, C.; Marsh, E.; Lei, S.B.; Koenig, A.; Lowers, H. Paragenesis and geochemistry of ore minerals in the epizonal gold deposits of the Yangshan gold belt, West Qinling, China. *Miner. Depos.* **2014**, *49*, 427–449. [CrossRef]

37. Li, N.; Deng, J.; Yang, L.-Q.; Groves, D.I.; Liu, X.-W.; Dai, W.-G. Constraints on depositional conditions and ore-fluid source for orogenic gold districts in the West Qinling Orogen, China: Implications from sulfide assemblages and their trace-element geochemistry. *Ore Geol. Rev.* **2018**, *102*, 204–219. [CrossRef]

38. Zhang, G.W.; Meng, Q.R.; Lai, S.C. Tectonics and structure of Qinling orogenic belt. *Sci. China Ser. B* **1995**, *38*, 1379–1394.

39. Meng, Q.R.; Zhang, G.W. Timing of collision of the North and South China blocks: Controversy and reconciliation. *Geology* **1999**, *27*, 123. [CrossRef]

40. Dong, Y.P.; Zhang, X.N.; Liu, X.M.; Li, W.; Chen, Q.; Zhang, G.W.; Zhang, H.F.; Yang, Z.; Sun, S.S.; Zhang, F.F. Propagation tectonics and multiple accretionary processes of the Qinling Orogen. *J. Asian Earth Sci.* **2015**, *104*, 84–98. [CrossRef]

41. Deng, J.; Wang, Q.F.; Li, G.J. Tectonic evolution, superimposed orogeny, and composite metallogenic system in China. *Gondwana Res.* **2017**, *50*, 216–266. [CrossRef]

42. Yan, Q.; Wang, Z.; Yan, Z.; Hanson, A.D.; Druschke, P.A.; Liu, D.; Song, B.; Jian, P.; Wang, T. Geochronology of the Bikou Group volcanic rocks:Newest results from SHRIMP zircon U-Pb dating. *Geol. Bull. China* **2003**, *22*, 456–458.

43. Chen, Y.J.; Zhang, J.; Zhang, F.; Franco, P.; Li, C. Carlin and Carlin-like gold deposits in Western Qinling Mountains and their metallogenic time, tectonic setting and model. *Geol. Rev.* **2004**, *50*, 134–152, (In Chinese with English Abstract).

44. Yang, L.Q.; Ji, X.Z.; Santosh, M.; Li, N.; Zhang, Z.C.; Yu, J.Y. Detrital zircon U–Pb ages, Hf isotope, and geochemistry of Devonian chert from the Mianlue suture: Implications for tectonic evolution of the Qinling orogen. *J. Asian Earth Sci.* **2015**, *113 Pt 2*, 589–609. [CrossRef]

45. Yang, L.Q.; Deng, J.; Qiu, K.F.; Ji, X.Z.; Santosh, M.; Song, K.R.; Song, Y.H.; Geng, J.Z.; Zhang, C.; Hua, B. Magma mixing and crust–mantle interaction in the Triassic monzogranites of Bikou Terrane, central China: Constraints from petrology, geochemistry, and zircon U–Pb–Hf isotopic systematics. *J. Asian Earth Sci.* **2015**, *98*, 320–341. [CrossRef]

46. Yang, L.Q.; Deng, J.; Dilek, Y.; Qiu, K.F.; Ji, X.Z.; Li, N.; Taylor, R.D.; Yu, J.Y. Structure, geochronology, and petrogenesis of the Late Triassic Puziba granitoid dikes in the Mianlue suture zone, Qinling orogen, China. *Geol. Soc. Am. Bull.* **2015**, *127*, 1831–1854. [CrossRef]

47. Dong, Y.P.; Zhang, G.W.; Neubauer, F.; Liu, X.M.; Genser, J.; Hauzenberger, C. Tectonic evolution of the Qinling orogen, China: Review and synthesis. *J. Asian Earth Sci.* **2011**, *41*, 213–237. [CrossRef]

48. Wang, X.X.; Wang, T.; Zhang, C.L. Neoproterozoic, Paleozoic, and Mesozoic granitoid magmatism in the Qinling Orogen, China: Constraints on orogenic process. *J. Asian Earth Sci.* **2013**, *72*, 129–151. [CrossRef]

49. Li, N.; Chen, Y.J.; Santosh, M.; Pirajno, F. Compositional polarity of Triassic granitoids in the Qinling Orogen, China: Implication for termination of the northernmost paleo-Tethys. *Gondwana Res.* **2015**, *27*, 244–257. [CrossRef]

50. Li, N.; Yang, L.Q.; Zhang, C.; Zhang, J.; Lei, S.B.; Wang, H.T.; Wang, H.W.; Gao, X. Sulfur isotope characteristics of the Yangshan gold belt, West Qinling: Constraints on ore-forming environment and material source. *Acta Petrol. Sin.* **2012**, *28*, 1577–1587, (In Chinese with English Abstract).

51. Li, N. Geochemistry of Ore-Forming Processes in the Yangshan Gold Belt, West Qinling, Central China. Ph.D. Thesis, China University of Geosciences, Beijing, China, 2013. (In Chinese with English Abstract).

52. Li, N.; Zhang, Z.C.; Liu, X.W.; Liu, J. The relationship between disseminated gold mineraliztion and vein-type gold-antimony mineralization: Example from the Yangshan gold belt, West Qinling. *Acta Petrol. Sin.* **2018**, *34*, 1312–1326, (In Chinese with English Abstract).

53. Zhao, H.X.; Frimmel, H.E.; Jiang, S.Y.; Dai, B.Z. LA-ICP-MS trace element analysis of pyrite from the Xiaoqinling gold district, China: Implications for ore genesis. *Ore Geol. Rev.* **2011**, *43*, 142–153. [CrossRef]

54. Longerich, H.P.; Jackson, S.E.; Günther, D. Inter-laboratory note. Laser ablation inductively coupled plasma mass spectrometric transient signal data acquisition and analyte concentration calculation. *J. Anal. At. Spectrom.* **1996**, *11*, 899–904. [CrossRef]

55. Zhao, J.; Liang, J.L.; Ni, S.J.; Xiang, Q.R. In situ sulfur isotopic composition analysis of Au-bearing pyrites by using Nanao-SIMS in Yangshan gold deposit, Gansu Province. *Miner. Depos.* **2016**, *35*, 653–662, (In Chinese with English Abstract).

56. Ward, J.; Mavrogenes, J.; Murray, A.; Holden, P. Trace element and sulfur isotopic evidence for redox changes during formation of the Wallaby Gold Deposit, Western Australia. *Ore Geol. Rev.* **2017**, *82*, 31–48. [CrossRef]

57. Zhang, J.; Deng, J.; Chen, H.Y.; Yang, L.Q.; Cooke, D.; Danyushevsky, L.; Gong, Q.J. LA-ICP-MS trace element analysis of pyrite from the Chang'an gold deposit, Sanjiang region, China: Implication for ore-forming process. *Gondwana Res.* **2014**, *26*, 557–575. [CrossRef]

58. Yang, L.Q.; Deng, J.; Li, N.; Zhang, C.; Ji, X.Z.; Yu, J.Y. Isotopic characteristics of gold deposits in the Yangshan gold belt, West Qinling, central China: Implications for fluid and metal sources and ore genesis. *J. Geochem. Explor.* **2016**, *168*, 103–118. [CrossRef]

59. Cook, N.J.; Ciobanu, C.L.; Mao, J.W. Textural control on gold distribution in As-free pyrite from the Dongping, Huangtuliang and Hougou gold deposits, North China Craton (Hebei Province, China). *Chem. Geol.* **2009**, *264*, 101–121. [CrossRef]

60. Li, Z.-K.; Li, J.-W.; Cooke, D.R.; Danyushevsky, L.; Zhang, L.; O'Brien, H.; Lahaye, Y.; Zhang, W.; Xu, H.-J. Textures, trace elements, and Pb isotopes of sulfides from the Haopinggou vein deposit, southern North China Craton: Implications for discrete Au and Ag–Pb–Zn mineralization. *Contrib. Mineral. Petrol.* **2016**, *171*, 99. [CrossRef]

61. Yue, H. *Flotation Test of Primary Ore from Yangshan Gold Deposit*; Changchun Gold Research Institute: Changchun, China, 2011; pp. 1–145, (In Chinese with English Abstract).

62. Yan, F.Z.; Qi, J.Z.; Guo, J.H. *Geology and Exploration of Yangshan Gold Deposit in Gansu Province*; Geological Publishing House: Beijing, China, 2010; p. 236.

63. Liang, J.L.; Sun, W.D.; Zhu, S.Y.; Li, H.; Liu, Y.L.; Zhai, W. Mineralogical study of sediment-hosted gold deposits in the Yangshan ore field, Western Qinling Orogen, Central China. *J. Asian Earth Sci.* **2014**, *85*, 40–50. [CrossRef]

64. Williams-Jones, A.E.; Norman, C. Controls of mineral parageneses in the system Fe-Sb-S-O. *Econ. Geol.* **1997**, *92*, 308–324. [CrossRef]

65. An, F.; Zhu, Y. Native antimony in the Baogutu gold deposit (west Junggar, NW China): Its occurrence and origin. *Ore Geol. Rev.* **2010**, *37*, 214–223. [CrossRef]

66. Xie, Z.J.; Xia, Y.; Cline, J.S.; Yan, B.W.; Wang, Z.P.; Tan, Q.P.; Wei, D.T. Comparison of the native antimony-bearing Paiting gold deposit, Guizhou Province, China, with Carlin-type gold deposits, Nevada, USA. *Miner. Depos.* **2017**, *52*, 69–84. [CrossRef]

67. Simon, G.; Kesler, S.E.; Chryssoulis, S. Geochemistry and textures of gold-bearing arsenian pyrite, Twin Creeks, Nevada; implications for deposition of gold in carlin-type deposits. *Econ. Geol.* **1999**, *94*, 405–421. [CrossRef]

 *minerals*

Article

Process and Mechanism of Gold Mineralization at the Zhengchong Gold Deposit, Jiangnan Orogenic Belt: Evidence from the Arsenopyrite and Chlorite Mineral Thermometers

Si-Chen Sun [1], Liang Zhang [1,*], Rong-Hua Li [1], Ting Wen [2], Hao Xu [2], Jiu-Yi Wang [1], Zhi-Qi Li [3], Fu Zhang [3], Xue-Jun Zhang [3] and Hu Guo [4]

[1] State Key Laboratory of Geological Process and Mineral Resources, China University of Geosciences, Beijing 100083, China; sunsichencugb@126.com (S.-C.S.); lironghuacugb@126.com (R.-H.L.); jywang1993@126.com (J.-Y.W.)

[2] The 214th Geological Brigade, Hunan Provincial Bureau of Non-Ferrous Metal and Geological Exploration, Zhuzhou 412007, China; wenting214@126.com (T.W.); sksxh@126.com (H.X.)

[3] The Liling Zhengchong Gold Mining Co., Ltd, Liling 412200, China; lizhiqi597762623@163.com (Z.-Q.L.); Dafuxing258@163.com (F.Z.); zhangxuejun1965@sohu.com (X.-J.Z.)

[4] Tianjin Center, China Geological Survey, Tianjin 300170, China; 028huguo@163.com

* Correspondence: zhang@cugb.edu.cn

Received: 12 January 2019; Accepted: 20 February 2019; Published: 25 February 2019

Abstract: The Zhengchong gold deposit, with a proven gold reserve of 19 t, is located in the central part of Jiangnan Orogenic Belt (JOB), South China. The orebodies are dominated by NNE- and NW- trending auriferous pyrite-arsenopyrite-quartz veins and disseminated pyrite-arsenopyrite-sericite-quartz alteration zone, structurally hosted in the Neoproterozoic epimetamorphic terranes. Three stages of hydrothermal alteration and mineralization have been defined at the Zhengchong deposit: (i) Quartz–auriferous arsenopyrite and pyrite; (ii) Quartz–polymetallic sulfides–native gold–minor chlorite; (iii) Barren quartz–calcite vein. Both invisible and native gold occurred at the deposit. Disseminated arsenopyrite and pyrite with invisible gold in them formed at an early stage in the alteration zones have generally undergone syn-mineralization plastic-brittle deformation. This resulted in the generation of hydrothermal quartz, chlorite and sulfides in pressure shadows around the arsenopyrite and the formation of fractures of the arsenopyrite. Meanwhile, the infiltration of the ore-forming fluid carrying Sb, Cu, Zn, As and Au resulted in the precipitation of polymetallic sulfides and free gold. The X-ray elements mapping of arsenopyrite and spot composition analysis of arsenopyrite and chlorite were carried out to constrain the ore-forming physicochemical conditions. The results show that the early arsenopyrite and invisible gold formed at 322–397 °C with $\lg f(S_2)$ ranging from -10.5 to -6.7. The crack-seal structure of the ores indicates cyclic pressure fluctuations controlled by fault-valve behavior. The dramatic drop of pressure resulted in the phase separation of ore-forming fluids. During the phase separation, the escape of H_2S gas caused the decomposition of the gold-hydrosulfide complex, which further resulted in the deposition of the native gold. With the weakening of the gold mineralization, the chlorite formed at 258–274 °C with $\lg f(O_2)$ of -50.9 to -40.1, as constrained by the results from mineral thermometer.

Keywords: orogenic-gold deposit; mineral geo-thermometry; physicochemical condition of mineralization; Zhengchong gold deposit; Jiangnan orogenic belt

---

1. Introduction

The process and mechanism of gold mineralization are essential for understanding the genesis of the deposit [1–4]. The evolution of P-T-X conditions of the ore-forming fluids provides unique insights into the process and mechanism of the mineralization [5–8]. The major and trace elements of hydrothermal minerals formed during mineralization contain important information on the ore-forming physicochemical conditions [9,10]. Many lode-gold deposits have been investigated in order to obtain the distribution of elements in the gold-bear fluids by in-situ analysis [11–13]. The principal elements, Fe, S and As, in the hydrothermal arsenopyrite can be used to constrain its temperature and sulfur fugacity combined with mineral assemblages [11,14]. Compared with the microthermometry of fluid inclusions commonly used in research on ore-forming P-T conditions of the orogenic gold deposit [15–22], the mineral thermometers can constrain a more accurate forming temperature of gold-bearing sulfides, such as arsenopyrite, resulting in more detailed analysis of the gold mineralization.

Jiangnan Orogenic Belt (JOB) is situated in the collisional and subduction zone between the Yangtze and Cathaysian Blocks (Figure 1a) [23], which have undergone long-term and multiple deformations and magmatism [24]. The belt is defined as a significant Neoproterozoic to Late Mesozoic polymetallic metallogenic zone in China, which hosts more than 250 deposits with ~970 t gold in total [25,26].

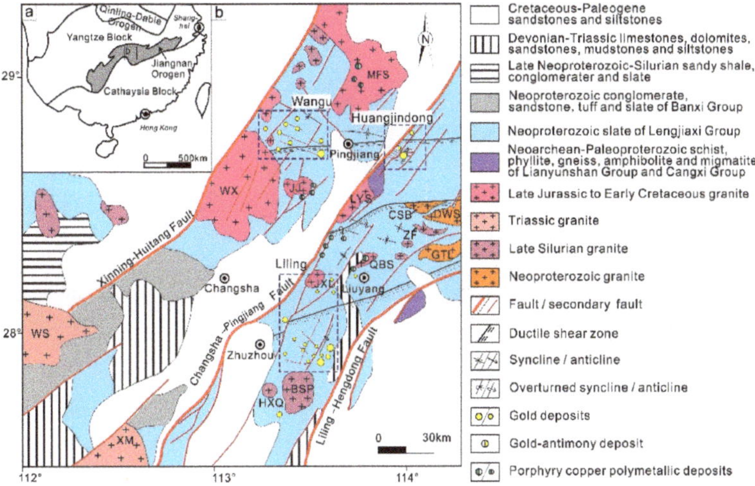

Figure 1. (**a**) Location of Jiangnan Orogenic Belt (JOB); (**b**) Geological map of the Changsha-Pingjiang metallogenic belt showing the distributions of gold deposits. Abbreviations for granites: BSP, Banshanpu; CSB, Changsanbei; DWS, Daweishan; GTL, Getengling; HXQ, Hongxiaqiao; JJ, Jinjing; JXL, Jiaoxiling; LYS, Lianyunshan; MFS, Mufushan; QBS, Qibaoshan; WS, Weishan; WX, Wangxiang; XM, Xiema; ZF, Zhangfang. Modified from references [26,27].

The Zhengchong gold deposit, with a proven gold reserve of 19 t and an average grade of 3.21 g/t [28], is located in the central part of JOB, South China. The gold deposit is defined as an orogenic gold deposit based upon mineral paragenesis, in-situ geochemistry of pyrites [29] and the characteristics of ore-forming fluids [28]. The lower limit of the ore-forming temperatures of the deposit were previously constrained by fluid inclusion microthermometer to 312–406 °C, 252–324 °C and 193–296 °C for a three-stage hydrothermal alteration and mineralization [28]. The deposition of gold was believed to result from the fluid-rock reaction [28]. However, the observations on the relationship of hydrothermal minerals with gold and constraints on the ore-forming chemical conditions are still lacking. Thus, the process and individual mechanism of the invisible and free gold are still unclear.

In this study, detailed field and laboratory works, including petrography and microstructural studies and electro probe micro-analyses (EPMA) on arsenopyrite and chlorite, spatially and temporally associated with gold have been carried out to constrain the evolution of the ore-forming physicochemical conditions. Finally, the process and individual mechanism of the invisible and free gold are summarized based on the ore-forming physicochemical data.

2. Regional and Local Geology

2.1. Regional Geology

The NE-NNE-trending Changsha-Pingjiang metallogenic belt, bounded by the Xinning-Huitang and Liling-Hengdong faults, is located in the central part of JOB, with 125 Au and polymetallic occurrences and deposits, and an estimated ~300 t of gold resources [26] (Figure 1b). Sedimentary rock successions mainly consist of Neoarchean-Paleoproterozoic schist, phyllite, gneiss, amphibolite and migmatite of Lianyunshan Group and Cangxi Group; Neoproterozoic slate of Lengjiaxi Group; Neoproterozoic conglomerate sandstone, tuff and slate of Banxi Group; Silurian-Sinian sandy shale conglomerate and slate; Devonian-Triassic limestones, dolomites, sandstones, mudstones and siltstones and Cretaceous-Paleogene sandstones and siltstones [27].

Three goldfields including Huangjindong, Liling and Wangu are aligned along the first-order NNE- to NE- trending long-live Chang-Ping Fault. The orebodies in these goldfields are mostly directly controlled by EW- to NW- trending faults [26,30], while a few thin orebodies are controlled by NNE-trending faults in the Liling gold field. Multiple intrusions are exposed in this belt, including the Neoproterozoic, Late Silurian, Triassic and Late Jurassic-Early Cretaceous granitoids.

2.2. Deposit Geology

The Zhengchong gold deposit, which is located in the Liling goldfield, is hosted by Neoproterozoic low-metamorphic sedimentary rock of Lengjiaxi Group (Figure 1b). The Lengjiaxi Group, whose protolith is mainly turbidite, is composed of greywacke and silty slate of Lower Huanghudong Formation, sandstone, siltstone and slate of Upper Huanghudong Formation and sandstone and silty slate of Lower Xiaomuping Formation. All these strata strike NE and dip to NW with dip angles of $26°{\sim}65°$ [31] (Figure 2a). Only small granitoid intrusion has been revealed by both geological mapping and drilling at the deposit [31] (Figure 2).

The dominated thick and less important thin orebodies are controlled by NW- and NE-NNE-trending faults, respectively. The NNE-trending orebodies have strike lengths of 1400 m and dip lengths of ~200 m with thicknesses of 0.38–5.4 m [28]. In plane graph, the orebodies and related hydrothermal alteration are lens-shaped with thickness changing greatly in a short distance. The alterations are mainly composed of the significant muscovite alteration, silicification and arsenopyrite, pyrite and other sulfide alterations with minor chloritization. The main orebodies can be divided into two types: (1) auriferous arsenopyrite-pyrite-quartz vein; (2) altered slates with disseminated pyrite, arsenopyrite, muscovite and quartz (Figure 3). Crack-seal, a typical structure at this deposit, developed in the laminated quartz vein (Figure 3d). This structural vein formed during the cyclic episodes of co-seismic extension fractures [32,33]. These cyclic processes typically result in high-grade ores with gold in the fractures of the veins [34].

2.3. Mineral Association and Paragenesis

Ore minerals are mainly composed of native gold, arsenopyrite and pyrite with minor galena, sphalerite, chalcopyrite, tetrahedrite and pyrrhotite. Based on detailed field mapping and observation under microscope, three mineralization stages were recognized at the Zhengchong deposit (Figure 4): (i) Quartz–auriferous arsenopyrite–pyrite; (ii) Quartz–polymetallic sulfides–native gold–minor chlorite;

and (iii) Barren quartz–calcite vein. Compared with a previous description [26], early quartz vein is barren including no sulfides. The gold content of ore-bearing quartz vein is more than 10 g/t. We observed both invisible gold evenly dispersed in arsenopyrite and pyrite in this type of deposit in NEHP [10,35,36], and native gold in fractures and curved dissolution edges of arsenopyrite and pyrite.

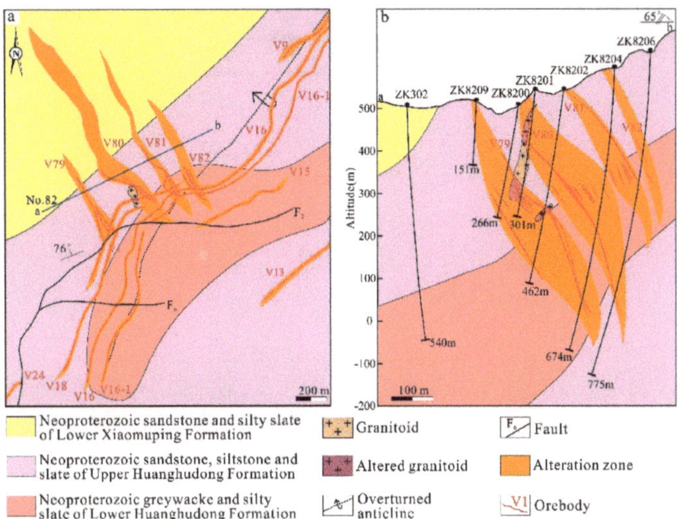

Figure 2. (**a**) Simplified geological map of Zhengchong gold deposit showing the major host rocks, orebodies and alteration zones; (**b**) Profile of the No. 82 cross section of the Zhengchong gold deposit.

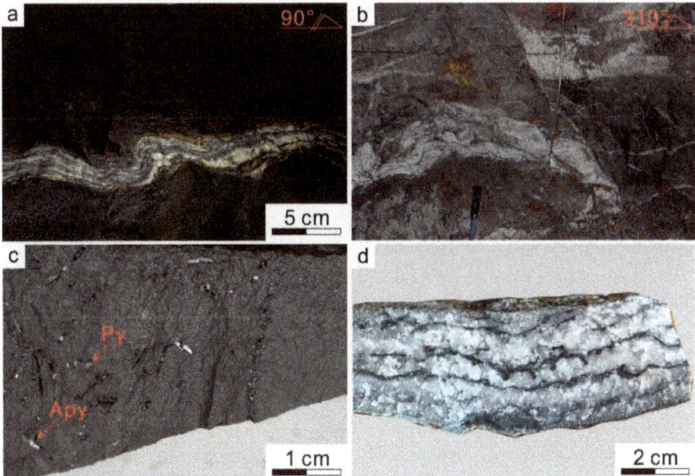

Figure 3. Photographs of representative ore styles at the Zhengchong gold deposit. (**a**) Auriferous arsenopyrite-pyrite-quartz vein folded in disseminated mineralized slate adjacent to the fault; (**b**) Auriferous arsenopyrite-pyrite-quartz vein cut by the post-fault; (**c**) Altered slate with disseminated arsenopyrite and pyrite; (**d**) A typical crack-seal structure showing white barren quartz vein with fractures infilled by the sulfides with fine-grained quartz and muscovite. Py-pyrite; Apy-arsenopyrite.

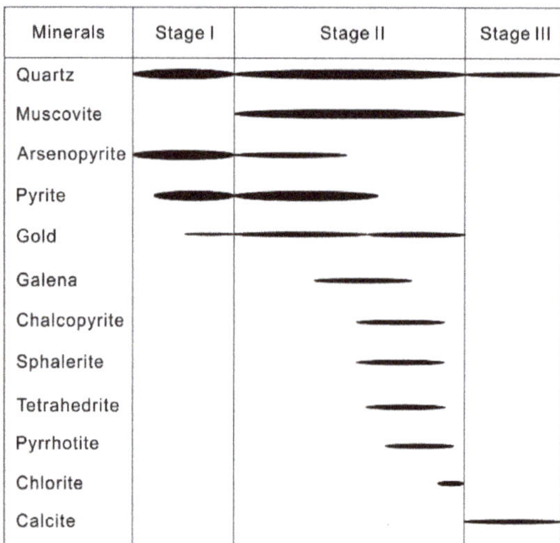

Figure 4. Paragenetic assemblages and sequences of minerals at the Zhengchong gold deposit.

In the altered slate with ~2 g/t gold, pressure shadows with gangue and metallic minerals around rigid arsenopyrite and pyrite in relatively weak matrix composed of fine-grained sericite and quartz are well developed (Figure 5b–e). The quartz fibers and chlorite are the major gangue minerals in the pressure shadow around the rigid sulfide grains (Figure 5b,d). Besides, polymetallic sulfides, such as sphalerite, chalcopyrite and tetrahedrite, which mainly occur in fractures of the pyrite and arsenopyrite, are also observed in these pressure shadows (Figure 5c,e). This indicates that the base metal sulfides formed contemporarily with the adjacent quartz fibers. In some cases, fractures of arsenopyrite were also filled with the fiber-shaped quartz (Figure 5d).

2.3.1. Arsenopyrite

Arsenopyrite is a significant and ubiquitous gold-bearing sulfide both in veins and altered rocks at the Zhengchong gold deposit. Disseminated euhedral arsenopyrite in an altered slate, generally, has acicular and granulous shapes, grain size of 100–600 μm, and straight boundaries (Figure 5a). Generally, the arsenopyrites disseminated in the altered rocks have been ductile-brittlely deformed, forming many cracks and a pressure shadow microstructure in low-stress areas (Figure 5a,b). The fractures of arsenopyrites have been filled by polymetallic sulfides (Figure 5c) and native gold (Figure 5h). The euhedral arsenopyrite in the veins is totally minor without deformation.

2.3.2. Polymetallic Sulfides

Polymetallic sulfides, such as galena, sphalerite, chalcopyrite, tetrahedrite and pyrrhotite, are the most relevant minerals associated with the native gold (Figure 5h). These sulfides formed at the second stage could be divided into several sub-stages. Chalcopyrite formed simultaneously with sphalerite on account of their exsolution-texture (Figure 5f). In places, the chalcopyrite was replaced by tetrahedrite in the fractures of the arsenopyrite (Figure 5c) and pressure shadows of arsenopyrite and pyrite (Figure 5e), or included by the late pyrrhotite. Galena predated the other polymetallic sulfides because of being cut by sphalerite (Figure 5i).

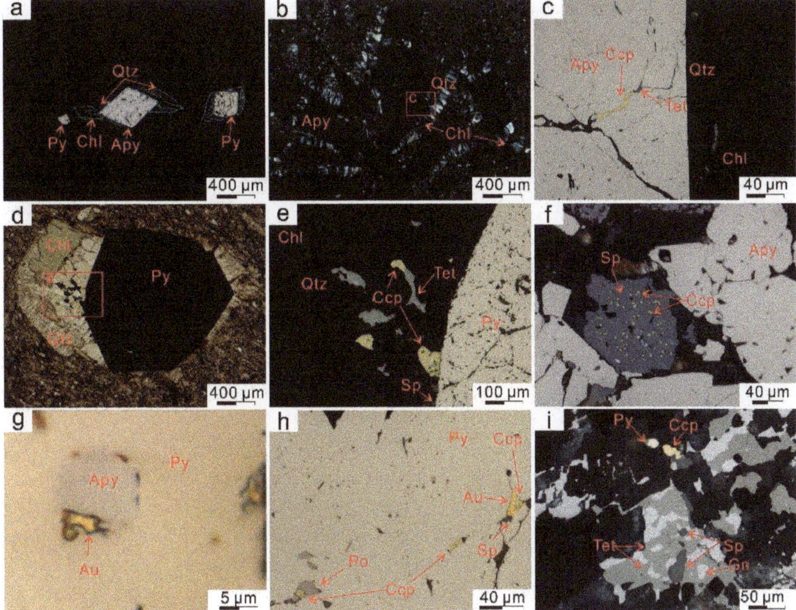

Figure 5. Photomicrographs showing the mineral assemblage. (**a**) Arsenopyrite and pyrite disseminated in the altered slates with pressure shadows infilling by the chlorite; (**b**) Typical pressure shadows filled by quartz and chlorite around the generation of arsenopyrite; (**c**) The fractures of arsenopyrite disseminated in the host-rock filled with the chalcopyrite, sphalerite and tetrahedrite, and later chlorite in the pressure shadows of the arsenopyrite; (**d**) Pyrite with the bulky pressure shadows filled by late quartz, sulfides, and chlorite; (**e**) Chalcopyrite, sphalerite and tetrahedrite in the pressure shadows; the tetrahedrite replaced the chalcopyrite; (**f**) Chalcopyrite and sphalerite exsolution-texture; (**g**) Gold along the grain edge between the pyrite and arsenopyrite; (**h**) Electrum, chalcopyrite, sphalerite and pyrrhotite in the healed and unhealed cracks of pyrite; chalcopyrite is included by the pyrrhotite; (**i**) The occurrence of polymetallic sulfides, where galena is cut by the sphalerite. Py—pyrite; Apy—arsenopyrite; Ccp—chalcopyrite; Gn—galena; Sp—sphalerite; Po—pyrrhotite; Tet—tetrahedrite; Au—native gold; Qtz—quartz; Chl—chlorite.

2.3.3. Chlorite

Late-stage chlorite is typically observed in the pressure shadows of early gold-bearing arsenopyrite and pyrite. Different from the fiber-shaped quartz in the pressure shadows, the chlorite is anhedral without specified shape.

3. Analytical Methods and Results

3.1. Sample Selection and Analytical Methods

Six gold-bearing sulfide-quartz veins and five altered slate samples were collected from the underground tunnel at 290 m level at the Zhengchong deposit. Elements-mapping for a deformed gold-bearing arsenopyrite, and spot element analysis on arsenopyrite, and chlorite in the pressure shadows were carried out by an Electron Probe Micro-Analyzer (EPMA).

Electron Probe Micro-Analyzer test was performed at the Laboratory of Isotope Geology, Tianjin Institute of Geology and Mineral Resources, Tian-jin, China. The operating conditions for the sulfide analysis were 15 kV acceleration voltages and 20 nA beam currents with 5 μm beam diameter.

The chemical compositions of chlorite were measured under the same operating condition but with a different beam diameter of 1–5 μm.

Content of major elements and its mineral association can indicate the forming environment of arsenopyrite. Its forming temperature and $\lg f(S_2)$ would be defined owing to the As content distribution in the diagram of sulfide symbiosis.

Chlorite is an important gangue mineral in the gold deposit, whose chemical formula is $(R_X^{2+}R_y^{3+}\square_{6-x-y})_6^{VI}(Si_zR_{4-z}^{3+})_4^{IV}O_{10}(OH)^8$, where R^{2+} and R^{3+} represent the divalent and trivalent cations, respectively, such as Fe^{2+}, Mg^{2+} and Al^{3+}, Fe^{3+}. The octahedral vacancy is expressed by $\square$. The chemical compositions of chlorite are controlled by the substitution between Fe^{2+} and Mg^{2+}, the Tschermark substitution between Al^{IV}-Al^{VI} and Si-(Mg, Fe^{2+}), as well as 3(Mg, Fe^{2+}) replacing $\square 2\ Al^{VI}$, which is due to the complexity of the formation environments. Analysis of the chemical compositions assists in deducing the temperature and oxygen fugacity of chlorites.

3.2. Results

The major and trace elements are not conformably distributed in the arsenopyrite deformed as shown in the elements map (Figure 6). Sulfur is a major element of the arsenopyrite with slightly higher content in the internal zone (Figure 6b), but this trending of value difference of other major and trace elements has not been observed. Cobalt, nickel, antimony and gold contents are uniformly distributed in the arsenopyrite with low values (Figure 6c,d,h,i). Similarly, copper and zinc also have uniform distribution in the arsenopyrite, but the contents of both elements are slightly high in the fractures and cracks (Figure 6e,f). Arsenic, a major element with high content, is homogeneously distributed in this kind of arsenopyrite (Figure 6g).

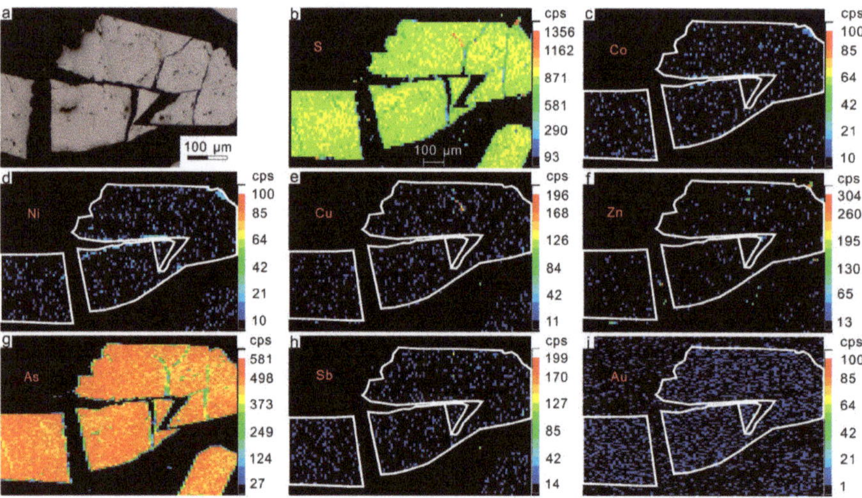

Figure 6. (a) Photomicrographs showing mineralogical features of the arsenopyrite selected to measure; (b–i) X-ray elements distribution maps of S, Co, Ni, Cu, Zn, As, Sb and Au.

Arsenic is the principal element in reflecting the forming temperature of different generations of arsenopyrite [37]. Twenty pieces of data of the core of ore-bearing arsenopyrite from the Zhengchong gold deposit yielded arsenic atomic percentages of 29.71%–31.66%, showing wide systematic variation within the same generation (Table 1). The yielded sulfur atomic percentages were 35.35%–37.44%. Moreover, the sulfur content has a negative correlation with arsenic values, with a regression coefficient of −0.97 and R^2 of 0.89 (Figure 7).

Table 1. Chemical compositions of arsenopyrite at the Zhengchong gold deposit.

Spot No.	Elements (wt %)											Chemical Formula
	As	Au	S	Ag	Sb	Fe	Co	Ni	Cu	Zn	Total	
ZC17D01B1-1	43.07	0	22.24	0	0	34.80	0.05	0.02	0	0	100.18	$FeAs_{0.92}S_{1.11}$
ZC17D01B1-2	42.43	0.04	22.80	0.01	0	34.64	0.01	0.07	0	0	99.99	$FeAs_{0.91}S_{1.15}$
ZC17D01B1-3	41.52	0	22.24	0.01	0.04	34.37	0.07	0.01	0.02	0	98.28	$FeAs_{0.90}S_{1.13}$
ZC17D01B1-4	43.22	0	22.43	0	0.05	34.45	0.02	0.01	0.01	0.12	100.30	$FeAs_{0.94}S_{1.13}$
ZC17D01B1-5	43.39	0.08	22.05	0.01	0.01	34.84	0.07	0.01	0.01	0.06	100.54	$FeAs_{0.93}S_{1.10}$
ZC17D01B1-6	43.85	0.01	22.02	0	0	35.03	0.06	0	0	0.03	100.99	$FeAs_{0.93}S_{1.10}$
ZC17D01B1-7	42.77	0.01	22.26	0	0.02	34.60	0.01	0.05	0.04	0	99.75	$FeAs_{0.92}S_{1.12}$
ZC17D01B1-8	44.72	0.07	21.37	0	0	34.65	0.01	0.02	0.01	0.04	100.89	$FeAs_{0.96}S_{1.07}$
ZC17D01B1-9	43.91	0.01	21.38	0	0	34.72	0.05	0	0.03	0.01	100.11	$FeAs_{0.94}S_{1.07}$
ZC17D01B1-10	44.04	0	21.66	0	0	34.58	0.02	0	0.07	0.01	100.37	$FeAs_{0.95}S_{1.09}$
ZC17D01B1-11	43.89	0	21.94	0.03	0	34.21	0.01	0	0	0	100.07	$FeAs_{0.96}S_{1.12}$
ZC17D01B1-12	44.21	0.09	21.88	0.02	0	34.71	0.01	0.05	0.06	0	101.02	$FeAs_{0.95}S_{1.10}$
ZC17D01B1-13	44.12	0	21.41	0.02	0	33.84	0.03	0	0.01	0	99.42	$FeAs_{0.97}S_{1.10}$
ZC17D01B1-14	43.91	0.01	21.46	0	0.01	34.29	0.02	0.03	0.08	0.02	99.82	$FeAs_{0.96}S_{1.09}$
ZC17D01B2-15	43.22	0	21.52	0	0	34.09	0.03	0.04	0.03	0	98.94	$FeAs_{0.95}S_{1.10}$
ZC17D01B2-16	43.40	0.06	21.60	0.03	0.01	34.67	0.07	0.05	0.06	0.05	100.00	$FeAs_{0.93}S_{1.09}$
ZC17D01B2-17	43.42	0.07	21.72	0	0.01	34.10	0.03	0.04	0	0	99.38	$FeAs_{0.95}S_{1.11}$
ZC17D01B2-18	42.20	0.03	22.42	0	0.02	34.55	0.04	0.01	0	0.04	99.29	$FeAs_{0.91}S_{1.13}$
ZC17D01B2-19	43.11	0.02	21.56	0	0.03	33.98	0.05	0.05	0.01	0.05	98.86	$FeAs_{0.95}S_{1.11}$
ZC17D01B2-20	43.44	0.01	21.29	0	0.03	34.28	0.03	9	0.07	0	99.14	$FeAs_{0.95}S_{1.08}$

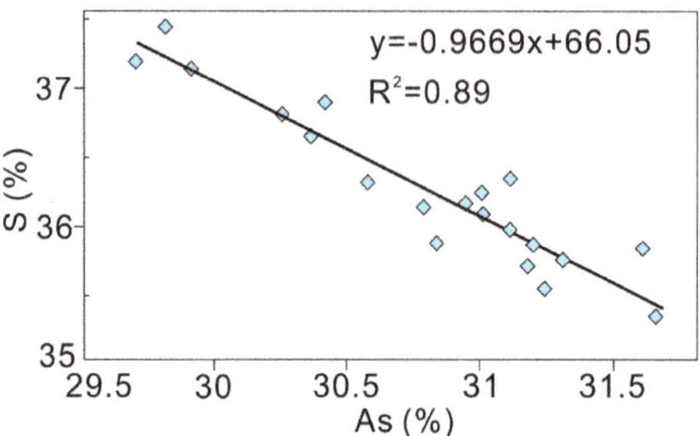

Figure 7. Distributions and relationship of As and S atomic percentages in the arsenopyrite [10].

Chlorite is widely distributed in the pressure shadows around rigid arsenopyrite and pyrite in ores at the Zhengchong deposit. Chlorite samples from different sites of the pressure shadows have similar compositions, such as SiO_2 (21.51–23.16 wt %), MgO (11.56–13.26 wt %), FeO (25.90–28.37 wt %) and Al_2O_3 (21.84–23.13 wt %) (Table 2). The contents of Cr_2O_3, CaO, K_2O and Na_2O are low. The total values of the Fe/(Fe + Mg) ratios range from 0.52 to 0.58. Owing to the extremely small amount of MnO in the chlorite, Fe/(Fe + Mg + Mn) is approximately equal to Fe/(Fe + Mg). Silicon values vary from 2.43 to 2.55 a.p.f.u. Hence, the chlorites in the pressure shadows were defined as ripidolite type (Figure 8), a kind of chlorite with relatively high Fe irons according to the classification of chlorites [38].

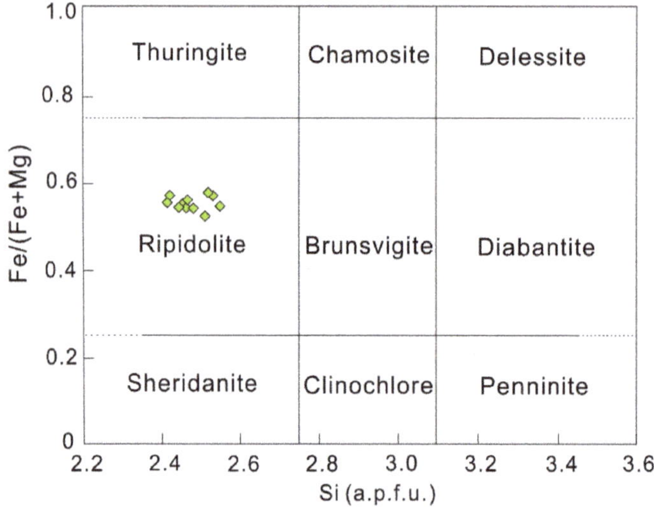

Figure 8. Chemical compositions of chlorite in the pressure shadows of arsenopyrite and pyrite at the Zhengchong gold deposit. Nomenclature and classification of chlorite [38].

4. Discussion

4.1. The Physicochemical Environment of Mineral Precipitations

Precipitation of minerals from the gold-bearing fluids is controlled by physicochemical conditions, including temperature and redox conditions. These conditions change with the evolution of the ore-forming fluids. Previous studies show that the major and trace elements of minerals provide important information of their forming physicochemical conditions which in turn could be used to constrain the P-T-X evolution of the fluids [9,10].

4.1.1. Physicochemical Conditions of Arsenopyrite

As shown above, arsenopyrites formed earlier than the polymetallic sulfides in the fractures or pressure shadows of arsenopyrite and pyrite (Figure 5a,b,d). Thus, the sulfide associated with arsenopyrites is only pyrite (Figure 4). Hence, the arsenic values from both internal and external zones of arsenopyrite fall in the "pyrite + arsenopyrite" zone of the buffered assemblages involving arsenopyrite in the Fe-S-As system [14,39]. The temperatures and $\lg f(S_2)$ of arsenopyrite are almost successive due to their similar arsenic contents (Figure 7). The temperature is constrained from 322 °C to 397 °C, and the yielded $\lg f(S_2)$ of the disseminated arsenopyrites is from -10.5 to -6.7 (Figure 9).

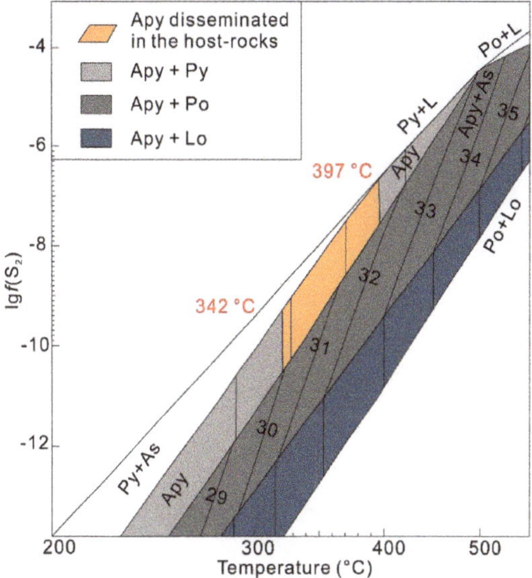

Figure 9. Activity of $\lg f(S_2)$ and temperature (T) projection of the stability field of arsenopyrite [14,39].

Table 2. Chemical compositions of chlorite in pressure shadows at the Zhengchong gold deposit.

Samples No.	zc17d01b1-1	zc17d01b1-2	zc17d01b1-3	zc17d01b1-4	zc17d01b1-5	zc17d01b1-6	zc17d01b1-7	zc17d01b1-8	zc17d01b1-9	zc17d01b1-10	zc17d01b1-11
W(B)/%											
Na_2O	0.08	0.05	0.09	0.06	0.00	0.05	0.07	0.03	0.03	0.02	0.04
MgO	12.11	12.47	12.57	12.30	12.61	11.56	11.85	13.26	11.84	12.06	12.58
Al_2O_3	22.61	23.09	23.13	22.92	22.16	21.97	22.50	22.37	22.74	22.96	21.84
SiO_2	22.22	22.07	22.39	22.15	23.16	22.57	23.02	22.79	21.74	21.51	21.96
K_2O	0.03	0.02	0.03	0.02	0.02	0.03	0.04	0.02	0.01	0.00	0.02
CaO	0.00	0.01	0.01	0.02	0.01	0.01	0.02	0.00	0.02	0.01	0.01
P_2O_5	0.00	0.00	0.00	0.00	0.01	0.03	0.01	0.00	0.01	0.00	0.01
FeO	27.59	26.65	26.70	25.90	27.19	28.37	28.18	26.06	28.14	26.86	27.29
V_2O_3	0.00	0.02	0.04	0.03	0.00	0.02	0.01	0.02	0.04	0.03	0.00
MnO	0.00	0.00	0.01	0.02	0.02	0.00	0.03	0.02	0.00	0.00	0.01
Cr_2O_3	0.00	0.00	0.00	0.01	0.01	0.00	0.00	0.00	0.00	0.00	0.00
Cations											
Si	2.48	2.45	2.47	2.48	2.55	2.53	2.53	2.52	2.43	2.43	2.48
Al^{IV}	1.52	1.55	1.53	1.52	1.45	1.47	1.47	1.48	1.57	1.57	1.52
$\sum R_t$	4.00	4.00	4.00	4.00	4.00	4.00	4.00	4.00	4.00	4.00	
Fe^{3+}	0.02	0.01	0.02	0.02	0.01	0.03	0.03	0.01	0.01	0.00	0.01
Fe^{2+}	2.55	2.46	2.44	2.41	2.50	2.63	2.57	2.40	2.62	2.53	2.56
Al^{VI}	1.45	1.48	1.48	1.51	1.43	1.42	1.45	1.43	1.44	1.48	1.38
Mg	2.01	2.07	2.07	2.05	2.07	1.93	1.94	2.18	1.98	2.03	2.11
$\sum R_o$	6.03	6.03	6.01	5.99	6.01	6.02	6.01	6.03	6.06	6.05	6.08
Component Activity											
Fe/(Fe + Mg)	0.56	0.55	0.54	0.54	0.55	0.58	0.57	0.52	0.57	0.56	0.55
Fe/(Fe + Mg + Mn)	0.56	0.55	0.54	0.54	0.55	0.58	0.57	0.52	0.57	0.56	0.55
d_{001}	14.1118	14.1109	14.1132	14.1152	14.1215	14.1159	14.1180	14.1195	14.1055	14.1063	14.1115
$t/°C$	267.23	268.13	265.79	263.84	257.54	263.10	261.04	259.48	273.52	272.65	267.54
lga_3	−1.34	−1.40	−1.42	−1.43	−1.39	−1.29	−1.32	−1.48	−1.29	−1.35	−1.37
lga_6	−2.32	−2.58	−2.34	−2.49	−2.83	−2.14	−2.24	−2.82	−2.52	−2.98	−2.54
lgK_1	9.77	9.74	9.81	9.87	10.05	9.89	9.95	10.00	9.58	9.61	9.76
$Lgf(O_2)$	−43.00	−43.66	−42.93	−43.71	−45.95	−42.98	−43.45	−45.32	−43.25	−44.97	−43.70

Calculations of Si, Al^{IV}, Al^{VI} and Mg were based on the fourteen oxygen atoms and on the assumption that iron was totally Fe^{2+}. $\sum R_t$ and and $\sum R_o$ are applied to the total number of the tetrahedral and octahedral coordination, respectively. The a_3 and a_6 in the table are the activity coefficient of the reaction between the C_3 and C_6 end members [40]. The K_1 and K_2 are the equilibrium coefficients of the reaction equation. $a_3 = 59.72(X_{Fe^{2+},o})^5(X_{Al,o})(X_{Si,t})(X_{Al,t})$, $a_6 = 729(X_{Fe^{2+},o})(X_{Fe^{3+},o})(X_{Al,o})(X_{Si,t})(X_{Al,t})$. $X_{i,j}$ stands for the mole fraction of the "i" ion in the "j" coordination.

4.1.2. Physicochemical Conditions of Chlorite

Chlorite thermometer was used in gold deposits to assist the constraints on the metallogenic temperature [9,10,41]. Typically, basal spacing d_{001} of chlorite shows a positive correlation with its crystallographic temperature [42]. The value of d_{001} can be calculated through chemical compositions of chlorite using EPMA. After that, calculation for the temperature was proceeded by empirical equation having a positive r-squared (r = 0.95) [43,44]. The results show that the forming temperatures of chlorite ranged from 258 °C to 274 °C, with an average of 265 °C.

Oxygen fugacity of chlorite was also calculated through its chemical compositions, in which six groups of chlorite solid solution models were divided into six thermodynamic end-members. Reactions of various end-members were established to deduce not only the temperature but also the $\lg f(O_2)$ through the calculation of activity quotient a_i and equilibrium coefficient of chemical equation K_j according to the paragenesis with chlorite [40].

The $\lg f(O_2)$ of chlorite can be calculated by the chemical reaction between C_3 and C_6 thermodynamic end-members [40],

$$\mathrm{Fe_5^{2+}Al_2Si_3O_{10}(OH)_8(C_3)} + \frac{1}{4}O_2(g) = \mathrm{Fe_4^{2+}Fe^{3+}Al_2Si_3O_{11}(OH)_7(C_6)} + \frac{1}{2}H_2O(l), \qquad (1)$$

and the $\lg f(O_2)$ is also inferred [45],

$$\lg f(O_2) = 4(\lg a_6 - \lg a_3 - \lg K_1). \qquad (2)$$

The K_1 in Equation (2) represents the equilibrium coefficient of Equation (1), the correlation between equilibrium and temperature was refitted after all $\lg K_1$ under the different temperatures were calculated by mathematical induction [46].

$$\lg K_1 = 21.77e^{-0.003t} \qquad (3)$$

The t and e represent the temperature and natural logarithm. Therefore, the temperature obtained is 258–274 °C using Equation (3). The yielded $\lg f(O_2)$ is -50.9–-40.1.

4.2. Process and Mechanism of the Gold Precipitation

At the Zhengchong gold deposit, uniform invisible gold distributions in the arsenopyrite and visible gold related to the polymetallic sulfides locating in the fractures of early sulfides were observed by microscope and elements mapping. The process and physicochemical conditions of the two-stage gold mineralization are constrained by geological evidences and results from mineral geothermometer.

At the early stage, free elements sulfur, arsenic and gold precipitated from the gold-bearing fluids when fluid-rock reaction occurred [28], which promoted the deposition of arsenopyrite and pyrite with the invisible gold, disseminated in the altered slates at a temperature of 322–397 °C and the $\lg f(S_2)$ from -8.5 to -6.7 (Figure 10a). Arsenic of the arsenopyrite has negative correlation with sulfur (Figure 7), indicating that S could be substituted by As [47]; therefore, Au could enter into the arsenopyrite by substituting Fe to maintain the electric balance [10,48,49]. Subsequently, because of shortening event, the euhedral arsenopyrite underwent minor crystal plastic deformation (Figure 10b). As the compressional deformation grew, the microfractures propagated in the arsenopyrite (Figure 10c). Then, the cracks generated due to the significant brittle deformation of the arsenopyrite [50–53]. This indicates that the hydrothermal alteration and mineralization occurred during the brittle-ductile transition [54]. The pressure shadow microstructures and quartz fibers in the fractures and adjacent extensional zones of arsenopyrite and pyrite (Figure 5b,d) indicate an ongoing shortening event [51].

With or following the brittle deformation of the arsenopyrite and pyrite, gold-, copper-, zinc- and antimony-bearing fluids filled the fractures forming gold, chalcopyrite, sphalerite and tetrahedrite (Figure 10d). The absence of late arsenopyrite and pyrite in the cracks of early arsenopyrite and pyrite reveal that fluid-rock reaction was not the major mechanism which induced the deposition of gold at this stage. Although gold precipitation associated with some low temperature polymetallic sulfides, temperature reduction is typically not the main factor for orogenic-gold deposits [3].

The low-pressure extensional fractures and pressure shadows of early sulfides (Figure 3d) indicate that the dilational site is an important control on gold precipitation [55]. Crack-seal structure formed by fault-valve behavior [33] also indicates a cyclic pressure drop which could result in phase separation of the ore-forming fluids [18,56]. Early fractures of the early veins and other dilational sites from arsenopyrite deformation will be sealed by the successive deposition of hydrothermal minerals [34,57,58]. A renewed period of incremental fluid pressure built up the new fractures with favorable orientation. Minor native gold deposited with the chlorite precipitated, following the quartz, in the pressure shadow microstructure (Figure 10e) at 258–274 °C and $-50.9 \sim -40.1$ for $\lg f(O_2)$. The presence of quartz-calcite veins marked the termination of the gold mineralization.

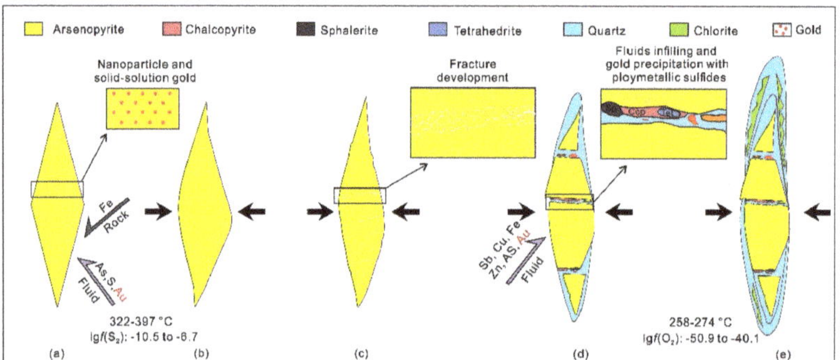

Figure 10. Simplified model interpreting the process and mechanism of deposition of minerals including gold. (**a**) Euhedral arsenopyrite with invisible gold disseminated in the altered slates; (**b**) Arsenopyrite underwent initial crystal plastic deformation; (**c**) Fractures development in arsenopyrite; (**d**) Fluids containing Sb, Cu, Zn, As and Au filled in the fractures and other dilational sites of arsenopyrite leading to gold and polymetallic sulfides precipitation. (**e**) Chlorite precipitated in the pressure shadow microstructure with minor gold.

5. Conclusions

Field and microscopic observations define three stages of hydrothermal alteration and mineralization at the Zhengchong gold deposit: (i) Quartz–auriferous arsenopyrite–pyrite; (ii) Quartz-polymetallic sulfides–native gold–minor chlorite; (iii) Barren quartz-calcite vein. Invisible and native gold both occurred in the deposit.

The invisible gold was distributed of the arsenopyrite generated from the fluid-rock reaction at about 322–397 °C and the $\lg f(S_2)$ from -8.5 to -6.7. Successively after crystal plastic and brittle deformations, the later fluid carrying gold, copper, zinc, antimony and sulfide was injected into the fractures and extensional zones. The fault-valve behavior of the ore-controlling fault resulted in cyclic pressure fluctuation. The dramatic drop in pressure induced the fluid phase separation, decomposition of gold-hydrosulfide complex and deposition of native gold. With the weakening of the gold mineralization, chlorite deposited in the pressure shadow microstructure following the quartz around arsenopyrite and pyrite at 258–274 °C and $-50.9 \sim -40.1$ for $\lg f(O_2)$. The formation of quartz-calcite veins indicates the end of the gold mineralization.

Author Contributions: S.-C.S. and L.Z. finished the main field work with T.W., H.X., Z.-Q.L., X.-J.Z. and F.Z.; S.-C.S. and H.G. performed the experiments; S.-C.S. and R.-H.L. analyzed the data; S.-C.S. wrote the paper; L.Z. and J.-Y.W. revised the paper.

Funding: This work was financially supported by National Natural Science Foundation of China (Grant No. 41702070), and the MOST Special Fund from the State Key Laboratory of Geological Processes and Mineral Resources, China University of Geosciences (Grant No. MSFGPMR201804) and the 111 Project under the Ministry of Education and the State Administration of Foreign Experts Affairs, China (Grant No. B07011), and Excellent Tutor Guidance Fund Project from China University of Geosciences (Grant No. 2652018146).

Acknowledgments: We thank Jun Deng and Li-Qiang Yang and Nan Li at China University of Geosciences (Beijing) and Ri Han of Key Lab. of Mineral Resource, Institute of Geology and Geophysics, Chinese Academy of Sciences for excellent comments and suggestions for the first draft of this paper.

Conflicts of Interest: The authors declare no conflict of interest.

References

1. Yang, L.Q.; Guo, L.N.; Wang, Z.L.; Zhao, R.X.; Song, M.C.; Zheng, X.L. Timing and mechanism of gold mineralization at the Wang'ershan gold deposit, Jiaodong Peninsula, eastern China. *Ore Geol. Rev.* **2017**, *88*, 491–510. [CrossRef]
2. Deng, J.; Wang, Q.F.; Li, G.J. Tectonic evolution, superimposed orogeny, and composite metallogenic system in China. *Gondwana Res.* **2017**, *50*, 216–266. [CrossRef]
3. Goldfarb, R.J.; Groves, D.I. Orogenic gold: Common or evolving fluid and metal sources through time. *Lithos* **2015**, *233*, 2–26. [CrossRef]
4. Qiu, K.F.; Taylor, R.D.; Song, Y.H.; Yu, H.C.; Song, K.R.; Li, N. Geologic and geochemical insights into the formation of the Taiyangshan porphyry copper–molybdenum deposit, Western Qinling Orogenic Belt, China. *Gondwana Res.* **2016**, *35*, 40–58. [CrossRef]
5. Hammond, N.Q.; Robb, L.; Foya, S.; Ishiyama, D. Mineralogical, fluid inclusion and stable isotope characteristics of Birimian orogenic gold mineralization at the Morila Mine, Mali, West Africa. *Ore Geol. Rev.* **2011**, *39*, 218–229. [CrossRef]
6. Wang, Z.L.; Yang, L.Q.; Guo, L.N.; Marsh, E.; Wang, J.P.; Liu, Y.; Zhang, C.; Li, R.H.; Zhang, L.; Zheng, X.L.; et al. Fluid immiscibility and gold deposition in the Xincheng deposit, Jiaodong Peninsula, China: A fluid inclusion study. *Ore Geol. Rev.* **2015**, *65*, 701–717. [CrossRef]
7. Yang, L.Q.; Deng, J.; Guo, L.N.; Wang, Z.L.; Li, X.Z.; Li, J.L. Origin and evolution of ore fluid and gold deposition processes at the giant Taishang gold deposit, Jiaodong Peninsula, eastern China. *Ore Geol. Rev.* **2016**, *72*, 585–602. [CrossRef]
8. Qiu, K.F.; Marsh, E.; Yu, H.C.; Pfaff, K.; Gulbransen, C.; Gou, Z.Y.; Li, N. Fluid and metal sources of the Wenquan porphyry molybdenum deposit, Western Qinling, NW China. *Ore Geol. Rev.* **2017**, *86*, 459–473. [CrossRef]
9. Chinnasamy, S.S.; Uken, R.; Reinhardt, J.; Selby, D.; Johnson, S. Pressure, temperature, and timing of mineralization of the sedimentary rock-hosted orogenic gold deposit at Klipwal, southeastern Kaapvaal Craton, South Africa. *Miner. Depos.* **2015**, *50*, 739–766. [CrossRef]
10. Sun, S.C.; Zhang, L.; Wu, S.G.; Peng, J.S.; Wen, T. Metallogenic mechanism of the Huangjindong gold deposit, Jiangnan Orogenic Belt: Constraints from mineral formation environment and physicochemical conditions of metallogenesis. *Acta Petrol. Sin.* **2018**, *34*, 1469–1483. (In Chinese with English Abstract)
11. Yang, L.Q.; Deng, J.; Wang, Z.L.; Guo, L.N.; Li, R.H.; Groves, D.I.; Danyushevsky, L.V.; Zhang, C.; Zheng, X.L.; Zhao, H. Relationships between gold and pyrite at the Xincheng gold deposit, Jiaodong, Peninsula, China: Implication for gold source and deposition in a brittle epizonal environment. *Econ. Geol.* **2016**, *111*, 105–126. [CrossRef]
12. Large, R.R.; Maslennikov, V.V.; Robert, F.; Danyushevsky, L.V.; Chang, Z.S. Multistage sedimentary and metamorphic origin of pyrite and gold in the giant Sukhoi Log Deposit, Lena gold province, Russia. *Econ. Geol.* **2007**, *102*, 1233–1267. [CrossRef]
13. Finch, E.G.; Tomkins, A.G. Pyrite-Pyrrhotite Stability in a Metamorphic Aureole: Implications for Orogenic Gold Genesis. *Econ. Geol.* **2017**, *112*, 661–674. [CrossRef]
14. Kretschmar, U.; Scott, S.D. Phase relations involving arsenopyrite in the system Fe-As-S and their application. *Can. Mineral.* **1976**, *14*, 364–386.
15. Yang, L.Q.; Deng, J.; Guo, R.P.; Guo, L.N.; Wang, Z.L.; Cheng, B.H.; Wang, B.H.; Wang, X.D. World-class Xincheng gold deposit: An example from the giant Jiaodong Gold Province. *Geosci. Front.* **2015**, *7*, 419–430. [CrossRef]
16. Zhang, L.; Liu, Y.; Li, R.H.; Huang, T.; Zhang, R.Z.; Chen, B.H.; Li, J.K. Lead isotope geochemistry of Dayingezhuang gold deposit, Jiaodong Peninsula, China. *Acta Petrol. Sin.* **2014**, *30*, 2468–2480. (In Chinese with English Abstract)

17. Zhang, L.; Yang, L.Q.; Wang, Y.; Weinberg, R.F.; An, P.; Chen, B.Y. Thermochronologic constrains on the processes of formation and exhumation of the Xinli orogenic gold deposit, Jiaodong Peninsula, eastern China. *Ore Geol. Rev.* **2017**, *81*, 140–153. [CrossRef]

18. Deng, J.; Yang, L.Q.; Gao, B.F.; Sun, Z.S.; Guo, C.Y.; Wang, Q.F.; Wang, J.P. Fluid evolution and metallogenic dynamics during tectonic regime transition: Example from the Jiapigou gold belt in Northeast China. *Resour. Geol.* **2009**, *59*, 140–152. [CrossRef]

19. Deng, J.; Wang, C.M.; Bagas, L.; Santosh, M.; Yao, E. Crustal architecture and metallogenesis in the south-eastern North China Craton. *Earth-Sci. Rev.* **2018**, *182*, 251–272. [CrossRef]

20. Groves, D.I.; Santosh, M.; Goldfarb, R.J.; Zhang, L. Structural geometry of orogenic gold deposits: Implications for exploration of world-class and giant deposits. *Geosci. Front.* **2018**, *9*, 1163–1177. [CrossRef]

21. Qiu, K.F.; Yu, H.C.; Gou, ZY.; Liang, Z.L.; Zhang, J.L.; Zhu, R. Nature and origin of Triassic igneous activity in the Western Qinling Orogen: The Wenquan composite pluton example. *Int. Geol. Rev.* **2018**, *60*, 242–266. [CrossRef]

22. Zhang, L.; Yang, L.Q.; Weinberg, R.F.; Groves, D.I.; Wang, Z.L.; Li, G.W.; Liu, Y.; Zhang, C.; Wang, Z.K. Anatomy of a world-class epizonal orogenic-gold system: A holistic thermochronological analysis of the Xincheng gold deposit, Jiaodong Peninsula, eastern China. *Gondwana Res.* **2019**, *70*, 50–70. [CrossRef]

23. Charvet, J.; Shu, L.; Shi, Y.; Guo, L.; Faure, M. The building of South china: Collision of Yangzi and Cathaysia blocks, problems and tentative answers. *J. Asian Earth Sci.* **1996**, *13*, 223–235. [CrossRef]

24. Xue, H.M.; Ma, F.; Song, Y.Q.; Xie, Y.P. Geochronology and geochemisty of the Neoproterozoic granitoid association from eastern segment of the Jiangnan orogeny, China: Constraints on the timing and process of amalgamation between the Yangtze and Cathaysia blocks. *Acta Petrol. Sin.* **2010**, *26*, 3215–3244. (In Chinese with English Abstract)

25. Deng, J.; Wang, Q.F. Gold mineralization in China: Metallogenic provinces, deposit types and tectonic framework. *Gondwana Res.* **2016**, *36*, 219–274. [CrossRef]

26. Xu, D.R.; Deng, T.; Chi, G.X.; Wang, Z.L.; Zou, F.H.; Zhang, J.L.; Zou, S.H. Gold mineralization in the Jiangnan Orogenic Belt of South China: Geological, geochemical and geochronological characteristics, ore deposit-type and geodynamic setting. *Ore Geol. Rev.* **2017**, *88*, 565–618. [CrossRef]

27. Zhang, L.; Yang, L.Q.; Groves, D.I.; Liu, Y.; Sun, S.C.; Qi, P.; Wu, S.G.; Peng, J.S. Geological and isotopic constraints on ore genesis, Huangjindong gold deposit, Jiangnan Orogen, southern China. *Ore Geol. Rev.* **2018**, *99*, 264–281. [CrossRef]

28. Liu, Q.Q.; Shao, Y.J.; Chen, M.; Algeo, T.J.; Li, H.; Dick, J.M.; Wang, C.; Wang, W.S.; Li, Z.Q.; Liu, Z.F. Insights into the genesis of orogenic gold deposits from the Zhengchong gold field, northeastern Hunan Province, China. *Ore Geol. Rev.* **2019**, *105*, 337–355. [CrossRef]

29. Shao, Y.J.; Wang, W.S.; Liu, Q.Q.; Zhang, Yu. Trace element analysis of pyrite from the Zhengchong gold deposit, Northeast Hunan Province, China: Implications for the ore-forming process. *Minerals* **2018**, *8*, 262. [CrossRef]

30. Wen, Z.L.; Deng, T.; Dong, G.J.; Zou, F.H.; Xu, D.R.; Wang, Z.L.; Lin, G.; Chen, W. Characteristics of ore-controlling structures of Wangu gold deposit in Northeastern Hunan Province. *Geotecton. Metallog.* **2016**, *40*, 281–294. (In Chinese with English Abstract)

31. Peng, Z.H. The Geological Characteristics and Metallogenic Prognosis of Xiaojiashan Gold Deposit in Liling, Hunan Province. Master's Thesis, Central South University, Changsha, China, 2012; pp. 1–143. (In Chinese with English Abstract)

32. Cox, S.F.; Sun, S.S.; Etheridge, M.A.; Wall, V.J. Structural and geochemical controls on the development of turbidite-hosted gold quartz vein deposits, Wattle Gully mine, central Victoria, Australia. *Econ. Geol.* **1995**, *90*, 1722–1786. [CrossRef]

33. Sibson, R.H.; Robert, F.; Poulsen, K.H. High-angle reverse faults, fluid-pressure cycling, and mesothermal gold-quartz deposits. *Geology* **1988**, *16*, 551–555. [CrossRef]

34. Wilson, C.J.L.; Schaubs, P.M.; Leader, L.D. Mineral Precipitation in the Quartz Reefs of the Bendigo Gold Deposit, Victoria, Australia. *Econ. Geol.* **2013**, *108*, 259–278. [CrossRef]

35. Zhang, W.L. EPMA of Au occurrence in gold, Huangjindong gold deposit. *Geol. J. China Univ.* **1997**, *3*, 256–262. (In Chinese)

36. Liu, Y.J.; Sun, C.Y.; Cui, W.D.; Ji, J.F. Study on the occurrence of gold in arsenopyrite of Huangjindong gold deposit in Hunan province. *Contrib. Geol. Miner. Resour. Res.* **1989**, *4*, 42–49. (In Chinese with English Abstract)

37. Large, R.; Thomas, H.; Craw, D.; Henne, A.; Henderson, S. Diagenetic pyrite as a source for metals in orogenic gold deposit, Otago schist, New Zealand. *N. Z. J. Geol. Geophys.* **2012**, *55*, 137–149. [CrossRef]

38. Hey, M.H. A new review of the chlorite. *Mineral. Mag.* **1954**, *30*, 277–292. [CrossRef]

39. Sharp, Z.D.; Essene, E.J.; Kelly, W.C. A re-examination of the arsenopyrite geobarometry: Pressure considerations and applications to natural assemblages. *Can. Mineral.* **1985**, *23*, 517–534.

40. Walshe, J.L. A six-component chlorite solid solution model and the conditions of chlorite formation in hydrothermal and geothermal systems. *Econ. Geol.* **1986**, *81*, 681–703. [CrossRef]

41. Dora, M.L.; Randive, K.R. Chloritisation along the Thanewasna shear zone, western Bastar Craton, Central India: Its genetic linkage to Cu–Au mineralization. *Ore Geol. Rev.* **2015**, *70*, 151–172. [CrossRef]

42. Battaglia, S. Applying X-ray geothermometer diffraction to a chlorite. *Clays Clay Miner.* **1999**, *47*, 54–63. [CrossRef]

43. Rausell-Colom, J.A.; Wiewiora, A.; Matesanz, E. Relation between composition and d_{001} for chlorite. *Am. Mineral.* **1991**, *76*, 1373–1379.

44. Nieto, F. Chemical composition of metapelitic chlorites: X-ray diffraction and optical property approach. *Eur. J. Mineral.* **1997**, *9*, 829–841. [CrossRef]

45. Zheng, Z.P.; Chen, F.R.; Yu, X.Y. Characteristics of chlorite and the mineralization significance of the gold deposit. *Acta Mineral. Sin.* **1997**, *17*, 100–106. (In Chinese with English Abstract)

46. Zhang, W.; Zhang, S.T.; Cao, H.W.; Wu, D.J.; Xiao, C.X.; Chen, H.J.; Tang, L. Mineral characteristics of chlorite and its significance in the tin deposit of Xiaolonghe in western Yunnan Province. *J. Chengdu Univ. Technol. (Sci. Technol. Ed.)* **2014**, *41*, 318–328. (In Chinese with English Abstract)

47. Velásquez, G.; Beziat, D.; Salvi, S.; Siebenaller, L.; Borisova, AY.; Pokrovski, G.S.; Parseval, P.D. Formation and deformation of pyrite and implications for gold mineralization in the El Callao district, Venezuela. *Econ. Geol.* **2014**, *109*, 457–486. [CrossRef]

48. Arehart, G.B. Gold and arsenic in iron sulfides from sediment-hosted disseminated gold deposits: Implications for depositional processes. *Econ. Geol.* **1993**, *88*, 171–185. [CrossRef]

49. Pokrovski, G.S.; Kokh, M.A.; Guillaume, D.; Borisova, A.Y.; Gisquet, P.; Hazemann, J.; Lahera, E.; Del Net, W.; Proux, O.; Testemale, D.; et al. Sulfur radical species from gold deposits on Earth. *Proc. Natl. Acad. Sci. USA* **2015**, *112*, 13484–13489. [CrossRef] [PubMed]

50. Bindi, L.; Moelo, Y.; Léone, P.; Suchaud, M. Stoichiometric arsenopyrite, FeAsS, from La Roche-Balue Quarry, Loire-Atlantique, France: Crystal structure and Mössbauer study. *Can. Mineral.* **2012**, *50*, 471–479. [CrossRef]

51. Fougerouse, D.; Micklethwaite, S.; Halfpenny, A.; Reddy, S.M.; Cliff, J.B.; Martin, L.A.J.; Kilburn, M.; Guagliardo, P.; Ultrich, S. The golden ark: Arsenopyrite crystal plasticity and the retention of gold through high strain and metamorphism. *Terr. Nova* **2016**, *28*, 181–187. [CrossRef]

52. Fougerouse, D.; Micklethwaite, S.; Tomkins, A.G.; Mei, Y.; Kilburn, M.; Guagliardo, P.; Fisher, L.A.; Halfpenny, A.; Gee, M.; Paterson, D.; et al. Gold remobilisation and formation of high grade ore shoots driven by dissolution-reprecipitation replacement and Ni substitution into auriferous arsenopyrite. *Geochim. Cosmochim. Acta* **2016**, *178*, 143–159. [CrossRef]

53. Fleck, N.A.; Muller, G.M.; Ashby, M.F.; Hutchinson, J.W. Strain gradient plasticity: Theory and experiment. *Acta Metall. Mater.* **1994**, *42*, 475–487. [CrossRef]

54. Dubosq, R.; Lawley, C.J.M.; Rogowitz, A.; Schneider, D.A.; Jackson, S. Pyrite deformation and connections to gold mobility: Insight from micro-structural analysis and trace element mapping. *Lithos* **2018**, *310–311*, 86–104. [CrossRef]

55. Weatherley, D.K.; Henley, R.W. Flash vaporization during earthquakes evidenced by gold deposits. *Nat. Geosci.* **2013**, *6*, 294–298. [CrossRef]

56. Beaudoin, G.; Chiaradia, M. Fluid mixing in orogenic gold deposits: Evidence from the H-O-Sr isotope composition of the Val-d'Or vein field (Abitibi, Canada). *Chem. Geol.* **2016**, *437*, 7–18. [CrossRef]

57. Pitcairn, I.K.; Olivo, G.R.; Teagle, D.A.H.; Craw, D. Sulfide evolution during prograde metamorphism of the Otago and Alpine schists, New Zealand. *Can. Mineral.* **2010**, *48*, 1267–1295. [CrossRef]

58. Williams-Jones, A.E.; Bowell, R.J.; Migdisov, A.A. Gold in Solution. *Elements* **2009**, *5*, 281–287. [CrossRef]

Article

Geological and Geochemical Characteristics of the Archean Basement-Hosted Gold Deposit in Pinglidian, Jiaodong Peninsula, Eastern China: Constraints on Auriferous Quartz-Vein Exploration

Ruihong Li [1,2,*], Ntwali Ntabira Albert [1], Menghe Yun [1], Yinsheng Meng [2] and Hao Du [3]

[1] State Key Laboratory of Geological Processes and Mineral Resources, China University of Geosciences, Beijing 100083, China; albertntnt@163.com (N.N.A.); ymmmmh@163.com (M.Y.)

[2] Institute of Geophysical and Geochemical Exploration, Chinese Academy of Geological Sciences, Langfang 065000, China; mengyinsheng@igge.cn

[3] Sanshandao Gold Company, Shandong Gold Mining Stock Co., Ltd., Laizhou 261438, China; duhaouu@163.com

* Correspondence: liruihong@igge.cn; Tel.: +86-(0316)-2267628

Received: 18 December 2018; Accepted: 16 January 2019; Published: 21 January 2019

Abstract: The gold deposits that are hosted in the Archean metamorphic rock, have yet to be explored beyond Pinglidian gold deposit in the northwestern Jiaodong Peninsula, eastern China. This kind of gold deposit differs from those that are hosted in Mesozoic granitoids, showing good potential for the prospecting of auriferous quartz-vein gold deposits controlled by the structures in greenfield Archean metamorphic rock. Pinglidian gold deposit is located in the hanging wall of the Jiaojia fault and consists of eight separated orebodies that are enveloped by altered rock in Archean biotite plagiogneiss. These orebodies and wall-rock alterations are strongly controlled by local structures that formed during the Mesozoic rotation and kink folding of the foliated and fissile Archean basement host. The major wall-rock alterations comprise sericitization, silicification, pyritization, and carbonation, which is up to 18 m in width and progressively increases in intensity towards the auriferous quartz vein. The visible gold is present as discrete native gold and electrum grains, which have basically filled in all manner of fractures or are adjacent to galena. We recognize two types of gold bearing quartz veins that are associated with mineral paragenetic sequences during hydrothermal alteration in the Pinglidian gold deposit. The petrological features and geochemical compositions in the reaction fronts of the alteration zone suggest variations in the physicochemical conditions during ore formation. These minerals in the wall rock, such as plagioclase, biotite, zircon, titanite, and magnetite, have been broken down to hydrothermal albite, sericite, and quartz in a K–Na–Al–Si–O–H system, and sulfides in a Fe–S–O–H system. The major and trace elements were calculated by the mass-balance method, showing gains during early alteration and losses during late alteration. The contents of K_2O, Na_2O, CaO, and LOI varied within the K–Na–Al–Si–O–H system during alteration, while Fe_2O_3 and MgO were relatively stable. Rare-earth elements (REE) changed from gains to losses alongside the breakdown of accessory minerals, such as large ion lithophile elements (LILE). The Sr and Ba contents exhibited high mobility during sericite-quartz alteration. Most of the low-mobility high-field strength elements (HFSE) were moderately depleted, except for Pb, which was extremely high in anomalous samples. The behavior of trans-transition elements (TRTE) was related to complicated sulfides in the Fe–S–O–H system and was constrained by the parameters of the mineral assemblages and geochemical compositions, temperature, pressure, pH, and fO_2. These factors during ore formation that were associated with the extents and intensity of sulfide alteration and gold precipitation can be utilized to evaluate the potential size and scale of an ore-forming hydrothermal system, and is an effective exploration tool for widespread auriferous quartz veins in Archean metamorphic basements.

Keywords: archean basement; quartz vein; hydrothermal alteration halo; Pinglidian gold deposit; Jiaodong

---

1. Introduction

The Precambrian terrane over the globe are important repositories of major gold deposits [1]. However, many of the world-class gold deposits in the North China Craton (NCC) are spatially associated with Mesozoic magmatic intrusions and related processes in Jiaodong [2–4], and their formation has been correlated with massive hydrothermal alterations [5–7]. Hydrothermal alterations are traditionally prevalent in the upper crust, especially in regional-scale fault zones, including K-feldspathization, albitization, silicification, sericitization, carbonatization, chloritization, and sulfidation [8–13]. Mesothermal gold deposits typically form during compressional or transpressional tectonic events [14,15], wherein regional-scale alteration requires abundant fluid and mass transport through channels by the reactivation of previous faults or hydrofractures [16,17]. However, the occurrence of large auriferous quartz veins and hydrothermal halos in low-permeability Archean metamorphic rocks results from the diffusive metasomatism of wall rock, invoking extremely large fluid fluxes and dense brittle microfracture pathways [18,19]. When the fluid contacts mineral assemblages in wall rock, the system is immediately no longer in equilibrium, and re-equilibration reactions in the presence of a fluid phase drive large-scale processes, such as dissolution and precipitation [20]. Thus, the formation and development of products, such as alteration halos and quartz veins, are related to fluid-rock interactions and mass transport, which strongly affect the petrology and geochemical composition of the original rock [21]. Numerous field evidence and experimental observations indicated that fluid-induced reaction fronts migrate from veins to fresh wall rock through pores or microfractures [12,20,22]. Metamorphism that is accompanied by a change in the rock composition is referred to as metasomatism and is normally assumed to be the response to fluid-controlled mass transfer rather than purely to changes in P–T conditions [23,24]. Accordingly, large-scale hydrothermal alteration zones depend on the development and maintenance of an efficient porosity-permeability system during fluid-rock interactions in natural rocks [25], emphasizing the importance of mineral-assemblage replacement reactions during ore formation.

Jiaodong Peninsula is the most important gold province in the NCC [13,26,27]. The majority of the gold resources (>95%) are hosted in Mesozoic granitoids and are currently recognized as one of the largest granitoid-hosted gold provinces in the world [9]. These gold deposits are generally classified as disseminated-/stockwork-type (Jiaojia-type) and auriferous quartz-vein-type (Linglong-type) deposits, and both are hosted in Mesozoic granitoids and prominently controlled by regional-scale NE- to NNE-trending fault zones (Figure 1) [13,28–32]. Li et al. detailed the significant alteration halos that envelop the low-grade disseminated-/stockwork-type ore [12], however, high-grade auriferous quartz vein-type ores have traditionally been conceptualized to have negligible wall-rock alteration. Recognizing such alteration zones increases the target mineralized zone and the volume of potential ore during exploration [10,33]. Actually, the original mineral assemblages and structure-controlled fluid flow are obscured by alteration reactions in centimeter- to decameter-scale alteration zones in Mesozoic granitoids. Meanwhile, the detailed hydrothermal alteration of host rocks, especially Precambrian metamorphic rock instead of Mesozoic rock in auriferous quartz-vein gold deposits, has rarely been described.

In this paper, we begin with observations at the underground-outcrop scale and then carefully examine changing mineral assemblages and textural relationships in thin sections, and conduct whole-rock geochemistry analysis and electron microprobe analysis on sulfides to interpret the ore formation in the hydrothermal alteration halos in the Pinglidian gold deposit in Archean metamorphic rock, instead of the well-known Mesozoic granitoids in Jiaodong.

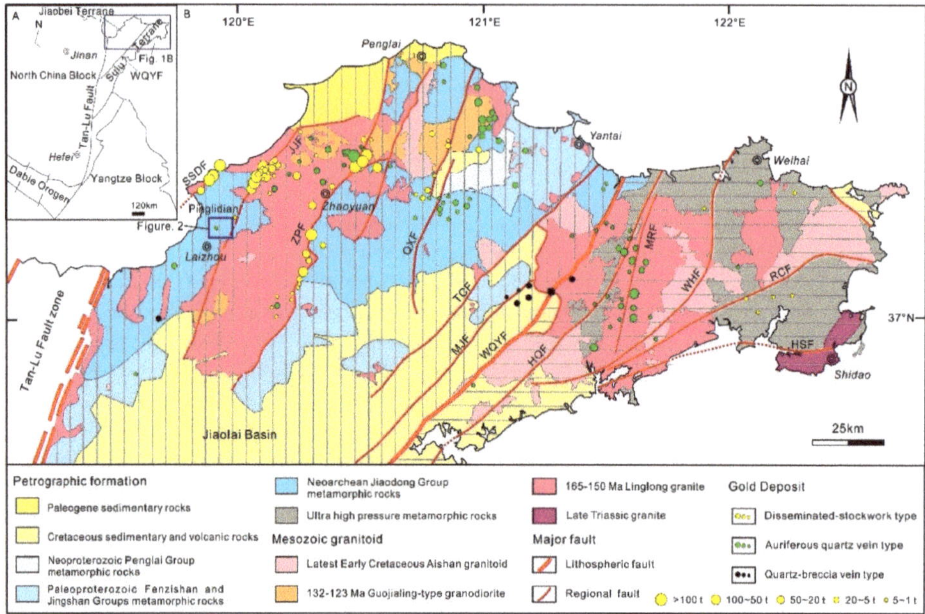

Figure 1. (**A**) Simplified tectonic map of the Jiaodong Peninsula. (**B**) Geological map of the Jiaodong gold province, showing the distribution of major fault zones, Precambrian metamorphic rocks, Mesozoic granitoid intrusions, sedimentary rocks, and gold deposits [26]. The different circle sizes represent five tonnage ranges for the gold deposits (>100 t, 100–50 t, 50–20 t, 20–5 t, and 5–1 t). The Jiaojia goldfield is situated in the northwestern Jiaodong Peninsula. HQF, Haiyang-Qingdao Fault; HSF, Haiyang-Shidao Fault; JJF, Jiaojia Fault; MJF, Muping-Jimo Fault; MRF, Muping-Rushan Fault; QXF, Qixia Fault; RCF, Rongcheng Fault; SSDF, Sanshandao Fault; TCF, Taocun Fault; WHF, Weihai Fault; WQYF, Wulian-Qingdao-Yantai Fault; ZPF, Zhaoping Fault.

2. Regional Geology

The Jiaodong Peninsula occupies the southeastern corner of the North China Block (NCB) and the northeastern corner of the South China block (SCB) [7,9,12,34]. These two blocks are separated by the Wulian-Qingdao-Yantai Fault (WQYF), associated with the Triassic collisional event [26]. The northwestern area of Jiaodong is called the Jiaobei Terrane, and the southeastern area is called the Sulu Terrane. The Jiaobei Terrane was divided into the Jiaobei uplift to the north and the Jiaolai Basin to the south by an E-W-trending normal fault during Early Cretaceous [35]. The Jiaobei uplift consists of Precambrian metamorphic basement and Mesozoic intrusions. More than 95% of the gold resources in the Jiaodong Peninsula are hosted by Late Jurassic and Early Cretaceous granitoids (Figure 1) [5,12]. Most of the gold deposits are controlled by NE- to NNE-trending fault systems, with a spacing of approximately 15–30 km from west to east (Figure 1) [36].

The Precambrian metamorphic basement is dominated by Mesoarchean to Neoarchean trondhjemite-tonalite-granodiorite (TTG) gneisses [37,38], Neoarchean Jiaodong Group metamorphic rocks, and Paleoproterozoic Fenzishan and Jingshan Group metamorphic rocks, which occupy approximately half of the northwestern Jiaodong Peninsula (Figures 1 and 2) [39,40]. The Jiaodong Group metamorphic rocks consist of biotite plagiogneiss, biotite granulite, amphibolite, and biotite-hornblende-plagioclase gneiss [41]. The Fenzishan Group is a Proterozoic high-grade metasedimentary sequence of calc-silicate rocks, marble, and minor amphibolite [42]. The Mesozoic intrusions can be clearly pinpointed to the Late Jurassic and middle Early Cretaceous, and these intrusions are associated with gold mineralization in close spatial and temporal relationships and are widely exposed in the Jiaobei uplift [34,43–45]. The Late Jurassic Linglong granitoids, emplaced in 165–150 Ma, host most of the gold deposits [46,47] and comprise biotite granite, quartz-diorite, and granodiorite. The middle Early Cretaceous Guojialing granitoids intruded the Linglong granitoids (Figures 1 and 2) [43,46] and consist of porphyritic quartz monzonite, granodiorite, and monzogranite, with large K-feldspar megacrysts. These granitoids contain Neoarchean, Paleoproterozoic, and Neoproterozoic xenocrystic zircons [34,43,44,47].

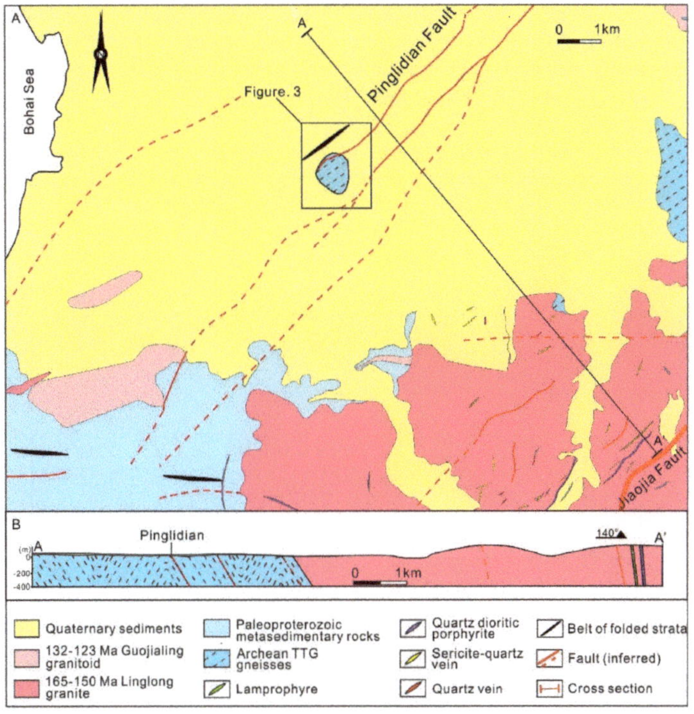

Figure 2. Sketch geological map of the northwestern Jiaodong gold province. (**A**) The Pinglidian gold deposit is situated within the Archean Jiaodong Group's metamorphic rocks in the hanging wall of Jiaojia Fault. (**B**) Inferred cross section of AA' in (A) to reveal the regional structure pattern (modified from unpublished internal material owned by Sanshandao Gold Company).

EW- to ENE-trending regional folding is well developed in the Precambrian metamorphic basement, such as Qixia anticlinorium and Laiyang anticlinoria [28,30]. NNE- to NE-trending fault systems rigidly controlled most of the gold mineralization that is hosted in the Mesozoic granitoids in the Jiaobei uplift [9,26]. From west to east, these fault systems are the Sanshandao Fault, Jiaojia Fault, and Zhaoping Fault, which are considered to be second-order faults of the lithospheric-scale Tan-Lu Fault [48–50]. These NNE- to NE-trending fault systems mainly follow the contact boundary between Mesozoic granitoid rocks in the footwall and Precambrian metamorphic rocks in the hanging wall, and partially cut through the Mesozoic granitoids and Precambrian metamorphic rocks. The lithological contacts along these faults, alongside fault jogs, exhibited the most control on the occurrence and distribution of the gold deposits (Figures 1 and 2). The gold deposits preferentially occurred in these jogs because of heterogeneous strain and increased fracture permeability, which focused ore fluid flow into their areas of influence [51–53], and fluid thus reacted with the wall rock. Second-order N- to NNE-trending faults, such as the Wang'ershan Fault in the footwall and Pinglidian Fault in the hanging wall, overprinted the Qixia anticlinorium in the Precambrian metamorphic basement [30], are broadly parallel to the Jiaojia Fault, and converge with the Jiaojia Fault at depth [13] (Figure 2).

3. Geology of the Pinglidian Gold Deposit

3.1. Ore Geology

The Pinglidian gold deposit is located approximately 13 km to the northeast of Laizhou City in Pinglidian, Shandong Province. The Pinglidian gold deposit, occupying an area of 1.2 km^2, is one of the few auriferous quartz-vein gold deposits that are hosted in Archean metamorphic rock instead of Mesozoic granite in the northwestern area of the Jiaobei uplift, with a proven reserve of >5 t of gold (Figure 3). This deposit has been exploited since 1996 by the 10th team of the Chinese Army, with the gold grade varying from 1.7 to 47.02 g/t.

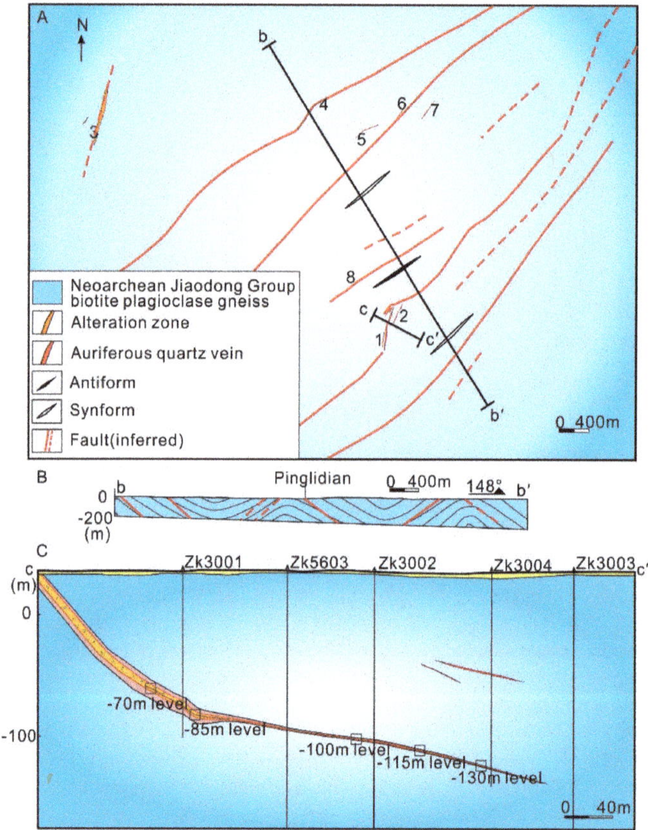

Figure 3. Simplified geological map and sketch map of the hydrothermal alteration zone and orebody in the Pinglidian gold deposit, showing the Archean basement fault's structural control on mineralization and hydrothermal alteration. (**A**) Plan view of an outcrop of the Pinglidian gold deposit. (**B**) Idealized regional scale ore-controlling structures of section bb' in Figure 3A. (**C**) The No. 1 orebodies controlled by local scale fault of section bb' in Figure 3A (modified from unpublished internal material owned by Sanshandao Gold Company).

Outcrops are rare in the ore district, covered by Paleogene sedimentary that is 1 m to 20 m thick. The host rock of the Pinglidian gold deposit is the Archean Jiaodong Group metamorphic rock (Figure 4), which comprises biotite plagiogneiss and amphibolite gneiss (Figure 4A,B). The biotite plagiogneiss is gray to dark-gray in color and medium- to fine-grained lepidoblastic texture. The major minerals are plagioclase (55–70%), biotite (15–25%), and quartz (10–15%), and accessory minerals include magnetite, apatite, and zircon. The plagioclase amphibolite is grayish in color and shows medium- to fine-grained crystalloblastic textures (Figure 4C). The major minerals are plagioclase (45–60%), amphibole (30–40%), and biotite (10–15%), and accessory minerals include magnetite, apatite, and zircon (Figure 4D,E,F).

Figure 4. Field photographs (**A,B**), hand specimen (**C**), and transmitted-light photomicrographs (**D–F**) of mineral assemblages in biotite plagiogneiss. (**A,B**) The biotite plagiogneiss also experienced processes such as regional metamorphism and migmatization, resulting in schistosity and boudinage. (**C**) The main mineral assemblages of the biotite plagiogneiss is plagioclase and biotite, and accessory minerals include magnetite, apatite, and zircon. (**D**) Subhedral to anhedral plagioclase and biotite form the preferred orientation distribution in the protolith. (**E,F**) The deformation of biotite and plagioclase is apparent, with local recrystallization along the margins of the plagioclase.

Auriferous quartz veins are predominantly enveloped by a hydrothermal alteration halo in cataclastic zones, structurally controlled by the NE-trending and SE-dipping Pinglidian Fault and subsidiary faults. The Pinglidian Fault, generally striking 50° and dipping 30° SE, is approximately

15 km in length and 5–20 m in width. These faults controlled the occurrence of the No. 1 to No. 8 orebodies, which are tabular or lenticular in shape. The Pinglidian Fault is filled with cataclastic and brecciated rocks and fault gouges, which are a few centimeters to tens of centimeters thick. Locally, the occurrence of the Pinglidian Fault has rapidly varied to accommodate the wavy foliation in the metamorphic wall rock. The auriferous quartz veins predominantly occur along these faults and are parallel to the foliation in the wall rock, indicating that the fault was deformed and overprinted the Archean metamorphic rock.

3.2. Characteristic of the Orebodies

Eight auriferous quartz vein orebodies are present in the Pinglidian gold deposit (Figure 3). The No. 1 and No. 2 orebodies comprise more than 80% of the proven reserves in the Pinglidian gold deposit and are believed to be controlled by secondary faults in the hanging wall of the Jiaojia Fault (Figures 2 and 3), as described below.

The No. 1 orebody is the largest orebody in the Pinglidian gold deposit and is characterized by massive auriferous quartz vein ores in hydrothermal breccias and cataclastic rocks, including 1.8-m- to 18-m-thick zones of reddish alteration and sericite and quartz, with both disseminated pyrite and pyrite-quartz stockworks. The major orebody is tabular in shape and approximately 550 m long, with a strike of 15–42° and dip of 5–25° SE. The orebody extends gently dips down for up to 250 m, limited between the −70 m and −130 m levels and ranging from 0.2 m to 1.2 m (average of 0.62 m) in thickness. The gold grade varies from 1.7 to 47.02 g/t, with an average of 6.03 g/t.

The No. 2 orebody is much smaller than the No. 1 orebody and also exhibits massive auriferous quartz veins. This orebody generally strikes 33° and dips 7–10° SE, with a length of approximately 130 m. The No. 2 orebody gently dips down from the −28 m to −45 m levels, ranging from 0.4 m to 0.6 m (average of 0.36 m) in thickness. The gold grade varies from 12.8 to 24.7 g/t, with an average of 21.51 g/t.

3.3. Wall-Rock Alteration and Gold Mineralization

Gold mineralization that is associated with intense wall-rock alteration is controlled by the fault zone in Jiaodong. Auriferous quartz-vein orebodies, enveloped by hydrothermal alteration halos that are up to 18 m in width, symmetrically occur along the Pinglidian Fault. The spatial zonation of the four major types of mineral assemblages reveal the alteration zone from protolith (biotite plagiogneiss) to orebody (auriferous quartz vein) in the Pinglidian gold deposit. The wall-rock alteration includes sericitization, silicification, and pyritization. All these alterations are related to quartz veins that occur in two main alteration halos (reddish alteration zones and sericite-quartz alteration zones) between the protolith rock and auriferous quartz veins (Figure 5).

3.3.1. Protolith Rock and Reddish Alteration

The fresh protolith rock originated from Archean biotite plagiogneiss (Figure 4A), which has experienced the regional metamorphism of amphibolite facies under medium-pressure deformation. According to field observations, the biotite plagiogneiss also experienced regional metamorphism to migmatization, as evidenced by the formation of schistosity and boudinage (Figure 4A,B). These schistosities strike 50–53° and dip 15–42° SE in Pinglidian. Mineral lineation and foliation are well developed in the biotite plagiogneiss, according to hand species and thin sections (Figure 4C,D). The biotites are euhedral to subhedral plate crystals and are distributed with preferential orientation (Figure 4D,F). The plagioclase is subhedral to anhedral tabular crystals and usually occurs as polysynthetic twins (Figure 4D–F). Accessory minerals include magnetite, zircon, monazite, apatite, and titanite.

Figure 5. Field photographs (**A,B**), hand specimen (**C**), and transmitted-light photomicrographs (**D–F**) of mineral assemblages that are related to reddish alteration. (**A**) Biotite plagiogneiss was replaced by iron stained alteration, displaying a pinkish color. (**B**) Reddish alteration occurred in the outermost alteration halo, changing the quartz-vein orebody to biotite plagioclase gneiss. (**C**) The main mineral assemblages of the altered K-feldspar rock are secondary K-feldspar, albite, and quartz, with minor amounts of muscovite and relict plagioclase. (**D**) Relict plagioclase is surrounded by quartz, muscovite, and secondary K-feldspar. Albite is present in the altered rock. (**E,F**) In the heavily altered area, K-feldspar and muscovite completely replaced plagioclase, while some remnant plagioclase is very dirty with massive sericite. Brittle fractures cut through the secondary K-feldspar, some of which were filled with sericite.

Reddish alteration in the outermost area of the alteration veins (Figures 3E,I and 5A,B) envelops the gold mineralization, which is related to earlier quartz veins and is characterized by scarce K-feldspar and albite that transformed from plagioclase or biotite in Archean metamorphic rock (Figure 5C–F). The thickness of this reddish alteration zone varies from tens of centimeters to meters because of diffusion and the intensity of fluid-rock reactions. The main mineral assemblages of this reddish alteration rock include secondary K-feldspar, albite, and quartz, with minor amounts of K-feldspar, muscovite, and relict plagioclase (Figure 5C). Biotite is completely absent, and relict plagioclase is surrounded by subhedral or anhedral quartz, muscovite, and secondary K-feldspar and albite in the altered rock (Figure 5D). In some local areas, plagioclase was almost completely replaced by albite and muscovite, while some remnant plagioclase is turbid with massive sericite (Figure 5E,F). Brittle fractures cut through the new secondary K-feldspar, some of which were filled with sericite. Accessory minerals in the biotite plagiogneiss, such as pyrite, magnetite, apatite, and zircon, were also replaced by albite and quartz in the weak alteration zone. Preferentially oriented foliation and lineation are present in the protolith instead of massive structures.

3.3.2. Sericite-Quartz Alteration

Adjacent to the auriferous quartz veins, relatively narrow sericite-quartz alteration zones vary from tens of centimeters to meters in thickness and are associated with broad or narrowing quartz veins. A sericite-quartz alteration zone closely envelops the quartz-vein orebody, pinched within the broad cataclastic sericite-quartz alteration zone (Figure 6A). Intensive sericitization, silicification, and pyritization overprinted previous reddish alteration along brittle cracks or open spaces in the cataclastic zone. Dense sericite-quartz alteration partially filled in brittle fractures in the quartz veins and enclosed earlier quartz-vein breccia (Figure 6B). The sericite-quartz alteration that is adjacent to earlier quartz vein consists of dense sericite, quartz, and minor pyrite (Figure 6C). The quartz and muscovite are distributed with preferential orientation in the dense sericite-quartz alteration zone (Figure 6D). Subhedral and anhedral quartz and muscovite are surrounded by a sericite matrix, and porphyroclasts show irregular shapes and various sizes (Figure 6E,F). Quartz is irregularly distributed in the sericite matrix in the sericite-quartz alteration zone (Figure 6G). Euhedral pyrite occurs in the sericite-quartz alteration zone, and anhedral galena filled in brittle fractures in pyrite (Figure 6H,I). A small amount of pyrite is disseminated within the sericite-quartz alteration rock. In addition, minor galena, which typically filled in cracks in pyrite, precipitated contemporaneously with the disseminated pyrite.

Figure 6. Field photographs (**A**,**B**), hand specimen (**C**), transmitted-light photomicrographs (**D–H**), and reflected-light photomicrographs of mineral assemblages that are associated with the sericite-quartz alteration zone. (**A**) The sericite-quartz alteration zone closely envelops the quartz vein orebody, which is pinched within the broad cataclastic sericite-quartz alteration zone. (**B**) Dense sericite-quartz alteration partially filled in brittle fractures in quartz veins and encloses the earlier quartz-vein breccia. (**C**) The sericite-quartz alteration that is adjacent to earlier quartz veins consists of dense sericite, quartz, and minor pyrite. (**D**) Quartz and muscovite are distributed along preferred orientations in the dense sericite-quartz alteration zone. (**E**,**F**) Subhedral and anhedral quartz and muscovite are surrounded by a sericite matrix, and porphyroclasts show irregular shapes and various sizes. (**G**) Quartz is irregularly distributed in the sericite matrix in the sericite-quartz alteration zone. (**H**,**I**) Euhedral pyrite occurred in the sericite-quartz alteration zone, and anhedral galena filled in brittle fractures in pyrite.

3.3.3. Quartz Veins and Gold Mineralization

Quartz veins occur in the center of the hydrothermal alteration halo. Two types of quartz veins could be identified from the cutting relationships and mineral assemblages. Type-I early quartz veins are milky in color and relatively broad and mainly consist of quartz (>90%) and minor muscovite and plagioclase (Figure 7A). These type-I quartz veins contain almost no gold, even if small pyrite veins or phenocrysts are present. Cataclastic texture is well developed in these early type-I quartz veins.

Younger brittle fractures overprinted the older broad quartz veins (Figure 7B). Type-II late quartz veins are gray to dark gray in color, have varying thickness, and commonly cut through the type-I quartz veins (Figure 7C). The type-II quartz veins predominantly consist of quartz (>80%), galena, and pyrite, with minor chalcopyrite and sphalerite (Figure 7D), and are closely associated with gold mineralization in the Pinglidian gold deposit. These type-II quartz veins have partially or completely replaced any type-I quartz veins.

Figure 7. Field photographs that show two types of quartz veins in the Pinglidian gold deposit. (**A**) This type-I broad milky quartz vein that contains almost no gold, with a few tiny pyrite veins or phenocrysts. (**B**) Late brittle fractures overprinted the early broad quartz vein. (**C**) This type-II quartz sulfide vein cut through the type-I quartz vein. (**D**) The type-II quartz sulfide vein predominantly consists of quartz, galena, pyrite, chalcopyrite, and sphalerite, which have partially or completely replaced the type-I quartz vein ore.

Based on the mineral assemblages and distributions, the ores could be divided into disseminated- (Figure 8A) and massive sulfide ores (Figure 8B,C). The former comprises sericite-quartz or milky quartz veins with disseminated or veinlet mineral assemblages in the center of alteration halos and represents early gold mineralization (stage I), associated with cataclasites and breccias. The quartz-pyrite veinlets (stage II) and massive quartz-sulfide (stage III) ores (Figure 8B,C) represent the main period of gold mineralization, and are associated with changing physicochemical conditions during ore formation in the hydrothermal alteration zones. Anhedral pyrite (stage II) experienced brittle deformation, with some fracture networks filled with younger material (Figure 8D). Some intergranular gold and galena filled in brittle fractures in the stage II pyrite (Figure 8E). The stage-III massive sulfide ore is associated with complex assemblages, including pyrite, galena, pyrrhotite, chalcopyrite, sphalerite, freibergite, and tetrahedrite (Figure 8F–I). Anhedral pyrite was replaced and enclosed by anhedral galena, which is the predominant mineral in the stage-III ores (Figure 8G). Euhedral pyrite coexists with pyrrhotite and galena (Figure 8H). Anhedral pyrite, chalcopyrite, and sphalerite are commonly present in the stage-III ore (Figure 8I). Quartz-carbonate ore is rare, and used to be considered as stage-IV ore and filled in any brittle fractures.

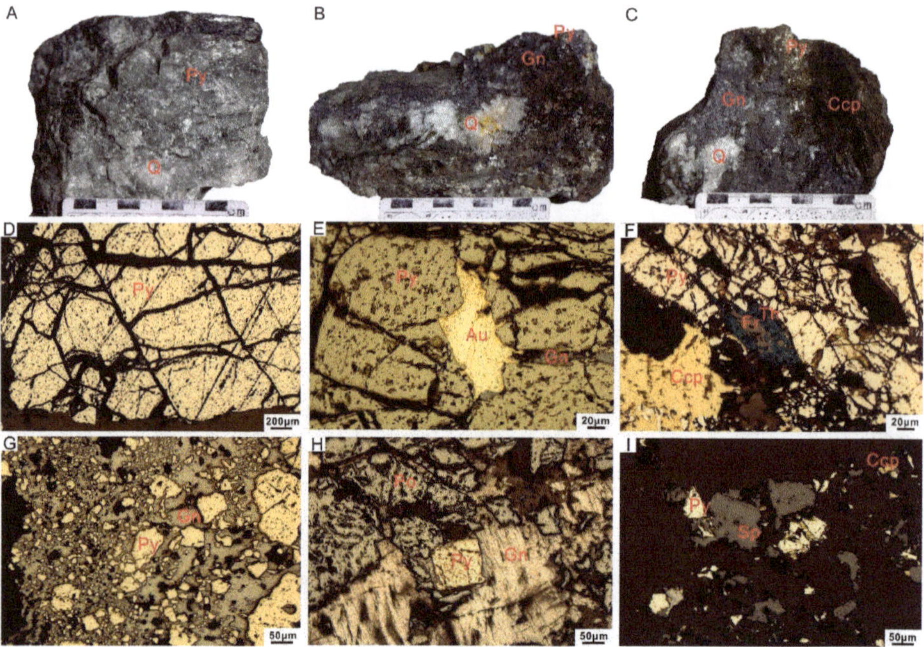

Figure 8. Ore and mineral assemblages in the Pinglidian gold deposit. (**A**) Quartz, pyrite, and sericite (Stage I). (**B**) Quartz, pyrite, galena, and chalcopyrite (Stages II and III). (**C**) Quartz and galena accumulations (Stage III). (**D**) Fracture network that was filled with anhedral pyrite (Stage III). (**E**) Intergranular gold and galena that were associated with brittle fractures in pyrite. (**F**) Pyrite, chalcopyrite, freibergite, and tetrahedrite. (**G**) Anhedral pyrite that was replaced and enclosed by anhedral galena. (**H**) Euhedral pyrite, pyrrhotite, and galena. (**I**) Anhedral pyrite, chalcopyrite, and sphalerite. Au, gold; Ccp, chalcopyrite; Fr, freibergite; Gn, galena; Py, pyrite; Po, pyrrhotite; Q, quartz; Ser, sericite; Sp, sphalerite.

Visible gold occurs as native gold and electrum; the latter comprises more than 90% of the observed visible gold. Visible gold is generally intergrown with anhedral pyrite and galena, and sometimes occurs in brittle fractures in the subhedral to anhedral pyrite or the matrix of the quartz veins (Figure 9A). Meanwhile, brittle cataclastic textures of pyrite are commonly filled with galena (Figure 9A). Electrum is elongated along anhedral galena-deformation fabrics (Figure 9A,B), and native gold shows exsolution texture within the electrum (Figure 9B). Visible gold mainly occurs as irregularly shaped electrum in pyrite and galena in the quartz veins (Figure 9C,D).

3.3.4. Paragenetic Sequences

Four paragenetic sequences have been identified based on the presence of alteration halos (Figure 10), cross-cutting relationships of the mineralized quartz veins or sulfide veinlets, and mineralogical and textural characteristics of the ores to interpret the gold mineralization of the Pinglidian gold deposit (Figures 4–9). As mentioned above, these sequences include quartz-sericite alteration (stage I), quartz-pyrite veinlets (stage II), massive quartz-sulfide veins (stage III), and carbonate (stage IV). The detailed paragenetic sequence of the Pinglidian gold deposit is summarized in Figure 10. Biotite, plagioclase, K-feldspar, and albite occurred before gold mineralization. Pyrite and quartz are the principal visible gold-hosting minerals throughout the entire gold mineralization, from stage I to IV. Native gold and electrum are rare in stages I and IV but are developed in stages II and III. In terms of sulfides, stage III consists of galena, followed by pyrite, chalcopyrite, sphalerite, pyrrhotite, freibergite, and tetrahedrite. In addition, sericite predominantly occurs in stage I, and calcite normally appears in late stage IV material.

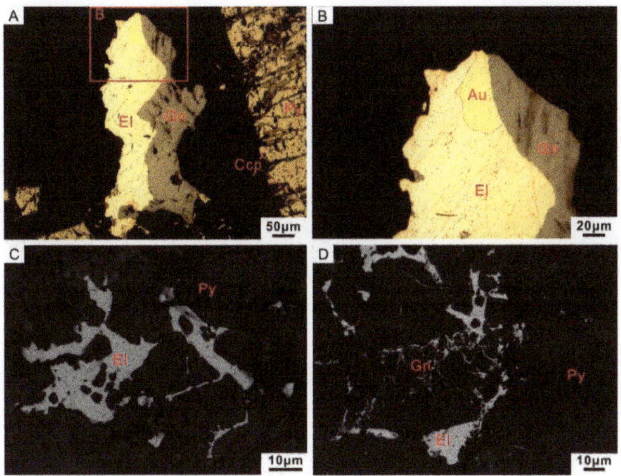

Figure 9. Reflected-light photomicrographs (**A**,**B**) and back-scattered electron (BSE) images (**C**,**D**) of gold in the ore. (**A**) Electrum intergrowth with galena. (**B**) Native gold that coexists with electrum. (**C**,**D**) Pyrite, galena, and electrum in a quartz vein.

Mineral	Stage I	Stage II	Stage III	Stage IV
Quartz				
Sericite				
Pyrite				
Pyrrhotite				
Native gold				
Electrum				
Galena				
Chalcopyrite				
Sphalerite				
Tetrahedrite				
Freibergite				
Calcite				

Figure 10. Paragenetic sequence of the Pinglidian gold deposit as interpreted from the field-cutting relationships, ore textures, and mineral assemblages that were associated with gold mineralization. The line thickness shows the relative abundance of minerals in the paragenetic sequence, as summarized from the above evidence.

4. Analysis Methods and Results

4.1. Electron Probe Microanalyses (EPMA)

Five samples (7 analyses on electrum, 4 analyses on pyrite, 4 analyses on galena, and 3 analyses on chalcopyrite) were analyzed. Quantitative chemical analyses of these 5 samples were conducted by using a JXA-8230 electron microprobe with an accelerating voltage of 15 kV and beam current of 20 nA at the Institute of Mineral Resources, Chinese Academy of Geological Sciences (CAGS), Beijing. Natural minerals and synthetic oxides were used as standards.

The EPMA results are shown in Table 1. Seven spot analyses of visible gold showed electrum contents (in wt %) of 55.92–72.15 for Au, 25.8–43.67 for Ag, 0.14–0.67 for Bi, and 0.03–0.07 for Te. Four spot analyses of pyrite showed elemental contents (in wt %) of 53.42–53.67 for S, 45.8–43.67 for Fe, 0.11–0.21 for Pb, and 0.04–0.06 for Au. Four spot analyses of galena showed elemental contents (in wt %) of 13.34–13.74 for S, 86.28–86.39 for Pb, 0.07–0.15 for Cd, and 0.21–0.3 for Bi. Three spot analyses of chalcopyrite showed elemental contents (in wt %) of 33.94–34.58 for S, 30.24–32.22 for Fe, 31.61–33.72 for Cu, 0.09–0.2 for Pb, 0.06–0.11 for Zn, and 0.11–0.15 for Bi.

Table 1. Representative electron probe microanalyses of metallic minerals from the Pinglidian gold deposit.

Samples	Au	Ag	S	As	Se	Fe	Zn	Pb	Cu	Sb	Cd	Bi	Te	Total (%)	Mineral
02B1-1	71.37	28.30	0.04	/	/	/	/	/	/	/	0.06	0.15	/	99.92	Electrum
02B1-2	62.19	36.89	0.05	/	/	/	/	/	/	/	0.07	0.67	0.05	99.92	Electrum
02B1-3	59.03	39.64	0.11	/	/	0.63	/	/	/	/	/	0.48	0.03	99.92	Electrum
02B1-4	72.86	25.80	0.03	/	/	0.62	/	/	/	/	/	0.61	/	99.92	Electrum
02B1-5	71.65	27.51	0.02	0.06	/	0.42	/	/	/	0.05	/	0.23	0.04	99.98	Electrum
02B1-6	55.92	43.67	/	/	0.06	0.14	/	/	/	/	/	0.14	0.05	99.98	Electrum
02B1-7	72.15	26.86	/	/	/	0.19	/	/	/	/	/	0.57	0.07	99.84	Electrum
14B2-1	/	/	34.42	/	/	30.44	0.06	0.14	34.23	/	/	/	/	99.29	Chalcopyrite
20B1-2	/	/	34.58	0.07	/	30.24	/	0.09	33.72	/	/	0.15	0.03	98.88	Chalcopyrite
28B1-2	0.08	0.72	33.94	/	/	32.22	0.11	0.20	31.61	/	0.06	0.11	/	99.05	Chalcopyrite
16B1-1	/	/	13.46	/	/	/	/	86.3	/	/	0.07	0.29	0.07	100.19	Galena
16B1-2	/	/	13.74	/	/	0.06	/	86.28	/	/	0.15	0.21	0.03	100.47	Galena
20B1-1	/	/	13.34	/	/	/	/	86.39	/	0.04	0.11	0.30	/	100.18	Galena
28B1-1	/	/	13.55	/	/	0.12	/	85.34	/	/	0.10	0.29	/	99.40	Galena
14B2-2	/	/	53.67	0.04	/	46.22	/	0.14	/	/	/	0.11	/	100.18	Pyrite
16B1-3	0.04	/	53.52	/	/	46.40	/	0.11	/	/	/	/	0.03	100.10	Pyrite
20B1-3	/	0.09	53.42	/	/	46.05	/	0.16	0.11	/	0.03	/	/	99.86	Pyrite
28B1-3	0.06	/	53.57	0.05	0.03	45.80	/	0.21	/	/	0.04	/	/	99.76	Pyrite

Note: "/" means the abundance is below the determination limits, 320 ppm for Au, 173 ppm for Ag, 99 ppm for S, 250 ppm for As, 227 ppm for Se, 199 ppm for Fe, 197 ppm for Zn, 421 ppm for Pb, 169 ppm for Cu, 350 ppm for Sb, 266 ppm for Cd, 460 ppm for Bi, and 297 ppm for Te.

4.2. Bulk Geochemistry

4.2.1. Major and Trace Elements

Thirteen samples (3 for biotite plagiogneiss, 4 for reddish alteration, and 6 for sericite-pyrite alteration) were collected to analyze the whole-rock major- and trace-element concentrations. Quantitative analysis was performed by using a Philips PW2404 X-ray Fluorescence Spectrometer (XRF) and an ELEMENT-1 plasma mass spectrometer (Finnigan-MAT Ltd.) at the Research Institute of Uranium Geology, Beijing, China. The analytical uncertainties were less than ±1% and ±5% for major and trace elements, respectively. The detailed analytical procedures are described in Li et al. [54].

4.2.2. Geochemical Characteristics of the Original and Altered Rock

Thirteen samples were classified into two groups according to their major-element concentrations, except for two abnormal samples (e.g., 17B1 is short for PLD15D017B1, and 24B2 is short for PLD15D024B2), as listed in Table 2 and Figure 11. The concentrations of SiO_2 (67.62–75.77%) and Al_2O_3 (12.8–18.8%) remained high and relatively stable (Table 2; Figure 11E). K_2O (1.02–5.94%) and LOI (1.01–4.37%) exhibited increasing concentrations from biotite plagiogneiss and albitized rock to sericite-quartz altered rock (Figure 11A,F), while Na_2O (0.213–5.49%) and CaO (0.154–3.02%) showed decreasing concentrations (Figure 11B,C). Fe_2O_3 (0.65–5.23%) showed a negative correlation with SiO_2 (Figure 11D).

The original and altered rock also exhibited variable chondrite-normalized rare-earth-element (REE) patterns and primitive mantle-normalized trace-element patterns [55] (Figure 12), including LILEs, HFSEs, and TRTEs. The whole-rock REE and trace-element patterns could be classified into four groups in accordance with biotite plagiogneiss, reddish alteration, sericite-quartz alteration, and quartz veins (Tables 2 and 3, Figures 12 and 13).

Table 2. Major oxides (wt %) and trace elements (ppm) for wall-rock samples from the Pinglidian gold deposit.

Samples	Biotite Plagioclase Gneiss			Reddish Alterated Rock				Sericite-Quartz Alterated Rock					
	07B3	13B1	18B5	04B1	17B1	18B4	27B1	02B2	03B2	07B2	14B1	19B2	24B2
SiO_2	70.96	67.62	74.34	75.77	58.24	72.84	72.09	72.47	75.02	67.57	68.18	69.16	58.67
Al_2O_3	15.24	16.11	14.34	13.64	16.46	13.82	13.55	12.80	15.30	15.42	18.80	14.49	19.60
Fe_2O_3	2.22	2.78	0.76	0.65	7.91	3.19	1.80	5.23	1.66	4.14	2.32	2.98	3.05
MgO	1.30	1.92	0.43	0.28	3.22	0.78	0.73	2.23	0.56	2.14	0.79	0.85	1.14
CaO	2.31	3.02	2.61	1.26	6.57	1.45	2.53	0.38	0.15	0.98	0.41	2.85	1.33
Na_2O	5.08	4.86	5.27	5.49	4.64	5.07	4.5	0.25	0.21	3.76	0.28	0.34	0.18
K_2O	1.35	1.54	1.02	1.29	0.84	1.29	1.65	3.30	4.57	2.55	5.94	4.40	6.11
MnO	0.03	0.04	0.02	0.03	0.21	0.04	0.04	0.08	0.02	0.03	0.01	0.09	0.08
TiO_2	0.27	0.39	0.06	0.06	0.80	0.21	0.19	0.25	0.12	0.44	0.12	0.29	0.20
P_2O_5	0.09	0.11	0.03	0.03	0.14	0.06	0.06	0.12	0.04	0.15	0.03	0.09	0.05
LOI*	1.04	1.48	1.01	1.49	0.86	1.12	2.81	2.81	2.24	2.75	3.02	4.37	4.55
TOTAL	99.91	99.89	99.90	100.00	99.90	99.89	99.96	99.94	99.91	99.94	99.93	99.93	94.99
B	9.42	10.50	10.80	10.90	7.69	8.59	10.30	27.20	34.60	11.40	32.50	14.20	24.10
Li	20.40	25.40	5.31	4.73	20.60	19.50	6.11	15.50	3.51	16.30	3.15	11.40	4.30
Be	0.84	1.27	0.99	0.63	1.21	0.97	0.61	1.30	1.45	0.86	1.69	1.14	1.61
Sc	3.96	5.09	1.35	1.29	16.80	3.35	2.70	3.43	1.63	5.18	2.03	4.23	2.67
V	28.00	46.60	10.50	10.80	127.00	28.10	14.30	28.80	12.80	41.80	30.30	25.90	28.50
Cr	5.03	8.19	2.72	3.03	23.10	7.46	2.96	4.86	2.98	8.87	2.25	4.93	2.24
Co	5.50	8.83	1.06	0.97	16.20	3.16	3.27	5.87	2.13	8.08	3.08	3.51	2.28
Ni	3.98	12.80	1.39	1.56	24.30	3.50	1.84	5.19	1.62	6.44	3.82	3.34	2.53
Cu	4.01	25.60	5.51	44.20	35.90	6.74	14.90	12.60	112.00	5.34	9.95	18.80	49.10
Zn	49.60	54.10	19.20	18.80	142.00	81.00	33.80	316.00	43.20	70.50	36.80	59.00	56.50
Ga	19.10	18.20	17.20	14.60	22.80	19.60	15.80	17.60	18.90	20.60	23.20	18.00	24.40
Rb	45.40	53.30	19.60	39.90	14.00	34.10	49.40	97.60	129.00	98.10	154.00	130.00	174.00
Sr	286	382	248	226	303	209	138	25.20	9.97	84.70	10.70	46.40	40.20
Y	5.49	5.60	3.09	2.23	19.80	6.91	4.32	6.78	5.24	8.44	1.52	6.69	3.93
Mo	0.14	0.13	0.14	0.11	0.49	0.15	0.18	0.20	0.15	0.21	1.09	0.13	0.05
Cd	0.15	0.13	0.16	0.15	0.44	0.34	0.14	4.16	0.44	0.25	0.29	0.16	8.51
In	0.02	0.02	0.01	0.01	0.07	0.02	0.01	0.02	0.02	0.03	0.05	0.03	0.11
Sb	0.08	0.11	0.07	0.18	0.08	0.16	0.20	0.41	0.46	0.26	0.27	0.45	4.73
Cs	1.93	1.44	1.30	1.17	0.41	1.66	1.88	1.42	1.90	2.10	2.04	2.94	2.54
Ba	252	374	133	220	102	171	228	207	270	779	723	387	571
La	19.70	17.00	9.59	4.69	14.70	23.30	7.14	16.10	24.40	28.50	9.39	22.50	18.70
Ce	34.70	32.10	17.80	8.40	31.10	43.60	12.60	30.30	45.50	53.20	17.20	39.10	31.70
Pr	3.81	3.78	2.03	0.94	3.89	4.90	1.40	3.46	5.12	5.94	1.90	4.23	3.27
Nd	13.70	14.90	7.53	3.38	15.70	17.90	5.49	13.20	19.50	21.40	7.07	15.20	11.40
Sm	2.22	2.36	1.37	0.66	3.29	3.23	1.04	2.28	3.46	3.37	1.06	2.66	1.67
Eu	0.78	0.78	0.57	0.42	1.41	0.68	0.44	0.51	0.46	0.85	0.26	0.77	0.47
Gd	1.92	2.01	1.16	0.55	2.92	2.67	0.95	1.89	2.82	2.96	0.91	2.29	1.48
Tb	0.27	0.28	0.16	0.09	0.61	0.40	0.17	0.29	0.37	0.41	0.11	0.33	0.18
Dy	1.22	1.28	0.66	0.43	3.26	1.61	0.79	1.40	1.47	1.87	0.39	1.49	0.78
Ho	0.21	0.21	0.12	0.08	0.67	0.27	0.15	0.25	0.20	0.31	0.06	0.24	0.14
Er	0.59	0.58	0.32	0.22	1.90	0.72	0.43	0.69	0.50	0.90	0.16	0.64	0.42
Tm	0.09	0.08	0.04	0.04	0.33	0.10	0.07	0.11	0.06	0.13	0.02	0.09	0.07
Yb	0.45	0.48	0.29	0.23	2.05	0.60	0.41	0.63	0.38	0.77	0.11	0.55	0.41
Lu	0.07	0.06	0.04	0.03	0.31	0.09	0.06	0.10	0.05	0.12	0.02	0.07	0.06
W	0.47	0.16	1.78	2.36	1.01	5.58	1.51	3.02	8.80	2.22	7.67	3.49	8.36
Re	0.005	0.004	0.003	<0.002	0.008	0.006	0.006	<0.002	0.002	0.002	0.007	<0.002	0.002
Tl	0.22	0.24	0.08	0.19	0.08	0.14	0.24	0.53	0.68	0.45	0.75	0.63	0.88
Pb	9.44	242	108	30.30	261	61.40	71.90	337	2081	10.70	165	85.90	35622
Bi	0.03	0.43	0.16	0.05	0.45	0.17	0.04	0.06	0.07	0.05	0.30	0.02	2.61
Th	2.39	1.88	1.87	0.76	2.03	4.94	1.41	2.07	4.37	3.43	2.03	2.88	2.44
U	0.41	1.96	0.36	0.49	1.24	0.42	0.41	1.17	1.07	0.96	1.26	0.29	0.37
Nb	4.49	6.40	1.94	2.62	10.50	5.07	3.39	5.53	4.40	8.22	1.54	7.73	3.75
Ta	0.29	0.78	0.23	0.18	0.81	0.28	0.23	0.44	0.38	0.50	0.11	0.37	0.35
Zr	23.50	27.70	31.50	27.00	11.80	49.10	42.60	53.40	31.00	54.00	21.30	46.10	51.70
Hf	0.70	0.69	0.97	0.94	0.63	1.49	1.25	1.44	1.12	1.24	0.61	1.15	1.43
Eu/Eu*	1.14	1.08	1.34	2.08	1.36	0.69	1.33	0.74	0.44	0.81	0.81	0.93	0.90
Ce/Ce*	0.92	0.94	0.94	0.92	0.99	0.95	0.92	0.95	0.95	0.95	0.94	0.92	0.91
ΣREE	79.77	75.94	41.71	20.2	82.16	100.09	31.17	71.23	104.32	120.75	38.67	90.2	70.78
LREE/HREE	15.45	14.17	13.79	10.88	5.81	14.46	9.19	12.26	16.75	15.13	20.65	14.73	18.85
$(La/Yb)_N$	31.13	25.30	23.48	14.38	5.14	27.86	12.34	18.27	45.94	26.51	59.08	29.08	32.48
$(Sm/Yb)_N$	5.43	5.44	5.20	3.15	1.78	5.98	2.78	4.01	10.09	4.86	10.33	5.33	4.49
$(La/Sm)_N$	5.73	4.65	4.52	4.57	2.88	4.66	4.43	4.56	4.55	5.46	5.72	5.46	7.23

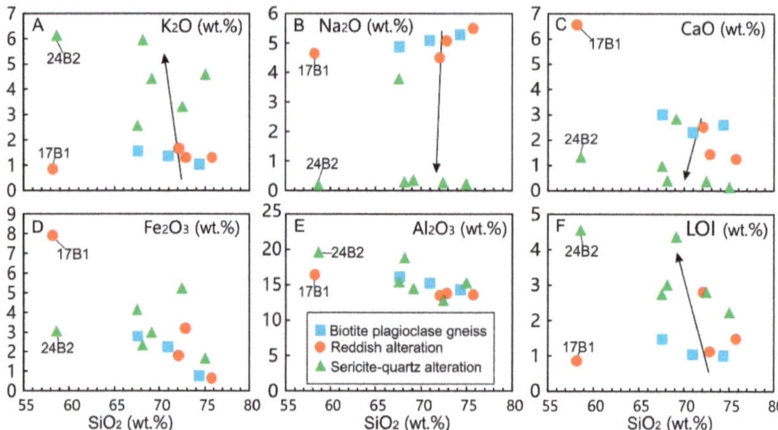

Figure 11. Selected plots of the wall rock in the Pinglidian gold deposit. Thirteen samples can be generally classified into two groups according to the major-element concentrations, except for two abnormal samples (17B1 and 24B2). (**A**,**F**) K_2O vs. SiO_2 and LOI vs. SiO_2 diagrams, which show increasing concentrations from the biotite plagioclase gneiss and albitized rock to sericite-quartz alteration rock. (**B**,**C**) Na_2O vs. SiO_2 and CaO vs. SiO_2 diagrams, which show decreasing concentrations. (**D**,**E**) Fe_2O_3 vs. SiO_2 and Al_2O_3 vs. SiO_2 diagrams, which show relatively stable concentrations.

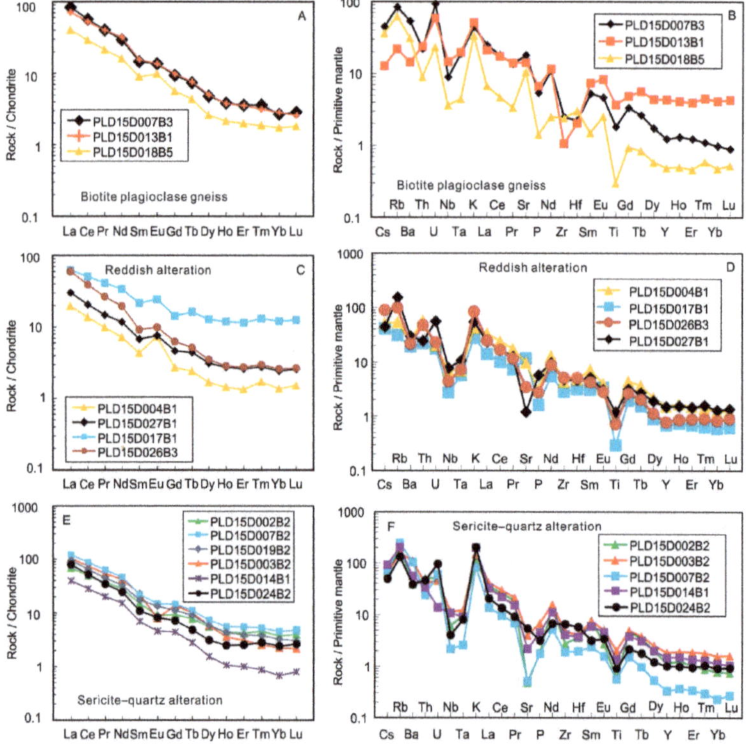

Figure 12. Chondrite-normalized REE patterns and primitive mantle-normalized trace-element patterns for biotite plagiogneiss, reddish alterations, and sericite-quartz alterations in the Pinglidian gold deposit. The chondrite- and primitive mantle-normalization data are from [55].

Table 3. Trace elements (ppm) for two types of quartz vein from the Pinglidian gold deposit.

Samples	Type I Quartz Veins					Type II Quartz Vein								
	14B3	18B1	19B1	20B2	23B2	02B1	03B1	08B1	11B1	14B2	23B1	26B1	26B2	27B3
Li	11.70	11.70	2.65	6.07	6.76	1.85	2.63	1.08	0.776	1.34	0.26	0.25	2.11	0.90
Be	0.87	0.07	0.08	0.04	0.04	0.08	0.17	0.01	0.01	0.12	0.01	0.04	0.05	0.002
Sc	8.66	0.96	0.39	0.15	0.62	2.04	0.39	0.35	0.73	3.72	0.80	0.31	0.57	0.15
V	70.60	2.97	0.58	0.97	3.58	4.32	5.18	5.93	0.83	5.76	0.04	2.41	0.33	1.07
Cr	7.22	3.14	7.05	1.73	10.80	2.17	3.89	2.01	1.61	1.39	1.36	1.70	2.93	1.88
Co	8.74	1.13	1.29	1.22	3.29	21.00	8.25	28.10	0.59	7.97	0.05	2.95	0.04	6.77
Ni	6.72	2.65	2.95	0.93	8.03	26.00	10.20	40.20	1.38	11.00	0.67	4.46	0.52	12.90
Cu	15.10	293	10.20	825	4780	57.50	2154	4356	4131	10647	1.67	321	1.35	28
Zn	70.40	445	18.90	124	12.50	87.80	5059	13.40	56.70	140	1.49	20.90	8.05	6.70
Ga	21.80	2.75	0.82	0.52	1.15	1.61	1.43	0.23	0.53	1.38	0.07	0.71	0.31	0.32
Rb	143	2.46	2.40	1.27	1.53	0.39	9.87	0.55	0.12	1.86	0.07	0.14	0.11	0.18
Sr	140	24.90	25.60	5.95	17.10	36.60	11.30	5.51	16.70	311	2.82	1.04	3.18	10.90
Y	9.85	3.21	2.05	0.56	0.27	4.07	1.40	0.49	2.43	11.90	0.18	0.29	0.37	0.56
Zr	17.60	1.56	2.49	1.35	2.00	0.95	13.80	0.53	0.41	0.67	0.50	0.21	0.59	0.57
Nb	9.68	0.25	0.07	0.03	0.39	0.10	0.83	0.04	0.02	0.31	0.02	0.01	0.01	0.01
Cs	4.93	0.11	0.07	0.04	0.09	0.03	0.31	0.01	0.01	0.03	0.01	0.02	0.01	0.01
Ba	565	26.70	125	13.80	473	7.69	23.20	22.90	2.86	30.50	4.33	3.14	91.50	4.80
La	17.20	4.71	2.26	0.91	0.54	2.10	2.22	0.16	0.78	3.10	0.06	0.11	0.12	0.19
Ce	34.30	7.50	3.94	1.47	0.95	3.60	3.62	0.32	1.44	6.52	0.12	0.17	0.22	0.36
Pr	4.22	0.89	0.45	0.15	0.10	0.43	0.39	0.03	0.17	0.90	0.01	0.01	0.02	0.03
Nd	17.30	3.51	1.78	0.65	0.44	1.93	1.45	0.16	0.86	4.32	0.06	0.09	0.11	0.18
Sm	3.14	0.71	0.38	0.12	0.08	0.53	0.26	0.07	0.24	1.61	0.01	0.03	0.02	0.05
Eu	0.98	0.46	0.24	0.05	0.03	0.57	0.12	0.03	0.21	1.28	0.01	0.02	0.01	0.04
Gd	2.64	0.66	0.38	0.11	0.08	0.56	0.24	0.07	0.26	1.63	0.01	0.03	0.04	0.05
Tb	0.45	0.11	0.07	0.02	0.01	0.13	0.04	0.01	0.06	0.41	0.01	0.01	0.01	0.02
Dy	2.15	0.57	0.35	0.08	0.05	0.78	0.22	0.08	0.38	2.27	0.02	0.05	0.07	0.07
Ho	0.38	0.10	0.06	0.02	0.01	0.14	0.04	0.01	0.08	0.44	0.01	0.01	0.01	0.01
Er	1.06	0.27	0.16	0.04	0.03	0.34	0.14	0.05	0.22	1.18	0.01	0.03	0.04	0.04
Tm	0.16	0.03	0.02	0.01	0.01	0.06	0.02	0.01	0.04	0.20	0.002	0.008	0.007	0.006
Yb	0.91	0.23	0.12	0.04	0.02	0.33	0.13	0.05	0.22	1.08	0.01	0.03	0.04	0.04
Lu	0.12	0.03	0.01	0.01	0.002	0.05	0.02	0.01	0.03	0.15	0.002	0.006	0.005	0.004
Hf	0.47	0.04	0.05	0.03	0.03	0.03	0.33	0.02	0.02	0.03	0.01	0.01	0.01	0.01
Ta	0.78	0.02	0.01	0.01	0.03	0.01	0.05	0.01	0.01	0.03	0.01	0.003	0.008	0.006
Pb	18.60	6092	1052	4689	11,871	592	/	/	165	579	251	/	544	/
Th	1.40	0.09	0.05	0.02	0.10	0.05	0.25	0.01	0.01	0.11	0.01	0.009	0.008	0.008
U	0.77	0.10	0.05	0.02	0.05	0.04	0.46	0.08	0.008	0.09	0.007	0.09	0.09	0.01
Eu/Eu*	1.02	2.05	1.94	1.51	1.05	3.14	1.53	1.58	2.61	2.90	2.19	1.60	2.26	2.39
Ce/Ce*	0.96	0.84	0.90	0.87	0.92	0.87	0.87	1.03	0.90	0.85	0.90	0.89	0.99	0.94
ΣREE	85.05	19.83	10.25	3.72	2.39	11.60	8.95	1.12	5.05	0.25	0.64	0.78	1.14	25.13
LREE	77.15	17.79	9.06	3.38	2.16	9.18	8.08	0.80	3.72	0.21	0.46	0.54	0.88	17.74
HREE	7.90	2.04	1.19	0.35	0.23	2.42	0.88	0.32	1.32	0.04	0.18	0.24	0.27	7.39
LREE/HREE	9.77	8.72	7.61	9.73	9.57	3.79	9.22	2.48	2.81	4.68	2.49	2.26	3.31	2.40
(La/Yb)$_N$	13.47	14.32	13.51	13.95	14.35	4.44	11.54	2.26	2.49	3.67	2.24	2.01	2.96	2.06
(La/Sm)$_N$	3.54	4.28	3.83	4.88	3.92	2.53	5.43	1.46	2.12	2.59	1.80	2.81	2.20	1.24
(Sm/Yb)$_N$	3.81	3.35	3.53	2.86	3.66	1.76	2.12	1.55	1.17	0.93	1.24	0.72	1.35	1.66

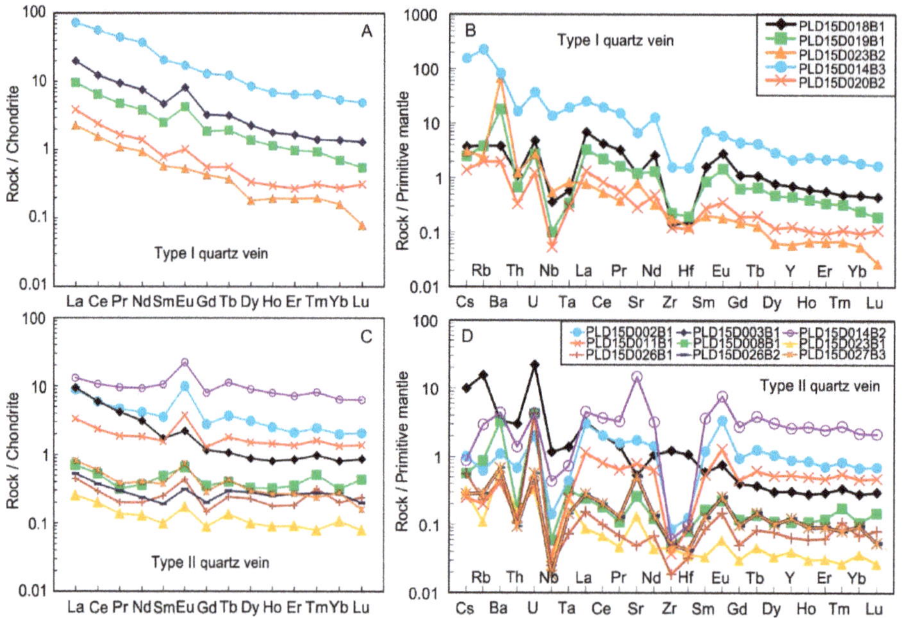

Figure 13. Chondrite-normalized REE patterns and primitive mantle-normalized trace-element patterns for the type-I and type-II quartz veins in the Pinglidian gold deposit. The chondrite- and primitive mantle-normalization data are from [55].

The ΣREE contents of biotite plagiogneiss, reddish alteration and sericite-quartz alteration were 41.71–79.77 ppm, 20.2–100.09 ppm, and 38.67–104.32 ppm (Table 2), respectively. These two types of rock were enriched in light rare earth elements (LREE), with LREE/HREE ratios of 13.79–15.45, 5.81–15.45 and 12.26–20.65, varying (La/Yb)$_N$ ratios of 23.48–31.13, 5.14–31.13 and 18.27–59.08 (Table 2), (Sm/Yb)$_N$ ratios of 5.2–5.44, 5.2–5.98 and 2.78–10.33, and (La/Sm)$_N$ ratios of 4.52–5.73, 2.88–5.73 and 4.43–7.23, respectively. Medium rare earth elements (MREE) were obviously depleted in these types of rock (Figure 12). These rocks showed moderately positive to weakly negative Eu anomalies (Table 2, Figure 12A,C,E), with Eu/Eu* ratios of 1.08–1.34, 0.69–2.08 and 0.44–0.93 for biotite plagiogneiss and sericite-quartz alteration, respectively. The large ion lithophile element (LILE) contents in the biotite plagiogneiss showed moderate variations in Rb (19.6–53.3 ppm), Sr (248–382 ppm), and Ba (133–374 ppm), that's similar to the reddish alteration rock having Rb (14–49.4 ppm), Sr (138–303 ppm), Ba (102–228 ppm). Meanwhile, the sericite-quartz alteration rock had higher contents of Rb (97.6–174 ppm), Sr (9.97–84.7 ppm), and Ba (207–779 ppm) (Table 2). The concentrations of high field strength elements (HFSE) in the biotite plagiogneiss, reddish alteration and sericite-quartz alteration were 3.09–5.6, 2.23–19.8 and 1.52–8.44 ppm for Y; 0.69–0.97, 0.63–1.49 and 0.51–1.47 ppm for Hf; 4.49–1.94, 1.94–10.5 and 1.54–7.73 ppm for Nb; 0.235–0.78, 0.184–0.81 and 0.106–0.502 ppm for Ta; and 1.87–2.39, 0.764–4.94 and 2.03–4.37 ppm for Th, respectively. Nb, Ta, and Th were obviously depleted, whereas Y and Hf were slightly depleted (Figure 12B,D,F). The abundances of transition trace elements (TRTE) in these two rock types were 10.5–127 and 12.8–41.8 ppm for V, 2.72–23.1 and 2.24–8.87 ppm for Cr, 0.976–16.6 and 2.13–8.08 ppm for Co, and 1.39–24.3 and 1.62–6.44 ppm for Ni. Overall, the chondrite-normalized REE patterns showed negative Sm and positive Eu anomalies; however, primitive mantle-normalized spidergrams were characterized by negative anomalies of Cs, Ba, Th, Nb, Ta, Nd, and Ti, and positive K, Eu, and Gd peaks for these types of rock (Figure 12).

4.2.3. Geochemical Characteristics of the Two Types Quartz Veins

Trace elements were not remarkable in the two types of quartz veins in the Pinglidian gold deposit, a huge difference from the enveloped alteration halos. The ΣREE contents of the early and late quartz veins were very low: 2.39–85.05 ppm and 0.25–25.13 ppm, respectively (Table 3). The type-I quartz veins were relatively enriched in LREEs, with LREE/HREE ratios of 7.61–9.77, $(La/Yb)_N$ ratios of 13.47–14.35, $(La/Sm)_N$ ratios of 3.54–4.88, and $(Sm/Yb)_N$ ratios of 2.86–3.81 (Table 3). The type-II quartz veins had low REE ratios, as mentioned above. Both types of quartz veins had similar MREE concentrations (i.e., Sm and Eu), although the chondrite-normalized REE patterns were quite different (Figure 13A,C), showing obvious depletion in all the rock types (Figure 13) and moderately positive to weakly negative Eu anomalies (Table 3, Figure 13A,C), with Eu/Eu* ratios of 1.02–2.05 and 1.53–3.14 for the early and late quartz veins, respectively. The LILE contents were extremely high in the type-I samples 14B3 (short for PLD15D014B3) and 23B2 (short for PLD15D023B2) and the type-II sample 14B2 (short for PLD15D014B2), including 143 ppm of Rb in 14B3, 311 ppm of Sr in 14B2, 140 ppm of Sr in 14B3, 565 ppm of Ba in 14B3, and 473 ppm of Ba in 23B2. Meanwhile, the other two types of samples had low contents of these elements (Table 3), and Cs and Rb were generally depleted in primitive mantle-normalized spidergrams (Figure 13B,D). The Zr content was relatively high in the type-I quartz veins, ranging from 1.35 to 17.6 ppm, while that in the type-II quartz veins ranged from 0.209 to 0.956 ppm, except for sample 03B1 (13.8 ppm). The other HFSEs generally had low contents, but the Pb content was extremely high in both types of quartz vein. Nb, Th, Zr, and Hf were depleted in primitive mantle-normalized spidergrams, whereas U was enriched (Figure 13B,D). The variations in the TRTE concentrations were very unstable; the Zn and Cu contents peaked at 5059 ppm and 10,647 ppm, respectively, but also reached minimums of 1.49 ppm and 1.35 ppm (Table 3), respectively. The other TRTEs had similar trends to Zn and Cu, showing relatively low contents.

4.3. Mass-Balance Calculations

The quantitative calculation of the geochemical mass-balance in alteration zones and metasomatically transformed rock is compatible with mineral assemblages in reaction fronts that are associated with the main alteration [56]. The elemental gains and losses of the altered rock were calculated by the isocon method [57] (Figure 14), which was used to reveal the relative geochemical compositional changes that accompanied the alteration of the original rock (Figure 14A,B). Relatively immobile components (e.g., Al_2O_3, TiO_2, Sc, Zr) always showed coherent behavior during alteration [58]. We applied Grant's isocon method with Al_2O_3 as the immobile element to produce the following isocon equation [57]:

$$\Delta C_i = \frac{C^O_{Al_2O_3}}{C^A_{Al_2O_3}} \times C^A_i - C^O_i$$

Here, C^O_i and C^A_i are the concentrations of element component (i) in the original (O) and altered (A) sample, respectively. Thus, the gain or loss in wt % for major elements or in ppm for trace elements could be calculated from ΔC. The results are listed in Table 4.

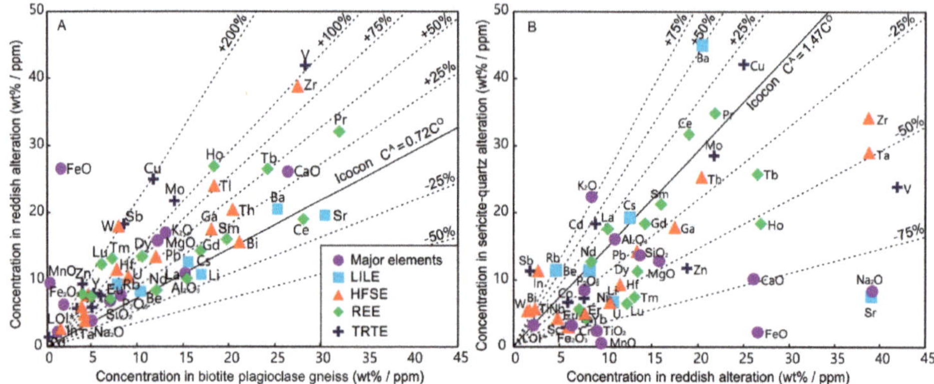

Figure 14. Isocon diagrams that show the elemental gains and losses of the altered rock (reddish alteration vs. the protolith biotite plagiogneiss and sericite-quartz alteration vs. the latest reddish alteration). The isocon lines are defined by the ratios of immobile elements (Al_2O_3), used to calculate the elemental gains and losses in the rock [57]. Enriched elements are above the isocon lines, while depleted elements are below the lines during alteration. Some elements are multiplied or divided by a constant (5, 10, 100, or 1000) to fit the diagram, which does not affect the validity of the mass-balance characteristics [57]. (**A**) Reddish alteration vs. biotite plagiogneiss. (**B**) Sericite-quartz alteration vs. reddish alteration.

Table 4. Calculated gains and losses from the isocon method for the average elemental contents in the two major alteration zones (except for abnormal samples 17B1 and 24B2).

$\triangle C_i$	Reddish Alteration	Sericite-Quartz Alteration	$\triangle C_i$	Reddish Alteration	Sericite-Quartz Alteration	$\triangle C_i$	Reddish Alteration	Sericite-Quartz Alteration
SiO_2	24.45	−21.94	Be	0.10	−0.04	Zr	26.34	−15.48
Fe_2O_3	6.86	−4.12	Rb	25.487	31.21	Hf	0.81	−0.51
MgO	0.97	−0.70	Sr	−33.70	−169.51	Y	6.02	−4.33
CaO	0.98	−1.91	Cs	0.19	0.05	Pb	66.43	−36.99
Na_2O	0.38	−3.34	Ba	31.48	106.51	U	0.53	−0.58
K_2O	1.05	1.36	Sc	4.53	−3.73	Bi	0.005	0.21
MnO	1.28	−0.91	V	29.88	−25.59	Th	0.78	−0.31
TiO_2	1.01	−0.73	Cr	5.77	−5.31	La	−1.28	1.82
P_2O_5	0.02	−0.01	Co	3.09	−1.36	Ce	−1.80	2.63
LOI *	1.94	−0.003	Ni	4.55	−2.66	Pr	−0.16	0.18
FeO	35.31	−25.03	Cu	23.02	3.73	Nd	−0.32	0.26
F	0.60	−0.41	Zn	89.49	−53.67	Sm	0.24	−0.15
Cl	6.49	−38.70	Mo	0.16	−0.02	Eu	0.27	−0.32
S	45.19	−32.29	Cd	1.08	0.36	Gd	0.29	−0.17
Au	0.009	−0.003	Sb	0.16	0.59	Tb	0.12	−0.08
Ag	0.01	1.40	In	0.02	0.05	Dy	0.81	−0.57
As	−0.30	0.24	Ga	6.11	−5.34	Ho	0.18	−0.14
Hg	−0.38	3.47	W	1.68	1.92	Er	0.54	−0.41
B	4.06	3.16	Re	0.004	−0.004	Tm	0.10	−0.08
Li	−2.08	−6.15	Tl	0.14	0.14			

* LOI, loss on ignition.

5. Discussion

5.1. Mineral Dissolution and Precipitation Processes

The details of ore formation are always preserved within alteration halos. Such systems are geochemically open on a small spatial scale but closed on a larger scale [59]. Some evidences have been found in the Pinglidian gold deposit based on observed changes in the mineralogy aggregation and the bulk-rock geochemical compositions (Figures 5–10). The alteration and metasomatic transformation of rocks are crucial characteristics of hydrothermal ore deposits [60], recording the details of ore formation through dissolution-precipitation mechanisms, with a sharp interface between the parent and product compositions in reaction fronts [20]. Reaction fronts where hydrothermal alteration occurs normal to fault cores in essentially unaltered wall rock can be easily recognized along regional-scale alteration zones of Mesozoic granitoids and Precambrian metamorphic rocks in Jiaodong. The compositions and textures in these reaction fronts before and after alteration represent the limits of ore-fluid infiltration and indicate the ore-formation process.

5.1.1. Early Alteration Stage

In the field, reaction fronts of reddish alteration can be easily recognized by the reddening of the rock from a hydrothermally altered granite because of the precipitation of hematite nano-platelets in pores [12,13,20,61] (Figure 3A,C,E,G,I). At the thin-section scale, the plagioclase and biotite in the Archean metamorphic rock were completely or partially replaced by K-feldspar, albite, muscovite, and quartz in the Pinglidian gold deposit (Figures 4 and 5). The textural relationships of the mineral assemblages intensively changed, as did their internal microstructures, with many brittle cracks forming. The original feldspars, generally clear and glassy in biotite plagiogneiss (Figure 4E,F), were replaced by turbid K-feldspar and albite, with massive porosity that was filled with nano-platelets of hematite and fluid inclusions [23] (Figure 5D–F). One of these replacement products, albite, had broader twinning striation than the original feldspar (Figures 4E and 5F). The presence of muscovite inclusions within feldspar is a significant indication that the replaced mineral overprinted another alteration (Figure 5D–F). The result of fluid interactions remained within the mineral assemblage, in which minor recrystallization with no significant changes in bulk−geochemical composition was also observed (Figure 5D–F).

The K_2O, Na_2O, and CaO contents were almost identical with the wall rock (Table 2, Figure 11A–C). The equilibration between the original and product feldspars likely accompanied cation exchange during fluid-induced phase separation, which strongly transformed the early feldspar's composition. The Fe_2O_3 and MgO contents decreased because of the consumption of biotite (Table 2, Figure 11D), except for the abnormal sample 17B1, which was enriched in Fe-bearing minerals. The variable TiO_2 and P_2O_5 contents were related to the breakdown or formation of accessory minerals, such as magnetite and apatite (Table 2). The LOI contents' increasing trend indicates that fluid flux increased. Thus, all the calculated major-element contents indicated gains, suggesting that the fluid was saturated in Fe and Si during the early alteration stage.

The ΣREE contents of reddish alteration were much lower than those of biotite plagiogneiss, except for two abnormal samples (17B1 and 18B4) (Table 2), which was a response to the mineral assemblage changing during the consumption of accessory titanite, monazite, zircon, and apatite. The abnormal data sets probably resulted from residual accessory minerals that were inherited from biotite plagiogneiss as inclusions in the feldspar. Most of the calculated REE contents indicated gains as high as 200% from the original to the product rock, except for La, Nd, and Ce loss (Figure 12). The biotite plagiogneiss was enriched in LREE (Table 2, Figure 12A), while the reddish alteration rock was relatively enriched in heavy rare earth elements (HREE, Table 2, Figure 12C), with lower LREE/HREE, $(La/Yb)_N$, $(Sm/Yb)_N$, and $(La/Sm)_N$ ratios than the biotite plagiogneiss. LREEs and MREEs were obviously depleted from the consumption of titanite and monazite (Figure 12). As mentioned above, LREEs and MREEs have higher mobility than HREEs during fluid−rock interaction, and the REE−rich minerals apatite and zircon gradually broke down because of increasing alteration intensity. The chondrite−normalized REE patterns showed obviously negative Sm and positive Eu anomalies. Moderately positive Eu anomalies (Table 2, Figure 12) with relatively high Eu/Eu* ratios in reddish alteration were associated with Eu−rich feldspar minerals.

LILEs, such as Rb, Sr, and Ba, are enriched in biotite, plagioclase, muscovite, albite, and K-feldspar because of their preferential substitution of K^+ and Na^+ in the lattice of the original and alteration rock. The LILE contents of muscovite, albite, and K-feldspar in the product rock influenced the consumption of biotite and plagioclase in the original rock, so the LILE contents exhibited gains and losses within ±25% of the original rock (Table 2, Figure 14A).

Y is found in apatite, while Hf, Ta, and Th are found in zircon and Nb in titanite. In the reddish alterations, Nb, Ta, and Th were depleted, and Y and Hf were slightly depleted in primitive mantle-normalized spidergrams (Figure 12). The consumption of these accessory minerals during accelerating alteration caused the gains in HFSE contents to contrast the losses in reddish alterations in the Mesozoic granite in the Sanshandao gold deposit [12] (Table 4, Figure 14A). Interpreting the exact reason for this abnormal situation is difficult. We suggest that the fluid flux rapidly dropped and the fluid dynamics became weak in the reaction fronts. Thus, these HFSEs remained in that area instead of migrating with the leaching fluid.

All the TRTEs increased in the reddish alteration rock (Table 4, Figure 14A). TRTEs were commonly concentrated in minerals such as biotite, magnetite, zircon, and hematite. Gains in Cu, Zn, and V exceeded the 100% isocon line because of the occurrence of minor disseminated sulfide during the early alteration stage (Table 4, Figure 14A). Cr, Co, and Ni preferentially appeared in nano-platelets of hematite when minerals such as biotite, magnetite, and zircon reacted with the fluid.

As mentioned above, the geochemical signatures of reddish alteration are consistent with changes in specific mineral assemblages as a response to various phase transitions from the relative roles of solid-state diffusion precipitation. A comparison of the Archean biotite plagiogneiss and reddish alteration rock compositions suggested that the bulk geochemistry was quite similar (Table 2, Figure 11), except for the abnormal sample 17B1, which consists of plagioclase or hematite phenocrysts. Even in a closed geochemical system, local mass transport must occur during the re-equilibration of rock [62]. Thus, the system may be chemically closed during reddish alteration, predominantly driven by changing P-T conditions during regional metamorphism. Minor fluid from metamorphic biotite in the parent gneiss accelerates the reaction.

5.1.2. Alteration that Was Related to Gold Mineralization

Representative sericite-quartz alterations were identified adjacent to auriferous quartz-vein orebodies in the field (Figure 3A–C,F–G,J, Figure 6A,B), which resulted from fluid-rock reactions overprinting the original reddish alteration rock. These alterations are characterized by reaction fronts with gray to dark-green color, consisting of massive sericite, chlorite, and epidote. Based on thin-section observations, the feldspar in the sericite-quartz alteration rock was definitely replaced by the sericite, quartz, muscovite, and sulfides that were observed in the typical hand specimens (Figure 6C–I). As evidence for the composition of the fluids during the reaction, the mineral assemblages and their textural relationships showed such reaction fronts (Figure 6E,F), which contained abundant porosity because of silica that was dissolved in the fluid. The reaction interface contains relict feldspars that are surrounded by disordered muscovite and fine laths of sericite mica (Figure 6E,F). The major minerals and their internal microstructures almost completely changed from the reddish alterations. Figure 6D shows typical cases, in which the crystallographic orientation was consistent with the foliation of the gneiss.

The bulk geochemistry significantly changed in the sericite-quartz alteration rock, related to the completely different alteration-mineral assemblage. Compared to the reddish alteration rock, the gains and losses are distinct compared to the original rock (Table 2, Figures 11, 12, and 14), including the abnormal sample 24B2, which is probably dominated by muscovite and sulfide phenocrysts.

The K_2O and LOI contents increased compared to those in the original rock (Table 2, Figure 11A,F), forming massive sericite and muscovite during the most intensive alteration and reaching equilibration with the fluid. However, the Na_2O and CaO contents decreased because of the complete consumption of feldspar, such as albite and K-feldspar (Table 2, Figure 6D,G, Figure 11B,C). Fe_2O_3 and Al_2O_3 had relatively stable contents compared to the Fe−bearing sulfide and Al-bearing clay (Table 2, Figure 6G–I, Figure 11D,E), indicating that Fe and Al were oversaturated in the ore-bearing fluid. The above accessory minerals were completely broken down, so the calculated TiO_2 and P_2O_5 contents indicated loss and transportation by the fluid (Table 4). Generally, most of the calculated contents of the major elements indicated losses, especially SiO_2 and FeO (Table 4), showing that the properties of the fluid had completely changed.

On the contrary, the ΣREE contents slightly increased with the consumption of REE-rich accessory minerals in the sericite-quartz alteration rock. In fact, the LREE contents, especially La and Ce, were relatively high in the product rock (Table 2), forming new or preserving relict La- and Ce-rich monazite in the alteration rock. Thus, the LREE/HREE and $(La/Yb)_N$ ratios were higher than those of the reddish alteration rock (Table 2), and the $(Sm/Yb)_N$ and $(La/Sm)_N$ ratios were relatively low. The chondrite-normalized REE patterns showed slightly negative Sm and positive Eu anomalies, except for sample 03B2 (Table 2, Figure 12). The Eu/Eu* ratios decreased because of the complete breakdown of feldspar in the sericite-quartz alterations. Most of the calculated REE contents indicated losses within -75%, while the LREE values indicated gains within +25% (Table 4; Figure 14B).

The LILE concentrations became complex in the sericite-quartz alteration rock because they are easily affected by the fluid behavior. The Sr content rapidly dropped because of the disappearance of feldspar (Table 2, Figure 12F), and Rb and Ba had extremely high contents because of their substitution of K, Na and Ca, exhibiting higher mobility during sericite-quartz alteration. Meanwhile, Cs, Ba and Sr were depleted in the chondrite-normalized REE patterns (Figure 12F), with an approximately 50% gain in Ba and over 75% loss in the calculated contents (Table 4; Figure 14B).

Most of the HFSEs were retained in the altered rock because of their low mobility. Nearly all the HFSEs, such as Nb, Ta, Ti, Th, Y, and Hf, were slightly to moderately depleted in the primitive mantle-normalized spidergrams (Table 2; Figure 12F). The Pb content of was closely associated with the gold-hosting mineral galena, and Hf increased in the sericite-quartz alteration rock. Most of the HFSEs' calculated contents decreased, except for W, Bi, Tl, and In (Table 4; Figure 14B).

Some of the TRTEs (Cu, Cd, and Sb) were enriched, while others (Sc, Co, Ni, Zn, Mo, and V) exhibited decreasing calculated contents (Table 4, Figure 14B) because of the dissolution and precipitation of disseminated sulfides in the sericite-quartz alteration rock.

A sequential study of stable mineral assemblages, textural relationships, microstructures, bulk geochemistry and mass balances indicated that the sericite-quartz alteration rock formed from dissolution-precipitation reactions, which occurred at the reaction front as reddish alteration rock reacted with $SiO_2{}^-$ and FeO^- bearing fluid (Table 4, Figure 6). Quartz, muscovite and minor pyrite precipitated with increasing metamorphic grades as fluid compositionally changed the affected system (Table 4; Figures 6 and 15).

5.1.3. Auriferous Quartz Veins and Gold Precipitation

The type II gray quartz veins in the Pinglidian gold deposit are virtual orebodies, in contrast to the type I milky quartz veins (Figure 7). As shown above, the type II auriferous veins precipitated at the center of the alteration zone and overprinted the type I quartz veins. The ore minerals were primarily pyrite and galena, followed by chalcopyrite and sphalerite, with minor pyrrhotite, freibergite, electrum, and native gold (Figure 8A–I).

Gold-hosting pyrites lack zoning in Pinglidian and are commonly cut by numerous brittle fractures (Figure 8D–F). Actually, these pyrites exhibited massive, ductile, single to interconnected, and linear to curvilinear distortions as lattice defects, and gold nanoparticles could also be present in the pyrite [13]. Pyrite and pyrrhotite corrosion textures are well developed because of replacement by galena (Figure 8G,H), including irregularly corroded interiors and concave margins. Pyrite precipitated slightly early, and then fluid-mineral reactions occurred, with galena changing the mineral textures and their assemblages during dissolution and precipitation.

The thin-section observations and electron probe microanalyses showed that the visible gold mainly occurred as inclusions in pyrite, microfracture fillings in pyrite, or inclusions along the margins of the sulfide grains [63] (Figure 8E, Figure 9A–D). SEM maps showed no zonation in terms of Au or Ag in electrum or gold, and other metals were not significantly present [13] (Table 1, Figure 9A,B). The fine native-gold intergrowths with electrum probably formed by sub-solid exsolution (Figure 9A,B) during reactions with aqueous fluid. We did not consider the original source of gold during the above fluid-mineral reactions, so supersaturated gold-bearing solution was transported by the increasing Fe- and Pb-bearing fluid. The loss of H_2S and the destabilization of gold-bearing complexes is normally the most likely mechanism for gold precipitation [13,64–66]. Obviously, gold precipitation was associated with significant hydrothermal alteration, especially sericite-quartz alteration in the Pinglidian gold deposit (Figure 3A–C,F,G,J, Figure 6A,B). The dissolution-precipitation reactions between gold-bearing fluids and wall rock gradually decreased the solubility of the gold complex $[Au(HS)^{2-}]$, causing gold deposition [67]. Fluid-immiscibility processes are also used to interpret the mechanism of gold precipitation [68].

5.2. Physicochemical Conditions

As mentioned above, principal physicochemical parameters, such as fluid compositions (H^+, Na^+, and K^+), temperature, pressure, pH, and fO_2, essentially control the processes of hydrothermal alteration and gold mineralization [12,69]. Although the details of these parameters are extremely complex and beyond the scope of the geologic problem in this paper, these simplified diagrams illustrate that this geochemical system corresponds to obvious changes in alteration mineralogy during hydrothermal processes. Numerous previous fluid inclusions of quartz showed homogeneous phase states at temperatures between 120 and 420 °C, particularly a range from 230 to 320 °C, with pressures of 78–300 MPa [49,68,70,71]. Brittle deformation occurred as fractures in the feldspar, and ductile deformation occurred as elongated quartz in the altered rocks (Figure 5B,C, Figure 6D–F), indicating that these microstructure deformations occurred at 300–400 °C [72–74]. We assume that local equilibrium was reached within the reaction front to precipitate the gangue−mineral assemblages because the temperatures varied (Figures 5D–F and 6D–F). Plümper and Putnis [69] calculated equilibrium constants to construct an activity diagram (H^+, Na^+ and K^+) that related the fluid composition to K-feldspar, albite, and sericite in K−Na−Al−Si−O−H systems at 200 MPa and various temperatures (Figure 15). This situation is similar to that in the Pinglidian gold deposit, which could be applied to the reaction of gangue−mineral assemblages from 200 to 300 °C. During the early stage, plagioclases and biotite $[K(Mg,Fe)_3AlSi_3O_{10}(F,OH)_2]$ were completely replaced by K-feldspar ($KAlSi_3O_8$), decreased aK^+, and increased aNa^+ in aqueous solutions, and the fluid compositions plotted in the stability field of K-feldspar and close to albite in the diagram (Figure 15). Then, both aK^+ and aNa^+ continuously decreased after K-feldspathization and albitization, during the fluid compositions entered the stability field of sericite (Figure 15). Textural observations, chemical analysis, and the calculated phase-equilibrium diagram suggested that albite and sericite predominantly formed intergrowths with each other, and occurred as contemporary alteration products during the early stage (Figure 5D,E, arrows 1 and 2). Sericite $[KAl_2(AlSi_3)O_{10}(OH)_2]$ began to precipitate in the externally derived fluid as the temperature decreased, which shrunk the stability field of albite, favoring sericitization [69] (Figure 15). Thus, all the feldspars had completely transformed into sericite and quartz [75] (Figure 6D,E, arrows 1 and 2).

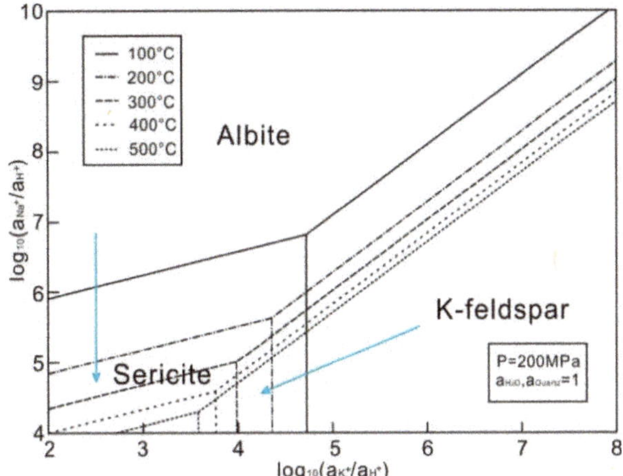

Figure 15. Log (aK$^+$/aH$^+$)−log (aNa$^+$/aH$^+$) diagram at various temperatures, 200 MPa and a(H$_2$O) × a(quartz) = 1, which shows the stability field of K−feldspar, albite, and sericite. The diagram was produced by using equilibrium constants, and the thermodynamic data are from the slop98.dat database, as calculated by the SUPCRT92 program [76]. The arrow represents a possible decrease in K$^+$ and Na$^+$, which caused the phase equilibrium of K−feldspar, albite, and sericite mineral assemblages.

Gold mineralization was closely related to sulfides and depended on the physicochemical conditions of the Fe-S-O-H system, as constrained by the fluid compositions (H$_2$S, HS$^-$, HSO$_4^-$, and SO$_4^{2-}$), temperature, pressure, pH, and fO_2 (Figure 16). Faure and Mensing [78] used sulfide-mineral pairs in equilibrium as a geothermometer, with fractionation errors of ±40–50 °C. The δ^{34}S values of mineral pairs (pyrite: +9.5‰, galena: +6.2‰) from sample PLD15D016B1 indicated isotopic equilibrium at temperatures between 400 and 300 °C. Thus, the diagram was calculated for a Fe-S-O-H system (Figure 16), assuming a closed system and equilibrium conditions of 350 °C, 300 MPa, and ΣS = 0.1 mol/kg (the thermodynamic properties for Au(HS)$_2^-$ during the precipitation of ore minerals, especially sulfides [12,66,76] (Figure 16). Sulfide minerals such as pyrite, galena, pyrrhotite, chalcopyrite, and sphalerite were directly associated with gold precipitation in the Pinglidian gold deposit (Figures 7–10). The coexistence of these sulfide minerals was further constrained by the ore-forming fluid's composition, fO_2, and pH [79] (Figure 16). Pyrite was the predominant gold-bearing mineral in the orebodies (Figure 8E, Figure 9C,D), and paragenetic pyrrhotite contacts pyrite and galena (Figure 8H). Furthermore, fO_2 was calculated to be between −23 and −35 in the stability field of the fluid compositions for pyrite and pyrrhotite (Figure 16). Gold was most likely transported as a gold−bisulfide complex below 350 °C rather than chloride [66]. Gold solubility contours [Au(HS)$_2^-$] were calculated in the diagram, which showed that the Au(HS)$_2^-$ concentrations occupied nearly the entire field of HS$^-$ and H$_2$S, predominantly overlapping the stability field of pyrite and pyrrhotite (Figure 16). However, the solubility of Au(HS)$_2^-$ in the ore-forming fluid within the same area was sensitive to fO_2 along arrow 1 and the pH along arrow 2. This result suggests that an increase in fO_2 and a decrease in pH triggered gold precipitation in the assumed situation for the Pinglidian gold deposit.

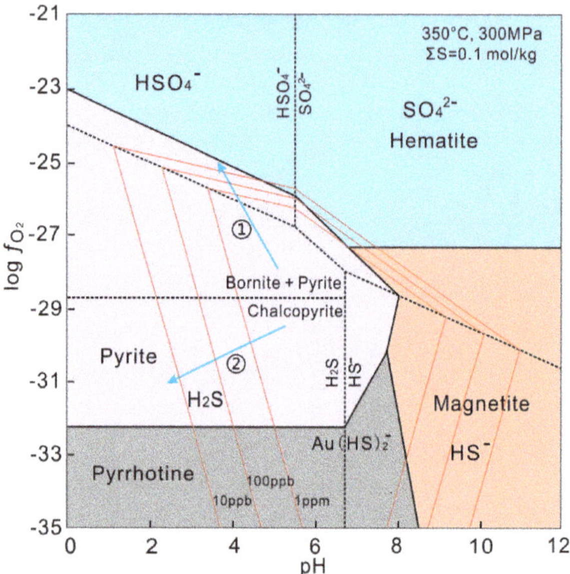

Figure 16. Log fO_2–pH diagram at 350°C, 300 MPa, and ΣS = 0.1 mol/kg. The thermodynamic properties for $Au(HS)_2^-$ are from [77], and other data are from the slop98.dat database [12]. The narrow and bold lines represent the boundaries between aqueous sulfur species and Fe-bearing minerals, respectively. The calculated solubility contours of $Au(HS)_2^-$ at 10, 100, and 1000 ppb (thin continuous lines) are shown in the predominant fields of various aqueous sulfur species. The arrow represents a possible decrease in fO_2 and pH, which caused gold precipitation and the coexistence of sulfide assemblages.

5.3. Implications for Exploration

Once hydrothermal alteration has been identified, it can be directly related to the potential ore-forming fluid fluxes and scale of the host structures [8,10]. The detailed petrological features and geochemical effects of the Pinglidian gold deposit were presented above, providing vast quantities of information regarding the alteration-mineral assemblages and physicochemical conditions of the mineralizing fluids, which revealed that these ore-forming fluids interacted with the surrounding rocks in conjunction with gold mineralization [80]. These indicators of alteration zones increased the prospecting target, revealing that the structural control of auriferous quartz veins was paralleling to the foliation of the Archean metamorphic basement rock.

As mentioned above, the existence of an apparent positive correlation between the size of quartz veins and width of the alteration envelope in the Pinglidian gold deposit indicates that (1) quartz veins and alteration halos were controlled by foliations in fold belts that were overprinted by brittle-deformation structures (Figure 17A); (2) fluid circulation and fluid-flow movement along the quartz veins were prominently restricted to narrow conduits because of factors such as the porosity and permeability of the wall rock; (3) silica was supersaturated in the ore-forming fluids and interacted with large volumes of host rock by dissolution and precipitation, proven by gain and loss of major elements (SiO_2, K_2O, Na_2O, and LOI), LREEs (La and Ce), MREEs (Eu), HREEs (Tm), LILEs (Rb, Ba, and Sr), HFSEs (Nb, Ta, Zr, and Hf) and TRTEs (Cu, Zn, Co, and Ni) (Figure 17B); and (4) equilibrium conditions existed in reaction fronts between the fluids and sub-solid rocks, with a sharp interface exchanging geochemical compositions (Figure 17B), as calculated above.

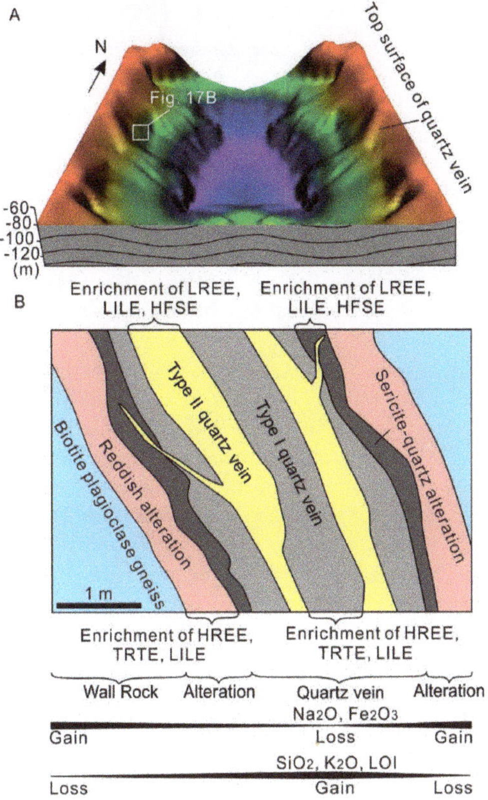

Wall Rock	Alteration	Quartz vein	Alteration
		Na2O, Fe2O3	
Gain		Loss	Gain
		SiO2, K2O, LOI	
Loss		Gain	Loss

Figure 17. (**A**) Simulation of the ore-controlling structural pattern of the Pinglidian gold deposit, revealing depth variation of the top surface of inferred gold bearing quartz vein. (**B**) The geochemical indicator of two types quartz veins and their relationship to alteration rock and wall rock.

In contrast, an intensive 200-m-wide alteration zone in the footwall of the regional fault zone, including the Jiaojia Fault, Zhaoping Fault and Sanshandao Fault, is hosted in Mesozoic granitoids [12,81]. Second- and third-order faults from these fault zones exhibit negligible wall—rock alterations the host structures, especially for the quartz-vein gold deposits (Linglong—type), but the hydrothermal reactions and gold mineralization are identical to the core of the alteration halo. Notable zonation is associated with the intensity of the alteration types from the fault core to wall rock, which depends on the permeability of the original wall rock. Fault cores or fault jogs are the pathways for fluid flow and the location for repeated deformation in fault zones [30], and these features have been proven to exhibit great potential for disseminated ores [7].

In summary, the geochemical loads of these altered mineral assemblages, alteration intensity, and range can be utilized to evaluate the potential size and scale of an ore-forming hydrothermal system under investigation, and thus is an effective exploration tool for estimating the structural deformation intensity decreased or increased in some degree in potential ore districts. These indicators reveal that the Archean metamorphic basement rock hosted auriferous quartz veins are structurally controlled by a local fault, which was parallel to the foliation of the biotite plagiogneiss.

6. Conclusions

(1) The Pinglidian gold deposit is hosted in Archean biotite plagiogneiss instead of Mesozoic granitoids. Gold mineralization was controlled by faults that are parallel to the foliations in the Archean biotite plagiogneiss. The orebodies consist of two types of quartz veins that are enveloped by intensive sericite-quartz alteration rock.

(2) Geological features and petrological observations showed clear spatial mineral-assemblage zonation from the infiltration of an external fluid during hydrothermal alteration, which produced auriferous quartz veins.

(3) The geochemical loads of these altered mineral assemblages displayed various concentrations of major elements (SiO_2, K_2O, Na_2O, and LOI), LREEs (La and Ce), MREEs (Eu), HREEs (Tm), LILEs (Rb, Ba, and Sr), HFSEs (Nb, Ta, Zr, and Hf) and TRTEs (Cu, Zn, Co, and Ni), which were preferentially enriched in specific minerals, especially sulfides and accessory minerals. The contents of these elements decreased as metasomatism increased during hydrothermal alteration.

(4) The re-enriched fluids triggered ore formation through a dissolution-precipitation mechanism, with a sharp interface between the parent and product compositions in the reaction fronts, causing the eclogitization of areas of the blueschist host that interacted with the passing fluid. Most of the original minerals were replaced, and the liberated Fe and Si were partitioned into the unsaturation fluid, thus precipitating the observed quartz veins and sulfides. More than 80% of the LILEs, REEs, and HFSEs were mobilized from the host rock as accessory minerals disappeared during intense alteration, with LILEs and LREEs mobilized more efficiently than HFSEs and HREEs. TRTEs mainly originated from the enriched fluid. The calculated sulfide-phase equilibrium showed that an increase in fO_2 and a decrease in pH caused gold precipitation in the assumed situation for the Pinglidian gold deposit.

(5) The petrological and geochemical indicators of the alteration zone increased the prospecting target and revealed the features of the structures that controlled the wall rock. The alteration intensity and range could be utilized to evaluate the potential size and scale of an ore-forming hydrothermal system, and thus represent an effective exploration tool for biotite plagiogneiss-hosted gold mineralization.

Author Contributions: Conceptualization, R.L.; methodology, R.L. and N.N.A.; formal analysis, M.Y.; investigation, Y.M.; data curation, M.Y.; writing—original draft preparation, R.L.; writing—review and editing, R.L. and M.Y.

Funding: This research was funded by the National Key Research and Development Program, Grant No. 2016YFC0600107−4; National Natural Science Foundation of China, Grant No. 41572069; National Key Research and Development Program of Deep-penetrating Geochemistry, Grant No. 2016YFC0600600 and National Nonprofit Institute Research Grant of IGGE, Grant No. AS2017J12.

Acknowledgments: We thank Li-Qiang Yang, Roberto Ferrez Weinberg, David Groves, and Richard Goldfarb for their assistance with this paper.

References

1.	Tang, L.; Santosh, M. Neoarchean granite-greenstone belts and related ore mineralization in the north china craton: An overview. *Geosci. Front.* **2018**, *9*, 751–768. [CrossRef]
2.	Li, S.R.; Santosh, M. Metallogeny and craton destruction: Records from the North China Craton. *Ore Geol. Rev.* **2014**, *56*, 376–414. [CrossRef]
3.	Li, S.R.; Santosh, M. Geodynamics of heterogeneous gold mineralization in the North China Craton and its relationship to lithospheric destruction. *Gondwana Res.* **2017**, *50*, 267–292. [CrossRef]
4.	Deng, J.; Wang, Q.F. Gold mineralization in China: Metallogenic provinces, deposit types and tectonic framework. *Gondwana Res.* **2016**, *36*, 219–274. [CrossRef]
5.	Goldfarb, R.J.; Santosh, M. The dilemma of the Jiaodong gold deposits: Are they unique? *Geosci. Front.* **2014**, *5*, 139–153. [CrossRef]

6. Deng, J.; Wang, Q.F.; Li, G.J. Tectonic evolution, superimposed orogeny, and composite metallogenic system in China. *Gondwana Res.* **2017**, *50*, 216–266. [CrossRef]

7. Groves, D.I.; Santosh, M. The giant Jiaodong gold province: The key to a unified model for orogenic gold deposits. *Geosci. Front.* **2016**, *7*, 409–418. [CrossRef]

8. Bierlein, F.P.; Arne, D.C.; McKnight, S.; Lu, J.; Reeves, S.; Besanko, J.; Marek, J.; Cooke, D.R. Wall-rock petrology and geochemistry in alteration halos associated with mesothermal gold mineralization, central Victoria, Australia. *Econ. Geol.* **2000**, *95*, 283–312. [CrossRef]

9. Qiu, Y.; Groves, D.I.; Mcnaughton, N.J.; Wang, L.G.; Zhou, T. Nature, age, and tectonic setting of granitoid-hosted, orogenic gold deposits of the Jiaodong Peninsula, Eastern North China Craton, China. *Miner. Depos.* **2002**, *37*, 283–305. [CrossRef]

10. Mackenzie, D.J.; Craw, D.; Begbie, M. Mineralogy, geochemistry, and structural controls of a disseminated gold-bearing alteration halo around the schist-hosted bullendale orogenic gold deposit, New Zealand. *J. Geochem. Explor.* **2007**, *93*, 160–176. [CrossRef]

11. Pirajno, F. *Hydrothermal Processes and Mineral Systems*; Springer: Dordrecht, The Netherlands, 2009; pp. 1–162.

12. Li, X.C.; Fan, H.R.; Santosh, M.; Hu, F.F.; Yang, K.F.; Lan, T.G. Hydrothermal alteration associated with Mesozoic granite-hosted gold mineralization at the Sanshandao deposit, Jiaodong gold province, China. *Ore Geol. Rev.* **2013**, *53*, 403–421. [CrossRef]

13. Yang, L.Q.; Deng, J.; Wang, Z.L.; Guo, L.N.; Li, R.H.; Groves, D.I.; Danyushevsky, L.V.; Zhang, C.; Zheng, X.L.; Zhao, H. Relationships between gold and pyrite at the xincheng gold deposit, jiaodong peninsula, china: Implications for gold source and deposition in a brittle epizonal environment. *Econ. Geol.* **2016**, *111*, 105–126. [CrossRef]

14. Goldfarb, R.J.; Baker, T.; Dube, B.; Groves, D.I.; Hart, C.J.; Gosselin, P. Distribution, character and genesis of gold deposits in metamorphic terranes. In *Economic Geology 100th Anniversary*; Hedenquist, J.W., Thompson, J.F.H., Eds.; Society of Economic Geologists: Littleton, CO, USA, 2005; pp. 407–450.

15. Goldfarb, R.J.; Taylor, R.D.; Collins, G.S.; Goryachev, N.A.; Orlandini, O.F. Phanerozoic continental growth and gold metallogeny of Asia. *Gondwana Res.* **2014**, *2014*. *25*, 48–102. [CrossRef]

16. Davies, J.H. The role of hydraulic fractures and intermediate depth earthquakes in generating subduction-zone magmatism. *Nature* **1999**, *398*, 142–145. [CrossRef]

17. Bons, P.D.; van Milligen, B.P. New experiment to model self-organize critical transport and accumulation of melt and hydrocarbons from their source rocks. *Geology* **2001**, *29*, 919–922. [CrossRef]

18. Faulkner, D.R.; Jackson, C.A.L.; Lunn, R.J.; Schlische, R.W.; Shipton, Z.K.; Wibberley, C.A.J.; Withjackd, M.O. A review of recent developments concerning the structure, mechanics and fluid flow properties of fault zones. *J. Struct. Geol.* **2010**, *32*, 1557–1575. [CrossRef]

19. Bons, P.D.; Elburg, M.A.; Gomez-Rivas, E. A review of the formation of tectonic veins and their microstructures. *J. Struct. Geol.* **2012**, *43*, 33–62. [CrossRef]

20. Putnis, A. Mineral replacement reactions. *Rev. Miner. Geochem.* **2009**, *70*, 87–124. [CrossRef]

21. Jamtveit, B.; Austrheim, H. Metamorphism: The role of fluids. *Elements* **2010**, *6*, 153–158. [CrossRef]

22. Arghe, F.; Skelton, A.; Pitcairn, I. Spatial coupling between spilitization and carbonation of basaltic sills in SW Scottish Highlands: Evidence of a mineralogical control of metamorphic fluid flow. *Geofluids* **2011**, *11*, 245–259. [CrossRef]

23. Parsons, I.; Lee, M.R. Mutual replacement reactions in alkali feldspars I: Microtextures and mechanisms. *Contrib. Miner. Petrol.* **2009**, *157*, 641–661. [CrossRef]

24. Skelton, A. Flux rate for water and carbon during greenschist facies metamorphism. *Geology* **2011**, *39*, 43–46. [CrossRef]

25. Jonas, L.; John, T.; King, H.E.; Geisler, T.; Putnis, A. The role of grain boundaries and transient porosity in rocks as fluid pathways for reaction front propagation. *Earth Planet. Sci. Lett.* **2014**, *386*, 64–74. [CrossRef]

26. Yang, L.Q.; Deng, J.; Wang, Z.L.; Zhang, L.; Guo, L.N.; Song, M.C.; Zheng, X.L. Mesozoic gold metallogenic system of the Jiaodong gold province, eastern China. *Acta Petrol. Sin.* **2014**, *30*, 2447–2467. (In Chinese)

27. Li, L.; Santosh, M.; Li, S.R. The 'Jiaodong type' gold deposits: Characteristics, origin and prospecting. *Ore Geol. Rev.* **2015**, *65*, 589–611. [CrossRef]

28. Lü, G.X.; Kong, Q.C. *Geology on the Linglong-Jiaojia-Type Gold Deposits in the Jiaodong Area*; Science Press: Beijing, China, 1993; pp. 1–253. (In Chinese)

29. Yang, L.Q.; Deng, J.; Wang, Q.F.; Zhou, Y.H. Coupling effects on gold mineralization of deep and shallow structures in the Northwestern Jiaodong Peninsula, Eastern China. *Acta Geol. Sin.* **2006**, *80*, 400–411.

30. Lu, H.Z.; Archambault, G.; Li, Y.; Wei, J. Structural geochemistry of gold mineralization in the Linglong-Jiaojia district, Shandong Province, China. *Chin. J. Geochem.* **2007**, *3*, 215–234. [CrossRef]

31. Deng, J.; Wang, Q.F.; Yang, L.Q.; Zhou, L.; Gong, Q.J.; Yuan, W.M.; Xu, H.; Guo, C.Y.; Liu, X.W. The structure of ore-controlling strain, stress fields in the Shangzhuang gold deposit in Shandong Province, China. *Acta Geol. Sin.* **2008**, *82*, 769–780.

32. Deng, J.; Yang, L.Q.; Gao, B.F.; Sun, Z.S.; Guo, C.Y.; Wang, Q.F.; Wang, P. Fluid Evolution and Metallogenic Dynamics during Tectonic Regime Transition: Example from the Jiapigou Gold Belt in Northeast China. *Resour. Geol.* **2009**, *59*, 140–152. [CrossRef]

33. Bierlein, F.P.; Crowe, D.E. Phanerozoic orogenic lode gold deposits. *Rev. Econ. Geol.* **2000**, *2000*, 103–139.

34. Wang, L.G.; Qiu, Y.M.; McNaughton, N.J.; Groves, D.I.; Luo, Z.K.; Huang, J.Z.; Miao, L.C.; Liu, Y.K. Constraints on crustal evolution and gold metallogeny in the northwestern Jiaodong Peninsula, China, from SHRIMP U–Pb zircon studies of granitoids. *Ore Geol. Rev.* **1998**, *13*, 275–291. [CrossRef]

35. Deng, J.; Wang, C.M.; Bagas, L.; Carranza, E.J.M.; Lu, Y.J. Cretaceous–Cenozoic tectonic history of the Jiaojia Fault and gold mineralization in the Jiaodong Peninsula, China: Constraints from zircon U–Pb, illite K–Ar, and apatite fission track thermochronometry. *Miner. Depos.* **2015**, *50*, 987–1006. [CrossRef]

36. Deng, J.; Yang, L.Q.; Li, R.H.; Groves, D.I.; Santosh, M.; Wang, Z.L.; Sai, S.X.; Wang, S.R. Regional structural control on the distribution of world-class gold deposits: An overview from the Giant Jiaodong Gold Province, China. *Geol. J.* **2018**, *3*, 1–14. [CrossRef]

37. Jahn, B.M.; Liu, D.Y.; Wan, Y.S.; Song, B.; Wu, J.S. Archean crustal evolution of the Jiaodong Peninsula, China, as revealed by zircon shrimp geochronology elemental and Nd-isotope geochemistry. *Am. J. Sci.* **2008**, *308*, 232–269. [CrossRef]

38. Wan, Y.S.; Dong, C.Y.; Xie, H.Q.; Wang, S.J.; Song, M.C.; Xu, Z.Y.; Wang, S.Y.; Zhou, H.Y.; Ma, M.Z.; Liu, D.Y. Formation Ages of Early Precambrian BIFs in the North China Craton: SHRIMP Zircon U-Pb Dating. *Acta Geol. Sin.* **2012**, *86*, 1447–1478. (In Chinese)

39. Wallis, S.; Enami, M.; Banno, S. The Sulu UHP Terrane: A review of the petrology and structural geology. *Int. Geol. Rev.* **1999**, *41*, 906–920. [CrossRef]

40. Tang, J.; Zheng, Y.F.; Wu, Y.B.; Gong, B.; Liu, X.M. Geochronology and Geochemistry of Metamorphic rocks in the Jiaobei terrane: Constraints on its tectonic affinity in the Sulu orogen. *Precambrian Res.* **2007**, *152*, 48–82. [CrossRef]

41. Zhou, X.W.; Zhao, G.C.; Wei, C.J.; Geng, Y.S.; Sun, M. EPMA U-Th-Pb monazite and SHRIMP U-Pb zircon geochronology of high-pressure pelitic granulites in the Jiaobei massif of the North China Craton. *Am. J. Sci.* **2008**, *308*, 328–350. [CrossRef]

42. Liu, P.H.; Liu, F.L.; Wang, F.; Liu, J.H.; and Cai, J. Petrological and geochronological preliminary study of the Xiliu ~2.1Ga meta-gabbro from the Jiaobei terrane, the southern segment of the Jiao-Liao-Ji Belt in the North China Craton. *Acta Petrol. Sin.* **2013**, *29*, 2371–2390. (In Chinese)

43. Deng, J.; Liu, X.F.; Wang, Q.F.; Dilek, K.; Yang, L.Q. Isotopic characterization and petrogenetic modeling of early cretaceous mafic diking-lithospheric extension in the North China craton, eastern Asia. *GSA Bull.* **2017**, *129*, 1379–1407. [CrossRef]

44. Yang, Q.Y.; Santosh, M. Early Cretaceous magma flare-up and its implications on gold mineralization in the Jiaodong Peninsula, China. *Ore Geol. Rev.* **2015**, *65*, 626–642. [CrossRef]

45. Yang, Q.Y.; Santosh, M.; Shen, J.F.; Li, S.R. Juvenile vs. recycled crust in NE China: Zircon U-Pb geochronology, Hf isotope and an integrated model for Mesozoic gold mineralization in the Jiaodong Peninsula. *Gondwana Res.* **2014**, *25*, 1445–1468. [CrossRef]

46. Zhang, J.; Zhao, Z.F.; Zheng, Y.F.; Dai, M.N. Postcollisional magmatism: Geochemical constraints on the petrogenesis of Mesozoic granitoids in the Sulu orogen, China. *Lithos* **2010**, *119*, 512–536. [CrossRef]

47. Ma, L.; Jiang, S.Y.; Dai, B.Z.; Jiang, Y.H.; Hou, M.L.; Pu, W.; Xu, B. Multiple sources for the origin of Late Jurassic Linglong adakitic granite in the Shandong Peninsula, eastern China: Zircon U-Pb geochronological, geochemical and Sr-Nd-Hf isotopic evidence. *Lithos* **2013**, *162*, 251–263. [CrossRef]

48. Yang, L.Q.; Deng, J.; Ge, L.S.; Wang, Q.F.; Zhang, J.; Gao, B.F.; Jiang, S.Q.; Xu, H. Metallogenic Age and Genesis of Gold Ore Deposits in Jiaodong Peninsula, Eastern China: A Regional Review. *Prog. Nat. Sci.* **2007**, *17*, 138–143.

49. Yang, L.Q.; Deng, J.; Zhang, J.; Guo, C.Y.; Gao, B.F.; Gong, Q.J.; Wang, Q.F.; Jiang, S.Q.; Yu, H.J. Decrepitation Thermometry and Compositions of Fluid Inclusions of the Damoqujia Gold Deposit, Jiaodong Gold Province, China: Implications for Metallogeny and Exploration. *J. Earth Sci.* **2008**, *19*, 378–390.

50. Zhao, Z.; Zhao, Z.X.; Xu, J.R. Velocity structure heterogeneity and tectonic motion in and around the Tan-Lu fault of China. J. *Asian Earth Sci.* **2012**, *57*, 6–14. [CrossRef]

51. Sibson, R.H. Brittle failure mode plots for compressional and extensional tectonic regimes. *J. Struct. Geol.* **1998**, *20*, 655–660. [CrossRef]

52. Cox, S.F. The application of failure mode diagrams for exploring the roles of fluid pressure and stress states in controlling styles of fracture-controlled permeability enhancement in faults and shear zones. *Geofluids* **2010**, *10*, 217–233.

53. Cox, S.F.; Munroe, S.M. Breccia formation by particle fluidization in fault zones: Implications for transitory, rupture-controlled fluid flow regimes in hydrothermal systems. *Am. J. Sci.* **2016**, *316*, 241–278. [CrossRef]

54. Li, X.H.; Qi, C.S.; Liu, Y.; Liang, X.R.; Tu, X.L.; Xie, L.W.; Yang, Y.H. Petrogenesis of the Neoproterozoic bimodal volcanic rocks along the western margin of the Yangtze Block: New constraints from Hf isotopes and Fe/Mn ratios. *Chin. Sci. Bull.* **2005**, *50*, 2481–2486. [CrossRef]

55. Sun, S.S.; McDonough, W.F. Chemical and isotopic systematics of oceanic basalts: Implications for mantle composition and processes. *Geol. Soc. Lond. Spec. Publ.* **1989**, *42*, 313–345. [CrossRef]

56. Gresens, R.L. Composition-volume relations of metasomatism. *Chem. Geol.* **1967**, *2*, 47–65. [CrossRef]

57. Grant, J.A. The isocon diagram—A simple solution to Gresens equation for metasomatic alteration. *Econ. Geol.* **1986**, *81*, 1976–1982. [CrossRef]

58. Kishida, A.; Kerrich, R. Hydrothermal alteration zoning and gold concentration at the kerr-Addison Archean lode gold deposit, Kirkland Lake, Ontario. *Econ. Geol.* **1987**, *82*, 649–690. [CrossRef]

59. Putnis, A.; John, T. Replacement processes in the Earth's Crust. *Elements* **2009**, *6*, 159–164. [CrossRef]

60. Cox, S.F.; Braun, J.; Knackstedt, M.A. Principles of structural control on permeability and fluid flow in hydrothermal systems. *Rev. Econ. Geol.* **2001**, *14*, 1–24.

61. Putnis, A.; Hinrichs, R.; Putnis, C.V.; Golla-Schindler, U.; Collins, L.G. Hematite in porous red-clouded feldspars: Evidence of large-scale crustal fluid-rock interaction. *Lithos* **2007**, *95*, 10–18. [CrossRef]

62. Carmichael, D.M. On the mechanism of prograde metamorphic reactions in quartz-bearing pelitic rocks. Contrib. *Miner. Petrol.* **1969**, *20*, 244–267. [CrossRef]

63. Du, X.J. Study on the technological mineralogy of the ores from the −145 level gold deposit of altered-rock type in Xincheng, Laizhou city. *Geol. Shandong* **1988**, *4*, 69–79. (In Chinese)

64. Huston, D.L.; Large, R.R. A chemical model for the concentration of gold in volcanogenic massive sulfide deposits. *Ore Geol. Rev.* **1989**, *4*, 174–200. [CrossRef]

65. Stefánsson, A.; Seward, T.M. Gold(I) complexing in aqueous sulphide solutions to 500 °C at 500 bar. *Geochim. Cosmochim. Acta* **2004**, *68*, 4121–4143. [CrossRef]

66. Williams-Jones, A.E.; Bowell, R.J.; Migdisov, A.A. Gold in solution. *Elements* **2009**, *5*, 281–287. [CrossRef]

67. Wilkinson, J.J.; Johnston, J.D. Pressure fluctuations, phase separation, and gold precipitation during seismic fracture propagation. *Geology* **1996**, *24*, 395–398. [CrossRef]

68. Wang, Z.L.; Yang, L.Q.; Guo, L.N.; Marsh, E.; Wang, J.P.; Liu, Y.; Zhang, C.; Li, R.H.; Zhang, L.; Zheng, X.L.; et al. Fluid immiscibility and gold deposition in the Xincheng deposit, Jiaodong Peninsula, China: A fluid inclusion study. *Ore Geol. Rev.* **2015**, *65*, 701–717. [CrossRef]

69. Plümper, O.; Putnis, A. The complex hydrothermal history of granitic rocks: Multiple feldspar replacement reactions under subsolidus conditions. *J. Petrol.* **2009**, *50*, 967–987. [CrossRef]

70. Yang, L.Q.; Deng, J.; Guo, L.N.; Wang, Z.L.; Li, X.Z.; Li, J.L. Origin and evolution of ore fluid, and Au deposition processes at the giant Taishang Au deposit, Jiaodong Peninsula, eastern China. *Ore Geol. Rev.* **2016**, *72*, 582–602. [CrossRef]

71. Guo, L.N.; Goldfarb, R.J.; Wang, Z.L.; Li, R.H.; Chen, B.H.; Li, J.L. A comparison of Jiaojia-and Linglong-type gold deposit ore-forming fluids: Do they differ? *Ore Geol. Rev.* **2017**, *88*, 511–533. [CrossRef]

72. Scholz, C.H. The brittle-plastic transition and the depth of seismic faulting. *Geol. Rundsch.* **1988**, *77*, 319–328. [CrossRef]

73. Passchier, C.W.; Trouw, R.A. *Microtectonics*, 2nd ed.; Springer: Berlin, Germany, 2005; pp. 1–289.

74. Li, R.H.; Liu, Y.; Li, H.L.; Zheng, X.L.; Zhao, H.; Lu, H.W. Ore-controlling structure deformation condition of Xincheng gold deposit, Jiaodong: The microstructure and EBSD fabrics analysis constrain. *Acta Petrol. Sin.* **2014**, *30*, 2546–2558. (In Chinese)

75. Sverjensky, D.A.; Hemley, J.J.; D'Angelo, W.M. Thermodynamic assessment of hydrothermal alkali feldspar-mica-aluminosilicate equilibria. *Geochim. Cosmochim. Acta* **1991**, *55*, 989–1004. [CrossRef]

76. Akinfiev, N.N.; Zotov, A.V. Thermodynamic description of aqueous species in the system Cu–Ag–Au–S–O–H at temperatures of 0–600 °C and pressures of 1–3000 bar. *Geochem. Int.* **2010**, *48*, 714–720. [CrossRef]

77. Johnson, J.W.; Oelkers, E.H.; Helgeson, H.C. SUPCRT92: A software package for calculating the standard molal thermodynamic properties of minerals, gases, aqueous species, and reactions from 1 to 5000 bars and 0 to 1000 °C. *Comput. Geosci.* **1992**, *18*, 899–947. [CrossRef]

78. Faure, G.; Mensing, T.M. *Isotopes Principles and Applications*, 2nd ed.; John Wiley & Sons: New York, NY, USA, 2005; pp. 1–343.

79. Yang, L.Q.; Guo, L.N.; Wang, Z.L.; Zhao, R.X.; Song, M.C.; Zheng, X.L. Timing and mechanism of gold mineralization at the Wang'ershan gold deposit, Jiaodong peninsula, eastern china. *Ore Geol. Rev.* **2016**, *88*, 491–510. [CrossRef]

80. Rollinson, H.R. *Using Geochemical Data: Evaluation, Presentation, Interpretation*; Longman Publishing Group: New York, NY, USA, 1993; pp. 1–441.

81. Zhang, L.; Yang, L.Q.; Wang, Y.; Weinberg, R.F.; An, P.; Chen, B.Y. Thermochronologic constrains on the processes of formation and exhumation of the Xinli orogenic gold deposit, Jiaodong Peninsula, eastern China. *Ore Geol. Rev.* **2017**, *81*, 140–153. [CrossRef]

Article

Geostatistical Determination of Ore Shoot Plunge and Structural Control of the Sizhuang World-Class Epizonal Orogenic Gold Deposit, Jiaodong Peninsula, China

Si-Rui Wang [1], Li-Qiang Yang [1,*], Jian-Gang Wang [2], En-Jing Wang [2] and Yong-Lin Xu [2]

1   State Key Laboratory of Geological Processes and Mineral Resources, China University of Geosciences, Beijing 100083, China; creedwang@163.com
2   Jiaojia Gold Mine, Shandong Gold Mining (Laizhou) Co., Ltd., Laizhou 261400, China; wbin920@sina.com (J.-G.W.); wangej123@163.com (E.-J.W.); xuyonglin@sd-gold.com (Y.-L.X.)
*   Correspondence: lqyang@cugb.edu.cn; Tel.: +86-(010)-82321937

Received: 18 January 2019; Accepted: 3 April 2019; Published: 4 April 2019

Abstract: The Jiaodong Peninsula in eastern China is the third largest gold-mining area and one of the most important orogenic gold provinces in the world. Ore shoots plunging in specific orientations are a ubiquitous feature of the Jiaodong lode deposits. The Sizhuang gold deposit, located in northwestern Jiaodong, is characterized by orebodies of different occurrences. The orientation of ore shoots has remained unresolved for a long time. In this paper, geostatistical tools were used to determine the plunge and structural control of ore shoots in the Sizhuang deposit. The ellipses determined by variogram modeling reveal the anisotropy of mineralization, plus the shape, size, and orientation of individual ore shoots. The long axes of the anisotropy ellipses trend NE or SEE and plunge 48° NE down the dip. However, individual ore shoots plunge almost perpendicular to the plunge of the ore deposit as a whole. This geometry is interpreted to have resulted from two periods of fluid flow parallel to two sets of striations that we identified on ore-controlling faults. Thrust-related lineations with a sinistral strike-slip component were associated with early-stage mineralization. This was overprinted by dextral and normal movement of the ore-controlling fault that controlled the late-stage mineralization. This kinematic switch caused a change in the upflow direction of ore-forming fluid, which in turn controlled the orientation of the large-scale orebodies and the subvertical plunge of individual ore shoots. Thus, a regional transition from NW-to-SE-trending compression to NW-to-SE-trending extension is interpreted as the geodynamic background of the ore-forming process. This research exemplifies an effective exploration strategy for studying the structural control of the geometry, orientation, and grade distribution of orebodies via the integration of geostatistical tools and structural analysis.

Keywords: ore shoots; structural control; geostatistics; Sizhuang gold deposit; Jiaodong Peninsula

---

1. Introduction

Ore shoots are discrete volumes of rock that contain high concentrations of mineralization (particularly high-grade ore). They are commonly hosted within particular structures or related to ore-controlling structures such as faults, shear zones, and folds [1]. The shape, orientation, and distribution of high-grade ore shoots are vital for accurate predictions in the gold mining industry. Ore shoots tend to be elongated in one direction (plunging direction) in lode gold deposits. In most cases, the ore shoot plunge of orogenic gold deposits is the result of mineral deposition and fluid flow direction controlled by structures. The possible structural controls on the localization of ore shoots include (1) the intersection of a fault or shear zone with a particular lithological unit [1]; (2) the intersection of two syn-metallogenic faults [2]; (3) dilation jogs, divergent bends in faults or zones of en echelon fault segmentation [3]; (4) fold hinge zones [4]; (5) flexures of lode or reef surface, with axes oblique to, and commonly at a high angle to, the movement direction [5]; and (6) zones that plunge subparallel to the striation [6].

More than one control can apply to an individual ore shoot and in which the specific controls on their localization are unclear. The exact plunge and structure control of ore shoots in most mines are hard to ascertain by geological methods alone, because ore shoots show little difference from the surrounding mineralization space in the macro- to mesoscopic view. Some geochemical studies have determined that high-grade ores may be developed with geochemical variations associated with hydrothermal events and enrichment by specific elements, such as Ag, Au, As, Bi, S, Sb, Te, and W. These geochemical indicators and their variations highlight hydrothermal alteration trends and define the plunge of ore shoots [7]. However, these geochemical data cannot access the exact scale and spatial distribution of ore shoots. Nowadays, a combination of structural geology and geostatistics is directly applied in the mining industry. They are integrated to create a method to model structurally controlled deposits numerically. By applying geostatistics, predictions of ore shoots and low-grade zones have been improved and a better understanding of the mechanism of mineralized fluid focus in deposit-scale structures and the results in different ore grades has also been obtained [8]. In this paper, this method is adopted to reveal the plunge of individual ore shoots and the whole deposit.

Sizhuang is a world-class epizonal orogenic gold deposit located in the Jiaodong Peninsula, Northeastern China. It is structurally controlled by the NE-to-NNE-trending and NW-dipping Jiaojia Fault [9]. In Sizhuang, the plunge of orebodies and ore shoots remains unresolved because of the complex architecture of auriferous lodes [10,11].

Thus, lineation measurement, geostatistical analysis, and ore grade data were used to investigate plunge regularity and structural control of high-grade ore shoots in the Sizhaung gold deposit. Finally, structural architecture, grade distribution, and an elaborate model for the formation of ore shoots were summarized.

2. Regional and Local Geology

2.1. Jiaodong Province and Jiaojia Belt

The Jiaodong gold province is located at the southeastern margin of the North China Craton. It is considered as perhaps the only world-class and giant gold accumulation on Earth, where relatively young gold ores (ca. 126–120 Ma [11–14]) occur within rocks that are at least 2 billion years older (ca. 2.9–1.9 Ga [15,16]) with proven gold resources over 4500 t. Although Jiaodong is commonly considered an orogenic gold province, the mineralization took place later, after regional metamorphism. During an anomalous lithospheric delamination event, the Archean-Paleoproterozoic metamorphic basement of the North China Block was intruded by multiple pulses of Mesozoic granitic magmas [17,18]. The driving force for widespread Late Jurassic and Early Cretaceous granitic magmatism, the switchover from a compressional to an extensional tectonic regime, and gold mineralization are considered to be plate subduction with lithospheric delamination and consequent asthenospheric upwelling [19–21]. Hundreds of gold deposits are widely distributed, but are concentrated in several fault systems with total resources of more than 4500 t gold. Three major NE- to NNE-trending gold belts in western Jiaodong, Sanshandao, Jiaojia, and Zhaoping contain more than 80% of the gold resources of the province and compose the broadly E-to-W-trending world-class gold corridor [22,23] (Figure 1).

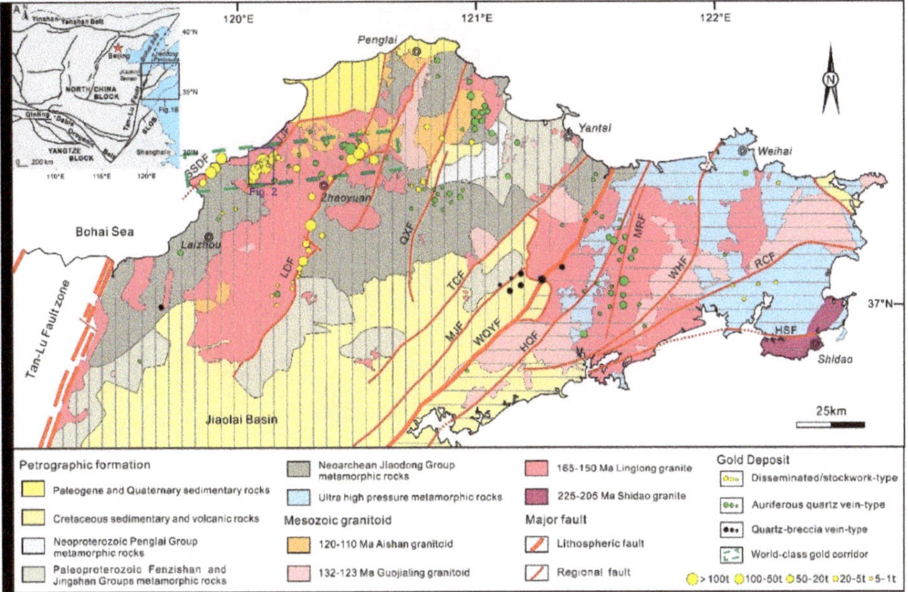

Figure 1. Geological map of Jiaodong gold province showing distribution of gold deposits. SSDF, Sanshandao Fault; JJF, Jiaojia Fault; LDF, Linglong Detachment Fault; QXF, Qixia Fault; TCF, Taocun Fault; MJF, Muping-Jimo Fault; WQYF, Wulian-Qingdao-Yantai Fault; HQF, Haiyang-Qingdao Fault; MRF, Muping-Rushan Fault; WHF, Weihai Fault; RCF, Rongcheng Fault; HSF, Haiyang-Shidao Fault [18].

Gold deposits in Jiaodong province have been mainly divided into the extensional massive quartz-vein type and the fracture-fill or disseminated type [24,25]. As the tectonic setting and origin of gold mineralization are different from typical conventional orogenic gold, some studies term these deposits "Jiaodong type" [26–28]. However, the Jiaodong-type gold deposits show no consistent spatial relationship to granitic intrusions of the same age or evidence of metal zonation related to thermal gradients surrounding anomalously hot intrusions in cooler host rocks [29–31]. They do show clear structural control along regional faults [32]. However, the ore and wall rock alteration mineralogy, fluid-inclusion composition, and stable-isotope chemistry are similar to the epizonal orogenic gold deposits [33–36].

The gold deposits in the Jiaojia gold belt, with total resources of >1400 t gold, are mainly controlled by the NE- to NNE-trending and gently NW-dipping Jiaojia and Wang'ershan Faults in an area 50 km from north to south and 1~2 km from east to west [37,38] (Figure 2). These faults underwent different tectonic activities and were normal-sinistral systems during mineralization. Major orebodies of these deposits occur at the footwall of the fault in the SE, characterized by a wide alteration zone and disseminated ores [39–41]. The Jiaojia gold belt is dominated by disseminated or fine-vein ores in the quartz–sericite alteration zone. In addition, gold-bearing quartz–pyrite veins can be found in the potassic alteration zone [42,43].

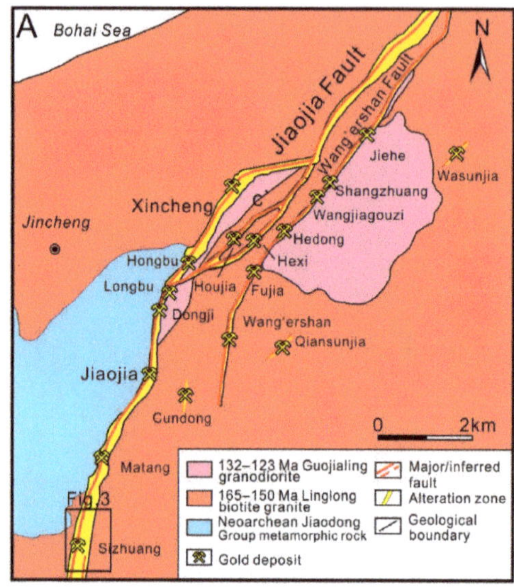

Figure 2. Simplified geological map of Jiaojia belt showing distribution of gold deposits [18].

2.2. Sizhuang Deposit Geology

The Sizhuang gold deposit (37°20′13″–37°23′28″ N, 120°04′44″–120°08′07″ E) is one of the largest gold deposits in the Jiaodong Peninsula, with a proven resource of >100 t and predicted resource of >200 t. The Sizhuang gold deposit is located about 28 km north of Laizhou City in western Jiaodong and near the southern extremity of the Jiaojia gold field. The Jiaodong group metamorphic rocks (Archean granite–greenstone belt, including TTG (Tonalite-Trondjemite-Granodiorite) gneiss, amphibolites, biotite leptynite, etc.) on the hanging wall of the Jiaojia Fault underwent high-grade metamorphism at about 2.5 Ga. The timing of gold mineralization at the Sizhuang gold deposit is likely to be between 126 and 115 Ma, while the LA-ICPMS (Laser Ablation Inductively Coupled Plasma Mass Spectrometry) zircon U-Pb age 132–123 Ma of the Guojialing granodiorites defines the earliest possible age of hydrothermal activity in the Jiaojia belt. [44,45]. Jurassic Linglong biotite granite is located on the footwall of the Jiaojia Fault to the east of the Sizhuang deposit [46].

The Linglong biotite granite was altered by hydrothermal activity, and altered rock in the footwall of the Jiaojia Fault displays a zoning distribution. Sericite–quartz–pyrite alteration close to the main fault in the footwall has an average width of 20 m. With increasing depth, the altered rocks become thinner. Sericite–quartz alteration is west of the sericite–quartz–pyrite alteration. The outermost part of the alteration zone is a potassic alteration zone, and in the west it is unaltered Linglong biotite granite [27,30,39].

According to their location and geological characteristics, orebodies in the Sizhuang deposit can be divided into three types. No. I orebodies are hosted by sericite–quartz–pyrite altered rock along a fault and are characterized by disseminated ore; No. II orebodies are hosted by sericite–quartz altered rock and are characterized by quartz–sulfide veinlets; and No. III orebodies are hosted by the potassic altered rock and are characterized by quartz–pyrite veins (Figure 3).

The orebodies are discontinuous in the shallow and the deep levels, and can be further divided into two enrichment zones accordingly. The first enrichment zone in the shallow levels is located at a depth from the ground surface to −260 m level, and the second enrichment zone is located at a depth from −310 m to −1000 m. The No. I orebodies strike 20–30°, dip 30–40° to NW overall and are the major components of the first enrichment zone. Ore in the No. I orebodies are characterized by disseminated gold in altered rocks, and tectonic breccias. The No. II orebodies are small, make up less than 10% of total gold resources in Sizhaung and totally distribute in the first enrichment zone. The second enrichment zone in the deep levels comprises the No. III orebodies and a small part of the No. I orebodies, the No. III orebodies mainly strike NNW-NNE (345°–30°) and dip slightly steeper than the No. I orebodies, with a mean dip angle of 35° [47]. The No. III orebodies are gold bearing pyrite-quartz veins hosted in the secondary en echelon structures in the potassic altered rock. The secondary en echelon fault system was the result of the slip movement of Jiaojia Fault during mineralization [48].

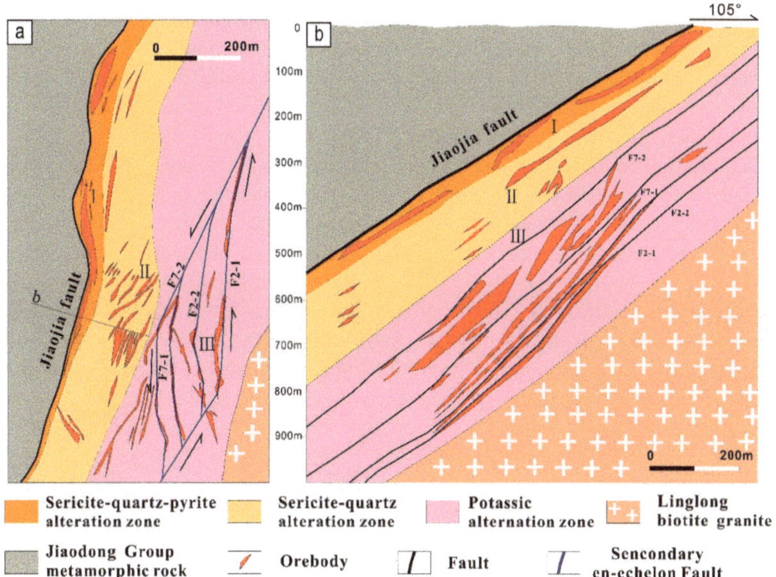

Figure 3. (**a**) Geological map of Sizhaung gold deposit, (**b**) NW–SE trending cross-section showing alteration and mineralization distribution. "I", "II" and "III" stand for No. I, No. II and No. III orebodies respectively.

3. Gold Grade and Ore-Shoot

3.1. Geostatistics Dataset and Analytical Methods

In order to determine the plunge of high-grade ore shoots in the Sizhuang deposit, the geostatistics method and semivariogram function were applied to gold grade distribution analysis.

Ore deposit structural analysis using a combination of structural geology and geostatistics has been directly applied in the mining industry to hydrothermal deposits. The main goal is to integrate structural measurements and assay data to reach the quantitative relation of mineralization and ore-controlling structure. Combined with fault kinematics, metallogenic dynamics can be revealed. Geostatistics is the study of phenomena that vary in space and/or time [49]. It is concerned with the spatial distribution of values, and was originally developed for the purpose of ore deposit modeling and ore reserve estimation [50].

However, the main goal of structural analysis in an ore deposit is to understand the behavior of grade distribution and correlate this information with the structural features of the deposit. The cornerstone of geostatistics is the semivariogram function, which geostatisticians generally refer to as a variogram. Given two locations, x and (x + h), a variogram is one-half the mean square error produced by assigning the value z(x + h) to the value z(x), or the variance of the increment.

The semivariogram function is expressed by the following equation:

$$\gamma(h) = \frac{1}{2N(h)} \sum_{i=i}^{i=N(h)} [z(xi) - z(xi + h)]^2$$

where N(h) is the number of experimental pairs [z(xi), z(xi + h)] of data, separated by the vector h [51].

A total of 8906 samples from bore holes were collected for the study. Most of them were sequential samples with approximately constant volume, taken in regularly spaced lines that cut through sections. Mine levels exist at 45 m intervals. The whole database representing No. I orebodies enclosed a volume of approximately 1000 m to the north, 150 m to the east, and 495 m in depth for −40 m to −535 m levels. The mineralization style of ore in No. I orebodies is disseminated/stockwork, but they have the best continuity as a whole. No. II orebodies are too small and No. III orebodies are too discrete to build up the database, so No. I orebodies were chosen for this study.

The database was multiplied by a constant value to maintain the mining company's confidentiality. This technique does not change the sample variance–range vector relationship and it still allows variographic structural analysis. The aim of variogram structural analysis is to partition the a priori variance and extract the range vectors to determine the anisotropy of the mineralization [52].

Summary statistics and frequency distribution plots were calculated using the moving-window technique to characterize the sample distribution and its general features through the deposit. This procedure indicated that the sample distribution closely followed the lognormal distribution. Therefore, a logarithmic transformation was applied to the dataset to avoid undesirable variations at the sills of the variograms. Frequency distribution plots totally follow normal distribution laws (Figure 4). These initial transformations confine directional variography dictated by the structural elements.

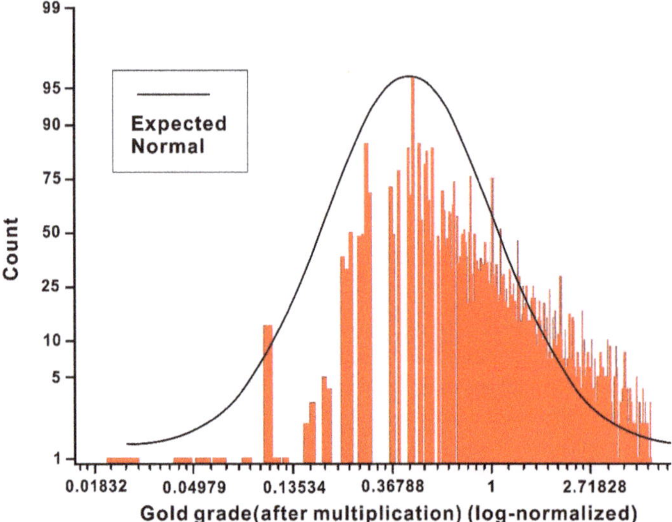

Figure 4. Distribution frequency of gold grades for samples from the Sizhuang deposit (logarithm of values). The gold grade presented here had been multiplied and log-normal transformed for confidentiality. The distribution of log-normalized gold grade resembles that of true gold grade, which can be used as the random variable to facilitate the analysis.

3.2. Ore-Grade Distribution

Because of the anisotropy of mineralization, gold grades change not only with distance but also direction. The semivariograms obtained by processing the semivariogram function demonstrate the anisotropy [53]. In the interpretation and modeling of variograms, variations in gold grade must be carefully evaluated in different directions, and the backbone of structural variographic analysis is to determine the mechanism by which mineralization spreads in different directions. If the semivariogram continues to climb steadily beyond the global variance value, this indicates a significant spatial trend in the variable. Within the distance of range, the lag distance at which the semivariogram levels off, all grades are autocorrelative. Mineralization anisotropy can be described as an ellipse with a preferred direction of maximum similarity obtained by variograms. The shape of the ellipse is determined by the range calculated from the semivariogram function in different directions. The ellipse that best fits the dataset was determined during variogram modeling. The distance of range shows the distance where gold grade is stable: where the direction range is the biggest, this indicates the best autocorrelation and the gold grade is most stable and changes at the lowest rate. The longer axis of the best-fitting ellipse indicates the orientation and distance where mineralization is stable, and the shorter axis of the ellipse indicates the orientation and distance where mineralization is unstable. In total, the range calculated by the semivariogram function will form an ellipse indicating the variation of gold grade in different directions. This further indicates the orientation and size of high-grade ore shoots as ore shoots plunge in the direction where mineralization is most stable and the range is the biggest [54].

Focusing on the structural geostatistical analysis of grade, general statistics and variograms were carried out. Several tests were conducted to define the parameters to calculate stable planar variograms. Several variographic directions (000°, 020°, 040°, 060°, 080°, 090°, 100°, 120°, 140°, and 160°) were tested. As a result, the directional dependency of the range value (range vector) was obtained during the modeling. The range was used to evaluate the shape of the ellipse that best describes the mineralization anisotropy. An example of variograms calculated for the data from the −40 m level is presented in Figure 5.

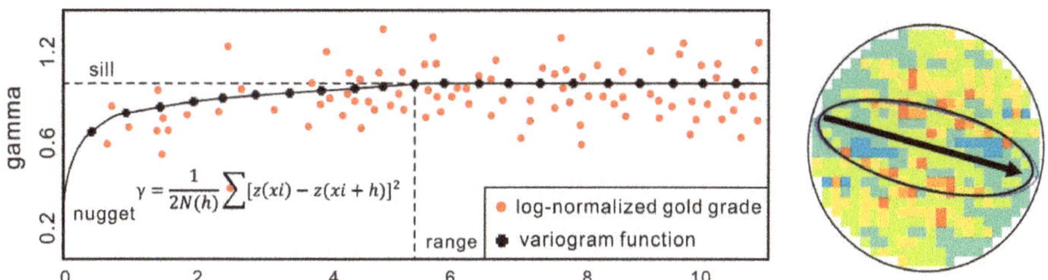

$$\gamma = \frac{1}{2N(h)} \sum [z(xi) - z(xi + h)]^2$$

Figure 5. Characteristic variogram longitudinal section for −40 m level. The horizontal axis is the sample distance of grade sample. The sill is the value at which the variogram levels off. Range is the lag distance where the semivariogram reaches the sill value and symbolizes the traditional concept of zone of influence of a sample, and represents the distance where the variogram function reaches the sill. The sill is the a priori variance of a random function, or simply the variance of the population. The ellipse on the right describes the mineralization anisotropy with preferred direction.

In addition, different ellipses describing the mineralization anisotropy for the −100 m to −535 m level are presented in Figure 6. In order to clarify the ore shape in three-dimensional space, a variogram in longitudinal section was calculated as well (Figure 7). The parameters determined during variogram modeling are presented in Table 1.

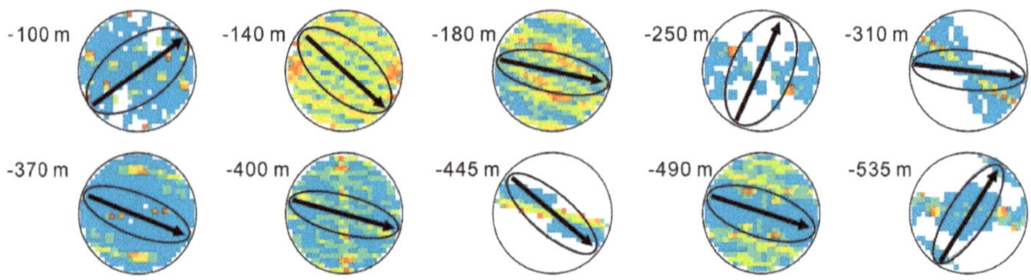

Figure 6. Mineralization ellipses to describe ore shape produced by variograms for −100 m to −535 m level.

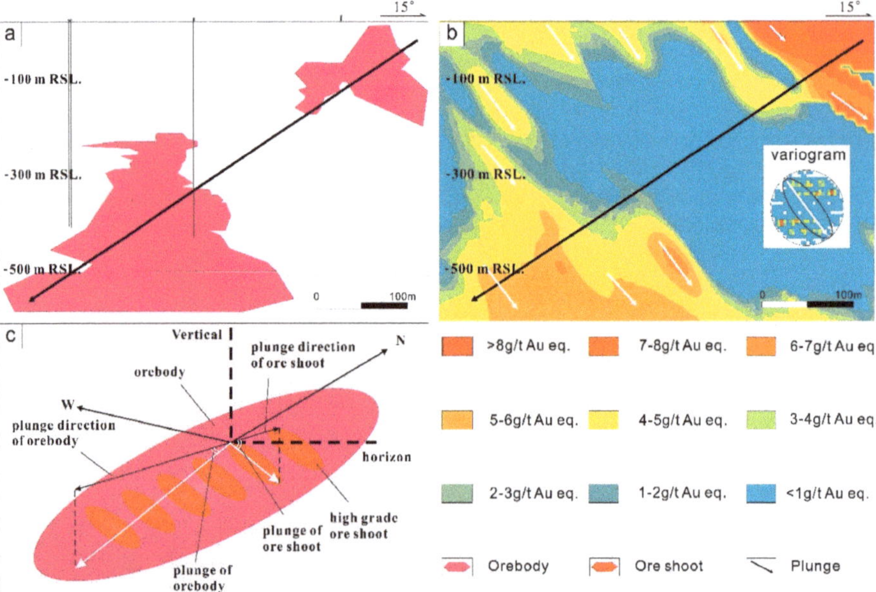

Figure 7. Mineralization anisotropy and ore shape in longitudinal section for Sizhuang deposit: (**a**) orebodies in longitudinal section; (**b**) mineralization anisotropy and best-fitting ellipse for ore shape in longitudinal section (black arrow signify plunge of orebodies as a whole, white arrows signify plunge of ore shoots); (**c**) schematic sketch of the geometry of orebodies and ore shoots.

Table 1. Parameters determined during variogram modeling.

Depth (m)	Nugget (g/t)	Maximum Range (m)	Minimum Range (m)	Partial Sill (g/t)	Step (m)	Sill (g/t)	Preferred Direction (°)
−40	0.59241	42.96557	14.38454	0.243249	4.941773	0.835659	105.6445
−100	0.777052	2.274927	1.241843	2.788349	0.189577	3.565401	56.07422
−140	2.021469	66.53855	31.10918	0.146952	5.544879	2.168421	129.5508
−180	1.111307	20.7399	6.931124	1.340726	1.728325	2.452033	103.3594
−250	0.005612	8.063627	4.352582	5.611534	0.061969	5.617146	22.14844
−310	0	5.251852	1.753606	95.27344	0.437654	95.27344	95.27344
−370	18.34196	5.702171	1.900767	15.11904	0.471809	33.461	112.8516
−400	0	56.14311	18.76465	16.14401	4.678593	16.14401	104.0625
−445	118.1267	12.01075	4.027388	15.78927	1.000896	133.916	130.0781
−490	3.8233	10.86118	4.2468	10.24774	0.905099	14.07104	107.2266
−535	7.714254	6.450409	2.151184	7.036681	0.537534	14.75094	33.7422
Longitudinal section	0	87.94258	58.94817	6.038677	7.328548	7.328548	138.6914

3.3. Plunge of Orebodies and Fault Kinematics

Orebodies in Sizhuang are generally considered as the products of ore-forming fluids, thus different fluid flow directions would lead to orebodies plunge in different orientations. According to the kinematic indicators of ore-controlling faults planes, the No. I orebodies are identified as the products of early fluid flow and overprinted by late fluid flow, while the No. II and No. III orebodies formed as the products of late fluid flow. In order to constrain the fluid flow that controlled the formation of ore shoots in the No. I orebodies, it was checked whether the plunge of ore shoots in the No. I orebodies matches that of the No. I, No. II and No. III orebodies. Individual orebodies of No. I, No. II, and No. III orebody groups have different orientations, shapes, and sizes (Figure 3). It used to be a problem to ascertain the plunge of individual orebodies. There is a simple and accurate way to determine the maximum, intermediate, and minimum shape axes of an ellipsoidal orebody by extracting and converting the eigenvectors and eigenvalues of the orientations of ellipsoids and put them with the boundaries in three-dimensional space [55]. The plunge and strike of the orebody can then be calculated using simple trigonometric functions and shape axes of ellipsoidal orebodies. The orientation and plunge were plotted in stereoplots (Figure 8). The orebodies in Sizhuang tend to strike to the NNE and plunge to the south on the whole. No. I orebodies strike N-to-NNE and plunge to the SW, and No. II and No. III orebodies strike to the NNW and plunge to the SE.

The ellipses determined by variograms indicate spaces where gold grade distribution is autocorrelative, and these elliptical spaces represent individual ore shoots [5]. The mineralization anisotropy diagrams in Figures 5 and 6 indicate two preferred directions, NE (22–56°) and SEE (95–130°), with an aspect ratio of about 2. This anisotropy reflects the preferential strike of high-grade ore shoots.

However, Figure 7 and Table 1 show that the high-grade ore shoots are general ellipsoid bodies that plunge 48° NE, almost perpendicular to the direction of the orebodies (SW) (Figure 7c). Combining the ellipses determined by variograms in sections with ellipses for each level, a three-dimensional ellipsoid can be determined to describe ore shoot plunge. Ore shoots can be regarded as ellipsoids with almost 100 m maximum axes plunging 48° to the NE, while 60–10 m intermediate axes trend to the NE or SEE and 30–5 m minimum axes vertical to them.

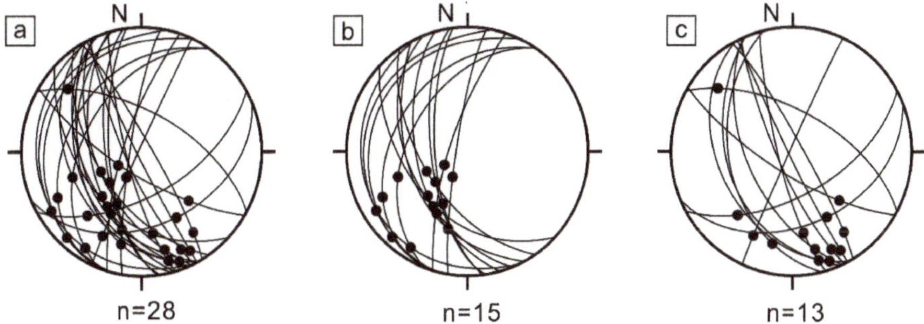

Figure 8. (**a**) Stereoplot of all attitudes and plunge calculated from shape axes of all orebodies in Sizhuang deposit. (**b**) Stereoplot of attitudes and plunge of No. I orebodies. (**c**) Stereoplot of attitudes and plunge of No. II and No. III orebodies. All stereoplots are low hemisphere, equal area.

The Sizhuang gold deposit is characterized by a NE- to NNE-trending major fault and secondary en echelon fault system in the footwall (Figure 3). Three concentrations of linear elements were recognized on these fault planes: one at 30°/40° (plunge/trend), one at 20°/190°, and one at 20°/157° (Figure 9c,f). The 30°/40° and 20°/190° planes are fairly close and defined by similar striations on the same rock. They may have been affected by later differential rotations, and we consider them to be one set, and the 20°/157° concentration is different from them. These two sets of linear elements are defined by striations caused by two different movement activities. The early set is only observed on the major fault and rocks within the sericite–quartz–pyrite alteration zone, which is dominated by No. I orebodies. This set is usually moderately steep and cuts through by later pyrite veins. The late set can be observed within No. I and No. III orebodies. Later lineations consistently trending 150° with shallow dips are usually found on quartz–pyrite veins and almost subparallel to the strike of the ore-controlling fault plane (Figure 9). The two sets of striations indicate early sinistral reverse movement on the ore-controlling faults of the No. I orebodies, followed by dextral normal fault movement on the ore-controlling faults of both the No. I and No. III orebodies.

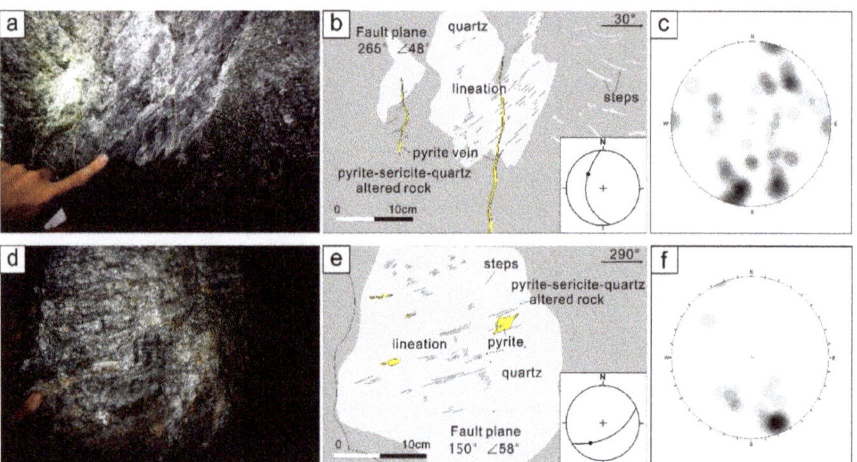

Figure 9. (**a,b**) Photograph and sketch of slickenfibers related to mineralization in No. I orebodies. (**c**) stereoplot of lineations and slickenfibers indicating movement directions on mineralization controlling fault planes in No. I orebodies. (**d,e**) Photograph and sketch of lineations and steps on fault plane. (**f**) Stereoplots of lineations and slickenfibers as movement vector on ore-controlling structure of No. III orebodies. All stereoplots are low hemisphere, equal area.

4. Discussion

Although the geostatistics dataset only contains gold grades from No. I orebodies, linear elements indicate that ore shoots of No. I orebodies, No. II and No. III orebodies were controlled by the same fluid flow. The plunge of ore shoots of No. I orebodies imply structural control on No. II and No. III orebodies. Structural control in gold mineralization (Figures 6 and 7) is the result of structurally controlled mineralizing fluid flow [56,57]. At the regional scale of western Jiaodong, the location of gold deposits is commonly controlled by jogs induced by deviations in the regional equivalent stresses. At these places, with resultant heterogeneous strain, rock permeability increased and ore-fluid ingress was focused [58,59]. At the scale of a single orebody, significant gold was deposited in dilational zones in faults. Such fault segments typically result in episodic pressure drop and provide favorable targets for large-tonnage gold ores in giant orogenic gold provinces [60,61].

In Sizhuang, significant gold was deposited in dilational zones in fault zones. Individual ore shoots plunge to the NE while the ore bodies plunge to the SW as a whole. The two main sets of linear elements on faults indicate the upflow of hydrothermal fluids migrating in different directions during two principal mineralization episodes. The directions of fluid flow were same as that of mineralization, and the sequence of episodes caused high-grade mineralization overprinting early wide-ranging mineralization. This led to the formation of SW-plunging orebodies and NE-plunging high-grade ore shoots, in a similar way as has been described for many hydrothermal environments [62,63].

Although hydrothermal flow is complex and poorly constrained, the fault kinematic data suggest a link between high-grade ore shoots and fluid flow controlled by these faults [64]. In addition, the fault movement matches the regional tectonic stress field during mineralization [33,65]. These faults are characterized by multistage events and extensive evolution from compression through transpression to transtension and finally extension, with the transitional phases of transpression and particularly transtension most closely related to gold mineralization.

We propose the following model for ore shoot formation: (1) an early mineralization episode with ore-forming fluids flowing up from SW to NE during sinistral thrusting due to NW-to-SE-trending regional compression that produced SW-plunging orebodies; and (2) a second mineralization episode with ore-forming fluids flowing up from NE to SW along a dextral normal ore-controlling fault as a result of a NW- to SE-trending extensional event that produced NE- plunging orebodies overprinting the early mineralization episode (Figure 10).

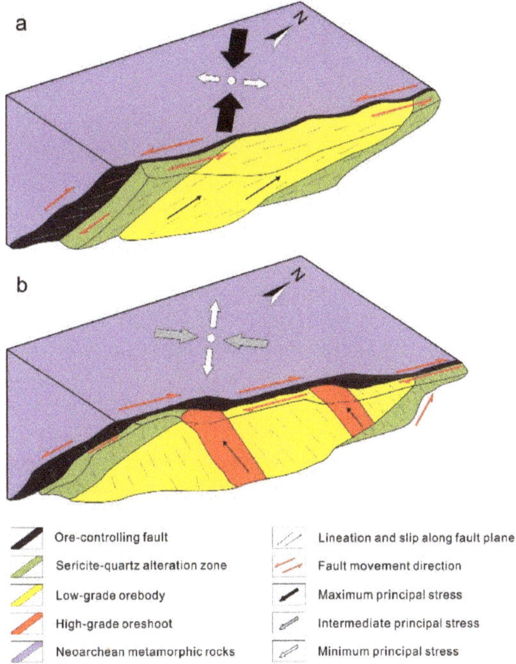

Figure 10. Model for structural control of mineralization showing ore-forming fluid flow for Sizhuang gold deposit: (**a**) early mineralization, (**b**) late mineralization.

5. Conclusions

The world-class Sizhuang epizonal orogenic gold deposit is strictly fault-controlled. Ellipses of mineralization anisotropy were determined by conducting a spatial analysis of gold grade distribution. Variographic structural analysis shows that high-grade ore shoots tend to plunge to the NE while orebodies plunge to the SW. The geometry and fault kinematics of No. I and No. III orebodies are quite different. No. I orebodies plunging to the SW underwent sinistral thrust and dextral normal movements during the mineralization epoch. No. III orebodies are more likely to plunge to the NE, the same as the plunge direction of high-grade ore shoots. No. III orebodies were controlled by syn-mineralization dextral normal movement. The late ore-forming fluid flow during the late mineralization episode overprinted upon previous alterations and mineralizations. It led to the phenomenon that high-grade ore shoots plunge nearly vertical to the major orebodies.

This study proves that structural analysis combined with geostatistical work is a useful way to determine the architecture of ore-controlling structures at deposit scale. The method could help to design more effective and economic exploration strategies for unknown ore shoots with an emphasis on structural control of the geometry, orientation, and grade distribution of the orebodies.

Author Contributions: L.-Q.Y. conceived and designed the ideas; J.-G.W., E.-J.W., and Y.-L.X. provided the geostatistics dataset; S.-R.W. performed the variogram function; L.-Q.Y. and S.-R.W. analyzed the data; S.-R.W. prepared the original draft; S.-R.W. and L.-Q.Y. reviewed and edited the draft.

Funding: This study was financially supported by the National Key Research Program of China (Grant No. 2016YFC0600107-4), the National Natural Science Foundation of China (Grant No. 41572069), the MOST Special Fund from the State Key Laboratory of Geological Processes and Mineral Resources, China University of Geosciences (Grant No. MSFGPMR201804), the Key Laboratory of Gold Mineralization Processes and Resource Utilization Subordinated to the Ministry of Natural Resources and Key Laboratory of Metallogenic Geological Process and Resources Utilization in Shandong Province, Shandong Institute of Geological Sciences (Grant No. KFKT201801), and the 111 Project under the Ministry of Education and the State Administration of Foreign Experts Affairs, China (Grant No. B07011).

Acknowledgments: We would like to thank Rongxin Zhao and Guangjun Guo at Shangdong Gold Mining for their help during field work, as well as David Groves, Jun Deng, Zhongliang Wang, and Liang Zhang for their help in improving the manuscript.

Conflicts of Interest: The authors declare no conflict of interest.

References

1. Hulin, C.D. Structural control of ore deposition. *Econ. Geol.* **1929**, *24*, 15–49. [CrossRef]

2. Blenkinsop, T.G. Thickness-displacement relationships for deformation zones: Discussion. *J. Struct. Geol.* **1989**, *11*, 1051–1052. [CrossRef]

3. Yang, L.Q.; Deng, J.; Guo, R.P.; Guo, L.N.; Wang, Z.L.; Chen, B.H.; Wang, X.D. World-class Xincheng gold deposit: An example from the giant Jiaodong gold province. *Geosci. Front.* **2016**, *7*, 419–430. [CrossRef]

4. Leader, L.D.; Wilson, C.J.L.; Robinson, J.A. Structural Constraints and Numerical Simulation of Strain Localization in the Bendigo Goldfield, Victoria, Australia. *Econ. Geol.* **2013**, *108*, 279–307. [CrossRef]

5. Blenkinsop, T.G. Orebody geometry in lode gold deposits from Zimbabwe: Implications for fluid flow, deformation and mineralization. *J. Struct. Geol.* **2004**, *26*, 1293–1301. [CrossRef]

6. Groves, D.I.; Goldfarb, R.J.; Knox-Robinson, C.M.; Ojala, J.; Gardoll, S.; Yun, G.; Holyland, P. Late-kinematic timing of orogenic gold deposits and significance for computer-based exploration techniques with emphasis on the Yilgarn Block, Western Australia. *Ore Geol. Rev.* **2000**, *17*, 1–38. [CrossRef]

7. Lebrun, E.; Thébaud, N.; Miller, J.; Roberts, M.; Evans, N. Mineralisation footprints and regional timing of the world-class Siguiri orogenic gold district (Guinea, West Africa). *Miner. Depos.* **2017**, *52*, 539–564. [CrossRef]

8. Páez, G.N.; Ruiz, R.; Guido, D.M.; Jovic, S.M.; Schalamuk, I.B. Structurally controlled fluid flow: High-grade silver ore-shoots at Martha epithermal mine, Deseado Massif, Argentina. *J. Struct. Geol.* **2011**, *33*, 985–999. [CrossRef]

9. Deng, J.; Liu, X.; Wang, Q.; Dilek, Y.; Liang, Y. Isotopic characterization and petrogenetic modeling of Early Cretaceous mafic diking—Lithospheric extension in the North China craton, eastern Asia. *Geol. Soc. Am. Bull.* **2017**, *129*, 1379–1407. [CrossRef]

10. Groves, D.I.; Santosh, M.; Goldfarb, R.J.; Zhang, L. Structural geometry of orogenic gold deposits: Implications for exploration for world-class and giant deposits. *Geosci. Front.* **2018**, *9*, 116–1177. [CrossRef]

11. Song, M.C.; Deng, J.; Yi, P.; Yang, L.Q.; Cui, S.X.; Xu, J.X.; Zhou, M.L.; Huang, T.L.; Song, G.Z.; Song, Y.X. The kiloton Jiaojia gold deposit in eastern Shandong Province and its genesis. *Acta Geosci. Sin.* **2014**, *88*, 801–824. [CrossRef]

12. Yang, L.Q.; Deng, J.; Wang, Z.L.; Zhang, L.; Goldfarb, R.J.; Yuan, W.M.; Weinberg, R.F.; Zhang, R.Z. Thermochronologic constraints on evolution of the Linglong Metamorphic Core Complex and implications for gold mineralization: A case study from the Xiadian gold deposit, Jiaodong Peninsula, eastern China. *Ore Geol. Rev.* **2016**, *72*, 165–178. [CrossRef]

13. Deng, J.; Wang, C.; Bagas, L.; Carranza, E.; Lu, Y. Cretaceous–Cenozoic tectonic history of the Jiaojia Fault and gold mineralization in the Jiaodong Peninsula, China: Constraints from zircon U–Pb, illite K–Ar, and apatite fission track thermochronometry. *Miner. Depos.* **2015**, *50*, 987–1006. [CrossRef]

14. Zhang, L.; Liu, Y.; Li, R.H.; Huang, T.; Zhang, R.Z.; Chen, B.H.; Li, J.K. Lead isotope geochemistry of Dayingezhuang gold deposit, Jiaodong Peninsula, China. *Acta Petrol. Sin.* **2014**, *30*, 2468–2480.

15. Deng, J.; Wang, Q.F. Gold mineralization in China: Metallogenic provinces, deposit types and tectonic framework. *Gondwana Res.* **2016**, *36*, 219–274. [CrossRef]

16. Shan, H.X.; Zhai, M.G.; Wang, F.; Zhou, Y.Y.; Santosh, M.; Zhu, X.Y.; Zhang, H.F.; Wang, W. Zircon U–Pb ages, geochemistry, and Nd–Hf isotopes of the TTG gneisses from the Jiaobei terrane: Implications for Neoarchean crustal evolution in the North China Craton. *J. Asian Earth Sci.* **2015**, *98*, 61–74. [CrossRef]

17. Wang, Z.L.; Yang, L.Q.; Deng, J.; Santosh, M.; Zhang, H.F.; Liu, Y.; Li, R.H.; Huang, T.; Zheng, X.L.; Zhao, H. Gold-hosting high Ba-Sr granitoids in the Xincheng gold deposit, Jiaodong Peninsula, East China: Petrogenesis and tectonic setting. *J. Asian Earth Sci.* **2014**, *95*, 274–299. [CrossRef]

18. Yang, L.Q.; Deng, J.; Wang, Z.L.; Guo, L.N.; Li, R.H.; Groves, D.I.; Danyushevkiy, J.V.; Zhang, C.; Zhang, X.L.; Zhao, H. Relationships between gold and pyrite at the Xincheng gold deposit, Jiaodong Peninsular, China: Implications for gold source and deposition in a brittle epizonal environment. *Econ. Geol.* **2016**, *111*, 105–126. [CrossRef]

19. Deng, J.; Wang, Q.F.; Xiao, C.H.; Yang, L.Q.; Liu, H.; Gong, Q.J.; Zhang, J. Tectonic-magmatic-metallogenic system, Tongling ore cluster region, Anhui Province, China. *Int. Geol. Rev.* **2011**, *53*, 449–476. [CrossRef]

20. Yang, L.Q.; Deng, J.; Ge, L.S.; Wang, Q.F.; Zhang, J.; Gao, B.F.; Jiang, S.Q.; Xu, H. Metallogenic age and genesis of gold ore deposits in Jiaodong Peninsula, Eastern China: A regional review. *Prog. Nat. Sci.* **2007**, *17*, 138–143.

21. Yang, L.Q.; Deng, J.; Qiu, K.F.; Ji, X.Z.; Santosh, M.; Song, K.R.; Song, Y.H.; Geng, J.Z.; Zhang, C.; Hua, B. Magma mixing and crust–mantle interaction in the Triassic monzogranites of Bikou Terrane, central China: Constraints from petrology, geochemistry, and zircon U–Pb–Hf isotopic systematic. *J. Asian Earth Sci.* **2015**, *98*, 320–341. [CrossRef]

22. Yang, L.Q.; Guo, L.N.; Wang, Z.L.; Zhao, R.X.; Song, M.C.; Zheng, X.L. Timing and mechanism of gold mineralization at the Wang'ershan gold deposit, Jiaodong Peninsula, eastern China. *Ore Geol. Rev.* **2017**, *88*, 491–510. [CrossRef]

23. Deng, J.; Wang, Q.F.; Li, G.J. Tectonic evolution, superimposed orogeny, and composite metallogenic system in China. *Gondwana Res.* **2017**, *50*, 216–266. [CrossRef]

24. Deng, J.; Yang, L.Q.; Li, R.H.; Groves, D.I.; Santosh, M.; Wang, Z.L.; Sai, S.X.; Wang, S.R. Regional structural control on the distribution of world-class gold deposits: An overview from the Giant Jiaodong Gold Province, China. *Geol. J.* **2019**, *54*, 378–391. [CrossRef]

25. Goldfarb, R.J.; Santosh, M. The dilemma of the Jiaodong gold deposits: Are they unique? *Geosci. Front.* **2014**, *5*, 139–153. [CrossRef]

26. Santosh, M.; Pirajno, F. The Jiaodong-type gold deposits: Introduction. *Ore Geol. Rev.* **2015**, *65*, 565–567. [CrossRef]

27. Li, L.; Santosh, M.; Li, S.R. The 'Jiaodong-type gold' deposits—Characteristics, origin and prospecting. *Ore Geol. Rev.* **2015**, *65*, 589–611. [CrossRef]

28. Deng, J.; Liu, X.F.; Wang, Q.F.; Pan, R. Origin of the Jiaodong-type Xinli gold deposit, Jiaodong peninsula, china: Constraints from fluid inclusion and C-D-O-S-Sr isotope compositions. *Ore Geol. Rev.* **2015**, *65*, 674–686. [CrossRef]

29. Charles, N.; Gumiauax, C.; Augier, R.; Chen, Y.; Zhu, R.; Lin, W. Metamorphic core complexes vs. synkinematic plutons in continental extension settings: Insights from key structures (Shandong Province, eastern China). *J. Asian Earth Sci.* **2001**, *40*, 261–278. [CrossRef]

30. Yang, L.Q.; Deng, J.; Goldfarb, R.J.; Zhang, J.; Gao, B.F.; Wang, Z.L. ^{40}Ar/^{39}Ar geochronological constraints on the formation of the Dayingezhuang gold deposit: New implications for timing and duration of hydrothermal activity in the Jiaodong gold province, China. *Gondwana Res.* **2014**, *25*, 1469–1483. [CrossRef]

31. Zhang, L.; Li, G.W.; Zheng, X.L.; An, P.; Chen, B.Y. ^{40}Ar/^{39}Ar and fission-track dating constraints on the tectonothermal history of the world-class Sanshandao gold deposit, Jiaodong Peninsula, Eastern China. *Acta Petrol. Sin.* **2016**, *32*, 2465–2476.

32. Lu, H.Z.; Archambault, G.; Li, Y.; Wei, J. Structural geochemistry of gold mineralization in the Linglong-Jiaojia district, Shandong Province, China. *Chin. J. Geochem.* **2007**, *26*, 215–234. [CrossRef]

33. Qiu, Y.M.; Groves, D.I.; McNaughton, N.J.; Wang, L.Z.; Zhou, T.H. Nature, age and tectonic setting of granitoid-hosted orogenic gold deposits of the Jiaodong Peninsula, eastern North China craton, China. *Miner. Depos.* **2002**, *37*, 283–305. [CrossRef]

34. Mao, J.W.; Wang, Y.T.; Li, H.M.; Piranjo, F.; Zhang, C.Q.; Wang, R.T. The relationship of mantle-derived fluids to gold metallogenesis in the Jiaodong Peninsula: Evidence from D-O-C-S isotope systematics. *Ore Geol. Rev.* **2008**, *33*, 361–381. [CrossRef]

35. Zhang, L.; Yang, L.Q.; Wang, Y.; Weinberg, R.F.; An, P.; Chen, B.Y. Thermochronologic constrains on the processes of formation and exhumation of the Xinli orogenic gold deposit, Jiaodong Peninsula, eastern China. *Ore Geol. Rev.* **2017**, *81*, 140–153. [CrossRef]

36. Zhang, L.; Yang, L.Q.; Weinberg, R.F.; Groves, D.I.; Wang, Z.L.; Li, G.W.; Liu, Y.; Zhang, C.; Wang, Z.K. Anatomy of a world-class epizonal orogenic-gold system: A holistic thermochronological analysis of the Xincheng gold deposit, Jiaodong Peninsula, eastern China. *Gondwana Res.* **2019**, *70*, 50–70. [CrossRef]

37. Yang, L.Q.; Deng, J.; Wang, Z.L.; Zhang, L.; Guo, L.N.; Song, M.C.; Zheng, X.L. Mesozoic gold metallogenic system of the Jiaodong gold province, eastern China. *Acta Geosci. Sin.* **2014**, *30*, 2447–2467.

38. Guo, L.N.; Marsh, E.; Goldfarb, R.J.; Wang, Z.L.; Li, R.H.; Chen, B.H.; Li, J.L. A comparison of Jiaojia- and Linglong-type gold deposit ore-forming fluids: Do they differ? *Ore Geol. Rev.* **2016**, *88*, 511–533. [CrossRef]

39. Gong, Q.J.; Deng, J.; Wang, C.M.; Wang, Z.L.; Zhou, L.Z. Element behaviors due to rock weathering and its implication to geochemical anomaly recognition: A case study on Linglong biotite granite in Jiaodong peninsula, China. *J. Geochem. Explor.* **2013**, *128*, 14–24. [CrossRef]

40. Liang, Y.Y.; Liu, X.F.; Chen, Q.; Li, Y.; Chen, J.; Jiang, J.Y. Petrogenesis of Early Cretaceous mafic dikes in southeastern Jiaolai basin, Jiaodong Peninsula, China. *Int. Geol. Rev.* **2017**, *59*, 131–150. [CrossRef]

41. Li, R.H. Ore-Controlling Model of the Jiaojia Gold Belt, Shandong Province, China. Ph.D. Thesis, China University of Geoscience, Beijing, China, 2017.

42. Yang, L.Q.; Deng, J.; Guo, L.N.; Wang, Z.L.; Li, X.Z.; Li, J.L. Origin and evolution of ore fluid, and gold-deposition processes at the giant Taishang gold deposit, Jiaodong Peninsula, eastern China. *Ore Geol. Rev.* **2016**, *72*, 585–602. [CrossRef]

43. Wang, Z.L.; Yang, L.Q.; Guo, L.N.; Marsh, E.; Wang, J.P.; Liu, Y.; Zhang, C.; Li, R.H.; Zhang, L.; Zheng, X.L.; et al. Fluid immiscibility and gold deposition in the Xincheng deposit, Jiaodong Peninsula, China: A fluid inclusion study. *Ore Geol. Rev.* **2015**, *65*, 701–717. [CrossRef]

44. Deng, J.; Wang, Q.F.; Wan, L.; Liu, H.; Yang, L.Q.; Zhang, J. A multifractal analysis of mineralization characteristics of the Dayingezhuang disseminated-veinlet gold deposit in the Jiaodong gold province of China. *Ore Geol. Rev.* **2011**, *40*, 54–64. [CrossRef]

45. Goldfarb, R.J.; Baker, T.; Dubé, B.; Groves, D.I.; Hart, C.J.R.; Gosselin, P. Distribution, character, and genesis of gold deposits in metamorphic terranes. In *Economic Geology One Hundredth Anniversary Volume*; Society of Economic Geologists: Littleton, CO, USA, 2005; pp. 407–450.

46. Sai, S.X.; Zhao, T.M.; Wang, Z.L.; Huang, S.Y.; Zhang, L. Petrogenesis of Linglong biotite granite: Constraints from mineralogical characteristics. *Acta Petrol. Sin.* **2016**, *32*, 2477–2493.

47. Qian, J.P.; Chen, H.Y.; Meng, Y. Geological characteristics of the Sizhaung gold deposit in the region of Jiaodong, Shandong Provience—A study on tectono-geochemical ore prospecting of ore deposits. *Chin. J. Geochem.* **2011**, *30*, 539–553. [CrossRef]

48. Wang, Z.L. Metallogenic System of Jiaojia Gold Orefield, Shandong Province, China. Ph.D. Thesis, China University of Geoscience, Beijing, China, 2012.

49. Deutsch, C.V. *Geostatistical Reservoir Modeling*, 1st ed.; Oxford University Press: Oxford, UK, 2002; pp. 27–40.

50. Monteiro, R.N.; Fyfe, W.S.; Chemale, F., Jr. The impact of the linkage between grade distribution and petrofabric on the understanding of structurally controlled mineral deposits: Ouro Fino gold mine, Brazil. *J. Struct. Geol.* **2004**, *26*, 1195–1214. [CrossRef]

51. Matheron, G. Principles of geostatistics. *Econ. Geol.* **1963**, *58*, 1246–1266. [CrossRef]

52. Matsumura, S. Three-dimensional expression of seismic particle motions by the trajectory ellipsoid and its application to the seismic data observed in the Kanto District, Japan. *J. Phys. Earth* **1981**, *29*, 221–239. [CrossRef]

53. Miller, J.M.; Wilson, C.J.L. Structural analysis of faults related to a heterogeneous stress history: Reconstruction of a dismembered gold deposit, stawell, western lachlan fold belt, Australia. *J. Struct. Geol.* **2004**, *26*, 1231–1256. [CrossRef]

54. Monteiro, R.N. Gold Mineralization at Ouro Fino Mine, Brazil: Structure and Alteration. Ph.D. Thesis, The University of Western Ontario, London, ON, Canada, 1996.

55. Blenkinsop, T. Comment on "Folding in high-grade rocks due to back-rotation between shear zones" by Lyal B. Harris. *J. Struct. Geol.* **2004**, *26*, 601–602. [CrossRef]

56. Yang, L.Q.; Deng, J.; Dilek, Y.; Qiu, K.F.; Ji, X.Z.; Li, N.; Taylor, R.D.; Yu, J.Y. Structure, Geochronology, and Petrogenesis of the Late Triassic Puziba Granitoid Dikes in the Mianlue Suture Zone, Qinling Orogen, China. *Geol. Soc. Am. Bull.* **2015**, *11*, 1831–1854. [CrossRef]

57. Goldfarb, R.J.; Groves, D.I.; Gardoll, S. Orogenic gold and geologic time; a global synthesis. *Ore Geol. Rev.* **2001**, *18*, 1–75. [CrossRef]

58. Sibson, R.H. Controls on maximum fluid overpressure defining conditions for mesozonal mineralisation. *J. Struct. Geol.* **2004**, *26*, 1127–1136. [CrossRef]

59. Cox, S.F. Injection-Driven Swarm Seismicity and Permeability Enhancement: Implications for the Dynamics of Hydrothermal Ore Systems in High Fluid-Flux, Overpressured Faulting Regimes—An Invited Paper. *Econ. Geol.* **2016**, *111*, 559–587. [CrossRef]

60. Goldfarb, R.J.; Groves, D.I. Orogenic gold: Common vs. evolving fluid and metal sources through time. *Lithos* **2015**, *237*, 2–26. [CrossRef]

61. Aerden, D. Foliation-boudinage control on the formation of the Rosebery Pb-Zn orebody, Tasmania. *J. Struct. Geol.* **1991**, *13*, 759–775. [CrossRef]

62. Guo, P.; Santosh, M.; Li, S. Geodynamics of gold metallogeny in the Shandong Province, NE China: An integrated geological, geophysical and geochemical perspective. *Gondwana Res.* **2013**, *24*, 807–1282. [CrossRef]

63. Deng, J.; Wang, Q.F.; Yang, L.Q.; Zhou, L.; Gong, Q.J.; Yuan, W.M.; Xu, H.; Guo, C.Y.; Liu, X.G. The Structure of Ore-controlling Strain and Stress Fields in the Shangzhuang Gold Deposit in Shandong Province, China. *Acta Geol. Sin.* **2008**, *82*, 769–780.

64. Deng, J.; Yang, L.Q.; Gao, B.F.; Sun, Z.S.; Guo, C.Y.; Wang, Q.F.; Wang, J.P. Fluid Evolution and Metallogenic Dynamics during Tectonic Regime Transition. *Resour. Geol.* **2009**, *59*, 140–153. [CrossRef]

65. Yang, L.Q.; Deng, J.; Wang, Q.F.; Zhou, Y.H. Coupling effects on gold mineralization of deep and shallow structures in the northwestern Jiaodong Peninsula, Eastern China. *Acta Geol. Sin.* **2006**, *80*, 400–411.

 *minerals*

Article

Geochronological and Geochemical Constraints on the Formation of the Giant Zaozigou Au-Sb Deposit, West Qinling, China

Hao-Cheng Yu [1,2], Chang-An Guo [1], Kun-Feng Qiu [1,2,3,*], Duncan McIntire [3], Gui-Peng Jiang [4], Zong-Yang Gou [1,2], Jian-Zhen Geng [1], Yao Pang [1,5], Rui Zhu [4] and Ning-Bo Li [2]

[1]   State Key Laboratory of Geological Processes and Mineral Resources, China University of Geosciences, Beijing 100083, China; haochengyu@cugb.edu.cn (H.-C.Y.); gca@cugb.edu.cn (C.-A.G.); zongygou@cugb.edu.cn (Z.-Y.G.); mumu1270@163.com (J.-Z.G.); pangyao1116@163.com (Y.P.)
[2]   Key Laboratory of Mineralogy and Metallogeny, Guangzhou Institute of Geochemistry, Chinese Academy of Sciences, Guangzhou 510640, China; liningbo@gig.ac.cn
[3]   Department of Geology and Geological Engineering, Colorado School of Mines, Golden, CO 80401, USA; dmcintire@mymail.mines.edu
[4]   Zhaojin Mining Industry CO., LTD., Zhaoyuan 265400, China; zzgjianggp@163.com (G.-P.J.); zzgzhurui@163.com (R.Z.)
[5]   Earthquake Monitoring Center, Sichuan Earthquake Administration, Chengdu 610041, China
*   Correspondence: kunfengqiu@cugb.edu.cn; Tel.: +86-10-82321383

Received: 8 December 2018; Accepted: 2 January 2019; Published: 11 January 2019

Abstract: The Zaozigou Au-Sb deposit has been controversial in its genesis and remains one of the most difficult ore systems to fully understand in West Qinling. The mineralization shows a broad spatial association with Triassic dikes and sills, which were previously thought to be genetically related to mineralization. Our U-Pb zircon dating in this contribution indicates that the ore-hosting porphyritic dacites were formed at 246.1 ± 5.2 Ma and 248.1 ± 3.8 Ma. The magmatic zircons yield $\varepsilon_{Hf}(t)$ values ranging from −12.5 to −8.9, with corresponding two-stage model ages of 2.08 to 1.83 Ga. The magma therefore could be derived from partial melting of Paleoproterozoic crustal materials. The ore-hosting porphyritic dacites have low oxygen fugacity, with ΔFMQ ranging from −4.61 to −2.56, indicating that magmas could have been sulfide-saturated during evolution in deep chambers and precluding the possibility that metals were released from the melt. Zaozigou exhibits characteristics widespread volcanics, massive sulfide mineralization, rare reduced mineral assemblage and discrete alteration zones which are not typical of reduced intrusion-related or porphyry gold systems. We propose that the spatially-related Triassic porphyritic dacite and dike swarm is not genetically related to the ore formation of Zaozigou Au-Sb deposit.

Keywords: zircon geochronology; magmatic oxygen fugacity; petrogenesis; Zaozigou deposit; West Qinling

1. Introduction

The Qinling Orogen linking the Dabie Orogen in the east with the Qilian and the Kunlun Orogens in the west can be separated into the West Qinling and East Qinling segments (Figure 1A) [1–5]. The West Qinling Orogen was assembled by subduction and closure of Palaeozoic Mianlue Ocean between the North China block and South China block during the Late Triassic [4–7]. Triassic granitoids and related mineralization are widespread in the West Qinling and constitute a major target for polymetallic exploration [6–11]. The Tongren–Xiahe–Hezuo district located in the northwest of the West Qinling can be further subdivided into northeastern and southwestern zones by the NNW-trending Tongren–Xiahe–Hezuo fault [6,8]. The northeastern part hosts several Triassic batholiths and Cu-Au-Fe deposits including Dewulu, Nanban, and Gangyi deposits. The southwestern part consists mainly of Triassic greenschist-facies metasedimentary rocks, which in places have been intruded by porphyry stocks or dikes. Au-Sb deposits such as at Zaozigou, Jiagantan, and Zaorendao can be found hosted in both (meta) sediments and dikes (Figure 1B). The batholiths in the northeastern zone have been dated at 250–235 Ma [6–8] and have been proven to be derived from partial melting of enriched sub-continental lithospheric mantle that had been previously modified by slab-derived melt during the continuous northward subduction of the Paleotethys oceanic slab [6,7,12–15]. The Cu-Au-Fe deposits in the northeastern zone have been recognized as porphyry-skarn systems genetically associated with Early Triassic granitoids [6]. However, a lack of research on felsic stocks and dikes in the southwestern zone makes understanding the dynamic processes governing the formation of the Triassic magmatism difficult. In addition, the genetic relationship between magmatism and ore formation remains controversial.

Successful exploration in the Tongren–Xiahe–Hezuo area relies on a robust understanding of ore deposition mechanisms. Some researchers have proposed that the mineralization is an expression of an intrusion-related gold system based on close spatial and temporal association between gold orebodies and dikes, and geochemical characteristics. Liu et al. [16] suggested that the Zaozigou deposit is a porphyry-type gold deposit related to diorite-porphyry intrusion. Sui et al. [8] and Sui and Li [17] have recently classified Zaozigou as a reduced intrusion-related gold system (RIRGS). Liang et al. [18] and Wei et al. [19] proposed that the activation of faults play important roles in the gold mineralization. Goldfarb et al. [20] believe that the lode gold deposits controlled by faults in metasedimentary rocks in this area are best classified as orogenic gold deposits. Porphyry deposits are spatially, temporally, and genetically related to hypabyssal dioritic to granitic intrusions. They are characterized by ore minerals in veinlets and disseminations in large volumes of hydrothermally altered rock [21]. The RIRGS model is as yet poorly defined. They are coeval ($\pm$2 Ma) with their associated, causative felsic and ilmenite-series pluton [22]. The most distinctive mineralization style of RIRGS is sheeted veins, which are composed of parallel, low-sulfide content, single stage quartz veins. Most orogenic gold deposits occur in greenschist facies rocks and are located adjacent to first-order, deep-crustal fault zones [23]. In Orogenic gold deposits ore minerals are deposited as vein fill of second- and third-order shears and faults, particularly at jogs or changes in strike along the crustal-scale fault zones.

The Zaozigou Au-Sb deposit is the largest gold deposit in the Tongren–Xiahe–Hezuo area (Figure 1B). Gold mineralization is controlled by fault systems, with primary mineralization manifesting as vein-hosted lode deposits. Secondary mineralization is primarily defined by wallrock alteration of intermediate to felsic dikes and Triassic slate host rocks, making it amenable to a comprehensive dike and ore genesis study. In this contribution, we performed LA-ICP-MS U-Pb dating, trace element analysis, and Lu-Hf isotope determination on zircons separated from ore-hosting porphyritic dacite at Zaozigou. Together with U-Pb ages, $\varepsilon_{Hf}(t)$, and oxygen fugacity values from the Triassic granitoids in the Tongren–Xiahe–Hezuo area, we evaluate their emplacement timing and petrogenesis in an attempt to elucidate the relationship between magmatism and mineralization.

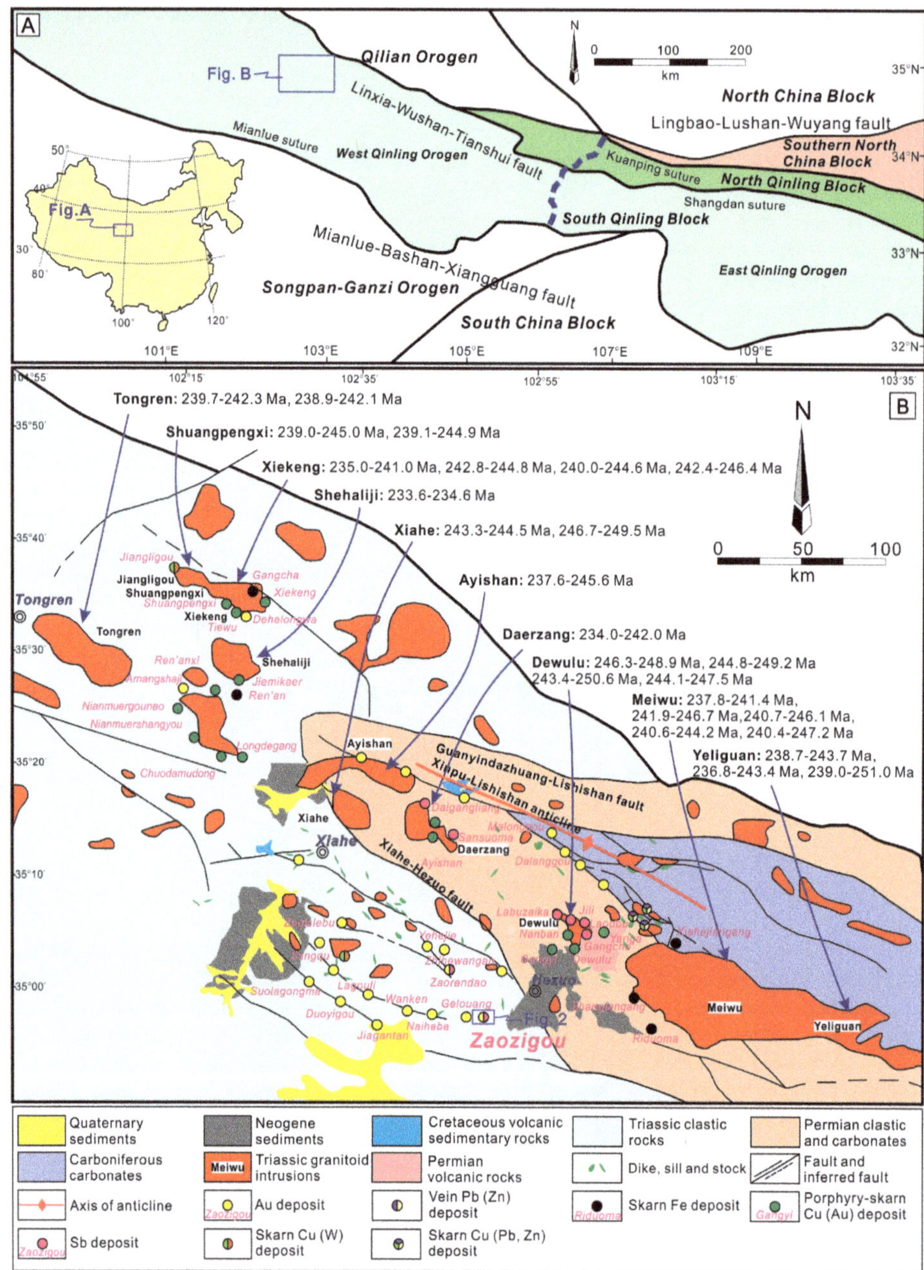

Figure 1. (**A**) Major tectonic domains of China and the location of the Qinling Orogenic Belt. Inset shows the location of the study area (modified after [6]). (**B**) Simplified geological map of the Tongren–Xiahe–Hezuo area of the Western Qinling Orogen, showing distributions and ages of early Triassic granitoids and ore deposits (modified after [6]).

2. Geological Background

2.1. Regional Geology

The Qinling Orogen in central China extends east–west for over 1500 km and is a major portion of the Central China Orogenic Belt [1,4,5,14,24,25]. It is tectonically bounded by the Qilian Orogen and North China block marked by the Lingbao–Lushan–Wuyang fault to the north, and Songpan–Ganzi Orogen and South China block marked by the Mianlue–Bashan–Xiangguang fault suture to the south (Figure 1A) [1,3,15]. The Qinling Orogen is generally further separated into the southern North China block, North Qinling block, South Qinling block and northern South China block from north to south by the Kuanping suture, Shangdan suture and Mianlue suture with related faults (Figure 1A) [2,3,10,24]. The Neoproterozoic Kuanping suture zone hosts greenschist and amphibolite facies rocks, which reflect remnants of an oceanic crust and oceanic island basalt of the Kuanping Ocean inferred to have existed at least during ca. 1.45–0.95 Ga [24]. The Paleozoic Shangdan suture zone mainly comprises ophiolitic assemblages, and subduction-related volcanic and sedimentary rocks, which are considered to have been associated with closure of the Shangdan Ocean and multistage amalgamation of the South China block-South Qinling block to the North China block during the Paleozoic [7,24]. The Triassic Mianlue suture zone is characterized by discontinuously exposed ophiolite sequences, ocean-island basalt, and island-arc volcanic rock units, marking the closure of a northern branch of the eastern Paleotethyan Ocean [1,7,24–28].

The West Qinling Orogen has traditionally been separated by the Huicheng basin or Foping dome roughly along the Baoji–Chengdu railway (shown as blue dash line in Figure 1A) to the East Qinling Orogen [7–10]. It is interpreted to have undergone a three-stage amalgamation process between the South and North China blocks, which include an early to middle Paleozoic accretionary orogen along the southern side of the North China block, a late Paleozoic to Triassic collisional orogen along the Mianlue suture zone, with amalgamation of the South China block to earlier accreted terranes, and Jurassic to Cretaceous intracontinental tectonism [7,24]. Sedimentary cover in West Qinling is dominated by Devonian to Cretaceous sediments and Precambrian basement is rarely exposed [5,8]. The pervasively folded and faulted strata record deformation related to the subduction and collision history [8,10]. The magmatism is widespread in West Qinling, including several dozens of Triassic granitoid intrusions in the eastern part and early to middle Triassic granitoids in the western part [5,8]. The intrusions in the eastern part corresponding to superimposed orogeny evolved from the northward subduction of Paleotethys Ocean through syn-collision to post-collision between the North China and South China blocks [7]. The granitoids in the western part were derived from arc magmatism related to an active continental margin setting [6].

The Tongren–Xiahe–Hezuo area is located to the northwestern segment of the West Qinling orogen (Figure 1A). This region hosts outcrops of late Paleozoic to early Mesozoic greenschist-facies slate, Triassic intrusions, with minor occurrences of Permian volcanic rocks and Cretaceous volcanic-sedimentary rocks (Figure 1B). The fine-grained foliated slate has been metamorphosed from original clastic rocks and volcanics. These rocks are mainly composed of quartz, feldspar and muscovite. Mesozoic igneous rocks, granitic stocks, dikes, and sills are widespread in the Xiahe–Hezuo district, and have been well documented to be closely-related spatially to hydrothermal ore deposits and occurrences (Figure 1B) [6,14].

2.2. Geology of the Zaozigou Deposit

The giant Zaozigou Au-Sb deposit (34°57′56″ N, 102°48′41″ E) is one of the largest gold deposits in the West Qinling. The estimated pre-mining resources included 106 t Au at an average grade of 3.34 g/t, and over 0.13 Mt Sb at an average grade of 1.34 g/t [8,16,29–33]. It is located about 9 km west of Hezuo City in Gansu Province (Figure 1B). The deposit was discovered in 1996 and has been mined since 2001 by the Zaozigou Gold Company. It was mined for oxidized ores by open pit prior to 2009 and subsequently by underground operations. Present annual production is about 3 t Au.

The Triassic Gulangdi Formation is the main host for the deposit (Figure 2). These rocks, which account for 70% of the outcrop area, are mainly composed of siliceous slate, calcareous slate, quartz sandstone, and siltstone [8,16–19]. Numerous NE- and NNE-trending intermediate to felsic sills and dikes intrude the regional greenschist facies meta-sedimentary rocks. These sills and dikes also host Au-Sb ore, and consist of porphyritic dacite, granodiorite, and porphyritic rhyolite. Previous age estimates suggest these igneous rocks vary from 250 to 215 Ma [8,16]. Four deformation events have been recognized. D1: NE-trending compression that produced NW-trending folds and thrusts, induced NS-, NE-, and nearly EW-trending fractures, and controlled emplacement of NE-trending dikes. D2: NE-, NNE-, NS-trending brittle faults. D3: Low-angle nearly EW-trending faults. D4: NNE-trending left-lateral transtensional faults that locally cut orebodies. The D2 and D3 stages are the main deformational events associated with the ores [8,16–19].

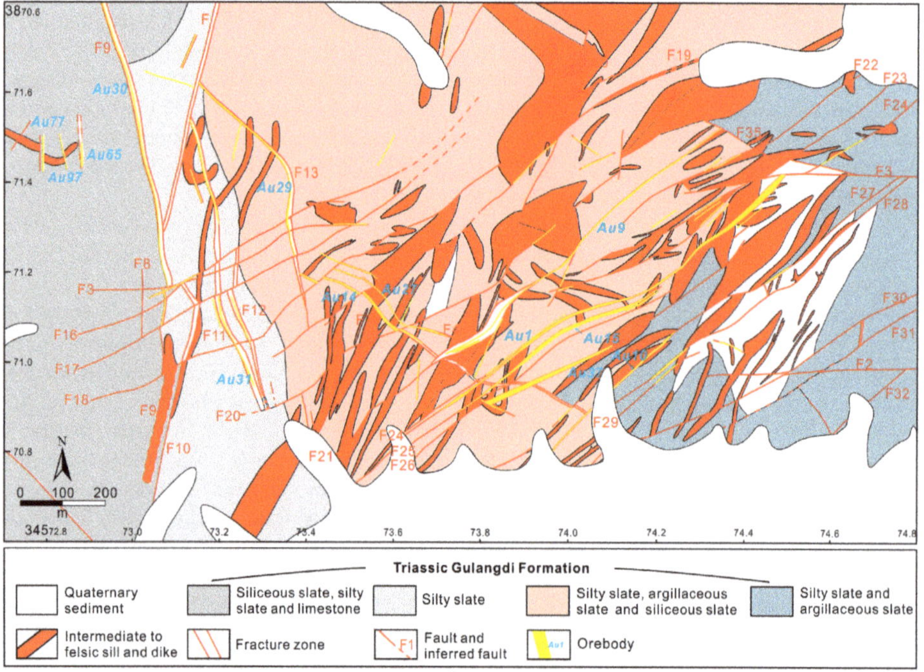

Figure 2. Sketch geologic map of the giant Zaozigou Au-Sb deposit in the West Qinling, China.

The ore-bearing zones comprise disseminated and veinlet ores within altered dikes (Figure 3A,B) and slates (Figure 3C,D), and gold- and/or stibnite-rich quartz ± calcite load vein mineralization (Figure 3E–H). They are structurally controlled by the NE-, NS-, and NW-trending faults, and are characterized by strong sulfidation (pyrite, arsenopyrite, stibnite), sericitization, silicification, and carbonatization in the wallrocks. Some of the gold- and stibnite-rich quartz±calcite veins are also structurally controlled by low-angle E–W faults. The iron-rich minerals or wallrock are important chemical traps for ore mineralization through wall rock sulfidation reactions (Figure 3A–D). The ore minerals in addition to the abundant stibnite, include lesser amounts of pyrite, arsenopyrite, sphalerite, chalcopyrite, tetrahedrite, galena, and native gold. The gangue minerals consist of quartz, biotite, calcite, sericite, feldspar, and epidote, with minor apatite, titanite, zircon, and monazite. Native gold grains occur in cracks or as inclusions within stibnite and quartz. Invisible gold is present in pyrite and arsenopyrite. Four paragenetic stages are identified based on crosscutting relationships and mineralogical and textural characteristics: stage 1 pyrite-chalcopyrite-tetrahedrite-quartz, stage 2 pyrite-arsenopyrite-quartz, stage 3 stibnite-sphalerite-quartz-native gold, and stage 4 stibnite-quartz-calcite (Figure 4).

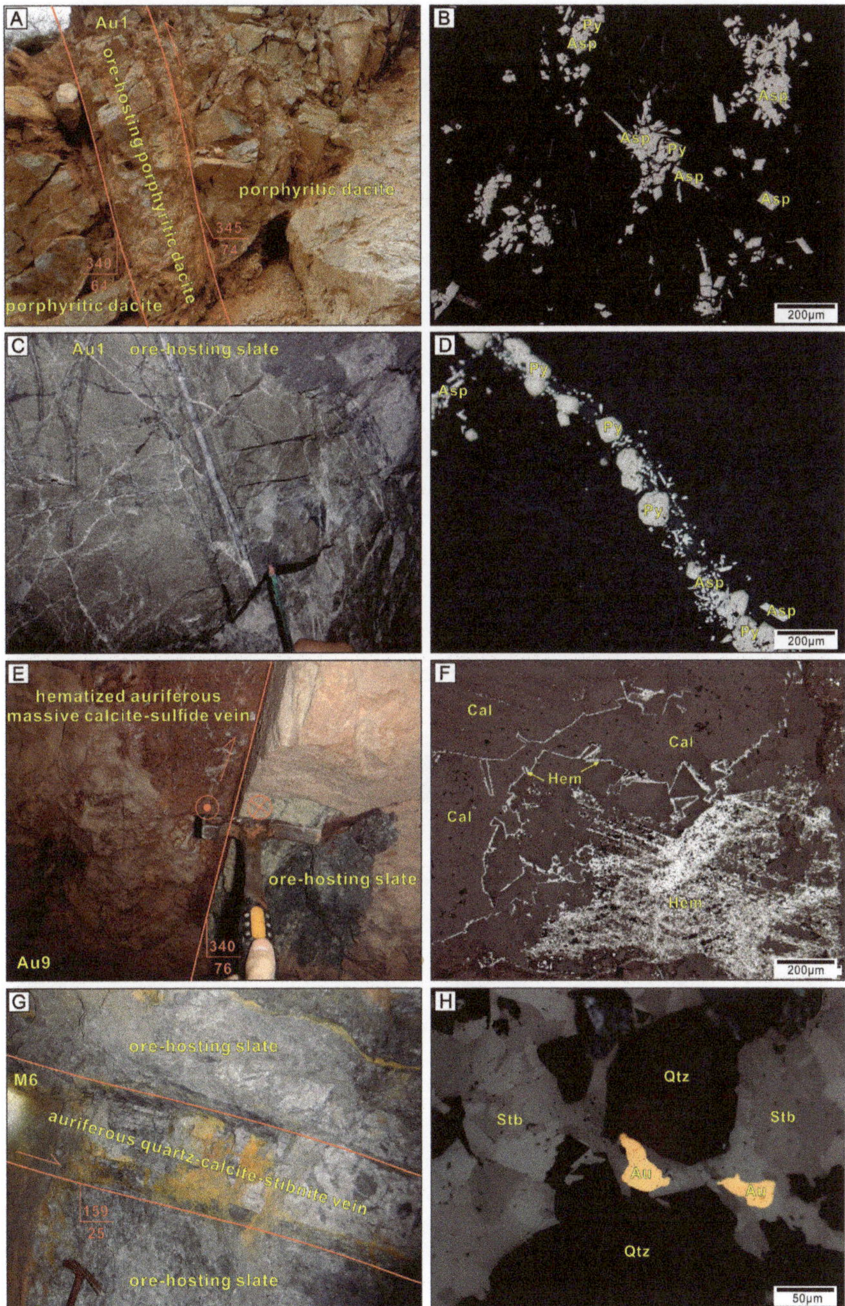

Figure 3. Exposure (**A,C,E,G**) and photomicrographs under reflected light (**B,D,F,H**) showing mineralization styles and mineral assemblages. (**A,B**) Pyrite and arsenopyrite as disseminated

aggregates in porphyritic dacite. (**C,D**) Pyrite and arsenopyrite veinlet in slate. (**E,F**) Hematized auriferous calcite veins. (**G,H**) Auriferous quartz-calcite-stibnite vein structurally controlled by NE-trending faults cutting slate and native gold grains occur in cracks and inclusions of stibnite and quartz. Py = pyrite, Asp = arsenopyrite, Stb = stibnite, Au = native gold, Qtz = quartz, Cal = calcite, Hem = hematite.

One hundred and forty-three orebodies have been delineated over the life of the mine, 16 of which have an individual reserves exceeding 1 t Au. The Au1, Au9 and M6 gold orebodies comprise over 40% of the Zaozigou deposit gold endowment. The Au1 and Au9 orebodies, which together comprise 33% of the gold endowment, are controlled by the NE-trending faults. The largest orebody, Au1 (pre-ore reserves ~20 t Au), is controlled by the F24 fault, which is characterized by multiphase activity and varying fault movement sense. The Au1 orebody consists of disseminated, veinlet, and auriferous quartz-stibnite ores (Figure 3A–D). The orebody developed in the porphyritic dacites dikes and slate, and known to extend along strike for at least 1240 m, with a trend of 55° and a dip of 80° to 85°. It ranges in thickness from 0.82 to 18.13 m (avg. 3.48 m) and extends downdip at least 1200 m. The gold grade varies from 1.00 to 11.20 g/t with an average of 3.47 g/t. The Au9 orebody hosts pre-ore reserves 15 t Au. Au9 is spatially controlled by the F21 fault, parallel with the Au1 orebody. It is hosted entirely within the slate and is characterized by hematized auriferous calcite veins. Au9 lode veins are mainly composed of coarse-grained calcite (1–8 cm interlocking crystals). The Au 9 orebody is 800 m long and 0.44 to 11.36 m thick (avg. 2.24 m), trending 55° and dipping ~80° (Figure 3E,F). The gold grade varies from 1.02 to 21.90 g/t with an average of 3.46 g/t. The M6 concealed orebody controlled by the EW-trending fault F3 contains ~15 t Au. M6 cross-cuts the Au1 and Au9 orebodies. The orebody is characterized by auriferous quartz-calcite-stibnite vein ores and appears as stratiform-like lenses that strike 165° to 180° and dip 10° to 26° (Figure 3G,H). It is 1160 m long, extending downdip from the 3226-m level to the 3030-m level, and ranges in thickness from 0.94 m to 15.87 m with an average thickness of 3.94 m. The gold grade varies from 1.01 to 12.97 g/t with an average of 4.01 g/t.

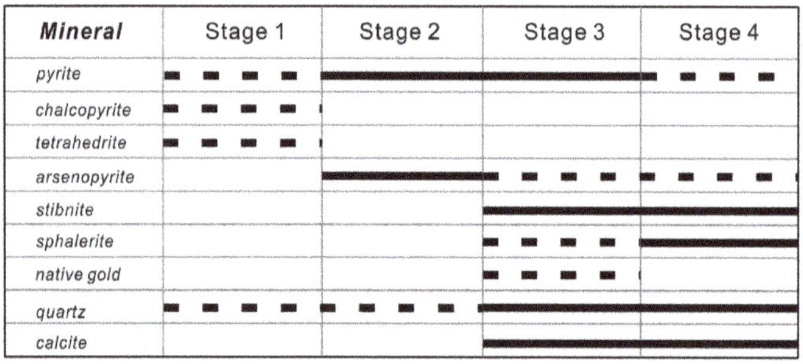

Mineral	Stage 1	Stage 2	Stage 3	Stage 4
pyrite				
chalcopyrite				
tetrahedrite				
arsenopyrite				
stibnite				
sphalerite				
native gold				
quartz				
calcite				

Figure 4. Paragenetic sequence of the Zaozigou deposit interpreted from cross-cutting relationships, ore textures and sulfide assemblages. The black full lines indicate high abundance and the black dashed lines represent minor amounts.

3. Sampling and Analytical Methods

3.1. Sampling

The detailed investigations of the present study focus on the porphyritic dacite dikes in the Zaozigou deposit. Representative samples were collected from the wallrock of the #Au1 orebody. The porphyritic dacite samples (ZZG11 and ZZG12) are light green to buff in color texture (Figure 5A,C). The phenocrysts include quartz (15–20 vol %, long dimension 0.1–1.5 mm), plagioclase (5–15 vol %, long dimension 0.1–2.5 mm), and biotite (5–15 vol %, long dimension 0.1–2 mm). Some quartz grains show rounded or angular-crystal habit, with dissolution textures. The plagioclase and biotite grains are variably altered to fine sericite and chlorite. The groundmass consists of plagioclase, quartz, and minor K-feldspar. Accessory minerals are mainly zircon, apatite, epidote, chlorite, and sericite (Figure 5B,D). Ore mineralogy consists mainly of pyrite and arsenopyrite, which are commonly sitting in iron-rich minerals, notably biotite which has been hydrothermally altered to sericite or chlorite (Figure 5B,D).

Figure 5. Hand specimen (**A,C**) and photomicrographs under transmitted light (**B,D**) of porphyritic dacite in this study. Py = pyrite, Asp = arsenopyrite, Bio = biotite, Pl = plagioclase, Qtz = quartz, Cal = calcite.

3.2. Analytical Methods

3.2.1. Zircon LA-ICP-MS U-Pb Dating and Trace Element Analyses

The porphyritic dacite samples were crushed to 40–60 mesh, and zircon crystals were separated through standard magnetic and density separation techniques. Zircon grains were carefully handpicked under a binocular microscope, mounted in epoxy, polished down to near half sections to expose internal structures, and then cleaned in an ultrasonic washer containing a 5% HNO_3 bath. Prior to analysis, polished sections of zircon were carbon coated for cathodoluminescence (CL) imaging, which were taken on a JXA-880 electron microscope and an image analysis software was used under operating conditions of 20 kV and 20 nA, at the Institute of Mineral Resources, Chinese Academy of Geological Sciences, Beijing, China, to identify the internal structure and texture of all zircon crystals. Zircon samples are checked carefully under the microscope and scanning electron microscope (SEM) to observe mineral and fluid inclusions and cracks.

Zircon U-Pb isotope and trace element analyses were simultaneously carried out using a LA-ICP-MS system in the Key Laboratory of Mineralogy and Metallogeny, Guangzhou Institute of Geochemistry, the Chinese Academy of Sciences. The LA-ICP-MS system includes an Agilent 7900 ICP-MS coupled with a Resonetics RESOlution S-155 ArF-Excimer laser source (λ = 193 nm). The operating conditions were 4 J/cm^2 of energy density, 29 μm of spot diameter, and 8 Hz of ablation frequency. Plesovice zircon, a new natural reference material for U-Pb isotopic microanalysis, was analyzed once every five analyses and TEMORA zircon was used as internal standards for U-Pb dating, and was analyzed twice every five analyses, in order to normalize isotopic fractionation during isotope analysis. The NIST610 glass standard was used as an external standard to normalize U, Th, and Pb concentrations of the unknowns. In addition, standard sample mud tank was used as an isotopic monitoring sample. The ICPMS DataCal program was used for processing analyses data. Common Pb was corrected according to the method proposed by [34]. The analytical results are reported with 1σ error. The weighted mean U-Pb ages (with 90% confidence) were calculated at 2σ level and Concordia plots were produced using ISOPLOT 3.23 V. The ^{29}Si was used as an internal standard for trace element analyses. The average analytical error ranges from 10% for light rare earth elements (LREE) to 5% for other trace elements. A detailed compilation of instrument and data acquisition parameters was presented in [35].

3.2.2. In Situ Zircon Lu-Hf Isotope Analyses

In situ zircon Lu-Hf isotopic analysis was carried out across samples on the analogous zircon zones where U-Pb age determinations were made. Hafnium isotopic compositions were determined with a Thermo Finnigan Neptune MC-ICP-MS system coupled to a New Wave UP193 nm laser ablation system at the Laboratory of Isotope Geology, Tianjin Institute of Geology and Mineral Resources, Tian-jin, China. A laser repetition rate of 11 Hz at 100 mJ was used for ablating zircons and the spot diameters were 50 μm. Helium was used as the carrier gas for the ablated aerosol. Isotopes, including ^{177}Hf, ^{178}Hf, ^{179}Hf, ^{180}Hf, ^{172}Yb, ^{173}Yb, ^{175}Lu, ^{176}(Hf + Yb + Lu), and ^{182}W, were measured during the analytical process. Isobaric interference of ^{176}Lu on ^{176}Hf was corrected based on the measured ^{175}Lu value and the recommended ^{176}Lu/^{175}Lu ratio of 0.02655. Similarly, the ^{176}Yb/^{172}Yb value of 0.5887 and mean βYb value obtained during Hf analysis on the same spot were used for interference correction of ^{176}Yb on ^{176}Hf. During the analyses, the GJ-1zircon standard yielded ^{176}Hf/^{177}Hf ratios of 0.282009 ± 24 (2σ, n = 13). These ratios are consistent with the recommended ^{176}Hf/^{177}Hf ratios of 0.282015 ± 19. The decay constant for ^{176}Lu of 1.865 × 10^{-11} year^{-1} and present-day chondritic ratios of ^{176}Hf/^{177}Hf = 0.282785 and ^{176}Lu/^{177}Hf = 0.0336 were used to calculate the ε$_{Hf}$(t) values [35–37].

3.2.3. Magmatic Oxygen Fugacity Estimation

Zou et al. [38] reviewed the underlying assumption of zircon REE oxy-barometers, the lattice strain model, systematically re-evaluated the common zircon REE oxy-barometers, and concluded that the $x^{melt}_{Ce^{4+}}/x^{melt}_{Ce^{3+}}$ oxy-barometer is a reliable accurate measurement. Oxygen fugacities in this study were estimated by algorithm based on MATLAB software using the formula (Computer Code S1 in Supplementary Materials) [39]

$$ln\left(\frac{x^{melt}_{Ce^{4+}}}{x^{melt}_{Ce^{3+}}}\right) = \frac{1}{4}lnfO_2 + \frac{13136\,(\pm591)}{T} - 2.604(\pm0.011)\frac{NBO}{T} - 8.878\,(\pm0.112)\cdot xH_2O - 8.955(\pm0.091)$$

where $x^{melt}_{Ce^{4+}}/x^{melt}_{Ce^{3+}}$ can be determined using the lattice strain model [40], T is the zircon crystallization temperature in K following the Ti-in-zircon thermometer and using a TiO$_2$ activity of 0.6 and SiO$_2$ activity of 1 [41], NBO/T is the proportion of non-bridging oxygen to tetrahedrally coordinated cations and can be determined on an anhydrous basis [42], and xH_2O is the mole fraction of water dissolved in the melt. In general, the water content of felsic magma is assumed to be 2.5–6.5 wt %. On the other hand, amphibole grain is rare in porphyritic dacite, indicating the water content is less than 4.5 wt %. Therefore, the water content is assumed to be 3 wt % in this paper [39,43].

4. Analytical Results

4.1. Zircon Morphology and U-Pb Ages

Most zircon grains are pristine, euhedral and display well-developed oscillatory growth zoning. Some crystals display inherited zircons with weak zoning or no zoning and some have mineral inclusions including apatite and zircon. The crystals have lengths of 100–200 µm and length/width ratios of 2:1 to 3:1. Several grains contain zircon and apatite inclusions, and have jagged edges, indicating that they may have been hydrothermally altered (Figure 6).

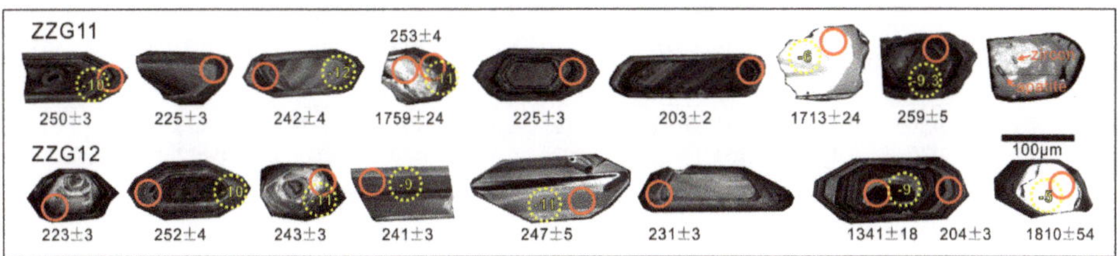

Figure 6. Representative cathodoluminescence images and photomicrographs of zircons derived from ZZG11 and ZZG12 with identified analytical spot, U-Pb age (Ma) and $\varepsilon_{Hf}(t)$ value. Red circle: U-Pb beam. Yellow dash circle: Hf beam.

Twenty-five analyses were carried out for U-Pb age dating of zircons from sample ZZG11 (Table 1, Figure 7A–C). Two analyses on the inherited zircons give $^{206}Pb/^{238}U$ ages of 1759 ± 24, and 1713 ± 24 Ma, respectively. Twenty analyses are concordant, yielding weighted average ages of 248.1 ± 3.8 Ma (MSWD = 3.6, n = 14), 222.1 ± 4.4 Ma (MSWD = 1.4, n = 5), and 203.5 ± 2.2 Ma (n = 1), respectively.

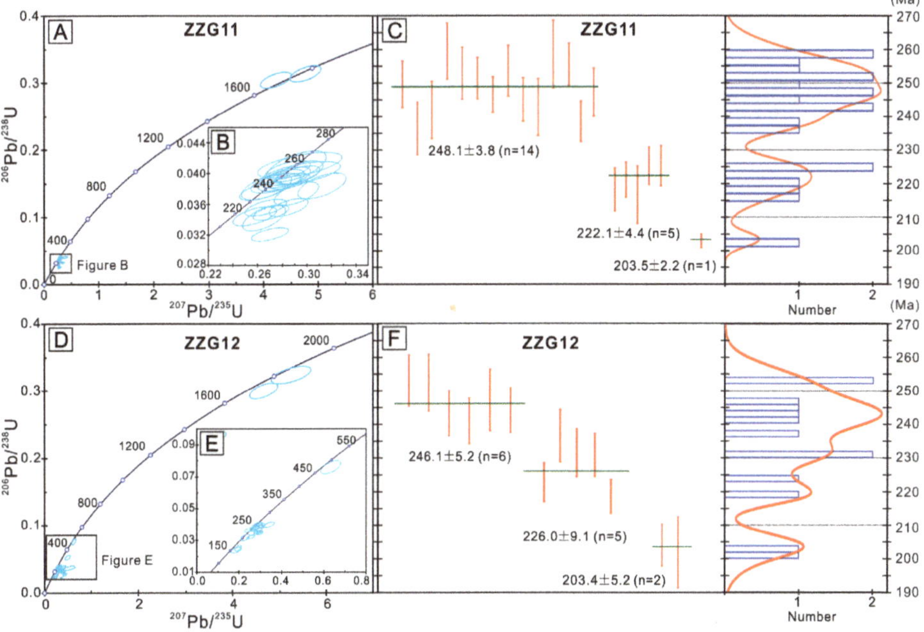

Figure 7. LA-ICP-MS zircon U-Pb concordia diagrams for ZZG11 (**A–C**) and ZZG12 (**D–F**).

Twenty analyses were carried out for U-Pb age dating of zircons from sample ZZG12 (Table 1, Figure 7D–F). Four analyses on the inherited zircons give ^{206}Pb/^{238}U ages of 1810 ± 26, 1693 ± 24, 1341 ± 18, and 472 ± 10, respectively. Thirteen analyses are concordant, yielding weighted average ages of 246.1 ± 5.2 Ma (MSWD = 1.8, n = 6), 226.0 ± 9.1 Ma (MSWD = 5.6, n = 5), and 203.4 ± 5.2 Ma (MSWD = 0.12, n = 2), respectively.

Table 1. LA-ICP-MS zircon U-Pb dating results of porphyritic dacite samples ZZG11 and ZZG12.

| Sample No. | Isotopic Ratios | | | | | | Ages (Ma) | | | | | |
| | ^{207}Pb/^{206}Pb | | ^{207}Pb/^{235}U | | ^{206}Pb/^{238}U | | ^{207}Pb/^{206}Pb | | ^{207}Pb/^{235}U | | ^{206}Pb/^{238}U | |
	Ratio	1σ	Ratio	1σ	Ratio	1σ	Age	1σ	Age	1σ	Age	1σ
ZZG11: Inherited Zircon												
ZZG11.9	0.1097	0.0020	4.7641	0.1176	0.3138	0.0050	1794.8	33.3	1778.6	20.8	1759.2	24.3
ZZG11.11	0.1009	0.0027	4.2254	0.1139	0.3044	0.0048	1640.4	54.6	1679.0	22.2	1713.2	24.0
ZZG11: 248.1 ± 3.8 Ma (MSWD = 3.6, n = 14)												
ZZG11.1	0.0516	0.0014	0.2831	0.0082	0.0395	0.0006	333.4	60.2	253.1	6.5	249.6	3.5
ZZG11.2	0.0522	0.0017	0.2919	0.0098	0.0409	0.0008	294.5	72.2	260.1	7.7	258.6	5.0
ZZG11.4	0.0507	0.0013	0.2639	0.0074	0.0377	0.0005	227.8	59.3	237.8	5.9	238.6	3.0
ZZG11.6	0.0518	0.0013	0.2794	0.0076	0.0391	0.0006	276.0	63.9	250.1	6.0	247.2	3.6
ZZG11.7	0.0511	0.0015	0.2675	0.0082	0.0382	0.0007	242.7	68.5	240.7	6.6	241.9	4.3
ZZG11.8	0.0519	0.0023	0.2946	0.0140	0.0411	0.0007	279.7	100.0	262.2	11.0	259.5	4.2
ZZG11.10	0.0535	0.0023	0.2956	0.0131	0.0400	0.0006	350.1	96.3	263.0	10.2	252.9	3.9
ZZG11.14	0.0510	0.0013	0.2845	0.0074	0.0404	0.0005	242.7	54.6	254.3	5.9	255.5	3.1
ZZG11.15	0.0533	0.0013	0.2925	0.0075	0.0398	0.0005	342.7	55.6	260.5	5.9	251.4	3.1
ZZG11.16	0.0532	0.0012	0.2865	0.0065	0.0390	0.0004	338.9	50.0	255.8	5.2	246.5	2.6
ZZG11.19	0.0532	0.0012	0.2944	0.0078	0.0401	0.0006	344.5	53.7	262.0	6.1	253.7	3.8
ZZG11.20	0.0567	0.0012	0.2917	0.0074	0.0374	0.0006	479.7	48.1	259.9	5.8	236.5	3.9
ZZG11.21	0.0512	0.0013	0.2750	0.0082	0.0387	0.0005	250.1	59.3	246.7	6.5	245.0	3.2
ZZG11.25	0.0597	0.0016	0.3124	0.0081	0.0384	0.0007	590.8	58.2	276.1	6.3	242.7	4.2
ZZG11: 222.1 ± 4.4 Ma (MSWD = 1.4, n = 5)												
ZZG11.3	0.0539	0.0013	0.2579	0.0072	0.0344	0.0005	368.6	55.6	233.0	5.8	218.3	3.2
ZZG11.5	0.0542	0.0013	0.2669	0.0068	0.0356	0.0004	388.9	51.8	240.2	5.5	225.3	2.7
ZZG11.22	0.0563	0.0011	0.2710	0.0059	0.0349	0.0004	464.9	44.4	243.5	4.7	221.2	2.5
ZZG11.23	0.0570	0.0015	0.2655	0.0070	0.0342	0.0007	500.0	57.4	239.1	5.6	216.8	4.2
ZZG11.24	0.0585	0.0014	0.2853	0.0064	0.0356	0.0005	550.0	51.8	254.8	5.1	225.3	3.0
ZZG11: 203.5 ± 2.2 Ma (n = 1)												
ZZG11.18	0.0612	0.0014	0.2702	0.0061	0.0321	0.0004	655.6	50.0	242.8	4.9	203.5	2.2
ZZG12: Inherited Zircon												
ZZG12.1	0.1130	0.0025	3.6467	0.0903	0.2312	0.0035	1850.0	39.7	1559.8	19.8	1341.0	18.3
ZZG12.2	0.1172	0.0039	5.2503	0.1720	0.3241	0.0053	1913.9	59.4	1860.8	28.0	1809.9	25.8
ZZG12.13	0.0601	0.0016	0.6331	0.0193	0.0759	0.0017	609.3	59.3	498.1	12.0	471.8	10.2
ZZG12.14	0.1117	0.0026	4.6682	0.1215	0.3003	0.0048	1827.8	42.0	1761.6	21.8	1692.9	23.7
ZZG12: 246.1 ± 5.2 Ma (MSWD = 1.8, n = 6)												
ZZG12.3	0.0541	0.0017	0.2895	0.0088	0.0385	0.0005	376.0	75.0	258.2	6.9	243.3	3.3
ZZG12.6	0.0522	0.0017	0.2756	0.0091	0.0381	0.0005	294.5	74.1	247.2	7.2	241.0	3.4
ZZG12.7	0.0541	0.0021	0.2935	0.0125	0.0391	0.0007	372.3	87.0	261.3	9.8	247.2	4.5
ZZG12.9	0.0556	0.0014	0.2948	0.0072	0.0386	0.0005	438.9	55.6	262.3	5.6	244.1	3.3
ZZG12.10	0.0521	0.0015	0.2889	0.0087	0.0400	0.0006	300.1	60.2	257.7	6.9	252.9	3.8
ZZG12.20	0.0614	0.0022	0.3404	0.0109	0.0399	0.0007	653.7	75.9	297.5	8.2	252.3	4.2
ZZG12: 226.0 ± 9.1 Ma (MSWD = 5.6, n = 5)												
ZZG12.4	0.0598	0.0017	0.2910	0.0080	0.0351	0.0005	594.5	63.0	259.3	6.3	222.6	2.9
ZZG12.5	0.0578	0.0018	0.2912	0.0081	0.0365	0.0006	524.1	68.5	259.5	6.4	231.4	3.5
ZZG12.12	0.0491	0.0016	0.2470	0.0080	0.0364	0.0005	153.8	77.8	224.1	6.5	230.7	3.2

Table 1. Cont.

Sample No.	Isotopic Ratios						Ages (Ma)					
	$^{207}Pb/^{206}Pb$		$^{207}Pb/^{235}U$		$^{206}Pb/^{238}U$		$^{207}Pb/^{206}Pb$		$^{207}Pb/^{235}U$		$^{206}Pb/^{238}U$	
	Ratio	1σ	Ratio	1σ	Ratio	1σ	Age	1σ	Age	1σ	Age	1σ
ZZG12.15	0.0624	0.0014	0.2984	0.0063	0.0345	0.0004	687.1	52.8	265.2	4.9	218.4	2.5
ZZG12.18	0.0531	0.0018	0.2755	0.0092	0.0374	0.0006	344.5	77.8	247.1	7.3	236.5	3.9
ZZG12: 203.4 ± 5.2 Ma (MSWD = 0.12, n = 2)												
ZZG12.8	0.0521	0.0015	0.2301	0.0086	0.0318	0.0008	300.1	60.2	210.3	7.1	201.8	5.2
ZZG12.16	0.0576	0.0019	0.2569	0.0078	0.0321	0.0005	522.3	70.4	232.2	6.3	204.0	3.1

4.2. Zircon Trace Element Composition and Oxygen Fugacity

Trace element compositions of zircons from sample ZZG11 and ZZG12 are listed in Table 2. The chondrite-normalized REE patterns are illustrated in Figure 8. The inherited zircons show strong positive Ce (Ce/Ce* ratios, where Ce* = $\sqrt{(La)_N \times (Pr)_N}$) and negative Eu (Eu/Eu* ratios, where Eu* = $\sqrt{(Sm)_N \times (Gd)_N}$) anomalies, ranging from 2.48 to 394.68, and 0.02 to 0.43, respectively. The ~248–246 Ma and ~226–222 Ma zircons show moderate positive Ce and negative Eu anomalies. The Ce/Ce* and Eu/Eu* of ~248–246 Ma zircons vary from 1.31 to 61.75 and 0.18 to 0.62, respectively. The Ce/Ce* and Eu/Eu* of ~226–222 Ma zircons vary from 1.24 to 4.90, and 0.46 to 0.63, respectively. The ~203 Ma zircons have high REE contents and weak positive Ce (Ce/Ce* = 1.48 to 4.37) and negative Eu anomalies (Eu/Eu* = 0.44 to 0.73).

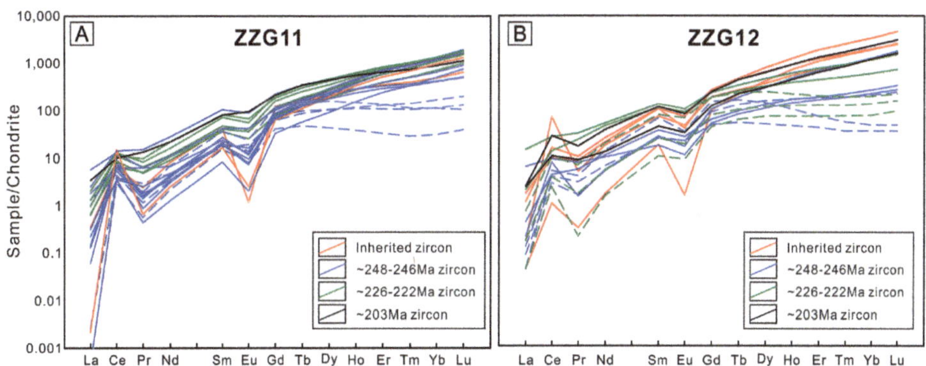

Figure 8. Chondrite-normalized REE patterns for ZZG11 (**A**) and ZZG12 (**B**). Note the HREE value anomalies (dashed lines) which may be indicative of contamination by inclusions or alteration.

The zircon crystallization temperatures and oxygen fugacities were calculated and listed in Table 2 and Figure 9. Aberrant REE values (dashed lines in Figure 8) which may be indicative of contamination by inclusions or alteration are excluded from the calculation (Figure 6). The logarithmic oxygen fugacities (fO_2) of the inherited zircons range from −10.58 to −1.39, with ΔFMQ ranging from −6.13 to 2.93. The logarithmic oxygen fugacities of the ~248–246 Ma zircons range from −9.17 to −7.16, with ΔFMQ ranging from −4.61 to −2.56. The logarithmic oxygen fugacities of the ~226–222 Ma zircons range from −8.08 to −6.09, with ΔFMQ ranging from −3.52 to −1.48. The logarithmic oxygen fugacities of the ~203 Ma zircons range from −7.99 to −6.45, with ΔFMQ ranging from −3.47 to −1.74.

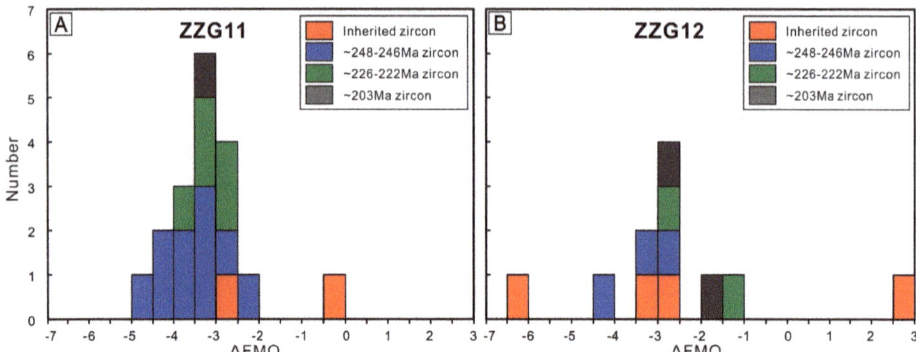

Figure 9. Magma oxygen fugacity for ZZG11 (**A**) and ZZG12 (**B**).

4.3. Zircon Lu-Hf Isotopic Composition

In situ zircon Lu-Hf isotopic data of magmatic zircons and inherited zircons are presented in Table 3 and shown in Figure 10. The single stage depleted-mantle model ages ($T_{DM}{}^1$) are determined for each sample by calculating the intersection of the zircon/parent-rock growth trajectory with the depleted-mantle evolution curve [6,9,35]. The two-stage model ages ($T_{DM}{}^2$) are calculated for the source rock of the magma by assuming a mean $^{176}Lu/^{177}Hf$ value of 0.015 for an average continental crust [6]. For sample ZZG11, the 14 magmatic zircon grains yield $^{176}Lu/^{177}Hf$ and $^{176}Hf/^{177}Hf$ ratios of 0.000031–0.000718 and 0.282263–0.282361, respectively. They show negative $\varepsilon_{Hf}(t)$ values (age corrected using U-Pb age for individual grains) in the range of −12.5 to −9.3, falling below the CHUR (chondrite uniform reservoir) line (Figure 10A,B). The corresponding calculated $T_{DM}{}^2$ values range from 1.86 to 2.08 Ga (Figure 10D). The Paleoproterozoic inherited zircons have $^{176}Lu/^{177}Hf$ ratio of 0.000424 and 0.000560, $^{176}Hf/^{177}Hf$ ratios of 0.281747 and 0.281540, and $\varepsilon_{Hf}(t)$ value of 2.4 and −6.1, with corresponding $T_{DM}{}^2$ of 2.28 and 2.78 Ga. For sample ZZG12, the six magmatic zircon grains have $^{176}Lu/^{177}Hf$ ratio of 0.000012 to 0.000542, $^{176}Hf/^{177}Hf$ ratios of 0.282323 to 0.282371, and negative $\varepsilon_{Hf}(t)$ value of −10.9 to −8.9, falling below the CHUR line (Figure 10A,B), with corresponding $T_{DM}{}^2$ of 1.97 to 1.84 Ga (Figure 10D). The 471.8 Ma zircon yielding negative $\varepsilon_{Hf}(t)$ values of −8.8 have corresponding $T_{DM}{}^2$ of 1.43 Ga. The Proterozoic inherited zircons have $\varepsilon_{Hf}(t)$ values of 2.5, −8.9, −5.2, with corresponding $T_{DM}{}^2$ of 2.02, 2.16, and 2.42 Ga.

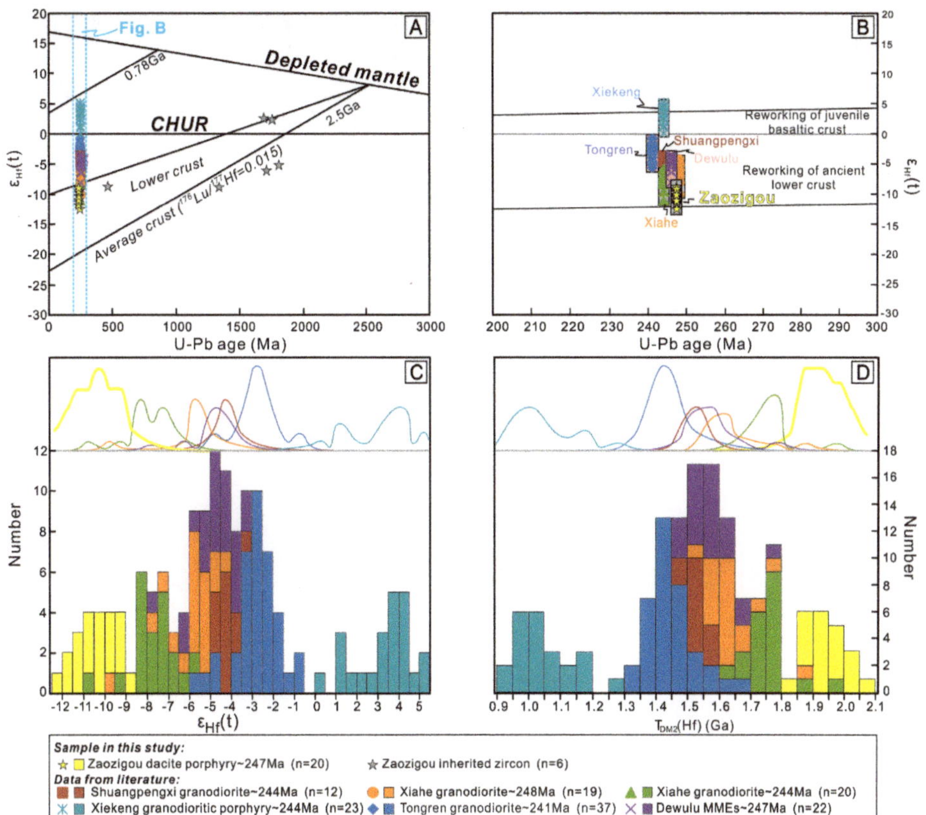

Figure 10. Variations of $\varepsilon_{Hf}(t)$ values versus U–Pb ages (Ma) of zircons from the Triassic granitoids in the Tongren–Xiahe–Hezuo area (**A**,**B**). (**C**) Histograms, and frequency curves of $\varepsilon_{Hf}(t)$ values. (**D**) Histograms, and frequency curves of corresponding two–stage Hf model ages ($T_{DM}{}^2$) (modified after [6]).

Table 2. Trace element concentration, and calculated Ti-in-zircon temperatures (T), logarithmic oxygen fugacity ($logfO_2$), and ΔFMQ of zircons from the porphyritic dacite samples ZZG11 and ZZG12.

Spot No.	Ti	Y	La	Ce	Pr	Nd	Sm	Eu	Gd	Tb	Dy	Ho	Er	Tm	Yb	Lu	Hf	Th	U	δEu	δCe	T(K)	$logfO_2$	ΔFMQ
									ZZG11: Inherited Zircon														
ZZG11.9	1088	464	0.08	3.91	0.23	3.85	6.52	0.07	25.5	6.11	53.7	16.2	59	10.3	84	16.8	10127	158	308	0.02	6.97	1957	−6.80	−2.87
ZZG11.11	916	564	0.00	9.07	0.06	1.18	2.57	0.15	13.1	4.15	50.4	19.4	86	18.0	167	34.2	10067	33	245	0.08	394.68	1899	−4.78	−0.41
									ZZG11: 248.1 ± 3.8 Ma (MSWD = 3.6, n = 14)														
ZZG11.1	867	714	0.24	4.93	0.17	1.64	2.98	0.42	15.8	5.23	61.7	22.7	102	21.1	196	42.2	9432	156	2962	0.19	5.88	1882	−7.78	−3.27
ZZG11.2 *	898	207	0.00	2.68	0.11	1.57	3.33	0.83	16.2	3.69	26.9	6.4	18	2.7	20	3.4	10066	118	1671	0.34	81.61	-	-	-
ZZG11.4 *	899	280	0.03	3.87	0.19	3.29	5.63	1.42	24.6	5.47	40.4	8.4	23	2.9	19	2.7	7966	174	2023	0.37	12.36	-	-	-
ZZG11.6	850	439	0.05	2.17	0.24	2.22	3.01	0.96	13.8	4.62	46.0	13.7	49	8.7	71	13.4	10943	49	3061	0.46	4.68	1875	−9.17	−4.61
ZZG11.7 *	861	220	0.33	2.35	0.28	2.24	2.64	1.12	12.8	3.26	26.7	6.6	22	3.6	29	5.2	10818	47	2023	0.59	1.90	-	-	-
ZZG11.8 *	823	73	0.00	1.85	0.05	1.01	2.40	0.47	8.5	1.79	11.1	2.2	6	0.7	5	1.0	9292	53	1118	0.32	61.75	-	-	-
ZZG11.10	871	282	0.00	2.21	0.04	0.60	1.27	0.12	6.9	2.17	23.2	8.9	40	8.8	86	19.6	11239	76	2031	0.12	83.58	1883	−8.88	−4.38
ZZG11.14	785	805	1.35	8.10	0.60	3.18	3.66	0.74	18.0	6.21	70.2	25.5	116	23.5	217	46.8	8666	183	3410	0.28	2.21	1850	−7.18	−2.42
ZZG11.15	874	514	0.07	3.34	0.14	2.09	4.19	0.69	17.8	5.07	51.9	16.5	66	12.9	113	23.6	9989	115	2672	0.24	8.16	1884	−8.37	−3.88
ZZG11.16	853	830	0.03	5.79	0.15	1.88	4.23	0.69	20.9	6.71	76.0	27.0	118	24.4	220	46.7	9246	264	3854	0.22	20.10	1876	−7.84	−3.28
ZZG11.19	811	797	0.01	4.81	0.08	1.46	3.22	0.46	17.9	5.75	68.1	25.6	114	23.9	227	49.6	8941	192	3348	0.18	34.23	1860	−8.05	−3.37
ZZG11.20	839	1033	0.60	8.81	1.45	13.33	16.15	5.17	46.2	13.15	118.0	33.6	127	24.3	208	41.6	9156	197	3884	0.58	2.32	1871	−7.16	−2.56
ZZG11.21	863	667	0.03	3.83	0.16	2.03	2.88	0.56	14.2	4.83	56.2	21.6	100	21.3	198	42.9	8769	126	2609	0.27	13.70	1880	−8.03	−3.51
ZZG11.25	871	523	0.43	3.83	0.59	5.29	6.50	2.43	22.4	6.79	61.1	16.6	56	9.5	72	12.9	10407	66	3051	0.62	1.85	1883	−8.57	−4.07
									ZZG11: 222.1 ± 4.4 Ma (MSWD = 1.4, n = 5)														
ZZG11.3	910	960	0.23	6.96	0.63	5.53	7.37	2.36	33.5	9.73	91.4	29.8	123	25.1	229	47.3	10302	213	4279	0.46	4.49	1897	−7.49	−3.10
ZZG11.5	851	496	0.16	4.01	0.49	5.10	5.57	1.51	16.6	4.89	49.0	16.0	66	13.4	123	25.9	10402	105	2595	0.48	3.50	1876	−8.08	−3.52
ZZG11.22	859	1139	0.49	8.03	0.80	7.71	11.12	3.25	42.8	12.98	119.9	36.2	140	26.8	230	46.8	9615	222	3925	0.46	3.14	1879	−7.31	−2.77
ZZG11.23	835	1027	0.31	6.93	0.94	10.39	12.67	3.93	38.9	11.59	110.1	33.2	131	25.7	228	46.4	9428	204	4037	0.54	3.15	1870	−7.58	−2.98
ZZG11.24	837	744	0.15	5.28	0.46	4.29	6.27	2.13	24.2	7.24	72.4	23.4	98	20.5	189	39.4	10041	168	3368	0.53	4.90	1870	−7.90	−3.30
									ZZG11: 203.5 ± 2.2 Ma (n = 1)														
ZZG11.18	863	1002	0.81	6.18	1.28	10.39	12.47	5.59	44.5	13.36	115.6	32.2	110	19.3	155	29.3	11251	59	4979	0.73	1.48	1880	−7.99	−3.47
									ZZG12: Inherited Zircon														
ZZG12.1	877	1508	0.27	9.66	0.97	11.88	16.37	2.35	54.4	16.81	166.8	52.5	207	39.3	336	63.3	9976	396	3372	0.24	4.66	1885	−7.06	−2.58
ZZG12.2	933	882	0.04	43.27	0.47	8.38	11.27	2.70	39.1	9.96	94.5	30.6	123	23.4	197	39.2	8268	42	94	0.39	73.75	1905	−1.39	2.93
ZZG12.13	803	2000	0.41	5.76	0.80	9.75	16.18	4.19	55.6	17.63	203.0	70.0	315	64.6	567	117.5	10164	210	2990	0.43	2.48	1857	−7.85	−3.15
ZZG12.14	889	1225	0.01	0.64	0.03	0.79	2.78	0.09	20.4	8.22	105.6	40.9	182	37.5	324	66.0	11673	37	1328	0.04	10.59	1890	−10.58	−6.13

Table 2. *Cont.*

Spot No.	Ti	Y	La	Ce	Pr	Nd	Sm	Eu	Gd	Tb	Dy	Ho	Er	Tm	Yb	Lu	Hf	Th	U	δEu	δCe	T(K)	logfO$_2$	ΔFMQ
ZZG12: 246.1 ± 5.2 Ma (MSWD = 1.8, n = 6)																								
ZZG12.3	834	290	0.10	2.59	0.52	5.12	5.67	1.55	12.9	3.54	33.2	9.4	31	5.6	45	8.4	10779	43	2614	0.55	2.72	1869	−8.89	−4.28
ZZG12.6 *	862	77	0.03	2.40	0.28	3.04	3.83	1.28	9.9	2.02	12.7	2.6	7	0.9	6	0.9	9121	103	1170	0.64	6.12	-	-	-
ZZG12.7 *	872	188	0.01	3.14	0.38	6.38	10.22	2.00	31.0	6.11	35.9	5.9	12	1.4	8	1.2	7997	94	989	0.34	14.85	-	-	-
ZZG12.9	844	248	1.49	5.59	0.73	4.63	2.70	0.64	10.6	3.06	27.7	8.0	27	4.7	38	6.9	11025	38	2051	0.37	1.31	1873	−7.2	−2.62
ZZG12.10 *	861	314	0.02	1.73	0.15	2.38	4.06	1.25	18.6	4.87	39.7	9.6	30	4.8	35	6.1	10190	61	2604	0.44	7.01	-	-	-
ZZG12.20	840	812	0.05	4.73	0.14	2.51	4.22	0.97	21.3	6.95	75.8	26.7	113	23.5	212	45.3	8990	154	2930	0.31	13.05	1871	−7.87	−3.28
ZZG12: 226.0 ± 9.1 Ma (MSWD = 5.6, n = 5)																								
ZZG12.4	771	672	0.48	8.16	2.24	21.62	17.29	5.25	38.0	9.49	82.1	21.8	73	12.9	101	18.6	9658	79	3721	0.63	1.94	1845	−7.6	−2.8
ZZG12.5 *	828	380	0.17	6.15	0.65	9.21	12.12	3.91	37.9	8.77	62.5	12.4	33	4.9	33	5.6	9227	256	3070	0.56	4.52	-	-	-
ZZG12.12 *	876	132	0.01	1.42	0.02	0.71	1.61	0.52	9.0	2.35	18.7	4.1	12	1.9	13	2.4	10659	52	1502	0.42	25.21	-	-	-
ZZG12.15	835	1064	3.41	16.16	3.01	24.46	20.45	6.02	48.7	12.61	117.2	34.4	127	23.6	200	38.8	10722	75	5414	0.58	1.24	1870	−6.09	−1.48
ZZG12.18 *	885	223	0.04	2.30	0.16	2.47	3.85	1.10	16.1	3.83	30.1	7.0	21	3.2	23	4.0	10274	68	1985	0.43	7.18	-	-	-
ZZG12: 203.4 ± 5.2 Ma (MSWD = 0.12, n = 2)																								
ZZG12.8	803	1532	0.58	17.32	1.62	17.87	18.07	4.78	50.4	16.62	161.1	52.6	219	43.9	393	78.8	10678	356	5174	0.48	4.37	1857	−6.45	−1.74
ZZG12.16	853	770	0.55	6.50	0.83	6.13	6.87	1.91	25.3	8.38	78.0	25.2	106	22.0	207	41.6	9977	99	3522	0.44	2.36	1876	−7.35	−2.8

Note the data marked with * are indicative of contamination by inclusions or alteration.

Table 3. Zircon Lu-Hf isotopic composition of the porphyritic dacite samples ZZG11 and ZZG12.

Sample No.	Age (Ma)	$^{176}Yb/^{177}Hf$	2σ	$^{176}Lu/^{177}Hf$	2σ	$^{176}Hf/^{177}Hf$	2σ	$\varepsilon_{Hf}(0)$	$\varepsilon_{Hf}(t)$	T_{DM1} (Ma)	T_{DM2} (Ma)	$f_{Lu/Hf}$
ZZG11: 248.1 $\pm$ 3.8 Ma												
ZZG11.1	249.6	0.024139	0.000326	0.000718	0.000008	0.282352	0.000018	−14.863592	−9.508981	1263.225048	1881.889668	−0.978380
ZZG11.2	258.6	0.001701	0.000022	0.000041	0.000001	0.282349	0.000018	−14.970353	−9.307346	1245.319464	1876.124379	−0.998762
ZZG11.4	238.6	0.001524	0.000016	0.000038	0.000001	0.282361	0.000019	−14.536947	−9.302832	1228.484312	1861.157809	−0.998852
ZZG11.6	247.2	0.001250	0.000048	0.000031	0.000001	0.282295	0.000016	−16.874484	−11.460123	1318.467147	2003.743246	−0.999062
ZZG11.7	241.9	0.013728	0.000178	0.000350	0.000003	0.282298	0.000017	−16.754266	−11.507066	1324.692857	2002.602403	−0.989470
ZZG11.8	259.5	0.001559	0.000108	0.000040	0.000002	0.282322	0.000018	−15.912184	−10.229071	1281.640849	1935.121937	−0.998798
ZZG11.10	252.9	0.009459	0.000060	0.000385	0.000002	0.282299	0.000025	−16.713617	−11.233104	1324.326437	1993.505528	−0.988407
ZZG11.14	255.5	0.021607	0.000541	0.000672	0.000017	0.282339	0.000021	−15.326172	−9.836895	1279.869410	1907.082223	−0.979755
ZZG11.15	251.4	0.004712	0.000219	0.000148	0.000006	0.282323	0.000017	−15.863755	−10.375831	1283.328974	1938.305577	−0.995557
ZZG11.16	246.5	0.018488	0.000432	0.000600	0.000012	0.282351	0.000021	−14.896557	−9.589016	1260.630563	1884.689690	−0.981924
ZZG11.19	253.7	0.019748	0.000071	0.000660	0.000005	0.282263	0.000024	−17.988048	−12.538169	1383.774218	2076.338657	−0.980125
ZZG11.20	236.5	0.019013	0.000292	0.000550	0.000006	0.282327	0.000019	−15.719942	−10.621397	1291.189494	1942.480958	−0.983433
ZZG11.21	245.0	0.019872	0.000146	0.000662	0.000003	0.282307	0.000024	−16.457244	−11.193214	1323.873187	1984.942448	−0.980068
ZZG11.25	242.7	0.008005	0.000139	0.000217	0.000003	0.282292	0.000018	−16.992240	−11.706242	1329.367364	2015.854097	−0.993465
ZZG11: Inherited Zircons												
ZZG11.9	1759.2	0.012876	0.000283	0.000424	0.000009	0.281747	0.000022	−36.261673	2.407467	2081.737912	2282.631032	−0.987219
ZZG11.11	1713.2	0.019195	0.000190	0.000560	0.000003	0.281540	0.000022	−43.572856	−6.121060	2370.087031	2775.472413	−0.983128
ZZG12: 246.1 $\pm$ 5.2 Ma												
ZZG12.3	243.3	0.000531	0.000008	0.000012	0.000000	0.282320	0.000019	−15.993589	−10.661623	1283.860523	1950.351280	−0.999637
ZZG12.6	241.0	0.001371	0.000036	0.000028	0.000001	0.282371	0.000017	−14.185466	−8.905572	1214.577203	1837.515937	−0.999168
ZZG12.7	247.2	0.000853	0.000011	0.000018	0.000000	0.282312	0.000016	−16.273862	−10.857290	1294.862014	1965.639935	−0.999462
ZZG12.9	244.1	0.000518	0.000009	0.000013	0.000000	0.282333	0.000016	−15.533076	−10.183396	1266.122384	1920.708045	−0.999616
ZZG12.10	252.9	0.006756	0.000341	0.000167	0.000008	0.282354	0.000020	−14.788333	−9.270597	1242.315778	1869.501545	−0.994974
ZZG12.20	252.3	0.018541	0.000319	0.000542	0.000007	0.282323	0.000016	−15.877285	−10.442352	1297.065034	1942.779033	−0.983679
ZZG12: Inherited Zircons												
ZZG12.1	1341.0	0.030050	0.001042	0.000808	0.000033	0.281702	0.000031	−37.845812	−8.861764	2163.964101	2663.095171	−0.975669
ZZG12.2	1809.9	0.019174	0.000179	0.000569	0.000003	0.281504	0.000027	−44.830641	−5.235327	2418.849006	2794.388577	−0.982856
ZZG12.13	471.8	0.031474	0.000291	0.001039	0.000012	0.282240	0.000019	−18.804106	−8.761496	1429.898743	2001.515941	−0.968699
ZZG12.14	1692.9	0.015942	0.000239	0.000436	0.000004	0.281792	0.000028	−34.661773	2.513847	2020.856289	2224.977230	−0.986874

5. Discussion

5.1. Triassic Magmatism in Tongren–Xiahe–Hezuo Area

Several detailed field observations and geochronological studies have been carried out on extensive pulse of Triassic magmatism in the Tongren–Xiahe–Hezuo area. As shown in Figure 1B, Jin et al. [44] preliminarily reported ages of Xiahe, Daerzang, and Yeliguan granitoids at 243–250 Ma, 234–242 Ma, and 237–251 Ma. Luo et al. [14] reported ages of Shuangpengxi granodiorite and Xiekeng gabbro diorite at ca. 239–245 Ma and 235–246 Ma. Huang et al. [45] subsequently gave the age of Shehaliji quartz monzonite at ca. 234–235 Ma. Li et al. [15] measured ages of emplacement of the Tongren granodiorite (238–242) Ma, Ayishan granitoid (238–242 Ma), Meiwu granitoid (243–251 Ma) and Dewulu granitoid (238–247 Ma). Qiu and Deng [6] reported that the dioritic MME of Dewulu intrusive complex yields an age of 247.0 ± 2.2 Ma. Sui et al. [8] also reported ages of granodiorite, quartz diorite porphyry, and diorite porphyry dike in Zaozigou deposit at 248.9 ± 1.4 Ma, 244.8 ± 1.4 Ma, and 237.5 ± 1.4 Ma, respectively. Combining the geochronology of the ore-hosting porphyritic dacite emplacement at 246.1 ± 5.2 Ma and 248.1 ± 3.8 Ma in the Zaozigou deposit presented in this study, in conjunction with the data from previously published articles, we propose that Triassic magmatism widespread in the Tongren–Xiahe–Hezuo area, including batholiths, stocks, sills, and dikes, mainly formed at ca. 248–235 Ma.

Several geochronological data focused on the age of metamorphism in the Qinling orogen have been published in the past decade. The earliest precise estimated ages came from $^{39}\mathrm{Ar}/^{40}\mathrm{Ar}$ of phengites and riebeckites, giving well-defined age plateaus of 236 ± 5 Ma and 217 ± 8 Ma, respectively [46]. Li et al. [47] employed the whole rock Sm-Nd and Rb-Sr isochron on schist and reported ages of 242 ± 21 Ma and 221 ± 13 Ma. Another estimate of 214 ± 11 Ma was made from U-Pb zircon analysis on granulites, representing the age of retrograde metamorphism in amphibolite facies rocks [48]. Consequently, the age of peak metamorphism in the studied area is ~220 Ma. Sui et al. [8] and Sui and Li [17] reported $^{40}\mathrm{Ar}/^{39}\mathrm{Ar}$ plateau ages of 245.6 ± 1.0 Ma, 242.1 ± 1.0 Ma, 230 ± 2.3 Ma, and 219.4 ± 1.1 Ma for sericite from porphyritic dacite and quartz diorite porphyry. The first three ages are consistent with zircon U-Pb ages, representing emplacement of Triassic magmatism. The latter age is close to the metamorphic peak and may represent the mineralization age. Therefore, the ages of ~225 Ma and ~203 Ma in the ore-hosting porphyritic dacite are interpreted to be influenced by hydrothermal events related to metamorphism.

5.2. Petrogenesis of Early Triassic Magmatism in Tongren–Xiahe–Hezuo Area

The Lu-Hf isotope system is a very sensitive geochemical tracer to detect the evolutionary history of crustal and mantle material [35–37,49]. Hafnium is partitioned more strongly into melts than Lu during partial melting; therefore the crust generally has lower $^{176}\mathrm{Lu}/^{177}\mathrm{Hf}$ and $^{176}\mathrm{Hf}/^{177}\mathrm{Hf}$ ratios than the mantle. Accordingly, high values of $^{176}\mathrm{Hf}/^{177}\mathrm{Hf}$ (i.e., positive $\varepsilon_{Hf}(t)$ values) are considered to be sourced from the partial melting of juvenile crustal materials, or directly via mantle-derived mafic melts [49–51]. Low values of $^{176}\mathrm{Hf}/^{177}\mathrm{Hf}$ (i.e., negative $\varepsilon_{Hf}(t)$ values) indicate old crust input [35,49–51]. The $\varepsilon_{Hf}(t)$ values of magmatic zircons from Zaozigou ore-hosting porphyritic dacite range from -12.5 to -8.9, with corresponding the zircon two-stage model ages ($T_{DM}{}^2$) of 2.08–1.84 Ga. These analyses plot below the chondrite uniform reservoir (CHUR) line (Figure 10A), and within the area of reworked ancient lower crust (Figure 10B), which indicates that the magma could be derived from partial melting of Paleoproterozoic lower crustal material [6]. This interpretation is further supported by data presented in this article indicating that most of the inherited zircons being crystallized in Paleoproterozoic (Table 1).

The $\varepsilon_{Hf}(t)$ and $T_{DM}{}^2$ values are similar to the Shuangpengxi granodiorites ($\varepsilon_{Hf}(t)$ = -4.7~-3.6; $T_{DM}{}^2$ = 1.49~1.57 Ga) [14], Xiahe granodiorites ($\varepsilon_{Hf}(t)$ = -11.0~-4.0; $T_{DM}{}^2$ = 1.53~1.97 Ga) [52], Tongren granodiorites ($\varepsilon_{Hf}(t)$ = -5.8~-0.6; $T_{DM}{}^2$ = 1.32~1.64 Ga) [15], and Dewulu intrusive complex

($\varepsilon_{Hf}(t) = -8.0 \sim -3.3$; $T_{DM}{}^2 = 1.48 \sim 1.78$ Ga) [6], but different from the Xiekeng high-Mg and high-Al diorites ($\varepsilon_{Hf}(t) = 0.2 \sim 5.3$; $T_{DM}{}^2 = 0.93 \sim 1.26$ Ga) [14]. Qiu and Deng [6] thus suggested that the magmas of Early Triassic magmatism were probably derived from a heterogeneous source that included both crustal and mantle components.

This hypothesis was further supported by the bulk geochemical signature of widespread magmatism coeval with the ore-hosting porphyritic dacite in the Tongren–Xiahe–Hezuo area. Luo et al. [14] reported the geochemical and Sr–Nd–Hf isotopic compositions of the Shuangpengxi granodiorite and proposed that the magma was derived from partial melting of crustal materials. Li et al. [15] suggest that the magma of the Tongren granodiorite was generated by dehydration melting of a mafic lower crustal component with additional input of a mafic component derived from the subcontinental lithospheric mantle. These geochemical traits suggest that the Early Triassic magmatism in Tongren–Xiahe–Hezuo area originated from the reworking of Mesoproterozoic to Paleoproterozoic ancient crust and partial melting of Neoproterozoic juvenile crust [6,35].

5.3. Implications on Links between Magmatism and Mineralization

The Zaozigou deposit has been controversial in its classification and remains one of the more difficult ore systems to fully understand in the West Qinling. The deposit shows a spatial association with Triassic dikes and sills. The broad spatial association between gold and magmatism has been argued as genetically important by some workers. Liu et al. [16] concluded that the ore-forming fluids were of a magmatic-hydrothermal origin based upon hydrogen, oxygen, and sulfur isotopic compositions. They further proposed that the Zaozigou deposit is a porphyry-type gold deposit related to a diorite porphyry, and therefore a ca. 216 Ma date for magmatic zircon records the age of the Au-Sb ore formation. However these 'magmatic' zircons yield CL textures and compositions that could correspond to overprints from a younger hydrothermal event. Dai and Chen [53] argued for a genetic relationship between Triassic magmatism and Au-Sb mineralization based on geological and geochemical characteristics. A recent ^{40}Ar/^{39}Ar plateau ages of 245.6 ± 1.0 Ma and 242.1 ± 1.0 Ma for hydrothermal sericite, which are bracketed by zircon U-Pb ages on pre- and post-ore dikes, were used to argue for a significantly older mineralizing event [8]. Sui et al. [8] further defined Zaozigou as a reduced intrusion-related gold system (RIRGS) on the basis of interpreted overlapping formation ages of igneous host rocks and gold mineralization.

Magmatic oxygen fugacity is a key factor that controls the formation of porphyry deposits [39,54]. Most researchers agree that oxidized magmas can hold high metal contents and are favorable for the generation of porphyry deposits [54]. In general, oxygen fugacities of >FMQ + 2 are necessary for the formation of economic porphyry Cu (Au) deposits and FMQ + 1.5 is a threshold for any porphyry deposit [54]. As shown in this study, the ore-bearing porphyritic dacites have very low oxygen fugacity, with ΔFMQ ranging from -4.61 to -2.56. Under the low oxygen fugacities, it would have been difficult to release metals out of the melt and instead magmas would have become sulfide-saturated during evolution in deep magma chambers [54]. A dozen of porphyry-skarn Cu (Au) deposits are hosted in the granitic intrusions and adjacent rock in northeastern part of Tongren–Xiahe–Hezuo area (Figure 1B). Oxygen fugacity from ore-bearing porphyries shows FMQ $\pm$ 3.3 [6], significantly higher than the porphyritic dacites at Zaozigou. In addition, hydrothermal alteration in porphyry deposits typically shows distinct temporal and spatial evolution and zonation from early, proximal, high-temperature potassic alteration to sericitic alteration to low-temperature, distal, advanced argillic and intermediate argillic alteration. However, the orebodies at Zaozigou are characterized by strong wallrock sulfidation (pyrite, arsenopyrite, and stibnite) and discrete proximal sericitized, silicified, and carbonatized alteration haloes (Figure 3). Such alteration haloes are easily distinguished from typical porphyry alteration profiles by their size and distinct alteration boundaries. Additionally, ^{40}Ar/^{39}Ar geochronology on hydrothermal sericite indicates the gold event may have formed at ~219 Ma [17], ~17–28 Ma later than the emplacement of Triassic porphyritic dacite.

Reduced Intrusion-Related Gold Systems have become a new exploration target deposit model for low-grade, large-tonnage gold deposits [22,55,56]. The best recognized examples of such deposits occur in Fort Knox in Alaska, Dublin Gulch, and Scheelite Dome in Yukon [22]. Comparisons of Zaozigou with these deposits reveals some similarities but critical differences. Important similarities include that the plutons are typically small and are dominated by felsic magmatic phases, the metal assemblages are Au-Sb-As-Hg, and the oxidation state of igneous rock are mostly at or below the FMQ oxide buffer [55,56]. Despite these similarities, significant characteristics that contrast with those of RIRGS include: (1) Typical RIRGS form in igneous rocks emplaced into a deformed continental margin backstop but are not products of arc magmatism and volcanic rocks are rare. The Zaozigou intrusions are products of arc magmatism and volcanic rocks are well developed. (2) The sulfide content of RIRGS is extremely low, commonly <1 vol. %, but Zaozigou is characterized by massive and economic stibnite (Figure 3G,H). (3) The RIRGS are characterized by reduced mineral assemblage containing pyrrhotite and minor loellingite [22]. In contrast, these reduced minerals are rare in Zaozigou. (4) The distinctive mineralization style of RIRGS is sheeted veins, which are composed of parallel, low grades (<1 g/t Au), single stage quartz veins. Nevertheless, the ore-bearing zone at Zaoaigou comprises massive auriferous calcite vein (Figure 3E,F) and stibnite-quartz-native gold veins (Figure 3G,H). (5) Exsolution of hydrothermal fluids directly from the emplacement of the porphyritic dacites at Zaozigou is unlikely given the ~219 Ma gold metallogenic events occurred 20 Ma or so later than the emplacement of the intrusions. As a result, the Zaozigou deposit is not likely to be genetically related to the reduced intrusions, and therefore should not be considered a Reduced Intrusion-Related Gold System.

6. Conclusions

(1) Zircon U-Pb dating indicates that the ore-hosting porphyritic dacite was formed at 246–248 Ma. The magma of Early to Middle Triassic porphyritic dacite could be derived from partial melting of Paleoproterozoic crustal materials.

(2) The emplacement of the Triassic porphyritic dacite was approximately 20 Ma earlier than the economic mineralization at Zaozigou. Oxygen fugacity of porphyritic dacite lower than FMQ and undeveloped typical porphyry alteration zones indicate that Zaozigou is not a porphyry-type deposit. In addition, massive sulfide minerals preclude the possibility that the Zaozigou is a reduced intrusion-related gold system. The new finding that the Zaozigou deposit is not likely to be genetically related to the magmatism will provide us with new ideas for prospecting. It allows explorers to concentrate prospecting on local and regional structures to vector prospective targets rather than focusing on magmatic rocks.

Supplementary Materials: The following are available online at http://www.mdpi.com/2075-163X/9/1/37/s1, Computer Code S1: OFnum1.0. Algorithms based on Matlab software for calculating fO_2.

Author Contributions: Conceptualization, K.-F.Q.; Data curation, H.-C.Y.; Funding acquisition, N.-B.L.; Investigation, H.-C.Y., K.-F.Q., Z.-Y.G., D.M., G.-P.J., and R.Z.; Methodology, H.-C.Y., K.-F.Q., Z.-Y.G., and J.-Z.G.; Project administration, K.-F.Q. and N.-B.L.; Resources, G.-P.J. and R.Z.; Software, C.-A.G. and Y.P.; Writing—original draft, H.-C.Y. and K.-F.Q.; Writing—review & editing, H.-C.Y., K.-F.Q., and D.M.

Funding: This research was funded by the National Natural Science Foundation of China (41702069), China Postdoctoral Science Foundation (2016M591221, 20170108), CAS Key Laboratory of Mineralogy and Metallogeny (KLMM20170201), and State Key Laboratory of Geological Processes and Mineral Resources (MSFGPMR201804).

Acknowledgments: The authors are very grateful for the logistic assistance from the management of Zaozigou Gold Company. The authors also would like to thank Richard Goldfarb, Xue Gao, and Yun-Chuan Zeng at the China University of Geoscience in Beijing for constructive discussion on the formatting and preparation of this manuscript.

References

1. Deng, J.; Wang, Q.F. Gold mineralization in China: Metallogenic provinces, deposit types and tectonic framework. *Gondwana Res.* **2016**, *36*, 219–274. [CrossRef]
2. Yang, L.Q.; Deng, J.; Dilek, Y.; Qiu, K.F.; Ji, X.Z.; Li, N.; Taylor, R.D.; Yu, J.Y. Structure, geochronology, and petrogenesis of the Late Triassic Puziba granitoid dikes in the Mianlue suture zone, Qinling Orogen, China. *Geol. Soc. Am. Bull.* **2015**, *11/12*, 1831–1854. [CrossRef]
3. Yang, L.Q.; Deng, J.; Qiu, K.F.; Ji, X.Z.; Santosh, M.; Song, K.R.; Song, Y.H.; Geng, J.Z.; Zhang, C.; Hua, B. Magma mixing and crust–mantle interaction in the Triassic monzogranites of Bikou Terrane, central China: Constraints from petrology, geochemistry, and zircon U-Pb-Hf isotopic systematics. *J. Asian Earth Sci.* **2015**, *98*, 320–341. [CrossRef]
4. Dong, Y.P.; Santosh, M. Tectonic architecture and multiple orogeny of the Qinling Orogenic Belt, Central China. *Gondwana Res.* **2016**, *29*, 1–40. [CrossRef]
5. Duan, M.; Niu, Y.L.; Kong, J.J.; Sun, P.; Hu, Y.; Zhang, Y.; Chen, S.; Li, J.Y. Zircon U-Pb geochronology, Sr-Nd-Hf isotopic composition and geological significance of the Late-Triassic Baijiazhuang and Lvjing granitic plutons in West Qinling Orogen. *Lithos* **2016**, *260*, 443–456. [CrossRef]
6. Qiu, K.F.; Deng, J. Petrogenesis of granitoids in the Dewulu skarn copper deposit: Implications for the evolution of the Paleotethys ocean and mineralization in Western Qinling, China. *Ore Geol. Rev.* **2017**, *90*, 1078–1098. [CrossRef]
7. Qiu, K.F.; Yu, H.C.; Gou, Z.Y.; Liang, Z.L.; Zhang, J.L.; Zhu, R. Nature and origin of Triassic igneous activity in the Western Qinling Orogen: The Wenquan composite pluton example. *Int. Geol. Rev.* **2018**, *60*, 242–266. [CrossRef]
8. Sui, J.X.; Li, J.W.; Jin, X.Y.; Vasconcelos, P.; Zhu, R. ^{40}Ar/^{39}Ar and U-Pb constraints on the age of the Zaozigou disseminated gold deposit, Xiahe-Hezuo district, West Qinling orogen, China: Implications for the early Triassic reduced intrusion-related gold metallogeny. *Ore Geol. Rev.* **2018**, in press. [CrossRef]
9. Qiu, K.F.; Deng, J.; Taylor, R.D.; Song, K.R.; Song, Y.H.; Li, Q.Z.; Goldfarb, R.J. Paleozoic magmatism and porphyry Cu-mineralization in an evolving tectonic setting in the North Qilian Orogenic Belt, NW China. *J. Asian Earth Sci.* **2016**, *122*, 20–40. [CrossRef]
10. Qiu, K.F.; Marsh, E.; Yu, H.C.; Pfaff, K.; Gulbransen, C.; Gou, Z.Y.; Li, N. Fluid and metal sources for the Wenquan porphyry molybdenite deposit, Western Qinling, NW China. *Ore Geol. Rev.* **2017**, *86*, 459–473. [CrossRef]
11. Qiu, K.F.; Taylor, R.D.; Song, Y.H.; Yu, H.C.; Song, K.R.; Li, N. Geologic and geochemical insights into the formation of the Taiyangshan porphyry copper-molybdenum deposit, Western Qinling Orogenic Belt, China. *Gondwana Res.* **2016**, *35*, 40–58. [CrossRef]
12. Li, N.; Chen, Y.J.; Santosh, M.; Pirajno, F. Compositional polarity of Triassic granitoids in the Qinling orogen, China: Implication for termination of the northernmost paleo-Tethys. *Gondwana Res.* **2016**, *27*, 244–257. [CrossRef]
13. Zhang, H.F.; Jin, L.L.; Zhang, L.; Harris, N. Geochemical and Pb-Sr-Nd isotopic compositions of granitoids from western Qinling belt: Constraints on basement nature and tectonic affinity. *Sci. China (Ser. D): Earth Sci.* **2007**, *50*, 184–196. [CrossRef]
14. Luo, B.J.; Zhang, H.F.; Lü, X.B. U–Pb zircon dating, geochemical and Sr-Nd-Hf isotopic compositions of Early Indosinian intrusive rocks in West Qinling, central China: Petrogenesis and tectonic implications. *Contrib. Mineral. Petrol.* **2012**, *164*, 551–569. [CrossRef]
15. Li, X.W.; Mo, X.X.; Huang, X.F.; Dong, G.C.; Yu, X.H.; Luo, M.F.; Liu, Y.B. U-Pb zircon geochronology, geochemical and Sr-Nd-Hf isotopic compositions of the Early Indosinian Tongren Pluton in West Qinling: Petrogenesis and geodynamic implications. *J. Asian Earth Sci.* **2015**, *97*, 38–50. [CrossRef]
16. Liu, Y.; Liu, Y.H.; Dong, F.C.; Li, Z.H.; Yu, J.K.; Ma, X.P. Accurate dating of mineralogenetic epoch and its geological significance in Zaozigou gold deposit, Gansu Province. *Gold* **2012**, *33*, 10–17. (In Chinese with English Abstract)
17. Sui, J.Z.; Li, J.W. Geochronology and genesis of the Zaozigou gold deposit, Xiahe-Hezuo district, West Qinling. *Acta Mineral. Sin.* **2013**, 346–347. (In Chinese)
18. Liang, Z.L.; Chen, G.Z.; Ma, H.S.; Zhang, Y.N. Evolution of Ore-controlling Faults in the Zaozigou Gold Deposit, Western Qinling. *Geotectonica et Metallogenia* **2016**, *40*, 354–366.

19. Wei, L.X.; Chen, Z.L.; Pang, Z.S.; Han, F.B.; Xiao, C.H. An Analysis of the Tectonic Stress Field in the Zaozigou Gold Deposit, Hezuo Area, Gansu Province. *Aeta Geosci. Sin.* **2018**, *39*, 79–93. (In Chinese with English Abstract)

20. Goldfarb, R.J.; Qiu, K.F.; Deng, J.; Chen, Y.J.; Yang, L.Q. Orogenic Gold Deposits of China. *Econ. Geol.* **2019**, unpublished.

21. Seedorff, E.; Dilles, J.H.; Proffett, J.M., Jr.; Einaudi, M.T.; Zurcher, L.; Stavast, W.J.A.; Johnson, D.A.; Barton, M.D. Porphyry Deposits: Characteristics and Origin of Hypogene Features. *Econ. Geol.* **2005**, *100*, 251–298.

22. Hart, C.J.R.; Goldfarb, R.J. Distinguishing intrusion-related from orogenic gold systems. In Proceedings of the New Zealand Minerals Conference, Auckland, New Zealand, 13–16 November 2005; pp. 125–133.

23. Goldfarb, R.J.; Baker, T.; Dube, B.; Groves, D.I.; Hart, C.J.R.; Gosselin, P. Distribution, character and genesis of gold deposits in metamorphic terranes. *Econ. Geol.* **2005**, *100*, 407–450.

24. Dong, Y.P.; Yang, Z.; Liu, X.M.; Sun, S.S.; Li, W.; Cheng, B.; Zhang, F.F.; Zhang, X.N.; He, D.F.; Zhang, G.W. Mesozoic intracontinental orogeny in the Qinling Mountains, central China. *Gondwana Res.* **2016**, *30*, 144–158. [CrossRef]

25. Deng, J.; Wang, Q.F.; Li, G.J.; Santosh, M. Cenozoic tectono-magmatic and metallogenic processes in the Sanjiang region, southwestern China. *Earth–Sci. Rev.* **2014**, *138*, 268–299. [CrossRef]

26. Deng, J.; Wang, C.M.; Bagas, L.; Carranza, E.J.M.; Lu, Y.J. Cretaceous-Cenozoic tectonic history of the Jiaojia Fault and gold mineralization in the Jiaodong Peninsula, China: Constraints from zircon U-Pb, illite K-Ar, and apatite fission track thermochronometry. *Miner. Depos.* **2015**, *50*, 987–1006. [CrossRef]

27. Yang, L.Q.; Deng, J.; Wang, Z.L.; Guo, L.N.; Li, R.H.; Groves, D.I.; Danyushevsky, L.V.; Zhang, C.; Zheng, X.L.; Zhao, H. Relationships between gold and pyrite at the Xincheng gold deposit, Jiaodong Peninsula, China: Implications for gold source and deposition in a brittle epizonal environment. *Econ. Geol.* **2016**, *111*, 105–126. [CrossRef]

28. Yang, L.Q.; Deng, J.; Goldfarb, R.J.; Zhang, J.; Gao, B.F.; Wang, Z.L. ^{40}Ar/^{39}Ar geochronological constraints on the formation of the Dayingezhuang gold deposit: New implications for timing and duration of hydrothermal activity in the Jiaodong gold province, China. *Gondwana Res.* **2014**, *25*, 1469–1483. [CrossRef]

29. Cao, X.F.; Mohaed, L.S.S.; Lü, X.B.; He, M.C.; Chen, C.; Zhu, J.; Tang, R.K.; Liu, Z.; Zhang, B. Ore-Forming Process of the Zaozigou Gold Deposit: Constraints from Geological characteristics, gold occurrence and stable isotope compositions. *J. Jilin Univ. (Earth Sci. Ed.)* **2012**, *42*, 1040–1054. (In Chinese with English Abstract)

30. Dai, W.J.; Chen, Y.Y.; Liu, D.X.; Ma, X.Y. Wallrock alteration and gold mineralization in zaozigou gold mine of gansu province. *Gansu Geol.* **2011**, *20*, 31–36. (In Chinese with English Abstract)

31. Liang, Z.L.; Zhang, Y.N.; Wang, J.L.; Wang, H.T. Geology feature and deep prospection analysis of Zaozigou gold deposit. *Gansu Metall.* **2012**, *34*, 64–69.

32. Zhang, T.; Chen, Z.L.; Xiao, C.H.; Liang, Z.L.; Zhou, Z.J.; Han, F.B.; Wei, L.X.; Hou, C.T. The characteristics of mesozonic magmatic rocks and its relation with gold mineralization in Xiahe-Hezuo area, Gansu province. *J. Mineral. Petrol.* **2017**, *37*, 40–51. (In Chinese with English Abstract)

33. Yin, Y. Relation of dike rock and gold mineralization in West Qinling region. *Gansu Geol.* **2011**, *20*, 28–37. (In Chinese with English Abstract)

34. Anderson, T. Correction of common lead in U-Pb analyses that do not report ^{204}Pb. *Chem. Geol.* **2002**, *192*, 59–79. [CrossRef]

35. Geng, J.Z.; Qiu, K.F.; Gou, Z.Y.; Yu, H.C. Tectonic regime switchover of Triassic Western Qinling Orogen: Constraints from LA-ICP-MS zircon U-Pb geochronology and Lu-Hf isotope of Dangchuan intrusive complex in Gansu, China. *Chemie der Erde-Geochem.* **2017**, *77*, 637–651. [CrossRef]

36. Li, C.Y.; Zhang, R.Q.; Ding, X.; Ling, M.X.; Fan, W.M.; Sun, W.D. Dating cassiterite using laser ablation Icp-Ms. *Ore Geol. Rev.* **2016**, *72*, 313–322. [CrossRef]

37. Zhang, R.Q.; Lu, J.J.; Lehmann, B.; Li, C.Y.; Li, G.L.; Zhang, L.P.; Guo, J.; Sun, W.D. Combined zircon and cassiterite U–Pb dating of the Piaotang granite-related tungsten–tin deposit, southern Jiangxi tungsten district, China. *Ore Geol. Rev.* **2017**, *82*, 268–284. [CrossRef]

38. Zou, X.Y.; Qin, K.Z.; Han, X.L.; Li, G.M.; Evans, N.J.; Li, Z.Z.; Yang, W. Insight into zircon REE oxy-barometers: A lattice strain model perspective. *Earth Planet. Sci. Lett.* **2019**, *56*, 87–96. [CrossRef]

39. Smythe, D.J.; Brenan, J.M. Magmatic oxygen fugacity estimated using zircon-melt partitioning of cerium. *Earth Planet. Sci. Lett.* **2016**, *453*, 260–266. [CrossRef]

40. Ballard, J.R.; Palin, M.J.; Campbell, I.H. Relative oxidation states of magmas inferred from Ce(IV)/Ce(III) in zircon: Application to porphyry copper deposits of northern Chile. *Contrib. Mineral. Petrol.* **2002**, *144*, 347–364. [CrossRef]

41. Ferry, J.M.; Watson, E.B. New thermodynamic models and revised calibrations for the Ti-in-zircon and Zr-in-rutile thermometers. *Contrib. Mineral. Petrol.* **2007**, *154*, 429–437. [CrossRef]

42. Virgo, D.; Mysen, B.O.; Kushiro, I. Anionic constitution of 1-atmosphere silicate melts: Implications for the structure of igneous melts. *Science* **1980**, *208*, 1371–1373. [CrossRef]

43. Hou, Z.Q.; Yang, Z.M.; Lu, Y.J.; Kemp, A.; Li, Q.Y.; Tang, J.X.; Yang, Z.S.; Duan, L.F. A genetic linkage between subduction- and collision-related porphyry Cu deposits in continental collision zones. *Geology* **2015**, *43*, 643–650. [CrossRef]

44. Jin, W.J.; Zhang, Q.; He, D.F.; Jia, X.Q. SHRIMP dating of adakites in western Qinling and their implications. *Acta Petrol. Sin.* **2005**, *21*, 959–966. (In Chinese with English Abstract)

45. Huang, X.F.; Mo, X.X.; Yu, X.H.; Li, X.W.; Ding, Y.; Wei, P.; He, W.Y. Zircon U-Pb chronology, geochemistry of the Late Triassic acid volcanic rocks in Tanchang area, West Qinling and their geological significance. *Acta Petrol. Sin.* **2013**, *29*, 3968–3980. (In Chinese with English Abstract)

46. Mattauer, M.; Matte, P.; Malavieille, J.; Tapponnier, P.; Maluski, H.; Qin, X.Z.; Lun, L.Y.; Qin, T.Y. Tectonics of Qinling Belt: Build-up and evolution of eastern Asia. *Nature* **1985**, *317*, 496–500. [CrossRef]

47. Li, S.G.; Sun, W.D.; Zhang, G.W.; Chen, J.Y.; Yang, S.C. Chronology and geochemistry of metamorphic volcanic rocks in Heigouxia, South Qinling. *Sci. Chine Earth Sci.* **1996**, *26*, 223–230. (In Chinese)

48. Liang, S.; Liu, L.; Zhang, C.L.; Yang, Y.C.; Yang, W.Q.; Kang, L.; Cao, Y.T. Metamorphism and zircon U-Pb age of high-pressure mafic granulites in in Mian-Lüe suture zone, South Qinling orogen. *Acta Petrol. Sin.* **2013**, *29*, 1657–1674.

49. Blichert-Toft, J.; Albarède, F. The Lu-Hf isotope geochemistry of chondrites and the evolution of the mantle-crust system. *Earth Planet. Sci. Lett.* **1997**, *148*, 243–258. [CrossRef]

50. Belousova, E.A.; Griffin, W.L.; O'Reilly, S.Y. Zircon crystal morphology, trace element signatures and hf isotope composition as a tool for petrogenetic modelling: Examples from eastern australian granitoids. *J. Petrol.* **2005**, *47*, 329–353. [CrossRef]

51. Mahdavi, A.; Karimpour, M.H.; Mao, J.W.; Shahri, M.R.H.; Shafaroudi, A.M.; Li, H.Y. Zircon U-Pb geochronology, Hf isotopes and geochemistry of intrusive rocks in the gazu copper deposit, Iran: Petrogenesis and geological implications. *Ore Geol. Rev.* **2016**, *72*, 818–837. [CrossRef]

52. Wei, P.; Mo, X.X.; Yu, X.H.; Huang, X.F.; Ding, Y.; Li, X.W. Geochemistry, chronology and geological significance of the granitoids in Xiahe, West Qinling. *Acta Geol. Sin.* **2013**, *29*, 3981–3992. (In Chinese with English Abstract)

53. Dai, W.J.; Chen, Y.Y. Relationship between neutral rock vine and mineralization in Zaozigou gold deposit, Gansu. *Gold* **2012**, *33*, 19–23. (In Chinese with English Abstract)

54. Zhang, C.C.; Sun, W.D.; Wang, J.T.; Zhang, L.P.; Sun, S.J.; Wu, K. Oxygen fugacity and porphyry mineralization: A zircon perspective of Dexing porphyry Cu deposit, China. *Geochim. Cosmochim. Acta* **2017**, *206*, 342–363. [CrossRef]

55. Qiu, K.F. Geological and geochemical characteristics of reduced intrusion-related gold system. *Geol. Rev.* **2013**, *59*, 590–591. (In Chinese)

56. Mériaud, N.; Jébrak, M. From intrusion-related to orogenic mineralization: The Wasamac deposit, Abitibi Greenstone Belt, Canada. *Ore Geol. Rev.* **2017**, *84*, 289–308. [CrossRef]

Article

Geochronology of Magmatism and Mineralization in the Dongbulage Mo-Polymetallic Deposit, Northeast China: Implications for the Timing of Mineralization and Ore Genesis

Fengxiang Wang [1,2], Qiangfeng Li [2], Yifei Liu [2], Sihong Jiang [2] and Chao Chen [1,*]

[1] Key Laboratory of Regional Geology and Mineralization, College of Resources, Hebei GEO University, Shijiazhuang 050031, China; fengxiang@hgu.edu.cn
[2] MNR Key Laboratory of Metallogeny and Mineral Assessment, Institute of Mineral Resources, Chinese Academy of Geological Sciences, Beijing 100037, China; liqiangfeng1989@163.com (Q.L.); lyfsky@126.com (Y.L.); jiangsihong@163.com (S.J.)
* Correspondence: chenchao@hgu.edu.cn; Tel.: +86-0311-8720-8285

Received: 5 March 2019; Accepted: 21 April 2019; Published: 26 April 2019

Abstract: The recently discovered Dongbulage Mo-polymetallic deposit is located in the southern part of the Great Xing'an Range, northeast China. Mineralization is closely related to the emplacement of Middle–Late Jurassic granitoids. In order to understand the petrogenetic link between mineralization and host granitoids, this study presents new zircon U–Pb ages, bulk-rock geochemistry, and molybdenite Re–Os ages for the Dongbulage deposits. LA-ICP-MS zircon U–Pb dating of the monzogranite and syenogranite intrusions yielded two weighted mean $^{206}Pb/^{238}U$ ages: of 164 ± 2 Ma and 165 ± 3 Ma, respectively. The subvolcanic rocks (red porphyritic granite and rhyolite) yielded a time interval between 161 ± 2 and 162 ± 3 Ma. In addition, molybdenite from the Dongbulage deposit gave a Re–Os isochron age of 162.6 ± 1.5 Ma, which was interpreted as the age of the mineralization. The new geochronology has established the close temporal and genetic relationships between the mineralization event and the emplacement of the Middle–Late Jurassic granitoids. Bulk-rock geochemistry shows that the Dongbulage granitoids are characterized by high SiO_2, K_2O, and A/CNK [Al_2O_3/(CaO + Na_2O + K_2O)(molar ratio)] values, and low TiO_2, CaO, and MgO values, indicating a metaluminous to peraluminous, high-K calc-alkaline affinity. The granitoids also featured enrichments of large ion lithophile elements and light rare earth elements (LREE), and a relative depletion of high field strength elements (HFSE), along with an increasing negative δEu anomaly. The high differentiation index (DI), ranging from 81.75 to 94.76, and obvious fractionation between LREE and HREE, indicate that the Dongbulage granitoids are highly fractionated, metaluminous–peraluminous, and high-K calc-alkaline I-type granites. Combined with the regional geology, the Dongbulage granitoids may have formed during post-orogenic extension that followed the Mongol–Okhotsk Ocean closure coeval with subduction of the paleo-Pacific plate.

Keywords: zircon U–Pb dating; Re–Os dating; middle–late Jurassic; Mo-polymetallic deposit; Great Xing'an Range; Dongbulage

1. Introduction

Porphyry Mo-polymetallic deposits are the world's most important source of Mo on account of their large tonnages, and currently account for <95% of the global Mo production [1]. The Great Xing'an Range mineral province with a Mo reservoir of >4 million tonnes (Mt), which is located at the eastern segment of the Central Asia Orogenic Belt (CAOB), and ranks as the second greatest Mo source in China. This region is characterized by abundant highly evolved Mesozoic igneous rocks [2], and contains large-scale Ag-, Sn-, Mo-, and Cu-polymetallic vein deposits (Figure 1a) [3,4]. In this region, seven large base metal (≥0.5 Mt) and rare metal (≥0.2 Mt) deposits, and 25 smaller (0.1–0.5 Mt) base metal deposits have been discovered [5]. Three newly recognized metallogenic belts are located within the Great Xing'an Range and its adjacent areas. These are: the northern Mo ± Cu ore-forming belt; the middle Sn ± W ± Ag metallogenic belt; and the southern Mo ore-forming belt (Figure 1b) [6]. The recently discovered Dongbulage deposit is located where the northern Mo ± Cu and middle Sn ± W ± Ag mineralization belts come together (Figure 1b). The deposit represents an unusual example of Mo-polymetallic mineralization dominated by abundant Mo–Pb–Zn with minor Ag–Cu mineralization. The polymetallic features of the Dongbulage deposit represent a change in the style of mineralization characteristics between the Sn ± W ± Ag and northern Mo ± Cu mineralization systems in this area.

The Dongbulage deposit shows a close spatial and genetic relationship with the enclosing granitic rocks, and displays a conspicuous feature of porphyry–epithermal ore-forming systems in which Mo orebodies are hosted within the subvolcanic rocks and adjacent breccia pipes, whereas the Ag-base metal orebodies are hosted in the contact zone between the subvolcanic rocks and host rocks. Based on fluid inclusion studies, Li et al. (2017) suggested that the ore-forming fluid is mainly derived from magmatic water [7]. However, the ages and petrogenesis of the ore-related magmatic rocks are still unclear, thus limits our understanding of the relationships between mineralization and magmatic activity.

Determining the age of porphyry magmatic–hydrothermal events is critical for understanding ore genesis and related geological processes. The most commonly used tools are U–Pb zircon geochronology to date intrusive and high-temperature hydrothermal events, and Re–Os molybdenite geochronology to directly date sulfide mineralization stages. Combining the two techniques allows for clarification of the duration of magmatic–hydrothermal events associated with porphyry systems. In this manuscript, the genesis of the Dongbulage deposit is discussed based on the geology, petrology, and geochemistry of ore-related magmatic rocks, in conjunction with the molybdenite Re–Os ages of the ores and the zircon LA-ICP-MS U–Pb ages for the magmatic rocks. The aim of this study is to accurately constrain the ore-forming age, discuss the petrogenesis of the ore-related granites, and decipher ore genesis in the context of reconstructing the genetic evolution of porphyry systems and evaluating their duration.

2. Geological Setting

The Dongbulage Mo-polymetallic deposit is located in the southern part of the Great Xing'an Range, in a tectonic unit known as the eastern section of the CAOB, which is often referred to as the Xing'an–Mongolian Orogenic Belt (XMOB) (Figure 1a) [8–11]. This region is considered to have experienced a complex geodynamic evolution under three distinct tectonic regimes: closure of the paleo-Asian Ocean during the Paleozoic, subduction and closure of the Mongol–Okhotsk Ocean during the Mesozoic, and subduction of the Pacific Plate during the Mesozoic–Cenozoic, which is associated with widespread magmatic and metamorphic events [9,10].

During the Paleozoic, this area recorded the migration and final closure of the paleo-Asian Ocean between the China Craton (NCC) and the Siberian Craton (SC). The XMOB is interpreted to have consisted of Phanerozoic juvenile crust formed by the amalgamation of several blocks, including, from northwest to southeast, the Erguna (EB), Xing'an–Airgin Sum (XAB), Songliao–Hunshandake (SHB), and Jiamusi blocks (JB), separated by the Xinlin–Xiguitu (XXS), Xilinhot–Heihe (XHS), Mudanjiang (MS), and Ondor Sum–Yongji sutures (OYS) (Figure 1b) [10]. Although final closure time and tectonic evolution are still debated, the XMOB was a united continent by the Mesozoic [12–16]. Hence, there was a significant transition since the Mesozoic, with the tectonic development being controlled by the northwest (NW)-dipping subduction of the paleo-Pacific plate and the south (S) to southeast (SE)-dipping subduction of the Mongol–Okhotsk Ocean plate [9,17–22]. During this tectonic regime, a NE-trending stratigraphy and magmatism developed that controlled the location of variously aged strata and magmatic intrusions (Figure 1b,c).

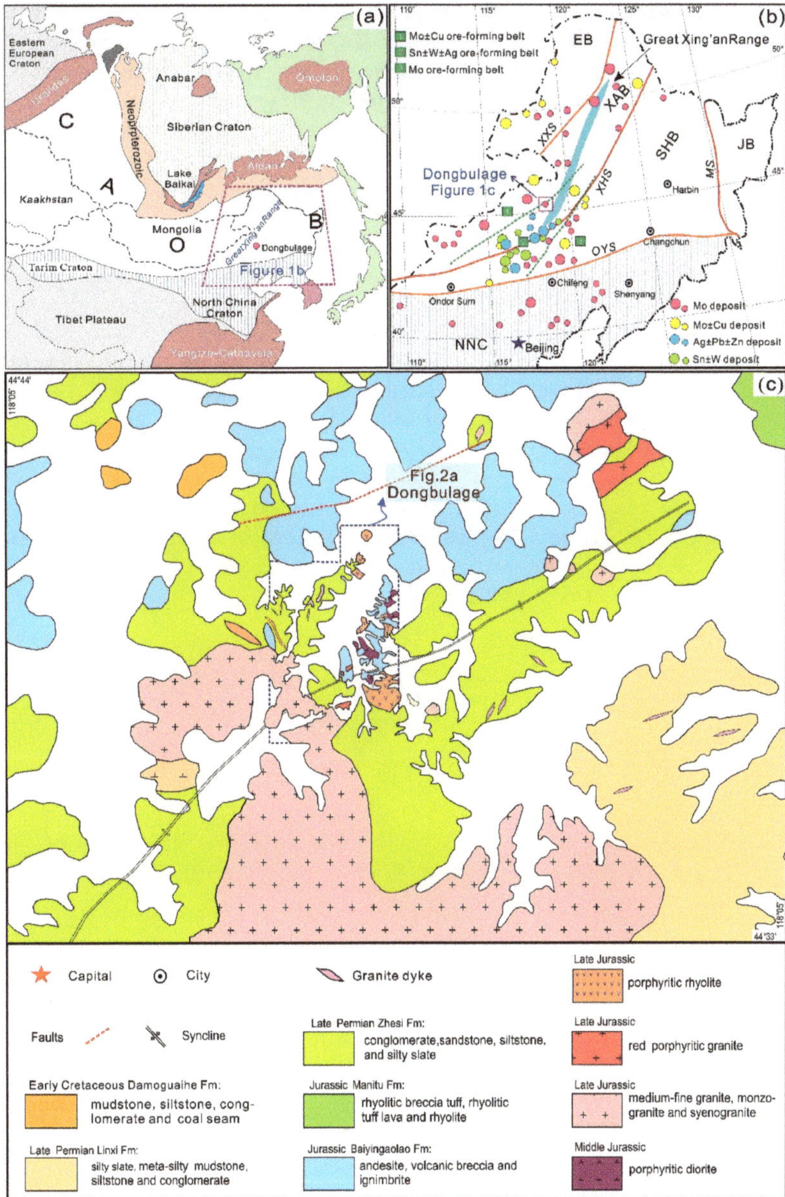

Figure 1. Geological maps: (**a**) location of the Central Asian Orogenic Belt (CAOB, modified after Jahn et al., 2000 [9]). (**b**) Map showing the regional tectonic framework of the Xing'an–Mongolian Orogenic Belt (XMOB, modified after Xu et al., 2015 [10]), NCC = North China Craton, EB = Erguna block, XAB = Xing'an–Airgin Sum block, SHB = Songliao–Hunshandake block, JB = Jiamusi block, XXS = Xinlin–Xiguitu Suture, XHS = Xilinhot–Heihe Suture, MS = Mudanjiang Suture, OYS = Ondor Sum–Yongji Suture. The large and small circles in deposit type represent large and smaller base metal deposit, respectively. White areas show Quaternary rock. (**c**) Simplified regional geological map of the study area (modified after BGMRNMAR, 1991 [11]). White areas show Quaternary rock.

The Dongbulage Mo-polymetallic deposit is located in the southeastern part of the XAB and the northern section of the deep-seated Erlian–Hougenshan Fault (Figure 1b). Two major tectonostratigraphic units have been recognized in the region: Permian sedimentary rocks and Jurassic–Cretaceous felsic volcano-sedimentary sequences (Figure 1c) [11,23]. The Permian sedimentary rocks show discontinuous upward facies change from shallow sea to terrestrial, and can be subdivided into the Zhesi and Linxi formations. They are unconformably overlain by Cretaceous lacustrine sedimentary rocks in the Damoguaihe Formation and Jurassic continental felsic volcanic rocks in the Manitu and Baiyingaolao formations (Figure 1c) [11].

Mesozoic granites are common in the region, and have intruded Permian sedimentary rocks and mid-Jurassic terrestrial volcano-sedimentary sequences (Figure 1c). According to the intrusive relationships (and our new ages), the Mesozoic granitoids in the region are mainly composed of Middle–Late Jurassic plutonic rocks and relatively young subvolcanic intrusive complexes. The plutonic rocks are mostly located in the southern part of region, and occur as batholith and stocks (Figure 1c). The plutonic rocks were considered to be a part of the Buerhaotu granitic batholith and were classified into fine-grained to medium-grained monzogranite and syenogranite (Figure 1c). At the northern marginal facies of the Buerhaotu granitic batholith, some rocks of the subvolcanic complex outcrop as dykes and small stocks, consisting mainly of porphyritic diorite, red porphyritic granite, and porphyritic rhyolite (Figures 1c and 2a). It should be noted that many polymetallic hydrothermal deposits (e.g., large Baiyinnuoer Pb–Zn–Ag, giant Shuangjianzishan Ag–Pb–Zn, and large Haobugao Fe–Zn–Cu, and Aonaodaba Cu–Sn–Ag deposits) are present in this region; these displays have a temporal and spatial relationship with the Mesozoic granites.

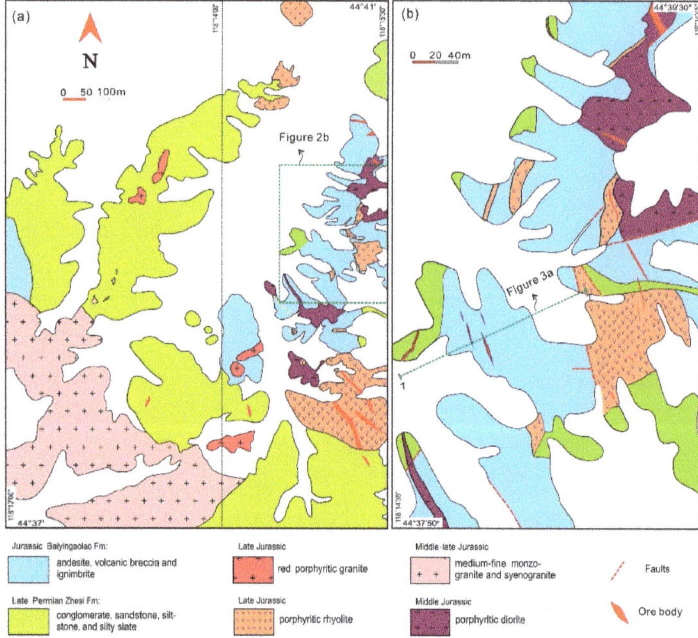

Figure 2. Geological maps of: (**a**) the Dongbulage ore district and (**b**) proximal Mo-bearing ore blocks in the mining exploration project area. White areas show quaternary (modified after ZXGE, 2010 [24]).

3. Ore Deposit Geology

The Dongbulage Mo-polymetallic deposit is located in the northern margin of the Buerhaotu granitic batholith and the axis of the NE-trending East Ujimqin Banner anticlinorium (Figures 1c and 2).

The Mo-polymetallic mineralization occurs in a small area of approximately 1–2 km^2, and is mainly hosted by the NNW, and roughly EW-trending faults and joints (Figures 2 and 3).

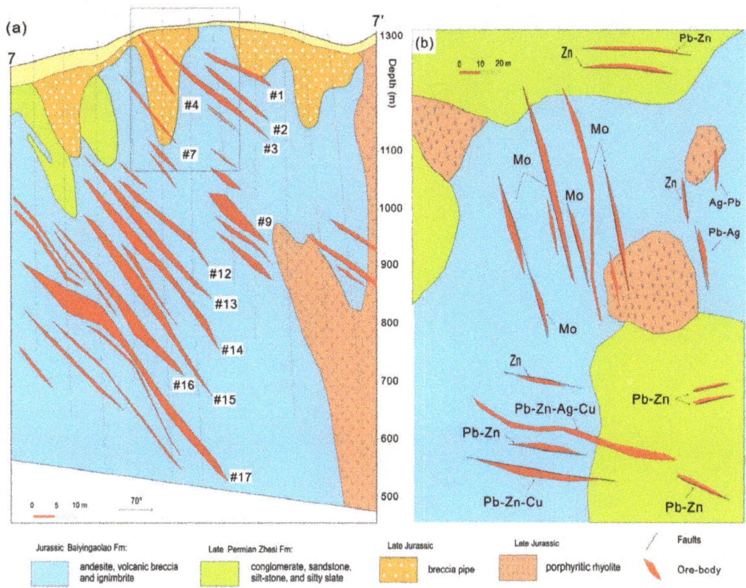

Figure 3. Diagrams showing: (**a**) simplified cross-section along No. 7 exploration line (7–7′) at Dongbulage showing the distribution of Mo orebodies, and the spatial relationship between orebodies and main host rocks (modified after ZXGE, 2010 [24]); (**b**) plan view of the Dongbulage deposit (1200 m above the see level) showing well-developed horizontal zoning outward from the subvolcanic stocks (modified after ZXGE, 2010 [24]).

3.1. Host Rocks and Structures

Host rocks at Dongbulage mainly consist of: (1) the NE-trending Late Permian Zhesi sedimentary succession (Figure 2a); (2) the Jurassic volcano-sedimentary sequences in the Baiyingaolao Formation and Porphyritic diorite (Figure 2a,b); (3) the Late Jurassic subvolcanic, together with the coeval concealed hydrothermal breccias (Figures 2 and 3a).

The major fractures and faults that developed in the inner and outer contact zones between the small stocks and adjacent host rocks are predominantly NNW-trending, and subordinately nearly EW-trending; these serve as major ore-controlling structures by providing conduits for the intrusion of magmatic rocks and the subsequent migration of the hydrothermal fluids that are responsible for mineralization (Figure 3).

3.2. Mineralization and Zoning

More than 100 orebodies had been outlined at Dongbulage by 2010, with 31 Mt of proved ore consisting of approximately 0.107% Mo, approximately 0.13% Zn, approximately 0.11% Pb, approximately 0.19% Cu, and approximately 4.53 ppm Ag [24]. Three types of mineralization have been recognized at Dongbulage. Disseminated and stockwork mineralization is mainly hosted by porphyritic rhyolite and adjacent country rock, the vein-type Mo-polymetallic mineralization is hosted by fractures and faults, and breccia mineralization occurs within cryptoexplosive breccia pipes. Of these, vein-type mineralization is the most abundant and economically important, and is essentially restricted to thick quartz-dominant veins (1 m in thickness), and small decimeter-scale (dm-scale) veins (0.1–1 m thick) and veinlets (0.05–0.1 m thick) (Figure 4a,b).

Figure 4. Photographs and photomicrographs of mineralization showing: (**a**) and (**b**) altered volcanic rocks with vein and veinlet molybdenite (Mol); (**c**) and (**d**) altered porphyritic granite with crumby, flaked, and veinlet molybdenite (Mol); (**e**) irregularly-shaped pyrite (Py) with chalcopyrite (Ccp) in ore-bearing quartz vein (under reflected light); and (**f**) xenomorphic sphalerite (Sp), pyrite (Pyr), chalcopyrite (Ccp), and pyrrhotite (Po) intergrowths (under reflected light).

The mineralization shows well-developed horizontal zoning outward from the subvolcanic stocks. Based on metal assemblages, two main mineralized vein systems are recognized at Dongbulage: proximal Mo-bearing veins, which occur in the central part of the ore district, and distal Zn ± Pb ± Cu ± Ag mineralization, which is found at the edge of the ore district (Figure 3b).

Mo mineralization is primarily restricted to the NNW-trending extensional faults and fractures that developed in the inner and the outer contact zones between small stocks and adjacent host rocks (Figure 3). Individual Mo orebodies are lens-shaped and vein-shaped. Lengths along the strike direction range from 20 to 650 m, and vein widths range from 1.2 to 17 m. Orebodies #3, #12, #17 are the largest (Figure 3a). The Mo-bearing vein-type ores consist of quartz, subhedral pyrite (<0.025 mm in size), and euhedral sheeted molybdenite (approximately 0.01–0.437 mm), with associated muscovite, fluorite, and K-feldspar (Figure 4c,d). Bonanza horizons in the vein are lenticular in shape with a thickness of <2 m. The grade of bonanzas in this vein reaches 1.793%, although the average grade is 0.107%.

Distal Zn ± Pb ± Cu ± Ag mineralization is associated with shallow-level, nearly EW-trending faults and fractures within the Zhesi and Baiyingaolao formations (Figure 3b). The individual Zn ± Pb ± Cu ± Ag orebodies are approximately 25 to 150 m in length, and approximately 1.03 to 9.80 m in thickness. The high concentrations of sulfides generally coincide with a high frequency of quartz veins. Economic Zn ± Pb ± Cu ± Ag vein-type mineralization mainly contains pyrite, sphalerite, and galena, with minor chalcopyrite, pyrrhotite, and argentite (Figure 4e,f). Gangue minerals are primarily quartz and K-feldspar, with minor amounts of chlorite, muscovite, fluorite, calcite, and kaolin.

3.3. Stages of Alterations and Ore Formation

Alteration is well widespread within the ore-bearing porphyry and the contact host rocks, and is characterized by an early stage of K-silicate alteration at deeper levels of the mine, fracture-controlled silicification and sericite alteration, widespread propylitic alteration, and later argillic alteration [7]. According to the alteration–mineralization patterns and cross-cutting relationship of the veins, five stages of alteration–mineralization have been recognized: (1) the K-silicate stage; (2) the quartz + sericite + molybdenite stage; (3) the chlorite + sericite + sulfides stage; (4) the quartz + calcite stage; and (5) the anhydrite stage (Figure 5) [7].

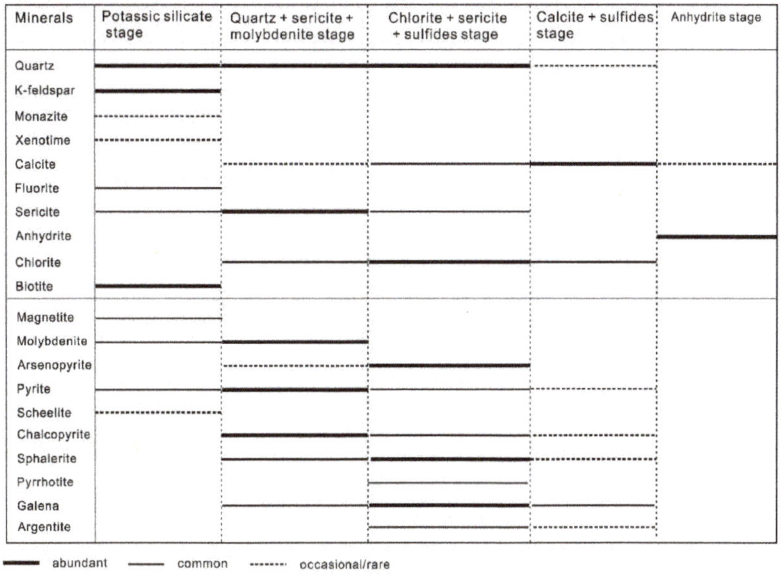

Minerals	Potassic silicate stage	Quartz + sericite + molybdenite stage	Chlorite + sericite + sulfides stage	Calcite + sulfides stage	Anhydrite stage
Quartz					
K-feldspar					
Monazite					
Xenotime					
Calcite					
Fluorite					
Sericite					
Anhydrite					
Chlorite					
Biotite					
Magnetite					
Molybdenite					
Arsenopyrite					
Pyrite					
Scheelite					
Chalcopyrite					
Sphalerite					
Pyrrhotite					
Galena					
Argentite					

——— abundant ——— common ········ occasional/rare

Figure 5. Mineral paragenesis of the Dongbulage Mo-polymetallic deposit (modified after Li et al., 2017 [7]).

The K-silicate stage represents the late magmatic and the initial hydrothermal stage, which is generally tightly confined to the volume occupied by deep porphyry stocks. Early K-silicate can be established, mainly consisting of quartz, K-feldspar, and biotite, with minor amounts of fluorite, pyrite, molybdenite, and magnetite, as well as traces of monazite, xenotime, and scheelite (Figure 5). During the K-silicate stage, K-feldspar and the biotite replacement of primary plagioclase in the porphyritic rhyolite is commonly observed. Moreover, in areas where the degree of K-silicate alteration is higher, disseminated and stockwork mineralization can be observed during this stage.

The quartz + sericite + molybdenite stage is the major alteration style associated with Mo mineralization, producing Mo-rich quartz veins hosted by NNW-trending faults and fractures. These veins are quite common in the porphyritic rhyolite and contact host rocks, and contain a medium-grained to coarse-grained quartz, sericite, calcite, and sulfide assemblage composed of chalcopyrite, pyrite, and molybdenite, with rare occurrences of euhedral arsenopyrite, sphalerite, and galena (Figure 5). Notably, economic vein-type Mo-mineralization occurs during this stage.

The chlorite + sericite + sulfides stage is third and the most important Zn ± Pb ± Cu ± Ag mineralizing event. It features veins controlled by the nearly EW-trending cracks and fault fractures developed primarily in the upper and outer parts of the porphyry stocks. The mineral assemblage in this stage is dominated by an intermediate–sulfidation combination composed of quartz + chlorite + muscovite + sphalerite + galena ± chalcopyrite ± arsenopyrite ± argentite ± argentite (Figure 5).

The quartz + calcite stage is subeconomic and characterized by vein, veinlets, and the dissemination of quartz and calcite, as well as accessory amounts of sulfides (Figure 5). The anhydrite stage is the final phase, and exhibits fewer mineralized sulfides and cross and cutting minerals from earlier stages. It is represented by anhydrite veins and veinlets that are commonly approximately between 15–50 mm width (Figure 5).

4. Sample Description and Analytical Methods

A total of 12 whole-rock samples were systematically collected from major magmatic rocks for geochemical analyses. The samples mainly included monzogranite, syenogranite, red porphyritic granite, and porphyritic rhyolite. The petrographic characteristics of the samples are described below.

The monzogranite (YP08, 10, and 11) and syenogranite samples (YP23, 25, and 26) were collected from the Buerhaotu granitic batholith in the southern section of the Dongbulage ore district (Figure 1c). The monzogranite consists of approximately 0.2 to 5-mm K-feldspar (30–35 vol. %), plagioclase (25–30 vol. %), quartz (20 vol. %), and biotite (5–10 vol. %), with accessory amounts of magnetite, zircon, and apatite (Figure 6a). The syenogranite is composed of approximately 2 to 7-mm long phenocrysts (30–40 vol. %) in an approximate 0.5 to 1-mm matrix (60–70 vol. %). It consists of quartz (~20 vol. %), plagioclase (~25 vol. %), K-feldspar (~45 vol. %), biotite (~3 vol. %), and accessory minerals (~2 vol. %), including zircon and fluorapatite (Figure 6b).

Figure 6. *Cont.*

Figure 6. Representative microphotographs for intrusive rocks around Dongbulage: (**a**) monzogranite collected from Buerhaotu granitic batholith, mainly comprising plagioclase, quartz, K-feldspar, and biotite phenocrysts, and a felsic matrix with a microgranitic texture (under polarized light); (**b**) syenogranite collected from Buerhaotu granitic batholith, mainly consisting of K-feldspar, quartz phenocrysts, and felsic matrix (under polarized light), (**c**) red porphyritic granite collected from drill cores, light red in color, which contains quartz and K-feldspar phenocryst in a cryptocrystalline matrix (under polarized light), and (**d**) porphyritic rhyolite collected from drill cores, moderately porphyritic and rhotaxitic texture, and phenocrysts that are solely grayish quartz and altered potassium feldspar, and the matrix is hyalopilitic, comprising intertwined quartz and feldspar crystallites (under polarized light). Abbreviations: Bi = biotite, Kfs = K-feldspars, Pl = plagioclase, and Qt = quartz.

The samples of red porphyritic granite (YP32–34) and the samples of porphyritic rhyolite (YP6, 7, 11, and 28) were selected from drill cores at Dongbulage. Red porphyritic granite is light red in color and mainly contains quartz and K-feldspar phenocrysts (20–25 vol. %) in a cryptocrystalline matrix (75–80 vol. %), consisting of quartz, feldspar, and minor amounts of chlorite and zircon (Figure 6c). Porphyritic rhyolite is characterized by a moderately porphyritic texture, with phenocrysts (20 vol. %) that are solely grayish quartz and altered potassium feldspar, and the matrix (80 vol. %) is hyalopilitic, mainly comprising intertwined quartz and feldspar crystallites (Figure 6d).

In order to constrain the ages of major magmatic rocks, whole-rock samples were selected for LA-ICP-MS zircon U–Pb dating. Furthermore, considering that the age for the mineralization is poorly constrained using Re–Os model ages, two samples of molybdenite were also collected from Dongbulage. The molybdenite in Sample YP-5 formed aggregates in the quartz vein, while the molybdenite in Sample YP-6 was disseminated and occurred in the Jurassic volcano-sedimentary sequences.

4.1. Major and Trace Element Analyses

A total of 12 whole-rock samples were crushed and powdered to less than ~200 mesh size. Chemical analyses were performed at the National Research Center for Geoanalysis, Chinese Academy of Geological Sciences (CAGS), Beijing. Oxide concentrations were analyzed using a using an X-ray fluorescence spectrometer (Instrument model: PE 300D) with an analytical error of <1.4%, and trace elements were measured using an inductively coupled plasma mass spectrometry (ICP-MS). The analytical precision and accuracy of the analyses was better than 5% for the major elements and 10% for the trace elements. The analytical uncertainties are based on the US Geological Survey rock standards BCR-1 and AVG-2, and the Chinese national rock standard GSR-3 [4].

4.2. LA-MC-ICP-MS Zircon U–Pb Dating

Zircon grains were extracted from samples using density and magnetic separation techniques. Representative zircons were handpicked using a binocular microscope and photographed using orthogonal polarization, reflected light, and cathodoluminescence (CL) imaging. Approximately 24 grains were selected for U–Pb isotope analyses, based on their morphology and internal structure. In situ zircon U–Pb dating was completed using a Thermo Finnigan Neptune MC-ICP-MS instrument at the Key Laboratory of Crust–Mantle Materials and Environments, University of Science and Technology of China, Chinese Academy Sciences, Anhui Province, China. Laser ablation was performed using a

New Wave UP 213 laser ablation system, with a laser ablation spot diameter of 32 μm. The carrier gas was He, and an external zircon standard (91,500) was analyzed after five sample points. The NIST SRM610 standard was used for calibrating element concentrations. The operating conditions for the laser ablation system were provided by Liu et al. (2007) [25]. Since ^{204}Pb could not be measured due to a low signal and interference from ^{204}Hg in the gas supply, a common lead correction was performed using the LaDating@Zrn Excel VBA program (Version 2, University of Oslo, Oslo, Norway) and Com Pb corr#3_18 [26]. Errors for individual analyses by LA-ICP-MS are quoted at the 1σ confidence level, while errors on pooled ages are quoted at the 95% (2σ) level. Isoplot (ver. 3.0) was used to calculate the weighted mean ages and generate Concordia plots [27].

4.3. Re–Os Dating

Samples were crushed, separated, and handpicked using a binocular microscope to obtain non-oxidized molybdenite. Sample pretreatment and analysis were completed at the Re–Os Laboratory of the National Research Center of Geoanalysis, CAGS. The molybdenite samples were dissolved in an HCl–HNO$_3$–H$_2$O solution in a Carius tube. Osmium was separated by distillation and rhenium was separated by extraction, as described by Du et al. (1995, 2001, 2004) [28–30]. The Re and Os isotope ratios were determined using a TJA X-series ICP-MS manufactured by the Thermo Electron Corporation, Waltham, MA, USA. Shirey and Walker (1995), Stein et al. (1997), and Du et al. (2004) [30–32] describe the additional details of the analytical procedures used in this analysis. The ^{187}Re/^{188}Os and ^{187}Os/^{188}Os were calculated using the decay constant of ^{187}Re, and λ = equaled approximately 1.666×10^{-11} a^{-1}, with an absolute uncertainty of approximately 0.017×10^{-11} [33].

The uncertainty for the Re and Os contents was comprised of the weighting error for samples and diluent, the calibration error for the diluent, the fractionation correction error for mass measurements, and the measurement error of the isotopic ratio for the samples. The uncertainty for model ages consisted of the uncertainty of the decay constant. The Re–Os age uncertainties for the molybdenite analyses reported here are given as 2σ.

5. Results

5.1. Whole-Rock Geochemistry

Whole-rock geochemical data for the main magmatic rocks of Dongbulage are listed in Table 1. A brief summary is presented here. Before plotting, the assay results were normalized to 100% after accounting for the loss on ignition (LOI).

5.1.1. Major Elements

The samples of monzogranite and syenogranite selected from the Buerhaotu granitic batholith had a narrow range of SiO$_2$ (65.09–66.22 wt. % and 74.41–75.31 wt. %), Al$_2$O$_3$ (15.99–16.78 wt. % and 12.79–13.10 wt. %), K$_2$O (3.22–4.22 wt. % and 3.62–4.48 wt. %), and Na$_2$O contents (4.96–5.62 wt. % and 4.07–4.11 wt. %), respectively. They are also characterized by high differentiation indices (DI; sum of normative Q + Or + Ab) [34], ranging from 81.75 to 86.92 and 87.60 to 92.42, and high Rittman indices [σ = (Na$_2$O + K$_2$O)^2/(SiO$_2$ − 43)] (oxides in wt. %), ranging from approximately 3.54 to 3.62 and approximately 1.98 to 2.30, respectively (Table 1) [34,35]. On the SiO$_2$ versus K$_2$O + Na$_2$O diagram, the monzogranite samples plot in the quartz monzonite field, the syenogranite samples plot in the granite field (Figure 7a), and all the samples fall within the high-K calc-alkaline series of granitic rocks (Figure 7b). On the A/CNK [A/CNK = Al$_2$O$_3$/(CaO + Na$_2$O + K$_2$O)(molar ratio)] versus A/NK [A/NK = Al$_2$O$_3$/(Na$_2$O + K$_2$O) (molar ratio)] diagram, the monzogranite samples are moderately peraluminous, with A/CNK values of approximately 1.02 to 1.06, and all the syenogranite samples fall within the metaluminous peraluminous transitional field (Figure 7c). One notable feature is that monzogranite and syenogranite both have low ratios of [Fe$_2$O$_3$/(Fe$_2$O$_3$ + FeO)] (0.19 to 0.27 and 0.26 to 0.37, respectively), suggesting that all are reduced intrusions (<0.4) [36].

Table 1. Major and trace elements compositions for the rocks in the Dongbulage Mo-polymetallic deposit.

Rock Type	Monzogranite			Syenogranite			Red Porphyritic Granite			Porphyritic Rhyolite		
Sample No.	YP8	YP10	YP11	YP23	YP25	YP26	YP32	YP33	YP34	YP28-11	YP6	YP7
Major Element (wt. %)												
SiO_2	65.99	66.22	65.09	74.41	75.15	75.31	76.16	75.97	75.41	76.35	76.05	76.39
TiO_2	0.43	0.43	0.40	0.16	0.14	0.14	0.09	0.09	0.11	0.08	0.08	0.08
Al_2O_3	16.10	15.99	16.78	12.79	13.10	13.05	13.05	13.17	13.30	11.96	12.15	11.93
Fe_2O_3	1.17	0.95	1.53	0.32	0.33	0.39	0.76	0.80	0.95	0.06	0.31	0.31
FeO	2.83	2.71	2.61	1.41	1.13	1.08	0.17	0.12	0.13	0.60	0.74	0.74
MgO	0.96	0.92	0.94	0.24	0.17	0.17	0.11	0.10	0.09	0.16	0.15	0.19
CaO	1.36	1.65	1.97	1.12	0.72	0.70	0.53	0.53	0.48	1.11	0.70	0.92
Na_2O	5.26	4.96	5.62	4.28	4.11	4.07	4.00	4.12	3.90	1.25	1.70	1.01
K_2O	3.77	4.22	3.22	3.62	4.48	4.47	4.51	4.47	4.84	6.49	6.54	6.44
P_2O_5	0.14	0.13	0.13	0.05	0.03	0.03	0.03	0.03	0.03	0.01	0.01	0.01
MnO	0.07	0.07	0.08	0.07	0.06	0.06	0.03	0.03	0.03	0.04	0.04	0.04
LOI	1.76	1.57	1.45	1.48	0.52	0.48	0.5	0.5	0.66	1.77	1.41	1.82
Total	99.85	99.83	99.82	99.94	99.95	99.95	99.95	99.94	99.93	99.88	99.88	99.88
$K_2O + Na_2O$	9.04	9.17	8.84	7.89	8.60	8.54	8.51	8.60	8.73	7.74	8.24	7.46
K_2O/Na_2O	0.72	0.85	0.57	0.85	1.09	1.10	1.13	1.09	1.24	5.19	3.85	6.36
σ	3.55	3.62	3.54	1.98	2.30	2.26	2.19	2.24	2.35	1.80	2.05	1.67
A/CNK	1.06	1.02	1.03	0.99	1.01	1.02	1.05	1.05	1.06	1.08	1.09	1.16
A/NK	1.26	1.26	1.32	1.17	1.13	1.13	1.14	1.14	1.14	1.32	1.23	1.38
DI	86.92	84.72	81.75	87.60	92.27	92.42	94.14	94.24	94.76	87.43	92.14	89.02
$Fe_2O_3/Fe_2O_3 + FeO$	0.29	0.26	0.37	0.18	0.23	0.27	0.82	0.87	0.88	0.09	0.23	0.26
Trace Element (ppm)												
Li	24.73	27.86	22.63	20.43	21.82	21.54	9.94	9.37	9.78	12.13	21.37	11.44
Be	5.12	5.50	3.15	3.80	5.76	5.57	5.76	6.47	5.20	2.39	4.76	2.49
Sc	6.41	6.75	5.62	3.22	3.43	3.47	2.63	2.63	3.31	1.67	1.75	1.77
V	57.91	55.45	50.59	33.58	32.86	33.15	35.43	37.12	36.54	32.48	31.35	31.15
Cr	5.85	5.92	5.84	5.08	5.14	5.40	5.96	5.80	5.45	5.39	5.33	5.32
Co	5.51	3.78	5.22	1.11	0.59	0.54	0.73	0.75	0.35	0.39	1.27	1.32
Ni	2.13	2.32	1.86	0.56	2.98	0.46	1.71	2.29	1.15	0.43	0.36	0.64
Cu	26.15	16.12	25.91	3.27	2.81	2.69	15.50	17.26	19.69	3.48	26.78	19.96
Zn	60.78	67.84	57.13	40.32	31.17	31.87	29.13	26.58	15.57	78.85	66.84	112.40
Ga	22.85	22.31	20.13	19.30	20.75	20.35	21.34	22.32	21.12	15.34	16.10	16.27
Rb	127.20	167.20	113.60	115.30	155.70	155.3	148.80	152.90	161.30	208.60	175.90	203.10
Sr	171.70	222.10	241.40	83.19	60.27	59.72	55.24	57.85	54.95	76.66	87.89	65.75
Y	27.40	26.13	22.81	23.83	38.57	32.36	25.52	23.12	17.43	32.86	35.47	33.06
Zr	379.67	398.37	413.42	159.30	176.46	157.48	141.28	142.25	146.92	117.18	119.05	117.95
Nb	9.39	9.67	7.76	9.11	21.29	21.29	11.70	14.14	20.79	10.56	11.76	10.39
Ba	547.62	855.74	658.85	188.16	156.70	147.2	173.56	181.50	172.19	579.87	569.09	514.89
Cs	4.08	6.62	6.88	2.94	7.06	7.05	3.79	3.84	4.82	5.40	6.13	5.85
Hf	10.44	11.41	12.03	6.58	8.77	8.37	7.22	7.08	7.52	6.09	6.17	6.12
Ta	0.61	0.67	0.54	0.60	1.79	1.74	1.05	1.24	1.65	1.00	1.19	1.07
Tl	0.96	1.00	0.90	0.84	0.91	0.92	1.11	1.12	0.98	2.10	2.04	2.11
Pb	11.55	16.73	8.99	11.90	11.91	11.63	14.18	14.29	12.20	15.89	28.98	10.89
Th	7.25	7.97	6.15	14.00	19.99	20.03	20.83	20.90	18.95	16.48	16.84	16.94
U	1.99	2.40	1.65	2.47	7.01	7.71	4.26	4.72	3.64	4.77	5.14	5.25
Mo	190.60	938.00	13.09	3.65	6.66	1.53	7.68	7.95	1.36	2.15	2.18	2.47
W	66.77	33.64	2.41	2.20	11.48	12.36	0.51	0.61	0.81	0.89	0.70	1.01
In	0.07	0.09	0.10	0.09	0.02	0.03	0.02	0.03	0.02	0.05	0.07	0.05
Y/Nb	2.92	2.70	2.94	2.62	1.81	1.52	2.18	1.64	0.84	3.11	3.02	3.18
Rb/Sr	0.74	0.75	0.47	1.39	2.58	2.60	2.69	2.64	2.94	2.72	2.00	3.09
Th/U	3.64	3.33	3.72	5.67	2.85	2.60	4.89	4.43	5.21	3.45	3.27	3.22
Nb/Pb	0.81	0.58	0.86	0.77	1.79	1.83	0.83	0.99	1.70	0.66	0.41	0.95
Th/Yb	2.68	3.04	2.71	6.19	5.06	5.76	7.58	8.41	9.33	4.56	4.62	4.92
Ta/Yb	0.23	0.25	0.24	0.27	0.45	0.50	0.38	0.50	0.81	0.28	0.33	0.31
Rare Earth Elements (ppm)												
La	30.61	29.22	25.25	36.92	44.22	31.63	25.01	22.43	20.86	19.59	22.07	22.09
Ce	60.28	58.83	49.93	72.86	89.58	64.18	52.35	48.14	39.67	45.26	48.89	49.51
Pr	7.11	6.90	6.01	8.21	10.24	7.64	6.38	5.78	4.69	5.74	5.99	6.17
Nd	27.09	26.35	22.85	28.83	35.83	27.64	23.34	21.46	17.49	21.16	22.56	22.91
Sm	6.02	5.74	4.95	5.94	8.02	6.38	5.50	5.02	3.83	5.34	5.92	5.63
Eu	0.86	0.94	1.12	0.43	0.43	0.40	0.36	0.33	0.34	0.36	0.39	0.36
Gd	5.15	5.04	4.31	5.08	6.89	5.74	4.40	3.94	3.09	4.72	5.13	4.95
Tb	0.84	0.82	0.70	0.76	1.14	0.96	0.73	0.67	0.48	0.88	0.91	0.87
Dy	5.12	4.96	4.53	4.47	7.11	5.94	4.58	4.15	3.22	5.67	6.16	5.74
Ho	0.98	0.96	0.84	0.82	1.39	1.12	0.87	0.81	0.62	1.12	1.21	1.14
Er	2.58	2.47	2.25	2.14	3.66	3.02	2.40	2.15	1.71	3.13	3.23	3.06
Tm	0.44	0.43	0.37	0.36	0.63	0.54	0.43	0.40	0.32	0.57	0.57	0.54
Yb	2.71	2.62	2.27	2.26	3.95	3.48	2.75	2.49	2.03	3.61	3.65	3.44
Lu	0.56	0.55	0.47	0.41	0.69	0.61	0.49	0.45	0.42	0.67	0.66	0.65
ΣREE	150.35	145.84	125.84	169.48	213.77	159.28	129.59	118.22	98.76	117.84	127.35	127.06
LREE/HREE	7.18	7.17	7.00	9.41	7.40	6.44	6.78	6.85	7.31	4.78	4.92	5.23
δEu	0.46	0.52	0.72	0.23	0.17	0.20	0.22	0.22	0.30	0.21	0.21	0.20
$(La/Yb)_N$	7.62	7.52	7.49	11.02	7.55	6.13	6.14	6.08	6.93	3.65	4.08	4.32

Notes: LOI is loss on ignition; TREE, total rare earth elements; LREE, light rare earth elements; HREE, heavy rare earth elements. A/CNK = $Al_2O_3/(CaO + Na_2O + K_2O)$(molar ratio), A/NK = $Al_2O_3/(Na_2O+K_2O)$ (molar ratio), DI = sum of normative Q + Or + Ab [34], $(La/Yb)_N = (La/La^*)/(Yb/Yb^*)$; δEu = $(Eu/Eu^*)/\{0.5 \times [(Sm/Sm^*) + (Gd/Gd^*)]\}$; σ (The Rittmann Index) = $(Na_2O + K_2O)_2/(SiO_2 - 43)$ (the oxide in wt. %); $(La/Yb)_N$ is normalized using values. Several rare earths are shown as chondrite-normalized using values from [35].

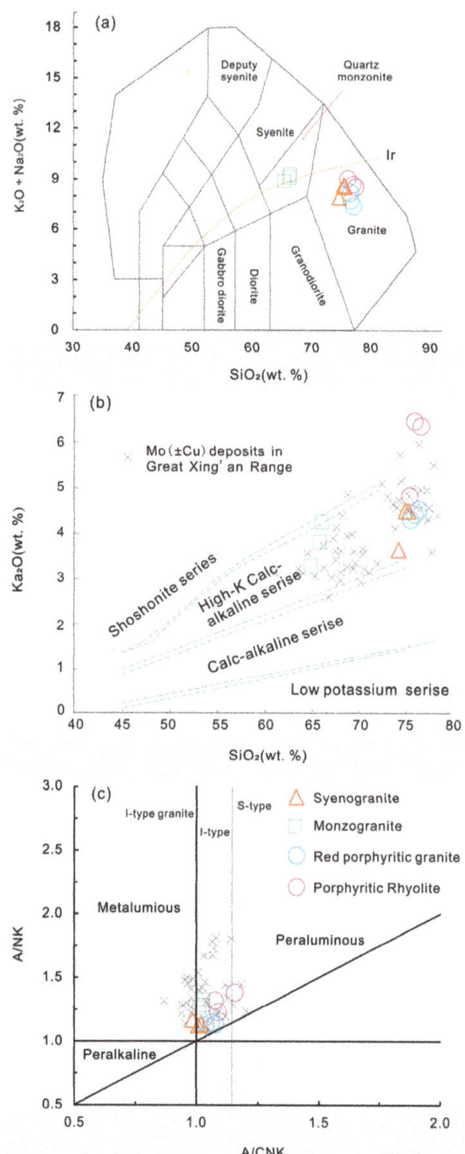

Figure 7. Geochemical plots for samples from Dongbulage of: (**a**) total alkali versus silica [37], (**b**) SiO$_2$ versus K$_2$O, and (**c**) A/CNK versus A/NK [38].

Compared to the plutons (monzogranite and syenogranite), the red porphyritic granites showed highly evolved features, with SiO$_2$ contents ranging from approximately 75.41 to 76.16 wt. % and DI values ranging between 94.14–94.76 (Table 1). The samples from the porphyritic rhyolite also exhibited high felsic compositions, with SiO$_2$ contents ranging between 76.05–76.39 wt. % and DI ranging between 87.43–92.14. Both red porphyritic granite and porphyritic rhyolite had high concentrations of total alkalis (8.51–8.73 wt. % and 7.46–8.24 wt. %, respectively) as well as Al$_2$O$_3$ (13.05–13.30 wt. % and 15.9–16.78 wt. %), with Rittman indices between 2.19–2.35 and 1.67–2.05, respectively (Table 1).

On the SiO_2 versus $K_2O + Na_2O$ diagram, all the samples from porphyritic granite and rhyolite plot in the granite field (Figure 7a) are weak peraluminous, with A/CNK ratios of 1.05–1.16, and belong to the high-K calc-alkaline series of granitic rocks (Figure 7b,c). Although samples of rhyolitic and granitic porphyry have similar major compositions, they differ in [Fe_2O_3/(Fe_2O_3 + FeO)] values. Samples from porphyritic rhyolite are more oxidized than samples of granitic porphyry [36].

5.1.2. Trace Elements

Chondrite normalized rare earth element (REE) patterns for monzogranite, syenogranite, porphyritic rhyolite, and porphyritic granite were roughly similar; they show high levels contents of total rare earth elements ($\sum$REE) and are enriched in light rare earth elements (LREEs), and depleted in heavy rare earth elements (HREEs), with negative Eu/Eu* (δEu) compared to common magmatic rocks (Table 1 and Figure 8a) [39]. Specifically, all the samples contained high levels of total rare earth elements ($\sum$REE), ranging from 125.84 to 150.35 ppm for monzogranite, 117.84 to 213.77 ppm for syenogranite, 98.76 to 129.59 ppm for granitic porphyry, and 117.84 to 127.35 ppm for porphyritic rhyolite. Monzogranite exhibited negative δEu values between 0.46–0.72, LREE/HREE values between 6.44–9.41, and $(La/Yb)_N$ values between 7.49–7.62 (Table 1), whereas syenogranite yielded moderately negative δEu values of approximately 0.17 to 0.23, LREE/HREE values of approximately 6.13 to 11.02, and $(La/Yb)_N$ values of approximately 7.00 to 7.18 (Table 1). Compared with pluton, porphyritic rhyolite and porphyritic granite both record negative δEu values between 0.20–0.21 and 0.22–0.30, LREE/HREE values between 4.78–5.23 and 6.78–7.31, and $(La/Yb)_N$ values between 3.65–4.32 and approximately 6.08–6.93, respectively (Table 1).

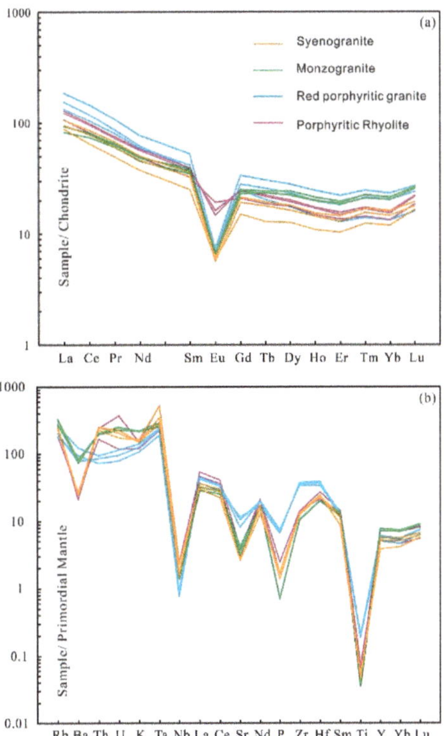

Figure 8. Geochemical plots for samples from Dongbulage of (**a**) chondrite-normalized rare earth element (REE) patterns; and (**b**) primitive-mantle normalized spider diagram (chondrite and primitive mantle values are from [35,39]).

The primitive mantle normalized spider diagram indicates that the trace element characteristics for all the samples had broadly similar patterns, with elevated large-ion lithophile elements (LILE, such as U, K, Rb, and Hf) and depleted high field strength elements (HFSE, such as Nb and Ta) (Figure 8).

5.2. Zircon U–Pb Dating

Zircons extracted from the main magmatic rocks at Dongbulage are a light to dark brown color, transparent, and dominated by typically short-prismatic, long-prismatic, and equigranular shapes, with width-to-length ratios of approximately 1:3.7 (Figure 9). All the zircon grains analyzed have a well-preserved oscillatory zoning indicative of a magmatic origin (Figure 9). Zircon ^{206}Pb/^{238}U ages for the main magmatic rocks have been summarized in Table 2 and are displayed in Figure 10.

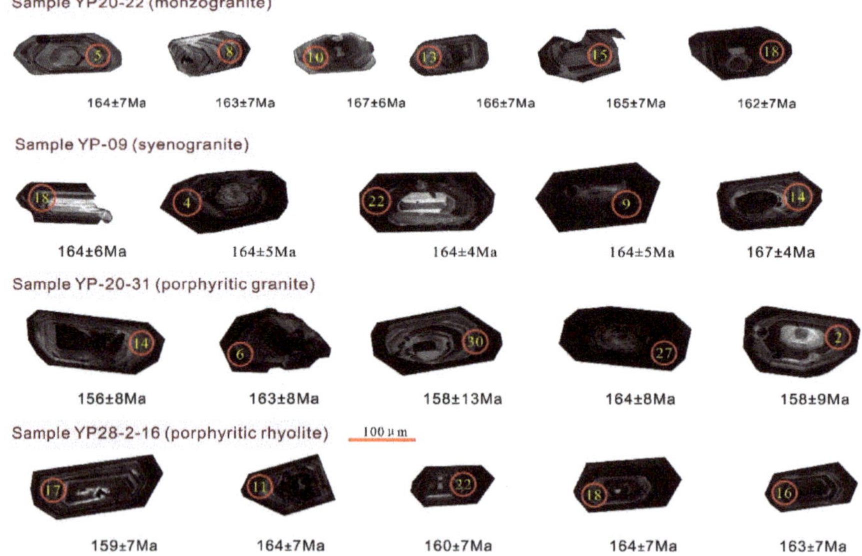

Figure 9. Representative cathodoluminescence (CL) images of zircons from samples. Note: the numbers in circles are test points.

A total of 24 zircon grains extracted from Sample YP20-22 (monzogranite) are analyzed, and they exhibit a wide range of ^{232}Th (55.39 to 1333.6 ppm) and ^{238}U (90.89 to 2288.74 ppm) concentrations and associated Th/U ratios of approximately 0.45 to 1.36 (Table 2). Seventeen zircon analyses yielded concordant ^{206}Pb/^{238}U dates ranging from 168 ± 6 to 159 ± 6 Ma, with a weighted mean date of 165 ± 3 Ma [MSWD (mean square of weighted deviates) = 0.13]; Table 2 and Figure 10a,b). This mean date was interpreted to be the crystallization age for the fine-grained to medium-grained monzogranite. Five points were not concordant and were excluded from the age calculation.

Table 2. LA-MC-ICP-MS zircon U–Pb data for samples from the Dongbulage Mo-polymetallic deposit. ICP-MS: inductively coupled plasma mass spectrometry.

Spot #	Total Pb	^{232}Th	^{238}U	^{232}Th/^{238}U	Corrected Isotopic Ratios						Corrected Age (Ma)					
		10^{-6}			^{207}Pb/^{206}Pb	±1σ	^{207}Pb/^{235}U	±1σ	^{206}Pb/^{238}U	±1σ	^{207}Pb/^{206}Pb	±1σ	^{207}Pb/^{235}U	±1σ	^{206}Pb/^{238}U	±1σ
YP20-22																
1	35.36	1241.6	934.87	1.33	0.04997	0.00248	0.17851	0.01084	0.02591	0.00092	194	116	167	9	165	6
4	18.59	467.34	542.68	0.86	0.05754	0.0033	0.20656	0.01375	0.02603	0.00098	512	129	191	12	166	6
5	20.53	523.71	570.89	0.92	0.04754	0.00326	0.17177	0.01302	0.0262	0.00099	77	152	161	11	167	6
7	7	136.57	221.77	0.62	0.05163	0.00499	0.18198	0.01813	0.02556	0.00104	269	206	170	16	163	7
9	72.07	1029.7	2288.74	0.45	0.04879	0.0011	0.17701	0.0072	0.02631	0.00089	138	50	165	6	167	6
10	11.48	264.61	338.01	0.78	0.05195	0.00233	0.18758	0.01086	0.02619	0.00094	283	97	175	9	167	6
12	6.2	113.71	192.55	0.59	0.05287	0.00385	0.19272	0.0156	0.02644	0.00102	323	161	179	13	168	6
14	5	95.3	156.4	0.61	0.05552	0.00534	0.19687	0.01958	0.02572	0.00114	433	214	182	17	164	7
16	6.38	136.99	201.09	0.68	0.05408	0.00602	0.18987	0.02098	0.02547	0.00112	374	237	177	18	162	7
17	8.46	233.86	257.07	0.91	0.04793	0.00307	0.16496	0.01193	0.02496	0.00092	96	137	155	10	159	6
18	5.82	104.26	184.02	0.57	0.05829	0.00565	0.20832	0.02044	0.02592	0.00105	541	213	192	17	165	7
19	7.27	183.33	213.82	0.86	0.04694	0.00367	0.16938	0.01413	0.02617	0.00101	46	168	159	12	167	6
20	3.22	55.39	102.65	0.54	0.04375	0.00459	0.15631	0.01695	0.02591	0.00116	-84	186	147	15	165	7
21	4.54	123.73	134.75	0.92	0.04973	0.00459	0.17623	0.01647	0.0257	0.00106	182	210	165	14	164	7
22	2.83	44.46	90.89	0.49	0.06012	0.00673	0.21371	0.02427	0.02578	0.00117	608	257	197	20	164	7
23	36.52	1333.6	980.7	1.36	0.05274	0.00233	0.1862	0.01028	0.0256	0.00094	318	103	173	9	163	6
24	7.36	193.43	212.5	0.91	0.05433	0.00397	0.19344	0.01552	0.02582	0.00096	385	170	180	13	164	6
YP-09																
2	8.89	261.72	250.22	1.05	0.04383	0.00413	0.15543	0.01483	0.02572	0.00087	82	102	158	7	163	5
3	35.94	498.95	1181.54	0.42	0.04766	0.00222	0.1682	0.00847	0.0256	0.00074	52	93	157	7	164	5
4	43.48	613.45	1435.61	0.43	0.04705	0.00204	0.16666	0.00793	0.02569	0.00077	283	153	169	11	161	5
5	13.72	264.63	436.01	0.61	0.05194	0.00356	0.18088	0.0124	0.02525	0.0008	308	134	173	10	164	5
6	17.92	238.3	598.49	0.40	0.05251	0.00315	0.18612	0.01164	0.02571	0.0008	814	304	212	26	161	6
7	5.62	194.09	144.2	1.35	0.06625	0.00941	0.23171	0.03203	0.02537	0.00101	521	178	194	15	168	6
8	21.84	283.93	689.22	0.41	0.05777	0.00478	0.21039	0.01745	0.02641	0.00088	170	99	164	7	164	5
10	4.25	76.4	136.88	0.56	0.03812	0.00625	0.13929	0.02229	0.0265	0.00112	763	126	207	11	162	5
11	14.33	347.12	420.49	0.83	0.06464	0.00398	0.22618	0.01352	0.02538	0.00073	338	94	176	8	164	4
12	30.72	818.64	888.25	0.92	0.05321	0.00243	0.18907	0.00949	0.02577	0.00071	474	163	185	13	163	5
13	8.44	251.76	237.23	1.06	0.05656	0.0046	0.20007	0.01549	0.02566	0.00084	338	89	179	7	167	4
14	37.59	1135.9	1054.6	1.08	0.05322	0.0023	0.19291	0.00861	0.02629	0.00071	208	77	166	6	163	4
15	62.59	945.5	2061.53	0.46	0.05029	0.00185	0.17798	0.00731	0.02567	0.0007	205	120	164	10	161	5
16	35.2	831.24	1088.36	0.76	0.05021	0.00291	0.17558	0.01178	0.02536	0.00083	136	166	164	12	166	5
17	22.5	306.95	704.98	0.44	0.04876	0.00405	0.17578	0.01406	0.02615	0.00086	427	129	182	11	164	6
18	11.62	298.73	338.83	0.88	0.05536	0.00355	0.19635	0.01329	0.02572	0.00092	424	201	182	17	164	7
19	6.4	89.64	211.38	0.42	0.0553	0.00557	0.19634	0.01961	0.02575	0.00119	508	173	187	15	163	6
21	16.67	604.9	448.85	1.35	0.04256	0.00312	0.14943	0.01073	0.02547	0.00079	238	114	168	8	164	4
22	24.67	776.52	692.41	1.12	0.05095	0.00268	0.18048	0.00954	0.02569	0.0007	315	104	174	8	164	4

Table 2. *Cont.*

| Spot # | Total Pb | ^{232}Th | ^{238}U | ^{232}Th/^{238}U | Corrected Isotopic Ratios | | | | | | Corrected Age (Ma) | | | | | |
		10^{-6}			^{207}Pb/^{206}Pb	±1σ	^{207}Pb/^{235}U	±1σ	^{206}Pb/^{238}U	±1σ	^{207}Pb/^{206}Pb	±1σ	^{207}Pb/^{235}U	±1σ	^{206}Pb/^{238}U	±1σ
YP20-31																
1	6.08	151.88	185.24	0.82	0.04357	0.00524	0.15557	0.01995	0.0259	0.00139	-93	191	147	18	165	6
2	11.68	494.06	299.72	1.65	0.04676	0.00586	0.1598	0.02066	0.02479	0.00142	37	231	151	18	158	6
4	14.09	227.63	469.44	0.48	0.05042	0.00352	0.17477	0.01468	0.02514	0.00126	214	153	164	13	160	8
7	13.43	388.05	413.16	0.94	0.05889	0.00421	0.20129	0.01757	0.02479	0.00127	563	154	186	15	158	6
9	30.32	394.77	1018.82	0.39	0.05549	0.00343	0.19611	0.01455	0.02563	0.00124	432	129	182	12	163	8
10	8.1	262.99	226.44	1.16	0.04746	0.00454	0.16428	0.01673	0.0251	0.0013	72	195	154	15	160	4
11	22.04	266.6	724.72	0.37	0.04884	0.00238	0.17788	0.01196	0.02642	0.00126	140	102	166	10	168	8
13	27.95	448.81	917.88	0.49	0.04952	0.00247	0.1733	0.01182	0.02538	0.00123	173	113	162	10	162	5
15	26.5	431.82	863.88	0.5	0.04803	0.00227	0.17132	0.01162	0.02587	0.00125	101	104	161	10	165	5
16	6.78	105.62	227.81	0.46	0.04767	0.00444	0.16884	0.01743	0.02569	0.00137	83	201	158	15	163	6
17	51.28	828.74	1707.37	0.49	0.0541	0.00214	0.19067	0.01161	0.02556	0.00121	375	89	177	10	163	4
18	26.17	429.5	877.47	0.49	0.05596	0.0024	0.19354	0.01235	0.02509	0.0012	451	96	180	11	160	8
19	13.26	211.65	429.66	0.49	0.05584	0.00441	0.19521	0.01797	0.02535	0.00133	446	168	181	15	161	4
22	24.56	367.41	844.86	0.43	0.04878	0.00245	0.16722	0.01145	0.02486	0.00121	137	107	157	10	158	5
23	9.82	163.25	325.98	0.50	0.05232	0.00355	0.18296	0.01447	0.02536	0.00126	299	146	171	12	161	5
24	27.32	351.97	902.32	0.39	0.0515	0.00206	0.18541	0.01128	0.02611	0.00123	263	95	173	10	166	8
25	61.86	749.89	2107.83	0.36	0.05229	0.00156	0.18216	0.0101	0.02527	0.0012	298	71	170	9	161	3
28	12.68	173.15	430.87	0.4	0.05921	0.00523	0.20254	0.01961	0.02481	0.00133	575	202	187	17	158	8
32	21.72	300.88	718.05	0.42	0.05122	0.00256	0.18355	0.01254	0.02599	0.00127	251	115	171	11	165	8
YP28-2-16																
2	38.76	529.35	1285.9	0.41	0.05424	0.00174	0.19166	0.01012	0.02563	0.00108	381	73	178	9	163	7
3	58.72	970.02	1947.29	0.5	0.04824	0.00146	0.16695	0.0086	0.0251	0.00106	111	69	157	7	160	7
6	23.97	330.9	773.69	0.43	0.05205	0.00259	0.1847	0.01203	0.02574	0.00111	288	121	172	10	164	4
7	30.93	735.09	992.41	0.74	0.0489	0.00231	0.16567	0.01047	0.02457	0.00107	143	112	156	9	156	7
9	25.43	472.51	846.98	0.56	0.05143	0.00205	0.17576	0.01028	0.02478	0.00109	260	97	164	9	158	5
10	25.03	374.35	846.73	0.44	0.04754	0.00254	0.16411	0.01096	0.02503	0.00108	77	122	154	10	159	7
11	31.23	674.5	969.95	0.7	0.05237	0.00197	0.1858	0.01041	0.02573	0.00111	302	77	173	9	164	7
12	30.37	469.11	995.36	0.47	0.04817	0.00165	0.17009	0.00938	0.02561	0.00109	107	71	159	8	163	7
13	53.97	878.18	1782.61	0.49	0.05038	0.00133	0.17523	0.00872	0.02523	0.00106	212	55	164	8	161	3
17	34.31	829.25	1086.9	0.76	0.05007	0.002	0.17229	0.01013	0.02496	0.00108	198	95	161	9	159	7
18	44.36	1085.2	1339.04	0.81	0.04956	0.00202	0.1765	0.01034	0.02583	0.00112	174	96	165	9	164	2
19	43.44	718.98	1415.64	0.51	0.05381	0.00196	0.1887	0.01051	0.02543	0.00109	363	85	176	9	162	7
20	29.25	453.45	974.32	0.47	0.05078	0.00178	0.17584	0.00967	0.02512	0.00109	231	78	164	8	160	7
21	26.99	404.76	903.86	0.45	0.05141	0.0021	0.18217	0.0107	0.0257	0.00111	259	90	170	9	164	4
22	45.97	750.09	1497.59	0.5	0.05443	0.00197	0.18803	0.0105	0.02506	0.00107	389	78	175	9	160	7
23	60.34	1008.4	2062.34	0.49	0.05166	0.00158	0.17432	0.00919	0.02447	0.00105	270	67	163	8	156	4
24	46.04	1176.2	1390.86	0.85	0.04997	0.00166	0.17282	0.00926	0.02508	0.00106	194	75	162	8	160	7
25	22.28	309.04	730.91	0.42	0.05014	0.00184	0.17786	0.01002	0.02573	0.0011	201	81	166	9	164	6
27	60.25	2007.1	1733.54	1.16	0.04958	0.0012	0.17172	0.00835	0.02512	0.00107	176	58	161	7	160	7
29	31.38	535.36	1009.31	0.53	0.05255	0.00177	0.18457	0.00995	0.02547	0.00109	309	80	172	9	162	7
30	38.08	484.21	1289.54	0.38	0.04965	0.00189	0.17296	0.00983	0.02527	0.00108	179	91	162	9	161	5
32	20.31	264.97	668.73	0.4	0.05217	0.00246	0.18645	0.0117	0.02592	0.00113	293	94	174	10	165	7
33	37.73	713.3	1195.59	0.6	0.05149	0.00175	0.18244	0.00996	0.0257	0.0011	263	68	170	9	164	6
34	65.63	1634.4	2123.01	0.77	0.05131	0.00153	0.18014	0.00938	0.02546	0.00108	255	60	168	8	162	7

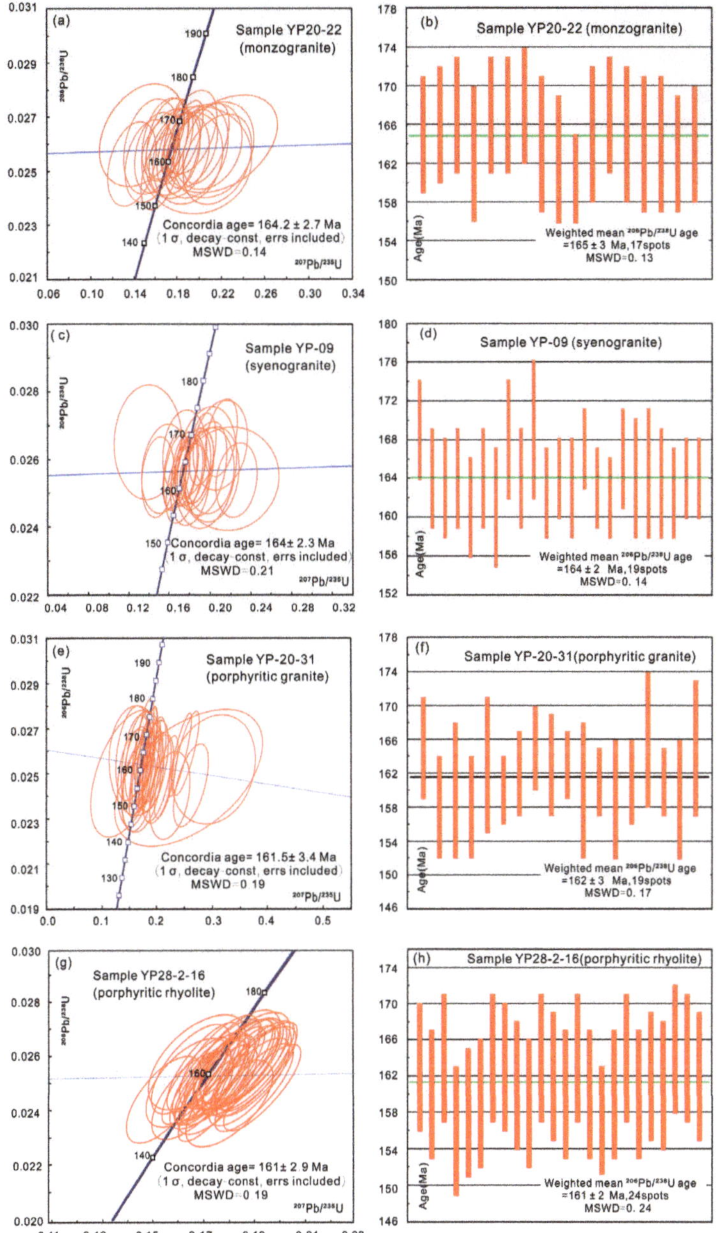

Figure 10. Zircon U–Pb concordia diagrams for main magmatic rocks: (**a**,**b**) monzogranite, (**c**,**d**) syenogranite, (**e**,**f**) porphyritic granite and (**g**,**h**) porphyritic rhyolite.

The Th and U concentrations of zircons extracted from Sample YP-09 (syenogranite) varied from 76.4 to 945.5 ppm and 136.88 to 1435.6 ppm, respectively, with associated Th/U ratios of 0.40 to 1.35 (Table 2). A total of 23 points were analyzed in the syenogranite, and three others were not on the concordance line. Excluding the points that were not on the concordance line, 20 zircons yielded concordant ^{206}Pb/^{238}U ages between 161 ± 5 Ma and 168 ± 8 Ma, with a weighted average age of 164 ± 2 Ma (MSWD = 0.14; Table 2 and Figure 10c,d). This average was interpreted to be the emplacement age of the syenogranite.

Zircon grains from Sample YP-20-31 (porphyritic granite) had U and Th contents between 105.62–749.89 and 185.24–2107.83 ppm, respectively, with associated Th/U ratios of approximately 0.37–1.65 (Table 2 and Figure 10c). A total of 32 points in the sample were analyzed; the 19 dated grains yielded concordant ^{206}Pb/^{238}U ages between 158 ± 6 Ma and 168 ± 8 Ma, with a mean ^{206}Pb/^{238}U age of 162 ± 3 Ma and an MSWD value of 0.17 (Table 2 and Figure 10e,f). We interpreted this mean age as the emplacement age of the granite porphyries. A total of 13 points were not concordant and were excluded from the age calculation (Table 2).

Zircon grains selected from Sample YP28-2-16 (porphyritic rhyolite) exhibited variable ^{232}Th (220–507 ppm) and ^{238}U (668.73–2123.01 ppm) concentrations, with typical magmatic Th/U ratios between 0.40–0.77 (Table 2). A total of 34 points were selected from the porphyritic rhyolite, of which the 24 that were analyzed yielded ^{206}Pb/^{238}U ages ranging from 156 ± 7 to 165 ± 7 Ma. Excluding the 10 points that were not on the concordant line, the weighted mean age of the remaining 24 samples was 161 ± 2 Ma (MSWD = 0.24; Table 2 and Figure 10h,f), which was interpreted as the emplacement age of porphyritic rhyolite.

5.3. Re-Os Age

The concentrations of Re and Os, as well as the Os isotopic composition of molybdenite from Dongbulage, are listed in Table 3. Since the molybdenite samples were extracted from different types of ores, the Re contents ranged widely, from 0.4896 to 101.6 µg/g. The ^{187}Os and common Os concentrations of the dated samples ranged between 0.843–172.8 ng/g and 0.0003–1.0195 ng/g, respectively (Table 3). Compared to the relatively high ^{187}Os values (0.843–172.8 ng/g) for the dated samples, the common Os values (0.0003–1.0195 ng/g) were negligible (Table 3), indicating that the measured Os is monoisotopic (^{187}Os) and the product of ^{187}Re decay [40]. Therefore, the Re–Os chronometer is an efficient method for dating.

Table 3. Re–Os data of molybdenite from the Dongbulage deposit.

Sample No.	Mineralization Types	Weight (g)	Re/µg·g^{-1}		Common Os/ng·g^{-1}		^{187}Re/µg·g^{-1}		^{187}Os/ng·g^{-1}		Model Age (Ma)	
			Measured	2σ	Measured	2σ	Measured	2σ	Measured	2σ	Measured	2σ
YP-5	VM	0.02280	101.6	0.7	1.0195	0.2459	63.86	0.45	172.8	1.1	162.2	2.2
YP-6	VM	0.01047	18.22	0.16	0.0022	0.0718	11.45	0.10	31.06	0.20	162.6	2.4
DBLG-8 *	DSM and BM	0.04999	10.786	0.11	0.0052	0.0292	6.78	0.07	18.61	0.19	164.7	2.7
DBLG-6 *	DSM and BM	0.30009	0.4896	0.0046	0.0026	0.0015	0.3077	0.0029	0.843	0.0071	164.2	2.5
DBLG-35 *	DSM and BM	0.03708	2.741	0.023	0.0065	0.0069	1.723	0.015	4.771	0.044	166.0	2.5
DBLG-36 *	DSM and BM	0.05048	6.344	0.056	0.0822	0.0108	3.988	0.035	11.14	0.10	167.5	2.5
DBLG-38 *	DSM and BM	0.05053	6.355	0.053	0.0031	0.0035	3.995	0.033	10.93	0.09	164.0	2.3
DBLG-39 *	DSM and BM	0.04992	0.6918	0.081	0.0003	0.0003	0.4348	0.051	1.216	0.019	167.6	3.5

Notes: * Data from Li et al. (2017) [7]; VM = vein-type mineralization; DSM = disseminated and stockwork mineralization; BM = breccia mineralization.

The Re–Os dating of molybdenite extracted from disseminated and breccia mineralization yielded Re–Os model ages ranging from 164.2 to 167.6 Ma, and yielded an identical Re–Os model age of 162.2 Ma for ore-bearing vein-type mineralization hosted by fractures and faults (Figure 11a and Table 3). Model ages of molybdenites overlapped with each other within the measurement uncertainty, and Mo mineralization occurred for a short period of time (Table 3). Based on the combination of the molybdenite Re–Os data above, the ^{187}Re and ^{187}Os data for molybdenite define a well-constrained isochron, yielding a regression age of 162.6 ± 1.5 Ma (MSWD = 3.70; Figure 11b), which was virtually identical to the weighted mean age of 164.5 ± 1.7 Ma (MSWD = 2.50, n = 8; Figure 11a).

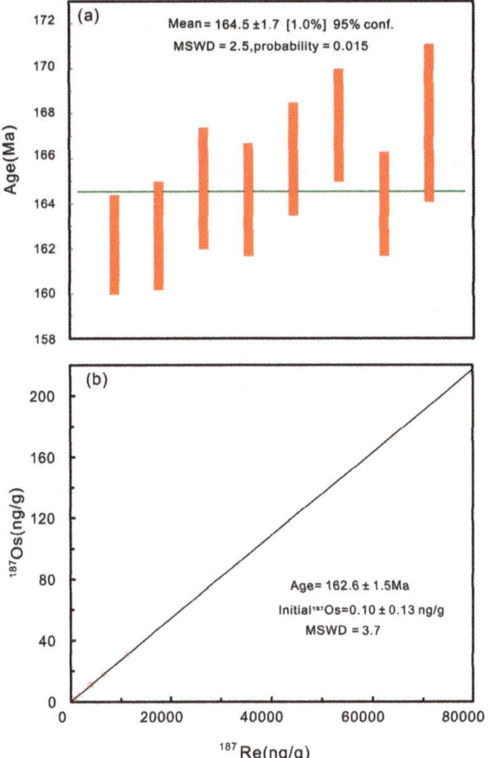

Figure 11. Re–Os isochron (**a**) and weighted average age (**b**) of molybdenite from Dongbulage Mo-polymetallic deposit.

6. Discussion

6.1. Magma Evolution and Petrogenesis

6.1.1. Fractional Crystallization

The Dongbulage Mo-polymetal deposit is temporally, spatially, and genetically related to the emplacement of the granitoids. Differentiations could play an important role in the Dongbulage Mo-rich magmatic evolution, and the process of fractional crystallization is recorded by the systematic distribution patterns of major and trace elements.

The Dongbulage granitoids exhibit a variation in chemical composition, and many major oxide elements (e.g., TiO_2, FeO_T, Al_2O_3, MgO, and CaO) show good negative correlations with SiO_2, with the exception of K_2O, Na_2O, and MnO (Figure 12). This suggests that fractional crystallization may have occurred during magma evolution, and complex processes influenced the compositions of the rocks. Fractional crystallization was also confirmed by the high differentiation index (DI) ranging from 81.75 to 94.76 (Table 1) and the obvious fractionation between LREE and HREE.

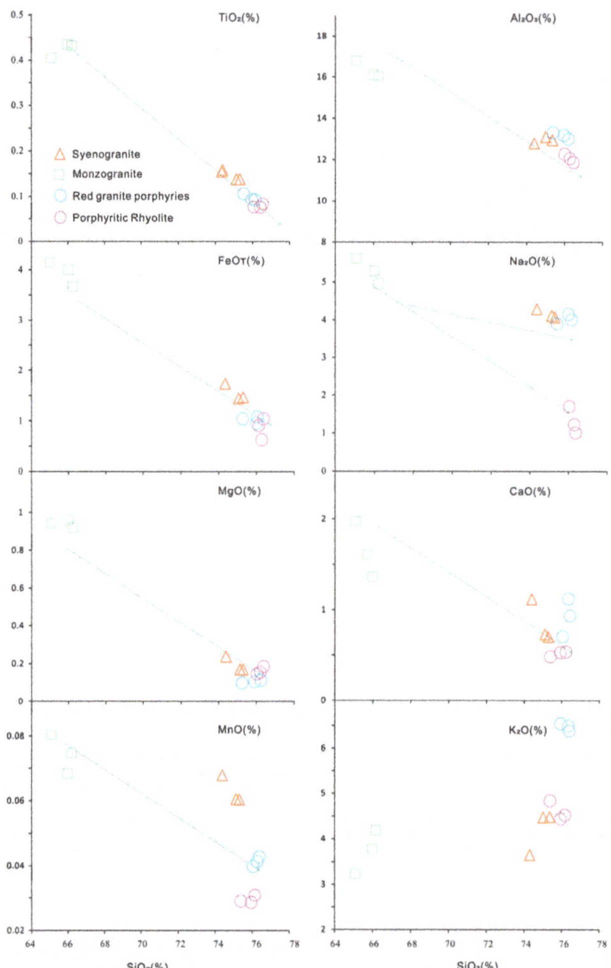

Figure 12. Harker diagrams for the granitoids at Dongbulage.

De Paolo (1981), Davidson et al. (1988), and Schiano et al. (2010) suggested that processes involved in partial melting and fractional crystallization can be identified from systematic changes in incompatible element concentrations and ratios [41–43]. The Ba (147.2–855.74 ppm) and Sr (83.19–241.10 ppm) concentrations of these rocks vary significantly, suggesting that the trace elements were obviously affected when hydrothermal processes are involved. Hence, all the dated samples only generally plot along a fractional crystallization trend in a Ba/Zr versus Ba diagram (Figure 13a), and the same conclusion can be achieved using a Ba–Sr diagram, in which the linear correlation between Sr and Ba almost passes through the origin (Figure 13b), suggesting that the magmas that formed the Dongbulage granitoids underwent significant fractional crystallization, but the assimilation in hydrothermal processes also play a significant role in the petrogenesis.

The separation of feldspars played a significant role during the evolution of the Dongbulage stocks (Figure 13c–e). Plagioclase fractionation led to the depletions of CaO, Sr, and Eu. On the other hand, negative Eu anomalies could be caused by the separation of K-feldspar, which can also explain the Ba depletion. Furthermore, the decrease in Ba with increase in Rb, and with decreasing Eu/Eu*, the decrease in Sr and increase in Rb/Sr. Trace element modeling also shows that biotite, minor apatite, and allanite are fractional phases (Figure 13c–f). Many researchers have suggested that peraluminous granitic rocks can be generated by the fractional crystallization of mafic parental magmas [44–46]; hence, we suggest that the peraluminous magmatic rocks around Dongbulage were successive products during the fractionation process of mafic parental magmas.

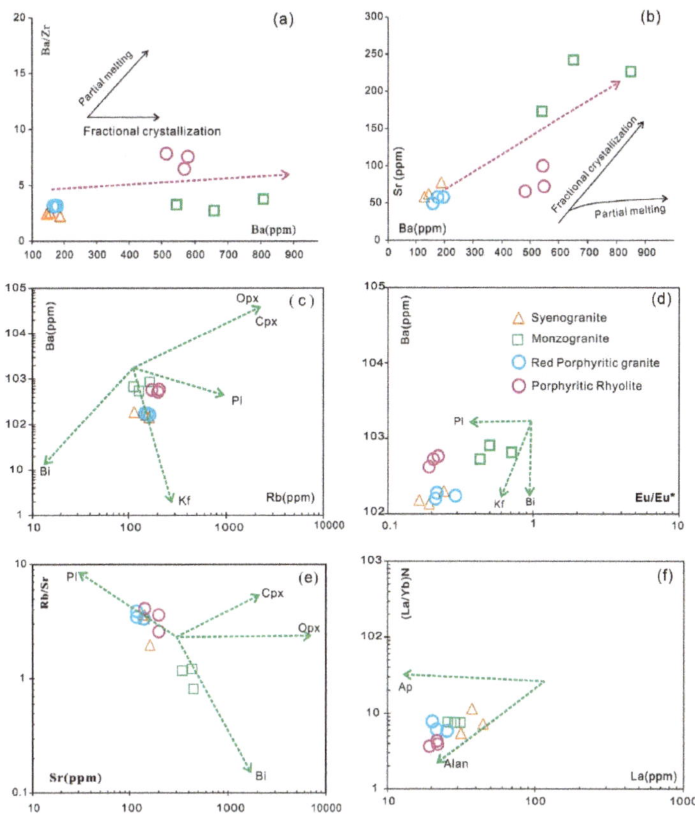

Figure 13. Geochemical plots for samples from Dongbulage showing variations in Ba/Zr versus Ba (**a**) and Sr versus Ba, (**b**) [43]; and Ba versus Rb, (**c**) Ba versus Eu/Eu* (**d**), Rb/Sr versus Sr, (**e**) (La/Yb)$_N$ versus La, and (**f**) diagrams for the Dongbulage granitoids illustrating the fractional crystallization of plagioclase, K-feldspar, biotite, and minor allanite. Partition coefficients are from [37].

6.1.2. Magma Types

In the (Zr + Nb + Ce + Y) versus FeO$_T$/MgO and the (Zr + Nb + Ce + Y) versus (K$_2$O + Na$_2$O)/CaO diagrams, with the exception of the monzogranite samples, many of the magmatic rock samples around Dongbulage fell within the fractionated granite field (Figure 14a,b). However, monzogranites showed slightly high (Zr + Nb + Ce + Y) values ranging from approximately 476.73 to 493.92, which is indicative of an A-type granite affinity (Figure 14a,b). Given that all of the magmatic rocks around Dongbulage had low 10,000 Ga/Al ratios (1.52–2.14) and low Y values (17.43–38.57 ppm), this actually indicates that these rocks were not A-type granites, but rather fractionated I or S-type granites (Figure 14c) [47].

The aluminum saturation index (A/CNK) of the dated samples was <1.1, with the exception of one sample (YP-7), suggesting that Dongbulage magmatic rocks are I-type granite (Figure 14c). Chappell et al. (1998) found that the P-content of S-type granites increases along with excess Al$_2$O$_3$ during fractional crystallization, while the P-content decreases to very low values in I-type granites [48].

The samples in our study had significantly low P-contents (approximately 0.01–0.14%), which showed a negative correlation with SiO_2, and was consistent with the typical evolution of I-type granites (Figure 14d). The geochemical affinity of I-type magmatic rocks was also further supported by low Y contents (17.43–38.57 ppm; Table 1) and Rb/Sr ratios (1.39–2.94; Table 1), compared to A-type granites (75 for Y and a3.42 for Rb/Sr ratios [47]). Thus, we conclude that the Dongbulage granitoids are a typical I-type, rather than an A-type or S-type intrusion.

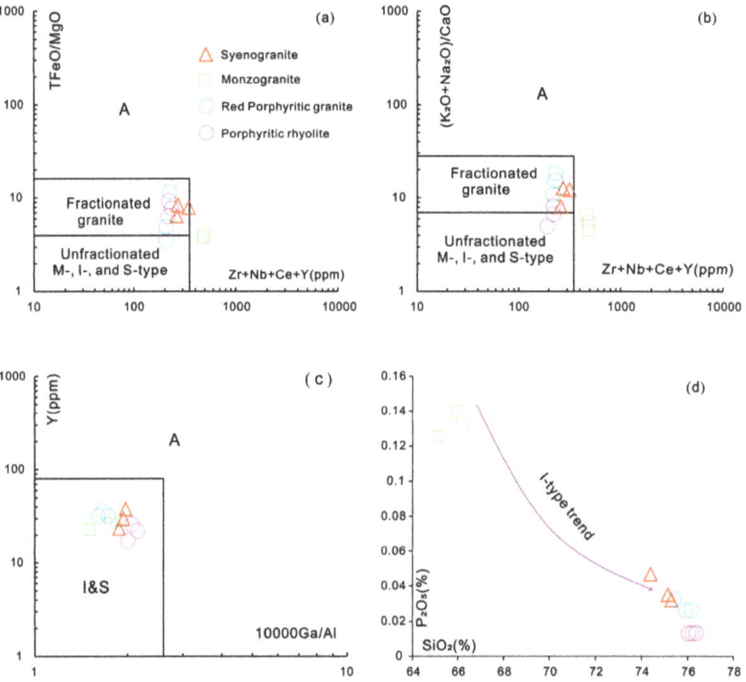

Figure 14. Dongbulage granites plot on diagrams (**a**) and (**b**) discriminating FG (fractionated granite) and OGT (unfractionated M-, I-, and S-type granites) granites from A-type granites [47]; (**c**) discriminating I-type and S-type granites from A-type granite [47] and (**d**) SiO_2 vs. P_2O_5 diagram showing P variation vs. silica.

6.1.3. Possible Sources

Most scholars proposed that most of the highly fractionated granites in the CAOB are derived from a juvenile crust mixed with a minor proportion of Precambrian crust in their source rocks, and the magma source of granitoids may contain two member components that are mantle-derived and crust-derived [4,49–51]. Dongbulage granitoids have high contents of SiO_2 and K_2O, and low abundances of Cr and Ni (Table 2), suggesting that they are mainly crust-derived magmas or highly evolved magmas. Generally, K-enriched calc-alkali granites are thought to be associated with mature subduction zones, which are confirmed by the Th/Yb versus Ta/Yb diagram (Figure 15a). Here, it can be seen that the Dongbulage granitoids plot near subduction zone compositions. In the Rb/Y versus Nb/Y and Nb/Y versus Th/Y diagrams (Figure 15b,c), the samples from magmatic rocks at Dongbulage plot close to the lower part of the continental crust field, indicating that the source magmas were of deeper origin. In contrast, the Th/U ratios for the Dongbulage granitoids were abnormally low, partly because granitic melts interact with primary mantle melts (1.20 to 4.05 [35]). These geochemical characteristics suggest that the source magmas of the Dongbulage granitoids had a deeper origin.

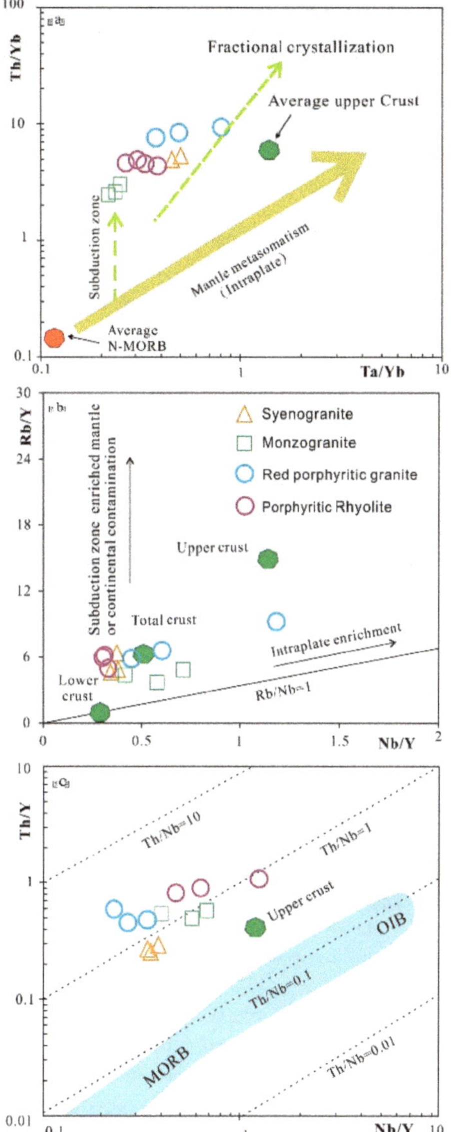

Figure 15. Ta/Yb versus Th/Yb diagram (**a**), Nb/Y versus Rb/Y diagram (**b**), and Nb/Y versus Th/Y diagram (**c**) of the samples from Dongbulage Mo-polymetallic deposit.

6.2. Rhenium Concentrations in Molybdenite

Berzina et al. (2005) suggested that the variations of Re content in molybdenites may be related to the composition of parent magmas, the concentration of Re in the ore-forming fluid, and variations of the physical and chemical crystallization conditions [52]. The Re contents in the dated molybdenites that were extracted from the Dongbulage deposit varied greatly, ranging from 0.50 to 101.6 ppm, indicating remarkable differences in metal sources. Mao et al. (1999) suggested that the Re contents of molybdenite might reflect the source of the deposits, with Re content decreasing from mantle ($n \times 10^{-4}$)

to I-type (n × 10^{-5}) to S-type (n × 10^{-6}) granite-related deposits [53]. Stein et al. (2001) hypothesized that the Re concentrations in molybdenite provide clues as to the origin of a deposit, and that the deposits involving mantle contributions usually have higher Re contents than deposits derived from crust [54]. Therefore, the molybdenite Re contents at Dongbulage vary widely, implying a mixed crust–mantle origin for the ore materials (Figure 16).

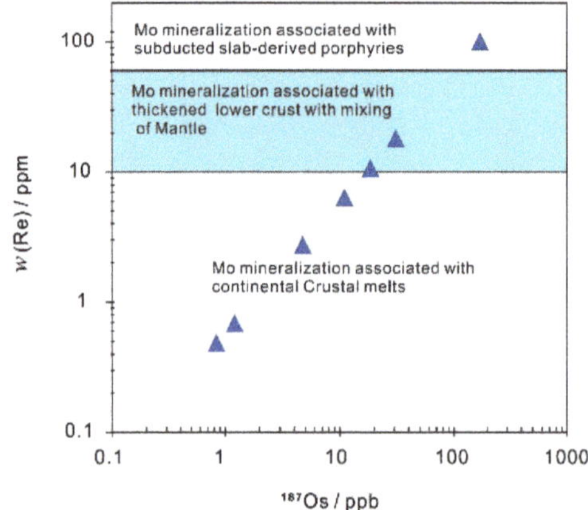

Figure 16. *w* (Re) vs. ^{187}Os diagram (modified from Chen et al., 2017 [55]) for Dongbulage Mo-polymetallic deposit.

6.3. Ages of Magmatism and Metallogenesis

Based on the geological field evidence, mineralization and alteration displayed horizontal zonal distributions centered on the porphyritic stocks at Dongbulage (Figure 3b), suggesting that mineralization in this area has a direct genetic relationship with porphyritic stocks. The accurate dating of magmatic–hydrothermal events is of fundamental importance when reconstructing the genetic evolution of porphyry systems and evaluating their duration.

According to the results of LA-ICP-MS zircon U–Pb dating, the ages of the felsic units at Dongbulage are 165 ± 3 Ma for monzogranite, 164 ± 2 Ma for syenogranite, 162 ± 2 Ma for red porphyritic granite, and 161 ± 2 Ma for porphyritic rhyolite (Figure 11). These dating results confirm a temporal correlation between the Buerhaotu granitic batholith (monzogranite and syenogranite) and adjacent small porphyritic stocks (porphyritic granite and porphyritic rhyolite), and indicate that the magmatism that occurred at Dongbulage continued for a protracted period ranging from approximately 165 ± 3 to 164 ± 2 Ma, with granitic pluton intruding at approximately 165 ± 3 to 164 ± 2 Ma and late phase felsic subvolcanic–volcanic complexes intruding at approximately 162 ± 3 to 161 ± 2 Ma. Hence, we suggest that the ore-hosting subvolcanic complexes represent an apophysis of a larger intrusion at depth.

Molybdenite (MoS$_2$) is the main economic mineral in Dongbulage, and can be dated using the Re–Os method. The stability of Re–Os in molybdenite is thought to be highly robust through post-ore geological processes [56,57]. The dated samples yielded model ages of approximately 162.2–167.6 Ma, with a weighted average age of 164.5 ± 1.7 Ma, initial ^{187}Os values ranging from approximately 0.4348 to 63.86 μg/g, and MSWD = 2.50 (Figure 11a). In addition, the dated samples yielded a Re–Os isochron age of 162.6 ± 1.5 Ma (2σ), with the initial ^{187}Os = 0.10 ± 0.13 ng/g and MSWD = 3.70 (Figure 11b). This ca. 162.6 ± 1.5 Ma age is coeval with the zircon U–Pb ages of ore-hosting subvolcanic complexes, suggesting that the mineralization may be related to the later emplacement of subvolcanic complexes, as indicated in geological evidence described above. Hence, this new molybdenite Re–Os isochron age was interpreted as the ore-forming age, and this also indicates that the Mo-polymetallic mineralization at Dongbulage was formed during the Middle-Late Jurassic stage.

Geochronological data published for the Mo (Pb–Zn–Ag) mineralization in the southern part of the Great Xing'an Range [55,58–72] are shown that this region records four stages for Mo (Pb–Zn–Ag) mineralization: (1) Late Permian (265 Ma), (2) Middle Triassic (244–235 Ma), (3) Middle–Late Jurassic (179–161 Ma), and (4) Late Jurassic–Early Cretaceous (147–132 Ma). The age of 162.6 ± 1.5 Ma obtained in this study is also consistent with the Middle–Late Jurassic stage of Mo-polymetallic mineralization in the southern part of the Great Xing'an Range, such as the Lianhuashan Cu + Ag + Mo deposit with a zircon U–Pb age of 161.8 Ma and the Shuangjianzishan deposit with a pyrite Re–Os isochron age of 165 ± 4 Ma, and slightly younger than the Meng'enTolgoi deposit, which had a muscovite Ar–Ar isochron age of 179 ± 2 Ma [58–60] (Figure 17).

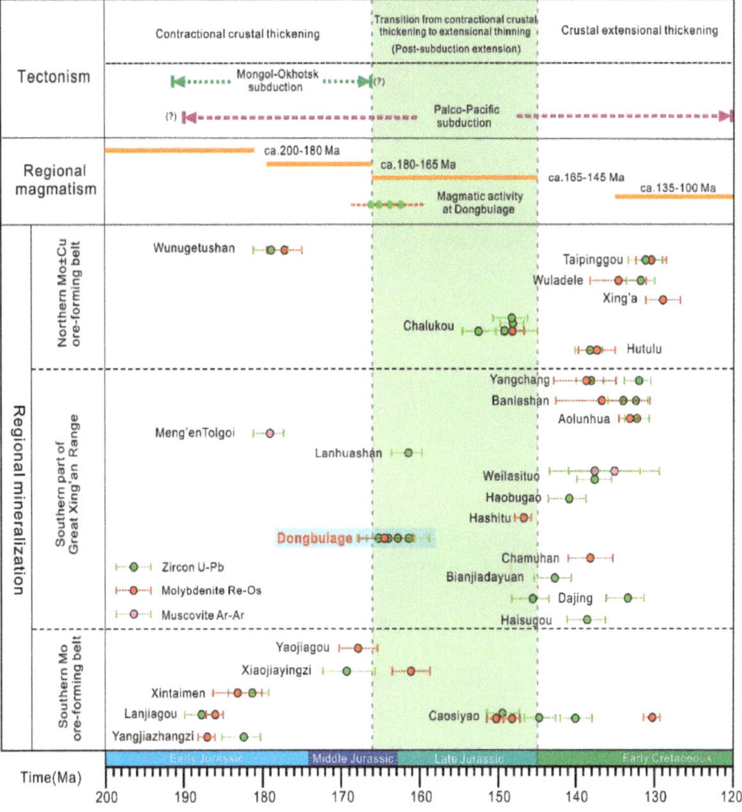

Figure 17. Time-scale showing the main plate tectonic events in XMOB, the ages for references to magmatism in XMOB, the ages for the main magmatic rocks around Dongbulage, and the Mo-polymetallic mineralizing events in XMOB. References to tectonism and magmatism: this study and [18,55,58–75]. References to mineralizing events: [4,55,58–72,75–85] and this study. The green areas show the post-subduction extension environment; the dark yellow lines represent five major periods of magmatic activity, and the dots show age determinations obtained in previous research.

Combined with the Mo-dominant deposits in the northern and southern Mo (±Cu) ore-forming belts, examples of Middle–Late Jurassic deposits are the Wunugetushan deposit, with an Re–Os isochron age of 177.6 ± 4.5 Ma, the Chalukou deposit, with an Re–Os isochron age of 148 ± 1 Ma, the Yaojiagou deposit, with an Re–Os isochron age of 168.8 ± 3.9 Ma, the Xiaojiayingzi deposit, with an Re–Os isochron age of 161.3 ± 2.4 Ma, the Xintaimen deposit, with an Re–Os isochron age of

183 ± 3 Ma, the Lanjiagou deposit, with an Re–Os isochron age of 186.5 ± 0.7 Ma, the Yangjiazhangzi deposit, with an Re–Os isochron age of 187 ± 2 Ma, and the Caosiyao deposit, with an Re–Os isochron age of 148.5 ± 1.1 Ma (Figure 17) [76–83]. These dates indicate that the Middle–Late Jurassic is an important Mo metallogenic stage in the region, although the peak of metallogenesis actually took place during Late Jurassic–Early Cretaceous periods (Figure 17). These ore-related granites are typically characterized by high degrees of fractionation, high silica content, and formations that were enriched by volatile elements such as F and Li [84–86].

6.4. Tectonic Significance

Even if the final closure time and location of the Paleo-Asian Ocean are still debated, the consensus is that the XMOB was a united continent during the Mesozoic period [87–91]. The geochronological data suggest that the Mesozoic magmatic activity in the XMOB have consisted of five major periods: 200–180 Ma related to a subduction–collision of the Mongol–Okhotsk Ocean, 180–165 Ma and 165–145 Ma with a post-collisional extensional setting, 145–135 Ma and 135–100 Ma formed during orogenic collapse coupled with back arc extension related to the subduction of the paleo-Pacific plate (Figure 17) [75]. Ouyang et al. (2015) pointed out that the magmatism and associated mineralization from the Late Jurassic to Early Cretaceous took place during the lithospheric extension resulting from the break-off of the south-dipping Mongol–Okhotsk oceanic slab at depth with the closure of the Mongol–Okhotsk Ocean, which also restricted the westward movement of the paleo-Pacific plate [5]. Based on our new results, the magmatic activities and related ore-forming events at Dongbulage took place between approximately 165 ± 3 and 161 ± 2 Ma (Figure 17), which is consistent with the period of the regional Middle–Late Jurassic magmatic activities (165–145 Ma). However, the tectonic setting of the Middle–Late Jurassic has controversial characteristics including: (1) Jurassic subduction of the paleo-Pacific plate [19,75]; (2) post-orogenic extension following the closure of the Mongolia–Okhotsk Ocean during the Late Jurassic–Early Cretaceous [21,62]; and (3) mantle plumes and mantle branches in an intraplate anorogenic environment [20,92,93].

Given that the Middle–Late Jurassic tectonic setting in the XMOB is still unclear, the new geochemical and isotopic chronology of the magmatic rocks around Dongbulage can provide us with potential information regarding the geodynamic setting. According to the granitoid tectonic classification scheme based on major elements (the SiO_2 versus Al_2O_3 and R_1–R_2 diagrams), most of the studied samples fell in the domains of post-orogenic granite and anorogenic granite (Figure 18a,b). The same conclusion is reached using the trace element discrimination diagrams: all the studied samples fell within the post-collisional granite (post-COLG) and volcanic arc granites (VAG) fields (Figure 18c,d). These tectonic setting discrimination diagrams suggest that either a post-collisional or post-orogenic geodynamic setting could represent the dominant tectonic environment from approximately 165 ± 3 to 161 ± 2 Ma at Dongbulage.

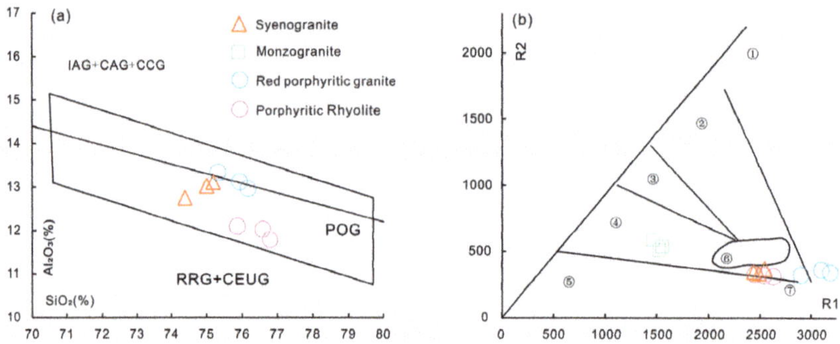

Figure 18. *Cont.*

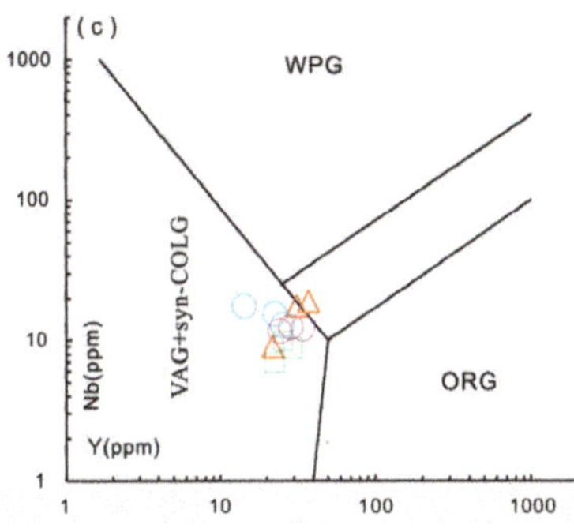

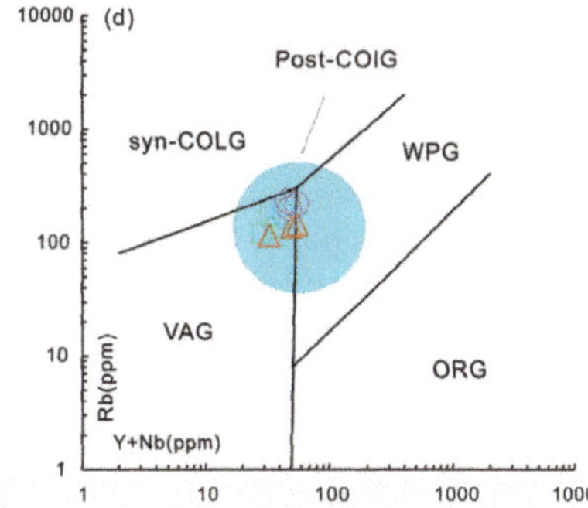

Figure 18. Major elements tectonic discrimination using (**a**) Al_2O_3 vs. SiO_2 diagram and (**b**) R1–R2 multicationic variation diagram R1 = 4Si – 11(Na + K) – 2(Fe + Ti) and R2 = 6Ca + 2Mg + Al after [94]; (**c**) and (**d**) show rare elements tectonic discrimination diagrams after [95] and that of post-COLG in the Rb–Y + Nb) diagram is adopted by [96]. CAG, continental arc granitoids; CCG, continental collision grantoids; CEUG, continental epeirogenic uplift granitoids; IAG, island-arc granitoids; ORG, oceanic ridge granites; POG, post-orogenic granitoids; post-COLG, post-collisional granites; RRG, rift-related granitoids; syn-COLG, syn-collisional granites; VAG, volcanic arc granites; WPG, within-plate granites; ①, mantle-differentiated granites; ②, pre-collisional granites; ③, post-collisional uplifting granites; ④, late-orogenic granites; ⑤, anorogenic granites; ⑥, syn-collisional granites; ⑦, post-orogenic granites.

Kravchinsky et al. (2012) proposed that the Mongol–Okhotsk Ocean may have closed gradually from west to east during the Permian–Early Cretaceous [97]. Recent studies have shown that the Mongol–Okhotsk Ocean was closed during the Middle Jurassic, and some Middle Jurassic–Early Cretaceous granitoids in the XMOB may be related to the post-orogenic collapse of the Mongol–Okhotsk orogen (Figure 17) [21,98–102]. According to Wang et al.(2015) [75], Middle–Late Jurassic (166–155 Ma) volcanic rocks west of the Songliao Basin belong to the high-K calc-alkaline series, which was formed during the collapse or delamination of a thickened continental crust associated with the evolution of the Mongol–Okhotsk suture belt. The evidence implies that the Middle–Late Jurassic magmatism in the XMOB may have been genetically related to the Mongol–Okhotsk orogenic event. However, much previous research data have indicated that magmatism and metallogeny are closely related to the subduction of the paleo-Pacific plates under the Eurasian continent [103], although these data overestimate the importance of the relationship between the subduction of the paleo-Pacific plate and magmatism in the XMOB. Even if the Middle–Late Jurassic magmatism in the region is active under different tectonic regimes, the consensus is that the Dongbulage granites and the associated Mo–Pb–Zn mineralization were formed in an extensional tectonic setting (Figure 17).

Therefore, the Middle–Late Jurassic Dongbulage magmatic rocks that most likely formed in a post-orogenic setting, which may have resulted from the combined influence of post-orogenic gravitational collapse and paleo-Pacific plate subduction, and perhaps been dominated by the former.

6.5. Implications for the Formation of the Dongbulage Deposit

Fluid inclusion studies from the Dongbulage deposit show that ores associated with Mo mineralization were deposited from an aqueous, high saline (63.9 eq. wt. % NaCl), high-temperature (approximately 534–312 °C) fluid, indicating that the ore-forming fluid was mainly derived from magmatic water [7]. New Re–Os model ages are similar to the crystallization age of the ore-bearing porphyritic stocks around Dongbulage; therefore, the Mo-polymetallic temporally and genetically linked with a porphyry–cryptoexplosive breccia igneous system.

The concentrations of Mo in siliceous melts of a rhyolitic composition are low, and the enrichment of Mo in these melts is possibly achieved through fractional crystallization and the interaction and movement of magmatic–hydrothermal fluids [104–106]. Furthermore, Audétat (2010) suggested that

Mo enrichment in ore-forming magma seems to occur by fractional crystallization rather than fluid transfer from associated mafic magmas [102]. The Dongbulage magmatic rocks have high differentiation indices (81.75–94.76), together with increasingly negative δEu anomalies (0.72–0.17), high values of incompatible trace elements (e.g., Zr, Ce, and Rb) and low values of compatible trace elements (e.g., Sr and CaO). This evidence indicates a fractional crystallization process during the evolution of ore-forming magma. Hence, we conclude that Mo content may have gradually increased in conjunction with the differentiation and evolution of magma at Dongbulage.

Most granitoids hosted by Mo-polymetallic deposits in the XMOB are highly fractionated. Examples include porphyritic granites hosting the giant Wunugetushan porphyry Cu–Mo deposit and Chalukou porphyry Mo deposit in the northern Great Xing'an Range; porphyritic monzonitic granite and granodiorite hosting the Aolunhua, Haolibao porphyry Cu–Mo deposit and Banlashan porphyry in the southern Great Xing'an Range; and monzonitic granite and granitic porphyries hosting the Caosiyao, Chaganhua porphyry Mo deposit in the northern margin of the North China Craton [22,47,50,58,65,66,107]. These granitoids are highly fractionated high-K calc-alkaline I-type and/or A-type granites, and they are geochemically characterized by enrichment of the LILEs (Ba, Rb, U, and Th) and Pb, and depletion of the HFSEs (Nb, Ta, and Ti) and HREE [22,47]. It should be noted that the highly fractionated I-type magmatic rocks at Dongbulage are members of the Mo-hosting granitoids in the XMOB.

Based on the available data, porphyry Mo deposits are directly associated with small volumes of porphyry emplaced as stocks, which is a part of relatively long-lived multiphase magmatic intrusion. The size of the magma chamber is a key factor forming an economic porphyry Mo deposit; due to the low saturation concentration of Mo in granitic melt, at least several tens of km^3 of magma are required to concentrate large amounts of metals to produce an economic porphyry deposit. It can be calculated that at least several tens of km^3 of magma were required to form the intermediate to large Mo deposits [108]. However, Mo-polymetallic mineralization at Dongbulage was mainly hosted by subvolcanic rocks (porphyritic granite and porphyritic rhyolite) and coeval breccia pipes (Figure 3). Such small stocks could not provide the total volume of Mo-polymetallic metal found at Dongbulage. Based on the essentially identical bulk-rock geochemical and isotopic data, we deduced that the highly fractionated I-type porphyritic granite and porphyritic rhyolite are part of the Buerhaotu granitic batholith, which provided much of the metal in the process of ore-forming hydrothermal evolution. In other words, magmatic rocks associated with Dongbulage Mo-polymetallic mineralization make up precursor pluton and ore-hosted subvolcanic complexes, and were emplaced over an interval of at least approximately four million years, between 165 ± 3 and 161 ± 2 Ma. Multiple pulses of intrusions can extend the life of a magmatic–hydrothermal system to a few million years. Harris et al. (2008) suggested that episodic replenishment is essential to ensure the longevity of a magmatic–hydrothermal system or its repeated reactivation beyond approximately one to two million years [109]. Therefore, the highly fractionated magma, along with the prolonged magmatic–hydrothermal interaction, jointly contributed to the formation of the Dongbulage deposit.

7. Conclusions

The Dongbulage is a porphyry deposit hosted by and genetically associated with the multiphase composite igneous body around Dongbulage. Geochemical studies show that porphyritic stocks at the Dongbulage deposit are highly fractionated I-types granites with negative Eu anomalies.

The new zircon U–Pb data indicate that the ages of the magmatism at Dongbulage are identical within error and lasted for a period of time, ranging from 165 ± 3 to 161 ± 2 Ma, with granitic pluton intruding at approximately 165 ± 3 to 164 ± 2 Ma and late-phase felsic subvolcanic complexes intruding at approximately 162 ± 3 to 161 ± 2 Ma. The molybdenite Re–Os isochron age for different types of ores is 162.6 ± 1.5 Ma, which is in accordance with previous zircon U–Pb age determinations of magmatic rocks in the area.

We propose that ore-hosting subvolcanic complexes represent an apophysis of a larger metal-rich intrusion at depth, which recorded a Middle–Late Jurassic Mo-rich felsic magma event during post-orogenic extension that followed the Mongol–Okhotsk Ocean closure coeval with the paleo-Pacific plate subduction.

Author Contributions: Conceptualization, S.J. and Y.L.; Data curation, Q.L.; Formal analysis, Q.L.; Funding acquisition, S.J., F.W. and C.C.; Investigation, F.W., and C.C.; Project administration, S.J.; Writing–original draft, F.W and C.C.

Funding: This work was jointly supported by National Key R&D Program of China (2017YFC0601303), National Natural Science Foundation of China (41873051), Graduate demonstration course of Hebei province (KCJSX2018089) and Doctoral Scientific Research Foundation of Hebei GEO University (BQ2018032).

Acknowledgments: I sincerely appreciate the great support and assistance provided by Xiang Guo.

Conflicts of Interest: The authors declare no conflict of interest.

References

1. Sinclair, W.D. Porphyry deposits. In *Mineral Deposits of Canada: A Synthesis of Major Deposit-Types, District Metallogeny, the Evolution of Geological Provinces, and Exploration Methods*; Geological Association of Canada, Mineral Deposits Division, Special Publication: St. John's, NL, Canada, 2007; pp. 223–243.

2. Sengör, A.M.C.; Natal'In, B.A. Turkic-type orogeny and its role in the making of the continental crust. *Annu. Rev. Earth Planet. Sci.* **1996**, *24*, 263–337. [CrossRef]

3. Chen, Y.C.; Zhu, Y.S.; Xiao, K.Y.; Zhang, X.H.; Mei, Y.X.; Yan, S.H.; Liu, Y.L.; Song, G.G.; Li, C.J.; Wang, Y.Y.; et al. Division of minerogenic provinces (belts) in China. *Miner. Depos.* **2006**, *S1*, 1–6. (In Chinese)

4. Wang, F.X.; Bagas, L.; Jiang, S.H.; Liu, Y.F. Geological, geochemical, and geochronological characteristics of Weilasituo Sn-polymetal deposit, Inner Mongolia, China. *Ore Geol. Rev.* **2017**, *80*, 1206–1229. [CrossRef]

5. Ouyang, H.G.; Mao, J.W.; Zhou, Z.H.; Su, H.M. Late Mesozoic metallogeny and intracontinental magmatism, southern Great Xing'an Range, northeastern China. *Gondwana Res.* **2015**, *27*, 1153–1172. [CrossRef]

6. Liu, Y.F.; Jiang, S.H. Mo mineralization in Xing'an-Mongolian orogen and north margin of China Craton: Review, question and a preliminary genetic model. *Miner. Depos.* **2017**, *36*, 557–594. (In Chinese)

7. Li, L.M.; Xie, Y.L.; Li, F.G.; Jia, L.; Chen, W.; Wang, Y.; Li, Z. Molybdenite Re-Os age and characteristics of ore-forming fluid of the Dongbulage Mo-Pb-Zn deposit, Inner Mongolia. *Geotecton. Metallog.* **2017**, *41*, 108–121. (In Chinese)

8. Wang, H.Z.; Mo, X.X. An outline of the tectonic evolution of China. *Episodes* **1995**, *18*, 6–16.

9. Jahn, B.M.; Wu, F.Y.; Chen, B. Granitoids of the Central Asian Orogenic Belt and continental growth in the Phanerozoic. *Tran. Royal Soc. Edinburgh Earth Sci.* **2000**, *350*, 181–193. [CrossRef]

10. Xu, B.; Zhao, P.; Wang, Y.; Liao, W.; Luo, Z.; Bao, Q.; Zhou, Y. The pre-Devonian tectonic framework of Xing'an-Mongolia orogenic belt (XMOB) in north China. *J. Asian Earth Sci.* **2015**, *97*, 183–196. [CrossRef]

11. Bureau of Geology and Mineral Resources of Inner Mongolia (BGMRIM). *Regional Geology of Nei Mongol (Inner Mongolia) Autonomous Region*; The memoirs of geological servey 25; Geological Publishing House: Beijing, China, 1991; p. 725. (In Chinese)

12. Wu, F.Y.; Zhao, G.C.; Sun, D.Y.; Wilde, S.A.; Yang, J.H. The Hulan Group: Its role in the evolution of the Central Asian Orogenic Belt of NE China. *Asian Earth Sci.* **2007**, *30*, 542–556. [CrossRef]

13. Zhang, X.Z.; Ma, Y.X.; Chi, X.G.; Zhang, F.X.; Sun, Y.W.; Guo, Y.; Zeng, Z. Discussion on Phanerozoic tectonic evolution in Northeastern China. *Jilin Univ. (Earth Sci. Ed.)* **2012**, *42*, 1269–1285. (In Chinese)

14. Pei, F.P.; Xu, W.L.; Yang, D.B.; Zhao, Q.G.; Liu, X.M.; Hu, Z.C. Zircon U-Pb geochronology of basement metamorphic rocks in the Songliao Basin. *Chin. Sci. Bull.* **2007**, *52*, 942–948. [CrossRef]

15. Zhou, X.H.; Ying, J.F.; Zhang, L.C.; Zhang, Y.T. The petrogenesis of Late Mesozoic volcanic rock and the contributions from ancient micro-continents: Constraints from the zircon U-Pb dating and Sr-Nd-Pb-Hf isotopic systematics. *Earth Sci. J. China Univ. Geosci.* **2009**, *34*, 1–10. (In Chinese)

16. Zhou, J.B.; Wilde, S.A.; Zhang, X.Z.; Ren, S.M.; Zheng, C.Q. Early Paleozoic metamorphic rocks of the Erguna block in the Great Xing'an Range, NE China: Evidence for the timing of magmatic and metamorphic events and their tectonic implications. *Tectonophysics* **2011**, *499*, 105–117. [CrossRef]

17. Mao, J.W.; Wang, Y.T.; Zhang, Z.H.; Yu, J.J.; Niu, B.G. Geodynamic settings of Mesozoic large-scale mineralization in the North China and adjacent areas: Implication from the highly precise and accurate ages of metal deposits. *Sci. China Ser. D.* **2003**, *33*, 838–851. [CrossRef]

18. Zhao, Y.; Xu, G.; Zhang, S.H.; Yang, Z.Y.; Zhang, Y.Q.; Hu, J.M. Yanshanian movement and conversion of tectonic regimes in East Asia. *Earth Sci. Front.* **2004**, *11*, 319–328. (In Chinese)

19. Ge, W.C.; Wu, F.Y.; Zhou, C.Y.; Zhang, J.H. Porphyry Cu–Mo deposits in the eastern Xing'an-Mongolian Orogenic Belt: Mineralization ages and their geodynamic implications. *Chin. Sci. Bull.* **2007**, *52*, 3416–3427. [CrossRef]

20. Pirajno, F.; Ernst, R.E.; Borisenko, A.S.; Fedoseev, G.; Naumov, E.A. Intraplate magmatism in Central Asia and China and associated metallogeny. *Ore Geol. Rev.* **2009**, *35*, 114–131. [CrossRef]

21. Xu, W.L.; Pei, F.P.; Wang, F.; Meng, E.; Ji, W.Q.; Yang, D.B.; Wang, W. Spatial-temporal relationships of Mesozoic volcanic rocks in NE China: Constraints on tectonic overprinting and transformations between multiple tectonic regimes. *Asian Earth Sci.* **2013**, *74*, 167–193. [CrossRef]

22. Wang, Y.H.; Zhao, C.B.; Zhang, F.F.; Liu, J.J.; Wang, J.P.; Peng, R.M.; Liu, B. SIMS zircon U-Pb and molybdenite Re-Os geochronology, Hf isotope, and whole-rock geochemistry of the Wunugetushan porphyry Cu–Mo deposit and granitoids in NE China and their geological significance. *Gondwana Res.* **2015**, *28*, 1228–1245. [CrossRef]

23. Nie, F.J.; Li, Q.F.; Liu, C.H.; Ding, C.W. Geology and origin of Ag-Pb-Zn deposits occurring in the Ulaan-Jiawula metallogenic province, northeast Asia. *Asian Earth Sci.* **2015**, *97*, 424–441. [CrossRef]

24. Zhongxing Geological Exploration Co. Ltd. (ZXGE). *Prospecting Report of the Dongbulage Mo-Pb-Zn Deposit in the Wulan Mining Area, West Ujimqin, Inner Mongolia*; Mining prospecting report; Zhongxing Geological Exploration Co. Ltd. (ZXGE): Wulan, China, 2010; pp. 1–103. Unpublished Report. (In Chinese)

25. Liu, X.; Gao, S.; Diwu, C.; Yuan, H.; Hu, Z. Simultaneous in-situ determination of U-Pb age and trace elements in zircon by LA-ICP-MS in 20 μm spot size. *Chin. Sci. Bull.* **2007**, *52*, 1257–1264. [CrossRef]

26. Andersen, T. Correction of common lead in U-Pb analyses that do not report ^{204}Pb. *Chem. Geol.* **2002**, *192*, 59–79. [CrossRef]

27. Ludwig, K.R. *User's Manual for Isoplot 3.00: A Geochronological Toolkit for Microsoft Excel*; Berkeley Geochronology Center Special Publication: Berkeley, CA, USA, 2003; pp. 1–70.

28. Du, A.D.; He, H.L.; Yin, N.W. A study of the rhenium-osmium geochronometry of molybdenites. *Acta Geol. Sin. English Edi.* **1995**, *8*, 171–181.

29. Du, A.D.; Wang, S.X.; Sun, D.; Zhao, D.M.; Liu, D.Y. Precise Re-Os dating of molybdenite using Carius tube, NTIMS and ICPMS. In *Mineral Deposits at the Beginning of the 21st Century, Proceedings of the Joint 6th Biennial SGA-SEG Meeting, Krakov, Poland, 26–29 August 2001*; Piestrzynski, A., Ed.; Swets and Zeitlinger Publishers: Lisse, The Netherlands, 2001; pp. 405–407.

30. Du, A.D.; Wu, S.Q.; Sun, D.Z.; Wang, S.X.; Qu, W.J.; Markey, R.; Stein, H.; Morgan, J.; Malinovskiy, D. Preparation and certification of Re-Os dating reference materials: Molybdenite HLP and JDC. *Geostand. Geoanal. Res.* **2004**, *28*, 41–52. [CrossRef]

31. Shirey, S.B.; Walker, R.J. Carius tube digestion for low-bank rhenium-osmium analysis. *Anal. Chem.* **1995**, *67*, 2136–2141. [CrossRef]

32. Stein, H.J.; Markey, R.J.; Morgan, J.W.; Du, A.; Sun, Y. Highly precise and accurate Re-Os ages for molybdenite from the East Qinling molybdenum belt, Shaanxi Province, China. *Econ Geol.* **1997**, *92*, 827–835. [CrossRef]

33. Smoliar, M.L.; Walker, R.J.; Morgan, J.W. Re-Os ages of group IA, IIA, IVA and IVB iron meteorites. *Science* **1996**, *271*, 1099–1102. [CrossRef]

34. Thornton, C.P.; Tuttle, O.F. Chemistry of igneous rocks. 1. Differentiation index. *Am. J. Sci.* **1960**, *258*, 664–684. [CrossRef]

35. Sun, S.S.; McDonough, W.F. Chemical and isotopic systematic of oceanic basalts: Implications for mantle composition and processes. In *Magmatism in the Ocean Basins*; Saunders, A.D., Norry, M.J., Eds.; Geological Society, London, Special Publications: London, UK, 1989; Volume 42, pp. 313–345.

36. Meinert, L.D. *Compositional Variation of Igneous Rocks Associated with Skarn Deposits—Chemical Evidence for a Genetic Connection between Petrogenesis and Mineralization*; Mineralogical Association of Canada: Québec, QC, Canada, 1995; pp. 401–418.

37. Rollinson, H.R. *Using Geochemical Data: Evaluation, Presentation, Interpretation*; Pearson Education: London, UK, 1993.

38. Maniar, P.D.; Piccoli, P.M. Tectonic discrimination of granitoids. *Geol. Soc. Am. Bull.* **1989**, *101*, 635–643. [CrossRef]

39. Taylor, S.R. The application of trace element data to problems in petrology. *Phys. Chem. Earth.* **1965**, *6*, 133–213. [CrossRef]

40. Raith, J.G.; Stein, H.J. Re-Os dating and sulfur isotope composition of molybdenite from tungsten deposits in western Namaqualand, South Africa: Implication for ore genesis and the timing of metamorphism. *Mine. Depos.* **2000**, *35*, 741–753.

41. Paolo, D.J. Trace element and isotopic effects of combined wall-rock assimilation and fractional crystallization. *Earth Planet. Sci. Lett.* **1981**, *53*, 189–202.

42. Davidson, J.P.; Dungan, M.A.; Ferguson, K.M.; Colucci, M.T. Crust-magma interactions and the evolution of arc magmas: The San Pedro-Pellado Volcanic complex, southern Chilean Andes. *Geology* **1988**, *15*, 443–446. [CrossRef]

43. Schiano, P.; Monzier, M.; Eissen, J.P.; Martin, H.; Koga, K.T. Simple mixing as the major control of the evolution of volcanic suites in the Ecuadorian Andes. *Contrib. Miner. Petrol.* **2010**, *160*, 297–312. [CrossRef]

44. Phillips, L.W. Assessing measurement error in key informant reports: A methodological note on organizational analysis in marketing. *J. Mark. Res.* **1981**, *18*, 395–415. [CrossRef]

45. Mittlefehldt, D.W.; Miller, C.F. Geochemistry of the Sweetwater Wash Pluton, California: Implications for "anomalous" trace element behavior during differentiation of felsic magmas. *Geochim. Cosmochim. Acta* **1983**, *47*, 109–124. [CrossRef]

46. Clemens, J.D. S-type granitic magmas—Petrogenetic issues, models and evidence. *Earth Sci. Rev.* **2003**, *61*, 1–18. [CrossRef]

47. Whalen, J.B.; Currie, K.L.; Chappell, B.W. A-type granites: Geochemical characteristics, discrimination and petrogenesis. *Contrib. Miner. Petrol.* **1987**, *95*, 407–419. [CrossRef]

48. Chappell, B.W.; Bryant, C.J.; Wyborn, D.; White, A.J.R.; Williams, I.S. High- and low temperature I-type granites. *Resour. Geol.* **1998**, *48*, 225–235. [CrossRef]

49. Jahn, B.M.; Wu, F.; Capdevila, R.; Martineau, F.; Zhao, Z.; Wang, Y. Highly evolved juvenile granites with tetrad REE patterns: The Woduhe and Baerzhe granites from the Great Xing'an Mountains in NE China. *Lithos* **2001**, *59*, 171–198. [CrossRef]

50. Jahn, B.M.; Capdevila, R.; Liu, D.Y.; Vernon, A.; Badarch, G. Sources of Phanerozoic granitoids in the transect Bayanhongor-Ulaan Baatar, Mongolia: Geochemical and Nd isotopic evidence, and implications for Phanerozoic crustal growth. *J. Asian Earth Sci.* **2004**, *23*, 629–653. [CrossRef]

51. Jiang, S.H.; Bagas, L.; Hu, P.; Han, N.; Chen, C.L.; Liu, Y.; Kang, H. Zircon U–Pb ages and Sr–Nd–Hf isotopes of the highly fractionated granite with tetrad REE patterns in the Shamai tungsten deposit in eastern Inner Mongolia, China: Implications for the timing of mineralization and ore genesis. *Lithos* **2016**, *261*, 322–339.

52. Berzina, A.N.; Sotnikov, V.I.; Economou-Eliopoulos, M.; Eliopoulos, D.G. Distribution of rhenium in molybdenite from porphyry Cu-Mo and Mo-Cu deposits of Russia (Siberia) and Mongolia. *Ore Geol. Rev.* **2005**, *26*, 91–113. [CrossRef]

53. Mao, J.W.; Zhang, Z.C.; Zhang, Z.H.; Du, A. Re-Os isotopic dating of molybdenites in the Xiaoliugou W (Mo) deposit in the northern Qinling mountains and its geological significance. *Geochim. Cosmochim. Acta* **1999**, *36*, 1815–1818.

54. Stein, H.J.; Markey, R.J.; Morgan, J.W. The remarkable Re-Os chronometer in molybdenite: How and why it works. *Terra Nova.* **2001**, *13*, 479–486. [CrossRef]

55. Chen, Y.J.; Zhang, C.; Wang, P.; Pirajno, F.; Li, N. The Mo deposits of Northeast China: A powerful indicator of tectonic settings and associated evolutionary trends. *Ore Geol. Rev.* **2017**, *81*, 602–640. [CrossRef]

56. Stein, H.J.; Markey, R.J.; Morgan, J.W.; Hannah, J.L.; Zák, K.; Sundblad, K. Re-Os dating of shear-hosted Au deposits using molybdenite, in Papunen. *Mine Depos.* **1997**, 313–317.

57. Frei, R.; Nägler, T.F.; Schöberg, R.; Kramers, J.D. Re-Os, Sm-Nd, U-Pb, and stepwise lead leaching isotope systematics in shear-zone hosted gold mineralization: Genetic tracing and age constraints of crustal hydrothermal activity. *Geochim. Cosmochim. Acta* **1998**, *62*, 1935–1936. [CrossRef]

58. Bai, L.A.; Sun, J.G.; Sun, Q.L.; Gu, A.L.; Zhao, K.Q.; Men, L.J. Ore-forming fluids and genesis of Lianhuashan Cu deposit in middle Da Hinggan Mountains. *Mine Depos.* **2012**, *31*, 1249–1258. (In Chinese)

59. Liu, C.H.; Bagas, L.; Wang, F.X. Isotopic analysis of the super-large Shuangjianzishan Pb-Zn-Ag deposit in Inner Mongolia, China: Constraints on magmatism, metallogenesis, and tectonic setting. *Ore Geol. Rev.* **2016**, *75*, 252–267. [CrossRef]

60. Zhu, X.Q.; Zhang, Q.; He, Y.L.; Zhu, C.H.; Huang, Y. Hydrothermal source rocks of the Meng'entaolegai Ag-Pb-Zn deposit in the granite batholith, Inner Mongolia, China: Constrained by isotopic geochemistry. *Geochem. J.* **2006**, *40*, 265–275. [CrossRef]

61. Zeng, Q.D.; Liu, J.M.; Zhang, Z.L. Re–Os geochronology of porphyry molybdenum deposit in south segment of Da Hinggan Mountains, Northeast China. *J. Earth Sci.* **2010**, *21*, 392–401. [CrossRef]

62. Ruan, B.X.; Lu, X.B.; Liu, S.T.; Yang, W. Genesis of Bianjiadayuan Pb-Zn-Ag deposit in Inner Mongolia: Constraints from U–Pb dating of zircon and multi-isotope geochemistry. *Miner. Depos.* **2013**, *32*, 501–514. (In Chinese)

63. Zhao, Y.M.; Wang, D.W.; Zhang, D.Q.; Fu, X.Z.; Bao, X.P.; Ai, Y.F. *Ore-Controlling Factors and Ore-Prospecting Models for Copper-Polymetallic Deposits in Southeastern Inner Mongolia*; Seismological Publishing House: Beijing, China, 1994; pp. 1–234. (In Chinese)

64. Zeng, Q.D.; Liu, J.M.; Zhang, Z.L.; Jia, C.S.; Yu, C.M.; Ye, J.; Liu, H.T. Geology and lead-isotope study of the Baiyinnuoer Zn-Pb-Ag deposit, south segment of the Da Hinggan Mountains, Northeastern China. *Resour. Geol.* **2009**, *59*, 170–180. [CrossRef]

65. Jiang, S.H.; Nie, F.J.; Bai, D.M.; Liu, Y.F.; Liu, Y. Geochronology evidence for Indosinian mineralization in Baiyinnuoer Pb-Zn deposit of Inner Mongolia. *Mine. Depo.* **2011**, *30*, 787–798. (In Chinese)

66. Wang, M.Y.; He, L. Re-Os dating of molybdenites from Chamuhan W-Mo deposit, Inner Mongolia and its geological implications. *Geotecton. Metallog.* **2013**, *37*, 49–56. (In Chinese)

67. Shu, Q.H.; Lai, Y.; Wang, C.; Xu, J.J.; Sun, Y. Geochronology, geochemistry and Sr-Nd-Hf isotopes of the Haisugou porphyry Mo deposit, northeast China, and their geological significance. *J. Asian Earth Sci.* **2013**, *79*, 777–791. [CrossRef]

68. Zhai, D.G.; Liu, J.J.; Wang, J.P.; Yang, Y.Q.; Zhang, H.Y.; Wang, X.L.; Zhang, Q.B.; Wang, G.W.; Liu, Z.J. Zircon U-Pb and molybdenite Re-Os geochronology, and whole-rock geochemistry of the Hashitu molybdenum deposit and host granitoids, Inner Mongolia, NE China. *J. Asian Earth Sci.* **2013**, *79*, 144–160. [CrossRef]

69. Zeng, Q.D.; Yang, J.H.; Zhang, Z.L.; Liu, J.M.; Duan, X.X. Petrogenesis of the Yangchang Mo-bearing granite in the Xilamulun metallogenic belt, NE China: Geochemistry, zircon U-Pb ages and Sr-Nd-Pb isotopes. *Geol. Mag.* **2013**, *49*, 1–14. [CrossRef]

70. Pan, X.F.; Guo, L.J.; Wang, S.; Xue, H.M.; Hou, Z.Q.; Tong, Y. Laser microprobe Ar -Ar dating of biotite from the Weilasituo Cu-Zn polymetallic deposit in Inner Mongolia. *Acta Petrol. Mineralog.* **2009**, *28*, 473–479. (In Chinese)

71. Chang, Y.; Lai, Y. Study of characteristics of ore-forming fluid and chronology in the Yindu Ag-Pb-Zn polymetallic ore deposit, Inner Mongolia. *Acta Scientiarum Naturalium Universitatis Pekinensis* **2010**, *46*, 581–593. (In Chinese)

72. Huang, D.H.; Du, A.D.; Wu, C.Y.; Liu, L.S.; Sun, Y.L.; Zou, X.Q. Metallochronology of molybdenum (–copper) deposits in the north China Platform: Re–Os age of molybdenite and its geological significance. *Miner. Depos.* **1996**, *15*, 365–373. (In Chinese)

73. Wang, F.X.; Sun, Z.J.; Pei, R.F.; Liu, Y.F.; Liu, C.H.; Jiang, S.H. The Geologic features and Genesis of Shuangjianzishan Ag-polymetallic Deposit, Balinzuo Qi, Inner Mongolia. *Geol. Rev.* **2016**, *62*, 1241–1256. (In Chinese)

74. Zhang, J.H.; Ge, W.C.; Wu, F.Y.; Wilde, S.A.; Yang, J.H.; Liu, X.M. Large-scale early Cretaceous volcanic events in the northern Great Xing'an Range, northeastern China. *Lithos* **2008**, *102*, 138–157. [CrossRef]

75. Wang, T.; Guo, L.; Zhang, L.; Yang, Q.; Zhang, J.; Tong, Y.; Ye, K. Timing and evolution of Jurassic-Cretaceous granitoid magmatism in the Mongol-Okhotsk belt and adjacent areas, NE Asia: Implications for transition from contractional crustal thickening to extensional thinning and geodynamic settings. *J. Asian Earth Sci.* **2015**, *97*, 365–392. [CrossRef]

76. Chen, Z.G.; Zhang, L.C.; Wan, B.; Wu, H.Y.; Cleven, N. Geochronology and geochemistry of the Wunugetushan porphyry Cu-Mo deposit in NE China, and their geological significance. *Ore. Geol. Rev.* **2011**, *43*, 92–105. [CrossRef]

77. Liu, J.; Mao, J.W.; Wu, G.; Wang, F.; Luo, D.F.; Hu, Y.Q. Zircon U-Pb and molybdenite Re-Os dating of the Chalukou porphyry Mo deposit in the northern Great Xing'an Range, China, and its geological significance. *J. Asian Earth Sci.* **2014**, *79*, 696–709. [CrossRef]

78. Fang, J.Q.; Nie, F.J.; Zhang, K. Re-Os isotopic dating on molybdenite separates and its geological significance from the Yaojiagou molybdenum deposit, Liaoning Province. *Acta Petrol. Sin.* **2012**, *28*, 372–378. (In Chinese)

79. Dai, J.Z.; Mao, J.W.; Zhao, C.S.; Xie, G.Q.; Yang, F.Q.; Wang, Y.T. New U-Pb and Re-Os age data and the geodynamic setting of the Xiaojiayingzi Mo (Fe) deposit, Western Liaoning province, Northeastern China. *Ore Geol. Rev.* **2009**, *35*, 235–244. [CrossRef]

80. Zhang, Z.Z.; Wu, C.Z.; Gu, L.X.; Feng, H.; Zheng, Y.C.; Huang, J.H.; Li, J.; Sun, Y.L. Molybdenite Re–Os dating of Xintaimen molybdenum deposit in Yanshan–Liaoning Metallogenic Belt, North China. *Miner. Depos.* **2009**, *28*, 313–320. (In Chinese)

81. Shu, Q.H.; Jiang, L.; Lai, Y.; Lu, Y.H. Geochronology and fluid inclusion study of the Aolunhua porphyry Cu-Mo deposit in Arhorqin Area, Inner Mongolia. *Acta Petrol. Sin.* **2009**, *25*, 2601–2614. (In Chinese)

82. Tao, J.X.; Chen, Z.H.; Luo, Z.Z.; Xu, L.Q.; Hao, X.Y.; Cui, L.W. The Re-Os isotopic dating of molybdenite from the Wulandele molybdenum–copper polymetallic deposit in Sonid Zuoqi of Inner Mongolia and its geological significance. *Rock Miner. Anal.* **2009**, *28*, 249–253. (In Chinese)

83. Wu, G.; Li, X.Z.; Xu, L.Q.; Wang, G.R.; Liu, J.; Zhang, T.; Quan, Z.X.; Wu, H.; Li, T.G.; Chen, Y.C. Age, geochemistry, and Sr-Nd-Hf-Pb isotopes of the Caosiyao porphyry Mo deposit in Inner Mongolia. China. *Ore Geol. Rev.* **2017**, *81*, 706–727. [CrossRef]

84. Yan, C.; Sun, Y.; Lai, Y. LA-ICP-MS zircon U-Pb and molybdenite Re-Os isotope ages and metallogenic geodynamic setting of Banlashan Mo deposit, Inner Mongolia. *Mine. Depo.* **2011**, *30*, 616–634. (In Chinese)

85. Zhang, C. Geology, Geochemistry and Metallogenic Evolution of Molybdenum Deposits in the Great Hinggan Range. Ph.D. Thesis, Peking University, Beijing, China, 2015. (In Chinese).

86. Zhang, C.; Li, N. Geochronology and zircon Hf isotope geochemistry of the granites at the giant Chalukou Mo deposit. NE China: Implication for tectonic setting. *Ore Geol. Rev.* **2017**, *81*, 780–793. [CrossRef]

87. Şengör, A.M.C.; Natal'in, B.A.; Burtman, V.S. Evolution of the Altaid tectonic collage and Palaeozoic crustal growth in Eurasia. *Nature* **1993**, *364*, 299–307. [CrossRef]

88. Xiao, W.J.; Windley, B.F.; Hao, J.; Zhai, M.G. Accretion leading to collision and the Permian Solonker suture, Inner Mongolia, China: Termination of the central Asian orogenic belt. *Tectonics* **2003**, *22*, 1069. [CrossRef]

89. Xiao, W.J.; Windley, B.F.; Huang, B.C.; Han, C.M.; Yuan, C.; Chen, H.L.; Sun, M.; Sun, S.; Li, J.L. End-Permian to mid-Triassic termination of the accretionary processes of the southern Altaids: Implications for the geodynamic evolution, Phanerozoic continental growth, and metallogeny of Central Asia. *Int. J. Earth Sci.* **2009**, *98*, 1189–1217. [CrossRef]

90. Wu, F.Y.; Sun, D.Y.; Ge, W.C.; Zhang, Y.B.; Grant, M.L.; Wilde, S.A.; Jahn, B.-M. Geochronology of the Phanerozoic granitoids in northeastern China. *J. Asian Earth Sci.* **2011**, *41*, 1–30. [CrossRef]

91. Wu, G.; Chen, Y.C.; Chen, Y.J.; Zeng, Q.T. Zircon U-Pb ages of the metamorphic supracrustal rocks of the Xinghuadukou Group and granitic complexes in the Argun massif of the northern Great Hinggan Range, NE China, and their tectonic implications. *J. Asian Earth Sci.* **2012**, *49*, 214–233. [CrossRef]

92. Niu, S.Y.; Nie, F.J.; Sun, A.Q.; Jiang, S.H. Mantle branch structure and silver polymetallic mineralization in the Da Hinggan Mountains, Inner Mongolia. *J. Jilin Univ.* **2011**, *41*, 1944–1958. (In Chinese)

93. Pirajno, F.; Zhou, T.F. Intracontinental porphyry and porphyry–skarn mineral systems in eastern China: Scrutiny of a special case "Made-in-China". *Econ. Geol.* **2015**, *110*, 603–629. [CrossRef]

94. Batchelor, R.A.; Bowden, P. Petrogenetic interpretation of granitoid rock series using multicationic parameters. *Chem. Geol.* **1985**, *48*, 43–55. [CrossRef]

95. Stern, R.J.; Gottfried, D. Discussion of the Paper 'Late Pan-African Magmatism and Crustal Development in Northeastern Egypt'. *Geol. J.* **1989**, *24*, 371–374. [CrossRef]

96. Pearce, J. Sources and settings of granitic rocks. *Episodes* **1996**, *19*, 120–125.

97. Kravchinsky, V.A.; Cogne, J.P.; Harbert, W.P.; Kuzmin, M.I. Evolution of the Mongol-Okhotsk Ocean as constrained by new palaeomagnetic data from the Mongol-Okhotsk suture zone, Siberia. *Geophys. J. Int.* **2002**, *148*, 34–57. [CrossRef]

98. Zorin, Y.A. Geodynamics of the western part of the Mongolia-Okhotsk collisional belt, Trans-Baikal region (Russia) and Mongolia. *Tectonophysics* **1999**, *306*, 33–56. [CrossRef]

99. Cogné, J.P.; Kravchinsky, V.A.; Halim, N.; Hankard, F. Late Jurassic-Early Cretaceous closure of the Mongol-Okhotsk Ocean demonstrated by new Mesozoic palaeomagnetic results from the Trans-Baïkal area (SE Siberia). *Geophys. J. Int.* **2005**, *163*, 813–832. [CrossRef]

100. Tomurtogoo, O.; Windley, B.F.; Kroner, A.; Badarch, G.; Liu, D.Y. Zircon age and occurrence of the Adaatsag ophiolite and Muron shear zone, central Mongolia: Constraints on the evolution of the Mongol-Okhotsk ocean, suture and orogeny. *J. Geol. Soc. Lond.* **2005**, *162*, 125–134. [CrossRef]

101. Deng, J.; Wang, Q.F. Gold mineralization in China: Metallogenic provinces, deposit types and tectonic framework. *Gondwana Res.* **2016**, *36*, 219–274. [CrossRef]

102. Yang, L.Q.; Deng, J.; Wang, Z.L.; Zhang, L.; Guo, L.N.; Song, M.C.; Zheng, X.L. Mesozoic gold metallogenic system of the Jiaodong gold province, eastern China. *Acta Geosci. Sin.* **2014**, *30*, 2447–2467. (In Chinese)

103. Dong, S.W.; Zhang, Y.Q.; Long, C.X.; Yang, Z.Y.; Ji, Q.; Wang, T.; Hu, J.M.; Chen, X.H. Jurassic tectonic revolution in China and new interpretation of the Yanshan movement. *Acta Geol. Sin.* **2007**, *81*, 1449–1461. (In Chinese)

104. Audétat, A. Source and Evolution of Molybdenum in the Porphyry Mo(-Nb) Deposit at Cave Peak, Texas. *J. Petrol.* **2010**, *51*, 1739–1760. [CrossRef]

105. Audétat, A.; Dolejs, D.; Lowenstern, J.B. Molybdenite saturation in silicic magmas: Occurrence and petrological implications. *Petrology* **2011**, *52*, 891–904. [CrossRef]

106. Christiansen, E.H.; Keith, J.D. Trace element systematics in silicic magmas: A metallogenic perspective. In *Trace Element Geochemistry of Volcanic Rocks: Applications for Massive Sulphide Exploration*; Geological Association of Canada, Short Course Notes; Wyman, D.A., Ed.; Geological Assn of Canada: St. John's, NL, Canada, 1996; pp. 115–151.

107. Liu, Y.F.; Nie, F.J.; Jiang, S.H.; Zhong, X.I.; Zhang, Z.G.; Xiao, W. Ore-forming granites from Chaganhua molybdenum deposit, central Inner Mongolia, China: Geochemistry, geochronology and petrogenesis. *Acta Petrol. Sin.* **2012**, *28*, 409–420. (In Chinese)

108. Lerchbaumer, L.; Audétat, A. The metal content of silicate melts and aqueous fluids in subeconomically Mo mineralized granites: Implications for porphyry Mo genesis. *Econ. Geol.* **2013**, *108*, 987–1013. [CrossRef]

109. Harris, A.C.; Dunlap, W.J.; Reiners, P.W.; Allen, C.M.; Cooke, D.R.; White, N.C. Multimillion-year thermal history of a porphyry copper deposit: Application of U-Pb, ^{40}Ar/^{39}Ar and (U-Th)/He chronometers, Bajo de la alumbrera copper-gold deposit, Argentina. *Miner. Depos.* **2008**, *43*, 295–314. [CrossRef]

Article

Geological, Geochronological, and Geochemical Insights into the Formation of the Giant Pulang Porphyry Cu (–Mo–Au) Deposit in Northwestern Yunnan Province, SW China

Qun Yang [1], Yun-Sheng Ren [1,2,*], Sheng-Bo Chen [3], Guo-Liang Zhang [4], Qing-Hong Zeng [3], Yu-Jie Hao [2], Jing-Mou Li [1], Zhong-Jie Yang [5], Xin-Hao Sun [1] and Zhen-Ming Sun [1]

[1] College of Earth Sciences, Jilin University, Changchun 130061, China; yangq2009jlu@163.com (Q.Y.); Lijmmy2019@163.com (J.-M.L.); xhsun17@mails.jlu.edu.cn (X.-H.S.); sunzhenming.9040@163.com (Z.-M.S.)
[2] Key Laboratory of Mineral Resources Evaluation in Northeast Asia, Ministry of Natural Resources of China, Changchun 130026, China; haoyujie@jlu.edu.cn
[3] College of Geo-Exploration Science and Technology, Jilin University, Changchun 130026, China; chensb@jlu.edu.cn (S.-B.C.); zengqh17@mails.jlu.edu.cn (Q.-H.Z.)
[4] Key Laboratory of Satellite Remote Sensing Application Technology of Jilin Province, Chang Guang Satellite Technology Co., Ltd., Changchun 130102, China; zhanggl0207@163.com
[5] Geological Survey Institute of Liaoning Porvice, Shenyang 110031, China; yangzj870227@163.com
* Correspondence: renys@jlu.edu.cn; Tel: +86-0431-88502708

Received: 7 March 2019; Accepted: 19 March 2019; Published: 21 March 2019

Abstract: The giant Pulang porphyry Cu (–Mo–Au) deposit in Northwestern Yunnan Province, China, is located in the southern part of the Triassic Yidun Arc. The Cu orebodies are mainly hosted in quartz monzonite porphyry (QMP) intruding quartz diorite porphyry (QDP) and cut by granodiorite porphyry (GP). New LA-ICP-MS zircon U–Pb ages indicate that QDP (227 ± 2 Ma), QMP (218 ± 1 Ma, 219 ± 1 Ma), and GP (209 ± 1 Ma) are significantly different in age; however, the molybdenite Re–Os isochron age (218 ± 2 Ma) indicates a close temporal and genetic relationship between Cu mineralization and QMP. Pulang porphyry intrusions are enriched in light rare-earth elements (LREEs) and large ion lithophile elements (LILEs), and depleted in heavy rare-earth elements (HREEs) and high field-strength elements (HFSEs), with moderately negative Eu anomalies. They are high in SiO_2, Al_2O_3, Sr, Na_2O/K_2O, Mg#, and Sr/Y, but low in Y, and Yb, suggesting a geochemical affinity to high-silica (HSA) adakitic rocks. These features are used to infer that the Pulang HSA porphyry intrusions were derived from the partial melting of a basaltic oceanic-slab. These magmas reacted with peridotite during their ascent through the mantle wedge. This is interpreted to indicate that the Pulang Cu deposit and associated magmatism can be linked to the synchronous westward subduction of the Ganzi–Litang oceanic lithosphere, which has been established as Late Triassic.

Keywords: Pulang porphyry Cu (–Mo–Au) deposit; LA-ICP-MS zircon U–Pb dating; molybdenite Re–Os dating; high-silica adakitic rocks (HSA); Northwestern Yunnan Province

---

1. Introduction

Two famous porphyry mineralization systems in Western Yunnan Province, Southwestern (SW) China, mainly compose the Yidun Arc metallogenic belt in the north and the Jinshajiang–Red River metallogenic belt in the south (Figure 1A). The southern Yidun Arc (Figure 1B), namely, the Zhongdian Arc, is important for its Cu resources. Numerous porphyry and skarn Cu deposits have been discovered in the area, and representative deposits include Pulang, Xuejiping, Lannitang, Songnuo, Langdu, and Chundu (Figure 1C). The geological characteristics and geochronological data indicate that these

Cu deposits have a temporal and spatial relationship with widespread Mesozoic intermediate-acid intrusions [1–9]. Recently, extensive isotopic geochronology studies have been carried out on these Cu deposits. Cu mineralization ages in the southern Yidun Arc have now been constrained to the Late Triassic period (210–230 Ma) by zircon U–Pb and molybdenite Re–Os dating [5–13].

The Pulang deposit, located ~36 km northeast of the Shangri-La County in Northwestern Yunnan Province, is one of the large-scale porphyry Cu (–Mo–Au) deposits in China. It is located in the Sanjiang Tethyan metallogenic domain in Yunnan Province, SW China, and contains reserves of 4.18 Mt Cu. The largest KT1 orebody contains reserves of 1.57 Mt Cu @ 0.52%, 54t Au @ 0.18g/t, and 84,800 t Mo @ 0.01% [14]. In the past ten years, because of its giant size, the deposit has attracted much attention, especially regarding its ore geology [15–17], geochronology, and petro-geochemistry [18–23]. Despite the existing geochronological and geochemical studies on the Pulang Cu deposit, there remains an insufficient understanding on the accurate diagenesis and mineralization ages, the relationships between Cu mineralization and associated magmatism, and the tectonic setting. To start with, the mineralization age of Pulang deposit is still a matter of debate, as the limited molybdenite Re–Os age ranging from 208 ± 15 Ma to 213 ± 4 Ma [3,18,24] is a substantial margin of error. Next, the published geochronological data for the quartz monzonite porphyry (QMP) and quartz diorite porphyry (QDP) in the Pulang deposit are in a broad range of 211–231 Ma and 214–228 Ma, respectively [4–6,10,18,19,22,23]. Some of these ages are younger than the Re–Os isochron ages of molybdenite [4,5,14,19], whereas some are ~15 Ma older than the Re–Os isochron ages of molybdenite [22,25]. Generally, a porphyry-type hydrothermal Cu mineralization system lasts for less than 10 Ma [26]. Therefore, most of these ages are not well explained in relation to the observed Cu mineralization. In addition, the relationship between Cu mineralization and associated magmatism remains unclear and controversial as a result of the ~20 Ma age interval (210–230 Ma) for the Pulang porphyry intrusions. Some scholars have considered the QMP to be the ore-causative intrusion [18,23], whereas others have deemed that all of the Pulang porphyry intrusions contributed to Cu mineralization [14,27,28]. Finally, there are also divergences in the understanding of the petrogenesis of these ore-bearing porphyries; some scholars have proposed that the Pulang porphyries associated with Cu mineralization are related to the partial melting of the Ganzi-Litang oceanic plate [10,13], while others have considered the formation of these porphyries to be related to the partial melting of enriched mantle and the assimilation and fractional crystallization (AFC) of mantle-derived basalt magma [14,27].

To address these issues, this study presents new zircon U–Pb geochronology and whole-rock major-trace elemental data for Pulang porphyry intrusions and molybdenite Re–Os ages for different types of ore. The aim of the study is to provide precise diagenesis and mineralization ages, explain the relationship between associated intrusions and Cu mineralization, evaluate the contribution of Pulang porphyry intrusions to the Cu mineralization, and discuss the geodynamic setting of the intrusions and associated Cu mineralization in the Pulang Cu (–Mo–Au) deposit.

2. Geological Background

The NNW-trending Yidun terrane is located between the Garze–Litang Suture and the Jinshajiang Suture (Figure 1B), which represent closure zones of two distinct Paleo-Tethys oceans during the Middle–Late Triassic and Late Paleozoic, respectively [29–31]. The Yidun Terrane is divided into the Zhongza Massif in the west and the Yidun Arc in the east by the NS-trending Xiangchang–Geza fault [32,33]. The subduction of the Garze–Litang oceanic plate beneath the Zhongza Block resulted in the accretion of the Zhongza Block to the Yangtze Craton via the Yidun Arc during the Late Triassic time. It is inferred that the Yidun Arc developed on the basement that was inherited from the Yangtze Craton [14].

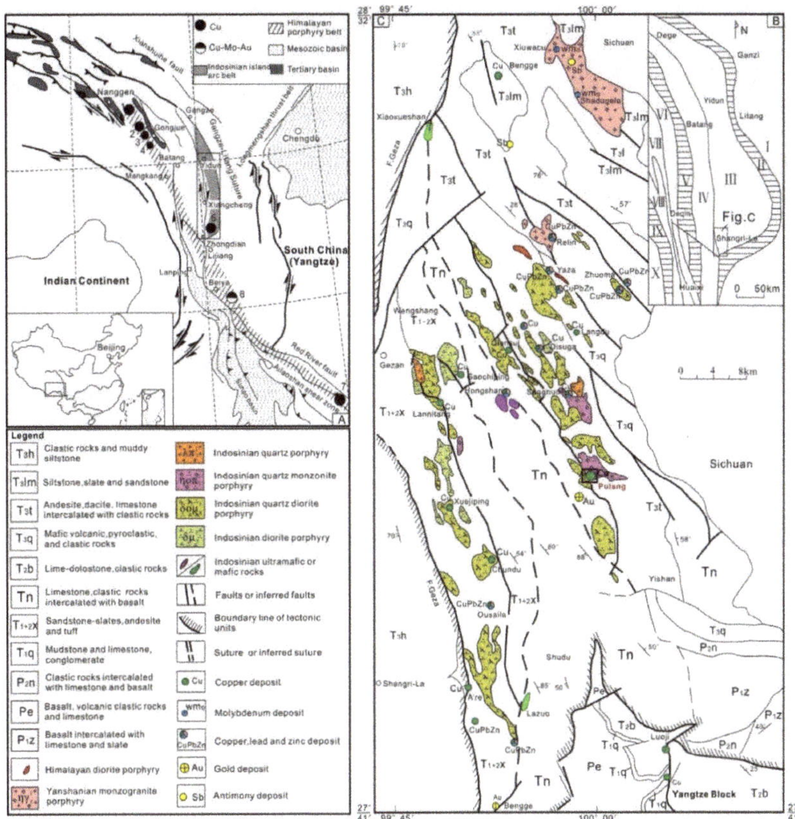

Figure 1. (**A**) Regional tectonic map of the Zhongdian island-arc zone (based on Li et al. [20]), showing the Indosinian and Himalayan porphyry Cu belts and deposits numbered as follows: 1 = Yulong; 2 = Malasongsuo; 3 = Duoxiasongsuo; 4 = Mamupu; 5 = Pulang; 6 = Machangqing; 7 = Tongchang. (**B**) Tectonic location map and (**C**) regional geology map of the Pulang Cu deposit (based on Wang et al. [34]). I = Yangtze block; II = Garze–Litang suture zone; III = Yidun arc belt; IV = Zhongza block; V = Jinsahjiang suture zone; VI = Jomda–Weixi magmatic arc; VII = Changdu–Lanping block; VIII = Sandashan–Jinghong volcanic arc; IX = Lancangjiang suture zone; X = Baoshan block.

The Yidun Arc is almost entirely overlain by the Triassic Yidun Group, which consists of slate, sandstone interbedded with minor limestone and flysch–volcanic succession, and intrusive rocks, which were produced by the westward subduction of the Garze–Litang oceanic plate (Figure 1C) [35,36]. Structures in the area are dominated by the NW–SE-trending or N–S-trending regional faults associated with the subduction, which channeled the NW–SE-trending Mesozoic intrusions (Figure 1C). The subduction-related magmatism resulted in abundant calc-alkaline arc-related volcanic rocks (such as andesite, rhyolite, and basalt), which developed at ~230 Ma in the northern segment of the Yidun Arc, and the voluminous Late Triassic (206–237 Ma) intermediate to felsic porphyritic rocks (such as quartz porphyry, quartz monzonite porphyry, quartz diorite porphyry, and diorite porphyry) in the southern segment of the Yidun Arc (Zhongdian Arc) [20,37].

Volcanic-associated massive sulfide (VMS) deposits, such as the Gayiqiong Zn–Pb–Cu, Shengmolong Zn–Pb, and Gacun giant Zn–Pb–Ag–Cu deposits, are distributed in the northern segment of the Yidun Arc, whereas abundant economic porphyry Cu deposits, such as the well-known Pulang, Xuejiping, Lannitang, and Chundu deposits, are mainly distributed in the southern segment of the

Yidun Arc (Figure 1C) [20,33]. The different types of deposits in the northern and southern segment of the Yidun Arc are probably attributable to the subduction angle of the Garze–Litang oceanic plate in the Middle–Late Triassic [20,33]. As a result of a steep subduction angle, the northern segment of the Yidun Arc was tensional, which induced magma spill over onto the surface and caused extensive volcanic rocks to accompany the formation of some volcanic massive sulfide-type deposits. By contrast, owing to a gentle angle of subduction, the southern segment of the Yidun Arc was compressional, which caused the magma to remain below the surface and formed hypabyssal porphyritic intrusions related to porphyry-type deposits.

3. Deposit Geology

The Pulang porphyry Cu (–Mo–Au) ore district (Location: 28°02′19″ N, 99°59′23″ E) is situated ~36 km northeast of the Shangri-La (Zhongdian) County in Northwestern Yunnan Province (Figure 1C). It contains the largest ore block with ~96% of the total ore reserves in the south and smaller ore blocks in the north and east. Besides minor Quaternary strata, the ore district mainly consists of the Late Triassic Tumugou Formation and large-scale porphyry intrusions over an area of 8.9 km^2 (Figure 2A). The Tumugou Formation consists of interlayered slates and sandstones, minor limestones, and volcanic rocks. NW-trending Heishuitang and NEE-trending Quanganlida faults controlled the porphyry intrusions and the major Cu orebodies [28].

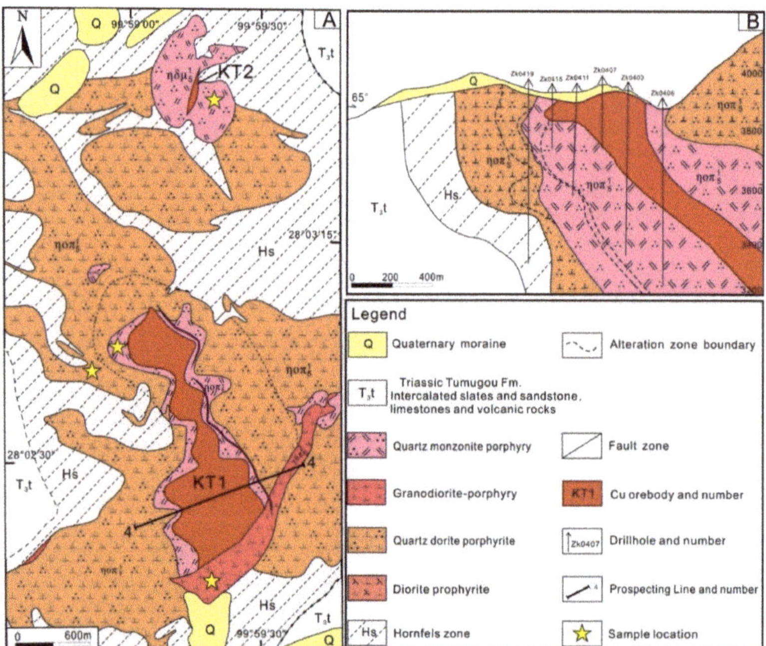

Figure 2. (**A**) Geological map of the Pulang porphyry Cu (–Mo–Au) deposit showing the biggest orebody KT1 and KT2 (based on Wang et al. [14]). (**B**) Geological section of the No. 4 exploration line in the Pulang porphyry Cu (–Mo–Au) deposit (based on Li et al. [20]).

The intrusive units in Pulang occurred mainly as NW-trending porphyry intrusions that can be grouped into three types on the basis of their cross-cutting relationship and petrographic characteristics. The quartz diorite porphyry (QDP) forms the largest intrusion and is widespread in the ore district. The QMP is the most important ore-hosting intrusion, and it trends mainly NNW–SSE within the QDP in the southern part of the ore district (Figure 2A,B). The NE–SW-trending GP is also distributed within the QDP, and it cuts the earlier QMP and the Cu orebody (Figure 2A). The QDP contains minor sulfides, such as chalcopyrite, pyrite, and molybdenite. Petrographic characteristics of the three types of porphyry intrusions are listed in Table 1 and shown in Figure 3.

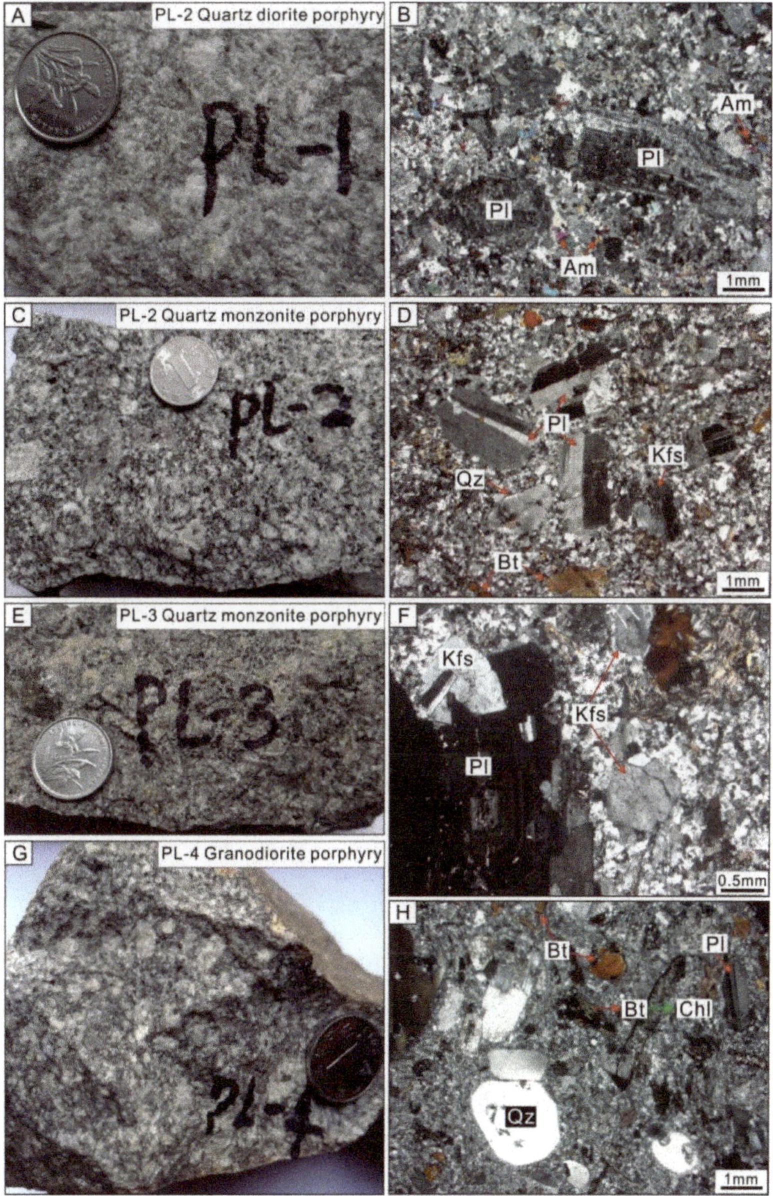

Figure 3. Hand specimen photographs and thin section photomicrographs of the porphyry intrusions in the Pulang ore district. (**A**,**B**) Quartz diorite porphyry (B under cross-polarized light). (**C–F**) Quartz monzonite porphyry (D and F under cross-polarized light). (**G**,**H**) Granodiorite porphyry (H under cross-polarized light). Abbreviations: Qz: quartz; Pl: Plagioclase; Kfs: K-feldspar; Bt: Biotite; Am: Amphibole; Chl: chlorite.

Table 1. Characteristics of the porphyry intrusions in the Pulang ore district.

Sample No.	Lithology	Location	Texture/Structure	Phenocrysts	Matrix
PL-1	Quartz diorite porphyry	N: 28°02′49″ E: 99°58′51″	Porphyritic texture, massive structure	Phenocrysts account for 20% of the rock mass and consist of plagioclase (~15%, 0.5–3.2 mm), amphibole (~3%, 0.3–1.8 mm), and quartz (~2%, 0.4–2.3 mm).	Matrix is primarily a fine-grained texture and is dominated by plagioclase with minor amphibole, K-feldspar, and quartz.
PL-2	Quartz monzonite porphyry	N: 28°03′44″ E: 99°59′19″	Porphyritic texture, massive structure	Phenocrysts account for 20% of the rock mass and consist of plagioclase (~9%, 0.6–3.8 mm), K-feldspar (~7%, 0.4–3.4 mm), quartz (~3%, 0.5–2.8 mm), and biotite (~1%, 0.2–1.6 mm).	Matrix shows a fine-grained texture and is composed of plagioclase and K-feldspar with minor quartz and biotite.
PL-3	Quartz monzonite porphyry	N: 28°02′54″ E: 99°58′57″	Porphyritic texture, massive structure	Phenocrysts account for 20% of the rock mass and consist of plagioclase (~9%, 0.6–3.8 mm), K-feldspar (~7%, 0.4–3.4 mm), quartz (~3%, 0.5–2.8 mm), and biotite (~1%, 0.2–1.7 mm).	Matrix shows a fine-grained texture and is composed of plagioclase and K-feldspar with minor quartz and biotite.
PL-4	Granodiorite porphyry	N: 28°02′07″ E: 99°58′18″	Porphyritic texture, massive structure	Phenocrysts account for 20% of the rock mass and consist of plagioclase (~8%, 0.4–2.3 mm), quartz (~5%, 0.5–2.6 mm), biotite (~3%, 0.3–1.4 mm), amphibole (~2%, 0.2–1.2 mm), and K-feldspar (~2%, 0.5–2.0 mm).	Matrix shows a fine-grained texture and is composed of plagioclase and quartz with minor biotite and amphibole.

Six Cu orebodies were explored in the initial mining area, and the "multi-nodal" KT1 that mainly occurs in the interior of the QMP is the largest orebody (Figure 2) controlled by the Heishuitang fault. It is about 1600 m in length and has a width of 360–600 m in the south and 120–300 m in the north. The thickness of this orebody varies from 52 m to 700 m [20]. The Cu grade varies from 0.20% to 3.74% with an average grade of 0.52% and a grade variation coefficient of 68.69%. From the center to the outer edge, the thickness of the orebody decreases, as does the grade of Cu and other associated metals (0.001% to 0.030% Mo, average 0.010%; 0.06 to 0.87 g/t Au, average 0.18 g/t; and 0.34 to 3.93 g/t Ag, average 1.27 g/t). Ore samples in the Pulang deposit display vein (Figure 4A), veinlet (Figure 4B), stockwork, disseminated, dense-disseminated (Figure 4C), and taxitic (Figure 4D) structures. The metallic minerals are dominated by chalcopyrite (Figure 4E), pyrite (Figure 4F–H), pyrrhotite, bornite, molybdenite (Figure 4I), and minor magnetite. Metal minerals display xenomorphic granular (Figure 4E), idiomorphic–hypidiomorphic granular (Figure 4F), filling metasomatic (Figure 4G), cataclastic (Figure 4G), and scaly textures (Figure 4I). The non-metallic minerals include quartz, plagioclase, K-feldspar, biotite, amphibole, chlorite, epidote, sericite, and calcite.

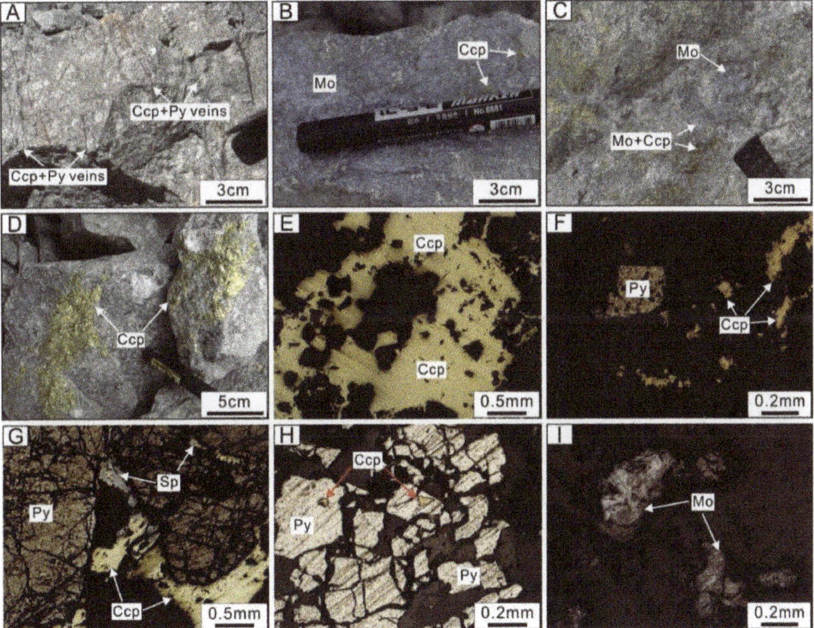

Figure 4. Photographs of ore fabrics and mineral assemblages of the Pulang porphyry Cu (–Mo–Au) deposit. (**A**) Chalcopyrite–pyrite veins and veinlets are distributed in the altered quartz monzonite porphyry. (**B**) Thin-film molybdenite veinlet. (**C**) Chalcopyrite intergrown with molybdenite that occurs as dense-disseminated in the altered quartz monzonite porphyry. (**D**) Nodular chalcopyrite. (**E**) Xenomorphic granular chalcopyrite. (**F**) Idiomorphic granular pyrite and xenomorphic granular chalcopyrite. (**G**) Chalcopyrite replaced by sphalerite, and they were filled into the fissures of pyrite. (**H**) Pyrite showing a cataclastic texture; and chalcopyrite filling into the holes of pyrite. (**I**) Scaly molybdenite. Abbreviations: Ccp: chalcopyrite; Py: pyrite; Mo: molybdenite; Sp: sphalerite.

The hydrothermal alteration in the Pulang deposit mainly includes silicification, K-feldspar alteration, sericitization, chloritization, epidotization, and carbonation. A typical porphyry alteration zoning occurs in this deposit from the center to the outer edge zone. An inner alteration zone is located at the core of the QMP and consists of extensive hydrothermal K-feldspar and quartz assemblages, which resulted from alteration of the plagioclase in the wall rock. Subsequently, the primary

K-feldspar, plagioclase, and biotite were altered to silicification and sericitization assemblages, with minor K-feldspathization forming the silicification–sericitization alteration zone, the latter of which is accompanied by economic Cu orebodies. The outer alteration zone (propylitization zone) is characterized by chloritization, epidotization with minor silicification, and carbonation.

On the basis of field research, cross-cutting relationships, mineral paragenetic associations, and associated hydrothermal alteration, the mineralization processes of the Pulang deposit can be divided into the following four stages: quartz + K-feldspar ± pyrite veins (Stage I); quartz + molybdenite ± chalcopyrite ± pyrite veins (Stage II); quartz + chalcopyrite + pyrite ± pyrrhotite ± molybdenite veins (Stage III); and quartz + calcite ± pyrite veins (Stage IV). Stage I is characterized by silicification and K-feldspathization with minor idiomorphic–hypidiomorphic disseminated pyrite mineralization distributed in the quartz–K-feldspar veins and host rocks. Stage II is characterized by the occurrence of molybdenite, with minor pyrite and chalcopyrite assemblages associated with silicification. The zone of Stage II silicification affected rocks that were slightly farther from the center of the orebody and around the margins of the zone of earlier K-feldspathization–silicification (Stage I). Stage III is the major Cu mineralization stage and is characterized by abundant precipitation of sulfide minerals accompanied by strong silicification and sericitization. Sulfides formed in this stage are dominated by chalcopyrite and pyrite (>85 vol.%), followed by bornite, pyrrhotite, and molybdenite. These sulfides generally replaced the earlier-formed pyrite and molybdenite, or filled in fractures of the earlier-formed euhedral pyrite (Figure 4G). The associated alteration including silicification and sericitization, as well as minor chloritization and epidotization, is distributed on both sides of the Stage III veins. Stage IV is the latest stage generating large numbers of quartz–calcite veins near the propylitization zone. Only minor disseminated pyrite was locally distributed in the veins and wall-rock. The abundant quartz and calcite minerals were observed in this stage.

4. Samples and Analytical Methods

4.1. LA-ICP-MS Zircon U–Pb Dating

Zircon grains were extracted from the quartz diorite porphyry, quartz monzonite porphyry, and granodiorite porphyry samples (Table 1) using magnetic and heavy liquid separation methods, and they were handpicked under a binocular microscope at the Integrity Geological Service Corporation in Langfang City, Heibei Province, China. The handpicked zircon grains were then mounted in epoxy resin and polished to expose crystal cores. All the zircon grains were examined under reflected and transmitted light with an optical microscope. Cathodoluminescence (CL) images, obtained using a JEOL scanning electron microscope (JEOL, Tokyo, Japan), were used to reveal their internal structures and to select spots for zircon U–Pb isotope analyses (Figure 5). These analysis spots were usually chosen near rims in areas with oscillatory CL response, which represents the final crystallization age of these intrusions.

Samples for LA-ICP-MS zircon U–Pb analyses were performed using a quadrupole ICP-MS (Agilent 7500c, Agilent Technologies, Santa Clara, CA, USA) coupled with an 193 nm ArF excimer laser (COMPexPro 102; Coherent, Santa Clara, CA, USA) equipped with an automatic positioning system, at the Key Laboratory of Mineral Resources Evaluation in Northeast Asia, Ministry of Natural Resources of China, Changchun, Jilin Province, China. The diameter of the laser spot was 32 μm. A zircon standard (91500) and the National Institute of Science and Technology (NIST) 610 reference standard were analyzed after each set of six unknown analyses. The external zircon standard was used to correct for isotope ratio fractionation. The NIST610 reference standard was used in calculations of element concentrations and Si was used as an internal standard. Uncertainties on isotope ratios and ages are presented as ±2σ. The detailed analytical process and data reduction methodology are described in detail by Hou et al. [38], and the isotope data were calculated using the GLITTER (Version 4.0) [39]. Concordia diagram and weighted-mean age calculations were produced using ISOPLOT (Version 3.0) [40]. Common Pb was corrected following the method of Anderson [41].

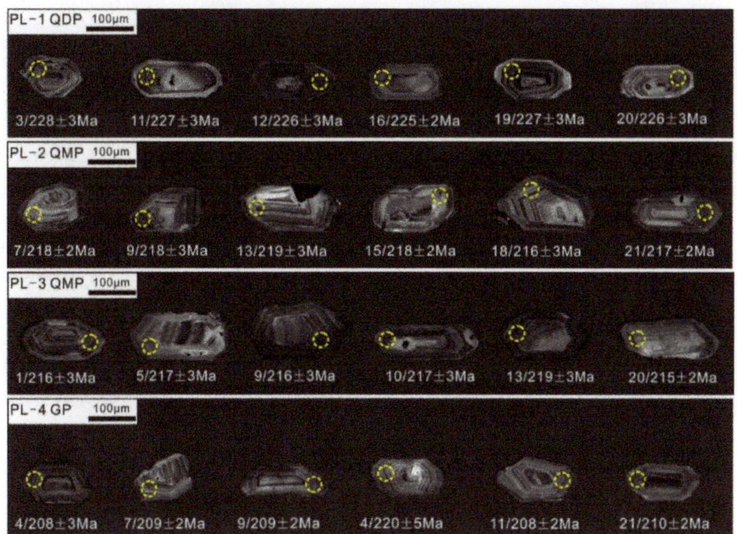

Figure 5. Cathodoluminescence (CL) images of respective zircon grains from the samples of Pulang porphyry intrusions. Yellow circles denote the locations of U–Pb dating. Abbreviations: QDP: Quartz diorite porphyry; QMP: Quartz monzonite porphyry; GP: Granodiorite porphyry.

4.2. Molybdenite Re–Os Dating

Minor disseminated molybdenite in the Pulang deposit were recognized in the QDP, and most of the molybdenite occurred as vein and dense-disseminated in the QMP. Six molybdenite samples from different ore-hosting porphyry intrusions in the Pulang Cu (–Mo–Au) deposit, including three QMP-hosted and three QDP-hosted samples, were collected for Re–Os dating. Molybdenite grains were magnetically separated and handpicked using a binocular microscope (purity > 99%). The Re–Os isotope analyses were performed at the Re–Os Isotope Laboratory, National Research Center of Geoanalysis (NRCG), Chinese Academy of Geological Sciences (CAGS), Beijing, China. Samples were dissolved and equilibrated with ^{185}Re- and ^{190}Os-enriched spikes using alkaline fusion. The separation of rhenium from matrix elements was achieved by solvent extraction and cation exchange resin chromatography, while osmium was distilled as OsO_4 from an H_2SO_4–$Ce(SO_4)_2$ solution. The Re and Os concentrations as well as their isotopic compositions were made by an inductively coupled plasma mass spectrometer (TJA X-series ICP-MS, Thermo Fisher Scientific, Waltham, MA, USA). Detailed chemical procedure of Re–Os analytical methods used in this study can be followed by Shirey and Walker [42] and Du et al. [43]. The analytical reliability was tested by analyses of the JDC standard certified reference material GBW04436. Molybdenite model ages were calculated by t = [ln (1 + ^{187}Os/^{187}Re)]/λ, where λ is the decay constant of ^{187}Re, 1.666×10^{-11} year^{-1} [44]. The Isoplot/Ex (Version 3.0) [40] was used for calculating the isochron age and drawing isochronological age diagrams. Absolute uncertainties of the Re–Os data are given at the 2σ level.

4.3. Major and Trace Element Concentrations

Following the removal of weathered surfaces, a total of 23 whole-rock samples were crushed and ground to ~200 mesh using an agate mill. Major and trace element compositions were analyzed by inductively coupled plasma optical emission spectroscopy (ICP-OES; Leeman Prodigy; precision better than ±1%) and inductively coupled plasma mass spectrometry (ICP-MS; Agilent—7500a; precision better than ±5%), respectively, at the Experiment Center, China University of Geosciences, Beijing, China.

5. Analytical Results

5.1. Zircon U–Pb Dating

Zircon U–Pb dating results of the quartz diorite porphyry (QDP, PL-1), and quartz monzonite porphyry (QMP, PL-2, PL-3) and granodiorite porphyry (GP, PL-4) in the Pulang deposit are listed in Table S1 and shown in Figure 6. All zircon grains from different types of porphyry samples are euhedral–subhedral in shape and transparent or pale brown in color, and exhibit marked oscillatory growth zoning in CL images (Figure 5). Most of them have high Th/U ratios (0.3–1.7; Table S1), indicating their magmatic origin [45–50]. For sample PL-1, the ^{206}Pb/^{238}U ages from 23 analytical spots range from 220 ± 3 Ma to 233 ± 3 Ma, with a concordant U–Pb age of 228 ± 1 Ma (2σ, MSWD = 0.30, N = 23; Figure 6A) and a weighted mean age of 227 ± 2 Ma (2σ, MSWD = 1.8, N = 23; Figure 6A). The ^{206}Pb/^{238}U ages of 21 analytical spots from sample PL-2 range from 214 ± 3 Ma to 225 ± 4 Ma, with concordant U–Pb and weighted mean ages of 218 ± 1 Ma (2σ, MSWD = 0.43, N = 21) and 218 ± 1 Ma (2σ, MSWD = 0.78, N = 21; Figure 6B), respectively. For sample PL-3, 21 analytical spots yield ^{206}Pb/^{238}U ages ranging from 215 ± 2 Ma to 224 ± 3 Ma, which give a concordant U–Pb age of 219 ± 1 Ma (2σ, MSWD = 0.25, N = 21; Figure 6C) and a weighted mean age of 219 ± 1 Ma (2σ, MSWD = 1.18, N = 21; Figure 6C). Twenty-one analytical spots from the sample PL-4 yield ^{206}Pb/^{238}U ages ranging from 204 ± 2 Ma to 211 ± 3 Ma, and yield a concordant U–Pb age of 209 ± 1 Ma (2σ, MSWD = 0.33, N = 21; Figure 6D) and a weighted mean age of 209 ± 1 Ma (2σ, MSWD = 0.98, N = 21; Figure 6D). These ages represent the emplacement ages of intrusions in the Pulang Cu (–Mo–Au) deposit.

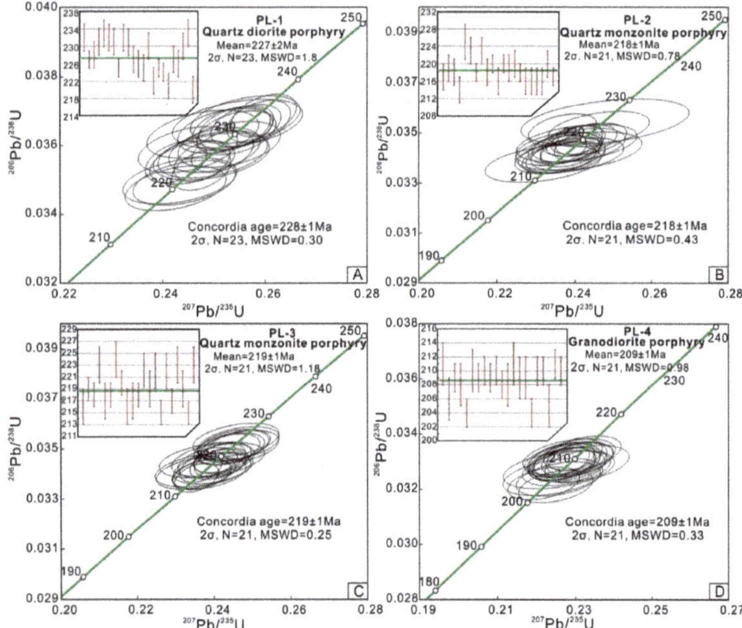

Figure 6. Zircon U–Pb concordia diagrams for samples of the quartz diorite porphyry (**A**), quartz monzonite porphyry (**B,C**), and granodiorite porphyry (**D**) in the Pulang Cu (–Mo–Au) deposit.

5.2. Molybdenite Re–Os Dating

The Re–Os isotopic compositions of six molybdenite samples obtained from the Pulang Cu deposit are presented in Table 2 and Figure 7. The Re and common Os concentrations are 101.22–933.39 µg/g and 0.0164–0.5393 ng/g, respectively, and the ^{187}Re and ^{187}Os compositions are 63.62–586.65 µg/g and 230.17–2129.61 ng/g, respectively. The six molybdenite samples yield a well-constrained Re–Os isochron age of 218 ± 2 Ma (2σ, MSWD = 0.23, N = 6; Figure 7A), and a weighted average age of 217 ± 1 Ma (2σ, MSWD = 0.23, N = 6; Figure 7B). The isochron age (218 ± 2 Ma) represents Cu mineralization age of the Pulang deposit.

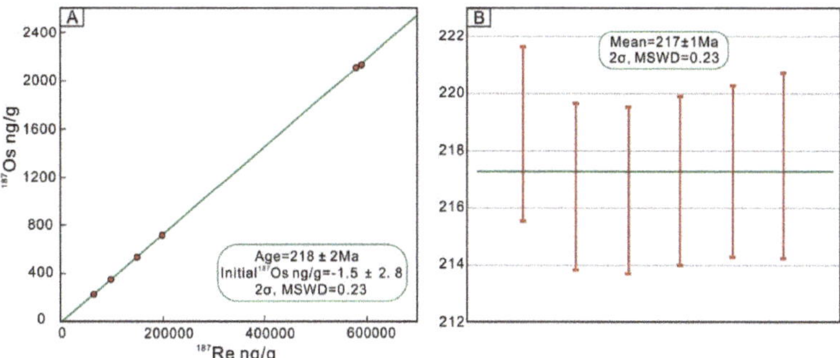

Figure 7. Diagrams of molybdenite Re–Os isochron age (**A**) and weighted mean age (**B**) in the Pulang Cu (–Mo–Au) deposit.

5.3. Major and Trace Element

The whole-rock major and trace elemental compositions for the QDP (PL-1), QMP (PL-2, PL-3), and GP (PL-4) in the Pulang deposit are listed in Table S2. Overall, the three different porphyry intrusions show wide-ranging compositional variation in terms of major elements. The QDP has an SiO_2 content of 61.90–64.24 wt %, a K_2O content of 1.10–1.14 wt %, a CaO content of 5.05–6.18 wt %, and an MgO content of 3.26–4.19 wt %, with Mg# [Mg# = 100 × Mg^{2+}/(Mg^{2+} + TFe^{2+})] values varying from 68 to 73. Compared with the QDP, the QMP and GP have higher SiO_2 (QMP = 64.32–67.69 wt %, GP = 69.95–74.10 wt %) and K_2O (QMP = 2.53–4.33 wt %, GP = 2.0–3.01 wt %), but relatively lower CaO (QMP = 2.60–3.85 wt %, GP = 1.49–2.14 wt %) and MgO (QMP = 1.69–2.06 wt %, GP = 1.80–3.23 wt %), with Mg# value (QMP = 46–60, GP = 47–71). In addition, the Al_2O_3 content of the GP (11.39–12.11 wt %) is lower than that of QDP (15.43–16.17 wt %) and QMP (14.32–16.07 wt %). The TiO_2, Fe_2O_3T, Na_2O, P_2O_5, and aluminum saturation index (A/CNK) of the three porphyry intrusions are similar, ranging from 0.40 to 0.77 wt %, from 2.17 to 4.56 wt %, from 2.46 to 4.98 wt %, from 0.23 to 0.45 wt %, and from 0.77 to 1.08, respectively. For SiO_2 vs. K_2O + Na_2O diagram (Figure 8A [51]), data for all samples are plotted within fields of the subalkaline series. In the SiO_2 vs. K_2O (Figure 8B [52]) and Na_2O vs. K_2O diagrams (Figure 8C [53]), data for QDP samples are plotted within fields of the middle-K calc-alkaline series and sodic suites, whereas data for the QMP and GP samples are plotted within fields of the high-K calc-alkaline series and potassic suites (Figure 8B,C). All samples have a narrow range of A/CNK (mole [Al_2O_3/(CaO + Na_2O + K_2O)]) ratio values (A/CNK = 0.77–1.08), indicating metaluminous to moderately peraluminous in an A/NK vs. A/CNK diagram (Figure 8D [54]).

In the chondrite-normalized REE diagram (Figure 9A), the three different petrographic porphyry intrusions are depleted of heavy rare-earth elements (HREEs) relative to light rare-earth elements (LREEs), with slightly negligible Eu anomalies (Eu/Eu* = 0.79–0.97). The QMP [(La/Yb)$_N$ = 17.0–25.9] are more fractionated than the GP [(La/Yb)$_N$ = 9.3–16.8] and QDP [(La/Yb)$_N$ = 7.6–9.0]. In the MORB-normalized geochemical pattern (Figure 9B), they display similar REE patterns, and all are characterized by a depletion of high field-strength elements (HFSEs; e.g., Nb, Ta, and Ti). They are characterized by high Sr (432–1150 ppm), and low Y (9–16 ppm) and Yb (0.8–1.5 ppm) concentrations, with high Sr/Y ratios (48–94).

Table 2. Molybdenite Re–Os dating results in the Pulang Cu (–Mo–Au) deposit.

Sample No.	Lithology	Ore-Hosted Rocks	Location	Re (µg/g)		Common Os (ng/g)		^{187}Re (µg/g)		^{187}Os (ng/g)		Model Age (Ma)	
				Measured	Error	Measured	Error	Measured	Error	Measured	Error	Measured	Error
PL-8-6	Mo-bearing quartz veinlet ore	QDP	N: 28°02′45″ E: 99°58′53″	921.25	8.08	0.0184	0.4222	579.03	5.08	2112.44	13.19	218.6	3.0
PL-11-2	Disseminated Mo-bearing chalcopyrite ore	QMP	N: 28°02′53″ E: 99°59′04″	101.22	0.65	0.2262	0.4283	63.62	0.41	230.17	1.42	216.8	2.9
PL-11-3	Disseminated Mo-bearing chalcopyrite ore	QMP	N: 28°02′53″ E: 99°59′04″	154.64	1.01	0.0164	0.3759	97.19	0.64	351.44	2.09	216.6	2.9
PL-10-4	Mo-bearing quartz veinlet ore	QDP	N: 28°02′56″ E: 99°58′52″	316.48	2.15	0.4501	0.4819	198.91	1.35	720.58	4.47	217.0	2.9
PL-11-5	Mo-bearing chalcopyrite quartz vein ore	QDP	N: 28°02′53″ E: 99°59′26″	236.07	1.64	0.5393	0.2190	148.37	1.03	538.16	3.42	217.3	3.0
PL-7-1	Mo-bearing quartz lump ore	QMP	N: 28°02′49″ E: 99°59′10″	933.39	8.19	0.0187	0.4278	586.65	5.15	2129.61	13.64	217.5	3.2

Note: QDP: Quartz diorite porphyry; QMP: Quartz monzonite porphyry.

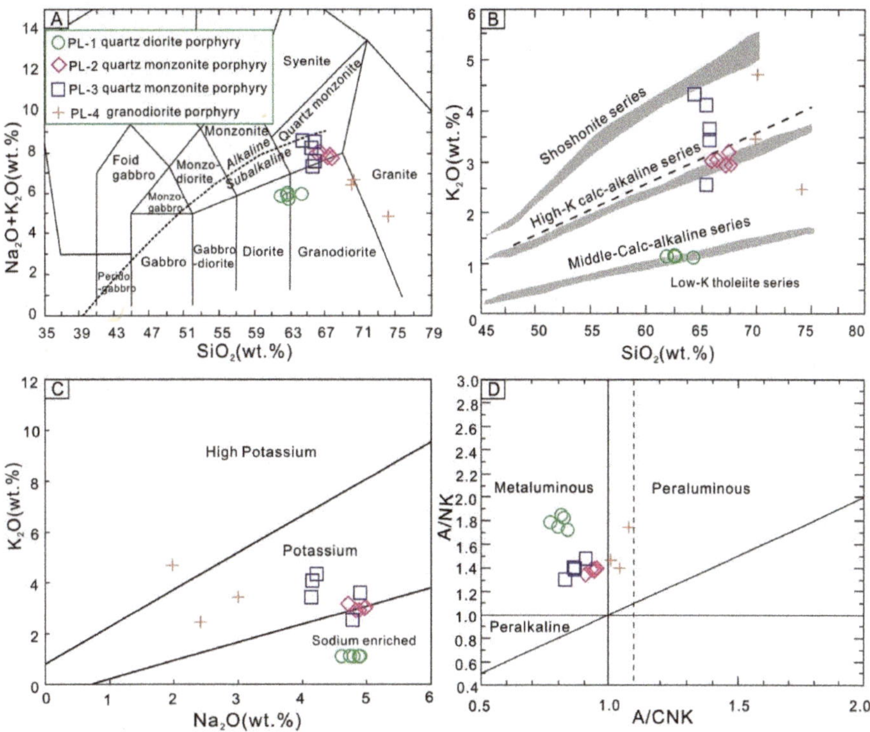

Figure 8. (**A**) Diagrams of SiO_2 vs. $Na_2O + K_2O$ (based on Irvine and Baragar [51]); (**B**) SiO_2 vs. K_2O (based on Peccerillo and Taylor [52]); (**C**) Na_2O vs. K_2O (based on Middlemost [53]); (**D**) A/CNK vs. A/NK for the Pulang porphyry intrusions (based on Maniar and Piccoli [54]).

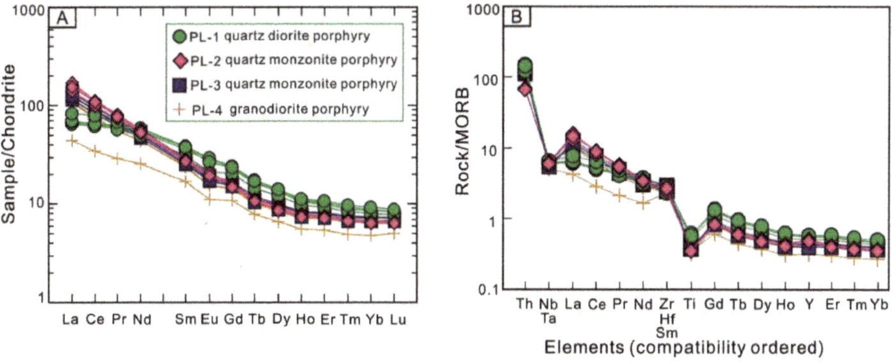

Figure 9. (**A**) Chondrite-normalized REE patterns and (**B**) MORB-normalized geochemical patterns for the Pulang porphyry intrusions (based on Pearce [55]). Chondrite and N-MORB values are from Boynton [56] and Sun and McDinough [57], respectively.

6. Discussion

6.1. Timing of Magmatism and Cu Mineralization

Some previous geochronological studies on the QDP, QMP, and GP in Pulang obtained a coeval age. For example, Wang et al. [14] obtained LA-ICP-MS zircon U–Pb ages of 214 ± 3 Ma, 214 ± 3 Ma, and 209 ± 4 Ma for QDP, QMP, and GP, respectively; Wang et al. [22] obtained SHRIMP zircon U–Pb ages of 228 ± 3 Ma and 226 ± 3 Ma for QDP and QMP, respectively. These isotope dating results have led to the conclusion that all the QDP, QMP, and GP contribute to the Pulang Cu mineralization. However, the ages of QDP (228 ± 3 Ma) and QMP (226 ± 3 Ma) are incompatible with the reported molybdenite Re–Os isochron ages (208 ± 15 Ma and 213 ± 4 Ma [3,18,24]). In this study, the U–Pb concordia age of the zircon coincides well with the weighted mean ^{206}Pb/^{238}U age in every porphyry intrusion sample (Figure 6), suggesting more accurate and reliable dating. Our new dating results for QMP (218 ± 1 Ma and 219 ± 1 Ma) show that it is younger than QDP (227 ± 2 Ma) by about 9 Ma, and it is older than GP (209 ± 1 Ma) by about 10 Ma, indicating that the three different porphyry intrusions were not formed coevally. This conclusion is further supported by the geological fact that the QMP and GP display as small dykes intruding into the QDP (Figure 2), and that the GP cuts both QMP and orebodies (Figure 2). The Pulang porphyry intrusions were not formed coevally, which makes it unlikely that the Pulang porphyry intrusions all contributed to Cu mineralization. To constrain the mineralization age accurately, we performed the molybdenite Re–Os dating, and the result is that the six molybdenite-bearing samples from the Pulang deposit yielded a well-constrained isochron age of 218 ± 2 Ma, and a weighted mean age of 217 ± 1 Ma (Figure 7). The precise zircon U–Pb and Re–Os ages indicate a close temporal relationship between Cu mineralization and QMP. In addition, Cu mineralization in the Pulang deposit mainly occurred within the QMP (Figure 7), and only minor mineralization developed in the QDP. From the center (QMP) to the outer edge (QDP), the thickness of the orebody decreases, as does the grade of Cu and other associated metals (0.030–0.001% Mo, average 0.010%; 0.87–0.06g/t Au, average 0.18 g/t; and 3.93–0.34 g/t Ag, average 1.27 g/t), which also suggests a spatial relationship between Cu mineralization and QMP. Moreover, the Pb isotope compositions of sulfur minerals (^{206}Pb/^{204}Pb = 18.079–18.694, ^{207}Pb/^{204}Pb = 15.603–15.632, and ^{208}Pb/^{204}Pb = 38.228–38.635 [58]) in Pulang are similar and homogeneous, and they coincide well with those of the QMP (^{206}Pb/^{204}Pb = 18.3005–18.5431, ^{207}Pb/^{204}Pb = 15.6061–15.6318 and ^{208}Pb/^{204}Pb = 38.4352–38.856 [58]), but significantly differ from those of the QDP (^{206}Pb/^{204}Pb = 18.8761–19.1700, ^{207}Pb/^{204}Pb = 15.6191–15.6413, and ^{208}Pb/^{204}Pb = 39.3458–39.5706 [23]). Furthermore, the Cu concentrations (Table S2) of QMP range from 343 to 5220 ppm (average = 1989 ppm), largely exceeding the Cu concentrations in QDP (113–117 ppm, average = 115 ppm) and GP (222–229 ppm, average = 225 ppm); this finding indicates that ore-forming materials in the Pulang deposit were mainly sourced from QMP. To summarize, geological, geochronological, isotopic, and geochemical data led to the conclusion that Cu mineralization in Pulang was spatially, temporally, and genetically related to the QMP rather than the QDP and GP.

The zircon U–Pb ages of QMP (218 ± 1 Ma and 219 ± 1 Ma) are consistent with molybdenite Re–Os age (218 ± 2 Ma), constraining the timing of emplacement and associated Cu mineralization in the Pulang Cu (–Mo–Au) deposit to the Late Triassic (~217 Ma). Moreover, these ages are broadly consistent with the zircon U–Pb and mineralization ages of some porphyry-type Cu deposits located in the same metallogenic belt of the southern Yidun Arc, such as Xuejiping, Lannitang, Songnuo, and Chundu (Table 3). Thus, the zircon U–Pb and Re–Os isochron ages, reported herein and previously, constrain the Late Triassic magmatism and associated Cu mineralization event to the southern Yidun Arc.

Table 3. Geological characteristics of the major Late Triassic Cu deposits in the southern Yidun Arc.

Deposit	Type	Commodity	Reserve/Grade	Ore-Hosting Rock	Metal Minerals	Alteration	Related Intrusion Age (Ma)	Mineralization Age (Ma)	Reference
Pulang	Porphyry	Cu–Mo–Au	Cu: 1.14 Mt @ 0.52% Mo: 6399 t @ 0.0004% Au: 54 t @ 0.18 g/t	QDP, QMP, GP	Ccp, Py, Mo, Bn, Mag, Sp, Pyr, Cv	Qz, Kfs, Ser, Chl, Ep, Cal	QMP (Zircon U–Pb:218 ± 1, 219 ± 1)	Molybdenite Re–Os: 219 ± 2	This paper [20]
Xuejiping	Porphyry	Cu	Cu: 0.30 Mt @ 0.52%	DP, QMP	Ccp, Py, Mo, Gn, Sp	Qz, Kfs, Ser, Chl, Ep, Cal	QMP (Zircon U–Pb:219 ± 2, 215 ± 3)	Molybdenite Re–Os 221 ± 2	[7,55]
Chundu	Porphyry	Cu	No data	DP, QMP, GP	Ccp, Py, Gn, Sp	Qz, Kfs, Ser, Chl, Ep, Cal	QMP (Zircon U–Pb: 220 ± 2)	Triassic	[2]
Lannitang	Porphyry	Cu	Cu: 0.18 Mt @ 0.5%	QDP, QMP	Ccp, Py, Clc, Mag, Hem	Qz, Kfs, Chl, Ep	QDP (Zircon U–Pb:225 ± 4)	Triassic	[33]
Songnuo	Porphyry	Cu	No data	QMP	Ccp, Py, Pyr	Qz, Chl, Ep, Cal	QMP (Zircon U–Pb:221 ± 4)	Triassic	[9]
Langdu	Skarn	Cu–Fe	Cu: 0.10 Mt @ 6%	Skarn and hornfel	Ccp, Py, Pyr, Mag	Qz, Grt, Tr, Chl, Ep, Cal	QMP (Zircon U–Pb:223 ± 1)	Triassic	[33]

Abbreviations: QDP: quartz diorite porphyry; QMP: quartz monzonite porphyry; GP: granodiorite porphyry; DP: diorite porphyry; Ccp: chalcopyrite; Py: pyrite; Mo: molybdenite; Bn: bornite; Mag: magnetite; Sp: sphalerite; Gn: galena; Pyr: pyrrhotite; Clc: chalcocite; Cv: covellite; Hem: hematite; Qz: quartz; Kfs: K-feldspar; Ser: sericite; Chl: chlorite; Grt: garnet; Tr: tremolite; Ep: epidote; Cal: calcite.

6.2. Petrogenesis and Tectonic Setting

A large number of geochronological data indicate that Late Triassic magmatism in the Yidun Arc lasted for at least 29 Ma from ~235 to ~206 Ma with a peak age at around 215 Ma [59–61]. According to the sequence of magma intrusion for the Pulang porphyry intrusions, it is clear that the SiO_2 content of the rocks increases gradually from QDP to QMP and then to GP, which indicates that the Late Triassic Pulang porphyry intrusions are the products of different evolution stages of the same magmatism in the same tectonic setting.

Pulang porphyry intrusions have high SiO_2 (62–74 wt %), Al_2O_3 (11.4–16.2 wt %, mostly >15.0 wt %), Na_2O (2.4–5.0 wt %, mostly >4.0 wt %), Sr (432–1150 ppm), Mg# values (46–73), and Sr/Y ratios (48–94), and they have low Y (9–16 ppm) and Yb values (0.8–1.5 ppm), as well as moderately negative Eu anomalies (Eu/Eu* = 0.79–0.97), suggesting a geochemical affinity to adakitic rocks [62–65]. In the $(La/Yb)_N$ vs. Yb_N and Sr/Y vs. Y geochemical classification diagrams (Figure 10A,B), most samples of the Pulang porphyry intrusions are plotted in the adakitic rocks field, and only a few samples are plotted in the transitional field between typical adakitic rocks and normal island arc rocks. Adakite was initially named for rocks with clear contributions from the partial melting of subducted young hot oceanic crust represented by mid-ocean ridge basalt (MORB) [62]. Some important geochemical characteristics of the adakites, such as high Sr and low Y and Yb, can be attributed to the presence of garnet $\pm$ amphibole and the absence of plagioclase within the melt residue [62,66–70]. To date, several genetic models have been proposed to explain the genesis of adakites, including (1) the assimilation of crustal material by a mantle-derived magma, and subsequent fractional crystallization (AFC model [64,71]), (2) the partial melting of the delaminated lower continental crust [72,73], (3) the partial melting of the thickened continental crust [74,75], and (4) the partial melting of a subducted oceanic slab [62–64].

To begin with, the fractionation of amphibole would result in high Sr/Y ratios, but there are no evident correlations between Sr/Y and SiO_2 (Figure 11A). The moderately negative Eu anomalies indicate little or no plagioclase fractionation. Fractional crystallization of a garnet-bearing assemblage from basaltic melts would result in a positive correlation between Sr/Y, La/Yb, and especially Dy/Yb ratios with increasing SiO_2 [10]. The Pulang porphyry intrusions do not show such correlation (Figure 11B). Instead, they display the clearly partial melting trends observed in the La/Yb vs. La and Rb/Nb vs. Rb diagrams (Figure 11C,D), indicating that the partial melting of the source plays a dominant role in the formation of the Pulang adakitic magmas. Moreover, adakitic magmas derived from the partial melting of thickened continental lower crust are characteristic of K-rich and low MgO magmas with low Mg# values (generally Mg# < 46) [66,71], which are not compatible with the geochemical features of the Pulang porphyry intrusions as described above. Moreover, adakitic magmas produced by the partial melting of delaminated lower crust generally have high MgO, Cr, Co, and Ni contents, which are similar to those of the Pulang porphyry intrusions. However, extensive isotope, geochemical, and geochronological data indicate that the delamination of the lower crust is inconsistent with the geological background of the Yidun Arc during the Late Triassic [10,14,33]. Finally, the geochemical characteristics of the Pulang adakitic porphyry intrusions and adakite are compared in detail (Table 4), and the comparison results indicate that Pulang adakitic porphyry intrusions are similar to adakite. Moreover, Pulang porphyry intrusions also have low TiO_2 (0.4–0.8 wt %), MgO (1.7–4.2 wt %), and CaO + Na_2O (4.6–10.9 wt %) contents and low Cr/Ni ratios (1.2–7.6, mostly 1.2–3.9). As shown in the detailed geochemical comparison in Table 4, Pulang porphyry intrusions can be classified as high silica adakites (HSA) [63]. In the discrimination diagrams of Martin et al. (2005) [63], Pulang porphyry intrusion samples are plotted in the fields of HSA (Figure 10C–F). It is generally believed that HSA are derived from oceanic-slab melts that may react with mantle wedges during the ascent [62–65]. Together with the isotopic geochronology and the tectonic setting discussed below, we propose that Pulang HSA porphyry intrusions were derived from the partial melting of the subducted Ganzi-Litang oceanic plate that reacted with peridotite during its ascent through the mantle wedge. Their high-Mg# features are caused by the interaction between slab melts and mantle peridotite [76,77].

It is generally accepted that the formation of the Yidun Arc can be attributed to the westward subduction of the Ganzi–Litang oceanic lithosphere during the Late Triassic period, and that magmatic activity in the Yidun Arc lasted for at least 29 Ma, from ~235 to ~206 Ma [59–61]. The initial westward subduction of the Ganzi–Litang oceanic lithosphere began at ~235 Ma (following the earlier closure of the Jinshajiang Ocean), and finally ended at ~206 Ma, accompanied by the occurrence of syn-collision granites. The Late Triassic volcanic and intrusive rocks in the Yidun Arc are characterized by middle- to high-K calc-alkaline rock series with an I-type affinity [10,11,14], together with enrichment in LILEs and LREEs, and depletion of HFSEs and HREEs. These geochemical characteristics are similar to those of QDP, QMP, and GP in Pulang, indicating an island arc setting [78]. Moreover, in the Rb vs. Y + Nb, Rb vs. Yb + Ta, Nb vs. Y, and Rb/10 vs. Hf vs. Ta × 3 diagrams (Figure 12A–D), all the QDP, QMP and GP samples from Pulang are plotted within the volcanic arc granite (VAG) field, also consistent with a subduction-related volcanic arc setting. In the Th/Yb vs. Ba/La and Ba/Th vs. Th/Nb diagrams (Figure 12E,F), the Pulang samples show a trend similar to that of the melt-related enrichment rather than the fluid-related enrichment, indicating a source associated with subducted plate melting. This further supports the conclusion discussed above that Pulang adakitic porphyry intrusions are related to oceanic-slab-derived melts. Importantly, these geochemical characteristics at the Pulang Cu deposit are similar to those of the porphyry intrusions associated with Cu mineralization at the Xuejiping, Chundu, Lannitang, Songnuo, and Langdu Cu deposits in the southern Yidun Arc. Combining similar geochemical characteristics with similar mineralization ages and geological characteristics of these porphyry-type Cu deposits in the southern Yidun Arc (Table 3), we conclude that all of these Cu deposits can be attributed to the Late Triassic tectonic–magmatic–hydrothermal event that was induced by the westward subduction of the Ganzi–Litang oceanic lithosphere.

Table 4. Comparisons of geochemical characteristics between the adakite, high-SiO_2 adakites (HSA), low-SiO_2 adakites (LSA), and Pulang porphyry intrusions.

	Adakite	HSA	LSA	Pulang Porphyry Intrusions
SiO_2	≥56 wt %	≥60 wt %	≤60 wt %	62–74 wt %
Al_2O_3	≥15 wt % (rarely lower)			11.4–16.2 wt % (mostly 15.1–16.2 wt %)
MgO	<3 wt % (rarely >6 wt %)	0.5–4.0 wt %	4.0–9.0 wt %	1.7–4.2 wt %
Na_2O	3.5–7.5 wt %			2.7–5.0 wt % (mostly 3.4–5.0 wt %)
K_2O/Na_2O	~0.42 (sodic)			0.2–1.0 wt % (sodic)
$CaO + Na_2O$		<11 wt %	>10 wt %	4.6–10.9 wt %
TiO_2		<0.9 wt %	>3.0 wt %	0.40–0.77 wt %
Sr	>400 ppm (rarely <400 ppm)	<1100 ppm	>1000 ppm	432–1150 ppm
Y	≤18 ppm	≤18 ppm	≤18 ppm	9–16 ppm
Yb	≤1.9 ppm	≤1.9 ppm	≤1.9 ppm	0.8–1.5 ppm
Sr/Y	40–100			48–94
Cr/Ni		0.5–4.5	1.0–2.5	1.2–7.6 (mostly 1.2–3.9 wt %)
Mg#	~50			46–73 (>46, High-Mg)
Reference	[62,65]	[79]	[79]	This paper

Note: HSA: High-SiO_2 adakites; LSA: Low-SiO_2 adakites.

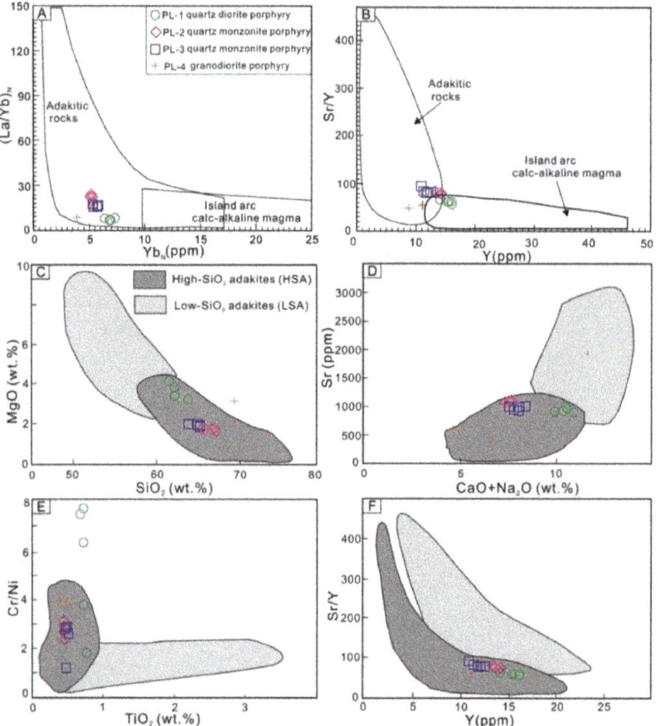

Figure 10. (**A**) (La/Yb)$_N$ vs. (Yb)$_N$ (based on Martin [79]), (**B**) Sr/Y vs. Y (Defant and Drummond [62]), (**C**) SiO$_2$ vs. MgO (based on Martin et al. [63]), (**D**) SiO$_2$ vs. Sr (based on Martin et al. [63]), (**E**) SiO$_2$ vs. Cr/Ni (based on Martin et al. [63]), and (**F**) SiO$_2$ vs. Sr/Y (based on Martin et al. [63]) discrimination diagrams for Pulang porphyry intrusions.

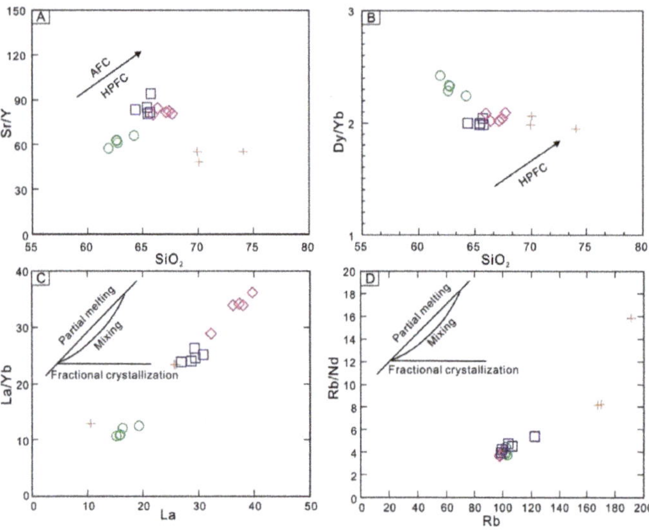

Figure 11. (**A**) Sr/Y vs. SiO$_2$, (**B**) Dy/Y vs. SiO$_2$, (**C**) La/Yb vs. La, and (**D**) Rb/Nd vs. Rb diagrams for the Pulang porphyry intrusions. Symbols are as in Figure 10.

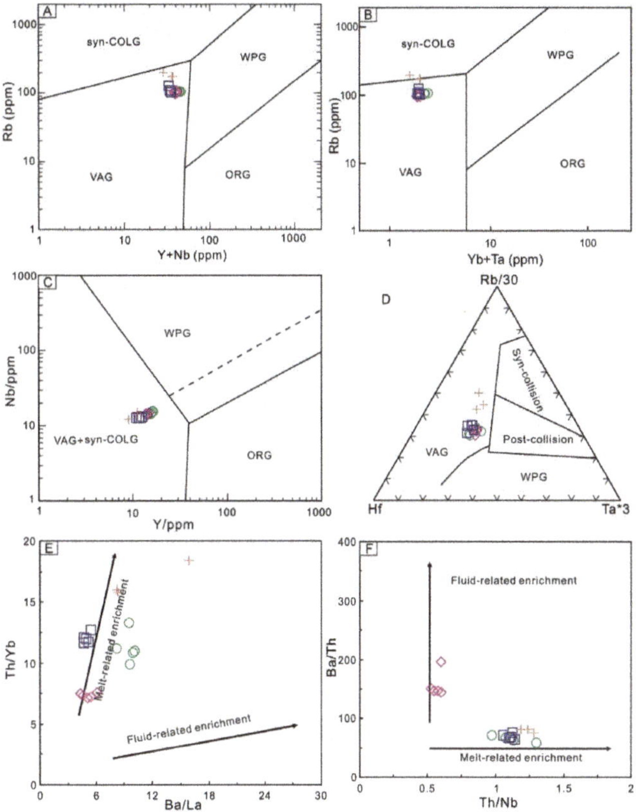

Figure 12. (**A**) Rb vs. Y + Nb (based on Pearce et al. [80]), (**B**) Rb vs. Yb + Ta (based on Pearce et al. [80]), (**C**) Y vs. Nb (based on Pearce [81]), and (**D**) Rb/30-Hf-Ta × 3 (based on Harris et al. [82]) tectonic discrimination diagrams for Pulang porphyry intrusions. (**E**) Th/Yb vs. Ba/La and (**F**) Ba/Th vs. Th/Nb diagrams showing the roles of the fluid- and melt-related enrichments of the source. Symbols are as in Figure 10.

6.3. Implications for the Cu Mineralization

The Yidun arc was formed during the Late Triassic in response to the westward subduction of the Ganzi–Litang oceanic plate. The Late Triassic bimodal volcanic suites and associated VMS-type deposits developed in the northern Yidun Arc, which can be attributed to a steep subduction. In contrast, the flat subduction in the southern Yidun Arc produced abundant adakitic porphyry intrusions and associated porphyry-type Cu deposits. This flat subduction is the most favorable tectonic setting for slab melting [83,84]. It has been recently proposed that oceanic-slab-derived adakites are genetically related to porphyry-type Cu–Au deposits [83–91]. Oceanic-slab-derived melts are considered to have high oxygen fugacity and high Cu concentrations to form adakitic magmas that may potentially produce porphyry Cu deposit [83–85]. At an oxygen fugacity above FMQ + 2 (FMQ refers to fayalite–magnetite–quartz oxygen buffer), sulfur and chalcophile elements (Cu, Au, etc.) easily form oxidized ionic compounds (SO_4^{2-}) that cause the Cu to remain in magmas in the form of sulfate during the magma ascent. By contrast, at an oxygen fugacity below FMQ, sulfur presents mainly in its reduced form (S^{2-}), which is less soluble in magmas. This leads to a premature precipitation of metallic elements in the mantle or deep crust, which is not good for the formation of porphyry Cu deposit [83]. For the Pulang deposit, the QMP associated with Cu mineralization has a high oxygen fugacity (fO_2 > FMQ + 5.5), and even the QDP of the barren mineralization has a high oxygen fugacity (fO_2 > FMQ + 5.1) [14], indicating that the Pulang Cu deposit formed in a high oxygen fugacity environment. For another, Cu concentrations in the oceanic crust range from 60 to 130 ppm [92], which is much higher than those of the primitive mantle (30 ppm [93]) and the continental crust (27 ppm [84]). Therefore, basaltic oceanic-slab-derived melts (oceanic-slab-derived adakites) have considerably higher Cu content than the lower continental crust melt or mantle-derived melt [83], which is supported by the higher Cu concentrations for the QDP (113–117 ppm), QMP (343–5220 ppm), and GP (222–229 ppm) in Pulang. The combination of these factors contributed to the formation of the subduction-related Pulang porphyry deposit (Figure 13).

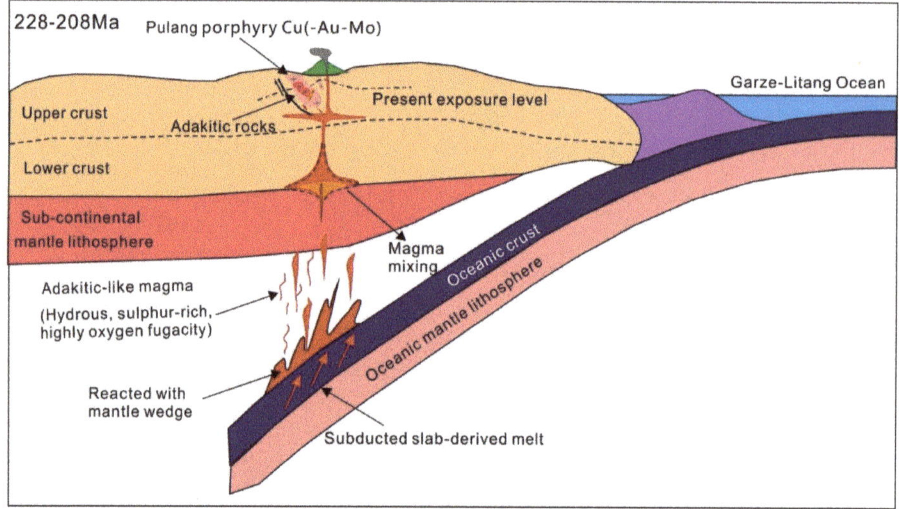

Figure 13. Tectonic setting and metallogenic model diagram of the Pulang porphyry Cu (–Mo–Au) deposit.

7. Conclusions

According to the LA-ICP-MS zircon U–Pb dating, molybdenite Re–Os dating and whole-rock geochemistry data, the following conclusions can be drawn:

1. Molybdenite from the Pulang porphyry Cu deposit yields the Re–Os age of 218 ± 2 Ma, which is in accordance with most of the porphyry-type Cu deposits in the southern Yidun Arc having mineralization ages of 217–221 Ma. Zircons from the quartz diorite porphyry, quartz monzonite porphyry, and granodiorite porphyry yield U–Pb ages of 227 ± 2 Ma, 218 ± 1 Ma, and 209 ± 1 Ma, respectively, suggesting a close spatial, temporal, and genetical relationship between Cu mineralization and quartz monzonite porphyry.

2. The Pulang porphyry intrusions geochemically belong to high silica (HSA) adakitic rocks. These intrusions derived from the partial melting of subducted Ganzi-Litang oceanic plate that reacted with peridotite during its ascent through the mantle wedge.

3. The Pulang deposit shows similar geological characteristics to the most other porphyry-type Cu deposits in the southern Yidun Arc. These porphyry-type Cu deposits are considered to be related to the westward subduction of the Ganzi-Litang oceanic lithosphere during the Late Triassic.

Supplementary Materials: The following are available online at http://www.mdpi.com/2075-163X/9/3/191/s1, Table S1: LA-ICP-MS zircon U–Pb dating data of the porphyry intrusions in the Pulang Cu (–Mo–Au) deposit, Table S2: Whole-rock major and trace element data of the porphyry intrusions in the Pulang Cu (–Mo–Au) deposit.

Author Contributions: Conceptualization: Q.Y., Y.-S.R., and S.-B.C.; field investigation: Q.Y., G.-L.Z., and Q.-H.Z.; experimental analysis: Y.-J.H.; software: J.-M.L., X.-H.S., Z.-J.Y., and Z.-M.S.; validation: Q.Y. and Y.-S.R.; resources: Y.-S.R. and S.-B.C.; data curation: Q.Y.; writing—original draft preparation: Q.Y.; writing—review & editing: Y.-S.R.; visualization: Q.Y.; funding acquisition: S.-B.C. and Y.-S.R.

Funding: This work was financially supported by The Program for Jilin University Science and Technology Innovative Research Team (JLUSTIRT, 2017TD-26), the Industrial Application Projects in Provinces (autonomous regions and cities) of the National High Resolution Earth Observation System Major Science and Technology Project: Black Soil Area Agricultural Information Industrialization Application of High Resolution Special Projects in the Central and Western Jilin (71-Y40G04-9001-15/18), and the Changbai Mountain Scholars Program (440050117009), Jilin Province, China.

Acknowledgments: We would like to thank the leaders and geologists of the Pulang ore district for their support of our fieldwork.

Conflicts of Interest: The authors declare no conflict of interest.

References

1. Leng, C.-B.; Zhang, X.-C.; Wang, S.-X.; Qin, C.-J.; Gou, T.-Z.; Wang, W.-Q. SHRIMP zircon U–Pb dating of the Songnuo ore–hosting porphyry, Zhongdian, Northwest Yunnan, China and its geological implication. *Geotecton. Metall.* **2008**, *32*, 124–130. (In Chinese)
2. Zhang, X.-C.; Leng, C.-B.; Yang, C.-Z.; Wang, W.-Q.; Qin, C.-J. Zircon SIMS U–Pb isotopic age and its significance of ore–bearing porphyry of Chundu porphyry copper deposit in Zhongdian, Northwestern Yunnan. *Acta Miner. Sin.* **2009**, *29*, 359–360. (In Chinese)
3. Cao, D.-H. Porphyry Copper Deposit Model and Exploration Technique in Zhongdian, Yunnan. Ph.D. Thesis, Chinese Academy of Geological Sciences, Beijing, China, 2007. (In Chinese)
4. Liu, J.-T. Late-Trassic Cu Mineralization in Porphyry Environment, Northwest Yunnan, China. Ph.D. Thesis, China University of Geosciences, Beijing, China, 2014. (In Chinese)
5. Pang, Z.-S.; Du, Y.-S.; Wang, G.-W.; Guo, X.; Cao, Y.; Li, Q. Single–grain zircon U–Pb isotopic ages, geochemistry and its implication of Pulang complex in Yunnan Province, China. *Acta Petrol. Sin.* **2009**, *25*, 159–165. (In Chinese)
6. Liu, X.-L.; Li, W.-C.; Yin, G.-H.; Zhang, N. The geochronology, mineralogy and geochemistry study of the Pulang porphyry copper deposits in Geza arc of Yunnan Province. *Acta Petrol. Sin.* **2013**, *29*, 3049–3064. (In Chinese)
7. Leng, C.-B.; Zhang, X.-C.; Hu, R.-Z.; Wang, S.-X.; Zhong, H.; Wang, W.-Q.; Bi, X.-W. Zircon U–Pb and molybdenite Re–Os geochronology and Sr–Nd–Pb–Hf isotopic constraints on the genesis of the Xuejiping porphyry copper deposit in Zhongdian, Northwest Yunnan, China. *J. Asian Earth Sci.* **2012**, *60*, 31–48. [CrossRef]
8. Cao, D.-H.; Wang, A.-J.; Huang, Y.-F.; Zhang, W.; Hou, K.-J.; Li, R.-P.; Li, Y.-K. SHRIMP geochronology and Hf isotope composition of zircons from Xuejiping porphyry copper deposit, Yunnan province. *Acta Geol. Sin.* **2009**, *83*, 1430–1435. (In Chinese)
9. Li, W.-C.; Yu, H.-J.; Gao, X.; Liu, X.-L.; Wang, J.-H. Review of Mesozoic multiple magmatism and porphyry Cu–Mo (W) mineralization in the Yidun Arc, eastern Tibet Plateau. *Ore Geol. Rev.* **2017**, *90*, 795–812. [CrossRef]
10. Wang, B.-Q.; Zhou, M.-F.; Li, J.-W.; Yan, D.-P. Late Triassic porphyritic instructions and associated volcanic rocks from the Shangri–La region, Yidun terrane, Eastern Tibetan Plateau: Adakitic magmatism and porphyry copper mineralization. *Lithos* **2011**, *127*, 24–38. [CrossRef]
11. Chen, J.-L.; Xu, J.-F.; Ren, J.-B.; Huang, X.-X.; Wang, B.-D. Geochronology and geochemical characteristics of Late Triassic porphyritic rocks from the Zhongdian arc, eastern Tibet, and their tectonic and metallogenic implications. *Gondwana Res.* **2014**, *26*, 492–504. [CrossRef]
12. Ren, J.-B.; Xu, J.-F.; Chen, J.-L. Zircon geochronology and geological implications of ore-bearing porphyries from Zhongdian arc. *Acta Petrol. Sin.* **2011**, *27*, 2591–2599. (In Chinese)
13. Ren, J.-B.; Xu, J.-F.; Chen, J.-L.; Zhang, S.Q.; Liang, H.Y. Geochemistry and petrogenesis of Pulang porphyries in Sanjiang region. *Acta Petrol. Miner.* **2011**, *30*, 581–592. (In Chinese)

14. Wang, P.; Dong, G.-C.; Zhao, G.-C.; Han, Y.-G.; Li, Y.-P. Petrogenesis of the Pulang porphyry complex, southwestern China: Implications for porphyry copper metallogenesis and subduction of the Paleo-Tethys Oceaincc lithosphere. *Lithos* **2018**. [CrossRef]

15. Zeng, P.-S.; Mo, X.-X.; Yu, X.-H.; Hou, Z.-Q.; Xu, Q.-D.; Wang, H.-P.; Li, H.; Yang, C.-Z. Porphyries and porphyry copper deposits in Zhongdian Area, Northwestern Yunnan. *Miner. Depos.* **2003**, *22*, 394–400. (In Chinese)

16. Li, W.-C.; Zeng, P.-S. Characteristics and metallogenic model of the Pulang superlarge porphyry copper deposit in Yunnan, China. *J. Chengdu Univ. Technol.* **2007**, *34*, 436–446. (In Chinese)

17. Guo, X.; Du, Y.-S.; Pang, Z.-S.; Li, S.-T.; Li, Q. Characteristics of the ore-forming fluids in alteration zones of the Pulang porphyry copper deposit in Yunnan Province and its metallogenic significance. *Geosciences* **2009**, *23*, 465–471.

18. Zeng, P.-S.; Li, W.-C.; Wang, H.-P.; Li, H. The Indoshinian Pulang superlarge porphyry copper deposit in Yunnan, China: Petrology and chronology. *Acta Petrol. Sin.* **2006**, *22*, 989–1000. (In Chinese)

19. Li, W.-C.; Yin, G.-H.; Lu, Y.-X.; Liu, X.-L.; Xu, D.; Zhang, S.-Q.; Zhang, N. The evolution and ^{40}Ar–^{39}Ar isotopic evidence of the Pulang Complex in Zhongdian. *Acta Geol. Sin.* **2009**, *83*, 1421–1429. (In Chinese)

20. Li, W.-C.; Zeng, P.-S.; Hou, Z.-Q.; Zeng, P.-S.; Noel, C.-W. The Pulang Porphyry Copper Deposit and Associated Felsic Intrusions in Yunnan Province, Southwest China. *Econ. Geol.* **2011**, *106*, 79–92.

21. Liu, H.; Zhang, C.-Q.; Jia, F.-D.; Zhou, Y.-M.; Lou, D.-B. Mineral and geochemical evidences of magma-mixing from Pulang porphyry copper deposit in SW Sanjiang, China. *Acta Petrol. Sin.* **2015**, *31*, 3189–3202. (In Chinese)

22. Wang, S.-X.; Zhang, X.-C.; Leng, C.-B.; Qing, J.-C.; Ma, D.-Y.; Wang, W.Q. Zircon SHRIMP U–Pb dating of the Pulang porphyry copper deposit, Northwestern Yunnan, China: The ore-forming time limitation and geological significance. *Acta Petrol. Sin.* **2008**, *24*, 2313–2321. (In Chinese)

23. Yang, Z. Late Triassic mineralization of the porphyry copper deposits in Yidun arc, Southwest China. Ph.D. Thesis, China University of Geosciences, Beijing, China, 2017. (In Chinese)

24. Fan, Y.-H.; Li, W.-C. Geological characteristics of the Pulang porphyry copper deposit, Yunnan. *Geol. China* **2006**, *33*, 352–362. (In Chinese)

25. Zeng, P.-S.; Wang, H.-P.; Mo, X.-X.; Yu, X.; Li, W. Tectonic setting and prospects of porphyry copper deposits in Zhongdian island-arc belt. *Acta Geosci. Sin.* **2004**, *25*, 535–540. (In Chinese)

26. Sun, H.-S.; Li, H.; Martin, D.; Xia, Q.-L.; Jiang, C.-L.; Wu, P.; Yang, H.; Fan, Q.-R.; Zhu, D.-S. U–Pb and Re–Os geochronology and geochemistry of the Donggebi Mo deposit, Eastern Tianshan, NW China: Insights into mineralization and tectonic setting. *Ore Geol. Rev.* **2017**, *86*, 584–599. [CrossRef]

27. Cao, K.; Xu, J.-F.; Chen, J.-L.; Huang, X.-X.; Ren, J.-B. Origin of porphyry intrusions hosting superlarge Pulang porphyry copper deposit in Yunnan Province: Implications for metallogenesis. *Miner. Depos.* **2014**, *33*, 307–322. (In Chinese)

28. Cao, K.; Yang, Z.-M.; Xu, J.-F.; Fu, B.; Li, W.-K.; Sun, M.-Y. Origin of dioritic magma and its contribution to porphyry Cu–Au mineralization at Pulang in the Yidun arc, eastern Tibet. *Lithos* **2018**. [CrossRef]

29. Xiao, L.; He, Q.; Pirajno, F.; Ni, P.; Du, J.-X.; Wei, Q.-R. Possible correlation between a mantle plume and the evolution of Paleo-Tethys Jinshajiang Ocean: Evidence from a volcanic rifted margin in the Xiaru-Tuoding area, Yunnan, SW China. *Lithos* **2008**, *100*, 112–126. [CrossRef]

30. Roger, F.; Jolivet, M.; Malavieille, J. The tectonic evolution of the Songpan-Garzê (North Tibet) and adjacent areas from Proterozoic to Present: A synthesis. *J. Asian Earth Sci.* **2010**, *39*, 254–269. [CrossRef]

31. Pullen, A.; Kapp, P.; Gehrels, G.E.; Vervoort, J.D.; Ding, L. Triassic continental subduction in central Tibet and Mediterranean-style closure of the Paleo-Tethys Ocean. *Geology* **2008**, *36*, 351–354. [CrossRef]

32. Hou, Z.-Q.; Yang, Y.-Q.; Qu, X.-M.; Huang, D.-H.; Lü, Q.-T.; Wang, H.-P.; Yu, J.-J.; Tang, S.-H. Tectonic evolution and mineralization systems of the Yidun Arc Orogen in Sanjiang Region, China. *Acta Geol. Sin.* **2004**, *78*, 109–120. (In Chinese)

33. Gao, X.; Yang, L.-Q.; Orovan, E.A. The lithospheric architecture of two subterranes in the eastern Yidun Terrane, East Tethys: Insights from Hf–Nd isotopic mapping. *Gondwana Res.* **2018**, *62*, 127–143. [CrossRef]

34. Wang, C.-Y.; Li, W.-C.; Wang, K.-Y.; Zhou, X.-B.; Yin, G.-H.; Yu, H.-J.; Xue, S.-R. Characteristics of fluid inclusions and genesis of Xuejiping copper deposit in Northwestern Yunnan Province. *Acta Petrol. Sin.* **2015**, *31*, 967–978. (In Chinese)

35. Reid, A.J.; Wilson, C.J.; Liu, S. Structural evidence for the Permo-Triassic tectonic evolution of the Yidun Arc, eastern Tibetan Plateau. *J. Struct. Geol.* **2005**, *27*, 119–137. [CrossRef]

36. Yang, T.-N.; Hou, Z.-Q.; Wang, Y.; Zhang, H.-R.; Wang, Z.-L. Late Paleozoic to Early Mesozoic tectonic evolution of northeast Tibet: Evidence from the Triassic composite western Jinsha–Garzê–Litang suture. *Tectonics* **2012**. [CrossRef]

37. Chen, J.-L.; Xu, J.-F.; Ren, J.-B.; Huang, X.-X. Late Triassic E-MORB-like basalts associated with porphyry Cu-deposits in the southern Yidun continental arc, eastern Tibet: Evidence of slab-tear during subduction? *Ore Geol. Rev.* **2017**, *90*, 1054–1062. [CrossRef]

38. Hou, K.-J.; Li, Y.-H.; Tian, Y.-R. In situ U−Pb zircon dating using, laser ablation–multiion counting−ICP−MS. *Miner. Depos.* **2009**, *28*, 481–492. (In Chinese)

39. Griffin, W.L.; Powell, W.J.; Pearson, N.J.; O'Reilly, S.Y. GLITTER: Data reduction software for laser ablation ICP–MS. In *Laser Ablation–ICP–MS in the Earth Sciences: Current Practices and Outstanding Issues*; Mineralogical Association Canada Short Course; Sylvester, P., Ed.; Mineralogical Association Canada: Quebec City, QC, Canada, 2008; Volume 40, pp. 308–311.

40. Ludwig, K.R. User's manual for Isoplot 3.0: A Geochronological Toolkit for Microsoft Excel. *Berkeley Geochronol. Cent. Spec. Publ.* **2003**, *4*, 25–32.

41. Anderson, T. Correction of common lead in U–Pb analyses that do not report ^{204}Pb. *Chem. Geol.* **2002**, *192*, 59–79. [CrossRef]

42. Shirey, S.B.; Walker, R.J. Carius tube digestion for low-blank rheniumosmium analysis. *Anal. Chem.* **1995**, *67*, 2136–2141. [CrossRef]

43. Du, A.-D.; Wu, S.-Q.; Sun, D.-Z.; Wang, S.-X.; Qu, W.-J.; Markey, R.; Stain, H.; Morgan, J.; Malinovskiy, D. Preparation and certification of Re–Os dating reference materials: Molybdenite HLP and JDC. *Geost. Geoanal. Res.* **2004**, *28*, 41–52. [CrossRef]

44. Smoliar, M.I.; Walker, R.J.; Morgan, J.W. Re–Os ages of group IIA, IIIA, IVA, and IVB iron meteorites. *Science* **1996**, *271*, 1099–1102. [CrossRef]

45. Hoskin, P.W.O.; Schaltegger, U. The composition of zircon and igneous and metamorphic petrogenesis. *Rev. Miner. Geochem.* **2003**, *53*, 27–62. [CrossRef]

46. Yang, Q.; Ren, Y.-S.; Sun, Z.-M.; Hao, Y.-J.; Zhang, B.; Sun, X.-H.; Lu, S.-Y. Geochronologic evidence of Late Paleozoic magmatic-hydrothermal mineralization in Tianbaoshan metallogenic region, Yanbian Area: A case study of Xinxing lead-zinc (silver) deposit. *Acta Petrol. Sin.* **2018**, *34*, 3153–3166. (In Chinese)

47. Qiu, K.-F.; Yu, H.-C.; Gou, Z.-Y.; Liang, Z.-L.; Zhang, J.-L.; Zhu, R. Nature and origin of Triassic igneous activity in the Western Qinling Orogen: The Wenquan composite pluton example. *Int. Geol. Rev.* **2018**, *60*, 242–266. [CrossRef]

48. Geng, J.-Z.; Qiu, K.-F.; Gou, Z.-Y.; Yu, H.-C. Tectonic regime switchover of Triassic Western Qinling Orogen: Constraints from LA-ICP-MS zircon U–Pb geochronology and Lu–Hf isotope of Dangchuan intrusive complex in Gansu, China. *Chem. Erde* **2017**, *77*, 637–651. [CrossRef]

49. Gou, Z.-Y.; Yu, H.-C.; Qiu, K.-F.; Geng, J.-Z.; Wu, M.-Q.; Wang, Y.-G.; Yu, M.-H.; Li, J. Petrogenesis of ore-hosting diorite in the Zaorendao gold deposit at the Tongren-Xiahe-Hezuo polymetallic district, West Qinling, China. *Minerals* **2019**, *9*, 76. [CrossRef]

50. Yu, H.-C.; Guo, C.-A.; Qiu, K.-F.; McIntire, D.; Jiang, G.-P.; Gou, Z.-Y.; Geng, J.-Z.; Pang, Y.; Zhu, R.; Li, N.-B. Geochronological and geochemical vonstraints on the formation of the giant Zaozigou Au-Sb deposit, West Qinling, China. *Minerals* **2019**, *9*, 37. [CrossRef]

51. Irvine, T.H.; Baragar, W.R.A. A guide to the chemical classification of the common volcanic rocks. *Can. J. Earth Sci.* **1971**, *8*, 523–548. [CrossRef]

52. Peccerillo, A.; Taylor, A.R. Geochemistry of Eocene calc-alkaline volcanic rocks from the Kastamonu area, Northern Turkey. *Contrib. Miner. Petrol.* **1976**, *58*, 63–81. [CrossRef]

53. Middlemost, E.A.K. A simple classification of volcanic rocks. *Bull. Volcanol.* **1972**, *36*, 382–397. [CrossRef]

54. Maniar, P.D.; Piccoli, P.M. Tectonic discrimination of granitoids. *Geol. Soc. Am. Bull.* **1989**, *101*, 635–643. [CrossRef]

55. Pearce, J.A. Geochemical fingerprinting of oceanic basalts with applications to ophiolite classification and the search for Archean oceanic crust. *Lithos* **2008**, *100*, 14–48. [CrossRef]

56. Boynton, W.V. Geochemistry of the rare earth elements: Meteorite studies. In *Rare Earth Element Geochemistry*; Henderson, P., Ed.; Elsevier: Amsterdam, The Netherlands, 1984; pp. 63–114.

57. Sun, S.-S.; McDonough, W.F. Chemical and isotopic systematics of oceanic basalts: Implications for mantle composition and processes. *Spec. Publ. Geol. Soc. Lond.* **1989**, *42*, 313–345. [CrossRef]

58. Yang, Q.; Ren, Y.-S.; Chen, S.-B.; Zhang, G.-L.; Zeng, Q.-H.; Hao, Y.-J.; Sun, X.-H.; Li, J.-M.; Yang, Z.-J.; Sun, Z.-M. Ore-forming fluid evolution and materiel sources of the giant Pulang porphyry Cu (–Mo–Au) deposit in Northwestern Yunnan, China: Evidence from fluid inclusions and C–H–O–S–Pb isotopes. *J. Geochem. Explor.* **2018**. under review.

59. Hou, Z.-Q.; Qu, X.-M.; Zhou, J.-R.; Yang, Y.-Q.; Huang, Q.-H.; Lv, Q.-T.; Tang, S.-H.; Yu, J.-J.; Wang, H.-P.; Zhao, J.-H. Collision-Orogenic Process of the Yidun Arc in the Sanjiang Region: Record of Granites. *Acta Geol. Sin.* **2001**, *4*, 484–497. (In Chinese)

60. Hou, Z.-Q.; Zaw, K.; Pan, G.-T.; Mo, X.-X.; Xu, Q.; Hu, Y.-Z.; Li, X.-Z. Sanjiang Tethyan metallogenesis in SW China: Tectonic setting, metallogenic epochs and deposit types. *Ore Geol. Rev.* **2007**, *31*, 48–87. [CrossRef]

61. Reid, A.; Wilson, C.J.L.; Shun, L.; Pearson, N.; Belousova, E. Mesozoic plutons of the Yidun Arc, SW China: U/Pb geochronology and Hf isotopic signature. *Ore Geol. Rev.* **2007**, *31*, 88–106. [CrossRef]

62. Defant, M.J.; Drummond, M.S. Derivation of some modern arc magmas by melting of young subducted lithosphere. *Nature* **1990**, *347*, 662–665. [CrossRef]

63. Martin, H.; Smithies, R.H.; Rapp, R.; Moyen, J.F.; Champion, D. An overview of adakite, tonalite-trondhjemite-granodiorite (TTG), and sanukitoid: Relationships and some implications for crustal evolution. *Lithos* **2005**, *79*, 1–24. [CrossRef]

64. Richards, J.P.; Kerrich, B. Adakite-like rocks: Their diverse origins and questionable role in metallogenesis. *Econ. Geol.* **2007**, *102*, 537–576. [CrossRef]

65. Moyen, J.F. High Sr/Y and La/Yb ratios: The meaning of the "adakitic signature". *Lithos* **2009**, *112*, 556–574. [CrossRef]

66. Rapp, R.P.; Watson, E.B. Dehydration melting of metabasalt at 8–32-kbar: Implications for continental growth and crust–mantle recycling. *J. Petrol.* **1995**, *36*, 891–931. [CrossRef]

67. Sen, C.; Dunn, T. Dehydration melting of a basaltic composition amphibolite at 1.5 and 2.0 GPa: Implications for the origin of adakites. *Contrib. Miner. Petrol.* **1994**, *117*, 394–409. [CrossRef]

68. Qiu, K.-F.; Deng, J.; Taylor, R.D.; Song, K.-R.; Song, Y.-H.; Li, Q.Z.; Goldfarb, R.J. Paleozoic magmatism and porphyry Cu-mineralization in an evolving tectonic setting in the North Qilian Orogenic Belt, NW China. *J. Asian Earth Sci.* **2016**, *122*, 20–40. [CrossRef]

69. Qiu, K.-F.; Deng, J. Petrogenesis of granitoids in the Dewulu skarn copper deposit: Implications for the evolution of the Paleotethys Ocean and mineralization in Western Qinling, China. *Ore Geol. Rev.* **2017**, *90*, 1078–1098. [CrossRef]

70. Stern, C.R.; Killian, R. Role of the subducted slab, mantle wedge and continental crust in the genesis of adakites from the Andean Austral Volcanic Zone. *Contrib. Miner. Petrol.* **1996**, *123*, 263–281. [CrossRef]

71. Castillo, P.R.; Janney, P.E.; Solidum, R.U. Petrology and geochemistry of Camiguin Island, southern Philippines: Insights to the source of adakites and other lavas in a complex arc setting. *Contrib. Miner. Petrol.* **1999**, *134*, 33–51. [CrossRef]

72. Kay, S.M.; Mpodozis, C. Magmatism as a probe to the Neogene shallowing of the Nazca plate beneath the modern Chilean flat-slab. *J. S. Am. Earth Sci.* **2002**, *15*, 39–57.

73. Pe-Piper, G.; Piper, J.W. Miocene magnesian andesites and dacites, Evia, Greece: Adakites associated with subducting slab detachment and extension. *Lithos* **1994**, *31*, 125–140. [CrossRef]

74. Atherton, M.P.; Petford, N. Generation of sodium-rich magmas from newly underplated basaltic crust. *Nature* **1993**, *362*, 144–146. [CrossRef]

75. Condie, K.C. TTGs and adakites: Are they both slab melts? *Lithos* **2005**, *80*, 33–44. [CrossRef]

76. Rapp, R.P.; Shimizu, N.; Norman, M.D.; Applegate, G.S. Reaction between slab-derived melts and peridotite in the mantle wedge: Experimental constraints at 3.8 GPa. *Chem. Geol.* **1999**, *160*, 335–356. [CrossRef]

77. Kelemen, P.B.; Hanghøj, K.; Greene, A.R. One view of the geochemistry of subduction-related magmatic arcs, with an emphasis on primitive andesite and lower crust. *Treatise Geochem.* **2014**, *4*, 749–806.

78. Gill, J.B. *Orogenic Andesites and Plate Tectonics*; Springer: New York, NY, USA, 1981; pp. 1–392.

79. Martin, H. Effect of steeper Archean geothermal gradient on geochemistry of subduction-zone magmas. *Geology* **1986**, *14*, 753–756. [CrossRef]

80. Pearce, J.A.; Harris, N.B.W.; Tindle, A.G. Trace element discrimination diagrams for the tectonic interpretation of granitic rocks. *J. Petrol.* **1984**, *25*, 956–983. [CrossRef]

81. Pearce, J.A. Sources and settings of granitic rocks. *Episodes* **1996**, *19*, 120–125.
82. Harris, N.B.W.; Pearce, J.A.; Tindle, A.G. Geochemical characteristics of collision-zone magmatism. *Geol. Soc. Lond. Spec. Publ.* **1986**, *19*, 67–81. [CrossRef]
83. Sun, W.-D.; Zhang, H.; Ling, M.-X.; Ding, X.; Chung, S.-L.; Zhou, J.-B.; Yang, X.-Y.; Fan, W.-M. The genetic association of adakites and Cu–Au ore deposits. *Int. Geol. Rev.* **2011**, *53*, 691–703. [CrossRef]
84. Sun, W.-D.; Ling, M.-X.; Chung, S.-L.; Ding, X.; Yang, X.-Y.; Liang, H.-Y.; Fan, W.-M.; Richard, G.; Yin, Q.-Z. Geochemical Constraints on Adakites of Different Origins and Copper Mineralization. *Geol. J.* **2012**, *120*, 105–120. [CrossRef]
85. Mungall, J.E. Roasting the mantle: Slab melting and the genesis of major Au and Au-rich Cu deposits. *Geology* **2002**, *30*, 915–918. [CrossRef]
86. Qiu, K.-F.; Yang, L.-Q. Genetic feature of monazite and its U–Th–Pb dating: Critical considerations on the tectonic evolution of Sanjiang Tethys. *Acta Petrol. Sin.* **2011**, *27*, 2721–2732. (In Chinese)
87. Qiu, K.-F.; Li, N.; Taylor, R.D.; Song, Y.-H.; Song, K.-R.; Han, W.-Z.; Zhang, D.-X. Timing and duration of metallogeny of the Wenquan deposit in the West Qinling, and its constraint on a proposed classification for porphyry molybdenum deposits. *Acta Petrol. Sin.* **2014**, *30*, 2631–2643. (In Chinese)
88. Qiu, K.-F.; Song, K.-R.; Song, Y.-H. Magmatic-hydrothermal fluid evolution of the Wenquan porphyry molybdenum deposit in the north margin of the Western Qinling, China. *Acta Petrol. Sin.* **2015**, *31*, 3391–3404. (In Chinese)
89. Qiu, K.-F.; Taylor, R.D.; Song, Y.-H.; Yu, H.-C.; Song, K.-R.; Li, N. Geologic and geochemical insights into the formation of the Taiyangshan porphyry copper–molybdenum deposit, Western Qinling Orogenic Belt, China. *Gondwana Res.* **2016**, *35*, 40–58. [CrossRef]
90. Qiu, K.-F.; Marsh, E.; Yu, H.-C.; Pfaff, K.; Gulbransen, C.; Gou, Z.-Y.; Li, N. Fluid and metal sources of the Wenquan porphyry molybdenum deposit, Western Qinling, NW China. *Ore Geol. Rev.* **2017**, *86*, 459–473. [CrossRef]
91. Oyarzun, R.; Marquez, A.; Lillo, J.; Lopez, I.; Rivera, S. Giant versus small porphyry copper deposits of Cenozoic age in northern Chile: Adakitic versus normal calcalkaline magmatism. *Miner. Depos.* **2001**, *36*, 794–798. [CrossRef]
92. Sun, W.-D.; Bennett, V.C.; Eggins, S.M.; Arculus, R.J.; Perfit, M.R. Rhenium systematics in submarine MORB and back-arc basin glasses: Laser ablation ICP-MS results. *Chem. Geol.* **2003**, *196*, 259–281. [CrossRef]
93. McDonough, W.F.; Sun, S.S. The composition of the earth. *Chem. Geol.* **1995**, *120*, 223–253. [CrossRef]

 *minerals*

Article

Geology, Geochronology and Geochemistry of Weilasituo Sn-Polymetallic Deposit in Inner Mongolia, China

Fan Yang [1,2], Jinggui Sun [1,*], Yan Wang [2], Junyu Fu [2], Fuchao Na [2], Zhiyong Fan [3] and Zhizhong Hu [4]

[1] College of Earth Sciences, Jilin University, Changchun 130061, China; fanyang19870509@163.com
[2] Shenyang Geological Survey Center, CGS, Shenyang 110034, China; wy68413@163.com (Y.W.); fjyzxy@163.com (J.F.); fuchaona1986@sina.com (F.N.)
[3] Inner Mongolia Weilasituo Mining Industry Co. Ltd., Chifeng 025350, China; brkyf98@163.com
[4] Chengdu Geological Survey Center, CGS, Chengdu 610000, China; hzz_pot@aliyun.com
* Correspondence: sunjinggui@jlu.edu.cn; Tel.: +86-024-86002927

Received: 2 December 2018; Accepted: 10 February 2019; Published: 12 February 2019

Abstract: The recently discovered Weilasituo Sn-polymetallic deposit in the Great Xing'an Range is an ultralarge porphyry-type deposit. The mineralization is closely associated with an Early Cretaceous quartz porphyry. Analysis of quartz porphyry samples, including zircon U-Pb dating and Hf isotopies, geochemical and molybdenite Re-Os isotopic testing, reveals a zircon U-Pb age of 138.6 ± 1.1 Ma and a molybdenite Re-Os isotopic age of 135 ± 7 Ma, suggesting the concurrence of the petrogenetic and metallogenic processes. The quartz porphyry has high concentrations of SiO_2 (71.57 wt %–78.60 wt %), Al_2O_3 (12.69 wt %–16.32 wt %), and $K_2O + Na_2O$ (8.85 wt %–10.44 wt %) and A/CNK ratios from 0.94–1.21, is mainly peraluminous, high-K calc-alkaline I-type granite and is relatively rich in LILEs (large ion lithophile elements, e.g., Th, Rb, U and K) and HFSEs (high field strength elements, e.g., Hf and Zr) and relatively poor in Sr, Ba, P, Ti and Nb. The zircon $\varepsilon_{Hf}(t)$ values range from 1.90 to 6.90, indicating that the magma source materials were mainly derived from the juvenile lower crust and experienced mixing with mantle materials. Given the regional structural evolution history, we conclude that the ore-forming magma originated from lower crust that had thickened and delaminated is the result of the subduction of the Paleo–Pacific Ocean. Following delamination, the lower crustal material entered the underlying mantle, where it was partially melted and reacted with mantle during ascent. The deposit formed at a time of transition from post-orogenic compression to extension following the subduction of the Paleo–Pacific Ocean.

Keywords: zircon U-Pb dating; molybdenite Re-Os dating; zircon Hf isotopes; Weilasituo Sn-polymetallic deposit; Inner Mongolia

1. Introduction

The southern Great Xing'an Range (SGXR) in southeastern Inner Mongolia is part of the eastern section of the Central Asian Orogenic Belt (Xing'anling Mongolian Orogenic belt). The SGXR is a superimposed compound structural region that has experienced processes associated with the evolution of the Paleo–Asian Ocean and the subduction of the Paleo–Pacific Ocean [1–3]. Multistage tectonic, magmatic, sedimentary and metamorphic events characterize the metallogenic background of the SGXR and have resulted in the development of extensive granitic plutonic rocks, intermediate-acidic volcanic-sedimentary rocks and Paleozoic strata [4–6]. Additionally, these events have given rise to associated porphyry-type molybdenum deposits, magmatic hydrothermal vein-type Sn-polymetallic (Pb-Zn-Ag) deposits, and skarn-type Pb-Zn-Ag deposits (Figure 1). The metallogenic ages of these deposits are mostly concentrated in the (I) Mid-Triassic (244–234 Ma; e.g., the Laojiagou Mo deposit and Baiyinnuoer Ag-Zn deposit; [7,8]), (II) Early to Mid-Jurassic (179 Ma; e.g., the Mengentaolegai Ag-Pb-Zn deposit; [9]), (III) Late Jurassic (150–153 Ma; e.g., the Budunhua Cu deposit, Hashitu Mo deposit, Jiguanshan Mo deposit, Zhanzigou Mo deposit, Maodeng Cu-Sn deposit and Aonaodaba Ag-Cu deposit, [10–13]), and (IV) the early stage of the Early Cretaceous (141–133 Ma; the Dajing Sn-polymetallic deposit, Weilasituo Sn-polymetallic deposit, Bairendaba Pb-Zn-Ag deposit, Huanggang Sn-Fe deposit, Haobugao Fe-Zn deposit, Chamuhan Sn-polymetallic deposit, Anle Ag-Sn deposit, Aolunhua Mo deposit, Banlashan Mo deposit, Yangchang Mo deposit, Hongshanzi Mo deposit, Xiaodonggou Mo deposit, and Gangzi Mo deposit; [1,7,14–24]).

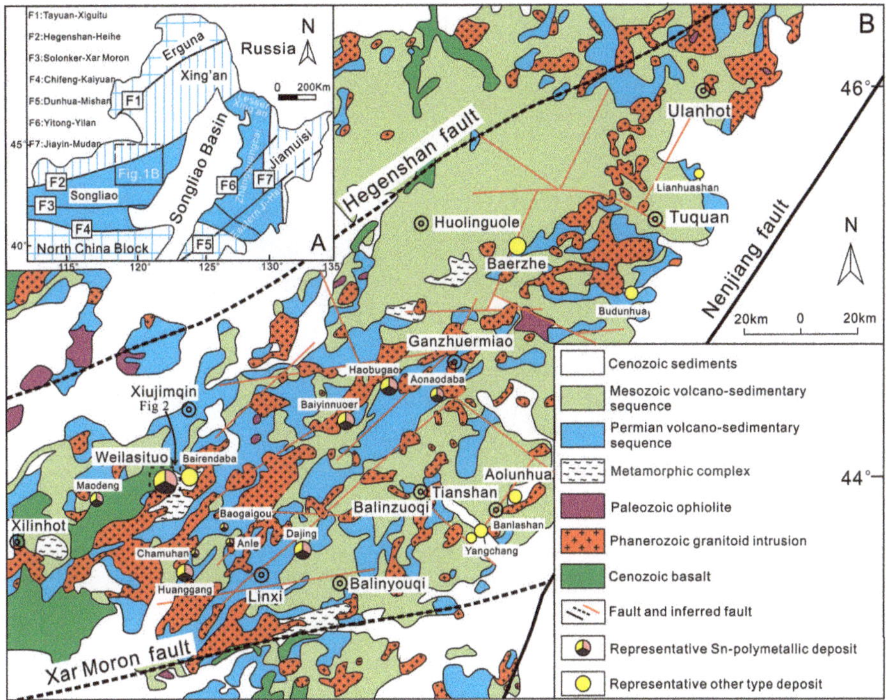

Figure 1. (**A**) Tectonic map of northeastern China (modified after [25]). (**B**) Simplified geologic map of the southern Great Xing'an Range (modified after [26]).

The Weilasituo deposit is a large recently discovered porphyry-type Sn-polymetallic deposit in the SGXR. Notably, most Sn-polymetallic deposits in this area are hydrothermal vein-type deposits and the porphyry-type Sn-polymetallic deposits are rare. The Weilasituo porphyry-type Sn-polymetallic deposit has estimated reserves of 93,127 t Sn at 1.35%, 15,800 t W at 0.44%, 32,540 t Zn at 1.63% and 9175 t Cu at 1.91% [16]. Recent studies have examined the geological characteristics, geochemistry, geochronology, and metallogenic mechanism of the Weilasituo Sn-polymetallic deposit [16,17,27,28], and have applied several geochronological methods to constrain the timing of petrogenesis and metallogenesis. The zircon U-Pb ages of the quartz porphyry are 138 ± 2 Ma and 135.7 ± 0.9 Ma [16,27]. The molybdenite Re-Os isochron ages are 135 ± 11 Ma and 125.7 ± 3.8 Ma [16,27]. The cassiterite U-Pb ages of the disseminated mineralization and vein-like mineralization are 138 ± 6 Ma and 135 ± 6 Ma, respectively [17]. Geochemical studies have determined the mineralized rock (i.e., the quartz porphyry) is a highly differentiated I-type granite by fractional crystallization [17].

In this contribution, we present LA-ICP-MS zircon U-Pb ages, molybdenite Re-Os ages, new whole-rock elemental compositions of the quartz porphyry and in situ zircon Hf isotopic data for the Weilasituo Sn-polymetallic deposit. We integrate these data with the results of previous research, enabling us to constrain the petrogenesis of the quartz porphyry and the mineralization timing of the deposit to evaluate the relationships between magmatism and metallogeny.

2. Regional Geology

The Weilasituo Sn-polymetallic deposit is located in the SGXR and is separated from the northern edge of the North China Block by the Xar Moron Fault to the south, from the Songliao Basin by the Nenjiang Fault to the east and from the Erguna-Xing'an Block by the Erlianhaote-Hegenshan Fault to the north (Figure 1B). Tectonically, this deposit forms part of the eastern side of the Central Asian Orogenic Belt (Figure 1A). From the Paleozoic through the Triassic, the area was influenced by the evolution of the Paleo–Asian Ocean [29], and experienced multiple collisions, accretion and regional extension events [30–33]. The SGXR was influenced by the subduction of the Paleo–Pacific Ocean during the Mesozoic [11,32], resulting in collision-related compression after extension, and also widespread late Mesozoic granitoids and volcanic rocks [11,33,34].

The oldest strata in the area consist of a Paleozoic, intermediate to high-grade metamorphic complex termed the "Xilin hot complex", which comprises amphibolite, plagioclase-bearing gneiss, mica schist and biotite-bearing granitic gneiss. In addition, Ordovician, Silurian, Devonian and Carboniferous clastic sedimentary, volcanic and carbonate rocks also crop out in the area [22]. The Permian volcanic-sedimentary strata are composed of carbonaceous clastic, volcanic and carbonate rocks [35].

The Paleozoic strata are intruded by Hercynian–Yanshanian granitic plutons and are overlain by Mesozoic volcanic-sedimentary sequences [36]. Paleozoic granitoids, tonalite, diorite and granodiorite, are chiefly located in the western part of the SGXR (Figure 1A), and the ages of zircon U-Pb range 321–237 Ma [28]. These granitoids comprise a post-subduction, high-potassium calc-alkaline magmatic suite and were generated under the geodynamic regime produced by the slab break-off of the Paleo–Asian Ocean [37]. The Mesozoic granitoids include monzogranite, granodiorite, and syenogranite, with ages ranging 150–131 Ma [28]. these granitoids show positive εNd values, suggesting that the pre-existing juvenile and ancient crustal components underwent remelting, recycling, and redistribution [38].

Mesozoic volcanic-sedimentary sequences are the stratigraphically highest rocks in the SGXR (Figure 1B). From bottom to top, these volcanic rocks can be subdivided into Manketouebo, Manitu, Baiyingaolao, and Meiletu Formations [39,40]. Zircon U-Pb ages show that Mesozoic volcanic activity occurred throughout the SGXR and began in the Late Jurassic and peaked in the Cretaceous [26]. Geochemically, these volcanic rocks are diverse, including trachyandesites, basaltic trachyandesites, rhyolites and trachytes, with smaller quantities of dacites and basalts [26]. The chemical properties of the mafic-intermediate rocks are similar to basalts formed in an intraplate extensional zone. The mafic-intermediate rocks have the characteristics of slightly enriched Sr isotope ratios and weakly depleted to slightly enriched Nd isotope ratios. These features show that the mafic-intermediate rocks were derived from weakly depleted to enriched continental lithospheric mantle [41].

3. Deposit Geology

In addition to the extensively exposed Holocene strata, the other rocks in the area include the Xilin hot complex, quartz diorite, cryptoexplosive breccia and buried quartz porphyry (Figure 2A,B). The Xilin hot complex is a combination of brown amphibole–plagioclase and biotite–plagioclase gneiss, both intensely metamorphosed and deformed. The quartz diorite is less extensively exposed and subject to NE-trending faults [42]. The cryptoexplosive breccia is spatially linked with the quartz porphyry in the upper part and it usually occurs at the contact between the Xilin hot complex and the quartz porphyry [17]. The buried quartz porphyry, whose overall morphology is unknown to date, lies approximately 400 m beneath the surface. The quartz porphyry samples recovered from a drillhole have an off-white color, a massive structure and a porphyritic texture (Figure 3A,B). The main minerals are K-feldspar (65%–70%), quartz (25%–27%) and biotite (3%–5%).

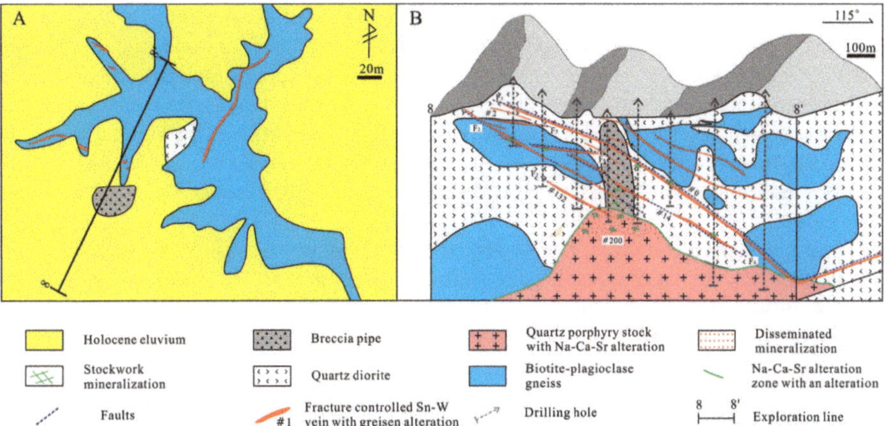

Figure 2. (**A**) Geological sketch map of the Weilasituo Sn-polymetallic deposit; (**B**) Schematic map of main orebodies of the Weilasituo Sn-polymetallic deposit (modified after [17]).

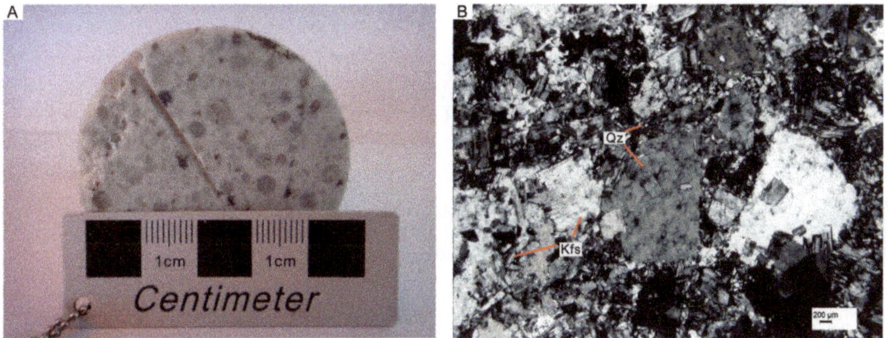

Figure 3. (**A**) Characteristic features of quartz porphyry; (**B**) Photomicrographs of quartz porphyry. Abbreviations: Qz—quartz; Kfs—K-feldspar.

The faults in the area trend primarily NE–SW and secondarily sub-E–W (Figure 2B). They are developed in the quartz diorite and, as the main ore-controlling structures, provided spaces for mineralization. The NE-trending faults F_1, F_2, F_3 and F_4 strike 20°–43°, dip from 18°–35°, and extend approximately 30–900 m with widths of 0.3–14 m. The E–W-trending faults (F_5 and F_6), located above or outside of the quartz porphyry, strike 85°–115°, dip 20°–25°, and extend approximately 20–800 m [17].

Mineralization and Alteration

The mining area is currently under mineral exploration, and ores have been encountered and verified by drilling. Ore bodies in the area come in three forms: (I) shallow quartz vein-type ore bodies, (II) moderate-depth cryptoexplosive breccia-type ore bodies, and (III) deep quartz porphyry-type ore bodies (Figure 2B).

Fifteen shallow quartz vein-type ore bodies of medium size or greater have been detected. These vein-like or stratiform subparallel ore bodies strike 25°–205°, dip towards 115° at 11°–54°, and exhibit a fairly simple morphology with basically continuous strike and some branching-reconnecting along dip. These orebodies extend 32–1400 m horizontally and 43–1300 m vertically, and their thicknesses range from 0.1 to 13 m. The cryptoexplosive breccia-type ore bodies are found in the cryptoexplosive breccia. Their morphology is associated with the internal structural morphology of the cryptoexplosive breccia. Local quartz vein-type ore bodies are found to cut the breccia-type ore bodies, adding to the complexity of these ore bodies. The quartz porphyry-type ore bodies found at the upper contact of the quartz porphyry bodies occur in lenticular form and are morphologically similar to the ore bodies above the quartz porphyry bodies, and those found inside the quartz porphyry bodies occur in a vein-like form. Seven ore bodies of industrial size, represented by Sn #200 (Figure 2B), represent more than 80% of the total quartz porphyry-type ore-body resources in the area. The ore bodies vary in size, extending dozens of meters to 400 m horizontally and 70 m vertically. Their thickness ranges from 1 to 10 m.

Three types of mineralization that are recognized in the Weilasituo deposit: stockwork mineralization, disseminated mineralization and vein-like mineralization. Stockwork mineralization exists both inside and outside the quartz porphyry and inside the cryptoexplosive breccia pipe. Disseminated mineralization occurs in the quartz porphyry. Vein-like mineralization occurs outside the quartz porphyry and inside the cryptoexplosive breccia pipe.

The mineralization level decreases with increasing depth. Stockwork and disseminated mineralization have resulted in disseminated and vein-like ores (Figure 4A–D). The ore minerals are mainly cassiterite + wolframite + sphalerite with lower amounts of chalcopyrite + arsenopyrite + tetrahedrite. Disseminated mineralization is the most important form of mineralization (Figure 4H–L). Vein-like mineralization is typically found in cracks and faults on top or outside of the quartz porphyry (Figure 2B) and is responsible for vein-like ores. The ore minerals include cassiterite + wolframite + sphalerite and less amounts of chalcopyrite + tetrahedrite + pyrrhotite + molybdenite + galena. Breccia pipe mineralization occurs in steep breccia pipes (Figures 2B and 4F), with no distinct boundary with the country rock. Disseminated and spots of cassiterite, chalcopyrite and dark sphalerite occur commonly as fragments in the breccia, suggesting that the breccia mineralization occurred after the disseminated mineralization. In addition, vein-like Sn-polymetallic mineralization was found to cut into the quartz porphyry-related breccia pipe mineralization (Figure 2B), suggesting that vein-like mineralization occurred after breccia mineralization.

The Sn-polymetallic metallogenic system in the Weilasituo deposit presents a high-to-low-temperature alteration zone from the quartz porphyry outward. The center of alteration is a Na-Ca-Rb alteration combination, surrounded by greisenization with high-level Na-Ca-Rb alteration. Stockwork and disseminated mineralization can be observed, typically in the quartz porphyry-type and breccia-type ore bodies. The alteration minerals are albite + amazonite + topaz + celestine + muscovite + epidote + fluorite (Figure 4A–C). Greisenization alteration mainly occurs in the vein-like orebodies on top or outside of the quartz porphyry, although some can also be found in the overlying breccia pipes. The alteration is controlled by fissures and faults [17]. The alteration minerals include muscovite + quartz + topaz + fluorite + zircon (Figure 4E).

According to field contacts, the ore mineral and gangue mineral combinations, and the interweaving pattern among veins, are chronologically divided into two stages of mineralization: early Na-Ca-Rb alteration-induced Sn-W-Rb mineralization and late greisenization-induced Sn-W-Rb mineralization. The mineral combination of the early stage is cassiterite + sphalerite + wolframite + stannite + tetrahedrite + chalcopyrite + arsenopyrite (Figure 4A–C) and that of the latter is quartz + cassiterite + muscovite + fluorite + wolframite + loellingite + sphalerite + chalcopyrite + galena + molybdenite (Figure 4H–L).

Figure 4. Photographs of mineralization features in the Weilasituo Sn-polymetallic deposit: (**A**) Disseminated cassiterite and sphalerite in quartz porphyry; (**B**) Amazonite in the quartz porphyry; (**C**) Cassiterite and galena in the quartz porphyry; (**D**) Euhedral cassiterite; (**E**) Typical greisen with wolframite; (**F**) Breccia collected from the surface; (**G**) Cassiterite and sphalerite intergrowths; (**H**) Sphalerite, arsenopyrite and tetrahedrite; (**I**) Cassiterite, sphaletire, arsenopyrite, pyrrhotite and chalcopyrite; (**J**) Greisen vein with sphaletire, tetrahedrite and chalcopyrite; (**K**) Greisen vein with molybdenite and cassiterite; (**L**) Greisen vein with cassiterite, arsenopyrite and chalcopyrite. Abbreviations: Cas—cassiterite, Am—amazonite, Apy—arsenopyrite, Ccp—chalcopyrite, Gn—galena, Po—pyrrhotite, Qz—quartz, Tet—tetrahedrite, Sp—sphalerite, Mol—molybdenite, Wol—wolframite, Mus—muscovite.

4. Sampling and Analytical Methods

Unaltered quartz porphyry samples for LA-ICP-MS U-Pb dating, zircon Lu-Hf isotope analysis and whole rock geochemistry analysis were collected from drillhole ZK809. Molybdenite samples for Re-Os isotopic analysis were also gathered from drillhole ZK809 of the porphyry Sn-W-Rb mineralization zone.

4.1. Zircon LA-ICP-MS U-Pb Dating

Two zircon U-Pb isotopic dating samples (ZK809-1 and ZK809-2) were analyzed at the Key Laboratory for Sedimentary Basin and Oil and Gas Resources, Ministry of Land and Resources. A GeoLasPro 193 nm laser system and an SF-ICPMS Element 2 were used for ablation and mass spectrometry, respectively. High-purity He gas was used as the carrier gas for the ablated material. The test was conducted with a laser wavelength of 193 nm, a beam spot of 32 μm, a pulse frequency of 6 Hz, and a laser energy of 6 J/cm^2. Before the test, the instrument was tuned to the optimal state using the NIST610 standard so that the ^{139}La and ^{232}Th signals were the highest and the oxide yield for ^{232}Th^{16}O/^{232}Th was less than 0.3%. GJ-1 zircon standard specimens were used as external criteria to calibrate the fractionation and mass discrimination of U-Pb isotopes, and Plěsovice zircon standard specimens were used as control specimens to monitor the stability of the analytical process. A set of standard specimens was inserted every five samples. Laser sampling of each sample included 20 s of background acquisition, 50 s of ablation sampling and 10 s of cell flushing. ICPMSDataCal was used to process the analysis data offline [43].

4.2. Molybdenite Re-Os Dating

Five molybdenite Re-Os isotope dating samples (ZK809-1 to ZK809-5) were obtained by separating the molybdenite samples collected from the drillhole (in basically the same way as for zircon) to a size far smaller than 2 mm to avoid decoupling. Re-Os isotopic testing was conducted at the Re-Os isotopic laboratory of the National Geological Test Center. A TJA X-series ICP-MS was used to determine isotopic ratios. For low Re-Os samples, a Thermo Fisher Scientific HR-ICP-MS Element 2 was used. For Re, mass numbers 187 and 185 were selected, and mass number 190 was used to monitor Os. For Os, mass numbers 186–190 and 192 were selected, and mass number 185 was used to monitor Re. Blank values of Os, Re and ^{187}Os measured by the TJA X-series ICP-MS were (0.0031 ± 0.0005) × 10^{-9}, (0.00023 ± 0.00008) × 10^{-9}, and (0.00039 ± 0.00005) × 10^{-9}, respectively, which were far smaller than the Os and Re measurements from the samples and the standard specimens and were therefore insignificant with respect to the test results. Common Os was calculated through the measured ^{192}Os/^{190}Os ratio according to the atomic and isotopic abundance charts. The uncertainties in Os and Re concentrations included weighing errors for the samples and thinner, calibration errors for the thinner, fractionation calibration errors for mass spectrometry, and isotopic ratio measurement errors for the samples in question. The confidence degree can reach 95%. The uncertainties in the model age also included this uncertainty (1.02%). The confidence degree was 95%. National standard materials (GBW04435 (JDC)) were used as standard specimens to monitor the reliability of the chemical process and analysis data. The principles and processes of Re-Os isotope analysis are detailed in Du et al. [44].

4.3. Whole Rock Geochemistry

Eight collected samples were crushed to 200-mesh powder. Major elements were analyzed on an Axios MaxX fluorescence spectrometer. FeO was titrated through a 50-mL burette. The sample was first placed in a Li$_2$B$_4$O$_7$ solution at a ratio of 1:5, melted at 1050–1250 °C, formed into thin glass sections and analyzed with approximate accuracies of 1% for SiO$_2$ and 2% for other oxides. Rare earth elements (REEs) and trace elements were analyzed on an X-series 2 instrument. A sample of whole-rock powder (200 mesh) weighing 50 mg was placed in a Teflon flask, dissolved with HNO$_3$ and HF solutions for 2 days, further dissolved by adding HClO$_4$, dried by distilling, and then diluted to 50 mL with a 5% HNO$_3$ solution before being analyzed for trace elements. During analysis, two GBW series specimens were measured: one for calibrating Li, Ba, Cr, V, Co, Ni, Zn, Cu, Ga, Sr, Rb, Cs, Ba, Pb, U, Th, Sc, Y and REEs and the other for calibrating Mo, W, Nb, Ta, Hf and Zr. The analysis errors were less than 5%. Major elements, trace elements and REEs were measured at the testing center of the Shenyang Geological Survey Center, Ministry of Land and Resources.

4.4. Zircon Lu-Hf Isotope

In situ analysis of the zircon Hf isotope composition of the quartz porphyry from the Weilasituo deposit was conducted at the State Key Laboratory for Mineral Deposits Research, Nanjing University using a Neptune Plus MC-ICP-MS equipped with a New Wave UP193 laser microprobe.

The data acquisition and instrumental conditions were similar with Wu et al. [45]. The analyses were mainly conducted with a repetition rate of 8 Hz and a beam diameter of 35 mm. Ar and He carrier gases were used to transfer the ablated sample from the laser ablation cell to the ICP-MS torch through the mixing chamber. Atomic masses 172, 173, 175–180 and 182 were measured simultaneously in the static collection mode. Corrections of isobaric interferences of ^{176}Lu and ^{176}Yb on ^{176}Hf were based on, Wu et al. [45], and Yang et al. [46]. The performance conditions and analytical accuracy were monitored by the 91500 and Mud Tank zircon standards (^{176}Hf/^{177}Hf = 0.282507 $\pm$ 6 [47]). The initial Hf values (ε_{Hf}) were calculated by using the decay constant of 1.865×10^{-11} per year [48] for ^{176}Lu and the chondritic model with ^{176}Lu/^{177}Hf = 0.0332 and ^{176}Hf/^{177}Hf = 0.282772 [49]. Depleted mantle Hf model ages (T$_{DM}$) were calculated by using the ^{176}Lu/^{177}Hf ratios of the zircon. These ages refer to a model of depleted mantle with a current ^{176}Hf/^{177}Hf ratio of 0.28325 which is similar to the average of MORB [50], which has a ^{176}Lu/^{177}Hf ratio of 0.0384 [51] and assumes an average continental crust ratio (f$_{CC}$) of -0.55 [52].

5. Results

5.1. Zircon U-Pb Dating

Pb isotopic age analysis was conducted on 52 zircon grains from the two quartz porphyry samples (ZK809-1 and ZK809-2). The analytical results are shown in Table 1. All the zircon grains analyzed are idiomorphic crystals that are 35–100 μm long. The aspect ratios are approximately 3:1–1:1 (Figure 5). As the U and Th concentrations are exceptionally high (for most of the analysis points, U is 8721–$48,802 \times 10^{-6}$ but can reach $165,623 \times 10^{-6}$; and Th is 819–$13,033 \times 10^{-6}$ but can reach $734,544 \times 10^{-6}$), most of the zircon grains from the two samples display weak zonal structures (Figure 5). Those showing obvious zones yield ages older than 189 Ma, suggesting they are inherited zircons. The Th/U ratio lies mostly in the 0.10–0.38 range, which also points to a magmatic origin. The zircon ^{206}Pb/^{238}U surface harmonic ages for sample ZK809-1 are 137–140 Ma, and the weighted average age is 138.0 ± 1.1 Ma (MSWD = 0.33; n = 12). The ^{206}Pb/^{238}U surface harmonic ages for sample ZK809-2 are 137–142 Ma, and the weighted average age is 138.6 ± 1.1 Ma (MSWD = 0.49; n = 15). Both ages represent the crystallization of the quartz porphyry. Both samples yield older ages (Figure 6A,C) for the inherited zircons.

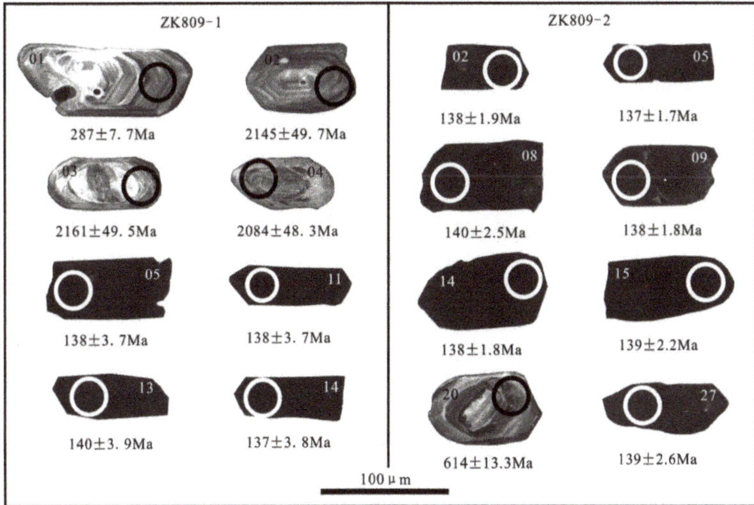

Figure 5. Typical CL images of zircons and analyzed spots from the quartz porphyry in the Weilasituo Sn-polymetallic deposit.

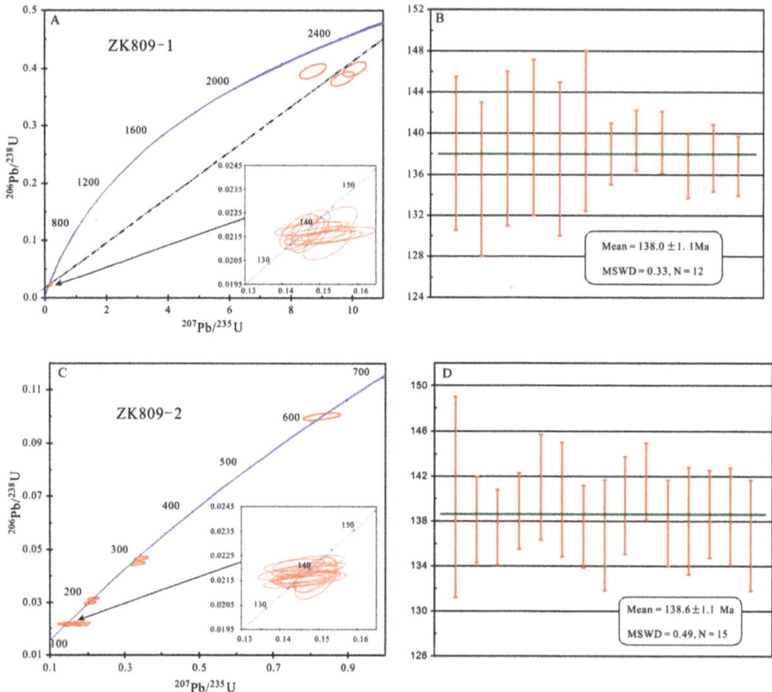

Figure 6. U-Pb Concordia diagram of zircon of quartz porphyry (**A,C**); weighted mean ages of zircon U-Pb dating for quartz porphyry (**B,D**).

5.2. Molybdenite Re-Os Geochronology

Table 2 presents the Re-Os isotopic results of nine molybdenite samples from the deposit. The test results indicate that w(Re) = 55.2–712.5 ng/g, w(Os) = 0.152–9.222 ng/g, w(^{187}Re) = 34.71–447.80 ng/g, and w(^{187}Os) = 0.0216–1.041 ng/g. As the relative concentration of common Os is higher than that of ^{187}Os, the isochron age (135 ± 7 Ma) was interpreted as the formation time of the molybdenite (Figure 7), which indicates that mineralization of the Weilasituo Sn-polymetallic deposit occurred in the early stage of the Early Cretaceous.

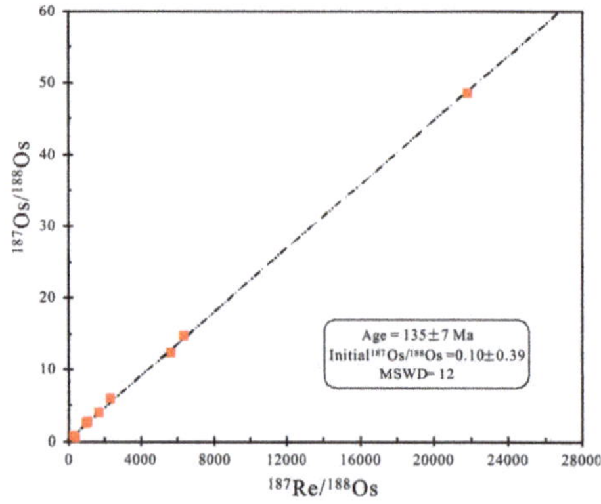

Figure 7. Re-Os isochron ages of quartz porphyry of the Weilasituo Sn-polymetallic deposit.

Table 1. LA-ICP-MS zircon U-Pb dating data of quartz porphyry of Weilasituo deposit.

Sample No.	Th (ppm)	U (ppm)	Th/U	Isotopic Ratios						Ages (Ma)					
				$^{207}Pb/^{206}Pb$		$^{207}Pb/^{235}U$		$^{206}Pb/^{238}U$		$^{207}Pb/^{206}Pb$		$^{207}Pb/^{235}U$		$^{206}Pb/^{238}U$	
				Ratio	1σ	Ratio	1σ	Ratio	1σ	Ages	1σ	Ages	1σ	Ages	1σ
ZK809-1															
ZK809-1-1	122	323	0.38	0.0508	0.0015	0.3419	0.0128	0.0455	0.0013	230	65	299	10	287	7.7
ZK809-1-2	261	337	0.77	0.16	0.0034	8.7234	0.3532	0.3947	0.0108	2455	36	2310	37	2145	49.7
ZK809-1-3	163	560	0.29	0.1619	0.0032	10.0799	0.3126	0.3982	0.0107	2475	33	2442	29	2161	49.5
ZK809-1-4	101	573	0.18	0.1666	0.0033	9.6905	0.3119	0.3816	0.0104	2524	33	2406	30	2084	48.3
ZK809-1-5	1699	11,150	0.15	0.0653	0.0015	0.1987	0.006	0.0216	0.0006	784	46	184	5	138	3.7
ZK809-1-6	2193	14,821	0.15	0.0482	0.001	0.1461	0.0042	0.0216	0.0006	109	48	138	4	138	3.7
ZK809-1-7	4465	18,362	0.24	0.0543	0.0015	0.1521	0.0052	0.0212	0.0006	384	59	144	5	136	3.7
ZK809-1-8	4961	23,997	0.21	0.052	0.0013	0.1619	0.0052	0.0218	0.0006	286	55	152	5	139	3.8
ZK809-1-9	13,033	47,595	0.27	0.0449	0.0009	0.1446	0.0042	0.0217	0.0006	0.1	0	137	4	139	3.8
ZK809-1-10	5151	27,486	0.19	0.0475	0.001	0.1477	0.0042	0.0219	0.0006	73	48	140	4	140	3.8
ZK809-1-11	12,609	48,802	0.26	0.0488	0.001	0.1474	0.0041	0.0216	0.0006	139	46	140	4	138	3.7
ZK809-1-12	3246	23,933	0.14	0.0556	0.0013	0.1745	0.0055	0.0218	0.0006	437	51	163	5	139	3.8
ZK809-1-13	1568	14,876	0.11	0.0492	0.0014	0.1535	0.0054	0.022	0.0006	158	65	145	5	140	3.9
ZK809-1-14	11,780	51,925	0.23	0.0521	0.0011	0.1579	0.0047	0.0214	0.0006	290	49	149	4	137	3.8
ZK809-1-15	3426	13,869	0.25	0.0487	0.0014	0.1503	0.0074	0.0216	0.0002	134	66	142	7	138	1.5
ZK809-1-16	8656	39,837	0.22	0.0475	0.0013	0.1488	0.0073	0.0218	0.0002	72	65	141	6	139	1.5
ZK809-1-17	8103	36,750	0.22	0.0472	0.0014	0.147	0.0073	0.0218	0.0002	59	67	139	6	139	1.5
ZK809-1-18	5164	25,663	0.2	0.0656	0.0021	0.204	0.0105	0.0211	0.0003	792	65	189	9	135	1.6
ZK809-1-19	2363	12,632	0.19	0.0497	0.0015	0.1501	0.0077	0.0215	0.0003	180	71	142	7	137	1.6
ZK809-1-20	4209	21,271	0.2	0.0528	0.0017	0.1551	0.0081	0.0216	0.0003	321	72	146	7	138	1.6
ZK809-1-21	7071	34,791	0.2	0.0481	0.0013	0.1523	0.0074	0.0215	0.0002	104	64	144	7	137	1.5
ZK809-1-22	1050	11,105	0.09	0.0721	0.0022	0.2256	0.0115	0.0215	0.0003	988	61	207	10	137	1.6
ZK809-2															
ZK809-2-1	1157	10,593	0.11	0.0489	0.0012	0.1487	0.0067	0.022	0.0002	144	57	141	6	140	4.5
ZK809-2-2	3862	21,422	0.18	0.048	0.0011	0.1476	0.0066	0.0217	0.0002	97	56	140	6	138	1.9
ZK809-2-3	7743	26,847	0.29	0.0524	0.0016	0.1685	0.0082	0.0217	0.0002	302	67	158	7	138	2.2
ZK809-2-4	236	694	0.34	0.0547	0.0016	0.3359	0.0163	0.0445	0.0005	400	62	294	12	280	2.7
ZK809-2-5	6779	12,142	0.56	0.0475	0.0014	0.1483	0.0073	0.0215	0.0002	72	71	140	6	137	1.7
ZK809-2-6	5844	34,485	0.17	0.0445	0.0011	0.1426	0.0065	0.0218	0.0002	0	0	135	6	139	1.7

Table 1. *Cont.*

Sample No.	Th (ppm)	U (ppm)	Th/U	$^{207}Pb/^{206}Pb$		$^{207}Pb/^{235}U$		$^{206}Pb/^{238}U$		$^{207}Pb/^{206}Pb$		$^{207}Pb/^{235}U$		$^{206}Pb/^{238}U$	
				Isotopic Ratios						**Ages (Ma)**					
				Ratio	1σ	Ratio	1σ	Ratio	1σ	Ages	1σ	Ages	1σ	Ages	1σ
						ZK809-2									
ZK809-2-7	4292	22,936	0.19	0.0464	0.0011	0.1438	0.0064	0.0221	0.0002	16	56	136	6	141	2.3
ZK809-2-8	3150	19,105	0.16	0.0481	0.0011	0.1459	0.0065	0.0219	0.0002	102	55	138	6	140	2.5
ZK809-2-9	3506	10,660	0.33	0.0496	0.0014	0.1471	0.0069	0.0216	0.0002	178	62	139	6	138	1.8
ZK809-2-10	5976	37,780	0.16	0.0438	0.0011	0.1333	0.006	0.0216	0.0002	0	0	127	5	138	1.7
ZK809-2-11	1160	4441	0.26	0.0532	0.0014	0.3462	0.0159	0.0464	0.0005	335	57	302	12	292	4.9
ZK809-2-12	734,544	165,623	4.44	0.4377	0.0102	1.8432	0.082	0.0369	0.0004	4044	35	1061	29	233	4.5
ZK809-2-13	6510	30,010	0.22	0.0463	0.0012	0.144	0.0066	0.0214	0.0002	12	59	137	6	137	2.5
ZK809-2-14	7371	35,849	0.21	0.042	0.001	0.1328	0.006	0.0216	0.0002	0	0	127	5	138	1.8
ZK809-2-15	4028	21,700	0.19	0.0488	0.0012	0.1477	0.0066	0.0219	0.0002	137	55	140	6	139	2.2
ZK809-2-16	1251	12,472	0.1	0.0461	0.0015	0.1475	0.0074	0.0222	0.0003	1.7	75	140	7	142	1.7
ZK809-2-17	5697	17,094	0.33	0.047	0.0011	0.1464	0.0066	0.0216	0.0002	50	57	139	6	138	1.9
ZK809-2-18	11.8	610	0.02	0.0462	0.0021	0.2071	0.0128	0.0297	0.0005	10	106	191	11	189	3.5
ZK809-2-19	7.2	242	0.03	0.049	0.0021	0.2164	0.0128	0.0313	0.0004	146	96	199	11	198	5.2
ZK809-2-20	172	802	0.21	0.0615	0.0017	0.8299	0.0411	0.0999	0.0011	656	58	614	23	614	13.3
ZK809-2-21	7783	35,035	0.22	0.0488	0.0012	0.1456	0.0066	0.0217	0.0002	140	56	138	6	138	2.4
ZK809-2-22	2588	15,834	0.16	0.0426	0.001	0.1292	0.0058	0.0218	0.0002	0	0	123	5	139	8.4
ZK809-2-23	1544	9814	0.16	0.0488	0.0011	0.162	0.005	0.0219	0.0006	139.7	51	153	4	139	2.3
ZK809-2-24	4594	11,160	0.41	0.0545	0.0013	0.1766	0.0057	0.0216	0.0006	390	52	165	5	138	2
ZK809-2-25	1478	10,299	0.14	0.0499	0.0011	0.1675	0.0052	0.0219	0.0006	189	52	157	5	140	2.1
ZK809-2-26	4561	25,814	0.18	0.0462	0.001	0.149	0.0044	0.0217	0.0006	9	49	141	4	139	1.9
ZK809-2-27	819	8721	0.09	0.0593	0.0013	0.1993	0.0061	0.0218	0.0006	579	47	185	5	139	2.6
ZK809-2-28	3215	20,693	0.16	0.0461	0.001	0.1494	0.0044	0.0217	0.0006	4	49	141	4	138	2.2
ZK809-2-29	4454	25,173	0.18	0.048	0.001	0.1493	0.0043	0.0214	0.0006	98	49	141	4	137	2.5
ZK809-2-30	5906	30,761	0.19	0.0558	0.0013	0.1846	0.0059	0.0217	0.0006	443	52	172	5	138	2.3

Table 2. Results of Re-Os isotopic analyses of greisen stage from the Weilasituo Sn-polymetallic deposit.

Sample No.	Weight (g)	Re (ppb)		Common Os (ppb)		^{187}Re (ppb)		^{187}Os (ppb)		^{187}Re/^{188}Os		^{187}Os/^{188}Os	
		Measured	2σ	Measured	2σ	Measured	2σ	Measured	2σ	Measured	2σ	Measured	2σ
ZK809-1	0.01951	552.7	1.6	1.136	0.0039	347.4	1	0.8724	0.0028	2337	23.6	5.852	0.01
ZK809-2	0.0168	491.6	1.4	2.155	0.0069	309	0.9	0.7616	0.0024	1092	11	2.682	0.004
ZK809-3	0.0106	495.8	1.4	9.222	0.2001	311.6	0.9	0.669	0.0125	256.8	6	0.5491	0.0036
ZK809-5	0.0116	306	0.9	1.477	0.0055	192.3	0.6	0.4799	0.0016	998	10.1	2.484	0.004
ZK809-4	0.0106	712.5	2.1	2.011	0.0066	447.8	1.3	1.041	0.0032	1704	17.2	3.954	0.006
IZK02401-2*	0.0497	55.2	0.38	0.698	0.0575	34.71	0.24	0.0216	0.0157	382.1	31.6	0.238	0.174
IZK0403-1*	0.0094	189.5	1.9	0.161	0.0032	119.1	1.2	0.2594	0.007	5677	127	12.36	0.42
IZK0403-3*	0.0289	43.31	0.42	0.033	0.0004	27.22	0.26	0.0627	0.0009	6382	96	14.71	0.28
IZK1501-1*	0.006	685.6	10.1	0.152	0.0018	430.9	6.4	0.9581	0.0122	21854	410	48.59	0.84

Note: * samples from [16].

5.3. Major and Trace Elements

Whole-rock geochemical measurements of quartz porphyry samples from the Weilasituo deposit are shown in Table 3. The samples are rich in silicon, alkalis and aluminum. The analyses indicate SiO_2, $Na_2O + K_2O$ and Na_2O/K_2O ranges of 71.57 wt %–78.60 wt %, 8.85 wt %–10.44 wt %, and 1.37–2.94. Therefore, these rocks belong to the high-K calc-alkaline series (Figure 8A, [53]). The Al_2O_3 values vary from 12.69 wt %–16.32 wt %, the A/CNK values range from 0.94 to 1.21, and the A/NK values range from 1.40 to 1.75, indicating that these samples are high-K calc-alkaline, weakly meta aluminous to peraluminous rocks (Figure 8B; [54]). In the TAS diagram (Figure 8C; [55]), the data points plot within the granite zone; in the Na_2O-K_2O diagram (Figure 8D; [56]), some of the samples plot within the potassium zone, while the remainder plot within the sodium zone. The quartz porphyry is relatively poor in Mg, with Mg concentrations ranging from 0.09 wt %–0.19 wt %.

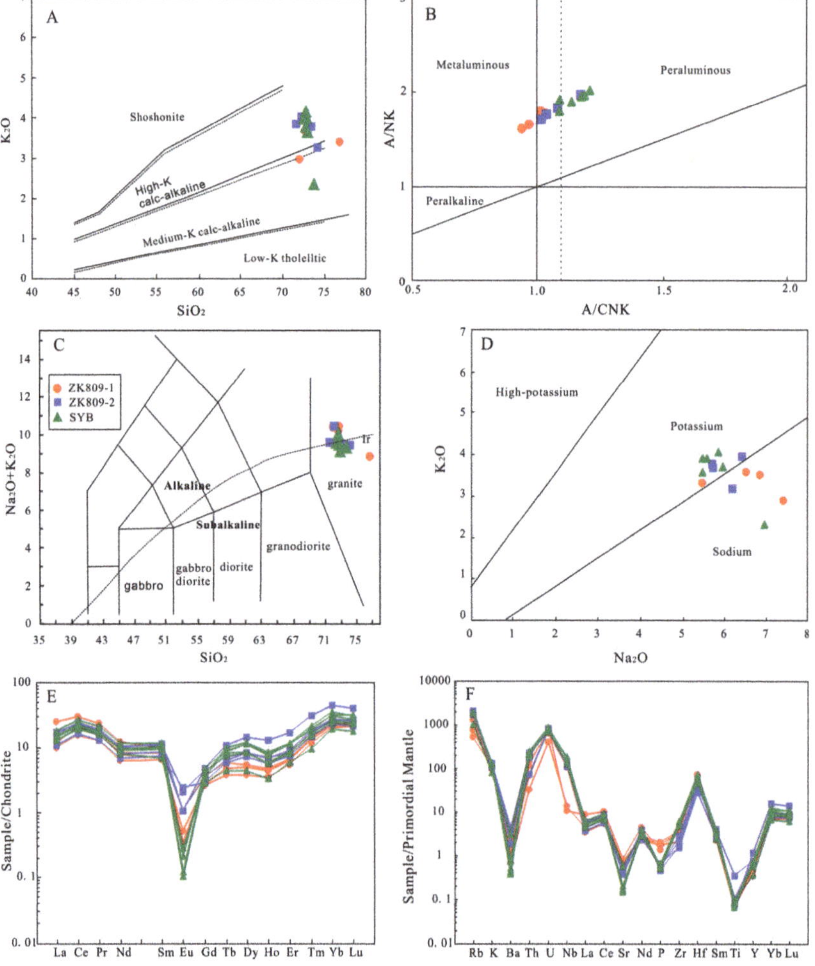

Figure 8. (**A**) SiO_2 versus (Na_2O+ K_2O) diagram (Middlemost, [55]); (**B**) Na_2O versus K_2O diagram (Middlemost, [56]); (**C**) SiO_2 versus K_2O diagram (Peccerillo et al. [53]); (**D**) A/CNK versus A/NK diagram (Manlar and Piccoli, [54]); (**E**) Chondrite-normalized REE variation (normalized values are from [57]); (**F**) Primitive mantle-normalized trace element patterns (normalized values are from [58]).

The REE analytical results (Table 3) indicate relatively low total REE contents in the quartz porphyry ($\sum$REE = 27.45–44.75 ppm). The standard REE chondrite-normalized diagram shows a relatively flat "seagull" pattern (Figure 8E), with light REE (LREE)/heavy REE (HREE) ratios of 1.48–3.91 and $(La/Yb)_N$ values of 0.37–1.09. The samples are rich in LREEs and poor in HREEs. The samples show distinct negative Eu anomalies (δEu = 0.01–0.46), suggesting that the magmatic evolution in the provenance involved long-term fractional crystallization with plagioclase as the main mineral phase [59].

On the standard trace element primitive mantle-normalized diagram (Figure 8F), the quartz porphyry samples display similar variations with very low Sr contents (3.09–17.45 ppm) and fairly high Yb contents (3.25–7.64 ppm), suggesting that the provenance pressure was less than 0.8–1.0 GPa [60]. The samples are relatively rich in large-ion lithophile elements (LILEs) such as Rb, Th, U, K as well as the high field strength elements (HFSEs) Hf and Zr and relatively poor in Ba, P, Sr, Ti and Nb, suggesting significant differentiation in the provenance of the quartz porphyry and implying the involvement of subduction zone processes.

5.4. Lu-Hf Isotopes

After zircon LA-ICP-MS U-Pb dating, microscale zircon Lu-Hf isotope analysis was conducted on the quartz porphyry samples from the Weilasituo deposit (ZK809-1 and ZK809-2). The analytical results are shown in Table 4. The zircons in the quartz porphyry have the similar Hf isotopic compositions. The ^{176}Hf/^{177}Hf and $\varepsilon_{Hf}(t)$ values at the 25 measuring points are 0.282742–0.282882 and 1.90–6.90, respectively. The $\varepsilon_{Hf}(t)$ isotopic calculations all plot between the chondritic meteorite and depleted mantle evolution lines (Figure 9B), as well as in the Phanerozoic igneous region of the eastern Central Asian Orogen (Figure 9A). The gray figures are quoted from [20,61,62]. At all the analysis points, T_{DM2} vary from 1069 to 747 Ma. As the zircon $f_{Lu/Hf}$ value is substantially lower than that in felsic crust (0.72, [63]) and mafic crust (0.34, [64]) and the T_{DM2} is close to the age of the source materials retained in the crust, the result indicates that the quartz porphyry magma originated from lower crustal materials.

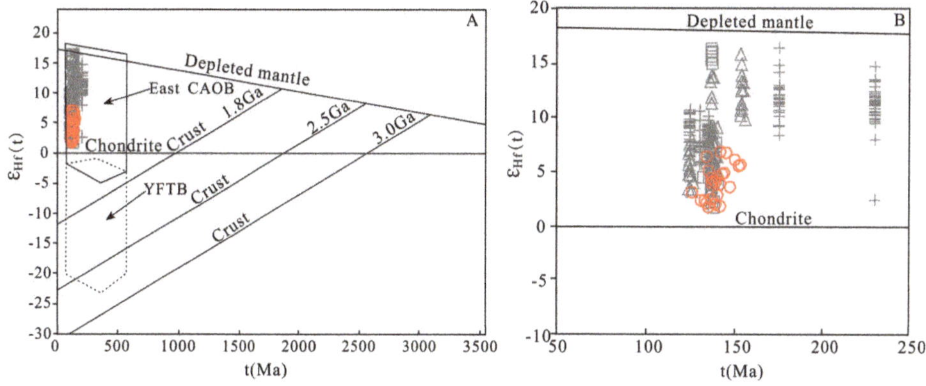

Figure 9. Variations of $\varepsilon_{Hf}(t)$ versus crystallization age of zircons (**A**,**B**) of the Weilasituo Sn-polymetallic deposit. The gray areas of granite is quoted from [20,61,62].

Table 3. Major (wt %) and trace elements (ppm) data of quartz porphyry from the Weilasituo Sn-polymetallic deposit.

Sample No.	ZK809-1-1	ZK809-1-2	ZK809-1-3	ZK809-1-4	ZK809-2-1	ZK809-2-2	ZK809-2-3	ZK809-2-4	SYB-01*	SYB-02*	SYB-03*	SYB-04*	SYB-05*	SYB-06*
Rock Type	Quartz Porphyry	Quartz Porphyry	Quartz Porphyry	Quartz Porphyry	Quartz Porphyry	Quartz Porphyry	Quartz Porphyry	Quartz Porphyry	Quartz Porphyry	Quartz Porphyry	Quartz Porphyry	Quartz Porphyry	Quartz Porphyry	Quartz Porphyry
SiO_2	72.84	76.8	72.67	72.01	73.4	72.15	71.57	74.15	73.69	72.68	72.91	72.79	72.52	72.76
Al_2O_3	14.63	12.69	15.63	16.16	15.09	15.42	16.32	14.43	15.54	16.04	15.93	16.12	16.15	15.63
TiO_2	0.016	0.018	0.015	0.015	0.076	0.023	0.016	0.023	0.020	0.019	0.014	0.018	0.017	0.02
Fe_2O_3	1.09	0.79	0.70	0.83	0.53	0.50	0.59	0.55	0.305	0.46	0.795	0.51	0.654	0.526
FeO	0.90	0.67	0.58	0.72	0.45	0.43	0.50	0.47	0.26	0.39	0.61	0.43	0.55	0.47
MnO	0.043	0.029	0.022	0.014	0.047	0.042	0.06	0.056	0.018	0.027	0.036	0.021	0.04	0.027
MgO	0.11	0.12	0.15	0.15	0.16	0.19	0.18	0.11	0.12	0.083	0.192	0.148	0.098	0.159
CaO	0.205	0.224	0.256	0.29	0.27	0.16	0.17	0.13	0.151	0.104	0.118	0.124	0.103	0.113
Na_2O	6.84	5.46	6.51	7.42	5.71	6.4	5.70	6.18	6.95	5.95	5.46	5.46	5.57	5.85
K_2O	3.6	3.39	3.66	2.96	3.78	4.04	3.86	3.26	2.36	3.79	3.65	4.00	3.99	4.16
P_2O_5	0.029	0.037	0.039	0.043	0.01	0.014	0.012	0.011	0.014	0.013	0.012	0.012	0.012	0.011
LOI	0.65	0.47	0.30	0.20	0.14	0.31	0.35	0.19	0.32	0.34	0.37	0.29	0.34	0.24
Total	100.94	100.7	100.53	100.8	99.66	99.68	99.33	99.6	99.75	99.9	100.1	99.92	100.04	99.97
$Na_2O + K_2O$	10.44	8.85	10.17	10.38	9.49	10.44	9.56	9.44	9.31	9.74	9.11	9.46	9.56	10.01
Na_2O/K_2O	1.90	1.61	1.78	2.51	1.51	1.58	1.48	1.90	2.94	1.57	1.50	1.37	1.40	1.41
La	6.02	2.93	2.37	2.77	3.75	2.57	4.13	2.98	3.10	3.63	3.58	3.08	4.30	3.43
Ce	18.28	11.72	9.60	12.86	14.98	10.10	14.27	12.00	11.80	13.70	13.50	12.10	16.30	13.20
Pr	2.28	1.52	1.23	1.65	1.83	1.24	1.79	1.48	1.51	1.71	1.71	1.59	2.06	1.66
Nd	5.83	3.72	2.99	3.86	4.97	3.20	4.73	3.79	3.78	4.50	4.45	4.29	5.43	4.26
Sm	1.55	1.06	1.00	1.27	1.65	1.15	1.75	1.30	1.06	1.47	1.57	1.60	1.79	1.46
Eu	0.06	0.03	0.03	0.02	0.12	0.14	0.06	0.06	0.016	0.012	0.007	0.006	0.012	0.006
Gd	0.75	0.53	0.54	0.57	0.91	0.59	0.99	0.72	0.58	0.839	0.806	0.785	0.975	0.73
Tb	0.22	0.14	0.18	0.18	0.27	0.22	0.41	0.24	0.165	0.282	0.338	0.37	0.329	0.31
Dy	1.40	0.97	1.28	1.28	2.10	1.84	3.64	2.12	1.12	2.09	2.87	3.05	2.89	2.15
Ho	0.27	0.20	0.26	0.25	0.40	0.33	0.73	0.39	0.189	0.334	0.431	0.483	0.453	0.322
Er	1.13	0.90	1.08	1.05	1.39	1.24	2.78	1.50	0.968	1.45	1.89	1.92	1.92	1.33
Tm	0.36	0.30	0.34	0.34	0.44	0.38	0.80	0.47	0.24	0.409	0.499	0.543	0.467	0.387
Yb	3.95	3.44	3.62	3.85	4.36	3.90	7.64	4.63	3.25	4.47	5.24	6.02	5.35	4.21
Lu	0.62	0.54	0.56	0.59	0.65	0.55	1.03	0.68	0.456	0.628	0.684	0.796	0.813	0.589
Y	2.45	1.61	2.05	2.00	3.50	2.70	5.22	3.48	1.68	2.75	3.33	3.35	3.18	2.60
$\sum$REE	42.72	28.00	25.08	30.54	37.82	27.45	44.75	32.36	28.23	35.524	37.575	36.633	43.089	34.044
LREE	34.02	20.98	17.22	22.43	27.30	18.40	26.73	21.61	21.27	25.02	24.82	22.67	29.89	24.02
HREE	8.70	7.02	7.86	8.11	10.52	9.05	18.02	10.75	6.97	10.5	12.76	13.97	13.2	10.03
LREE/HREE	3.91	2.99	2.19	2.77	2.60	2.03	1.48	2.01	3.05	2.38	1.95	1.62	2.27	2.39
δEu	0.15	0.11	0.11	0.06	0.27	0.46	0.13	0.17	0.06	0.03	0.02	0.01	0.03	0.02
$(La/Yb)_N$	1.09	0.61	0.47	0.52	0.62	0.47	0.39	0.46	0.68	0.58	0.49	0.37	0.58	0.58
Li	1200	811	556	405	388	347	650	496	287	374	512	274	569	365
Be	5.54	4.57	5.83	5.94	5.20	5.45	5.32	5.19	5.05	4.73	5.08	4.22	4.99	5.17
Sc	2.82	2.42	2.63	2.02	3.52	2.87	2.67	3.08	1.64	1.73	1.89	2.43	2.26	1.92

Table 3. *Cont.*

Sample No.	ZK809-1-1	ZK809-1-2	ZK809-1-3	ZK809-1-4	ZK809-2-1	ZK809-2-2	ZK809-2-3	ZK809-2-4	SYB-01*	SYB-02*	SYB-03*	SYB-04*	SYB-05*	SYB-06*
Rock Type	Quartz Porphyry	Quartz Porphyry	Quartz Porphyry	Quartz Porphyry	Quartz Porphyry	Quartz Porphyry	Quartz Porphyry	Quartz Porphyry	Quartz Porphyry	Quartz Porphyry	Quartz Porphyry	Quartz Porphyry	Quartz Porphyry	Quartz Porphyry
Cr	6.10	9.23	4.29	10.59	9.01	9.59	3.39	4.32	5.23	1.07	2.32	2.08	2.17	1.16
Co	0.08	0.40	0.78	4.81	2.03	1.18	1.40	1.65	0.45	0.14	0.33	0.14	0.19	0.14
Ni	2.59	1.67	5.02	6.02	3.39	2.85	3.07	2.88	4.11	0.39	1.50	0.90	0.99	0.30
Ga	68.2	70.4	69.5	67.6	64.0	68.7	65.0	64.6	76.2	73.9	72.4	75.0	76.1	75.7
Rb	476	339	1012	819	1200	1200	1300	1100	651	1157	1153	1233	1262	1192
Sr	17.45	14.77	13.95	14.36	9.92	9.43	9.47	7.81	12.40	3.34	4.22	3.35	3.09	3.63
Ba	8.37	4.80	8.54	16.65	13.10	30.7	26.78	26.62	21.70	3.62	3.12	2.68	5.31	2.84
U	17.3	8.7	15.5	13.4	16.6	15.9	14.6	17.5	14.8	16.2	15.8	18.6	17.1	15.1
Th	5.9	2.8	9.8	11.9	15.6	6.3	16.9	19.1	19.7	17.9	16.7	20.9	17.7	17.0
Nb	77.5	97.7	76.9	76.3	87.9	90.0	79.8	86.1	95.1	102	127	133	119	126
Ta	0.9	0.5	2.7	2.4	28.1	26.2	30.1	32.42	26.3	32.9	41.4	46.4	40.7	34.8
Zr	43.9	39.4	23.2	39.4	17.8	16.8	23.7	32.3	66.3	55.2	59.1	65.7	69.9	54.1
Hf	22.5	20.8	17.1	21.0	11.9	8.8	13.8	13.6	22.8	15.6	17.8	18.0	20.2	14.9
DI	95.06	97.41	96.39	96.19	94.91	96.22	93.68	97.07	96.97	96.26	94.47	95.28	95.27	96.50
A/CNK	0.943	0.972	1.033	1.014	1.08	1.015	1.177	1.036	1.09	1.139	1.212	1.191	1.181	1.091
A/NK	1.40	1.43	1.54	1.56	1.59	1.48	1.71	1.53	1.67	1.65	1.75	1.70	1.69	1.56
SI	0.88	1.15	1.29	1.24	1.47	2.45	2.51	1.03	1.20	0.78	1.79	1.40	0.90	1.42
AR	5.75	5.36	4.56	4.42	4.23	5.06	3.76	4.69	3.92	4.04	3.63	3.79	3.86	4.49
R1	1523	2339	1635	1419	1945	1562	1817	1961	1883	1827	2032	1956	1893	1774
R2	313	278	341	353	334	335	355	304	329	331	336	338	334	327
A/MF	4.96	5.61	7.46	6.57	6.99	6.64	6.42	8.00	14.63	11.87	6.73	9.85	8.67	8.98
C/MF	0.13	0.18	0.22	0.21	0.23	0.13	0.12	0.13	0.26	0.14	0.09	0.14	0.1	0.12
Tzr	672	674	638	670	626	616	649	662	717	706	717	723	727	701
Zr/Hf	1.95	1.89	1.36	1.88	1.50	1.92	1.71	2.38	2.91	3.54	3.32	3.65	3.46	3.63
Nb/Ta	86.11	195.40	28.48	31.80	3.13	3.43	2.65	2.65	3.62	3.10	3.07	2.87	2.92	3.62
Y/Ho	9.07	8.05	7.88	8.00	8.75	8.18	7.15	8.92	8.89	8.23	7.72	6.93	7.01	8.07

Note: * samples from [17].

Table 4. Zircon Hf isotopic composition for zircons from quartz porphyry from the Weilasituo Sn-polymetal deposit.

Sample No.	$\dfrac{^{176}\text{Yb}}{^{177}\text{Hf}}$	$\dfrac{^{176}\text{Lu}}{^{177}\text{Hf}}$	2σ	$\dfrac{^{176}\text{Hf}}{^{177}\text{Hf}}$	2σ	I_{Hf}	$\text{å}_{Hf}(0)$	$\text{å}_{Hf}(t)$	T_{DM}	T_{DMc}	$f_{Lu/Hf}$
				ZK809-1							
ZK809-1-5	0.14894	0.004678	0.000011	0.282894	0.00003	0.282882	4.32	7	562	747	−0.86
ZK809-1-6	0.070153	0.002193	0.000011	0.282827	0.000033	0.282822	1.96	4.91	621	883	−0.93
ZK809-1-8	0.094014	0.002688	0.00013	0.282759	0.000034	0.282753	−0.44	2.28	731	1045	−0.92
ZK809-1-11	0.119131	0.003901	0.00006	0.282757	0.000029	0.282747	−0.52	2.16	759	1053	−0.88
ZK809-1-12	0.099831	0.003076	0.00011	0.282794	0.000046	0.282787	0.78	3.3	687	973	−0.91
ZK809-1-13	0.058014	0.001766	0.000017	0.282821	0.000016	0.282816	1.73	4.57	624	899	−0.95
ZK809-1-14	0.23863	0.007394	0.000128	0.28278	0.000028	0.282762	0.29	2.53	806	1025	−0.78
ZK809-1-16	0.208726	0.006686	0.00006	0.282781	0.000027	0.282763	0.3	2.69	787	1018	−0.8
ZK809-1-17	0.187535	0.00576	0.000021	0.282887	0.000025	0.282873	4.07	6.52	592	773	−0.83
				ZK809-2							
ZK809-2-1	0.068173	0.002062	0.000087	0.282811	0.000039	0.282805	1.37	4.23	643	921	−0.94
ZK809-2-2	0.204887	0.00609	0.000197	0.282841	0.000021	0.282825	2.43	5.03	674	876	−0.82
ZK809-2-7	0.08257	0.002656	0.000066	0.282884	0.000022	0.282877	3.97	6.9	545	756	−0.92
ZK809-2-8	0.101062	0.003427	0.000041	0.28285	0.000019	0.28284	2.77	5.77	609	834	−0.9
ZK809-2-10	0.225517	0.007277	0.000082	0.282789	0.00003	0.28277	0.59	3	788	1002	−0.78
ZK809-2-13	0.089718	0.002885	0.000107	0.282749	0.000024	0.282742	−0.81	2.05	750	1065	−0.91
ZK809-2-14	0.151158	0.004783	0.000067	0.28283	0.000035	0.282817	2.04	4.67	666	896	−0.86
ZK809-2-16	0.055678	0.0018	0.000027	0.282792	0.000014	0.282787	0.7	3.75	666	959	−0.95
ZK809-2-17	0.11063	0.003649	0.000132	0.282849	0.000024	0.282838	2.71	5.67	615	840	−0.89
ZK809-2-23	0.062346	0.002142	0.000029	0.282862	0.000034	0.282856	3.18	6.27	570	801	−0.94
ZK809-2-24	0.055027	0.00173	0.000045	0.282813	0.000017	0.282808	1.44	4.31	635	917	−0.95
ZK809-2-25	0.060326	0.001936	0.000013	0.282804	0.000014	0.282799	1.15	4.06	651	935	−0.94
ZK809-2-26	0.054206	0.001611	0.000031	0.282765	0.00003	0.282761	−0.24	2.54	701	1026	−0.95
ZK809-2-27	0.0481	0.001494	0.000018	0.282807	0.000016	0.282803	1.24	4.12	639	928	−0.95
ZK809-2-28	0.089327	0.00277	0.000015	0.282844	0.000018	0.282837	2.54	5.27	607	854	−0.92
ZK809-2-29	0.116556	0.00372	0.000267	0.282751	0.00004	0.282742	−0.74	1.9	765	1069	−0.89

6. Discussion

6.1. Magmatic-Metallogenic Age Context

The determination of the precise age of mineralization and closely associated intrusions is essential for understanding genetic processes and locating economic deposits [65,66]. Presently, several high-precision and accurate isotopic dating methods exist to determine petrogenetic and metallogenic ages, such as zircon U-Pb [16,25], molybdenite Re-Os [16,25], and cassiterite U-Pb dating [17].

The zircon LA-ICP-MS U-Pb ages we obtained are 138.0 ± 1.1 Ma and 138.6 ± 1.1 Ma, which agree with the zircon U-Pb age of the quartz porphyry (138 ± 2 Ma and 135.7 ± 0.9 Ma) and the cassiterite age of the disseminated mineralization (138 ± 6 Ma) obtained by previous studies [16,17]. In addition, the molybdenite Re-Os isochron age (135 ± 7 Ma) we obtained also agrees with the molybdenite Re-Os isochron age (135 ± 11 Ma), and the cassiterite U-Pb age of the vein-like mineralization (135 ± 6 Ma) obtained by previous studies. Taking into account previous findings, we assume a petrogenetic age range of 138.0–138.6 Ma, which coincides with the age of the disseminated mineralization (138 ± 6 Ma). The vein-like mineralization formed at 135 ± 7 Ma, which was slightly later than the disseminated mineralization. These results indicate that the petrogenesis, as well as the metallogenesis, of the deposit occurred during the early Early Cretaceous. The petrogenetic ages of the various deposits found in the area have been widely examined and summarized over the past years. The results mostly correspond to the Mid-Triassic, Early to middle Jurassic and Early Cretaceous [67], with the Early Cretaceous becoming the most prevalent age. The Early Cretaceous deposits include more hydrothermal Pb-Zn-Ag (Sn-Cu), skarn-type Pb-Zn-Ag (Sn-Fe) and porphyry-type Cu-Mo deposits than porphyry-type Sn (W) deposits.

The study area lies at the eastern end of the Central Asian Orogen, which is sandwiched between the North China Block and the Siberian Plates. The study area was influenced by the evolution and closure of the Paleo–Asian Ocean during the Paleozoic and by the circum-Pacific tectonic domain from the Mesozoic to the Cenozoic [27,68]. Many studies have disconnected the magmatism of late Mesozoic in the SGXR from the subduction of the Pacific Plate [4,69], however, it is evident that the Early Cretaceous magmatism occurred throughout East China. This evidence suggests that the geodynamics underlying the formation of granite during this period must be taken into account. We believe the Pacific Plate played an important role in this magmatism [70]. During the Early Cretaceous, northern China is thought to have experienced a post-orogenic extensional collapse, an environment of post-arc extension closely followed by the subduction of the Paleo–Pacific Plate [26,71–73]. In the Nb-Y tectonic discrimination diagram, the sample points plot within the volcanic arc + syn-collisional granite fields (Figure 10C). In the Rb-Y + Nb diagram, the sample points are all distributed within the syn-collisional granite field (Figure 10D).

As discussed above, we have dated the petrogenesis and metallogenesis of the Weilasituo deposit to the early stage of the Early Cretaceous and conclude that the deposit was affected by the Early Cretaceous compression associated with the Paleo–Pacific Plate.

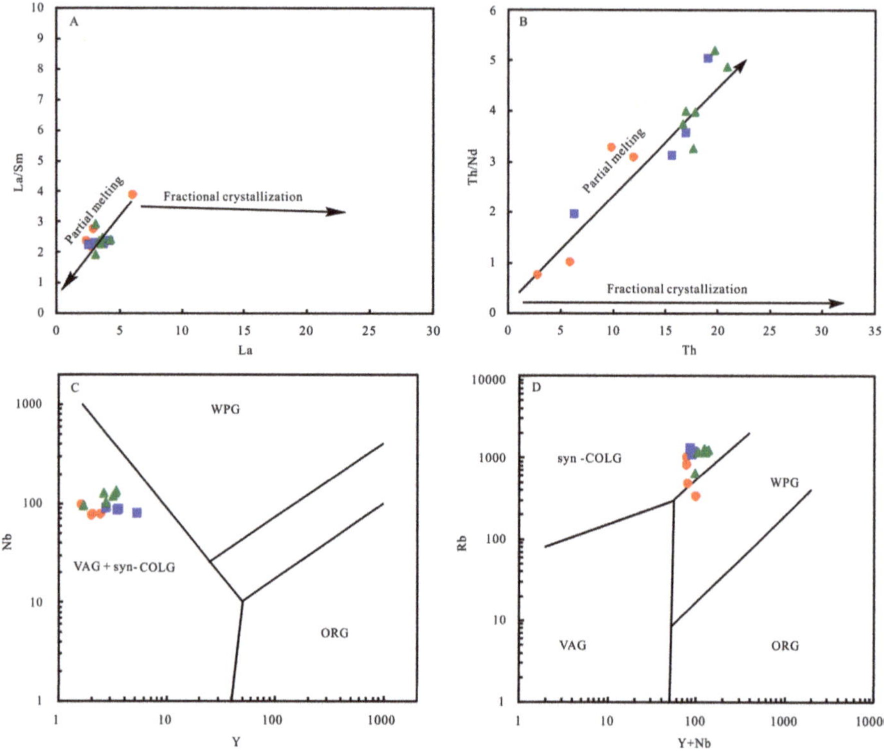

Figure 10. (**A**) La versus La/Sm diagram (Yang et al. [62]); (**B**) Th versus Th/Nd diagram (Schiano et al. [74]) (**C**) Y versus Nb diagram (Pearce et al. [75]); (**D**) Y + Nb versus Rb diagram (Pearce et al. [75]). Same symbols as Figure 8.

6.2. Petrogenesis and Magma Source

It is of great significance to determine the genetic type of granitoids so that one can understand the magma source, magmatism, and tectonic setting [75,76]. According to the (Zr + Nb + Ce + Y)-FeO^T/MgO diagram (Figure 11A) and the SiO$_2$-FeO^T/MgO diagram (Figure 11B), the quartz porphyry from the Weilasituo deposit exhibit no characteristics of A-type granites. In addition, the rocks are rich in silica (average SiO$_2$ = 73.07 wt %) and alkalis (average Na$_2$O + K$_2$O = 9.71 wt %) with Na$_2$O/K$_2$O ratios from 1.37 to 2.94, Al$_2$O$_3$ from 12.69 wt % to 16.32 wt %, and A/CNK ratios from 0.94 to 1.21. These features are typical of high-K calc-alkaline peraluminous granite. The low MgO content (<0.3 wt %) is also indicative of S-type granites [77].

In all the samples, the P-contents of quartz porphyry is very low (0.010%–0.043%) and P$_2$O$_5$ decreases with increasing SiO$_2$ (Figure 11C), while Th increases with Rb (Figure 11D), both of which are important gauges for discriminating I-type granites [77]. The Nb/Ta ratios of the quartz porphyry mostly range from 2.65 to 3.62 (<5) in Table 3, which is a sign of differentiation of peraluminous granite [78]. In the Nb-Nb/Ta diagram and Ta-Nb/Ta diagram (Figure 12A,B), the degree of differentiation of the quartz porphyry samples reaches more than 90%. The Zr/Hf ratios of quartz porphyry range from 1.36 to 3.65 (<25), and the U and Th concentrations are high in zircon, which are typical features of highly differentiated granite [79]. In the Zr-Zr/Hf diagram (Figure 12C), the ratio of Zr/Hf decreases with decreasing Zr content. In the Nb-Y/Ho diagram (Figure 12D), the Y/Ho ratio decreases with the increasing Nb content. The above two characteristics are typical of highly differentiated granite [80]. Additionally, the relatively low total REE contents, the relatively flat "seagull" pattern in the standard REE chondrite-normalized diagram (Figure 8E), the relatively abundant LREEs with LREE/HREE ratios of 1.48–3.91 (suggesting distinct differentiation), the very low contents of Sr (<400 ppm), the high contents of Yb (>2 ppm), $(La/Yb)_N = 0.37–1.09$, the intensely negative Eu anomalies ($\delta Eu = 0.01–0.46$, Figure 8F) and the relatively high abundances of the LILEs Rb, U, Th, and K compared to the relatively low abundances of Ba, P, Sr, Ti are all indicative of a highly differentiated I-type granite [81].

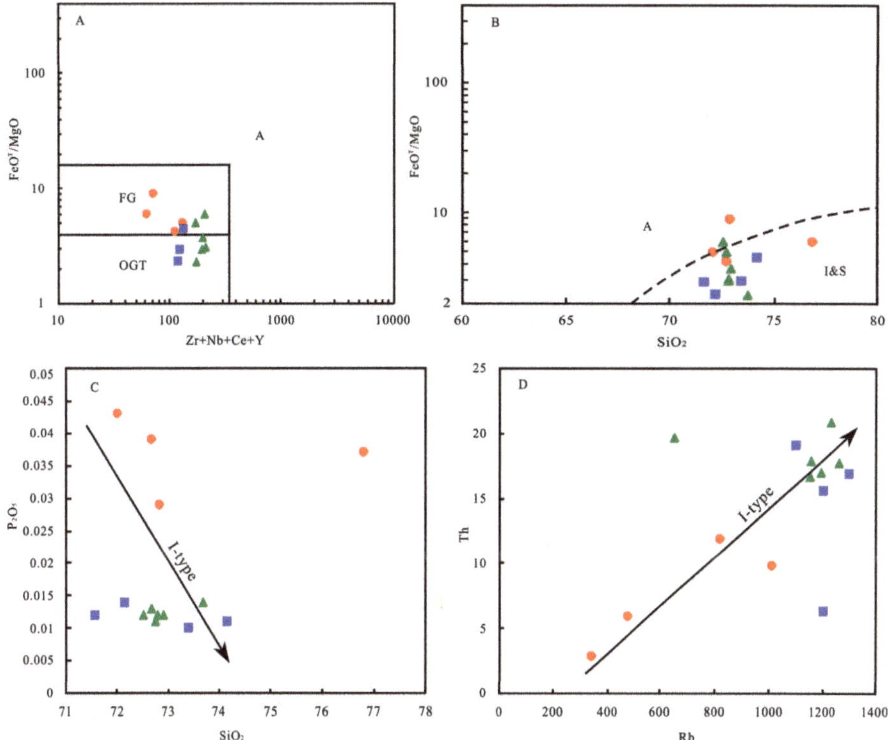

Figure 11. (**A**) Zr + Nb + Ce + Y versus FeO^T/MgO diagram (Whalen et al. [81]); (**B**) SiO_2 versus FeO^T/MgO diagram (Eby et al. [82]); (**C**) SiO_2 versus P_2O_5 diagram; (**D**) Rb versus Th diagram. Same symbols as Figure 8. FG—fractionated I-type granite; OGT—unfractionated I, S and M type granite; A—A-type granite; I—I-type granite; S—S-type granite.

Overall, we assume that the metallogenesis-related quartz porphyry in the deposit is a peraluminous, high-K calc-alkaline I-type granite as previously stated by Wang et al. [17].

Using the bulk rock Zr composition to represent the melt composition (Table 3) in the model of Watson and Harrison [83], zircon-saturation temperatures (T_{Zircon}) were calculated for the quartz porphyry of the Weilasituo deposit. The T_{Zircon} of the quartz porphyry ranges from 616° to 727° (averaging 678°), thus corresponding to a low T_{Zircon} value. In general, highly fractionated granitic magmas generally feature high primary temperatures or large amounts of various volatiles during the later stage. Although the T_{Zircon} of the quartz porphyry is low, the hydrothermal fluid of the Weilasituo depositwas characterized by high-Na, high-F contents [17], which would have been favorable for the differentiation of the quartz porphyry. In addition, volatiles in the residual melt, such as H_2O, Li, F, and Cl, tend to increase with crystallization [84], thus promoting the differentiation of magma.

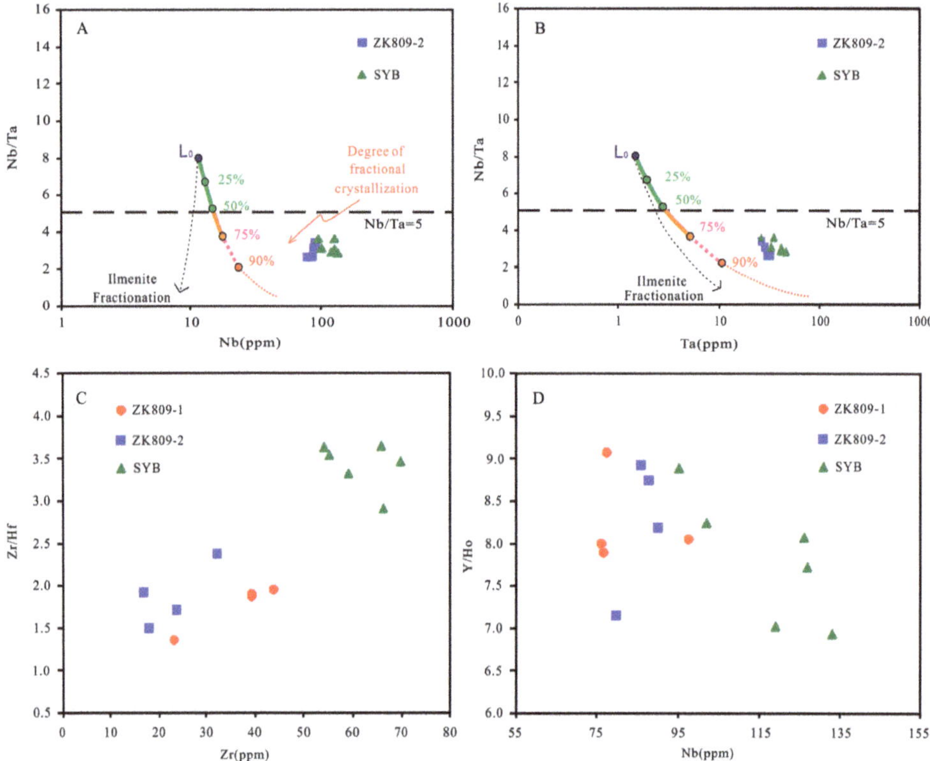

Figure 12. (**A**) Nb versus-Nb/Ta diagram (Ballouard et al. [78]); (**B**) Ta versus Nb/Ta diagram (Ballouard et al. [78]); (**C**) Zr versus Zr/Hf diagram; (**D**) Nb versus Y/Ho diagram.

Partial melting test results for common crustal rocks have revealed that this type of granite can be produced by the partial melting of water-bearing calc-alkaline to high-K calc-alkaline rocks, and mafic to intermediate metamorphic rocks in the earth's crust. In theory, granitoids can forme in many ways, including assimilation-fractional crystallization, magma mixing, or melting of the lower crust [85]. Magma mixing generates supercooled particles and magmatic inclusions [86]. However, no dioritic inclusions were found in the Weilasituo granites. Hence, the granites did not experience significant magma mixing. The zircons from the Weilasituo granites yield $^{206}Pb/^{238}U$ ages that are younger than 200 Ma and Th/U values that are greater than 0.1. These data suggest that the zircons were not inherited as a result of contamination by the surrounding host rocks or assimilation. Furthermore, the La-La/Sm diagram (Figure 10A) and the Th-Th/Nd diagram (Figure 10B) also illustrate the presence of partially melted magma in the quartz porphyry from the deposit. All these findings confirm that partial melting controlled the formation of the magma [87].

The Hf isotope composition of the Late Cretaceous granite in this area is summarized. The Hf isotope composition is the same as that of other rocks from the same era, suggesting that these rocks shared the same magma source. The $\varepsilon_{Hf}(t)$ value of the rock is positive, and the T_{DM2} model ages are 1069–747 Ma, and thus exhibit a large range. Therefore, the materials in the source region mainly came from the juvenile lower crust with some degree of mixing with mantle source materials, which is consistent with the conclusions based on the Sr-Nd isotope [17].

These data suggest that the subduction of the Paleo–Pacific Plate caused the lower crust to thicken and delaminate. The delaminated material then entered the underlying mantle, where it partially melted. As the resulting magma ascended, it reacted with the mantle before intruding into the shallow crust and solidifying in the form of the quartz porphyry.

6.3. Metallogenic Implications

Because the mineralization of the deposit occurred after the intrusion of the quartz porphyry, the mineralization had an inherently close genetic, spatial and temporal relationship with the quartz porphyry [17]. Thus, the ore-forming fluids of Sn-polymetallic mineralization were derived from the highly differentiated quartz porphyry.

Sn, W and Rb are highly incompatible elements. These elements are much more abundant in the crust than in the mantle [88]. Partial melting can lead to the enrichment of these elements in the presence of favorable complexing agents [89].

Chen et al. [90] and Chen and Fu [91] discussed the mechanism of granitic magma formation via the partial melting of thickened lower crust and established three models. (1) During a major period of continent orogenic movement, the strong collision between continental plates compresses and shortenes the crust, resulting in thickening and shearing of the lower crust. The temperature rise caused by the thickening of the lower crust and the heat generated by tectonic shearing may lead to partial melting of the lower crust, potentially assisted by the addition of external fluids. (2) After the main orogenic movement, the crust is in the state of tectonic decompression, which affects the thickening of the crust. This process leads to the dehydration of water-bearing minerals, and the addition of these fluids causes the partial melting of the crust. (3) After the main orogenic movement, gravitational instability results in lithospheric delamination of over-thickened crust results in mantle upwelling and the under-plating of the lower crust. These events cause partial melting of the lower crust.

At least in the late Mesozoic, the Great Xing'an Range had the characteristics of a post-collisional environment from the existing paleomagnetic evidence [92]. In recent years, studies in this region have found mafic granulite xenoliths and mafic-ultramafic cumulate xenoliths from the Early Mesozoic (based on ^{40}Ar-^{39}Ar ages, whole-rock K-Ar isochron ages, and Rb-Sr whole-rock isochron ages in a range of 220 to 237 Ma) and the late Mesozoic (based on zircon U-Pb ages in a range of 120 to 140 Ma). These rocks are believed to be the result of both upwelling and asthenospheric underplating [93]. Therefore, we believe that the granites associated with the Weilasituo are the result of the third mechanism discussed above.

Based on geochronological and geochemical research on the intermediate-acidic intrusive rocks exposed in the area and given the deposit geology and regional tectonics, we assume the following diagenetic and metallogenic dynamic processes. From the Late Jurassic to the Early Cretaceous, the subduction of the Paleo–Pacific Ocean caused the lower crust to thicken and delaminate. The delaminated material then entered the underlying mantle, where it partially melted. The roof of a magma chamber can become enriched in Sn, W, Mo and other elements by convection-driven thermogravimetric diffusion [94]. During the subsequent ascent of this magma, a portion mixed with remelted old crust and became calc-alkaline magma, which rose to the shallow crust and was emplaced in weak locations resulting from local extension, corresponding to the quartz porphyry emplacement. This process caused the ore minerals, Sn in particular, to leach out of the quartz porphyry (Figure 13A). During 138–135 Ma, due to the local extension, fractures and faults developed in the surrounding rocks above the quartz porphyry, and these fractures acted as conduits for the continuous upward migration of fluid. The increase in fluid pressure eventually exceeded the lithostatic pressure of the surrounding rocks, leading to cryptoexplosion and the formation of the breccia pipe. After the emplacement of the quartz porphyry, a large amount of metal-rich magmatic volatiles were released (Figure 13B). Because the magma was rich in Cl and F, Sn tendeds to migrate in the form of $SnCl_2$. The high F activity in the magma lowereds the viscosity and enableds the segregation of a less viscous, highly enriched residual fluid [95]. At approximately 135 Ma, ore-bearing hydrothermal fluid passed through the fractures, cleavages, and faults, and reacted with the surrounding rocks ("Xilin hot complex") as follows: $SnCl_2 + 3(Na,K)AlSi_3O_8 + 2H_2O \rightarrow SnO_2(cassiterite) + KAl_3Si_3O_{10}(OH)_2 + 6SiO_2 + 2NaCl$

+ H$_2$. Sn crystallized as cassiterite and was accompanied by the precipitation of other metallogenic elements. These processes formed the Weilasituo Sn-polymetallic deposit (Figure 13C).

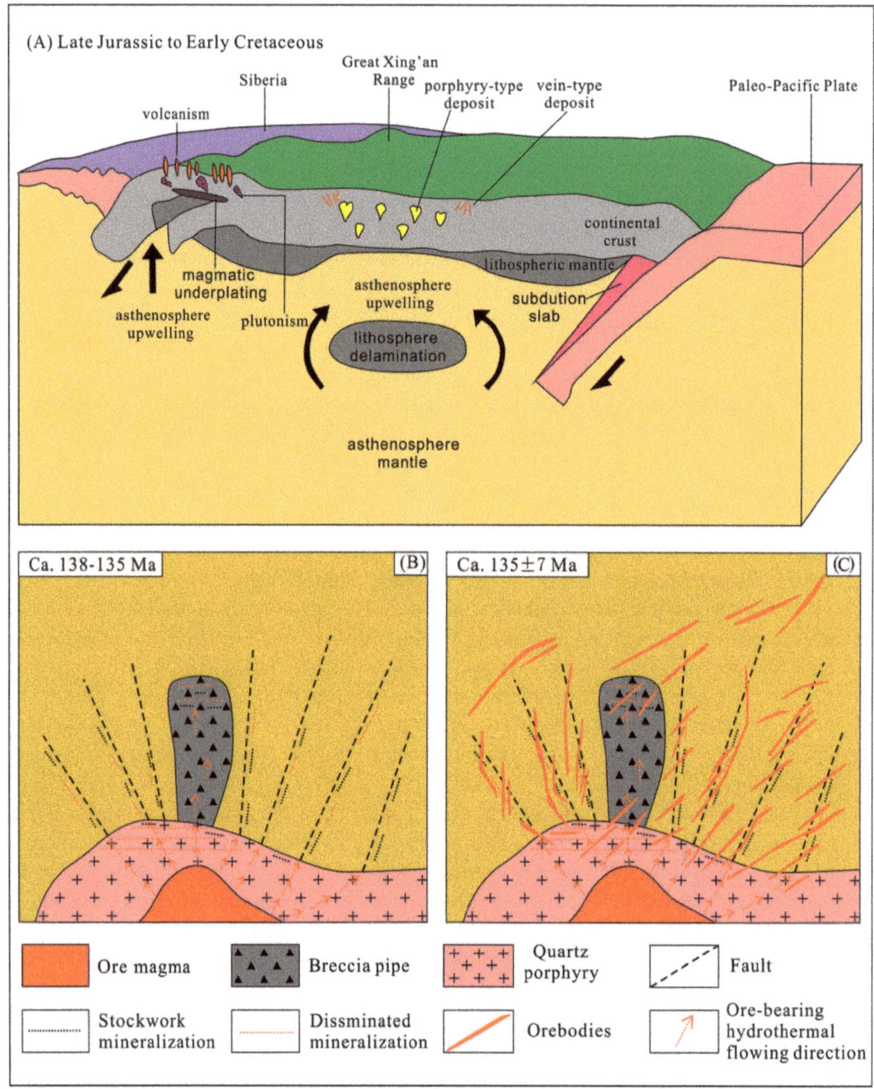

Figure 13. (**A**) Schematic model of the Late Jurassic to Early Cretaceous geodynamic evolution and associated ore deposits in the Great Xing'an Range. (**B**) Schematic model for the Weilasituo Sn-polymetallic deposit during 138–135 Ma. (**C**) Schematic model for the Weilasituo Sn-polymetallic deposit at 135 Ma.

7. Conclusions

Integrating zircon U-Pb and Re-Os dating, whole-rock geochemistry, and Hf isotopic data presented here with the results of previous research leads to the following conclusions:

(1)   The zircon U-Pb ages of the ore-forming porphyry (quartz porphyry) samples from the Weilasituo Sn-polymetallic deposit are 138.0 ± 1.1 Ma and 138.6 ± 1.1 Ma; the molybdenite Re-Os age is

135 ± 7 Ma. These ages date the metallogenesis of the deposit to the early stage of the Early Cretaceous at a time of transition from post-orogenic compression to extension following the subduction of the Paleo–Pacific Ocean.

(2) The geochemistry data show that the ore-forming porphyry (quartz porphyry) of the deposit has high concentrations of SiO_2, Al_2O_3, and K_2O+Na_2O, and is a peraluminous, high-K calc-alkaline highly differentiated I-type granite by partial melting. Quartz porphyry has $\varepsilon_{Hf}(t)$ values ranging from 1.90 to 6.90, corresponding to the T_{DM2} of 1069–747 Ma. The magmas were sourced from the juvenile lower crust mixed with mantle.

(3) The ore-forming fluids of Sn-polymetallic mineralization were derived from partial melting of highly differentiated quartz porphyry. The interaction between ore-bearing hydrothermal fluids and surrounding rocks ("Xilin hot complex") results in the precipitation of ore-forming elements.

Author Contributions: Conceptualization, F.Y. and J.S.; Investigation, F.Y., Y.W., J.F., F.N. and Z.F.; Data curation, F.Y.; Formal analysis, F.Y. and Z.H.; Funding acquisition, F.Y.; Project administration, J.S.; Writing—original draft preparation, F.Y.; Writing—review and editing, F.Y., J.S. and Y.W.

Funding: This research was funded by the National Natural Science Foundation of China (Grants No. 41172072) and the China Geological Survey (Grants No. DD20160048-01, DD20160048-16, DD20160048-05).

Acknowledgments: We thank the geologists from the Weilasituo Mining, Co. Ltd. with their assistance in the field work for the Weilasituo Sn-polymetallic deposit.

Conflicts of Interest: The authors declare no conflict of interest.

References

1. Zeng, Q.D.; Liu, J.M.; Qin, F.; Zhang, Z.L.; Chen, W.J. Preliminary research on Mesozoic four-stage Mo mineralization in Da Hinggan mountains of China. In Proceedings of the Gondwana 13th Program and Abstracts, Yunnan, China, 28 September 2008; pp. 251–252.

2. Meng, E.; Xu, W.L.; Yang, D.B.; Qiu, K.F.; Li, C.H.; Zhu, H.T. Zircon U-Pb chronology, geochemistry of Mesozoic volcanic rocks from the Lingquan basin in Maznhouli area, and its tectonics. *Acta Petrol. Sin.* **2011**, *27*, 1209–1226. (In Chinese)

3. Qin, K.Z.; Zhai, M.G.; Li, G.M.; Zhao, J.X.; Zeng, Q.D.; Gao, J.; Xiao, W.J.; Li, J.L.; Sun, S. Links of collage orogenesis of multiblocks and crust evolution to characteristic metallogeneses in China. *Acta Petrol. Sin.* **2017**, *33*, 305–325. (In Chinese)

4. Zhang, Z.D.; Wu, C.Z.; Gu, L.X.; Feng, H.; Zheng, Y.C.; Huang, J.H.; Li, J.; Sun, Y.L. Molybdenite Re-Os dating of Xintaimen molybdenum deposit in Yanshan-Liaoning metallogenic belt, North China. *Miner. Depos.* **2009**, *28*, 313–320. (In Chinese)

5. Chen, W.J.; Liu, J.M.; Liu, H.T.; Sun, X.G.; Zhang, R.B.; Zhang, Z.L.; Qin, F. Geochronology and fluid inclusion study of the Jiguanshan porphyry Mo deposit, Inner Mongolia. *Acta Petrol. Sin.* **2010**, *26*, 1423–1436. (In Chinese)

6. Deng, J.; Wang, C.M.; Bagas, L.; Carranza, E.J.M.; Lu, Y.J. Cretaceous-Cenozoic tectonic history of the Jiaojia Fault and gold mineralization in the Jiaodong Peninsula, China: Constraints from zircon U-Pb, illite K-Ar, and apatite fission track thermochronometry. *Miner. Depos.* **2015**, *50*, 987–1006. [CrossRef]

7. Zeng, Q.D.; Liu, J.M.; Chu, S.X.; Wang, Y.B.; Sun, Y.; Duan, X.X.; Zhou, L.L. Mesozoic molybdenum deposits in the East Xingmeng orogenic belt, northeast China: Characteristics and tectonic setting. *Inter. Geol. Rev.* **2012**, *54*, 1843–1869. [CrossRef]

8. Jiang, S.H.; Nie, F.J.; Bai, D.M.; Liu, Y.F.; Liu, Y. Geochronology evidence for Indosinian mineralization in Baiyinnuoer Pb-Zn deposit of Inner Mongolia. *Miner. Depos.* **2011**, *30*, 787–798. (In Chinese)

9. Zhu, X.Q.; Zhang, Q.; He, Y.L.; Zhu, C.H.; Huang, Y. Hydrothermal source rocks of the Meng'entaolegai Ag-Pb-Zn deposit in the granite batholith, Inner Mongolia, China. *Geochem. J.* **2006**, *40*, 265–275. [CrossRef]

10. Sheng, J.F.; Fu, X.Z. *The Geolocial Features of Copper-Polymetallic Deposit and the Ore-Forming Environment in Middle Part of the Greater Khingan Range*; Earthquake Press: Beijing, China, 1999; pp. 1–65. (In Chinese)

11. Zhang, K.; Nie, F.J.; Hou, W.R.; Li, C.; Liu, Y. Re-Os istopic age dating of molybdenite separates from Hashitu Mo deposit in Linxi County of Inner Mongolia and its geological significance. *Miner. Depos.* **2012**, *31*, 129–138. (In Chinese)

12. Zhao, Y.M.; Zhang, D.Q. *Metallogeny and Prospective Evaluation of Copperpolymetallic Deposits in the Da Hinggan Mountains and Its Adjacent Regions*; Seismological Publishing House: Beijing, China, 1997; pp. 1–318. (In Chinese)

13. Zhao, Y.M.; Wang, D.W.; Zhang, D.Q.; Fu, X.Z.; Bao, X.P.; Ai, Y.F. *Ore-Controlling Factors and Ore-Prospecting Models for Copper-Polymetallic Deposits in Southeastern Inner Mongolia*; Seismological Publishing House: Beijing, China, 1994; pp. 1–234. (In Chinese)

14. Ishiyama, D.; Sato, R.; Toshio, M.; Yohei, I.; Wang, J.B. Characteristic features of tiniron- coppermineralization in the Anle-Huanggangliangmining area, InnerMongolia, China. *Resour. Geol.* **2008**, *51*, 377–392. [CrossRef]

15. Jiang, S.H.; Liang, Q.L.; Liu, Y.F.; Liu, Y. Zircon U-Pb ages of themagmatic rocks occurring in and around the Dajing Cu-Ag-Sn polymetallic deposit of Inner Mongolia and constrains to the ore-forming age. *Acta Petrol. Sin.* **2012**, *28*, 495–513. (In Chinese)

16. Liu, Y.F.; Jiang, S.H.; Bagas, L. The genesis of metal zonation in the Weilasituo and Bairendaba Ag-Zn-Pb-Cu-(Sn-W) deposits in the shallow part of a porphyry Sn-polymetal system, Inner Mongolia, China. *Ore Geol. Rev.* **2016**, *75*, 150–173. [CrossRef]

17. Wang, F.X.; Bagas, L.; Jiang, S.H.; Liu, Y.F. Geological, geochemical, and geochronological characteristics of Weilasituo Sn-polymetal deposit, Inner Mongolia, China. *Ore Geol. Rev.* **2017**, *80*, 1206–1229. [CrossRef]

18. Pan, X.F.; Guo, L.J.; Wang, S.; Xue, H.M.; Hou, Z.Q.; Tong, Y.; Li, Z.M. Laser microprobe Ar-Ar dating of biotite from the Weilasituo Cu-Zn polymetallic deposit in Inner Mongolia. *Acta Petrol. Miner.* **2009**, *28*, 473–479. (In Chinese)

19. Chang, Y.; Lai, Y. Study on characteristics of ore-forming fluid and chronology in the Yindu Zn-Pb-Ag polymetallic ore deposit, Inner Mongolia. *Acta Sci. Natl. Univ. Pekinensis* **2010**, *46*, 581–593. (In Chinese)

20. Zhou, Z.H. Geology and Geochemistry of Huanggang Sn-Fe Deposit, Inner Mongolia. Ph.D. Thesis, Chinese Academy of Geological Sciences, Beijing, China, 2011; pp. 1–182. (In Chinese)

21. Wang, C.Y. Lead-Zinc Polymetallic Metallogenic Series and Prospecting Direction of Huanggangliang-Ganzhuermiao Metallogenic Belt, Inner Mongolia. Ph.D. Thesis, Jilin University, Changchun, China, 2014; pp. 1–178. (In Chinese)

22. Wang, J.B.; Wang, Y.W.; Wang, L.J.; Uemoto, T. Tin-polymetallic mineralization in the southern part of the Da Hinggan Mountains, China. *Resour. Geol.* **2001**, *51*, 283–291. [CrossRef]

23. Zeng, Q.D.; Liu, J.M.; Zhang, Z.L. Re-Os geochronology of porphyry molybdenum deposit in south segment of Da Hinggan Mountains, Northeast China. *J. Earth Sci.* **2010**, *21*, 392–401. [CrossRef]

24. Nie, F.J.; Zhang, W.Y.; Du, A.D.; Jiang, S.H.; Liu, Y. Re-Os isotopic dating on molybdenite separates from the xiaodonggou porphyry Mo deposit, Hexigten Qi, Inner Mongolia. *Acta Geol. Sin.* **2007**, *81*, 898–905. (In Chinese) [CrossRef]

25. Zhang, J.H.; Gao, S.; Ge, W.C.; Wu, F.Y.; Yang, J.H.; Wilde, S.A.; Li, M. Geochronology of the Mesozoic volcanic rocks in the Great Xing'an Range, northeastern China: Implications for subduction-induced delamination. *Chem. Geol.* **2010**, *276*, 144–165. [CrossRef]

26. Guo, F.; Fan, W.; Li, C.; Gao, X.; Miao, L. Early Cretaceous highly positive ε_{Nd} felsic volcanic rocks from the Hinggan Mountains, NE China: Origin and implications for Phanerozoic crustal growth. *Int. J. Earth Sci.* **2009**, *98*, 1395–1411. [CrossRef]

27. Zhai, D.G.; Liu, J.J.; Li, J.M.; Zhang, M.; Li, B.Y.; Fu, X.; Jiang, H.C.; Ma, L.J.; Qi, L. Geochronological study of Weilasituo porphyry type Sn deposit in Inner Mongolia and its geological significance. *Miner. Depos.* **2016**, *35*, 1011–1022. (In Chinese)

28. Liu, Y.F.; Fan, Z.Y.; Jiang, H.C.; Nie, F.J.; Jiang, S.H.; Ding, C.W.; Wang, F.Z. Genesis of the Weilasituo-Bairendaba Porphyry-Hydrothermal vein type system in Inner Mongolia, China. *Acta Geol. Sin.* **2014**, *88*, 2373–2385. (In Chinese)

29. Xu, W.L.; Wang, F.; Meng, E.; Gao, F.H.; Pei, F.P.; Yu, J.J.; Tang, J. Paleozoic-Early Mesozoic tectonic evolution in the Eastern Heilongjiang Province, NE China: Evidence from igneous rock association and U-Pb geochronology of detrital zircons. *J. Jilin Univ.* **2012**, *42*, 1378–1389. (In Chinese)

30. Sengör, A.M.C.; Natal'in, B.A.; Burtman, V.S. Evolution of the Altaid tectonic collage and Palaeozoic crustal growth in Eurasia. *Nature* **1993**, *364*, 299–307. [CrossRef]

31. Xiao, W.J.; Windley, B.F.; Hao, J.; Zhai, M.G. Accretion leading to collision and the Permian Solonker suture, Inner Mongolia, China: Termination of the central Asian orogenic belt. *Tectonics* **2003**, *22*, 1069. [CrossRef]

32. Xiao, W.J.; Li, S.Z.; Santosh, M.; Jahn, B.M. Orogenic belts in Central Asia: Correlations and connections. *J. Earth Sci.* **2012**, *49*, 1–6. [CrossRef]

33. Wu, F.Y.; Sun, D.Y.; Li, H.M.; Jahn, B.M.; Wilde, S.A. A-type granites in northeastern China: Age and geochemical constraints on their petrogenesis. *Chem. Geol.* **2002**, *187*, 143–173. [CrossRef]

34. Zhang, J.H.; Ge, W.C.; Wu, F.Y.; Wilde, S.A.; Yang, J.H.; Liu, X.M. Large-scale Early Cretaceous volcanic events in the northern Great Xing'an Range, northeastern China. *Lithos* **2008**, *102*, 138–157. [CrossRef]

35. Qin, G.J.; Kawachi, Y.; Zhao, L.Q.; Wang, Y.Z.; Ou, Q. The Upper Permian sedimentary facies and its role in the Dajing Cu-Sn deposit, Linxi County, Inner Mongolia, China. *Resour. Geol.* **2001**, *51*, 293–305. [CrossRef]

36. Shi, Y.R.; Liu, D.Y.; Zhang, Q.; Jian, P.; Zhang, F.Q.; Miao, L.C. SHRIMP geochronology of dioritic-granitic intrusions in Sunidzuoqi area, Inner Mongolia. *Acta Geol. Sin.* **2004**, *79*, 789–799. (In Chinese)

37. Zhang, X.H.; Wilde, S.A.; Zhang, H.F.; Zhai, M.G. Early Permian high-K calc-alkaline volcanic rocks from NW Inner Mongolia, North China: Geochemistry, origin and tectonic implications. *J. Geol. Soc.* **2011**, *168*, 525–543. [CrossRef]

38. Guo, F.; Fan, W.M.; Gao, X.F.; Li, C.W.; Miao, L.C.; Zhao, L.; Li, H.X. Sr-Nd-Pb isotope mapping of Mesozoic igneous rocks in NE China: Constraints on tectonic framework Phanerozoic crustal growth. *Lithos* **2010**, *120*, 563–578. [CrossRef]

39. Wang, Y.X.; Zhao, Z.H. Geochemistry and origin of the Baerzhe REE-Nb-Be-Zr super-large deposit. *Geochimica* **1997**, *26*, 24–35. (In Chinese)

40. Zhao, G.L.; Yang, G.L.; Fu, J.Y.; Wang, Z.; Fu, J.Y.; Yang, Y.Z. *Mesozoic Volcanic Rocks in Central and Southern Da Hinggan Ling Range*; Beijing Science and Technology Publishing House: Beijing, China, 1989; pp. 1–86. (In Chinese)

41. Deng, J.; Wang, Q.F.; Li, G.J.; Santosh, M. Cenozoic tectono-magmatic and metallogenic processes in the Sanjiang region, southwestern China. *Earth-Sci. Rev.* **2014**, 138268–138299. [CrossRef]

42. Zhong, R.C.; Yang, Y.F.; Shi, Y.X.; Li, W.B. Ore characters and ore genesis of the Bairendaba Ag polymetallic ore deposit in Keshiketeng banner, Inner Mongolia. *Geol. China* **2008**, *6*, 1–25. (In Chinese)

43. Yang, L.Q.; Deng, J.; Dilek, Y.; Qiu, K.F.; Ji, X.Z.; Li, N.; Taylor, R.D.; Yu, J.Y. Structure, Geochronology, and Petrogenesis of the Late Triassic Puziba Granitoid Dikes in the Mianlue Suture Zone, Qinling Orogen, China. *Geol. Soc. Am. Bull.* **2015**, *11–12*, 1831–1854. [CrossRef]

44. Du, A.D.; Wu, S.Q.; Sun, D.Z.; Wang, S.X.; Qu, W.J.; Markey, R.; Stain, H.; Morgan, J.; Malinovskiy, D. Preparation and certification of Re-Os dating reference materials: Molybdenites HLP and JDC. *Geostand. Geoanal. Res.* **2004**, *28*, 41–52. [CrossRef]

45. Wu, F.Y.; Yang, Y.H.; Xie, L.W.; Yang, J.H.; Xu, P. Hf isotopic compositions of the standard zircons and baddeleyites used in U-Pb geochronology. *Chem. Geol.* **2006**, *234*, 105–126. [CrossRef]

46. Yang, L.Q.; Deng, J.; Qiu, K.F.; Ji, X.Z.; Santosh, M.; Song, K.R.; Song, Y.H.; Geng, J.Z.; Zhang, C.; Hua, B. Magma mixing and crust-mantle interaction in the Triassic monzogranites of Bikou Terrane, central China: Constraints from petrology, geochemistry, and zircon U-Pb-Hf isotopic systematic. *J. Asian Earth Sci.* **2015**, *98*, 320–341. [CrossRef]

47. Woodhead, J.D.; Hergt, J.M. A preliminary appraisal of seven natural zircon reference materials for in situ Hf isotope determination. *Geostand. Geoanal. Res.* **2005**, *29*, 183–195. [CrossRef]

48. Scherer, E.; Munker, C.; Mezger, K. Calibration of the Lutetium-Hafnium clock. *Science* **2001**, *293*, 683–687. [CrossRef]

49. Blichert-Toft, J.; Albarède, F. The Lu-Hf geochemistry of chondrites and the evolution of the mantle-crust system. *Earth Planet. Sci. Lett.* **1997**, *148*, 243–258. [CrossRef]

50. Nowell, G.M.; Kempton, P.D.; Noble, S.R.; Fitton, J.G.; Saunders, A.D.; Mahoney, J.J.; Taylor, R.N. High precision Hf isotope measurements of MORB and OIB by thermal ionisation mass spectrometry: Insights into the depleted mantle. *Chem. Geol.* **1998**, *149*, 211–233. [CrossRef]

51. Griffin, W.L.; Pearson, N.J.; Belousova, E.; Jackson, S.E.; Van Achterbergh, E.; O'Reilly, S.Y.; Shee, S.R. The Hf isotope composition of cratonic mantle, LAM-MC-ICPMS analysis of zircon megacrysts in kimberlites. *Geochim. Cosmochim. Acta* **2000**, *64*, 133–147. [CrossRef]

52. Griffin, W.L.; Wang, X.; Jackson, S.E.; Pearson, N.J.; O'Reilly, S.Y.; Xu, X.S.; Zhou, X.M. Zircon chemistry and magma genesis, SE China: In-situ analysis of Hf isotopes, Tonglu and Pingtan igneous complexes. *Lithos* **2002**, *61*, 237–269. [CrossRef]

53. Peccerillo, A.; Taylor, S.R. Geochemistry of eocene calc-alkaline volcanic rocks from the Kastamonu area, Northern Turkey. *Contrib. Mineral. Petrol.* **1976**, *58*, 63–81. [CrossRef]

54. Maniar, P.D.; Piccoli, P.M. Tectonic discrimination of granitoids. *Geol. Soc. Am. Bull.* **1989**, *101*, 635–643. [CrossRef]

55. Middlemost, E.A.K. Naming materials in the magma/igneous rock system. *Earth Sci. Rev.* **1994**, *37*, 215–224. [CrossRef]

56. Middlemost, E.A.K. A simple classification of volcanic rocks. *Bull. Volcanol.* **1972**, *36*, 382–397. [CrossRef]

57. Boynton, W.V. Geochemistry of the rare earth elements, meteorite studies. In *Rare Earth Element Geochemistry*; Henderson, P., Ed.; Elsevier: Amsterdam, The Netherlands, 1984; pp. 63–114.

58. Sun, S.S.; McDonough, W.F. Chemical and isotopic systematics of oceanic basalts: Implications for mantle composition and processes. *Geol. Soc. Lond. Spec. Pub.* **1989**, *42*, 313–345. [CrossRef]

59. Lightfoot, P.C.; Hawkesworth, C.J.; Sethna, S.F. Petrogenesis of rhyolites and trachytes from the Deccan Trap: Sr, Nd and Pb isotope and trace element evidence. *Contrib. Mineral. Petrol.* **1987**, *95*, 44–54. [CrossRef]

60. Xiang, A.P.; Chen, Y.C.; Bagas, L. Molybdenite Re-Os and U-Pb zircon dating and genesis of the Dayana W-Mo deposit in eastern Ujumchin, Inner Mongolia. *Ore Geol. Rev.* **2016**, *78*, 268–280. [CrossRef]

61. Zhou, Z.H.; Mao, J.W. Geochronology and isotopic geochemistry of the A-type granites from the Huanggang Sn-Fe deposit, southern Great Hinggan Range, Ne China: Implication for their origin and tectonic setting. *J. Asian Earth Sci.* **2012**, *49*, 272–286. [CrossRef]

62. Yang, Q.D.; Guo, L.; Wang, T.; Zeng, T.; Zhang, L.; Tong, Y.; Shi, X.J.; Zhang, J.J. Geochronology, origin, sources and tectonic settings of Late Mesozoic two-stage granites in the Ganzhuermiao region, central and southern Da Hinggan Range, NE China. *Acta Petrol. Sin.* **2014**, *30*, 1961–1981. [CrossRef]

63. Amelin, Y.; Lee, D.C.; Halliday, A.N. Early-middle Archean crustal evolution deduced from Lu-Hf and U-Pb isotopic studies of single zircon grains. *Geochim. Cosmochim. Acta* **2000**, *64*, 4205–4225. [CrossRef]

64. Vervoort, J.D.; Pachelt, P.J.; Gehrels, G.E.; Nutman, A.P. Constraints on early Earth differentiation from hafnium and neodymium isotopes. *Nature* **1996**, *379*, 624–627. [CrossRef]

65. Yang, L.Q.; Deng, J.; Goldfarb, R.J.; Zhang, J.; Gao, B.F.; Wang, Z.L. 40Ar/39Ar geochronological constraints on the formation of the Dayingezhuang gold deposit: New implications for timing and duration of hydrothermal activity in the Jiaodong gold province, China. *Gondwana Res.* **2014**, *25*, 1469–1483. [CrossRef]

66. Yang, L.Q.; Deng, J.; Wang, Z.L.; Guo, L.N.; Li, R.H.; Groves, D.I.; Danyushevsky, L.V.; Zhang, C.; Zheng, X.L.; Zhao, H. Relationships between gold and pyrite at the Xincheng gold deposit, Jiaodong Peninsula, China: Implications for gold source and deposition in a brittle epizonal environment. *Econ. Geol.* **2016**, *111*, 105–126. [CrossRef]

67. Qiu, K.F.; Deng, J. Petrogenesis of granitoids in the Dewulu skarn copper deposit: Implication for the evolution of the Paleotethys ocean and mineralization in Western Qinling, China. *Ore Geol. Rev.* **2017**, *90*, 1078–1098. [CrossRef]

68. Sun, J.G.; Zhang, Y.; Han, S.J.; Men, L.J.; Li, Y.X.; Chai, P.; Yang, F. Timing of formation and geological setting of low-sulphidation epithermal gold deposits in the continental margin of NE China. *Int. Geol. Rev.* **2013**, *55*, 142–161. [CrossRef]

69. Qiu, K.F.; Deng, J.; Taylor, R.D.; Song, K.R.; Song, Y.H.; Li, Q.Z.; Goldfarb, R.J. Paleozoic magmatism and porphyry Cu-mineralization in an evolving tectonic setting in the North Qilian Orogenic Belt, NW China. *J. Asian Earth Sci.* **2016**, *122*, 20–40. [CrossRef]

70. Wu, F.Y.; Lin, J.Q.; Wilde, S.A.; Sun, D.Y.; Yang, J.H. Nature and significance of the Early Cretaceous giant igneous event in eastern China. *Earth Planet. Sci. Lett.* **2005**, *233*, 103–119. [CrossRef]

71. Mao, J.W.; Wang, Z.L. A preliminary study on time limits and geodynamic setting of largescalemetallogeny in East China. *Miner. Depos.* **1999**, *19*, 289–296. (In Chinese)

72. Mao, J.W.; Konopelko, D.; Seltmann, R.; Lehmann, B.; Chen, W.; Wang, Y.T.; Eklund, O.; Usubaliev, T. Post collisional age of the Kumtor gold deposit and timing of Hercynian events in the Tien Shan, Kyrgyzstan. *Econ. Geol.* **2004**, *99*, 1771–1780. [CrossRef]

73. Qiu, K.F.; Marsh, E.; Yu, H.C.; Pfaff, K.; Gulbransen, C.; Gou, Z.Y.; Li, N. Fluid and metal sources for the Wenquan porphyry molybdenite deposit, Western Qinling, NW China. *Ore Geol. Rev.* **2017**, *86*, 459–473. [CrossRef]

74. Schiano, P.; Monzier, M.; Eissen, J.P.; Martin, H.; Koga, K.T. Simple mixing as the major control of the evolution of volcanic suites in the Ecuadorian Andes. *Contrib. Mineral. Petrol.* **2010**, *160*, 297–312. [CrossRef]

75. Pearce, J.A.; Harris, N.B.W.; Tindle, A.G. Trace element discrimination diagrams for the tectonic interpretation of granitic rocks. *J. Petrol.* **1984**, *25*, 956–983. [CrossRef]

76. Sylvester, P.J. Post-collisional strongly peraluminous granite. *Lithos* **1998**, *45*, 29–44. [CrossRef]

77. Chappell, B.W.; White, A.J.R. I-type and S-type granites in the Lachlan Fold Belt. *Trans. R. Soc. Ed. Earth Sci.* **1992**, *83*, 1–26. [CrossRef]

78. Ballouard, C.; Poujol, M.; Boulvais, P.; Branquet, Y.; Tartese, R.; Vigneresse, J.L. Nb-Ta fractionation in peraluminous granites: A marker of the magmatic-hydrothermal transition. *Geology* **2016**, *44*, 231–234. [CrossRef]

79. Breiter, K.; Lamarao, C.N.; Borges, R.M.K.; Dall'Agnol, R. Chemical characteristics of zircon from A-type granites and comparison to zircon of S-type granites. *Lithos* **2014**, *192–195*, 208–225. [CrossRef]

80. Chen, J.Y.; Yang, J.H. Petrogenesis of the Fogang highly fractionated I-type granitoids: Constraints from Nb, Ta, Zr and Hf. *Acta Geol. Sin.* **2015**, *31*, 846–854. (In Chinese)

81. Whalen, J.B.; Currie, K.L.; Chappell, B.W. A-type granites: Geochemical characteristics, discrimination and petrogenesis. *Contrib. Mineral. Petrol.* **1987**, *95*, 407–419. [CrossRef]

82. Eby, G.N.; Woolley, A.R.; Din, V.; Platt, G. Geochemistry and petrogenesis of nepheline syenite. Kasungu-Chipala, Ilomba, and Ulindi nepheline syenite intrusions, North Nyasa Alkaline Province, Malawi. *J. Petrol.* **1990**, *39*, 1405–1424. [CrossRef]

83. Watson, E.B.; Harrison, T.M. Zircon saturation revisited: Temperature and composition effects in a variety of crustal magma types. *Earth Planet. Sci. Lett.* **1983**, *64*, 295–304. [CrossRef]

84. Zhu, J.C.; Rao, B.; Xiong, X.L.; Li, F.C.; Zhang, P.H. Comparison and genetic interpretation of Li-F rich, rare-metal bearing granitic rocks. *Geochimica* **2002**, *31*, 141–152. (In Chinese)

85. Liu, W.; Pan, X.F.; Xie, L.W.; Li, H. Sources of material for the Linxi granitoids, the southern segment of the Da Hinggan Mts.: When and how continental crust grew? *Acta Petrol. Sin.* **2007**, *23*, 441–460. (In Chinese)

86. Poli, G.E.; Tommasini, S. Model for the origin and significance of microgranular enclaves in calc-alkaline granitoids. *J. Petrol.* **1991**, *32*, 657–666. [CrossRef]

87. Qiu, K.F.; Yu, H.C.; Gou, Z.Y.; Liang, Z.L.; Zhang, J.L.; Zhu, R. Nature and origin of Triassic igneous activity in the Western Qingling Orogen: The Wenquan composite pluton example. *Int. Geol. Rev.* **2018**, *60*, 242–266. [CrossRef]

88. König, S.; Münker, C.; Hohl, S.; Paulick, H.; Barth, A.R.; Lagos, M.; Pfänder, J.; Büchl, A. The Earth's tungsten budget during mantle melting and crust formation. *Geochim. Cosmochim. Acta* **2011**, *75*, 2119–2136. [CrossRef]

89. Qiu, K.F.; Taylor, R.D.; Song, Y.H.; Yu, H.C.; Song, K.R.; Li, N. Geologic and geochemical insights into the formation of the Taiyangshan porphyry copper-molybdenum deposit, Western Qinling Orogenic Belt, China. *Gondwana Res.* **2016**, *35*, 40–58. [CrossRef]

90. Chen, Y.J.; Pirajno, F.; Qi, J.P. Origin of gold metallogeny and sources of oreforming fluids, Jiaodong province, eastern China. *Int. Geol. Rev.* **2005**, *47*, 530–549. [CrossRef]

91. Chen, Y.J.; Fu, S.G. *Mineralization of Gold Deposits in West Henan*; China Seismological Press: Beijing, China, 1992; pp. 1–234. (In Chinese)

92. Kuzmin, M.L.; Abramovich, G.Y.A.; Dril, S.L.; Kravchisky, V.Y.A. The Mongolian–Okhotsk suture as the evidence of late Paleozoic-Mesozoic collisional processes in Central Asia. In Proceedings of the Abstract of 30th International Geological Congress, Beijing, China, 4–14 August 1996; Volume 1, p. 261.

93. Deng, J.; Wang, Q.F. Gold mineralization in China: Metallogenic provinces, deposit types and tectonic framework. *Gondwana Res.* **2016**, *36*, 219–274. [CrossRef]

94. Eugster, H.P. Granites and hydrothermal ore deposits: A geochemical framework. *Mineral. Mag.* **1985**, *49*, 7–23. [CrossRef]

95. Dingwell, D.B. The structure and properties of fluorine-rich magmas: A review of experimental studies. *Can. Inst. Min. Metall.* **1985**, *39*, 1–12.

Article

Mineralogy and Garnet Sm–Nd Dating for the Hongshan Skarn Deposit in the Zhongdian Area, SW China

Bo Zu [1], Chunji Xue [2,*], Chen Dong [3] and Yi Zhao [4]

[1] School of Earth Resources, China University of Geosciences, Wuhan 4330074, China; zubo@cug.edu.cn
[2] State Key Laboratory of Geological Processes and Mineral Resources,
Faculty of Earth Sciences and Resources, China University of Geosciences, Beijing 100083, China
[3] Jiangxi Institute of Geological Exploration for Mineral Resources of Nonferrous Metals,
Nanchang 330025, China; chendcugb@163.com
[4] Yunnan Geological Survey, Kunming 650051, China; ttrainy@163.com
* Correspondence: chunji.xue@cugb.edu.cn; Tel.: +86-10-8232-1895

Received: 23 February 2019; Accepted: 17 April 2019; Published: 19 April 2019

Abstract: The Hongshan deposit is one of the largest Cu-polymetallic deposits in the Zhongdian area, southwest China. Two types of Cu–Mo ores, mainly developed in the skarns, have been recognized in the Hongshan deposit, i.e., massive or layered skarn and vein-type, with the former being dominant. The highly andraditic composition of garnet (Adr_{100} to $Adr_{64}Gr_{32}$) and diopsidic composition of pyroxene ($Di_{90}Hd_9$ to Di_1Hd_{99}) indicate the layered skarn ores are of magmatic-hydrothermal origin that formed under oxidized conditions. Sm–Nd dating of garnet yield a well-constrained isochron age of 76.48 ± 7.29 Ma (MSWD = 1.2) for the layered skarn ores. This age was consistent with the Re–Os age for the pyrrhotite from the layered skarn ores, and thereby indicated that the layered skarn mineralization was formed in the Late Cretaceous, rather than in the Triassic as was previously thought. The coincidence of the geochronology from the layered skarn ores and vein-type mineralization further indicated that both ores were the result of a single genetic event, rather than multiple events. The recognition of the Late Cretaceous post-collisional porphyry–skarn Cu–Mo–W belt in the Zhongdian area exhibited a promising prospecting potential.

Keywords: skarn mineralogy; high fugacity; garnet Sm–Nd dating; Late Cretaceous; Hongshan skarn deposit

---

1. Introduction

The Zhongdian Cu-polymetallic area, situated in the southern segment of the Yidun arc (Figure 1a,b), is commonly regarded as one of the most fertile regions for porphyry and skarn copper deposits in China [1–3]. Many of the porphyry copper deposits in the Zhongdian area (Figure 1c), such as Pulang (803.85 Mt with 0.52% Cu, 0.18 g/t Au, [4,5]), Xuejiping (54.15 Mt with 0.53% Cu, 0.06 g/t Au), Langdu (1.67 Mt with 6% Cu), Chundu, and Lannitang (36 Mt with 0.50% Cu, 0.45 g/t Au) deposits were formed in the Late Triassic as a result of westward subduction of the Garzê–Litang oceanic crust [4,5]. Recently, the recognition of some Late Cretaceous porphyry–skarn deposits, represented by the Hongshan Cu–Mo, Tongchanggou Mo–Cu, Xiuwacu W–Mo, and Relin W–Mo deposits, has drawn much attention on the collision-related metallogeny of the Zhongdian area [6–9].

The Hongshan deposit is one of the largest Cu-polymetallic deposit in the Zhongdian area (Figure 1c) with contained metals of 0.64 t Cu, 5769 t Mo, 7532 t W, 323 t Ag, and 25262 t Pb + Zn [10]. Generally, two types of ores, namely the major layered skarn Cu-polymetallic ores and later vein-type

Cu–Mo mineralization, have been formed [9,11]. Trace elements of pyrite and pyrrhotite from both types display distinct signatures and could plausibly indicate different origins [10]. Previous research proposed a two-stage model for the genesis of the Hongshan deposit with Triassic skarn Cu-polymetallic ores overprinted by Late Cretaceous vein-type Cu–Mo mineralization [10,11]. The age of later vein-type Cu–Mo mineralization has been well-constrained by molybdenite Re–Os and zircon U–Pb ages [9,11,12]. The mineralization age of the layered skarn Cu-polymetallic ores, however, has been poorly constrained. Zu et al. reported a Cretaceous age for the layered skarn ores using Re–Os dating on pyrrhotite [9]; however, the closure temperature of Re–Os systems for pyrrhotite is as low as 400 °C and might have undergone osmium diffusion or resetting in the presence of a later stage overprint event [13,14]. Thus, the timing of the layered skarn ores is still one key problem with the two-stage model.

A number of studies have indicated that garnets are useful geochemical tracers and can be used in geochronological studies [15,16]. In this contribution, we report systematic petrography and geochemistry of the main skarn mineral and Sm–Nd isotope systematics of garnet for layered skarn ores in the Hongshan deposit. The results, combined with the previous data, allow us to constrain the mineralization age of the skarns and to explain the genesis and geologic setting of the Hongshan Deposit. The implications for Cretaceous metallogeny in the Zhongdian area are also discussed.

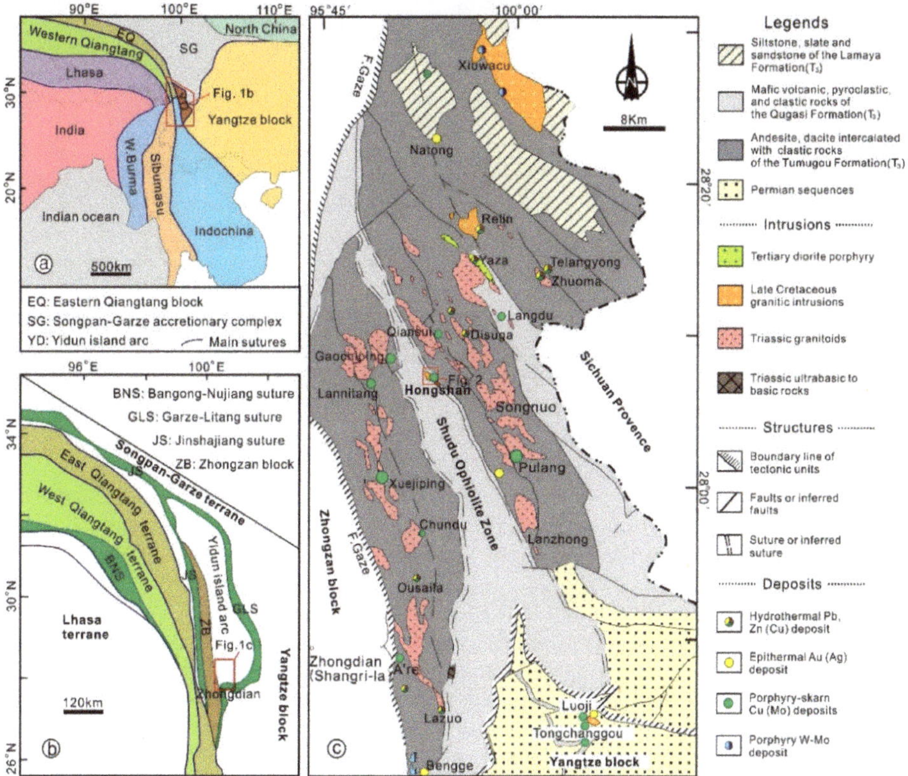

Figure 1. (**a**) Tectonic map sketch showing the distribution of principal continental blocks and sutures of Southeast Asia. The location and outline of the Yidun island arc are indicated. (**b**) Tectonic framework of the Sanjiang Tethyan domain and location of the southern Yidun arc. (**c**) Geological map showing the distribution of porphyry and skarn deposits in the southern Yidun arc. Modified from Zu et al. and Deng et al. [1,17].

2. Geological Setting

The Zhongdian area is located at the southern end of the Yidun island arc, forming a significant metallogeny belt in the Sanjiang Tethyan metallogenic domain (Figure 1b) [15,16]. The main tectonic framework of the Yidun arc is defined by the Garzê–Litang suture belt to the south and east and the major Dege–Xiangcheng fault to the west separates it from Zhongzan massif (Figure 1c) [9,18].

The southern Yidun arc witnessed a complex tectonic evolution history including Triassic oceanic crustal subduction, Jurassic–Cretaceous continental collision, and Cenozoic regional strike-slip faulting [18–20]. In the Triassic, westward subduction of the Garzê–Litang ocean led to the formation of the Yidun arc in the eastern part of the Zhongzan block, accompanied by the formation of voluminous volcanic and granitic rocks with ages mainly from 238 to 210 Ma (Figure 1c) [20,21]. These volcanic and granitic rocks host numerous mineral deposits, such as the Gacun volcanic massive sulfide (VMS) Ag-polymetallic deposit and porphyry–skarn deposits that were formed in a compressional setting (Figure 1c), in association with subduction of different angles [22]. When the Garzê–Litang ocean closed completely at the end of the Triassic, the Yidun island arc collided with the Yangtze block, accompanied by the emplacement of sparsely distributed granites from 138 to 75 Ma. Mineral deposits formed in this period of time, such as the Xiuwacu, Relin, Hongshan, and Tongchanggou deposits [8,23]. During the Cenozoic, a remote response to the Indian–Asian continental collision caused intracontinental strike-slip processes with the emplacement of alkaline granite intrusions and associated Au mineralization within the southern Yidun arc, including the Yaza and Bengge deposits (Figure 1c).

3. Geology of the Hongshan Deposit

The Hongshan Cu-polymetallic deposit is located in the central part of the Zhongdian area (Figure 1c). The main country rocks of the deposit comprise mainly a thick Triassic sequence of crystalline limestone, siltstone, gray to very dark gray argillaceous slate, and interlayered andesitic tuff of the Upper Triassic Qugasi formation (Figure 2). These rocks underwent contact thermal and replacement metamorphism with the development of extensive hornfels, marble and skarn throughout the deposit and surrounding areas (Figure 3).

Sparse felsic dikes and offshoots are exposed in the Hongshan deposit area. Diorite porphyries are the most widely distributed, and occur mainly in the southeastern and northern parts of the deposit area with SHRIMP U–Pb ages of 214 ± 2 Ma [17]. However, a direct contact between these porphyries and the orebodies has not been observed in the Hongshan deposit (Figure 2a). Quartz monzonite porphyry outcrops only as three small intrusions in the middle of the Hongshan area; a small part of the quartz monzonite porphyries have been altered to endoskarn near contact zones with marble (Figure 3a). Boreholes CK 17-2 and CK 17-4 have revealed concealed mineralized stocks that may be larger than the outcrops (Figure 2b,c) [17]. Previous dating suggests that the quartz monzonite porphyry and the concealed granite porphyry were formed in Late Cretaceous with zircon U–Pb ages of 77–81 Ma [12,17,23].

The copper orebodies in the study area are sheet-like, tabular, or lenticular; are subparallel to the main NNW-striking structures in the deposit area; and are hosted by skarn and hornfels zones. Individual orebodies generally strike NNW and range from 30 to 1223 m in length along strike, 20 to 420 m in width along the dip direction, and 3.9 to 19.6 m in thickness. Based on mineral assemblages and ore textures, two types of Cu–Mo ores, mainly developed in the skarns, have been recognized, i.e., massive or layered skarn (Figure 3a) and vein-type, with the former being dominant. Some other ore types are also present: hydrothermal Pb–Zn ores in the periphery of the Hongshan deposit (Figure 2a), and Mo–Cu porphyry ores associated with concealed intrusions at depth [17]. The alteration and mineralization in the Hongshan deposit can be divided into three alteration stages. The early stage (stage 1) is characterized by the alteration of wallrocks to albite–biotite–quartz–andalusite hornfels and massive, sage-green diopside hornfels (Figure 3d–f). In the skarn stage (stage 2), the interaction between the magmatic hydrothermal fluids and marble resulted in diverse assemblages of calc–silicate skarn minerals including garnet, diopside, wollastonite, and magnetite. The early retrograde alteration stage (stage 3a) is characterized by hydrosilicate minerals and sulfide minerals overprinting on the calcsilicate minerals, forming the massive and disseminated skarn ores (Figure 3b,c). The quartz–sulfide stage (stage 3b) featured a variety of quartz–sulfide veins overprinting skarns and hornfels. Marble-hosted hydrothermal Pb–Zn mineralization scattered on the periphery of the Cu–Mo skarn orebodies (Figure 2a), exhibiting a clear mineralization zoning.

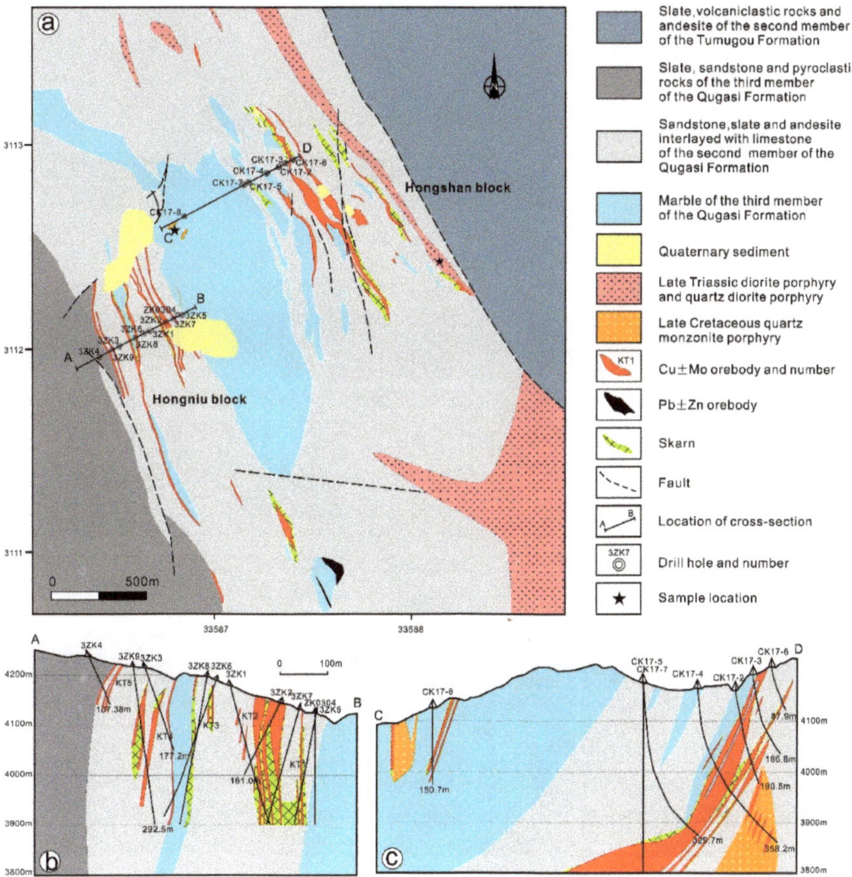

Figure 2. Simplified geological map (**a**) and cross-sections (**b,c**) of the Hongshan Cu–Mo deposit. Modified from Wang et al. [23].

The spatial zonation of the skarn minerals in the Hongshan deposits are also apparent with at least three associations. The endoskarns are only present within the quartz monzonite porphyry and have a mineral assemblage of garnet, diopside, and vesuvianite (Figure 3a) [17]. Exoskarns are more widely developed with a mineral assemblage of garnet, diopside, fluorite, and actinolite in the massive or layered skarn ores (Figure 3b,c). Minor distal skarns are scattered in the periphery of the main orebodies and associated with marble, hornfels, or wollastonite (Figure 3d–f). Garnet is the most abundant mineral in different assemblages and several different garnet generations can be optically distinguished. Garnets in the endoskarns are reddish, fine-grained, anisotropic, and have dodecahedral and/or polysynthetic twinning (Figure 3a). Garnets in the exoskarns are isotropic and euhedral, anisotropic (Figure 3j,k), and display evident oscillatory zoning (Figure 3g–i). The epitaxial growth with oscillatory zoning exoskarn garnets indicate these garnets might have formed during different stages (Figure 3g). Garnets from the distal skarns are greenish, brown, and rarely reddish in color without apparent oscillatory zonings (Figure 3l).

Figure 3. Photographs and photomicrographs showing characteristics of skarn ores in the Hongshan deposit. (**a**) Garnet in endoskarn in contact with residual quartz monzonite porphyry. (**b**) Massive or layered skarn ores with dense disseminated pyrrhotite in exoskarn. (**c**) Euhedral garnet in exoskarn ores with dense disseminated pyrrhotite and chalcopyrite. (**d**) Greenish-brown garnet in distal skarns along the marble. (**e**) Reddish-brown distal skarn in pyroxene hornfels. (**f**) Brown eduhedral garnet along with marble and wollastonite. (**g–k**) Garnets in the exoskarns are isotropic and euhedral, anisotropic, and display evident oscillatory zoning. (**l**) Garnet from distal skarn lacks of apparent oscillatory zonings. Ccp: chalcopyrite, Dio: diopside, Grt: garnet, Po: pyrrhotite, Qtz: quartz, Ser: sericite, Wo: wollastonite

4. Sampling and Analytical Methods

Polished sections were used to study skarn minerals following by electron microprobe analyses on representative garnet, pyroxene, and tremolite (actinolite). Major element analyses of these minerals were carried out at the China University of Geosciences (Beijing) using a Shimadzu EPMA-1600. An accelerating voltage of 25 kV, a beam current of 20 nA, and a beam diameter of 1 μm were applied.

Natural minerals and synthetic oxides were used as standards, and all data were corrected using standard ZAF correction procedures.

Ten skarn-bearing samples from the layered skarn ores in the Hongshan deposit were collected to pick pure garnet separates from high precision Sm–Nd isotope systematic analyses. Pure garnet separates were handpicked under a binocular microscope to select grains similar in appearance to the chips of the 60–80 mesh size crushed by agate mortar, and then powered to <200 mesh size. Sm–Nd isotopic analyses were performed on a ISOPROBE-T thermal ionization mass spectrometry (TIMS) at the Beijing Research Institute of Uranium Geology.

5. Analytical Results

5.1. Mineral Composition

Electron microprobe analyses show that garnets from the deposit form a grossular–andradite solid solution and dominated with andradite (Tables 1–3). Forty-nine spots on garnet grains from exoskarn from this study range from almost pure andradite (Adr_{100}) to $Adr_{64}Gr_{32}$ with less than 5% of other types of garnets (Figure 4a). They contain significant less Al contents than the garnets from endoskarn ($Adr_{22-57}Grs_{78-43}$) and from distal skarn ($Adr_{14-60}Grs_{86-40}$) [24]. Oscillatory zoned garnets are generally chemically zoned with fluctuation-ore repeated changes of FeO_T content and Al_2O_3 content (Figure 5). The invariably negative correlation between FeO_T and Al_2O_3 compositions also indicate the common existence of grossular–andradite solid solution (Figure 5).

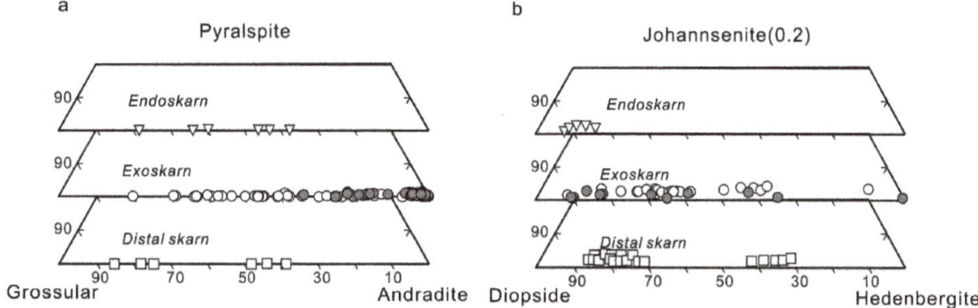

Figure 4. Major element compositions of garnets (**a**) and pyroxene (**b**) showing their dominance of end members from three selected skarn mineral deposits. Pyralspite–(pyrope + spessartine + almandine + uvarovite).

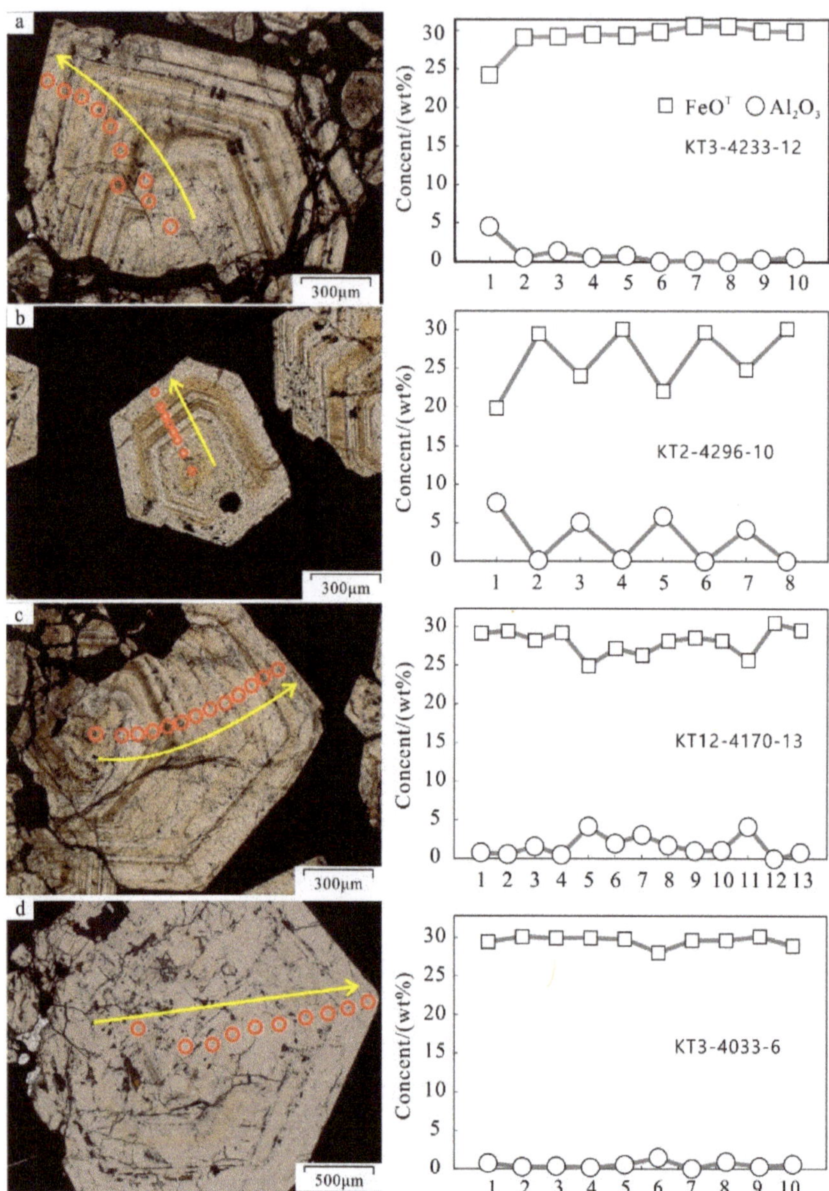

Figure 5. Variation of FeO$_T$ and Al$_2$O$_3$ content from core to margin in the garnet grains from the layered skarn ores.

The pyroxene (12 spots) from the Hongshan deposit are mainly diopside in composition, with minor hedenbergite and rarely johannsenite (Figure 4b). The composition of the pyroxene from the exoskarn has large variations with Di90Hd9 to Di1Hd99 (Table 4).

The prograde garnet, pyroxene, and wollastonite are commonly overprinted by retrograde tremolite, actinolite, epidote, and minor chlorite. The actinolite minerals (6 spots) in this study belongs to ferro-actinolite with high amounts of MgO (19–22 wt %) and FeO$_T$ (5–11 wt %) (Table 5).

5.2. Sm–Nd Dating Result

The Sm and Nd concentrations and isotopic ratios of garnets from the Hongshan deposit are listed in Table 6 and plotted in Figure 6. All the 10 analyzed samples showed a roughly linear correlation with an isochron age of 60.91 ± 4.94 Ma (MSWD = 26) (Figure 6a) using IsoplotR [25]. Three garnet samples (D21-5, D48-13, and KT2-4296-6) deviate the isochron line and could plausibly represent the incorporation of other minerals or/and affected by later perturbation despite detailed petrography study before analyzing (Figure 6a). The remaining 7 samples revealed a well-constrained isochron age of 76.48 ± 7.29 Ma (MSWD = 1.2) (Figure 6b), which represents the crystallization time of the garnet. The uncertainty of ±7.29 Ma is due to low ^{147}Sm/^{144}Nd ratios (<0.30) rather than low precision of the Nd isotope ratios.

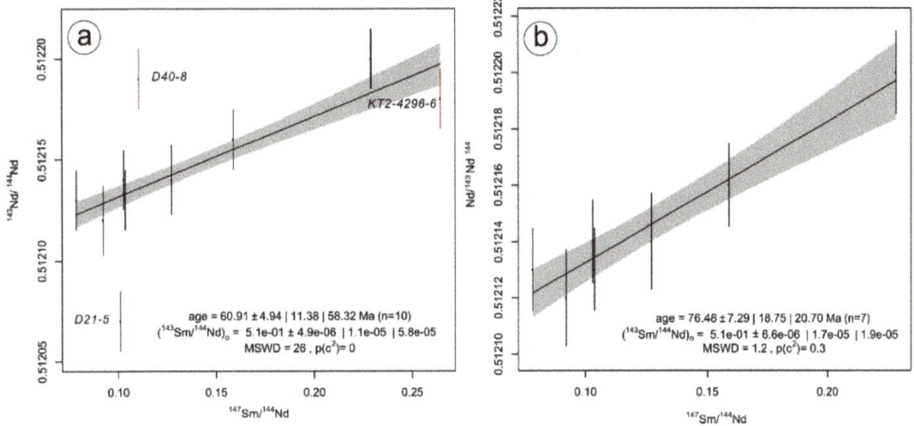

Figure 6. Sm–Nd garnet from the layered skarn ore in the Hongshan deposit. Age provided in (**a**) is calculated from all of the plotted data and age in (**b**) is calculated from seven garnet samples.

Table 1. Electron microprobe analyses of garnet from the Hongshan deposit (wt %).

Sample	Garnet–Diopside Skarn KT3-4033-12										Garnet–Diopside Skarn KT4-4078-1				
	Core				→		Rim				Core		→	Rim	
SiO_2	36.21	35.74	35.48	35.48	35.26	35.46	35.27	35.44	35.36	35.29	36.44	36.45	36.32	35.92	35.94
TiO_2	0.10	0	0	0	0.02	0	0	0	0.03	0.03	0.20	0.21	0.20	0.04	0.11
Al_2O_3	4.54	0.57	1.36	0.56	0.79	0.02	0.16	0	0.32	0.57	5.02	3.56	5.19	3.92	4.11
Cr_2O_3	0.04	0.09	0.10	0.07	0.06	0.17	0	0.05	0.25	0.09	0.17	0.12	0.12	0.16	0
FeO	24.12	29.01	29.11	29.39	29.27	29.72	30.51	30.45	29.83	29.78	22.74	23.95	23.03	24.53	24.40
MnO	0.28	0.47	0.38	0.38	0.45	0.36	0.44	0.40	0.29	0.37	0.36	0.46	0.34	0.19	0.22
MgO	0.14	0.19	0.15	0.16	0.03	0.20	0.11	0.11	0.10	0.11	0.22	0.21	0.33	0.21	0.10
CaO	33.01	32.21	32.06	32.82	32.41	32.31	32.12	32.36	32.30	32.63	33.00	33.06	33.06	33.34	33.81
Na_2O	0.15	0.14	0.05	0.08	0.03	0.06	0.05	0.15	0.17	0.08	0.20	0.20	0.23	0.11	0.24
K_2O	0	0	0	0	0	0	0	0	0	0	0	0	0	0	0
Total	98.59	98.42	98.69	98.94	98.32	98.30	98.66	98.96	98.65	98.95	98.35	98.22	98.82	98.42	98.93
Number of cations on the basis of 12 O															
Si	2.97	2.98	2.96	2.95	2.95	2.97	2.95	2.96	2.96	2.94	2.99	3.00	2.97	2.96	2.95
Ti	0.01	0	0	0	0	0	0	0	0	0	0.01	0.01	0.01	0	0.01
Al	0.44	0.06	0.13	0.05	0.08	0	0.02	0.00	0.03	0.06	0.48	0.35	0.50	0.38	0.40
Fe^{3+}	1.57	1.95	1.89	1.97	1.95	2.01	2.02	2.02	1.98	1.98	1.51	1.64	1.51	1.63	1.63
Fe^{2+}	0.08	0.08	0.14	0.07	0.10	0.08	0.12	0.10	0.11	0.10	0.05	0.01	0.07	0.06	0.04
Mn	0.02	0.03	0.03	0.03	0.03	0.03	0.03	0.03	0.02	0.03	0.02	0.03	0.02	0.01	0.02
Mg	0.02	0.02	0.02	0.02	0.00	0.02	0.01	0.01	0.01	0.01	0.03	0.03	0.04	0.03	0.01
Ca	2.90	2.88	2.86	2.93	2.91	2.90	2.88	2.90	2.90	2.91	2.90	2.92	2.89	2.94	2.97
And	78.07	96.91	93.10	97.07	95.96	99.34	99.22	99.84	97.63	96.96	75.23	82.25	74.82	80.69	80.41
Gro	17.85	1.66	0.55	1.25	0.67	4.11	4.66	4.77	3.14	1.86	20.74	15.02	20.52	15.67	17.27
Pyr	0.57	0.78	0.61	0.65	0.12	0.82	0.45	0.45	0.41	0.45	0.90	0.86	1.33	0.85	0.40

Table 2. Electron microprobe analyses of garnet from the Hongshan deposit (wt %).

Sample	Garnet–Diopside, Exoskarn													Diopside–Garnet Skarn		
	KT12-4170-13													KT2-4235-10		
	Core					→					Rim			Core	→	Rim
SiO_2	35.42	35.46	35.58	35.51	35.49	35.82	35.90	35.48	35.49	36.00	35.76	35.44	35.26	35.73	34.82	35.64
TiO_2	0	0.12	0	0	0.11	0.08	0.03	0	0.09	0	0.06	0.07	0.11	0	0.02	0
Al_2O_3	0.80	0.58	1.58	0.50	4.19	1.97	3.09	1.74	1.01	1.06	4.16	0	0.81	0.17	0.08	0.03
Cr_2O_3	0.02	0.25	0	0	0	0	0.06	0.01	0	0.07	0.12	0	0.05	0.09	0.00	0.05
FeO	29.26	29.54	28.32	29.31	25.03	27.28	26.41	28.26	28.68	28.29	25.77	30.61	29.66	30.12	30.26	30.15
MnO	0.23	0.18	0.09	0.08	0	0.37	0.15	0.02	0.23	0.26	0.26	0.32	0.30	0.43	0.66	0.28
MgO	0.28	0.23	0.10	0.16	0.09	0.26	0.17	0.15	0.20	0.35	0.25	0.26	0.05	0.09	0.13	0.12
CaO	31.87	31.55	32.60	33.16	33.76	32.75	32.38	32.48	32.32	32.45	32.44	31.92	32.13	32.18	31.92	32.72
Na_2O	0.19	0.44	0.09	0.02	0.12	0.13	0.22	0.01	0.19	0.11	0.02	0.07	0.16	0.05	0.10	0.21
K_2O	0	0.08	0	0	0	0	0	0	0	0	0	0	0.03	0	0.06	0
Total	98.07	98.43	98.36	98.74	98.79	98.66	98.41	98.15	98.21	98.59	98.84	98.69	98.56	98.86	98.05	99.20
Number of cations on the basis of 12 O																
Si	2.97	2.98	2.96	2.96	2.92	2.97	2.97	2.96	2.97	2.99	2.94	2.96	2.95	2.98	2.94	2.97
Ti	0	0.01	0	0	0.01	0	0	0	0.01	0	0	0	0.01	0	0	0
Al	0.08	0.06	0.16	0.05	0.41	0.19	0.30	0.17	0.10	0.10	0.40	0	0.08	0.02	0.01	0
Fe^{3+}	1.94	1.94	1.87	1.98	1.64	1.83	1.71	1.85	1.92	1.90	1.63	2.02	1.94	1.99	2.03	2.02
Fe^{2+}	0.11	0.14	0.11	0.06	0.08	0.06	0.12	0.12	0.09	0.06	0.15	0.12	0.13	0.11	0.11	0.08
Mn	0.02	0.01	0.01	0.01	0	0.03	0.01	0	0.02	0.02	0.02	0.02	0.02	0.03	0.05	0.02
Mg	0.04	0.03	0.01	0.02	0.01	0.03	0.02	0.02	0.02	0.04	0.03	0.03	0.01	0.01	0.02	0.01
Ca	2.86	2.84	2.91	2.96	2.98	2.91	2.87	2.90	2.90	2.89	2.86	2.86	2.88	2.87	2.89	2.92
And	96.02	96.32	92.33	97.58	80.14	90.47	84.84	91.53	95.06	94.61	79.82	99.67	95.90	98.88	99.61	99.69
Gro	1.53	3.07	3.57	0.41	16.81	5.52	9.99	3.94	0.65	0.99	13.39	5.72	1.31	4.09	5.17	3.70
Pyr	1.16	0.95	0.41	0.65	0.36	1.06	0.69	0.61	0.82	1.44	1.00	1.07	0.21	0.37	0.54	0.49

Table 3. Electron microprobe analyses of garnet from the Hongshan deposit (wt %).

Sample	Garnet–Diopside Skarn									Diopside–Garnet Skarn								
	KT2-4296-10									KT3-4033-6								
	Core			→		Rim				Core			→			Rim		
SiO_2	35.91	35.34	35.84	35.36	35.74	35.10	36.17	35.24	35.50	35.31	35.15	35.33	35.93	35.98	35.49	35.99	35.10	35.37
TiO_2	1.32	0	0	0.17	0.06	0.11	0	0	0.01	0.09	0.05	0	0	0.10	0.01	0	0.05	0.12
Al_2O_3	7.58	0.11	5.03	0.23	5.79	0	4.10	0.05	0.72	0.22	0.32	0.16	0.52	1.46	0	0.92	0.27	0.59
Cr_2O_3	0	0.07	0	0	0	0.13	0.05	0	0.14	0.14	0.09	0	0	0.12	0	0	0.07	
FeO	19.83	29.49	24.02	30.12	22.04	29.71	24.81	30.17	29.45	30.10	29.93	29.94	29.80	28.06	29.65	29.64	30.16	28.94
MnO	0.56	0.37	0.36	0.20	0.35	0.24	0.33	0.36	0.49	0.33	0.36	0.22	0.17	0.23	0.55	0.31	0.47	0.43
MgO	0.13	0.07	0	0.07	0.00	0.13	0.12	0.07	0.06	0.24	0.30	0.23	0.16	0.27	0.30	0.16	0.33	0.18
CaO	33.45	32.42	32.75	32.66	32.00	33.15	33.04	32.19	32.13	31.95	32.15	32.09	32.38	32.39	31.83	31.49	31.97	32.48
Na_2O	0	0.18	0.05	0.08	0.09	0.12	0.04	0.14	0.12	0.07	0.16	0.07	0.01	0.11	0.18	0.08	0	0.19
K_2O	0	0	0	0	0	0	0	0	0.01	0.01	0	0.02	0	0	00	0	0	0
Total	98.78	98.05	98.05	98.89	96.07	98.69	98.66	98.22	98.63	98.46	98.51	98.06	98.97	98.72	98.01	98.59	98.42	98.30
Number of cation on the basis of 12 O																		
Si	2.91	2.97	2.96	2.95	2.99	2.94	2.97	2.96	2.97	2.96	2.95	2.97	2.98	2.98	2.98	3.00	2.94	2.96
Ti	0.08	0	0	0.01	0	0.01	0	0	0	0.01	0	0	0	0.01	0	0	0	0.01
Al	0.72	0.01	0.49	0.02	0.57	0	0.40	0	0.07	0.02	0.03	0.02	0.05	0.14	0.00	0.09	0.03	0.06
Fe^{3+}	1.29	2.00	1.54	2.00	1.43	2.03	1.62	2.02	1.94	1.99	2.00	2.01	1.96	1.86	2.01	1.91	2.01	1.96
Fe^{2+}	0.06	0.07	0.12	0.10	0.11	0.05	0.08	0.10	0.12	0.12	0.10	0.10	0.11	0.09	0.07	0.15	0.11	0.06
Mn	0.04	0.03	0.03	0.01	0.02	0.02	0.02	0.03	0.03	0.02	0.03	0.02	0.01	0.02	0.04	0.02	0.03	0.03
Mg	0.02	0.01	0	0.01	0	0.02	0.01	0.01	0.01	0.03	0.04	0.03	0.02	0.03	0.04	0.02	0.04	0.02
Ca	2.90	2.92	2.90	2.92	2.87	2.97	2.91	2.90	2.88	2.87	2.89	2.89	2.88	2.88	2.87	2.81	2.87	2.91
And	63.99	99.23	75.87	98.88	71.49	99.58	80.19	99.76	96.04	98.47	98.15	99.22	97.47	92.51	99.85	95.49	98.46	97.12
Gro	32.31	2.97	19.37	2.85	24.01	2.74	15.62	4.25	1.69	4.49	3.83	3.93	2.07	2.57	4.97	1.95	4.70	0.98
Pyr	0.52	0.29	0	0.29	0	0.53	0.48	0.29	0.25	0.99	1.23	0.95	0.66	1.11	1.25	0.66	1.35	0.74

Table 4. Electron microprobe analyses of diopside from the Hongshan deposit (wt %).

Sample	Garnet–Diopside Skarn			Garnet–Diopside Skarn			Diopside Skarn			Diopside Skarn		
	KT12-4170-13			KT4-4078-1			KT3-4033-6			KT2-4235-10		
SiO_2	45.76	54.94	53.91	56.01	54.86	55.06	54.32	54.23	53.76	51.74	52.17	51.88
TiO_2	0.01	0.01	0.01	0.01	0.01	0.01	0.01	0.01	0	0.01	0	0
Al_2O_3	19.63	0.12	0.21	0.21	0.34	0.11	0.37	0.18	0.34	0.15	0.12	0.26
FeO	12.65	2.95	4.05	2.81	2.96	5.06	5.78	8.94	10.12	12.15	16.37	16.32
Cr_2O_3	0.03	0.02	0.01	0.03	0.05	0.04	0.02	0.01	0.06	0.01	0.02	0.05
MnO	0.06	0.21	0.82	0.15	0.15	0.43	0.52	0.42	0.11	0.56	0.17	0.63
MgO	0.05	16.72	15.91	15.72	15.34	14.37	15.62	11.53	10.64	9.99	4.99	6.89
CaO	20.61	24.32	23.97	24.21	25.37	24.83	22.30	24.18	23.91	23.94	25.01	22.93
Na_2O	0.01	0.00	0.02	0.02	0.01	0.01	0.00	0.01	0.04	0.00	0.03	0.05
K_2O	0.01	0.01	0.00	0.01	0.02	0.01	0.02	0.01	0.01	0.01	0.03	0.02
Total	98.82	99.30	98.91	99.18	99.11	99.93	98.96	99.52	98.99	98.56	98.91	99.03
Number of cations on the basis of 6 O												
Si	1.73	2.01	2.00	2.05	2.02	2.03	2.01	2.04	2.04	2.00	2.05	2.03
Al(IV)	0.27	0	0	0	0	0	0	0	0	0	0	0
Al(VI)	0.60	0.01	0.01	0.01	0.01	0	0.02	0.01	0.02	0.01	0.01	0.01
Ti	0	0	0	0	0	0	0	0	0	0	0	0
Cr	0	0	0	0	0	0	0	0	0	0	0	0
Fe^{3+}	0	0	0	0	0	0	0	0	0	0	0	0
Fe^{2+}	0.42	0.09	0.13	0.09	0.09	0.16	0.18	0.28	0.32	0.39	0.54	0.54
Mn	0	0.01	0.03	0	0	0.01	0.02	0.01	0	0.02	0.01	0.02
Mg	0	0.91	0.88	0.86	0.84	0.79	0.86	0.65	0.60	0.58	0.29	0.40
Ca	0.83	0.95	0.95	0.95	1.00	0.98	0.89	0.97	0.97	0.99	1.05	0.96
Na	0	0	0		0	0	0	0	0	0	0	0
K	0	0	0	0	0	0	0	0	0	0	0	0
Di	0.67	90.37	85.31	90.34	89.73	82.24	81.45	68.49	64.70	58.30	34.68	41.80
Hd	98.88	8.98	12.20	9.17	9.78	16.36	17.00	30.10	34.92	39.85	64.65	56.02
Jo	0.46	0.65	2.50	0.49	0.50	1.40	1.54	1.42	0.38	1.86	0.67	2.17

Table 5. Electron microprobe analyses of actinolite from the Hongshan deposit (wt %).

Samples	SiO_2	TiO_2	Al_2O_3	FeO	Cr_2O_3	MnO	MgO	CaO	Na_2O	K_2O	Total
Diopside–actinolite	55.08	0.02	1.11	8.03	0.13	0.11	20.08	12.53	0.30	0.04	97.43
skarn	54.43	0.22	1.46	7.98	0.12	0.15	19.86	12.90	0.30	0	97.42
KT12-4170-19	54.76	0	1.32	7.07	0	0.09	20.32	12.88	0.33	0	96.77
Actinolite skarn	55.03	0.46	2.12	4.72	0.06	0.13	21.48	12.92	0.45	0.01	97.38
KT2-4296-11	56.23	0.03	0.93	4.95	0.09	0.13	21.95	13.09	0.25	0.02	97.67
	53.91	0.17	0.55	10.89	0.13	0.30	18.44	12.16	0.35	0.12	97.02
Cations on the basis of 23 oxygens	Si	Ti	Al	Fe^{3+}	Fe^{2+}	Cr^{3+}	Mn	Mg	Ca	Na	K
KT12-4170-19	7.71	0	0.18	0.41	0.53	0.01	0.01	4.19	1.88	0.08	0.01
	7.64	0.02	0.24	0.40	0.54	0.01	0.02	4.16	1.94	0.08	0
	7.70	0	0.22	0.35	0.49	0	0.01	4.26	1.94	0.09	0
KT2-4296-11	7.63	0.02	0.35	0.25	0.30	0.01	0.02	4.44	1.92	0.12	0
	7.76	0	0.15	0.31	0.27	0.01	0.02	4.52	1.94	0.07	0
	7.69	0.02	0.09	0.52	0.78	0.02	0.04	3.92	1.86	0.10	0.02

Table 6. Sm–Nd concentrations and isotopic ratios of the garnet from the Hongshan deposit.

Sample	Sm (µg/g)	Nd (µg/g)	147Sm/144Nd	143Nd/144Nd	Standard Error	εNd (t)
KT3-4033-7	5.760	44.50	0.0784	0.512127	0.000006	−8.78
WY-4296-1	0.192	1.12	0.1040	0.512128	0.000006	−8.81
KT2-4296-6	1.780	9.73	0.1105	0.512194	0.000006	−6.64
7ZK16-1	0.480	2.82	0.1030	0.512138	0.000006	−9.02
D40-6	12.900	49.10	0.1592	0.512155	0.000006	−8.84
D40-8	12.500	33.00	0.2300	0.512202	0.000006	−9.05
D21-5	0.873	5.22	0.1012	0.512073	0.000006	−0.97
D48-13	3.060	6.99	0.2648	0.512178	0.000006	−8.52
KT2-4235-10	0.503	2.39	0.1269	0.512135	0.000007	−9.11
KT2-4170-21	1.890	12.40	0.0922	0.512120	0.000007	−9.04

6. Discussion

6.1. Mineralogy and Geochemistry of Skarn Minerals

The identification and classification of the skarn deposits were based on their mineralogy [26]. The Hongshan deposit is featured with layered or stratabound Cu ores with their occurrence in accordance with the host volcano-sedimentary wall rocks of the Qugasi Formation, which has led to the conception of sedimentary or SEDEX origin for these ores [10,11]. However, the pervasively distributed skarn minerals, e.g., andraditic garnet (Figure 4a), diopside–hedenbergite pyroxene (Figure 4a), actinolite, in the layered Cu ores are similar to the typical skarn Cu deposits as a result of interaction between felsic intrusion and surrounding carbonate rocks [26]. The lithologic or bedding contacts may provide highly permeable and/or reactive layers for the infiltration and lateral movement of the skarn-forming fluids [27]. Other evidences including the vertical zonation of skarn minerals, from top to bottom, are biotite hornfels–pyroxene hornfels, marble, wollastonite skarn (Wo > Pyx–Grt), pyroxene skarn (Pyx > Grt), garnet skarn (Grt > Pyx), pyroxene skarn, pyroxene hornfels, and biotite hornfels [6].

The chemical composition of the skarn minerals can also be used to evaluate the oxidation state of the mineralization fluids [27]. Many skarn Cu deposits contain a high abundance of chalcopyrite, bornite, pyrite, and magnetite ± hematite, which are believed to have formed from a high oxygen fugacity environment [26]. However, the large proportions of pyrrhotite (>60% sulfide) in the Hongshan deposit tends to be much more reduced than other skarn Cu deposits [28]. The highly andraditic composition of garnet (Adr_{100} to $Adr_{64}Gr_{32}$, Figure 4a) and diopsidic composition of pyroxene ($Di_{90}Hd_9$ to Di_1Hd_{99}, Figure 4b) indicate the prograde skarn minerals are formed under oxidized conditions as other skarn Cu deposits [26]. This suggest that the initial fluids during the prograde stage are inherently oxidized as other skarn Cu deposits and the rapid descent of the fugacity in the retrograde stage could be led by other effective factors.

6.2. Timing of Mineralization

The layered skarn Cu ores and vein-type Mo–Cu ores are two dominant ores in the Hongshan deposit. The age of the vein-type Mo–Cu ores have been precisely constrained by molybdenite Re–Os ages, ranging from 77 to 81 Ma [9,11,12]. However, the age of the layered skarn Cu ores is still in dispute. The existence of Triassic intrusions in the deposit area led some authors to argue the main layered skarn ores were formed in the Triassic [29], similar with the Pulang and Xuejiping porphyry Cu deposits in the Zhogndian area [4]. Re–Os dating with pyrrhotite, however, yielded a Late Cretaceous age with large error 79 ± 16 Ma [9]. Considering the low closure temperature of Re–Os systems in the pyrrhotite (as low as 400 °C) [30], a more robust and convincing method should be considered.

Garnet crystals grow between 400 °C and ~700 °C [31] and the primary growth ages can survive from wide range of P–T metamorphism, such as amphibolite metamorphism [15,16]. The strong preference of the garnet lattice for Sm over Nd makes Sm–Nd dating in garnet as a highly suitable chronometer for more than 30 years [16]. The garnets in the Hongshan deposit showed consistent or typical oscillatory zonations (Figure 5), indicating they were formed from a single hydrothermal system that underwent chemical fluctuation. The low $^{147}Sm/^{144}Nd$ ratios of 0.07 to 0.26 for the garnet samples (Table 6) in the Hongshan deposit indicated some silicate inclusions, such as epidote and pyroxene, might not be completely removed when compared with the inclusion-free garnet with high $^{147}Sm/^{144}Nd$ ratios (>1) [16,32]. Since these inclusions grew in isotopic equilibrium with the surrounding garnet and rock matrix during prograde metamorphism, the requirements of isochron dating are fulfilled. Therefore, the well-constrained isochron age of 76.48 ± 7.29 Ma (MSWD = 1.2) (Figure 6b) revealed from seven garnet separates can represent the crystallization time of the garnet. This age was consistent with the Re–Os age for the pyrrhotite from the layered skarn ores, and thereby indicated that the layered skarn mineralization was formed in the Late Cretaceous, rather than in the Triassic as previously thought [9]. The coincidence of the geochronology from the layered skarn ores and vein-type mineralization further indicated that both ores were the result of a single genetic event, rather than multiple events [10,11]. The development of endoskarn in the quartz monzonite porphyry (Figure 3a) and granite porphyry indicated the Cu–Mo mineralization in the Hongshan deposit was plausibly related to the syntectonic intrusions.

6.3. Implication for the Regional Metallogeny

The metallogenic age of the Hongshan deposit is constrained from the Sm–Nd dating of garnet and previous data to be Late Cretaceous. The Hongshan skarn Cu–Mo deposit, associated with the Xiuwacu quartz vein W–Mo, Relin quartz vein W–Mo, and Tongchanggou porphyry Mo–Cu deposit [8,33–35], thus defined a roughly N–S trending Late Cretaceous porphyry belts in the Zhongdian area, locally overlapping the Late Triassic porphyry belt (Figure 1c). Numerous geochemical research on the related intrusions revealed that the Zhongdian area in the Late Cretaceous underwent post-collisional extension event, which was also recently recognized as an optimal setting for the formation of porphyry–skarn deposits [36–38]. During post-collisional extension setting, remobilization of chalcophile metals from metasomatized mantle lithosphere have led to the generation of fertile magmas capable of forming porphyry–skarn Cu–Mo–W deposits in the Zhongdian area [17,33].

7. Conclusions

Two types of Cu–Mo ores, mainly developed in the skarns, have been recognized in the Hongshan deposit, i.e., massive or layered skarn (Figure 3a) and vein-type, with the former being dominant. The highly andraditic composition of garnet (Adr_{100} to $Adr_{64}Gr_{32}$) and diopsidic composition of pyroxene ($Di_{90}Hd_9$ to Di_1Hd_{99}) indicate the layered skarn ores are magmatic-hydrothermal origin that formed under oxidized conditions. Sm–Nd dating of garnet yielded an isochron age of 76.48 ± 7.29 Ma for the layered skarn ores, indicating the Hongshan deposit was formed in the Late Cretaceous in a unified system rather than from multiple mineralization events. The recognition of the Late Cretaceous post-collisional porphyry–skarn Cu–Mo–W belt in the Zhongdian area exhibits a promising prospecting potential.

Author Contributions: Data curation, B.Z. and Y.Z.; funding acquisition, B.Z. and C.X.; project administration, B.Z.; software, C.D.; supervision, C.X.; writing—original draft, B.Z.

Funding: This work was supported by National Key Research and Development Program of China (2017YFC0601202), National Science Foundation of China (41802108), and Fundamental Research Funds for the Central Universities (CUGL170812).

Acknowledgments: We would like to thank Sinacili, Baimakangzhu, He Jianwen, and Lurongcili from the Hongshan deposit for the help in the field. The editor and two reviewers are thanked for their constructive and detailed review of this manuscript.

Conflicts of Interest: The authors declare no conflict of interest.

References

1. Deng, J.; Wang, Q.; Li, G.; Santosh, M. Cenozoic tectono-magmatic and metallogenic processes in the Sanjiang region, southwestern China. *Earth-Sci. Rev.* **2014**, *138*, 268–299. [CrossRef]

2. Hou, Z.; Zaw, K.; Pan, G.; Mo, X.; Xu, Q.; Hu, Y.; Li, X. Sanjiang Tethyan metallogenesis in S.W. China: Tectonic setting, metallogenic epochs and deposit types. *Ore Geol. Rev.* **2007**, *31*, 48–87. [CrossRef]

3. Mao, J.; Pirajno, F.; Lehmann, B.; Luo, M.; Berzina, A. Distribution of porphyry deposits in the Eurasian continent and their corresponding tectonic settings. *J. Asian Earth Sci.* **2014**, *79*, 576–584. [CrossRef]

4. Li, W.; Zeng, P.; Hou, Z.; White, N.C. The Pulang porphyry copper deposit and associated felsic intrusions in Yunnan Province, Southwest China. *Econ. Geol.* **2011**, *106*, 79–92.

5. Leng, C.; Cooke, D.R.; Hou, Z.; Evans, N.J.; Zhang, X.; Chen, W.T.; Danišík, M.; McInnes, B.I.A.; Yang, J. Quantifying Exhumation at the Giant Pulang Porphyry Cu–Au Deposit Using U-Pb-He Dating. *Econ. Geol.* **2018**, *113*, 1077–1092. [CrossRef]

6. Peng, H.; Mao, J.; Hou, L.; Shu, Q.; Zhang, C.; Liu, H.; Zhou, Y. Stable Isotope and Fluid Inclusion Constraints on the Source and Evolution of Ore Fluids in the Hongniu-Hongshan Cu Skarn Deposit, Yunnan Province, China. *Econ. Geol.* **2016**, *111*, 1369–1396. [CrossRef]

7. Wang, X.; Hu, R.; Bi, X.; Leng, C.; Pan, L.; Zhu, J.; Chen, Y. Petrogenesis of Late Cretaceous I-type granites in the southern Yidun Terrane: New constraints on the Late Mesozoic tectonic evolution of the eastern Tibetan Plateau. *Lithos* **2014**, *208*, 202–219. [CrossRef]

8. Yang, L.; Deng, J.; Gao, X.; He, W.; Meng, J.; Santosh, M.; Yu, H.; Yang, Z.; Wang, D. Timing of formation and origin of the Tongchanggou porphyry–skarn deposit: Implications for Late Cretaceous Mo–Cu metallogenesis in the southern Yidun Terrane, SE Tibetan Plateau. *Ore Geol. Rev.* **2016**, *81*, 1015–1032. [CrossRef]

9. Zu, B.; Xue, C.; Zhao, Y.; Qu, W.; Li, C.; Symons, D.T.A.; Du, A. Late Cretaceous metallogeny in the Zhongdian area: Constraints from Re–Os dating of molybdenite and pyrrhotite from the Hongshan Cu deposit, Yunnan, China. *Ore Geol. Rev.* **2015**, *64*, 1–12. [CrossRef]

10. Leng, C. Genesis of Hongshan Cu polymetallic large deposit in the Zhongdian area, NW Yunnan: Constraints from LA-ICP-MS trace elements of pyrite and pyrrhotite. *Earth Sci. Front.* **2017**, *24*, 162–175.

11. Xu, X.; Cai, X.; Qu, W.; Song, B.; Qin, K.; Zhang, B. Later Cretaceous granitic porphyritic Cu-Mo mineralization system in the Hongshan area, northwestern Yunnan and its significances for tectonics. *Acta Geol. Sin.* **2006**, *80*, 1422–1433.

12. Peng, H.; Mao, J.; Pei, R.; Zhang, C.; Tian, G.; Zhou, Y.; Li, J.; Hou, L. Geochronology of the Hongniu-Hongshan porphyry and skarn Cu deposit, northwestern Yunnan province, China: Implications for mineralization of the Zhongdian arc. *J. Asian Earth Sci.* **2014**, *79*, 682–695. [CrossRef]

13. Stein, H.J.; Morgan, J.W.; Scherstén, A. Re-Os dating of low-level highly radiogenic (LLHR) sulfides: The Harnäs gold deposit, southwest Sweden, records continental-scale tectonic events. *Econ. Geol.* **2000**, *95*, 1657–1671. [CrossRef]

14. Li, Y.; Li, J.; Li, X.; Selby, D.; Huang, G.; Chen, L.; Zheng, K. An Early Cretaceous carbonate replacement origin for the Xinqiao stratabound massive sulfide deposit, Middle-Lower Yangtze Metallogenic Belt, China. *Ore Geol. Rev.* **2017**, *80*, 985–1003. [CrossRef]

15. Baxter, E.F.; Scherer, E.E. Garnet Geochronology: Timekeeper of Tectonometamorphic Processes. *Elements* **2013**, *9*, 433–438. [CrossRef]

16. Baxter, E.F. Garnet: A Rock-Forming Mineral Petrochronometer. *Rev. Miner. Geochem.* **2017**, *83*, 469–553. [CrossRef]

17. Zu, B.; Xue, C.; Chi, G.; Zhao, X.; Li, C.; Zhao, Y.; Yalikun, Y.; Zhang, G.; Zhao, Y. Geology, geochronology and geochemistry of granitic intrusions and the related ores at the Hongshan Cu-polymetallic deposit: Insights into the Late Cretaceous post-collisional porphyry-related mineralization systems in the southern Yidun arc, SW China. *Ore Geol. Rev.* **2016**, *77*, 25–42. [CrossRef]

18. Li, W.; Yu, H.; Gao, X.; Liu, X.; Wang, J. Review of Mesozoic multiple magmatism and porphyry Cu–Mo (W) mineralization in the Yidun Arc, eastern Tibet Plateau. *Ore Geol. Rev.* **2017**, *90*, 795–812. [CrossRef]

19. Hou, Z.; Yang, Y.; Qu, X.; Huang, D.; Lv, Q.; Wang, H.; Yu, J.; Tang, S. Tectonic evolution and mineralization systems of the Yidun arc orogen in Sanjiang region. *Acta Geol. Sin.* **2004**, *78*, 109–120.

20. Yang, L.; He, W.; Gao, X.; Xie, S.; Yang, Z. Mesozoic multiple magmatism and porphyry–skarn Cu–polymetallic systems of the Yidun Terrane, Eastern Tethys: Implications for subduction- and transtension-related metallogeny. *Gondwana Res.* **2018**, *62*, 144–162. [CrossRef]

21. Chen, J.; Xu, J.; Ren, J.; Huang, X.; Wang, B. Geochronology and geochemical characteristics of Late Triassic porphyritic rocks from the Zhongdian arc, eastern Tibet, and their tectonic and metallogenic implications. *Gondwana Res.* **2014**, *26*, 492–504. [CrossRef]

22. Gao, X.; Yang, L.; Orovan, E.A. The lithospheric architecture of two subterranes in the eastern Yidun Terrane, East Tethys: Insights from Hf–Nd isotopic mapping. *Gondwana Res.* **2018**, *62*, 127–143. [CrossRef]

23. Wang, X.; Bi, X.; Leng, C.; Zhong, H.; Tang, H.; Chen, Y.; Yin, G.; Huang, D.; Zhou, M. Geochronology and geochemistry of Late Cretaceous igneous intrusions and Mo–Cu–(W) mineralization in the southern Yidun Arc, SW China: Implications for metallogenesis and geodynamic setting. *Ore Geol. Rev.* **2014**, *61*, 73–95. [CrossRef]

24. Peng, H.; Zhang, C.; Mao, J.; Santosh, M.; Zhou, Y.; Hou, L. Garnets in porphyry–skarn systems: A LA–ICP–MS, fluid inclusion, and stable isotope study of garnets from the Hongniu–Hongshan copper deposit, Zhongdian area, NW Yunnan Province, China. *J. Asian Earth Sci.* **2015**, *103*, 229–251. [CrossRef]

25. Vermeesch, P. IsoplotR: A free and open toolbox for geochronology. *Geosci. Front.* **2018**, *9*, 1479–1493. [CrossRef]

26. Meinert, L.D.; Dipple, G.M.; Nicolescu, S. World Skarn Deposits. *Econ. Geol.* **2005**, *100*, 299–336.

27. Maher, K.C. Skarn alteration and mineralization at coroccohuayco, Tintaya District, Peru. *Econ. Geol.* **2010**, *105*, 263–283. [CrossRef]

28. Zu, B.; Xue, C.; Yaxiaer, Y.; Wang, Q.; Liang, H.; Zhao, Y.; Liu, M. Sulfide zonal texture and its geological significance of ores from the Hongshan copper deposit in Shangri-la, Yunnan Province, China. *Aata Petrol. Sin.* **2013**, *29*, 1203–1213.

29. Wang, S.; Zhang, X.; Leng, C.; Qin, C.; Wang, W.; Zhao, M. Stable isotopic compositions of the Hongshan skarn copper deposit in the Zhongdian area and its implication for the copper mineralization process. *Acta Petrol. Sin.* **2008**, *24*, 480–488.

30. Brenan, J.M.; Cherniak, D.J.; Rose, L.A. Diffusion of osmium in pyrrhotite and pyrite: Implications for closure of the Re-Os isotopic system. *Earth Planet. Sci. Lett.* **2000**, *180*, 399–413. [CrossRef]

31. Caddick, M.J.; Kohn, M.J. Garnet: Witness to the Evolution of Destructive Plate Boundaries. *Elements* **2013**, *9*, 427–432. [CrossRef]

32. Mühlberg, M.; Hegner, E.; Klemd, R.; Pfänder, J.A.; Kaliwoda, M.; Biske, Y.S. Late Carboniferous high-pressure metamorphism of the Kassan Metamorphic Complex (Kyrgyz Tianshan) and assembly of the SW Central Asian Orogenic Belt. *Lithos* **2016**, *264*, 41–55.

33. Yang, L.; Gao, X.; Shu, Q. Multiple Mesozoic porphyry-skarn Cu (Mo–W) systems in Yidun Terrane, east Tethys: Constraints from zircon U–Pb and molybdenite Re–Os geochronology. *Ore Geol. Rev.* **2017**, *90*, 813–826. [CrossRef]

34. Gao, X.; Yang, L.; Meng, J.; Zhang, L. Zircon U–Pb, molybdenite Re–Os geochronology and Sr–Nd–Pb–Hf–O–S isotopic constraints on the genesis of Relin Cu–Mo deposit in Zhongdian, Northwest Yunnan, China. *Ore Geol. Rev.* **2017**, *91*, 945–962. [CrossRef]

35. He, J.; Wang, B.; Wang, L.; Wang, Q.; Yan, G. Geochemistry and geochronology of the Late Cretaceous Tongchanggou Mo–Cu deposit, Yidun Terrane, SE Tibet; implications for post-collisional metallogenesis. *J. Asian Earth Sci.* **2018**, *172*, 308–327. [CrossRef]
36. Hou, Z.; Zhang, H.; Pan, X.; Yang, Z. Porphyry Cu (–Mo–Au) deposits related to melting of thickened mafic lower crust: Examples from the eastern Tethyan metallogenic domain. *Ore Geol. Rev.* **2011**, *39*, 21–45. [CrossRef]
37. Richards, J.P. Postsubduction porphyry Cu-Au and epithermal Au deposits: Products of remelting of subduction-modified lithosphere. *Geology* **2009**, *37*, 247–250. [CrossRef]
38. Richards, J.P. Magmatic to hydrothermal metal fluxes in convergent and collided margins. *Ore Geol. Rev.* **2011**, *40*, 1–26. [CrossRef]

Article

Geochemical Characteristics of A-Type Granite near the Hongyan Cu-Polymetallic Deposit in the Eastern Hegenshan-Heihe Suture Zone, NE China: Implications for Petrogenesis, Mineralization and Tectonic Setting

Chen Mao [1], Xinbiao Lü [1,2,*] and Chao Chen [2]

[1] Faculty of Earth Resources, China University of Geosciences, Wuhan 430074, China; maochen_110@cug.edu.cn

[2] Institute of Geological Survey, China University of Geosciences, Wuhan 430074, China; chenchao_109@163.com

* Correspondence: luxb@cug.edu.cn; Tel.: +86-159-7293-4906

Received: 11 February 2019; Accepted: 16 April 2019; Published: 18 May 2019

Abstract: In the eastern Hegenshan-Heihe suture zone (HHSZ) of NE China, Cu-Au hydrothermal mineralization at the newly discovered Hongyan deposit is associated with the Shanshenfu alkali-feldspar granite (SAFG). Zircon U-Pb dating showed that the inner phase and outer phase of the SAFG were formed at 298.8 ± 1.0 Ma and 298.5 ± 1.0 Ma, respectively. Whole rock geochemistry suggests that the SAFG can be classified as an A-type granite. Halfnium isotopes and trace elements in zircon suggest that the SAFG has high Ti-in-zircon crystallization temperature (721–990 °C), high magmatic oxygen fugacity and largely positive $\varepsilon_{Hf}(t)$ (from +6.0 to +9.9). We proposed that the SAFG was derived from crustal assimilation and fractional crystallization of juvenile crust metasomatized by subducting oceanic crust. The high oxygen fugacity of the SAFG suggests the chalcophile elements (e.g., Cu, Au) remained in the magma as opposed to the magma source. An arc-related juvenile source favors enrichment of Cu and Au in the resulting magma. Combined, these magmatic characteristics suggest Cu ± Au exploration potential for magmatic-hydrothermal mineralization related to the SAFG, and similar bodies along the HHSZ. The results obtained combined with regional geological background suggest that the Permian A-type granites and related mineralization along the HHSZ were formed in a post-collisional slab break-off process.

Keywords: Hegenshan-Heihe suture zone; Cu-Au hydrothermal mineralization; Hongyan deposit; Permian A-type granite; granite petrogenesis; magmatic oxygen fugacity; post-collisional slab break-off

---

1. Introduction

Since the introduction of the A-type granite classification by Loiselle and Wones [1], rocks of this type have been studied extensively due to their important geodynamic significance, complicated petrogenesis and economic potential [2–13]. It has been recognized that A-type granites can be formed in a variety of extensional regimes, such as continental back-arc extension, post-collisional extension or within-plate settings [2,14]. In recent years, there have been many studies on the origin and oxidation state of A-type granites and their relationships to mineralization [7–10,12,13]. In general, previous workers have related oxygen fugacity and a degree of magma contamination to different types of mineralization [7–10]. High magmatic oxygen fugacity and juvenile magma sources such as mantle-derived crust or partial melting of an oceanic slab have been related to Cu-Mo-Au-Pb-Zn mineralization [7,8,10]. Low magmatic oxygen fugacity, and wall-rock interaction or assimilation of continental crust are associated with W-Sn mineralization [7–9]. The oxidation state and petrogenesis of A-type granites has important geological significance, not only for regional tectonic evolution, but also for understanding the metallogenic process related to A-type granites. Decoding this petrogenetic intricacy, inherent geodynamic background and the related metallogenic process requires the use of an integrated whole rock geochemical and zircon chemical approach. Zircon chemistry relates to crystallization temperature and oxygen fugacity (fo_2) of magma [15–18], the composition of parental melts [19,20], source-rock type and crystallization environment [21–23]. The refractory nature and low elemental diffusion [24] mean that zircon chemistry remains unchanged through most geological processes, such as hydrothermal alteration, providing a powerful tool for documenting magmatic conditions (e.g., magma source and fo_2) and the resulting behavior of elements such as Cu [25–28].

In northeastern China (NE China), numerous Permian A-type granitoids (260–300 Ma) have been identified along the Hegenshan-Heihe suture zone (HHSZ) in the past two decades (Figure 1a,b) [3,4,29–32]. These A-type granites are on the eastern side of the larger Central Asian Orogenic Belt (CAOB, Figure 1a). Previous studies have provided mineralogical and geochemical constraints on the types (A2 type), sources (crust-mantle mixing or crustal sources) and tectonic settings (post-orogenic extension or post-collisional extension) of the Permian A-type granites along the HHSZ [3,4,29–32]. Newly discovered hydrothermal deposits are associated with these Permian A-type granites, including Aoyoute Cu-polymetallic deposit (287 Ma), Bayandulan Cu-polymetallic deposit (284 Ma) and Ataiwula Cu-polymetallic deposit (276 Ma) [32]. Despite a clear relationship between Permian A-type granitoids and mineralization, there is debate about the petrogenesis and tectonic setting of these rocks, and influence of the magma source and magmatic oxygen fugacity on the related metallogenic process needs further research. The newly discovered Hongyan Cu-polymetallic deposit is located in the northeastern part of the HHSZ (Figure 1c), and contains Cu-polymetallic hydrothermal mineralization clearly associated with a Permian A-type granitoid. The deposit and surrounding rocks provide a good opportunity to study the petrogenesis and tectonic setting of Permian A-type granitoids and their possible relationship with hydrothermal mineralization.

In this paper, we present zircon U-Pb ages, trace elements data and Hf isotope data with whole rock major and trace element data from Permian A-type granite in the Hongyan Cu-polymetallic deposit. We then use these data to discuss the oxidation state, petrogenesis and tectonic setting of these rocks. The results provide new insights into the petrogenesis and tectonic setting of Permian A-type granites along the HHSZ, and highlight the exploration potential of these rocks.

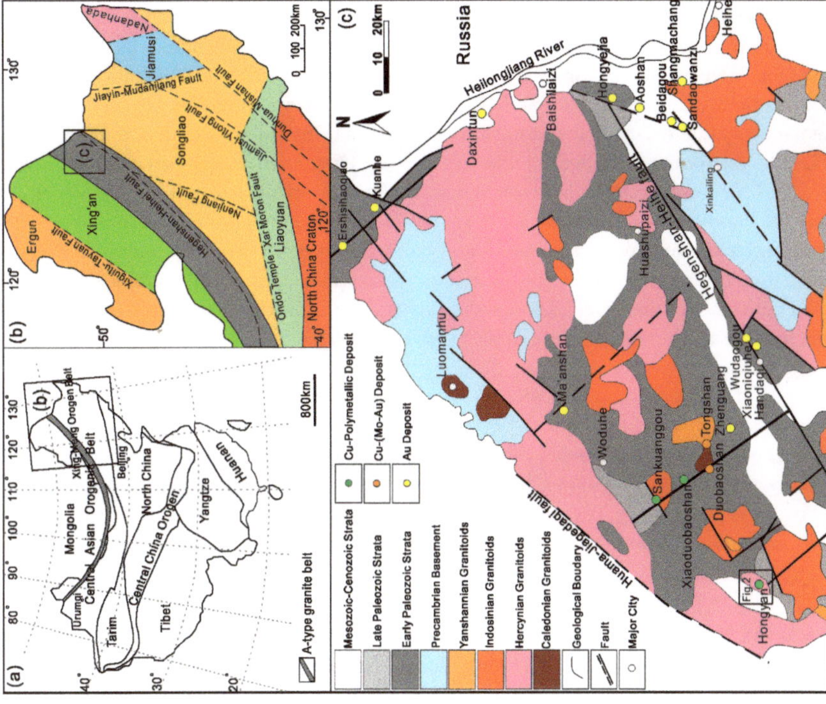

Figure 1. (**a**) Geotectonic division of China (after Mao et al. [33]); (**b**) tectonic subdivisions of northeast China (after Wu et al. [34]); (**c**) regional geological map of the northern Xing'an Block (after Gao et al. [35]).

2. Regional Geology

The Xing-Meng Orogen Belt (XMOB) is located in the eastern part of the CAOB, and was formed in the Permian through the collision between the Siberia block and North China block (Figure 1a) [34,36,37]. The XMOB is further subdivided by several NE-striking faults into different microcontinental massifs or terranes, including the Ergun massif, the Xing'an terrane, the Songliao terrane, the Jiamusi terrane and the Liaoyuan terrane (Figure 1b). These terranes have undergone EW-trending tectonic evolution of the Paleo-Asian Ocean during the Paleozoic and NNE-trending tectonic evolution of the western Pacific Ocean during the Mesozoic and Cenozoic.

The HHSZ is a suture zone resulting from the collision of the Xing'an and Songliao terranes (Figure 1b). The outcropping strata in the northern Xing'an terrane and its adjacent area can be approximately divided into four units (Figure 1c) [38,39]: (1) Sporadic distribution of Precambrian crystalline basement rocks composed of granulites, gneisses and schists; (2) extensive exposure of early Paleozoic metamorphosed volcanic and sedimentary rocks consisting of schist, sandy slate, marble and andesite, which are formed in continental margin and arc accretion settings; (3) the late Paleozoic units are similar to the early Paleozoic units but with lower metamorphic grades, such as the extensive distribution of the Baoligaomiao volcanic succession that represents Carboniferous subduction-related, mature, continental arc volcanism [40]; and (4) Mesozoic continental intermediate-felsic volcanic and sedimentary rocks, and Cenozoic sedimentary basin strata. Intense magmatic activity occurred along the HHSZ, which can be divided into the Caledonian, Hercynian, Indosinian and Yanshanian periods [41]. The Caledonian granitoids, consisting of granodiorite and quartz diorite, are sporadically distributed in the Luomahu and Duobaoshan areas. The Hercynian granitoids including granodiorite, monzogranite and alkali-feldspar granite are widely distributed on the northern side of the HHSZ. The Indosinian granitoids, consisting of monzogranite, syenogranite and two-mica granite, are widely distributed on both sides of the HHSZ. The Yanshanian granitoids, being composed of quartz diorite, granodiorite, tonalite and biotite granite, exhibit a sporadic distribution (Figure 1c) [4,42].

Intense hydrothermal activity was associated with these periods of magmatism, and resulted in various types of hydrothermal deposits, primarily including skarn Cu-polymetallic deposits (e.g., Sankuanggou and Xiaoduobaoshan), porphyry Cu-(Mo-Au) deposits (e.g., Duobaoshan and Tongshan) and epithermal Au deposits (e.g., Shangmachang, Beidagou, Sandaowanzi, Tianwangtaishan and Zhenguang, Figure 1c) [34,43,44], which constitute the northeastern segment of the Xing'an Cu-Mo-Fe-Pb-Zn-Au belt [37,45]. These deposits are located in a nearly NE-trending band along the HHSZ and its secondary NE- and NW-trending faults (Figure 1c).

3. Deposit Geology

The Hongyan Cu-polymetallic deposit comprises three orebodies (from E125°01′28″ to E125°06′30″, from N49°33′30″ to N49°42′30″; Figure 2), located in the northeastern part of the HHSZ (Figure 1c). Paleozoic and Mesozoic volcanic rocks are widespread near the deposit (Figure 2). Paleozoic volcanic rocks mainly include the Upper Carboniferous–Lower Permian Baoligaomiao Formation, which is a suite of continental volcanic rocks composed of rhyolitic tuff, andesitic tuff and basalt [46]. Mesozoic volcanic rocks are subdivided into the Lower Jurassic Manitu Formation, the Upper Cretaceous Baiyingaolao Formation and Damoguaihe Formation. The Manitu Formation comprises dacitic ignimbrite and dacite. The Baiyingaolao Formation is mainly composed of rhyolite, rhyolitic volcanic breccias and rhyolitic ignimbrite. The Damoguaihe Formation is mainly composed of conglomerate, fine sandstone and tuff breccias [46]. Steeply dipping brittle faults are dominantly NE-trending (e.g., F1 and F2), NWW-trending (e.g., F3, F4, F5, F6, F7 and F9) and E-trending faults (e.g., F8). The Shanshenfu alkali-feldspar granite (SAFG) with an exposed area of ~75 km^2 dominates plutonic rocks. The SAFG intrudes Paleozoic volcanic rocks with an irregular contact and is covered by Mesozoic volcanic rocks and Quaternary sediments (Figure 2).

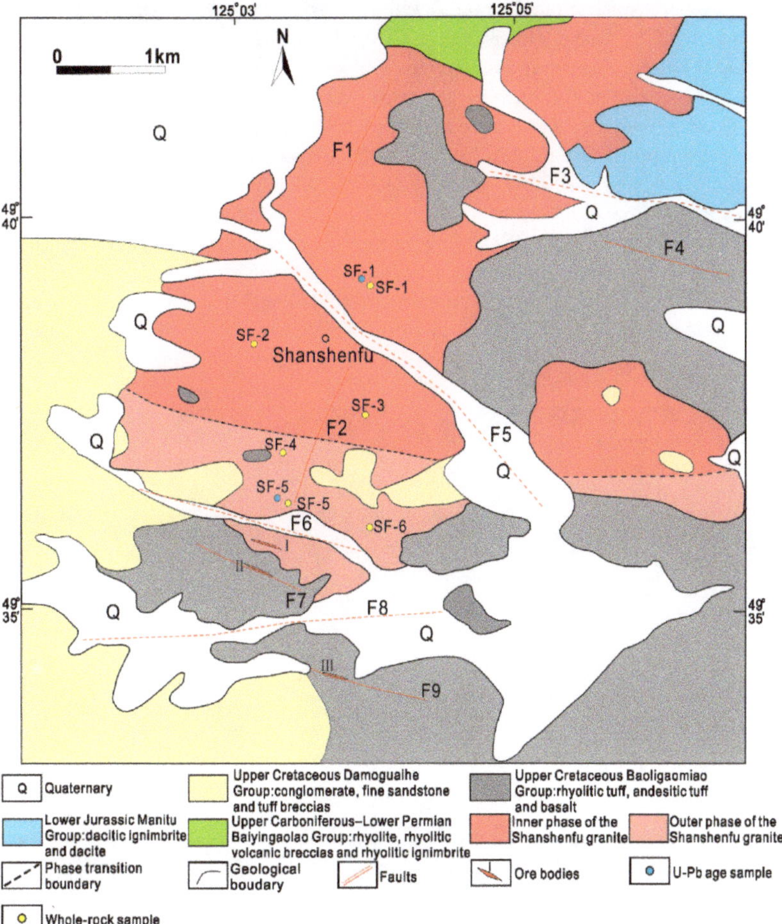

Figure 2. Geological sketch of the Hongyan Cu-polymetallic deposit [46].

The SAFG comprises an inner medium-grained phase and outer fine-grained phase (Figure 2). The boundary between the phases is gradational. The inner phase contains subhedral perthite (50%–55%), anhedral quartz (30%–35%), plagioclase (5%–10%), hornblende (0%–5%) and muscovite (0%–5%, Figures 3a and 4a). The outer phase comprises subhedral perthite (55%–60%), anhedral quartz (25%–30%), plagioclase (5%–10%), hornblende (0%–5%) and muscovite (0%–5%, Figures 3b and 4b). Accessory minerals are magnetite (Figure 4d), zircon and apatite.

The SAFG is pervasively altered by potassic (Figure 3c) and quartz-sericite alteration (Figure 4c), and is accompanied by some disseminated Cu mineralization (Figure 4d). Cu-polymetallic mineralization generally occurs as quartz veins and veinlets in the SAFG and the Baoligaomiao Formation, which is structurally controlled by tensile cracks and NWW-striking faults, respectively (Figure 2). Three vein-shaped mineralized bodies are identified in the Hongyan Cu-polymetallic deposit, which from north to south are No. I, II and III, respectively [46] (Figure 2): The No. I ore body hosted in the outer phase of the SAFG develops NWW-trending ore veins with a length of 250 m, an average thickness of 3.1 m and Cu grades of 0.5%–1.5%; the No. II ore body hosted in the Baoligaomiao Formation develops NWW-trending and N-dipping ore veins with a length of 220 m, an average thickness of 2.3 m and Cu, Au and Ag grades of 0.4%–1.2%, 0.34–2.21 g/t and 0.76–21.6 g/t, respectively; the No. III ore body hosted in the Baoligaomiao Formation develops NWW-trending and N-dipping ore veins with a length of 215 m, an average thickness of 1.8 m and Cu, Au and Ag grades of 0.5%–1.35%, 0.27–2.62 g/t and 0.64–16.2 g/t, respectively.

Figure 3. Hand specimen photos of the Shanshenfu alkali-feldspar granite and Stage I–III mineralization from the Hongyan Cu-polymetallic deposit. (**a**) Medium-grained alkali-feldspar granite (inner phase); (**b**) fine-grained alkali-feldspar granite (outer phase); (**c**) Stage I K-feldspar–quartz–pyrite vein; (**d**) Stage II disseminated chalcopyrite–pyrite–quartz mineralization; (**e**) Stage II quartz–pyrite–chalcopyrite–bornite vein; (**f**) Stage II quartz–pyrite–bornite vein; (**g**) Stage II quartz–bornite–galena vein; (**h**) Stage III quartz–calcite vein; (**i**) Stage III quartz–calcite–fluorite vein. Qz—quartz, Kf—K-feldspar, Py—pyrite, Ccp—chalcopyrite, Bn—bornite, Gn—galena, Cc—calcite, Fl—fluorite.

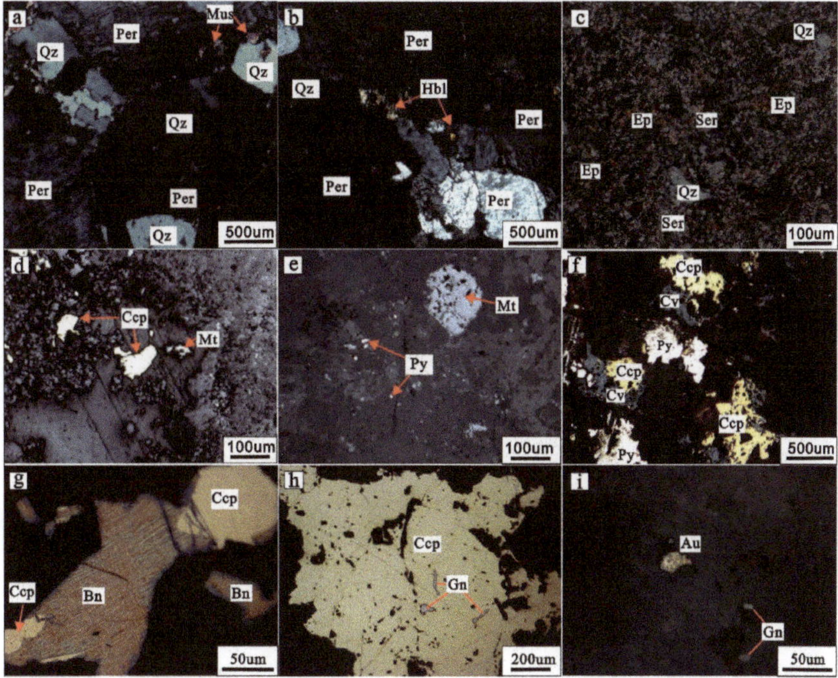

Figure 4. Transmitted light (**a**–**c**) and reflected light (**d**–**i**) microphotographs of the Shanshenfu alkali-feldspar granite, hydrothermal alteration features and metal mineral assemblages. (**a**) Medium-grained alkali-feldspar granite (inner phase); (**b**) fine-grained alkali-feldspar granite (outer phase); (**c**) widespread alteration of quartz, sericite and epidote in the granite; (**d**) chalcopyrite and magnetite in fine-grained alkali-feldspar granite (outer phase); (**e**) pyrite and magnetite in Stage I K-feldspar–quartz–pyrite vein; (**f**) chalcopyrite intergrown with pyrite replaced by the covellite in Stage II quartz vein; (**g**) chalcopyrite replace bornite in Stage II quartz vein; (**h**) chalcopyrite intergrown with galena in Stage II quartz vein; (**i**) visible gold intergrown with galena in Stage II quartz vein. Py—pyrite, Ccp—chalcopyrite, Bn—bornite, Gn—galena, Mt—magnetite, Cv—covellite, Au—native gold, Qz—quartz, Per—Perthite, Mus—Muscovite, Hbl—Hornblende, Ser—sericite, Ep—epidote.

4. Mineralization and Alteration in Ore Bodies

Based on our field and petrographic observations, the main ore minerals are chalcopyrite and pyrite, with minor bornite, covellite, galena, sphalerite, magnetite and native gold. The most common gangue minerals are quartz, K-feldspar, sericite, muscovite, chlorite, epidote, calcite and fluorite. Silicification is well developed and closely related to sulfide mineralization. Based on hydrothermal alteration mineral assemblages, we identify three stages of mineralization (Figure 5): Quartz ± K-feldspar ± pyrite (Stage I), quartz + chalcopyrite ± pyrite ± bornite ± sphalerite ± galena (Stage II) and quartz + carbonate ± fluorite (Stage III). Stage I is widely distributed in No. I and II ore bodies, which is defined by the occurrence of quartz and K-feldspar with minor pyrite and magnetite as veins (Figures 3c and 4e). Stage II is the main ore-forming stage and mainly distributed in the three ore bodies, which is represented by the widespread occurrence of quartz, chalcopyrite and pyrite minerals as veins or veinlets with minor bornite, covellite, galena and sphalerite (Figures 3d–g and 4f–h). Chalcopyrite replaces early bornite and is commonly intergrown with pyrite and galena (Figure 4g,h). Chalcopyrite and pyrite are partly replaced by covellite (Figure 4f). Gold mineralization is most commonly hosted within sulfide minerals in this stage, and visible gold can be found in the quartz veins (Figure 4i). Stage III is mainly distributed in No. II and III ore bodies, which is characterized by the appearance of calcite and quartz with minor fluorite as veins or veinlets (Figure 3h–i). Small amounts of pyrite occur in quartz-calcite veins, and gold mineralization is less in this stage. The presence of fluorite mineralization has been reported to be closely related to A-type granites [47,48]. Combined, the development of Cu-polymetallic mineralization in the SAFG and widespread K-feldspar alteration and fluorite mineralization in ore bodies suggest that these ore-forming hydrothermal fluids are mainly derived from the SAFG.

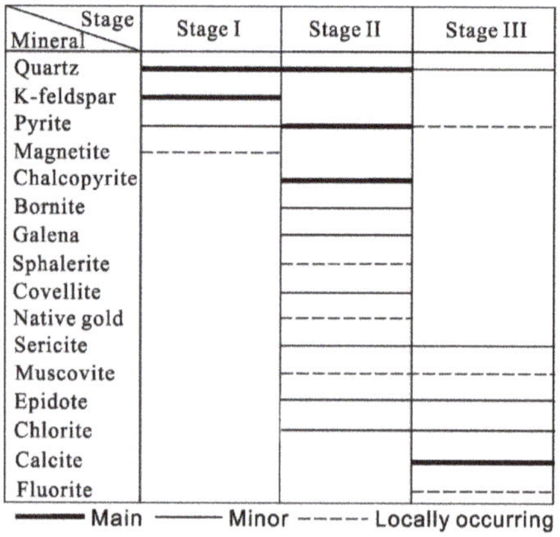

Mineral \ Stage	Stage I	Stage II	Stage III
Quartz			
K-feldspar			
Pyrite			
Magnetite			
Chalcopyrite			
Bornite			
Galena			
Sphalerite			
Covellite			
Native gold			
Sericite			
Muscovite			
Epidote			
Chlorite			
Calcite			
Fluorite			

Figure 5. Paragenetic sequence for the Hongyan Cu-polymetallic deposit.

5. Analytical Methods

5.1. Zircon U-Pb Dating Analysis

The sample SF-1 from the inner phase and the sample SF-5 from the outer phase were collected for geochronology (Figure 2). Zircon grains were separated using conventional heavy liquid and magnetic techniques, and then handpicked under a binocular microscope. Handpicked zircon grains were mounted on adhesive tape, enclosed in epoxy resin and polished so that approximately half of each zircon is exposed. To study interior textures of zircons cathodoluminescence (CL) images were collected with a JEOL JXA-8100 electron microprobe (TESCAN MIRA 3 LMH FE-SEM, TESCAN, Brno, Czech Republic) at the State Key Laboratory of Geological Processes and Mineral Resources, China University of Geosciences, Wuhan. Zircon U-Th-Pb measurements were done under a 32 μm diameter laser beam at the same laboratory using a Geo-Las 2005 System. An Agilent 7500a ICP-MS (Agilent, Santa Clara, CA, USA) instrument was used to acquire ion-signal intensities with a 50 mJ/pulse 193 nm ArF-excimer laser at 10 Hz and a homogenizing, imaging optical system (MicroLas, Göttingen, Germany). The data acquisition for each analysis took 100 s (40 s on background and 60 s on signal). Detailed instrumentation and analytical accuracy description were given by Liu et al. [49,50]. Time-dependent drifts of U-Th-Pb isotopic ratios were corrected using a linear interpolation (with time) for every six analyses according to the variations of external standard zircon 91,500 (2 zircon 91,500 + 6 samples + 2 zircon 91,500) [50]. The ages are calculated by inhouse software ICPMSDataCal (version 6.9, China University of Geosciences) [49] and concordia diagrams were made by Isoplot/Ex ver3.0 [51]. Errors on individual analyses are quoted at the 1σ level, whereas errors on weighted mean ages are quoted at the 2σ (95% confidence) level. Trace element compositions of zircon are calibrated against GSE-1G combined with internal standardization ^{29}Si [50].

Zircon Ce and Eu anomalies, Ti-in-zircon temperatures and magmatic oxygen fugacity (f_{O_2}) were calculated using the trace element compositions of zircons collected during the same analysis interval as U-Pb dating. The method for calculating these parameters has been described in Ferry and Watson [16] and Trail et al. [17,18]. CGDK software [52] was used for plotting data.

5.2. Zircon Lu-Hf Isotopes Analysis

In-situ zircon Hf isotopic analyses were conducted using a Neptune Plus MC-ICP-MS (Thermo Fisher Scientific, Germany) equipped with a Geolas 2005 excimer ArF laser ablation system (Lambda Physik, Göttingon, Germany) at the state Key Laboratory of Geological Processes and Mineral Resources, China University of Geosciences (Wuhan). All data were acquired on zircon in single spot ablation mode at a pulse rate of 20 Hz at 200 mJ with a spot size of 44 µm in this study. Each measurement consisted of 20 s of acquisition of the background signal followed by 50 s of ablation signal acquisition. Detailed operating conditions for the laser ablation system and the analytical method are the same as description by Hu et al. [53]. The $^{176}Hf/^{177}Hf$ ratios of the standard zircon (GJ1) were 0.282013 ± 0.000022 (2σ, n = 276), in agreement with recommended values within 2σ error [54,55]. Offline selection and integration of analyte signals, and mass bias calibrations were performed using ICPMSDataCal [50].

5.3. Whole-Rock Major and Trace Element Analyses

Three samples from the inner phase (from SF-1 to SF-3) and three samples from the outer phase (from SF-4 to SF-6) were collected for major and trace element determinations. All the samples are unaltered and unweathered. Geochemical analyses were carried out at the Guangzhou Institute of Geochemistry, Chinese Academy of Sciences (GIG-CAS). Major element oxides were analyzed using a Rigaku RIX 2000 X-ray fluorescence spectrometer (XRF, Panalytical, Almelo, The Netherlands), and analytical uncertainties are mostly between 1% and 5% [56]. Trace elements were obtained by inductively coupled plasma-mass spectrometry (ICP-MS, Perkin Elmer Elan 9000, Perkin, Waltham, MA, USA) after acid digestion of samples in high-pressure Teflon vessels, and detailed procedures are described by Li et al. [56]. The USGS and Chinese National standards AGV-2, GSR-1, GSR-2, MRG-1, BCR-1, W-2 and G-2 were chosen for calibrating element concentrations of the analyzed samples. Analytical precision of REE and other incompatible element analyses is typically 1%–5%.

6. Results

6.1. Zircon U-Pb Ages

The zircon U-Pb ages for 23 zircons are shown in Table 1. Most of the zircon grains from two samples (SF-1 and SF-5) are euhedral and prismatic, and are relatively transparent and colorless. They have a length between 61 µm and 179 µm, with length to width ratios between approximately 1:1–3:1 (Figure 6a,b). CL images show that most zircons have magmatic oscillatory overgrowth rims, although a few zircons with high uranium contents are dark brown and turbid (Figure 6a,b). Most zircons have high Th/U ratios >0.5, which are typical of an igneous origin. Twelve spot analyses of zircons from sample SF-1 yield $^{206}Pb/^{238}U$ ages of 297 Ma to 300 Ma (Figure 6c) with a weighted mean age of 298.8 ± 1.0 Ma (MSWD = 0.27) for all 12 analyses. Eleven spot analyses from sample SF-5 yield $^{206}Pb/^{238}U$ ages between 297 Ma and 300 Ma (Figure 6d) with a weighted mean age of 298.5 ± 1.0 Ma (MSWD = 0.36) for all 11 analyses. These ages are coeval and are interpreted as the crystallization age for the SAFG.

Table 1. Zircon LA-ICP-MS U-Pb data of the Shanshenfu alkali-feldspar granite.

Spot	Trace Elements (ppm)			Isotopic Ratio						Apparent Age (Ma)					
	U	Th	Th/U	$^{207}Pb/^{235}U$	1σ	$^{206}Pb/^{238}U$	1σ	$^{207}Pb/^{206}Pb$	1σ	$^{206}Pb/^{238}U$	1σ	$^{207}Pb/^{235}U$	1σ	$^{207}Pb/^{206}Pb$	1σ
Sample SF-1															
SF1-1	635	337	0.53	0.3362	0.0039	0.0472	0.0002	0.0517	0.0006	297	1.7	294	4.0	272	21.3
SF1-2	591	349	0.59	0.3425	0.0046	0.0475	0.0002	0.0523	0.0007	299	1.9	299	4.6	298	−2.8
SF1-3	817	683	0.84	0.3341	0.0035	0.0473	0.0002	0.0512	0.0005	298	1.8	293	3.8	250	24.1
SF1-4	656	345	0.53	0.3413	0.0046	0.0475	0.0001	0.0521	0.0007	299	1.6	298	4.6	300	31.5
SF1-5	1413	809	0.57	0.3434	0.0024	0.0473	0.0001	0.0526	0.0004	298	1.6	300	3.0	322	14.8
SF1-6	495	262	0.53	0.3403	0.0052	0.0475	0.0002	0.0519	0.0008	299	1.8	297	5.1	283	30.6
SF1-7	1130	738	0.65	0.3468	0.0029	0.0476	0.0002	0.0528	0.0004	300	2.0	302	3.4	320	16.7
SF1-8	988	482	0.49	0.3463	0.0031	0.0476	0.0002	0.0528	0.0005	300	1.8	302	3.5	320	20.4
SF1-9	822	430	0.52	0.3465	0.0033	0.0476	0.0002	0.0528	0.0005	300	1.9	302	3.7	320	20.4
SF1-10	1202	673	0.56	0.3418	0.0024	0.0475	0.0002	0.0522	0.0004	299	2.0	299	3.0	295	14.8
SF1-11	191	144	0.76	0.3417	0.0124	0.0473	0.0002	0.0523	0.0019	298	2.0	298	11.1	298	79.6
SF1-12	719	283	0.39	0.3428	0.0046	0.0475	0.0002	0.0523	0.0007	299	1.8	299	4.6	298	−2.8
Sample SF-5															
SF5-1	613	326	0.53	0.3459	0.0037	0.0476	0.0002	0.0527	0.0005	300	1.9	302	3.9	317	22.2
SF5-2	215	148	0.69	0.3457	0.0103	0.0475	0.0002	0.0527	0.0015	299	1.8	302	9.2	317	69.4
SF5-3	698	481	0.69	0.3456	0.0044	0.0475	0.0002	0.0528	0.0006	299	2.0	301	4.4	320	25.9
SF5-4	680	214	0.31	0.3413	0.004	0.0473	0.0001	0.0524	0.0006	298	1.6	298	4.1	302	25.9
SF5-5	1062	490	0.46	0.3397	0.0027	0.0475	0.0002	0.0519	0.0004	299	1.7	297	3.2	280	18.5
SF5-6	239	144	0.6	0.3451	0.0087	0.0475	0.0002	0.0526	0.0013	299	1.8	301	7.9	309	55.6
SF5-7	1077	67	0.06	0.3433	0.0025	0.0472	0.0001	0.0527	0.0004	298	1.6	300	3.1	317	23.1
SF5-8	201	95	0.47	0.3472	0.0142	0.0476	0.0001	0.0531	0.0022	300	1.7	303	12.5	332	92.6
SF5-9	1407	651	0.46	0.3501	0.0024	0.0472	0.0001	0.0538	0.0003	297	1.6	305	3.1	365	13.0
SF5-10	583	243	0.42	0.3407	0.0058	0.0472	0.0001	0.0523	0.0009	297	1.7	298	5.5	298	41.7
SF5-11	1213	646	0.53	0.3422	0.0027	0.0473	0.0001	0.0525	0.0004	298	1.6	299	3.2	306	12.0

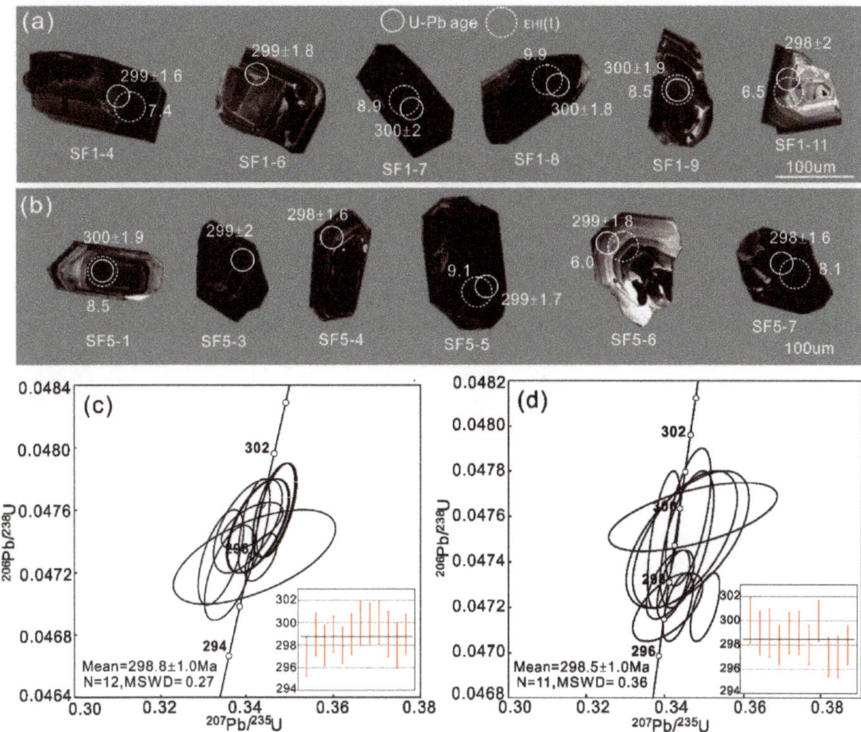

Figure 6. Zircon cathodoluminescence (CL) images and zircon U-Pb concordant curves for SF-1 from the inner phase (**a,c**) and SF-5 from the outer phase (**b,d**) of the Shanshenfu alkali-feldspar granite.

6.2. Zircon Trace Element, Ti-in-Zircon Temperature and Oxygen Fugacity

Trace element compositions of zircons are listed in Table 2, and the corresponding chondrite-normalized REE patterns are plotted in Figure 7. The zircons of sample SF-1 and SF-5 have ΣREE contents ranging from 3815 ppm to 10,727 ppm and 1411 ppm to 8447 ppm, respectively, Ce anomalies ranging from 1.2 to 276 (average of 86) and from 1.7 to 215 (average of 44), respectively, Eu anomalies ranging from 0.01 to 0.09 and from 0 to 0.38, respectively and Ti contents ranging from 8.6 ppm to 58.3 ppm and from 4.8 ppm to 173 ppm, respectively. Typical igneous zircon has Ti values ≤75 ppm [21]. One grain in our study has high Ti value (SF5-9 = 173 ppm). A value this high is unusual for igneous zircon and may be the result of inclusions (e.g., ilmenite) in the analysis; this analysis is excluded from the following discussion. The temperature of the melt during zircon crystallization was calculated by the Ti-in-zircon thermometer [16] as shown in the following Equation (1):

$$T_{zircon}(^{\circ}\mathrm{C}) = 4800/(5.711 - \lg(Ti_{zircon}) - \lg(Si_a) + \lg(Ti_a)) - 273 \tag{1}$$

As quartz is one of the major mineral phases in the SAFG, the activity of silica (Si_a) is set to 1. Due to the absence of rutile in the SAFG, the activity of titanium (Ti_a) is conservatively estimated to be 0.6 [15,57]. The calculated Ti-in-zircon temperatures for the inner phase (sample SF-1) and the outer phase (sample SF-5) are in the range of 773 °C to 990 °C (average of 854 °C) and 721–919 °C (average of 841 °C, Table 2), respectively. A new calibration has been presented by Trail et al. [17,18] to determine the oxygen fugacity of magmatic melt based on the incorporation of cerium into zircon and Ti-in-zircon temperature, which can be expressed by the following empirical equation:

$$\ln\left(\frac{Ce}{Ce^*}\right)_D = (0.1156 \pm 0.0050) \times \ln(f_{o2}) + \frac{13860 \pm 708}{T(K)} - 6.125 \pm 0.484 \tag{2}$$

where f_{o2} represents oxygen fugacity, and T is absolute temperature calculated by revised Ti-in-zircon thermometry. The Ce anomaly can be estimated by the following equation:

$$\left(\frac{Ce}{Ce^*}\right)_D \approx \left(\frac{Ce}{Ce^*}\right)_{CHUR} = \frac{Ce_N}{\sqrt{La_N \cdot Pr_N}} \tag{3}$$

where Ce_N, La_N and Pr_N are chondrite-normalized values for Ce, La and Pr in zircon, respectively.

The $\lg(f_{o2})$ values for the zircons of the inner phase and the outer phase are in the range of −21.3 to −3.4 (average of −10.3) and from −24.2 to −2.7 (average of −13.1), respectively, with corresponding ΔFMQ values of −8.4 to +10.6 (average of +3.1) and from −9.4 to +10.6 (average of +0.7), respectively (Table 2).

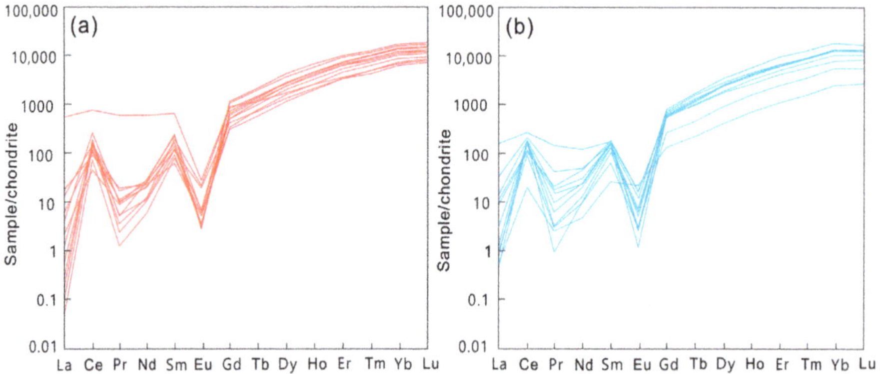

Figure 7. Chondrite-normalized REE patterns for the zircons from the Shanshenfu alkali-feldspar granite. (**a**) SF-1 (inner phase); (**b**) SF-5 (outer phase). Normalization values for chondrite are from McDonough and Sun [58].

6.3. Zircon Lu–Hf Isotopes

The analytical results on Hf isotopes for zircons are listed in Table 3. The $\varepsilon_{Hf}(t)$ values are calculated using their U-Pb ages. Seven spots from the inner phase show a range of initial ^{176}Hf/^{177}Hf ratios from 0.282955 to 0.283052 and $\varepsilon_{Hf}(t)$ values from +6.5 to +9.9, with the two-stage Hf model ages (T_{DM2}) from 394 Ma to 587 Ma. Nine spots from the outer phase show a range of initial ^{176}Hf/^{177}Hf ratios from 0.282941 to 0.283029 and $\varepsilon_{Hf}(t)$ values from +6.0 to +9.1, with T_{DM2} from 441 Ma to 614 Ma.

Table 2. Zircon trace element abundance, Ce and Eu anomalies, Ti-in-zircon temperatures (T) and magmatic oxygen fugacity (fo_2) of the Shanshenfu alkali-feldspar granite.

Spot	La ppm	Ce ppm	Pr ppm	Nd ppm	Sm ppm	Eu ppm	Gd ppm	Tb ppm	Dy ppm	Ho ppm	Er ppm	Tm ppm	Yb ppm	Lu ppm	ΣREE ppm	Ti ppm	δCe	δEu	T_{Ti} °C	lg(fo_2)	FMQ Buffer	ΔFMQ
									Sample SF-1													
SF1-1	1.57	268	1.39	21.3	53.9	0.6	333	119	1451	510	2244	417	3937	663	10,019	16.3	39.2	0.01	837	−10.1	−13.7	3.6
SF1-2	2.23	103	1.63	18.4	39.1	0.32	224	82.2	1050	395	1808	346	3297	566	7931	33.2	11.7	0.01	917	−11.5	−12.2	0.7
SF1-3	0.88	44.6	1.23	15.5	27.3	1.78	135	43.3	506	182	829	159	1586	284	3815	58.3	9.3	0.09	990	−9.9	−11.1	1.2
SF1-4	6.60	92.0	2.77	15.9	15.0	0.47	95.2	33.9	450	176	854	172	1798	321	4026	22.6	4.7	0.04	872	−16.7	−13.1	−3.6
SF1-5	0.11	161	0.74	19.8	58.6	0.26	348	126	1590	570	2490	457	4227	679	10,727	19.1	120.8	0.01	854	−5.2	−13.4	8.2
SF1-6	210	750	84.6	433	154	2.44	280	67.4	674	217	934	179	1774	312	5861	24.6	1.2	0.04	882	−21.3	−12.9	−8.4
SF1-7	0.24	116	0.5	9.2	44.4	1.84	251	86.8	1097	385	1700	316	2933	475	7416	19.2	73.2	0.05	855	−7.0	−13.4	6.4
SF1-8	0.04	137	0.75	7.9	27.8	0.59	186	70.9	947	350	1602	306	2990	504	7130	8.6	174.8	0.03	773	−7.4	−15.1	7.7
SF1-9	0.49	152	1.45	14.3	27.2	0.61	163	59.1	757	281	1317	261	2633	443	6110	9.3	39.0	0.03	781	−12.6	−14.9	2.3
SF1-10	0.07	191	0.34	7.6	21.2	0.33	162	67.4	904	338	1576	297	2838	452	6855	13.9	270.5	0.02	820	−3.6	−14.1	10.5
SF1-11	5.14	147	2.31	18.6	33.7	0.28	208	81.2	1073	406	1859	362	3546	585	8322	16.7	9.2	0.01	840	−15.4	−13.7	−1.7
SF1-12	0.02	72.7	0.17	4.3	17.9	0.53	110	43.5	589	229	1105	224	2242	367	5005	14.2	276.0	0.04	823	−3.4	−14.0	10.6
									Sample SF-5													
SF5-1	3.49	99.6	2.87	32.7	42.1	1.53	198	60.2	707	237	1038	204	1995	334	4952	33.6	6.8	0.05	919	−13.5	−12.2	−1.3
SF5-2	0.50	184	0.49	9.6	30.9	0.11	180	69.9	936	358	1607	321	3265	500	7462	32.5	80.7	0.00	915	−4.3	−12.3	8.0
SF5-3	12.2	212	6.04	35.3	36.1	0.24	199	75.3	987	369	1676	336	3362	528	7822	12.1	5.3	0.01	807	−18.9	−14.4	−4.5
SF5-4	5.38	183	2.59	22.3	36.9	0.49	232	92.7	1142	382	1685	338	3349	488	7954	9.2	10.6	0.02	780	−17.6	−14.9	−2.7
SF5-5	0.16	135	0.11	6.7	20.7	0.51	168	63.6	876	320	1488	290	2977	431	6777	20.4	215.0	0.03	861	−2.7	−13.3	10.6
SF5-6	0.31	161	1.34	17.9	38.1	0.27	216	76.6	962	347	1561	320	3247	502	7450	18.9	53.8	0.01	853	−8.3	−13.4	5.1
SF5-7	57.0	260	19.2	84.4	38.8	1.45	171	57.9	748	271	1258	255	2667	409	6241	9.8	1.7	0.05	786	−24.2	−14.8	−9.4
SF5-8	1.14	156	0.83	13.9	30.1	0.69	188	74.1	970	357	1595	324	3276	473	7459	18.4	34.8	0.03	850	−10.0	−13.5	3.5
SF5-9	0.36	87.4	0.42	6.8	15.1	0.27	82.6	27.8	370	134	627	135	1424	225	3135	173	48.3	0.02	1158	-	-	-
SF5-10	0.20	20.0	0.35	3.4	6.19	1.98	42.1	13.2	160	60.9	285	59.2	651	107	1410	33.5	16.4	0.38	918	−10.2	−12.2	2.0
SF5-11	4.03	109	1.65	14.6	30.1	0.78	204	78.5	1068	403	1890	376	3683	585	8444	4.8	9.2	0.03	721	−21.0	−16.3	−4.7

Note: T_{Ti} (°C) is the Ti temperature of zircon, $\delta Ce = Ce_N/(La_N \times Pr_N)^{0.5}$, $\delta Eu = Eu_N/(Sm_N \times Gd_N)^{0.5}$; FMQ buffer is the buffer of the fayalite-magnetite-quartz, which is calculated using the formula FMQ buffer $= -24,441.9/(T + 273) + 8.290 \ (\pm0.167)$ [59]; $\Delta FMQ = lg fo_2 -$ FMQ buffer.

Table 3. LA-MC-ICP-MS zircon Lu-Hf isotope data for the Shanshenfu alkali-feldspar granite.

Spot	^{176}Hf/^{177}Hf [a]	1σ	^{176}Lu/^{177}Hf [a]	1σ	^{176}Yb/^{177}Hf [a]	1σ	Age (Ma)	ε_{Hf} (t) [b]	1σ	T_{DM1} (Ma) [c]	T_{DM2} (Ma) [d]	$f_{Lu/Hf}$
					Sample SF-1							
SF1-1	0.282995	0.000015	0.002996	0.000008	0.086319	0.000235	297	7.9	0.74	384	507	−0.91
SF1-4	0.282982	0.000014	0.002209	0.000051	0.061855	0.001585	299	7.4	0.70	395	533	−0.93
SF1-7	0.283024	0.000015	0.004243	0.000064	0.115732	0.00164	300	8.9	0.75	353	450	−0.87
SF1-8	0.283052	0.000016	0.003983	0.000026	0.117902	0.00081	300	9.9	0.76	307	394	−0.88
SF1-9	0.283011	0.000014	0.003086	0.000051	0.089502	0.001556	300	8.4	0.72	362	476	−0.91
SF1-11	0.282955	0.000013	0.00128	0.000005	0.032593	0.000132	298	6.5	0.70	424	587	−0.96
SF1-12	0.282967	0.000014	0.002444	0.000009	0.068065	0.000324	299	6.9	0.73	420	563	−0.93
					Sample SF-5							
SF5-1	0.283011	0.000017	0.004065	0.000056	0.11864	0.001618	300	8.4	0.79	372	476	−0.88
SF5-2	0.283020	0.000014	0.002091	0.000018	0.055445	0.0004	299	8.8	0.72	338	458	−0.94
SF5-5	0.283029	0.000017	0.003066	0.000016	0.090641	0.000562	299	9.1	0.79	335	441	−0.91
SF5-6	0.282941	0.000014	0.000794	0.000014	0.020234	0.000404	299	6.0	0.70	438	614	−0.98
SF5-7	0.283002	0.000014	0.003232	0.000016	0.093923	0.000628	298	8.1	0.72	377	494	−0.90
SF5-8	0.282948	0.000015	0.001153	0.000078	0.029802	0.002258	300	6.2	0.73	433	602	−0.97
SF5-9	0.283015	0.000014	0.003446	0.00001	0.099972	0.000247	297	8.6	0.72	359	468	−0.90
SF5-10	0.283025	0.000014	0.002295	0.000037	0.066671	0.001076	297	8.9	0.72	333	448	−0.93
SF5-11	0.283028	0.000015	0.003929	0.00002	0.113707	0.000769	298	9.0	0.75	344	443	−0.88

[a] The measured values; [b] Initial ^{176}Hf/^{177}Hf ratios were calculated using the measured ^{176}Lu/^{177}Hf ratios and the ^{176}Lu decay constant of 1.867×10^{-11} year^{-1} [60], and ε_{Hf} values were calculated using the chondritic ratios of ^{176}Hf/^{177}Hf (=0.282772) and ^{176}Lu/^{177}Hf (=0.0332) [61]; [c] The single-stage model age (T_{DM1}) was calculated using the present-day ratios ^{176}Hf/^{177}Hf = 0.28325 and ^{176}Lu/^{177}Hf = 0.0384 [62]; [d] Two-stage model age (T_{DM2}) was calculated by projecting the initial ^{176}Hf/^{177}Hf ratios of the zircon back to the depleted mantle model growth curve, assuming a ^{176}Lu/^{177}Hf value of 0.015 for the average continent crust [63].

6.4. Whole-Rock Major and Trace Element Analyses

The major- and trace-element compositions of the SAFG are shown in Table 4 as weight percent (wt %) for major oxides and as parts per million (ppm) for all trace elements. According to the QAP ternary diagram [64] (Figure 8a), all samples classify as alkali-feldspar granite. They are characterized by high SiO_2 (74.89–76.83 wt %), Al_2O_3 (12.34–12.85 wt %) and alkali earths ($Na_2O + K_2O$ = 8.17–8.81 wt %) with low MgO (0.09–0.16 wt %) and CaO (0.08–0.24 wt %). They exhibit a high-K calc-alkaline, peraluminous and high-silica content, with A/CNK (molar Al_2O_3/CaO + Na_2O + K_2O) ranging from 1.04 to 1.12 and A/NK (molar Al_2O_3/Na_2O + K_2O) from 1.06 to 1.14 (Figure 8b,c). The inner phase has slightly higher SiO_2 content with lower Fe_2O_3, TiO_2, CaO and P_2O_5 than the outer phase (Table 4). This may indicate that SAFG cooled from the margin into the centre and that this cooling process resulted in a slightly more evolved inner phase. Both phases exhibit enriched REEs (except Eu) with total REE ranging from 147 ppm to 357 ppm. They are characterized by slightly enriched light REE (LREE, $(La/Yb)_N$ = 2.02–6.57) and flat heavy REE (HREE, $(Gd/Yb)_N$ = 0.6–1.27) with strong negative Eu anomalies (δEu = 0.05–0.12, Figure 9b). All samples are enriched in large-ion lithophile elements (LILEs, e.g., Rb, Th, U and K), high-field-strength elements (HFSEs, e.g., Nb, Ta, Zr and Hf), and exhibit strong depletion of Ba, Sr, P and Ti (Figure 9a).

Table 4. Major oxides and trace elements of the Shanshenfu alkali-feldspar granite.

Sample	The Inner Phase			The Outer Phase		
	SF-1	SF-2	SF-3	SF-4	SF-5	SF-6
SiO_2	76.83	75.23	76.02	76.22	74.89	75.36
Al_2O_3	12.34	12.56	12.78	12.67	12.85	12.54
Fe_2O_3	1.51	1.48	1.37	1.64	2.13	1.95
CaO	0.08	0.11	0.09	0.19	0.21	0.24
MgO	0.14	0.12	0.11	0.16	0.09	0.12
Na_2O	3.71	4.08	3.87	3.67	4.23	4.13
K_2O	4.46	4.31	4.42	4.87	4.53	4.68
TiO_2	0.15	0.27	0.19	0.17	0.25	0.31
MnO	0.06	0.08	0.04	0.11	0.05	0.06
P_2O_5	0.03	0.08	0.04	0.06	0.11	0.08
LOI	0.45	0.67	0.54	0.24	0.48	0.37
Total	99.98	99.56	99.71	100.32	99.67	99.48
La	52.3	22.6	45.5	58.8	27.2	40.1
Ce	161	54.2	101.3	174.5	72.3	89.6
Pr	14.75	6.58	13.2	15.1	7.22	12.6
Nd	53.6	20.8	44.2	51.3	29.1	45.5
Sm	12.8	5.13	9.35	11.26	7.32	9.81
Eu	0.34	0.15	0.13	0.42	0.17	0.21
Gd	10.72	5.82	7.62	10.11	9.61	6.97
Tb	1.91	1.28	1.18	1.83	2.36	1.11
Dy	12.3	10.44	7.61	12.28	13.12	6.5
Ho	2.54	2.35	1.51	2.57	2.68	1.27
Er	8.57	7.42	4.83	7.93	6.68	5.24
Tm	1.25	1.11	0.68	1.17	0.97	0.72
Yb	8.65	8.03	4.97	8.21	7.32	5.16
Lu	1.4	1.28	0.83	1.32	1.15	0.88
Ga	21.2	19.7	22.6	24.3	21.8	18.9
Rb	177	153.5	172	181.5	167	148
Ba	60.2	107	171	58.6	123	86
Th	16.3	17.51	14.2	17.1	18.2	13.6
U	3.79	3.65	3.48	2.87	3.67	3.13
Ta	2.23	2.65	2.31	2.95	2.46	2.33
Nb	43.7	49.1	44.3	48.2	46.2	44.7
Sr	8.3	14.8	22.4	24.1	15.3	18.7
Zr	358	435	647	248	487	276
Hf	11.9	13.1	18.4	10.8	13.9	11.2
Y	70.4	75.1	81.8	54.2	78.3	63.1
ΣREE	342	147	243	357	187	226
ΔEu	0.09	0.08	0.05	0.12	0.06	0.08

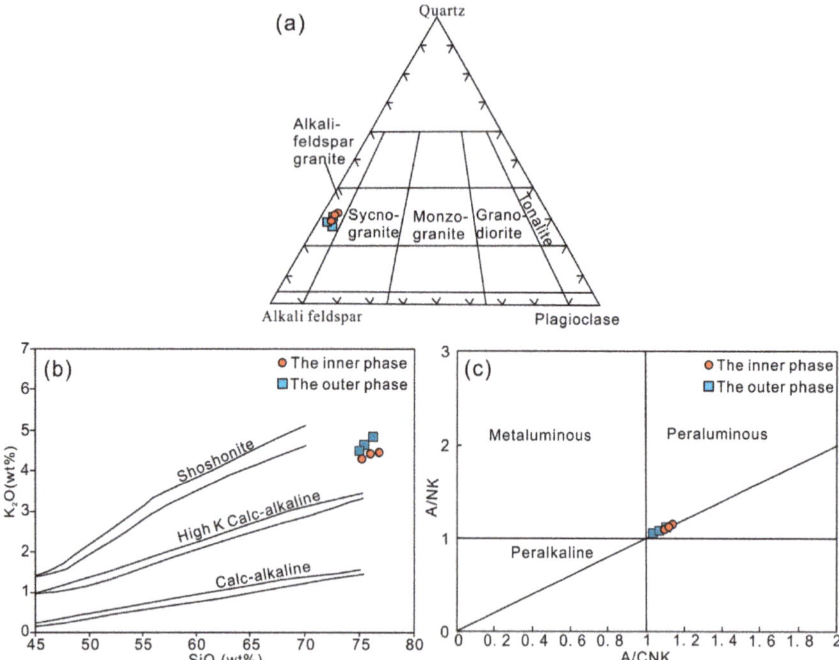

Figure 8. (**a**) QAP ternary diagram [64]; (**b**) SiO$_2$ vs. K$_2$O diagram of the Shanshenfu alkali-feldspar granite [65]; (**c**) A/NK vs. A/CNK diagram, where A/NK = Al$_2$O$_3$/(Na$_2$O + K$_2$O) and A/CNK = Al$_2$O$_3$/(CaO + Na$_2$O + K$_2$O) (all oxides on molar basis).

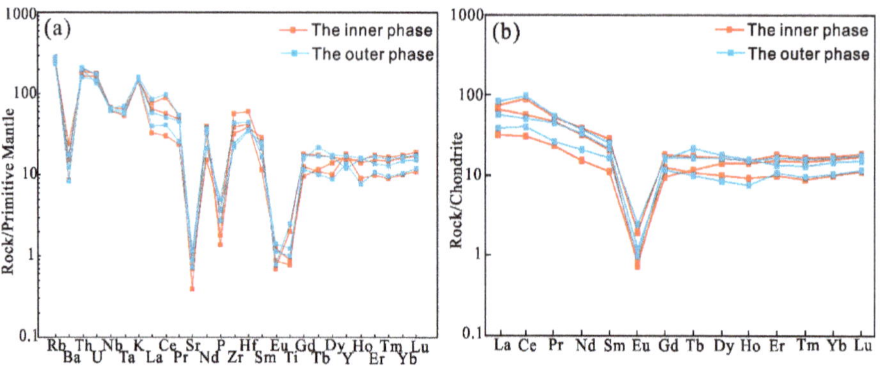

Figure 9. (**a**) Primitive mantle-normalized trace element diagram; (**b**) chondrite-normalized REE pattern. Normalization values are from Sun and McDonough [66].

7. Discussion

7.1. Genetic Type and Magmatic Oxygen Fugacity

The SAFG has high SiO_2 (74.89–76.83 wt %), total alkalis (K_2O + Na_2O = 8.17–8.81 wt %) and low P_2O_5 (0.03–0.11 wt %) contents (Table 4). It is enriched in HFSE (e.g., Nb, Ta, Zr and Hf) and REEs (except Eu), and is depleted in Sr, Ba, P, Ti and Eu (Figure 8c). These geochemical features are common for A-type granites [4]. An A-type granite classification is also supported by the elevated Ga content of all samples such that they plot in the A-type field on the discrimination diagram of Whalen et al. [67] (Figure 10a) and the within-plate-granite field on the diagram Y vs. Nb of Pearce et al. [68] (Figure 10b). Chondrite normalized REE patterns of zircons show they are depleted in LREE and enriched in HREE, and exhibit strong positive Ce anomalies (δCe = 1.22–275.99) and deep negative Eu anomalies (δEu = 0.1–0.38, Figure 7), which are typical of unaltered magmatic zircons [19]. Positive Ce anomalies may be related to a relatively oxidized melt with favorable incorporation of Ce^{4+} while negative Eu anomalies may be resulted from plagioclase fractionation in magma composition [17,18]. Due to the close relationship between the Ce anomaly in zircon and the oxidation state of the melt from which the zircon crystallizes [17,18], zircon can indicate the fo_2 of magmas. By plotting the Ti-in-zircon temperatures and the logarithmic oxygen fugacities on T vs. lg(fo_2) diagram (Figure 10c) that can be divided by the curves of some specific mineral oxidation buffers such as magnetite-hematite (MH), fayalite-magnetite-quartz (FMQ) and iron-wustite (IW) into several oxygen fugacity fields [17], nearly two thirds of the data points plot above the FMQ buffer (Figure 10c), which indicates oxidized magmatic conditions. The presence of magnetite (Figure 4d) and absence of significant amounts of ilmenite also suggest the oxidized magmatic conditions for the SAFG [12].

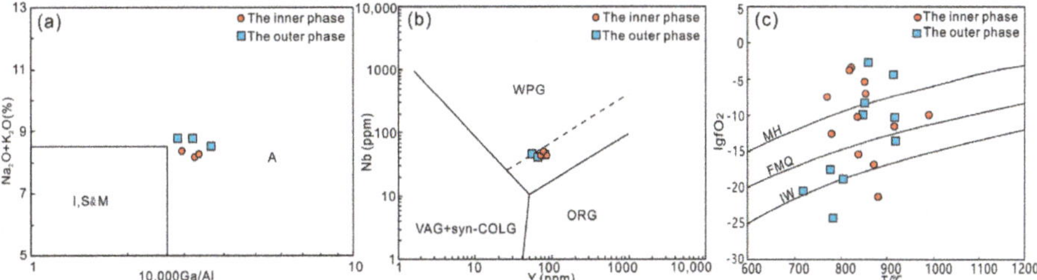

Figure 10. (a) K_2O + Na_2O vs. 10,000 Ga/Al discrimination diagram of Whalen et al. [67]. I, S and M—I-, S- and M-type granites; (b) Y vs. Nb tectonic discrimination diagram of Pearce et al. [68]. VAG—volcanic arc granites, WPG—within plate granites, COLG—collisional granites, ORG—oceanic ridge granites; (c) magmatic oxygen fugacity (fo_2) of the Shanshenfu alkali-feldspar granite, MH, FMQ and IW curves are from Trail et al. [17].

7.2. Petrogenesis and Implications for Mineralization

Several petrogenetic models have been proposed for the origin of A-type granites, including (1) extreme fractional crystallization of mantle-derived tholeiitic or alkaline basaltic magmas [6,69]; (2) low-degree partial melting of lower-crustal granulites with depletion of incompatible elements [70]; (3) anatexis of underplated I-type tonalitic crustal source [12,71,72]; and (4) hybridization between anatectic crustal and mantle-derived magmas, such as crustal assimilation and fractional crystallization of mantle-derived magmas, or of mixing between mantle-derived and crustal magmas [72–74]. The zircons of the SAFG exhibit high ε_{Hf}(t) values (from +6.0 to +9.9) and young T_{DM2} (394–614 Ma, Table 3), which is consistent with juvenile crustal sources. Thus, we can rule out the possibility of partial melting of lower-crustal granulites, which would have a more isotopically evolved signature. The Ti-in-zircon temperatures of the SAFG are between 721–990 °C, which suggests that anatexis of underplated I-type tonalitic crust is unlikely to be the source because that process cannot provide the required high temperature needed for the formation of A-type granites [6,75]. In addition, this magma source is not compatible with the observed elemental variations (e.g., Nb, Ta, Ti, P, Sr and Eu) in the SAFG [66]. Due to the absence of spatially or temporally associated alkaline mafic rocks in the study area [76], alkaline basaltic juvenile crust is not a suitable source for the SAFG. Thus, Two petrogenetic models that may explain the characteristics reported here for the SAFG are discussed further: (1) Extreme fractional crystallization from mantle-derived tholeiitic magma, and (2) hybridization between anatectic crustal and mantle-derived magmas.

Numerous studies suggest that the high magmatic oxygen fugacity is closely related to the nature of magma source regions [30,77,78]. The SAFG not only has similar U-Pb age and $\varepsilon_{Hf}(t)$ values to late Carboniferous arc intrusions from northern inner Mongolia (Figure 11a,b), but also has the high oxygen fugacity similar to arc magmas that can range from FMQ to FMQ + 6 [77]. The chemistry of zircon is related to the magma from which the zircon crystallized, two recent discrimination diagrams proposed by Grimes et al. [22] suggest zircon with relatively higher U or lower Yb or Y are sourced from continental crust, whereas zircon with lower U or higher Yb or Y are sourced from ocean crust. In Figure 11c,d, most of the studied zircons plot in the field of ocean crust zircon or in the intersection between continental zircon and ocean crust zircon. The evidence above suggest that arc-related juvenile crust is a possible source for the SAFG. This magma source of Permian A-type granites in the CAOB has been proposed by previous studies [76,79]. The depletion of Nb, Ta, Ti, P, Sr and Eu in the SAFG indicates contamination and metasomatism by subducting oceanic crust [80]. This is also consistent with peraluminous nature and muscovite reported in mineralogy. Therefore, we propose that the SAFG was derived from the crustal assimilation and fractional crystallization of juvenile crust metasomatized by subducting oceanic crust. This magma source is similar to charnockitized juvenile crust proposed by previous studies [79,81–83].

Based on related studies conducted on Cu-Au-Mo hydrothermal deposits, the link between oxidized felsic magmas and mineralization is well recognized [13,27,73,78,84]. Sun et al. [27,78] documented that sulfur in melt exists mainly as SO and SO_2 at high fo_2 and as S^{2-} at low fo_2. Numerous experiments suggested that the partition coefficients of chalcophile elements (e.g., Cu, Au and Mo) between silicate melt and pyrrhotite are positively correlated with fo_2 [84–86]. Therefore, high fo_2 of the SAFG favors transmission of the chalcophile elements (e.g., Cu and Au) of the source region into the melt together with oxidized sulfur, rather than into the sulfides (reduced sulfur) [17,25,26,38,78,87]. With the decrease of temperature and pressure, hydrous magma can release a water-rich volatile phase to form a magmatic-hydrothermal ore-forming system [88]. Mao et al. [89] revealed the magmatic-hydrothermal mineralization process of the SAFG in detail through the study of melt-fluid inclusions. In this process metals such as Cu can be effectively enriched in the fluid phases after magmatic fractional crystallization due to high fo_2 and high exsolution temperature of the SAFG [89], consequently facilitating subsequent Cu-polymetallic hydrothermal mineralization.

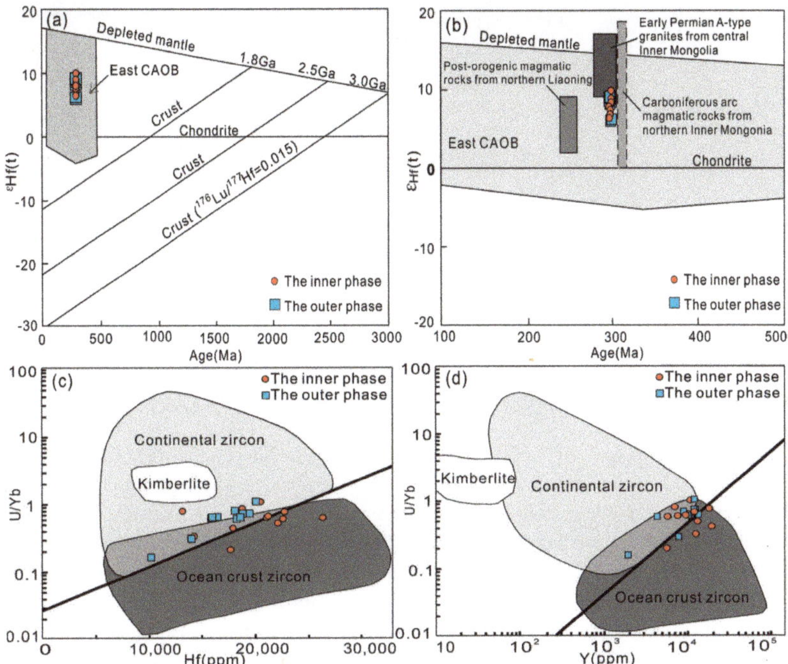

Figure 11. (**a**,**b**) Compilation diagram of ε_{Hf} (t) vs. U-Pb age of the Shanshenfu alkali-feldspar granite. The Hf isotopic evolution line of the Archean average crust with $^{176}Lu/^{177}Hf = 0.015$ is after Griffin et al. [62]. The fields for Carboniferous arc intrusions from northern inner Mongolia, early Permian A-type granites from central inner Mongolia and late Permian to early Triassic post-orogenic melts from northern Liaoning are from Chen et al. [36], Zhang et al. [79] and Zhang et al. [90], respectively. (**c**,**d**) U/Yb ratio vs. Y and Hf content diagrams of Grimes et al. [22] to discriminate between continental and oceanic crust zircon. Heavy lines indicate the lower limit of zircons from continental crust.

7.3. Extensional Setting for Cu-Polymetallic Mineralization

Recent studies suggest that Cu-polymetallic mineralization associated with extensional setting in the XMOB was mainly formed in the early Cretaceous [10,91–93]. The extensional setting for Cu-polymetallic mineralization in the Permian is rarely reported. Therefore, the clear tectonic setting of the Permian A-type granites associated with Cu-polymetallic mineralization along the HHSZ is of great significance for regional tectonic evolution and prospecting. The SAFG with an age of 298 Ma is not only contemporary with immense alkali granites from southern Mongolia [94], but also coincident with the equivalents in the central inner Mongolia [79] and Erguna-Xing'an terrane of NE China [4,34]. Eby [2] divided the A-type granites into A1 and A2 groups. The A1 group is emplaced in anorogenic settings such as plumes, hotspots or continental rift zones. The A2 group is related to a cycle of subduction-zone or continent-continent collision magmatism in crust and emplaced in a variety of tectonic settings. In our study, the SAFG plots in the A2 field on the Nb-Y-3Ga diagram of Eby [2] (Figure 12a), and plot in the post-collisional setting on the Y + Nb versus Rb plot of Pearce et al. [68] (Figure 12b). Such an environment is also supported by the previous studies for other Permian A-type granitoids along the HHSZ [4,78,95]. The SAFG together with the other Permian A2-type granites [3,4,29–31] are commonly found with Carboniferous calc-alkaline I-type plutons along the HHSZ [36]. The increase in granitoid alkalinity with time indicates the tectonic transition from an earlier arc setting to a later extensional setting [96]. Similar tendency also occurs in southern Mongolia [94,95] and central inner Mongolia [79].

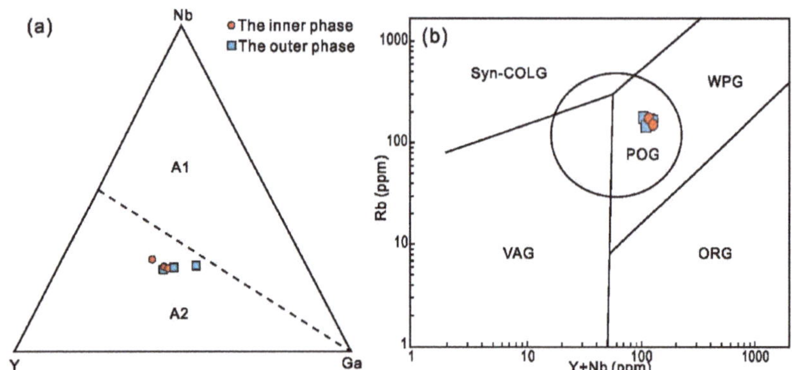

Figure 12. (**a**) Nb-Y-Ga ternary diagram of Eby [2]; (**b**) Y + Nb vs. Rb plot of Pearce et al. [68].

Previous studies suggest that the ocean along the Hegenshan-Heihe was closed during the late Carboniferous, which could be proved by the following evidences: (1) The Hegenshan ophiolite was emplaced before the early Permian, most likely between 300 Ma and 335 Ma [97,98]; (2) Permian and Carboniferous submarine turbidite strata with predominant magmatic arc source exists along the HHSZ [99]. However, the last Paleo-Asian ocean was not closed along the HHSZ during the late Carboniferous but closed along the Solonker-Xra Moron-Changchun suture during the late Permian–early Triassic [100]. This is well evidenced by late Permian to early Triassic post-orogenic magmatic rocks form belts along the Solonker-Xra Moron-Changchun suture [73,75]. Taking into account the fact that the Permian A2-type granites (260–300 Ma) are strikingly sparse and small in volume, and show a NE-trending migration together with the widespread occurrence of Permian volcanic rocks of island arc affinity and late Permian terrestrial sediments along the HHSZ [4], the post-collisional slab break-off process along the HHSZ during the Permian was generally proposed by previous studies [74,101]. Such tectonic regime is also evidenced by contemporaneous thermal metamorphism recorded by the Xilinhot metamorphic complexes in northern inner Mongolia [36], which once represented a segment of fore-arc basin [36]. In addition, the corresponding change can be observed in sedimentary style and clastic provenance of the early Permian strata from central inner Mongolia [102] and southern Mongolia [103]. The hallmark features of post-collisional slab break-off event is well recorded in numerous Phanerozoic cases [72,101,104], including transition in magmatic affinity from I-type to A-type, a narrow linear zone of magmatism, thermal metamorphismand change in uplift rate and concomitant sedimentary style [105]. All these features fit relatively well with the corresponding expressions along the HHSZ during Permian [4,36,102,103]. This unusual tectonic regime not only terminates a prolonged northward subduction of the Paleo-Asian ocean, but also heralds the amalgamation between the Songliao terrane and the Liaoyuan terrane along the Ondor Temple-Xar Moron suture zone.

The age of ore-related A-type granite in the Hongyan Cu-polymetallic deposit is consistent with the mineralized A-type granites in the southwestern part of the HHSZ [32], which suggests that these A-type granites associated with mineralization are formed in the same extension setting. During the Permian, the HHSZ was characterized by post-collisional extensional tectonism with the slab break-off, which not only caused upwelling of asthenospheric mantle and resulted in partial melting of the overlying lithospheric mantle and the juvenile arc crust [79], but also provided the thermal flux to form some narrow extensional magmatic channels, subsequently followed by A-type magmatism and related mineralization deposition along the HHSZ. Thus, the Permian A-type granites with arc-related juvenile crustal source and high fo_2 along the HHSZ have great potential for prospecting Cu-polymetallic hydrothermal deposits, and deserve more attention in future exploration.

8. Conclusions

Based upon geochronological and geochemical data from the SAFG, we suggest the following:

(1) LA-ICP-MS U-Pb zircon dating shows the inner phase and outer phase of the SAFG were formed at 298.8 ± 1.0 Ma and 298.5 ± 1.0 Ma, respectively. This makes the SAFG an early Permian pluton.

(2) The SAFG is a typical A-type granite with high magmatic oxygen fugacity. The source for the SAFG magma includes both crustal assimilation and fractional crystallization of juvenile crust metasomatized by subducting oceanic crust.

(3) The oxidized magmatic conditions favor transport and enrichment of chalcophile elements (e.g., Cu and Au) while the arc-related juvenile source favors Cu and Au enrichment. Combined, these magmatic characteristics suggest related magmatic-hydrothermal mineralization may be restricted to Cu ± Au.

(4) Combined with regional geological background, the Permian A-type granites and related mineralization along the HHSZ were formed in post-collisional extensional tectonism with the slab break-off.

Author Contributions: Data curation, C.M.; Formal analysis, C.M.; Funding acquisition, X.L.; Investigation, C.M., X.L. and C.C.; Project administration, X.L. and C.C.; Writing—original draft, C.M.; Writing—review & editing, C.M., X.L. and C.C.

Funding: This research was supported by the Geological Survey Project of China Geological Survey Bureau (NMKD2010-3).

Acknowledgments: We gratefully thank Ruan Banxiao and Yang Yongsheng from Institute of Geological Survey, China University of Geosciences, Wuhan, for their important guidance and assistance in geochemical analysis.

Conflicts of Interest: The authors declare no conflict of interest.

References

1. Loiselle, M.C.; Wones, D.R. Characteristics and origin of anorogenic granites. *Geol. Soc. Am. Abstr. Prog.* **1979**, *11*, 468.

2. Eby, G.N. Chemical subdivision of the A-type granitoids: Petrogenetic and tectonic implications. *Geology* **1992**, *20*, 641–644. [CrossRef]

3. Hong, D.W.; Huang, H.Z.; Xiao, Y.J.; Xu, H.M. Permian alkaline granites in center Inner Mongolia and their geodynamic significance. *Acta Geol. Sin.* **1994**, *68*, 219–230. (In Chinese) [CrossRef]

4. Wu, F.Y.; Sun, D.Y.; Li, H.M.; Jahn, B.M.; Wilde, S. A-type granites in northeastern China: Age and geochemical constraints on their petrogenesis. *Chem. Geol.* **2002**, *187*, 143–173. [CrossRef]

5. Wu, S.P.; Wang, M.Y.; Qi, K.J. Present situation of researches on A-type granites: A review. *Acta Petrol. Mineral.* **2007**, *26*, 57–66, (In Chinese with English Abstract). [CrossRef]

6. Bonin, B. A-type granites and related rocks: Evolution of a concept, problems and prospects. *Lithos* **2007**, *97*, 1–29. [CrossRef]

7. Li, H.; Watanabe, K.; Yonezu, K. Geochemistry of A-type granites in the Huangshaping polymetallic deposit (South Hunan, China): Implications for granite evolution and associated mineralization. *J. Asian Earth Sci.* **2014**, *88*, 149–167. [CrossRef]

8. Li, H.; Palinkaš, L.A.; Watanabe, K.; Xi, X.S. Petrogenesis of Jurassic A-type granites associated with Cu-Mo and W-Sn deposits in the central Nanling region, South China: Relation to mantle upwelling and intra-continental extension. *Ore Geol. Rev.* **2018**, *92*, 449–462. [CrossRef]

9. Liu, P.; Mao, J.W.; Santosh, M.; Bao, Z.A.; Zeng, X.Z.; Jia, L.H. Geochronology and petrogenesis of the Early Cretaceous A-type granite from the Feie'shan W-Sn deposit in the eastern Guangdong Province, SE China: Implications for W-Sn mineralization and geodynamic setting. *Lithos* **2018**, *300*, 330–347. [CrossRef]

10. Wang, X.D.; Xu, D.M.; Lv, X.B.; Wei, W.; Mei, W.; Fan, X.J.; Sun, B.K. Origin of the Haobugao skarn Fe-Zn polymetallic deposit, Southern Great Xing'an range, NE China: Geochronological, geochemical, and Sr-Nd-Pb isotopic constraints. *Ore Geol. Rev.* **2018**, *94*, 58–72. [CrossRef]

11. Cisse, M.; Lü, X.B.; Algeo, T.J.; Cao, X.F.; Li, H.; Wei, M.; Yuan, Q.; Chen, M. Geochronology and geochemical characteristics of the Dongping ore-bearing granite, North China: Sources and implications for its tectonic setting. *Ore Geol. Rev.* **2017**, *89*, 1091–1106. [CrossRef]

12. Dall'Agnol, R.; Oliveira, D.C. Oxidized, magnetite-series, rapakivi-type granites of Carajás, Brazil: Implications for classification and petrogenesis of A-type granites. *Lithos* **2007**, *93*, 215–233. [CrossRef]

13. Qiu, J.T.; Yu, X.Q.; Santosh, M.; Zhang, D.H.; Chen, S.Q.; Li, J.P. Geochronology and magmatic oxygen fugacity of the Tongcun molybdenum deposit, northwest Zhejiang, SE China. *Miner. Depos.* **2013**, *48*, 545–556. [CrossRef]

14. Frost, C.D.; Frost, B.R. Reduced rapakivi-type granites: The tholeiite connection. *Geology* **1997**, *25*, 647–650. [CrossRef]

15. Watson, E.B.; Wark, D.A.; Thomas, J.B. Crystallization thermometers for zircon and rutile. *Contrib. Mineral. Petrol.* **2006**, *151*, 413–433. [CrossRef]

16. Ferry, J.M.; Waston, E.B. New thermodynamic models and revised calibrations for the Ti-in-zircon and Zr-in-rutile thermometers. *Contrib. Mineral. Petrol.* **2007**, *154*, 429–437. [CrossRef]

17. Trail, D.; Watson, E.B.; Tailby, N.D. The oxidation state of Hadean magmas and implications for early Earth's atmosphere. *Nature* **2011**, *480*, 79–82. [CrossRef]

18. Trail, D.; Watson, E.B.; Tailby, N.D. Ce and Eu anomales in zircon as proxies for oxidation state of magmas. *Geochim. Cosmochim. Acta* **2012**, *97*, 70–87. [CrossRef]

19. Hoskin, P.W.O.; Schaltegger, U. The composition of zircon and igneous and metamorphic petrogenesis. *Rev. Mineral. Geochem.* **2003**, *53*, 27–62. [CrossRef]

20. Hanchar, J.M.; VanWestrenen, W. Rare earth element behavior in zircon-melt systems. *Elements* **2007**, *3*, 37–42. [CrossRef]

21. Hoskin, P.W.O. Trace-element composition of hydrothermal zircon and the alteration of Hadean zircon from the Jack Hills, Australia. *Geochim. Cosmochim. Acta* **2005**, *69*, 637–648. [CrossRef]

22. Grimes, C.B.; John, B.E.; Kelemen, P.B.; Mazdab, F.K.; Wooden, J.L.; Cheadle, M.J.; Hanghøj, K.; Schwartz, J.J. Trace element chemistry of zircons from oceanic crust: A method for distinguishing detrital zircon provenance. *Geology* **2007**, *35*, 643–646. [CrossRef]

23. Yang, J.H.; Wu, F.Y.; Wilde, S.A.; Xie, L.W.; Yang, Y.H.; Liu, X.M. Tracing magma mixing in granite genesis: In situ U–Pb dating and Hf-isotope analysis of zircons. *Contrib. Mineral. Petrol.* **2007**, *153*, 177–190. [CrossRef]

24. Cherniak, D.J.; Watson, E.B. Diffusion in zircon. *Rev. Mineral. Geochem.* **2003**, *53*, 113–143. [CrossRef]

25. Maughan, D.T.; Keith, J.D.; Christiansen, E.H.; Pulsipher, T.; Hattori, K.; Evans, N.J. Contributions from mafic alkaline magmas to the Bingham porphyry Cu-Au-Mo deposit, Utah, USA. *Miner. Depos.* **2002**, *37*, 14–37. [CrossRef]

26. Ma, X.H.; Chen, B.; Yang, M.C. Magma mixing origin for the Aolunhua porphyry related to Mo-Cu mineralization, eastern Central Asian Orogenic belt. *Gondwana Res.* **2013**, *24*, 1152–1171. [CrossRef]

27. Sun, W.D.; Liang, H.Y.; Ling, M.X.; Zhan, M.Z.; Ding, X.; Zhang, H.; Yang, X.Y.; Li, Y.L.; Ireland, T.R.; Wei, Q.R.; et al. The link between reduced porphyry copper deposits and oxidized magmas. *Geochim. Cosmochim. Acta* **2013**, *103*, 263–275. [CrossRef]

28. Lu, Y.J.; Loucks, R.R.; Fiorentini, M.L.; McCuaig, T.C.; Evans, N.J.; Yang, Z.M.; Hou, Z.Q.; Kirkland, C.L.; Parra-Avila, L.A.; Kobussen, A. Zircon compositions as a pathfinder for porphyry Cu ± Mo ± Au deposits. *Soc. Econ. Geol. Spec. Publ.* **2016**, *19*, 329–347.

29. Shi, G.H.; Miao, L.C.; Zhang, F.Q.; Jian, P.; Fan, W.M.; Liu, D.Y. The age and its district tectonic implications on the Xilinhaote A-type granites, Inner Mongolia. *Chin. Sci. Bull.* **2004**, *49*, 384–389. (In Chinese) [CrossRef]

30. Zhang, Y.Q.; Xu, L.Q.; Kang, X.L. Age dating of alkali granite in Jingesitai area of Dong Ujingqin Banner, Inner Mongolia, and its significance. *Geol. Chin.* **2009**, *36*, 988–995. (In Chinese) [CrossRef]

31. Guo, K.C.; Zhang, W.L.; Yang, X.P.; Wang, L.; Shi, D.Y.; Yu, H.T.; Su, H. Origin of early Permian A-type granite in the Wudaogou area, Heihe City. *J. Jilin Univ. (Earth Sci. Ed.)* **2011**, *41*, 1077–1083, (In Chinese with English abstract). [CrossRef]

32. Yang, K. Ore-forming Characters and Depositing Orientation of Metal Deposit in the Dong Ujimqin, Inner Mongolia. Master's Thesis, China University of Geosciences in Wuhan, Wuhan, China, 2013.

33. Mao, J.W.; Pirajno, F.; Cook, N. Mesozoic metallogeny in East China and corresponding geodynamic settings—An introduction to the special issue. *Ore Geol. Rev.* **2011**, *43*, 1–7. [CrossRef]

34. Wu, F.Y.; Sun, D.Y.; Ge, W.C.; Zhang, Y.B.; Grant, M.L.; Wilde, S.A.; Jahn, B.M. Geochronology of the Phanerozoic granitoids in northeastern China. *J. Asian Earth Sci.* **2011**, *41*, 1–30. [CrossRef]

35. Gao, S.; Xu, H.; Zang, Y.; Wang, T. Mineralogy, ore-forming fluids and geochronology of the Shangmachang and Beidagou gold deposits, Heilongjiang province, NE China. *J. Geochem. Explor.* **2018**, *188*, 137–155. [CrossRef]

36. Chen, B.; Jahn, B.M.; Tian, W. Evolution of the Solonker suture zone: Constraints from zircon U-Pb ages, Hf isotopic ratios and whole-rock Nd–Sr isotope compositions of subduction and collision-related magmas and forearc sediments. *J. Asian Earth Sci.* **2009**, *34*, 245–257. [CrossRef]

37. Zhang, X.H.; Xue, F.H.; Yuan, L.L.; Wilde, S.A. Late Permian appinite–granite complex from northwestern Liaoning, North China craton: Petrogenesis and tectonic implications. *Lithos* **2012**, *155*, 201–217. [CrossRef]

38. Wang, Y.H.; Zhao, C.B.; Zhang, F.F.; Liu, J.J.; Wang, J.P.; Peng, R.M.; Liu, B. SIMS zircon U-Pb and molybdenite Re-Os geochronology, Hf isotope, and whole-rock geochemistry of the Wunugetushan porphyry Cu-Mo deposit and granitoids in NE China and their geological significance. *Gondwana Res.* **2015**, *28*, 1228–1245. [CrossRef]

39. Zhou, Z.H.; Mao, J.W.; Wu, X.L.; Ouyang, H.G. Geochronology and geochemistry constraints of the Early Cretaceous Taibudai porphyry Cu deposit, northeast China, and its tectonic significance. *J. Asian Earth Sci.* **2015**, *103*, 212–228. [CrossRef]

40. Wei, R.H.; Gao, Y.F.; Xu, S.C.; Xin, H.T.; Santosh, M.; Liu, Y.F.; Lei, S.H. The volcanic succession of Baoligaomiao, central Inner Mongolia: Evidence for Carboniferous continental arc in the central Asian orogenic belt. *Gondwana Res.* **2017**, *51*, 234–254. [CrossRef]

41. Wu, G.; Chen, Y.C.; Sun, F.Y.; Liu, J.; Wang, G.R.; Xu, B. Geochronology, geochemistry, and Sr–Nd–Hf isotopes of the early Paleozoic igneous rocks in the Duobaoshan area, NE China, and their geological significance. *J. Asian Earth Sci.* **2015**, *97*, 229–250. [CrossRef]

42. Liu, W.; Liu, X.J.; Liu, L.Q. Underplating generated A- and I-type granitoids of the East Junggar from the lower and the upper oceanic crust with mixing of mafic magma: Insights from integrated zircon U-Pb ages, pertrography, geochemistry and Nd-Sr-Hf isotopes. *Lithos* **2013**, *179*, 293–319. [CrossRef]

43. Mao, J.W.; Wang, Y.T.; Zhang, Z.H.; Yu, J.J.; Niu, B.G. Geodynamic settings of Mesozoic large-scale mineralization in North China adjacent areas. *Sci. Chin. Ser. D* **2003**, *46*, 838–851. [CrossRef]

44. Zhang, F.F.; Wang, Y.H.; Liu, J.J.; Wang, J.P.; Zhao, C.B.; Song, Z.W. Origin of the Wunugetushan porphyry Cu–Mo deposit, Inner Mongolia, NE China: Constraints from geology, geochronology, geochemistry, and isotopic compositions. *J. Asian Earth Sci.* **2016**, *117*, 208–224. [CrossRef]

45. Zeng, Q.D.; Liu, J.M.; Zhang, L.C. Re-Os geochronology of porphyry molybdenum deposit in southern segment of Da Hinggan Mountains, Northeast China. *J. Earth Sci.* **2010**, *21*, 390–401. [CrossRef]

46. Yang, Y.S. Metallogenic Specialization of Gold, Copper and Molybdenum Mineralized Igneous Rocks in Middle-Northern Great Xing'an Range and Metallogenic Prediction in Hongyan Area. Ph.D. Thesis, China University of Geosciences in Wuhan, Wuhan, China, 2017.

47. Hassan, M.H.; Reinhard, K.; Tomoyuki, S. Genetically related Mo-Bi-Ag and U-F mineralization in A-type granite, Gabal Gattar, Eastern Desert, Egypt. *Ore Geol. Rev.* **2014**, *62*, 181–190. [CrossRef]

48. Bunyamin, A. Geochemical associations between fluorite mineralization and A-type shoshonitic magmatism in the Keban-Elazig area, East Anatolia, Turkey. *J. Afr. Earth Sci.* **2015**, *11*, 222–230. [CrossRef]

49. Liu, Y.S.; Hu, Z.C.; Gao, S.; Günther, D.; Xu, J.; Gao, C.G.; Chen, H.H. In situ analysis of major and trace elements of anhydrous minerals by LA-ICP-MS without applying at internal standard. *Chem. Geol.* **2008**, *257*, 34–43. [CrossRef]

50. Liu, Y.S.; Gao, S.; Hu, Z.C.; Gao, C.G.; Zong, K.Q.; Wang, D.B. Continental and oceanic crust recycling-induced melt-peridotite interactions in the Trans-North China Orogen: U-Pb dating, Hf isotopes and trace elements in zircons of mantle xenoliths. *J. Petrol.* **2010**, *51*, 537–571. [CrossRef]

51. Ludwig, K.R. Isoplot 3.00. A Geochronological toolkit for microsoft excel. *Berkeley Geochronol. Center Spec. Publ.* **2003**, *3*, 1–70.

52. Qiu, J.T.; Song, W.J.; Jiang, C.X.; Wu, H.; Dong, R.M. CGDK: An extensible CorelDRAW VBA program for geological drafting. *Comput. Geosci.* **2013**, *51*, 34–48. [CrossRef]

53. Hu, Z.C.; Liu, Y.S.; Gao, S.; Liu, W.G.; Yang, L.; Zhang, W.; Tong, X.R.; Lin, L.; Zong, K.Q.; Li, M. Improved in situ Hf isotope ratio analysis of zircon using newly designed X skimmer cone and Jet sample cone in combination with the addition of nitrogen by laser ablation multiple collector ICP-MS. *J. Anal. At. Spectrom.* **2012**, *27*, 1391–1399. [CrossRef]

54. Woodhead, J.; Herget, J.; Shelley, M.; Eggins, S.; Kemp, R. Zircon Hf-isotope analysis with an excimer laser, depth profiling, ablation of complex geometries, and concomitant age estimation. *Chem. Geol.* **2004**, *209*, 121–135. [CrossRef]

55. Wu, F.Y.; Yang, Y.H.; Xie, L.W.; Yang, J.H.; Xu, P. Hf isotopic compositions of the standard zircons and baddeleyites used in U-Pb geochronology. *Chem. Geol.* **2006**, *234*, 105–126. [CrossRef]

56. Li, X.H.; Li, Z.X.; Wingate, M.T.D.; Chung, S.L.; Liu, Y.; Lin, G.C.; Li, W.X. Geochemistry of the 755 Ma Mundine Well dyke swarm, northwestern Australia: Part of a Neoproterozoic mantle superplume beneath Rodinia? *Precambr. Res.* **2006**, *146*, 1–15. [CrossRef]

57. Hiess, J.; Nutman, A.P.; Bennett, V.C.; Holden, P. Ti-in-zircon thermometry applied to contrasting Archean metamorphic and igneous systems. *Chem. Geol.* **2008**, *247*, 323–338. [CrossRef]

58. McDonough, W.F.; Sun, S.S. The composition of the Earth. *Chem. Geol.* **1995**, *120*, 223–253. [CrossRef]

59. Myers, J.; Eugster, H.P. The system Fe-Si-O: Oxygen buffer calibrations to 1500 K. *Contrib. Mineral. Petrol.* **1983**, *82*, 75–90. [CrossRef]

60. Soderlund, U.; Patchett, P.J.; Vervoort, J.D.; Isachsen, C.E. The ^{176}Lu decay constant determined by Lu-Hf and U-Pb isotope systematic of Precambrian mafic intrusions. *Earth Planet. Sci. Lett.* **2004**, *219*, 311–324. [CrossRef]

61. Blichert-Toft, J.; Chauvel, C.; Albarede, F. Separation of Hf and Lu for high-precision isotope analysis of rock samples by magnetic sector-multiple collector ICP-MS. *Contrib. Mineral. Petrol.* **1997**, *127*, 248–260. [CrossRef]

62. Griffin, W.L.; Pearson, N.J.; Belousova, E.; Jackson, S.E.; Achterbergh, E.V.; O'Reilly, S.Y.; Shee, S.R. The Hf isotope composition of cratonicmantle: LAM-MC-ICPMS analysis of zircon megacrysts in kimberlites. *Geochim. Cosmochim. Acta* **2000**, *64*, 133–147. [CrossRef]

63. Griffin, W.L.; Wang, X.; Jackson, S.E.; Pearson, N.J.; O'Reilly, S.Y.; Xu, X.; Zhou, X. Zircons chemistry and magma genesis in SE China: In situ analysis of Hf isotopes, Pingtan and Tonglu igneous complexes. *Lithos* **2002**, *61*, 237–269. [CrossRef]

64. Strecheissen, A. To each plutonic rock its proper name. *Earth Sci. Rev.* **1976**, *12*, 1–33. [CrossRef]

65. Peccerillo, A.; Taylor, S.R. Geochemistry of Eocen calc-alkaline volcanic rocks from the Kastamounu area, northern Turkey. *Contrib. Miner. Petrol.* **1976**, *58*, 63–81. [CrossRef]

66. Sun, S.S.; McDonough, W.F. Chemical and isotopic systematics of oceanic basalts: Implications for mantle composition and processes. *Geol. Soc. Lond. Spec. Publ.* **1989**, *42*, 313–345. [CrossRef]

67. Whalen, J.B.; Currie, K.L.; Chappell, B.W. A-type granites: Geochemical characteristics, discrimination and petrogenesis. *Contrib. Mineral. Petrol.* **1987**, *95*, 407–419. [CrossRef]

68. Pearce, J.A.; Harris, N.B.W.; Tindle, A.G. Trace-element discrimination diagrams for the tectonic interpretation of granitic-rocks. *J. Petrol.* **1984**, *25*, 956–983. [CrossRef]

69. Patino Douce, A.E. What do experiments tell us about the relative contributions of crust and mantle to the origin of granitic magmas? *Geol. Soc. Lond. Spec. Publ.* **1999**, *168*, 55–75. [CrossRef]

70. Huang, H.Q.; Li, X.H.; Li, W.X.; Li, Z.X. Formation of high δ^{18}O fayalite-bearing A-type granite by high-temperature melting of granulitic metasedimentary rocks, southern China. *Geology* **2011**, *39*, 903–906. [CrossRef]

71. Patino Douce, A.E. Generation of metaluminous A-type granites by low-pressure melting of calc-alkaline granitoids. *Geology* **1997**, *25*, 743–746. [CrossRef]

72. Dall'Agnol, R.; Frost, C.D.; Ramo, O.T. IGCP Project 510 "A-type granites and related rocks through time": Project vita, results, and contribution to granite research. *Lithos* **2012**, *151*, 1–16. [CrossRef]

73. Han, Y.G.; Zhang, S.H.; Pirajno, F.; Zhou, X.W.; Zhao, G.C.; Qu, W.J.; Liu, S.H.; Zhang, J.M.; Liang, H.B.; Yang, K. U-Pb and Re-Os isotopic systematics and zircon Ce^{4+}/Ce^{3+} ratios in the Shiyaogou Mo deposit in eastern Qinling, central China: Insights into the oxidation state of granitoids and Mo (Au) mineralization. *Ore Geol. Rev.* **2013**, *55*, 29–47. [CrossRef]

74. Zhang, X.H.; Yuan, L.L.; Xue, F.H.; Zhang, Y. Contrasting Triassic ferroan granitoids from northwestern Liaoning, North China: Magmatic monitor of Mesozoic decratonization and craton–orogen boundary. *Lithos* **2012**, *144*, 12–23. [CrossRef]

75. Touret, J.L.R. Fluid inclusions and pressure-temperature estimates in deep-seated rocks. In *Chemical Transport in Metasomatic Processes*; Helgeson, H.C., Ed.; NATO ASI Series: Brussels, Belgium, 1987; Volume 218, pp. 91–121.

76. Wei, R.H.; Gao, Y.F.; Xu, S.C.; Santosh, M.; Xin, H.T.; Zhang, Z.M.; Li, W.L.; Liu, Y.F. Carboniferous continental arc in the Hegenshan accretionary belt: Constrains from plutonic complex in central Inner Mongolia. *Lithos* **2018**, *308*, 242–261. [CrossRef]

77. Lee, C.T.A.; Leeman, W.P.; Canil, D.; Canil, D.; Li, Z.X.A. Similar V/Sc systematics MORB and arc basalts: Implications for the oxygen fugacities of their mantle source regions. *J. Petrol.* **2005**, *46*, 2313–2336. [CrossRef]

78. Sun, W.D.; Huang, R.F.; Li, H.; Hu, Y.B.; Zhang, C.C.; Sun, S.J.; Zhang, L.P.; Ding, X.; Li, C.Y.; Zartman, R.E. Porphyry deposits and oxidized magmas. *Ore Geol. Rev.* **2015**, *65*, 97–131. [CrossRef]

79. Zhang, X.H.; Yuan, L.L.; Xue, F.H.; Yan, X.; Mao, Q. Early Permian A-type granites from central Inner Mongolia, North China: Magmatic tracer of post-collisional tectonics and oceanic crustal recycling. *Gondwana Res.* **2015**, *28*, 311–327. [CrossRef]

80. Fitton, J.G.; James, D.; Lccman, W.P. Basic magmatism associated with the late Cenozoic extension in the western United States: Compositional variations in space and time. *Geophys. Res.* **1991**, *13*, 693–711. [CrossRef]

81. Yuan, L.; Zhang, X.; Xue, F.; Liu, F. Juvenile crustal recycling in an accretionary orogen: Insights from contrasting early permian granites from central Inner Mongolia, North China. *Lithos* **2016**, *264*, 524–539. [CrossRef]

82. Landenberger, B.; Collins, W.J. Derivation of A-type granites from a dehydrated charnockitic lower crust: Evidence from the Chaelundi Complex, Eastern Australia. *J. Petrol.* **1996**, *37*, 145–170. [CrossRef]

83. Zhao, X.F.; Zhou, M.F.; Li, J.W.; Wu, F.Y. Association of neoproterozoic A-and I-type granites in South China: Implications for generation of A-type granites in a subduction-related environment. *Chem. Geol.* **2008**, *257*, 1–15. [CrossRef]

84. Mengason, M.J.; Candela, P.A.; Piccoli, P.M. Molybdenum, tungsten and manganese partitioning in the system pyrrhotite–Fe–S–O melt–rhyolite melt: Impact of sulfide segregation on arc magma evolution. *Geochim. Cosmochim. Acta* **2011**, *75*, 7018–7030. [CrossRef]

85. Jugo, P.; Candela, P.; Piccoli, P. Magmatic sulfides and Au: Cu ratios in porphyry deposits: An experimental study of copper and gold partitioning at 850 °C, 100 MPa in a haplogranitic melt–pyrrhotite–intermediate solidsolution–gold metal assemblage, at gas saturation. *Lithos* **1999**, *46*, 573–589. [CrossRef]

86. Yang, X.M.; Lentz, D.R.; Sylvester, P.J. Gold contents of sulfide minerals in granitoids from southwestern New Brunswick, Canada. *Miner. Depos.* **2006**, *41*, 369–386. [CrossRef]

87. Thompson, J.F.H.; Sillitoe, R.H.; Baker, T.; Lang, J.R.; Mortensen, J. Intrusion-related gold deposits associated with tungsten-tin provinces. *Miner. Depos.* **1999**, *34*, 323–334. [CrossRef]

88. Burnham, C.W. Magmas and hydrothermal fluids. In *Geochemistry of Hydrothermal Ore Deposits*, 3rd ed.; Barnes, H.L., Ed.; JohnWiley and Sons: New York, NY, USA, 1997; pp. 63–123.

89. Mao, C.; Lu, X.B.; Chen, C.; Cao, M.Y.; Gun, M.S.; Liao, P.C.; Jia, Q.Y. Melt-fluid inclusions study of Shanshenfu granite and its mineralization significance in Hongyan Town, Inner Mongolia. *Earth Sci.* **2016**, *41*, 139–152, (In Chinese with English Abstract).

90. Zhang, J.H.; Gao, S.; Ge, W.C.; Wu, F.Y.; Yang, J.H.; Wilde, S.A.; Li, M. Geochronology of the mesozoic volcanic rocks in the Great Xing'an Range, northeastern China: Implications for subduction-induced delamination. *Chem. Geol.* **2010**, *276*, 144–165. [CrossRef]

91. Mei, W.; Lv, X.B.; Liu, Z.; Tang, R.K.; Ai, Z.L.; Wang, X.D.; Cisse, M. Geochronological and geochemical constraints on the ore-related granites in Huanggang deposit, Southern Great Xing'an Range, NE China and its tectonic significance. *Geosci. J.* **2015**, *19*, 53–67. [CrossRef]

92. Zhai, D.G.; Liu, J.J.; Zhang, H.Y.; Yao, M.J.; Wang, J.P.; Yang, Y.Q. S-Pb isotopic geochemistry, U-Pb and Re-Os geochronology of the Huanggangliang Fe–Sn deposit, Inner Mongolia, NE China. *Ore Geol. Rev.* **2014**, *59*, 109–122. [CrossRef]

93. Ruan, B.X.; Lv, X.B.; Yang, W.; Liu, S.T.; Yu, Y.M.; Wu, C.M.; Munir, M.A.A. Geology, geochemistry and fluid inclusions of the Bianjiadayuan Pb-Zn-Ag deposit, Inner Monglia, NE China: Implications for tectonic setting and metallogeny. *Ore Geol. Rev.* **2015**, *71*, 121–137. [CrossRef]

94. Blight, J.H.S.; Crowley, Q.G.; Petterson, M.G.; Cunningham, D. Granites of the southern Mongolia Carboniferous arc: New geochronological and geochemical constraints. *Lithos* **2010**, *116*, 35–52. [CrossRef]

95. Blight, J.H.S.; Petterson, M.G.; Crowley, Q.G.; Cunningham, D. The Oyut Ulaan volcanic group: Stratigraphy, magmatic evolution and timing of Carboniferous arc development in SE Mongolia. *J. Geol. Soc. Lond.* **2010**, *167*, 491–509. [CrossRef]

96. Litvinovsky, B.A.; Tsygankov, A.A.; Jahn, B.M.; Katzir, Y.; Be'eri-Shlevin, Y. Origin and evolution of overlapping calc-alkaline and alkalinemagmas: The late palaeozoic post-collisional igneous province of transbaikalia (Russia). *Lithos* **2011**, *125*, 845–874. [CrossRef]

97. Zhou, J.B.; Han, J.; Zhao, G.C.; Zhang, X.Z.; Cao, J.L.; Pei, S.H.; Wang, B. The emplacement time of the Hegenshan Ophiolite: Constraints from the unconformably overlying paleozoic strata. *Tectonophysics* **2015**, *622*, 398–415. [CrossRef]

98. Pei, S.H.; Zhou, J.B.; Li, L. U–Pb ages of detrital zircon of the Paleozoic sedimentary rocks: New constraints on the emplacement time of the Hegenshan ophiolite, NE China. *J. Asian Earth Sci.* **2016**, *130*, 75–87. [CrossRef]

99. Eizenhöfer, P.R.; Zhao, G.; Sun, M.; Zhang, J.; Han, Y.G.; Hou, W. Geochronological and Hf isotopic variability of detrital zircons in Paleozoic strata across the accretionary collision zone between the North China craton and Mongolian arcs and tectonic implications. *Geol. Soc. Am. Bull.* **2015**, *127*, 1422–1436. [CrossRef]

100. Han, J.; Zhou, J.B.; Wang, B.; Cao, J.L. The final collision of the CAOB: Constraint from the zircon U–Pb dating of the Linxi Formation, Inner Mongolia. *Geosci. Front.* **2015**, *6*, 211–225. [CrossRef]

101. Janasi, V.; Vlach, S.R.F.; Campos Neto, M.; Ulbrich, H.H.G.J. Associated A-type subalkaline and high-K calc-alkaline granites in the Itu granite province, southeastern Brazil: Petrological and tectonic significance. *Can. Mineral.* **2009**, *47*, 1505–1526. [CrossRef]

102. Li, D.P.; Chen, Y.L.; Wang, Z.; Hou, K.; Liu, C.Z. Detrital zircon U–Pb ages, Hf isotopes and tectonic implications for Palaeozoic sedimentary rocks from the Xing-Meng Orogenic Belt, Middle-East part of Inner Mongolia, China. *Geol. J.* **2011**, *46*, 63–81. [CrossRef]

103. Lamb, M.A.; Badarch, G. Paleozoic sedimentary basins and volcanic arc systems of southern Mongolia; new geochemical and petrographic constraints. *Geol. Soc. Am. Mem.* **2001**, *194*, 117–149.

104. Whalen, J.B.; McNicoll, V.J.; van Staal, C.R.; Lissenberg, C.J.; Longstaffe, F.J.; Jenner, G.A.; van Breeman, O. Spatial, temporal and geochemical characteristics of Silurian collision-zone magmatism, Newfoundland Appalachians: An example of a rapidly evolving magmatic system related to slab break-off. *Lithos* **2006**, *89*, 377–404. [CrossRef]

105. Bonin, B. Do coeval mafic and felsic magmas in post-collisional to within-plate regimes necessarily imply two contrasting, mantle and crustal, sources? A review. *Lithos* **2004**, *78*, 1–24. [CrossRef]

Article

Geochemistry, Zircon U–Pb Geochronology, and Lu–Hf Isotopes of the Chishan Alkaline Complex, Western Shandong, China

Pengfei Wei [1,2,3], Xuefeng Yu [1,*], Dapeng Li [1,*], Qiang Liu [1], Lidong Yu [3], Zengsheng Li [1], Ke Geng [1], Yan Zhang [1], Yuqin Sun [1] and Naijie Chi [1]

[1] Shandong Institute of Geological Sciences, Shandong Key Laboratory of Geological Processes and Resource Utilization in Metallic Minerals, Key Laboratory of Gold Mineralization Processes and Resources Utilization Subordinated to the Ministry of Land and Resources, Jinan 250013, China; pfeiwei@126.com (P.W.); lqiang715@163.com (Q.L.); lizengsheng@126.com (Z.L.); gengke@126.com (K.G.); sddkyzy@163.com (Y.Z.); sunyuqin86@163.com (Y.S.); chinaijie@163.com (N.C.)

[2] Shandong Provincial Key Laboratory of Depositional Mineralization & Sedimentary Mineral, Shandong University of Science and Technology, Qingdao 266590, China

[3] School of Earth Sciences and Resources, China University of Geosciences, Beijing 100083, China; yldongy@126.com

[*] Correspondence: xfengy@sohu.com (X.Y.); dpengli@126.com (D.L.); Tel.: +86-0531-8655-6925 (X.Y.); +86-0531-8655-8802 (D.L.)

Received: 26 March 2019; Accepted: 5 May 2019; Published: 13 May 2019

Abstract: Mass alkaline magmatic activities in Western Shandong during the late Mesozoic controlled the mineralization processes of gold and rare earth element (REE) polymetallic deposits in the region. The Chishan alkaline complex is closely associated with the mineralization of the Chishan REE deposit, which, as the third largest light REE deposit in China following the Baiyenebo (Inner Mongolia) and Mianning (Sichuan) deposits, is considered a typical example of alkaline rock mineralization throughout the North China Craton. To determine how the Chishan alkaline complex and REE deposit interact with each other, a systematic study was conducted on the petrology, rock geochemistry, zircon U–Pb geochronology, Lu–Hf isotopes of the quartz syenite, and alkali granite contained in the Chishan alkaline complex. The results reveal that the deposits feature similar geochemical characteristics typical of an alkaline rock series—both are rich in alkali, high in potassium, metaluminous, and poor in Ti, Fe, Mg, and Mn. In terms of REEs, the deposits are strongly rich in light REEs but poor in heavy REEs, with weak negative Eu anomalies. In terms of trace elements, they are rich in large ion lithophile elements Ba, Sr, and Rb but poor in high field-strength elements Nb, Ta, and Hf. Zircon LA-ICP-MS U–Pb dating indicated that the quartz syenite and alkali granite formed in Early Cretaceous at 125.8 ± 1.2 Ma and 127.3 ± 1.0 Ma, respectively; their εHf(t) values are −22.67 to −13.19, with depleted model ages (T_{DM}) ranging from 1296 Ma to 1675 Ma and crustal model ages (T_{DM}^{C}) of 2036–2617 Ma. The Chishan alkaline complex originated from partial of the EM I-type (enriched mantle I) lithospheric mantle with assimilation of ancient crustal materials. The complex is of the same origin as the REE deposit, and developed in an extensional setting that resulted from plate subduction and lithospheric thinning and upwelling in the eastern area of the North China Craton.

Keywords: Chishan alkaline complex; rock geochemistry; zircon U–Pb dating; Lu–Hf isotopes

---

1. Introduction

In late Mesozoic, the tectonic regime of the North China Craton changed from an extrusional to an extensional environment. The mass lithospheric thinning [1–8] and extensive uptrusion of asthenospheric mantle materials led to violent magmatic activities, accompanied by the mineralization of gold and rare earth elements (REE) polymetallic deposits [9–13]. Intense and frequent magmatic activities dating back to 130–120 Ma, in particular, controlled the formation of a large number of deposits [14–16]. In East China's Shandong Province, the acidic magmas to the east of the Yishu fault zone, such as the Guojialing-style granite, controlled the formation of large-sized gold deposits in Eastern Shandong (e.g., the Linglong, Rushan, and Wulongshan gold deposits) and the polymetallic deposits in the eastern part of Eastern Shandong (e.g., Meishan and Taocun iron deposits) [11,14,17–22]. The alkaline magmas to the west of the fault zone controlled the formation of gold and REE deposits in Western Shandong [14,20,22]. Examples include the Longbaoshan alkaline complex, which controlled the mineralization of the Longbaoshan gold deposit [23,24], and the Chishan alkaline complex, which controlled the mineralization of the Chishan REE Deposit [25–30]. These alkaline magmas usually originate from the upper mantle and have experienced complicated underplating, contamination, and evolution, recording the tectonic evolution and lithospheric thinning of the North China Craton during late Mesozoic.

The Chishan alkaline complex in Western Shandong is closely associated with the mineralization of the Chishan REE deposit which, as the third largest light REE (LREE) deposit in China following the Baiyenebo (Inner Mongolia) and Mianning (Sichuan) deposits, is considered as a typical example of alkaline rock mineralization throughout the North China Craton [25], and valued by scientific research institutes and mining enterprises [25,27,28,31–35]. In particular, the recent discovery of the REE mineralization bodies from the neighboring Shagou (Xuecheng) and Longbaoshan (Zaozhuang) alkaline complexes has triggered a considerable interest in this alkaline magmatic rock series. However, despite the understandings on the formation age [25,35] and geochemistry [28,32,33] of this complex, the lack of systematic research still limits further identification of the local tectonic magma evolution and REE accumulation processes of the area.

In this study, we attempted to discuss the magmatic origin, evolution, and tectonic setting of these alkaline magmas through a systematic study on the petrology, geochemistry, zircon U–Pb geochronology, and Lu–Hf isotopes of the Chishan alkaline complex, with a view to further gain insights into the mechanisms behind the lithospheric thinning of the North China Craton and the genetic connections between alkaline rocks and polymetallic deposits.

2. Regional Geology

The Chishan alkaline complex lies near the Chishan village about 18 km southeast of Weishan County of Zaozhuang City, Shandong Province. Tectonically, this complex belongs to the Luxi Block at the southeastern margin of the North China Craton and the west side of the Tanlu fault zone (Figure 1a) [28,29]. The Luxi Block is bounded by the Tanlu fault zone to the east, the Liaocheng-Lankao fault zone to the west, the south of the Qihe-Guangrao fault to the north, and the Fengpei fault to the south (Figure 1b) [36–45]. The exposed formations in the study area include the Neoarchean Taishan rock group, Paleozoic Cambrian to Permian carbonate and clastic rocks, Mesozoic Triassic to Cretaceous clastic and volcanic rocks, and Cenozoic clastic rocks (Figure 1b), with the Neoarchean granodiorite constituting the crystalline basement. In this area, the Yanshanian intermediate-to-alkaline complexes consisting of diorite porphyrite, syenite, syenite porphyrite, diabase, and lamprophyre mostly occur in forms of batholiths, stocks, or dykes. The representative rock bodies include the Shagou, Tongshi, Chishan, and Longbaoshan volcanic–intrusive complexes composed of monzonitic–orthofelsic rocks.

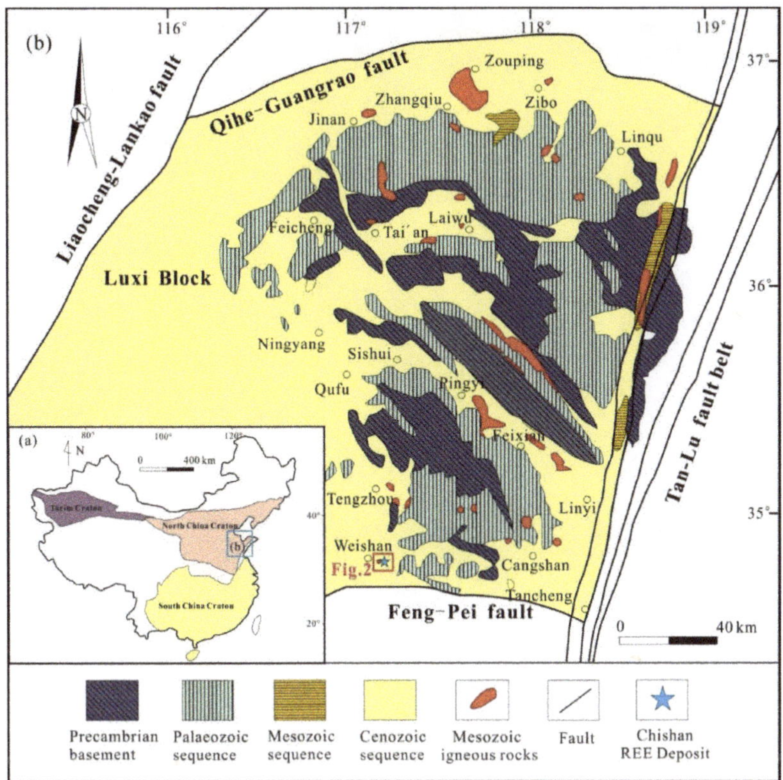

Figure 1. Simplified geological map showing the major tectonic units in China (**a**) and regional geological sketch of Western Shandong (**b**). (**a**) Geologic and tectonic map of China and the location of Luxi Block, modified after [40]. (**b**) Geological map of the Luxi Block, modified after [37].

The principal part of the Chishan alkaline complex comprises quartz syenite, aegirine quartz syenite, and alkali granite. The complex stretches in the NE–SE orientation, intruding into the early granodiorite and interfacing with the country rock like irregular branches. Varying degrees of alkaline metasomatism can be observed along most parts of the contact. The outcrop of this complex, occupying an area of an area of 0.5 km^2, is almost fully covered by Quaternary except for sporadic bedrock outcrops at the Chishan hilltop. The Chishan alkaline complex is primarily controlled by tectonic faults. Tectonic fissures in the rocks are often filled up by rare earth veins and occur in the NW orientation as single veins, spiderwebs, or disseminations (Figure 2) [46].

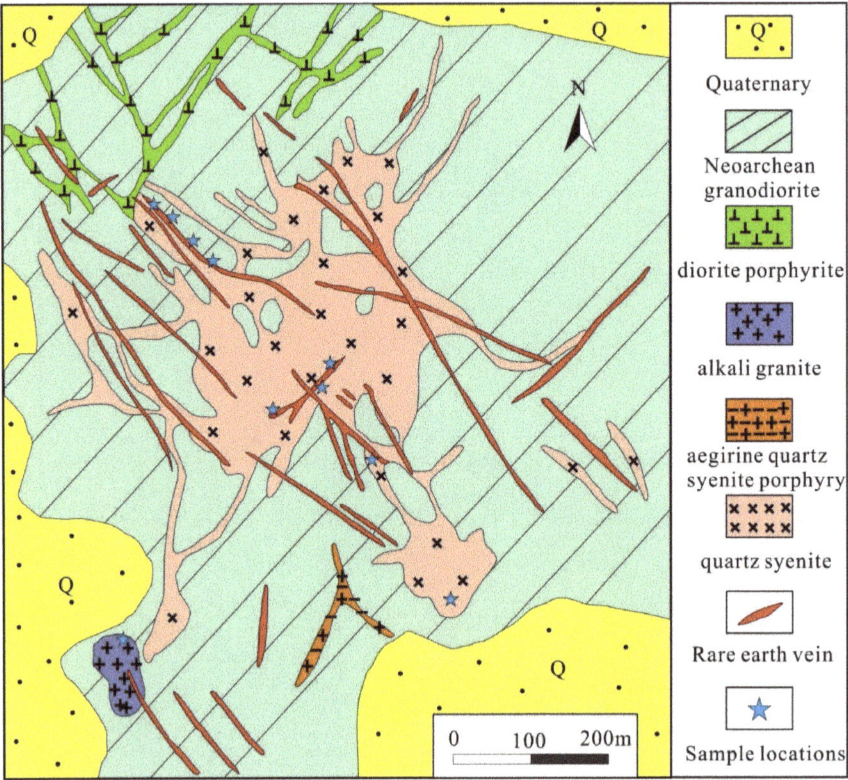

Figure 2. Regional geological map of the Chishan alkaline complex, modified after [46].

3. Petrology and Petrography

The quartz syenite is grayish white or light flesh pink in color (Figure 3a), porphyroid or porphyritic in texture, and massive in structure. The mineral components mainly include K-feldspar, plagioclase, quartz, with portions of hornblende. Microscopy revealed a fine-grained porphyritic texture. The phenocrysts primarily include K-feldspar (±10%) and a small amount of hornblende (±1%). The matrix, which is primarily composed of K-feldspar (±53%), plagioclase (±10%), quartz (±10%), and hornblende (±5%), exhibits a xenomorphic platy, columnar-granular, or irregular granular shape, and 0.01–1.0 mm grain size. The K-feldspar phenocrysts, which mainly consists of orthoclase, features a platy or irregular granular texture and 1.8–2.5 mm grain size, with detectable Carlsbad twinning (Figure 3c) and zonal structures, and are mostly clayized. The hornblende (±1%) phenocrysts are columnar in shape, approximate 1.2 mm in grain size, green to olive green in color (Figure 3c,e), and have been partly replaced by epidote and chlorite. The quartz occurs interstitially between feldspar grains (Figure 3e).

Figure 3. Petro-mineralogical diagram of the Chishan alkaline complex: (**a**) Quartz syenite; (**b**) alkali granite; (**c**) Carlsbad twins of the syenite; (**d**) crosshatch twins of the microcline; (**e**) hornblende occuring in columnar form; (**f**) quartz occurs interstitially between the feldspar grains. The (**c**) and (**e**) photos under the microscope refer to the sample present in (**a**); the (**d**) and (**f**) photos under the microscope refer to the sample present in (**b**). Qz-quartz; Or-orthoclase; Pl-plagioclase; Amp-amphibole; Mc-microcline.

The alkali granite features fine-grained granite texture and massive structure (Figure 3b). Microscopy revealed a fine-grained, subhedral granular texture. The main mineral components include orthoclase ($\pm$65%), microcline ($\pm$10%), quartz ($\pm$20%), and small fractions of biotite ($\pm$3%) and epidote ($\pm$1%). Accessory minerals include sphene and magnetite. The orthoclase is subhedral columnar or granular in shape with Carlsbad twinning (Figure 3d) and evident perthitic texture. The microcline is subhedral columnar in shape and 1–5 mm in grain size with crosshatch twinning (Figure 3d). The quartz is xenomorphic granular in shape and approximates 0.02–1.2 mm in grain size;

a portion of the quartz occurs interstitially between other minerals in fine granular forms (Figure 3f). The biotite is layered with about 0.1–1 mm by 0.02–0.2 mm in grain size, and notably pleochroism; the rock is reddish brown to sepia and appears as sandy beige or olive green to green. Some of the biotite is associated with epidote.

4. Samples and Analysis

All 12 samples used for our study originated from the fresh alkaline rocks of the Chishan REE deposit. Six samples were quartz syenite, and the other six were alkali granite. Ten samples were collected from the −160 m level of the deposit, and two were obtained from the surface outcrop. Figure 2 shows the exact sampling locations. Except for samples 18CS-01 (quartz syenite) and 18CS-27 (alkali granite) on which zircon dating analysis was performed, the total rock major and trace element analyses were performed on all other 10 samples.

4.1. Total Rock Major and Trace Element Analysis

The total rock major and trace element analyses were performed by Beijing Createch Testing Technology Co., Ltd. (Beijing, China). Ten fresh quartz syenite and alkali granite samples were crushed to smaller than 200 mesh. For the major elements, the lithium borate plus lithium nitrate melting method and X-ray fluorescence spectrometry (XRF-1800) with an accuracy higher than 1% were used. For the trace elements, the lithium borate melting method and quantitative inductively-coupled plasma (ICP) spectrometry (Agilent 7500ce) with an accuracy better than 5% were used. For REEs, the ICP–MS (Agilent 7500ce ICP-MS) with an accuracy of better than 5% was used. Table 1 presents the analysis results.

4.2. Zircon U–Pb Dating

Monomineral separation for zircon dating was completed by the Fengze Source Rock and Mineral Test Technology Co., Ltd. (Langfang, China); target fabrication, transmittance and reflectance measurement, and cathode luminescence (CL) photography and testing were all completed by Beijing Createch Testing Technology Co., Ltd. (Beijing, China). Further details about the target fabrication and testing are described in the work of Song et al. [47]. All of the zircons were studied with micrographs and cathode luminescence (CL) images to illustrate their microstructures. The zircon U–Pb compositions were determined using a sensitive high-resolution laser ablation multi-collector inductively-coupled plasma source mass spectrometer (AnalytikJena PlasmaQuant MS Elite ICP-MS) composed of an ESI NWR 193 nm FX laser and a Neptune mass spectrometer. Helium was used as the gas carrier for denudation material. The analytical spot sizes were 35 μm, but each spot was rastered over 120 μm for three minutes to remove common Pb on the zircon surfaces. Five consecutive scans were performed for each zircon spot. The analysis process is described in the work of Yuan et al. [48]. The final test data were processed with ICP–MS DataCal [49]. Finally, the U–Pb age concordia plot was developed, and the weighted average ages were calculated using ISOPLOT 3.70 [50]. Table 2 presents the analysis results.

4.3. Zircon Lu–Hf Isotopes

Zircon Hf isotopes were measured with a Neptune system in or near the same position used for U–Pb dating analysis. The laser and mass spectrometers used were the same as previously described for zircon U–Pb dating test. The analysis conditions, models of instruments, and analysis processes are described in the work of Geng et al. [51]. During the analysis, the $\varepsilon Hf(t)$ values were calculated by relying on the zircon U–Pb ages at the measuring points. For the purpose of this study, the ^{176}Lu decay constant of 1.867×10^{-11} year^{-1} [52], chondrite $^{176}Lu/^{177}Hf$ ratio of 0.0332, and $^{176}Hf/^{177}Hf$ ratio of 0.282772 were used [53]. The depleted mantle model age (T_{DM}) was calculated at the present $^{176}Lu/^{177}Hf$ ratio of 0.0384 and $^{176}Hf/^{177}Hf$ ratio of 0.28325 of the depleted mantle [54]. The formulas for calculating are shown below [55]. In the formulas, f_{cc}, f_s and f_{DM} respectively represent the $f_{Lu/Hf}$

of the large continental crust, sample, and depleted mantle, and T is the crystallization age of zircon. Table 3 provides the analysis results.

$$\varepsilon_{Hf}(0) = ((^{176}Hf/^{177}Hf)_S/(^{176}Hf/^{177}Hf)_{CHUR,0} - 1) \times 10,000; \tag{1}$$

$$\varepsilon_{Hf}(t) = ((^{176}Hf/^{177}Hf)_S - (^{176}Lu/^{177}Hf)_S \times (e^{\lambda t} - 1))/((^{176}Hf/^{177}Hf)_{CHUR,0} - (^{176}Lu/^{177}Hf)_{CHUR} \times (e^{\lambda t} - 1)) - 1) \times 10,000; \tag{2}$$

$$T_{Hf1} = 1/\lambda \times \ln[1 + ((^{176}Hf/^{177}Hf)_S - (^{176}Hf/^{177}Hf)_{DM})/((^{176}Lu/^{177}Hf)_S - (^{176}Lu/^{177}Hf)_{DM})]; \tag{3}$$

$$T_{Hf2} = T_{Hf1} - (T_{Hf1} - t)((f_{cc} - f_s)/(f_{cc} - f_{DM})); \tag{4}$$

$$f_{Lu/Hf} = (^{176}Lu/^{177}Hf)_S/(^{176}Lu/^{177}Hf)_{CHUR} - 1. \tag{5}$$

5. Results

5.1. Analysis Results of Total Rock Major and Trace Elements

Major element analysis (Table 1) indicated that the quartz syenite features a SiO_2 content of 69.02–71.72%, a $Na_2O + K_2O$ content of 8.87–10.94%, and MgO, CaO, and Fe_2O_3 contents of 0.15–0.49%, 0.73–1.59%, and 0.48–1.92%, respectively. The rock is relatively high in SiO_2, rich in alkali, low in Ti (with a TiO_2 content of 0.06–0.17%), and rich in Al (with an Al_2O_3 content of 13.51–16.12%). On the total alkali versus silica (TAS) diagram, the quartz syenite falls within either the quartz monzonite or granite zone (Figure 4a). The aluminum saturation index (A/CNK) of 0.80–1.00 suggests that the quartz syenite belongs to the metaluminous series. On the A/NK–A/CNK diagram, all the samples fall within the metaluminous zone (Figure 4b). The quartz syenite features an alkalinity rate (AR) of 2.58–9.02. On the AR–SiO_2 diagram (Figure 4c), the rock samples were projected into the alkaline rock zone, signifying that the quartz syenite belongs to the alkaline rock series.

The alkali granite features a similar major and trace element geochemistry as the quartz syenite, with a SiO_2 content of 64.07–71.66%, a $Na_2O + K_2O$ content of 7.62–8.35%, and MgO, CaO, and Fe_2O_3 contents of 0.62–1.55%, 2.32–4.14%, and 0.73–2.16%, respectively. The rock is relatively high in SiO_2, rich in alkali, low in Ti (with a TiO_2 content of 0.19–0.40%), and rich in Al (with an Al_2O_3 content of 14.07–16.75%). On the TAS diagram, the alkali granite typically falls within either the quartz syenite or granite zone (Figure 4a). The A/CNK value mostly falls within 0.75–1.00. On the A/NK–A/CNK diagram, the peralkaline contents are mostly found in the metaluminous zone (Figure 4b). The alkali granite exhibits an AR of 2.58–9.02. On the AR–SiO_2 diagram (Figure 4c), the rock samples were projected into the alkaline rock zone, except for a few that fall within the strongly alkaline rock zone. Hence, the rocks in this complex are alkaline rocks.

The trace element analysis of the Chishan alkaline complex indicated that the rocks are rich in large ion lithophile elements (LILEs) Ba, Sr, and Rb but poor in high field-strength elements (HFSEs) Nb, Ta, and Hf. The primitive mantle-normalized spidergram (Figure 5b) indicates a similarly incompatible element distribution pattern between the quartz syenite and alkaline granite of the Chishan alkaline complex. Both minerals exhibit remarkable positive k anomalies, strong negative Sr and Ti anomalies, and positive Ba and Zr anomalies.

The Chishan alkaline complex features a high total REE ($\sum$REE) value, with the quartz syenite featuring an average $\sum$REE value of 692.18×10^{-6} and the alkali granite yielding an average $\sum$REE of 247.61×10^{-6}. The complex is extremely rich in LREEs. The LREE/HREE ratio ranges from 11.00 to 51.17 with an average of 29.88. The $(La/Yb)_N$ value is 13.70–231.51; the δEu value is 0.84–1.24, exhibiting a negative Eu anomaly. The δCe value is 1.01–1.23, exhibiting an unremarkable Ce anomaly. On the chondrite-normalized REE diagram (Figure 5a), the REE curve displays a visible right-inclined separation, showing a strong LREE enrichment. The rocks are poor in HREEs (heavy rare earth elements) and display appreciable LREE/HREE fractionation. This finding suggests that the rocks have experienced strong LREE/HREE fractionation and are highly rich in LREEs.

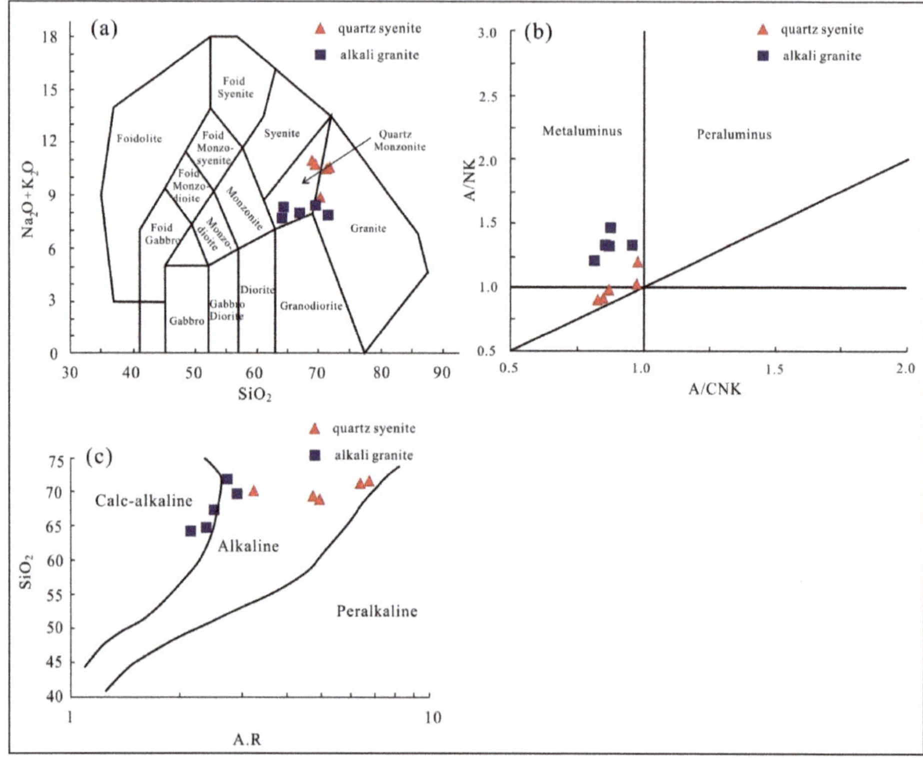

Figure 4. Plots of SiO$_2$ vs. Na$_2$O + K$_2$O (**a**), A/NK (molar ratio Al$_2$O$_3$/(Na$_2$O + K$_2$O)) vs. A/CNK (molar ratio Al$_2$O$_3$/(CaO + Na$_2$O + K$_2$O)) (**b**), and SiO$_2$ vs. A.R (Al$_2$O$_3$ + CaO + (Na$_2$O + K$_2$O))/(Al$_2$O$_3$ + CaO − (Na$_2$O + K$_2$O)) (**c**) of the Chishan alkalic complex. (**a**) is from [56], (**b**) is from [57], and (**c**) is from [58].

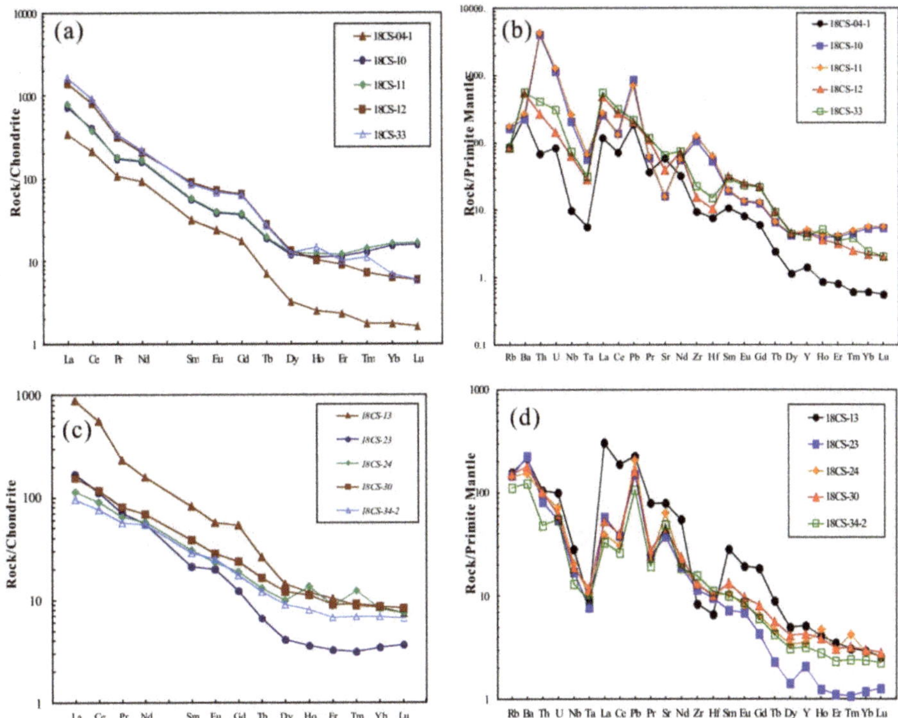

Figure 5. Chondrite-normalized REE patterns and primitive mantle (PM) normalized trace element spider diagram for quartz syenites (**a**,**b**) and alkali granites (**c**,**d**) from the Chishan alkalic complex (chondrite and PM values are from [59]).

Table 1. Analysis results of the major (wt %) and trace (ppm) elements of the Chishan alkaline complex.

Samples	18CS-04-1	18CS-10	18CS-11	18CS-12	18CS-33	18CS-13	18CS-23	18CS-24	18CS-30	18CS-34-2
			quartz syenite					alkali granite		
SiO_2	70.20	71.35	71.72	69.02	69.42	69.55	71.66	64.44	67.13	64.07
Al_2O_3	15.25	13.74	13.51	16.12	15.41	14.07	14.70	16.50	15.09	16.75
MgO	0.49	0.16	0.19	0.15	0.48	0.62	0.71	1.39	1.33	1.55
Na_2O	5.53	6.25	6.12	7.13	7.11	4.76	4.64	6.20	4.99	5.91
K_2O	3.34	4.29	4.43	3.82	3.67	3.59	3.20	2.08	2.95	1.70
P_2O_5	0.07	0.03	0.04	0.10	0.11	0.09	0.09	0.18	0.11	0.20
TiO_2	0.17	0.08	0.06	0.11	0.11	0.19	0.19	0.36	0.32	0.40
CaO	1.59	0.73	0.74	0.38	1.15	3.03	2.32	3.78	3.32	4.14
TFe_2O_3	1.32	2.35	2.31	1.38	1.50	1.74	1.93	3.75	3.19	4.19
FeO	0.76	0.39	0.39	0.12	0.37	0.89	1.08	1.70	1.50	1.83
Fe_2O_3	0.48	1.92	1.87	1.25	1.09	0.76	0.73	1.87	1.52	2.16
MnO	0.02	0.13	0.12	0.05	0.05	0.09	0.04	0.08	0.07	0.07
LOI	2.45	0.77	0.71	0.96	0.63	2.00	0.68	1.36	1.60	1.17
Total	100.43	99.87	99.94	99.21	99.65	99.73	100.15	100.13	100.10	100.14
$Na_2O + K_2O$	8.87	10.53	10.56	10.94	10.78	8.35	7.84	8.28	7.94	7.62
A/NK	1.199	0.921	0.908	1.017	0.983	1.2	1.325	1.325	1.324	1.447
A/CNK	0.977	0.846	0.833	0.974	0.867	0.816	0.96	0.854	0.865	0.877
AR	3.23	6.36	6.73	4.94	4.73	2.91	2.71	2.38	2.51	2.15
Rb	54.93	102.57	109.93	52.93	54.15	103.24	94.02	92.16	97.01	71.09
Sr	1235.82	339.67	337.60	830.67	1381.18	1670.70	793.43	1354.19	945.03	1027.48
Ba	1694.82	1572.33	1866.33	3825.37	3975.62	1539.17	1572.63	1075.93	1245.54	856.73
Th	5.80	342.33	362.58	23.00	34.86	9.64	6.94	8.37	8.63	4.06
U	1.77	23.87	26.92	2.99	6.55	1.87	1.13	1.50	1.30	1.16
Nb	7.09	148.94	187.67	44.96	53.76	17.12	12.23	14.60	13.35	9.38
Ta	0.23	2.30	2.75	1.16	1.29	0.39	0.32	0.43	0.49	0.37
Zr	106.19	1196.82	1413.12	175.74	255.58	87.80	128.83	143.37	146.91	174.67
Hf	2.35	16.56	19.43	3.16	4.67	1.96	2.97	3.16	3.09	3.46
Co	4.62	1.82	1.32	2.35	3.45	5.59	6.43	12.59	13.08	16.41
Ni	3.27	14.14	1.99	4.01	9.86	4.79	4.83	12.95	14.50	13.17
Cr	12.69	29.35	10.00	18.35	19.12	12.65	11.27	13.84	25.16	13.75
V	23.75	32.94	31.11	25.39	27.36	29.32	34.95	57.90	52.34	78.94
Sc	7.15	6.38	6.99	8.74	7.85	6.58	5.80	9.76	13.28	16.56
Cs	0.91	0.27	0.32	0.18	0.22	1.65	1.78	2.86	2.25	2.23

Table 1. Cont.

Samples	18CS-04-1	18CS-10	18CS-11	18CS-12	18CS-33	18CS-13	18CS-23	18CS-24	18CS-30	18CS-34-2
Ga	58.10	60.66	61.71	67.52	46.97	55.99	45.02	59.30	50.84	58.69
Cu	9.62	8.52	8.03	6.73	5.69	10.58	5.18	7.36	45.68	21.13
Pb	13.27	61.06	50.37	14.47	15.78	14.20	10.64	14.83	11.47	7.55
Zn	36.49	99.67	104.08	37.83	73.80	41.55	35.42	65.29	60.14	64.96
Be	2.30	17.92	18.15	3.89	4.89	1.98	1.90	2.37	2.34	2.12
La	81.50	174.62	187.46	330.85	383.69	219.65	39.33	26.67	36.75	22.78
Ce	128.85	244.16	231.99	489.15	560.84	347.44	67.59	54.80	70.24	46.35
Pr	10.20	16.42	16.97	30.48	32.84	22.83	6.62	6.13	7.57	5.41
Nd	43.37	74.72	76.91	97.28	101.77	77.68	25.97	27.17	32.22	25.25
Sm	4.78	8.64	8.78	13.91	13.26	13.28	3.22	4.66	5.93	4.44
Eu	1.38	2.23	2.29	4.15	3.94	3.46	1.16	1.36	1.64	1.45
Gd	3.58	7.52	7.73	13.43	13.04	11.43	2.53	3.85	4.85	3.58
Tb	0.26	0.70	0.73	1.03	1.00	1.02	0.25	0.49	0.61	0.45
Dy	0.83	3.08	3.21	3.36	3.25	3.84	1.05	2.51	3.06	2.28
Ho	0.14	0.64	0.70	0.58	0.84	0.72	0.20	0.77	0.63	0.46
Er	0.39	1.89	1.99	1.51	1.67	1.77	0.53	1.46	1.49	1.12
Tm	0.04	0.33	0.36	0.18	0.28	0.24	0.08	0.31	0.24	0.18
Yb	0.30	2.61	2.77	1.08	1.19	1.52	0.59	1.40	1.48	1.16
Lu	0.04	0.40	0.42	0.15	0.15	0.20	0.09	0.19	0.21	0.17
Y	6.44	21.15	22.93	20.51	18.35	24.60	9.42	15.97	19.39	14.44
$\sum$REE	275.68	537.98	542.32	987.16	1117.77	705.08	149.20	131.77	166.91	115.08
LREE/HREE	18.11	19.12	17.56	21.40	17.17	18.14	5.04	2.70	7.21	2.95
La$_N$/Yb$_N$	31.99	41.74	35.56	31.51	23.52	38.04	3.33	1.59	5.73	1.73
δEu	1.36	1.56	2.30	1.35	2.53	1.28	0.93	2.35	7.37	1.42

5.2. Zircon U–Pb Dating

In the quartz syenite (18CS-01, Figure 6a), zircons exist as automorphic columns or stumps with a complete crystal form and smooth surface. Most of the zircons contain zonal texture, suggesting a magmatic origin. The zircon grains in the sample are large, with clear internal texture, well-arranged zones, and an aspect ratio of 1:1.5–1:5, which agrees with the description of a magmatic zircon. In the alkali granite (18CS-27, Figure 6b), the zircons are well automorphic, mostly columnar in shape with a complete crystal form and smooth surface, and vary in size from 30 μm up to 100 μm. The zircons contain a clear visible zonal texture, whereas some zircon samples showed inherited zircon cores. The Th/U values are unexceptionally greater than 0.4. Almost all the zircon samples displayed the features typical of an alkaline magmatic zircon. Some of the zircons displayed a core-mantle texture, suggesting inherited growths. In the sample, the zircons feature relatively large grain sizes, clear internal texture, well-arranged zones, and an aspect ratio of 1:1.5–1:5, which agree with the description of a magmatic zircon. The measured U and Th contents spanned broadly from 249×10^{-6} to 2338×10^{-6} and from 20×10^{-6} to 133×10^{-6}, respectively; the measured Th/U values ranged narrowly from 0.01 to 0.06.

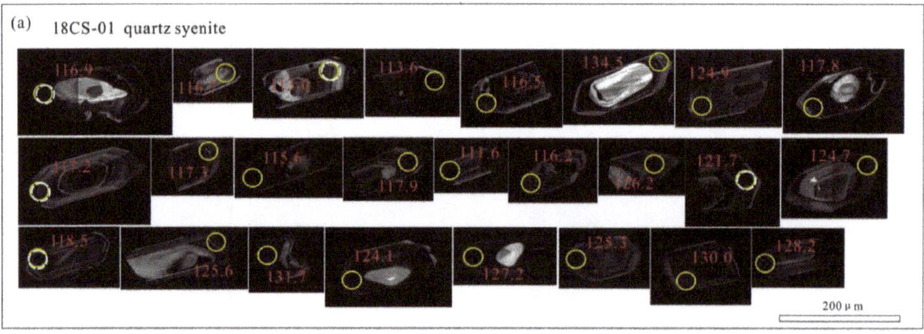

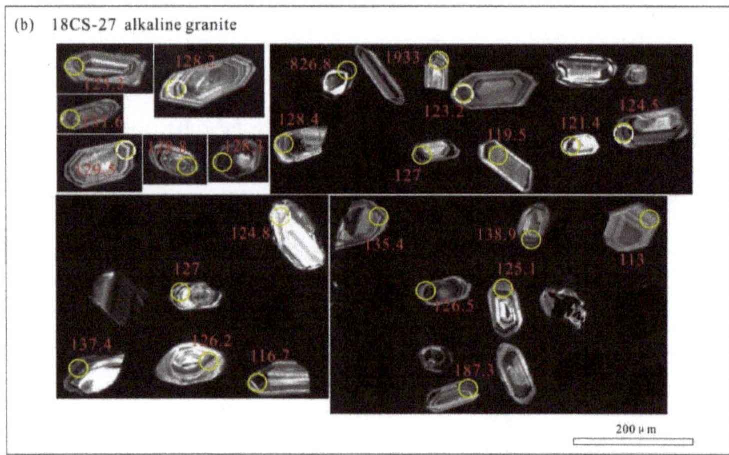

Figure 6. Zircon cathode luminescence (CL) diagram of the Chishan alkaline complex. (**a**) is the CL of quartz syenite (18CS-01); (**b**) is the CL of alkali granite (18CS-27).

Table 2 presents the valid data of the zircons in samples 18CS-01 and 18CS-27. The $^{206}\text{Pb}/^{238}\text{U}$–$^{207}\text{Pb}/^{235}\text{U}$ concordia curve projection was performed on 23 valid data of sample 18CS-01 and 17 valid data of the magmatic zircons in sample 18CS-27; the $^{206}\text{Pb}/^{238}\text{U}$ ages were weighted averaged (Figure 7a–d). The results indicated a concordia age of 125.8 ± 1.2 Ma (MSWD = 1.4) for zircons in the quartz syenite and 127.3 ± 1.0 Ma (MSWD = 1.13) for zircons in the alkali granite of the Chishan alkaline complex.

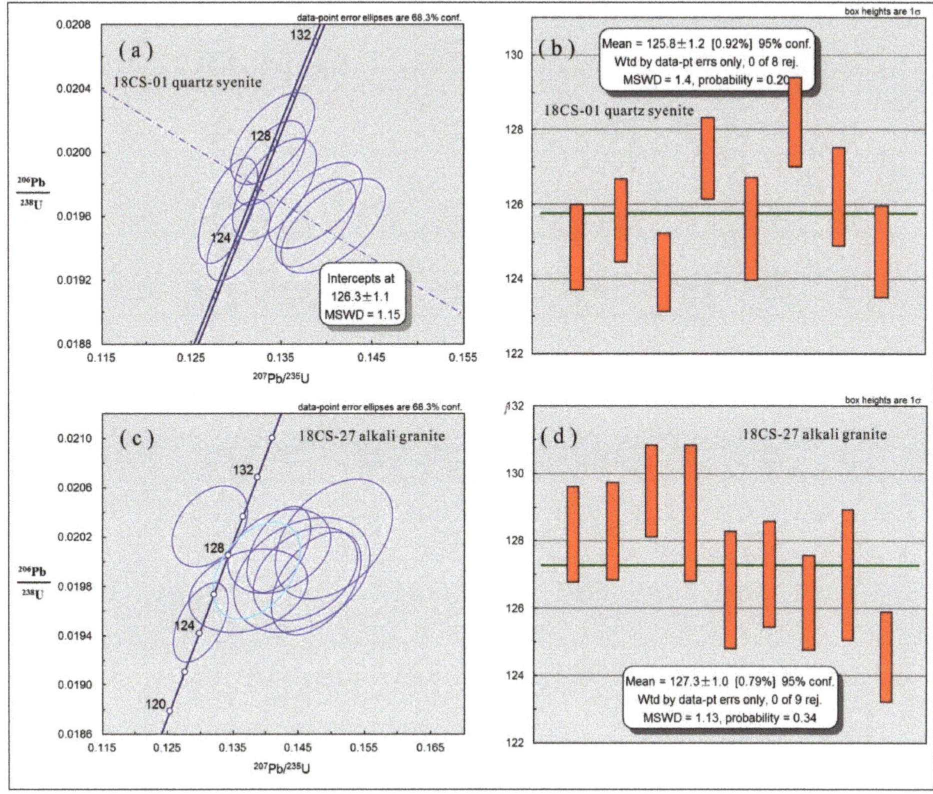

Figure 7. Zircon U–Pb age concordia plot of the Chishan alkaline complex. (**a**) is the concordia plot of quartz syenite (18CS-01); (**b**) is the average plot of quartz syenite (18CS-01); (**c**) is the concordia plot of alkali granite (18CS-27); (**d**) is the average plot of alkali granite (18CS-27).

5.3. Zircon Lu–Hf Dating

In situ Hf isotope analysis was conducted on five already-dated zircon samples from the quartz syenite and from the alkali granite, respectively, at roughly the same points as those used for zircon U–Pb dating. Table 3 presents the results. Then, the initial ratios and εHf(t) values of zircon Hf isotopes were calculated with the measured ages. From Table 3, the ^{176}Lu/^{177}Hf ratios were smaller than 0.002 at all points, suggesting that the zircons have gained no considerable radiogenic Hf after the complex formation. Consequently, the zircon ^{176}Hf/^{177}Hf ratios can be used to discuss the genetic information of the complex [55]. As the zircons in the samples show low $f_{Lu/Hf}$ ratios (-0.95–-0.98), the one-stage model ages of the zircon Lu–Hf isotopes can represent the time at which the source materials separated from the mantle [60]. Zircons from the quartz syenite (18CS-01) have variable ^{176}Hf/^{177}Hf ratios (0.28225–0.28232) with εHf(t) values ranging from -13.2 to -15.8. Their depleted model ages (T_{DM}) range from 1296 Ma to 1359 Ma and crustal model ages ($T_{DM}{}^C$) from 2074 Ma to 2186 Ma. Zircons from the alkali syenite (18CS-27) have relatively low and variable ^{176}Hf/^{177}Hf ratios (0.28206–0.28232) with εHf(t) values ranging from -13.4 to -22.7, depleted model ages (T_{DM}) ranging from 1324 Ma to 1675 Ma and crustal model ages ($T_{DM}{}^C$) of 2036–2617 Ma.

Table 2. Zircon U–Pb dating data of the Chishan alkaline complex.

Samples	Content(10^{-6})			$^{232}Th/^{238}U$	Ratio/1sigma						Age(Ma)/1sigma						Concord-Ance
	Th^{232}	U^{238}	Pb		$^{207}Pb/^{206}Pb$	(1σ)	$^{207}Pb/^{235}U$	(1σ)	$^{206}Pb/^{238}U$	(1σ)	$^{207}Pb/^{206}Pb$	(1σ)	$^{207}Pb/^{235}U$	(1σ)	$^{206}Pb/^{238}U$	(1σ)	
18CS-01-01	472.266	1574.096	72.42149	0.013808	0.047655	0.0008	0.120187	0.001995	0.018313	0.000146	83.425	42.5875	115.2415	1.807961	116.9831	0.926284	98%
18CS-01-02	141.3717	484.5583	22.00171	0.045451	0.048555	0.0015	0.121936	0.00375	0.018278	0.00019	127.865	77.7675	116.8258	3.394018	116.7661	1.200294	99%
18CS-01-03	112.2566	536.1725	19.61299	0.050986	0.049164	0.0013	0.119702	0.00321	0.017678	0.000192	166.75	60.175	114.8018	2.911336	112.9642	1.216271	98%
18CS-01-04	899.1592	2338.122	124.8586	0.008009	0.050982	0.0008	0.124976	0.00207	0.017779	0.000154	238.955	32.4025	119.5729	1.868731	113.6043	0.977245	94%
18CS-01-05	356.5362	1040.718	54.79529	0.018249	0.052719	0.0016	0.132715	0.004329	0.018234	0.000151	316.725	72.215	126.5343	3.880344	116.484	0.958009	91%
18CS-01-06	472.2965	1369.869	85.68922	0.011670	0.064813	0.0010	0.188885	0.003809	0.021084	0.000231	768.52	227.775	175.6776	3.252768	134.5061	1.460831	73%
18CS-01-07	319.7484	1300.715	55.01009	0.018178	0.050943	0.0010	0.129436	0.002559	0.018442	0.000168	238.955	46.2875	123.5911	2.300364	117.8035	1.064738	95%
18CS-01-08	318.2334	745.503	40.82856	0.024492	0.049643	0.0011	0.125423	0.00292	0.018346	0.000172	188.97	53.695	119.977	2.634657	117.1929	1.088303	97%
18CS-01-09	456.0166	1273.565	63.5378	0.015738	0.047779	0.0007	0.121213	0.002327	0.018359	0.000169	87.13	37.0325	116.1708	2.107511	117.278	1.069839	99%
18CS-01-10	1128.152	1975.129	133.8668	0.007470	0.053944	0.0008	0.134592	0.002222	0.01809	0.000142	368.57	67.585	128.2157	1.988704	115.5733	0.898444	89%
18CS-01-11	651.9653	1577.681	86.08203	0.011616	0.04734	0.0008	0.120433	0.002483	0.018459	0.000219	64.91	44.44	115.4646	2.249844	117.9099	1.387347	97%
18CS-01-12	898.3659	2328.23	124.3088	0.008044	0.053159	0.0012	0.128194	0.003227	0.01747	0.000168	344.5	51.8475	122.4733	2.904556	111.6455	1.061992	90%
18CS-01-13	788.5011	1671.612	103.3689	0.009674	0.050465	0.0009	0.132032	0.002546	0.019059	0.000271	216.74	44.435	125.922	2.283971	121.7033	1.71218	96%
18CS-01-14	457.0531	1327.671	73.53172	0.013599	0.0583	0.0013	0.14635	0.003806	0.018185	0.000228	542.63	45.365	138.6842	3.371065	116.1738	1.443957	82%
18CS-01-15	586.9446	1444.842	83.7469	0.011940	0.051583	0.0011	0.138845	0.002916	0.019556	0.000181	333.39	53.6975	132.0148	2.599895	124.8503	1.146076	94%
18CS-01-16	1575.995	1557.102	167.3218	0.005976	0.052591	0.0011	0.134521	0.00305	0.018556	0.000209	322.28	52.7725	128.1519	2.729537	118.5211	1.322501	92%
18CS-01-17	332.9182	1256.546	57.44716	0.017407	0.051117	0.0011	0.138803	0.003422	0.019668	0.000177	255.62	53.695	131.9771	3.05101	125.558	1.116436	95%
18CS-01-18	228.5987	1247.068	48.28645	0.020709	0.048046	0.0008	0.136866	0.002847	0.020647	0.000245	101.94	42.59	130.249	2.542711	131.7423	1.54961	98%
18CS-01-19	823.7941	1948.765	108.8157	0.009189	0.048503	0.0007	0.130254	0.002346	0.019449	0.000167	124.16	37.0325	124.3262	2.107497	124.1746	1.058586	99%
18CS-01-20	660.8823	1571.493	90.21077	0.011085	0.048625	0.0008	0.133771	0.002641	0.019933	0.000173	131.57	42.5875	127.4804	2.365375	127.2304	1.094315	99%
18CS-01-21	420.7892	1471.985	65.17453	0.015343	0.047701	0.0007	0.129111	0.002201	0.019633	0.000218	83.425	32.405	123.2988	1.97913	125.3364	1.377201	98%
18CS-01-22	156.5514	737.5803	28.84884	0.034663	0.049044	0.0011	0.137333	0.003077	0.02037	0.000237	150.085	58.325	130.6653	2.74676	129.9958	1.496093	99%
18CS-01-23	278.1643	755.6438	39.65799	0.025215	0.048441	0.0010	0.134069	0.003091	0.020085	0.00019	120.46	47.2175	127.7476	2.767645	128.1955	1.199058	99%
18CS-01-24	431.8319	1024.951	53.68663	0.018626	0.049214	0.0010	0.134265	0.003093	0.019769	0.000208	166.75	48.14	127.9231	2.768721	126.1937	1.315243	98%
18CS-01-25	542.2902	1066.892	70.0304	0.014279	0.052021	0.0011	0.140767	0.003797	0.019538	0.000195	287.1	54.625	133.7268	3.379782	124.7336	1.231627	93%
18CS-27-01	417.35	574.775	50.98	0.01962	0.0529	0.0015	0.1409	0.004	0.0193	0.0002	324.13	62.96	133.83	3.4989	123.3	1.0768	91%
18CS-27-02	206.17	391.128	28.38	0.03524	0.0512	0.0016	0.1421	0.005	0.02008	0.0002	255.62	74.06	134.91	4.6077	128.2	1.4283	94%
18CS-27-03	401.89	638.972	101	0.0099	0.1234	0.004	0.3537	0.013	0.02063	0.0002	2005.3	58.02	307.52	9.8711	131.6	1.3585	19%
18CS-27-04	1575.8	989.838	173.7	0.00576	0.0512	0.001	0.142	0.003	0.0201	0.0002	250.07	43.51	134.86	2.8708	128.3	1.4591	94%
18CS-27-05	280.19	391.704	37.03	0.027	0.0471	0.0014	0.1315	0.004	0.02029	0.0002	53.8	66.66	125.46	3.4071	129.5	1.3713	96%
18CS-27-06	686.96	449.545	74.51	0.01342	0.0541	0.002	0.1504	0.006	0.02018	0.0003	375.98	85.18	142.27	5.1808	128.8	2.026	90%
18CS-27-07	147.02	323.972	21.98	0.0455	0.0539	0.0025	0.1567	0.007	0.02123	0.0003	364.87	110.2	147.84	5.9398	135.4	1.7638	91%
18CS-27-08	406.1	671.044	56.57	0.01768	0.0542	0.0018	0.1477	0.004	0.01982	0.0003	388.94	75.92	139.88	3.9736	126.5	1.7396	89%
18CS-27-09	708.24	531.91	321.2	0.00311	0.3259	0.0103	1.3329	0.067	0.02948	0.0008	3598.1	48.22	860.14	29.004	187.3	4.9803	−29%
18CS-27-10	497.94	520.623	65.74	0.01521	0.0583	0.0022	0.1574	0.006	0.0196	0.0002	542.63	83.32	148.41	5.2574	125.1	1.4374	82%
18CS-27-11	495.14	480.37	92	0.01087	0.1088	0.0038	0.3258	0.011	0.02177	0.0003	1788.9	64.51	286.38	8.7345	138.9	1.5795	30%
18CS-27-12	344.03	316.666	36.02	0.02776	0.053	0.0023	0.1289	0.006	0.01769	0.0002	327.84	99.99	123.07	5.0352	113	1.1601	91%

Table 2. *Cont.*

Samples	Content(10^{-6})			^{232}Th/^{238}U	Ratio/1sigma							Age(Ma)/1sigma						Concord-Ance
	Th^{232}	U^{238}	Pb		^{207}Pb/^{206}Pb	(1σ)	^{207}Pb/^{235}U	(1σ)	^{206}Pb/^{238}U	(1σ)	^{207}Pb/^{206}Pb	(1σ)	^{207}Pb/^{235}U	(1σ)	^{206}Pb/^{238}U	(1σ)	
18CS-27-13	403.35	361.82	52.96	0.01888	0.0542	0.002	0.1606	0.006	0.02155	0.0003	388.94	81.47	151.21	4.9154	137.4	1.7486	90%
18CS-27-14	329.94	493.923	39.33	0.02542	0.0531	0.0018	0.146	0.005	0.0199	0.0003	344.5	74.07	138.41	4.6296	127	1.5806	91%
18CS-27-15	187.46	249.474	183.4	0.00545	0.0736	0.0012	1.6201	0.025	0.15983	0.0015	1031.5	31.95	978.03	9.6871	955.8	8.5274	97%
18CS-27-16	549.1	370.138	60.88	0.01643	0.0504	0.0021	0.1374	0.006	0.01976	0.0002	213.04	91.65	130.71	5.3468	126.2	1.4083	96%
18CS-27-17	234.19	784.505	35.75	0.02798	0.0493	0.0014	0.1238	0.003	0.01827	0.0002	161.2	66.66	118.49	3.1387	116.7	1.1585	98%
18CS-27-18	391.19	527.786	54.24	0.01844	0.0569	0.0016	0.1581	0.005	0.02012	0.0003	487.08	58.33	149	4.3331	128.4	1.6889	85%
18CS-27-19	82.005	878.631	237.1	0.00422	0.1403	0.0018	2.6566	0.068	0.13684	0.0026	2231.5	22.07	1316.5	18.848	826.8	14.47	54%
18CS-27-20	506.93	324.127	814.7	0.00123	0.1664	0.002	8.0428	0.221	0.3497	0.0078	2521.3	20.83	2235.8	24.777	1933	37.16	85%
18CS-27-21	380.77	680.57	49.65	0.02014	0.0495	0.0014	0.1316	0.004	0.01929	0.0002	172.31	66.66	125.52	3.4689	123.2	1.0877	98%
18CS-27-22	633.59	522.519	72.68	0.01376	0.053	0.0024	0.1455	0.007	0.01989	0.0003	327.84	97.21	137.93	6.069	127	1.9475	91%
18CS-27-23	960.9	598.913	99.37	0.01006	0.0504	0.0014	0.1299	0.003	0.01871	0.0002	213.04	66.66	124.04	3.129	119.5	1.102	96%
18CS-27-24	366.76	391.313	45.96	0.02176	0.066	0.0028	0.1724	0.007	0.01902	0.0002	809.26	88.89	161.52	5.6966	121.4	1.503	71%
18CS-27-25	1361.7	1031.35	149.1	0.00671	0.0483	0.0009	0.1301	0.003	0.01951	0.0002	122.31	44.44	124.18	2.4746	124.5	1.3435	99%

Table 3. Lu–Hf isotopic analysis result of the Chishan alkaline complex.

Samples	^{176}Yb/^{177}Hf	^{176}Lu/^{177}Hf	^{176}Hf/^{177}Hf	2σ	εHf(0)	εHf(t)	T$_{DM}$(Ma)	T$_{DM}$^{C}(Ma)	$f_{Lu/Hf}$
18CS01-01	0.013407	0.000586	0.282297	0.000018	−16.8	−14.1	1334	2078	−0.98
18CS01-03	0.015576	0.000718	0.282282	0.000029	−17.3	−14.6	1359	2112	−0.98
18CS01-08	0.013584	0.000573	0.282248	0.000025	−18.5	−15.8	1400	2186	−0.98
18CS01-13	0.011379	0.000519	0.282322	0.000019	−15.9	−13.2	1296	2020	−0.98
18CS01-16	0.022202	0.000943	0.282299	0.000021	−16.7	−14.0	1343	2074	−0.97
18CS-27-02	0.020686	0.001017	0.282241	0.000018	−18.8	−16.1	1426	2203	−0.97
18CS-27-05	0.022520	0.001098	0.282317	0.000023	−16.1	−13.4	1324	2036	−0.97
18CS-27-15	0.040275	0.001565	0.282200	0.000025	−20.2	−17.6	1505	2298	−0.95
18CS-27-21	0.017104	0.000765	0.282055	0.000022	−25.4	−22.7	1675	2617	−0.98
18CS-27-25	0.022610	0.000942	0.282062	0.000024	−25.1	−22.4	1672	2601	−0.97

The following parameters were applied to the calculation: (^{176}Lu/^{177}Hf)$_{CHUR}$ = 0.0332, (^{176}Hf/^{177}Hf)$_{CHUR,0}$ = 0.282772 [53]; (^{176}Lu/^{177}Hf)$_{DM}$ = 0.0384, (^{176}Hf/^{177}Hf)$_{DM,0}$ = 0.28325 [54]; ^{176}Lu decay constant λ = 1.867 × 10^{-11} a^{-1} [52].

6. Discussion

6.1. Rock Ages

Zircon U–Pb dating analysis demonstrated that in the Chishan alkaline complex, the quartz syenite formed at 125.8 ± 1.2 Ma, and the alkali granite formed at 127.3 ± 1.0 Ma, which concurs with the ages of the quartz syenite (122.4 ± 2.0 Ma) and aegirine quartz syenite porphyrite (130.1 ± 1.4 Ma) yielded by Liang et al. [35] through LA-MC-ICPMS zircon U–Pb measurement. This finding suggests that the rock complex formed in Early Cretaceous, which concurs with the ages of the late Mesozoic magmatic rocks extensively distributed in Eastern China [61]. According to Lan et al. [32], the dolomite Rb–Sr age of the Chishan REE deposit ores is 119 ± 1.4 Ma, which is a bit later than the emplacement of this alkaline rock complex, suggesting a close genetic connection between this alkaline rock system and mineralization of the REE ores in this area.

6.2. Magmatic Sources and Evolution

The Mesozoic Era witnessed three major geodynamic events in the North China lithosphere: collisional orogeny, tectonic regime transition, and mass lithospheric thinning [11]. Numerous alkaline intrusions emerged in the North China Craton [23], giving rise to various types of metal deposits. As the Chishan REE deposit represents a typical example of alkaline rock mineralization throughout the North China Craton [25], studying the material sources of alkaline rocks bears significance in the studies of both geodynamic mechanisms and metallogenic relationships.

The quartz syenite and alkali granite in the Chishan alkaline complex are rich in alkali (with a Na_2O + K_2O content of 7.62–10.94%), high in potassium (with a Na_2O/K_2O content of 0.28–0.89), metaluminous (with an aluminum saturation index (A/CNK) of 0.75–1.00), and poor in Ti, Fe, Mg, and Mn, which are typical features of an alkaline rock series. The rocks feature strong LREE/HREE fractionation; the REEs display a right-inclined distribution pattern with weak positive and negative Eu anomalies (the δEu is 1.16–4.15). For the similar compositions and close spatial relationship, we regard that quartz syenite and alkali granite are the members of one intrusion (Chishan alkaline complex).

On the Ta/Yb–Th/Yb diagram (Figure 8) [62], the absolute majority of the rock samples from the Chishan alkaline complex fall within or near the enriched mantle zone, suggesting a close connection between the material sources of the Chishan alkaline rocks and enriched mantle. As reported by Yan et al. [27], the aegirine syenite features a Rb content of 170.46×10^{-6}–4550.67×10^{-6}, a $^{87}Rb/^{86}Sr$ ratio of 2.6800–0.0500, and a $^{87}Sr/^{86}Sr$ ratio of 0.71176–0.70780. These findings agree with the geochemistry of mantle-derived magmas and are close to the features of an EMI (enriched mantle I)-type deposit. The rocks must have originated from highly enriched mantle-derived materials.

Zircons are well representative of the isotopic composition of magma sources as their high stability has guarded their Lu–Hf isotopes from magmatic differentiation and late weathering processes. For this reason, zircon Lu–Hf isotopes are often used for tracing magma sources and magmatic evolution [55]. Zircon εHf(t) reflects the composition of a magma source. A positive εHf(t) represents a depleted mantle or a young, new crust overgrowing a depleted mantle and indicates remelting from the new crust. The negative εHf(t) values of the studied samples might indicate Chishan alkaline complex was originated from an enriched mantle source or thickened lower crust or a mixed source region [55,63,64]. Given that the quartz syenite and alkali granite of the Chishan Alkaline Complex share the same zircon Hf isotopic composition with relatively low εHf(t) values (−17.60–−13.19), the Hf isotopic compositions fall above the 2.5 Ga continental crust evolutional line on the diagram for T-εHf(T) (Figure 9), and represent a mixing of mantle sources with remobilized old continental crust. Also, the results are consistent with those of the contemporary Longbaoshan alkaline complex (εHf(t) = −19.20 to −14.00) [65], which is believed to be produced by partial melting of the enriched lithospheric mantle (EMI) [66–69]. Further, there showed some inherited zircons in the alkali granite. Specially, the ages and Hf isotopes composition of inherited zircon cores could be an important and critical clue to identify the mantle source of rocks. Although, the ages and Hf isotopes of the inherited zircons cores in Chishan complex were not measured this time, the late Archean ages (2.51–2.64 Ga) with positive εHf(t) values (−0.2 to −6.2) derived from inherited zircon cores of the Longbaoshan complex [65] suggest that the ancient crust of the North China Craton was also involved in the formation of those alkaline complex. Thus, it is reasonable to deduce that the Chishan alkaline complex originated from partial of the EMI-type lithospheric mantle with assimilation of ancient crustal materials.

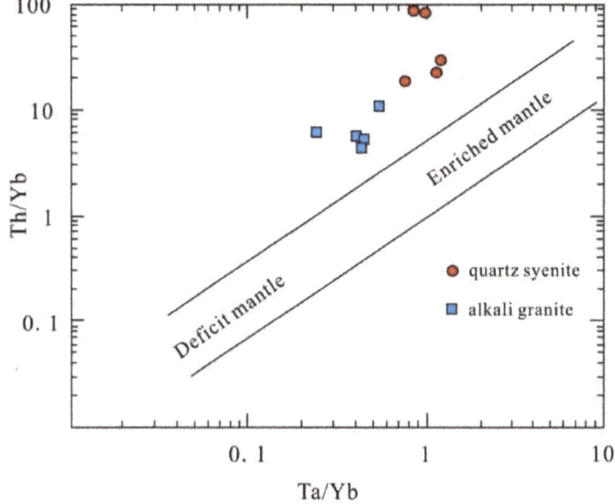

Figure 8. Ta/Yb–Th/Yb diagram of the Chishan alkaline complex. The diagram is from [62].

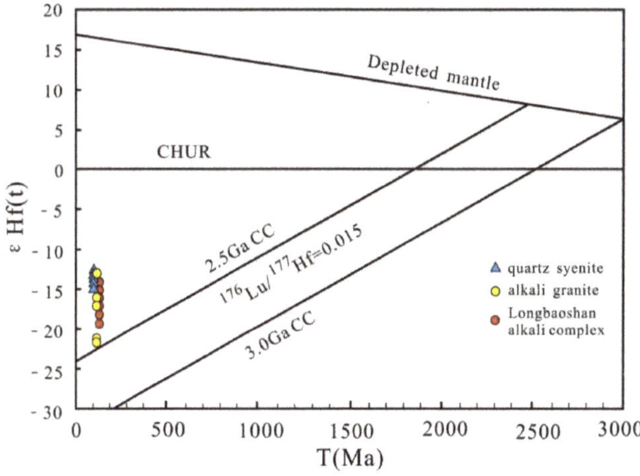

Figure 9. Zircon Hf isotope diagram of Chishan alkali complex. Other data sources: Longbaoshan alkaline complex [65].

Furthermore, in terms of Nd isotopic composition, the εNd(t) value of the Chishan alkaline rocks is −8.7 to −8.11 [70], the εNd(t) value of the Neoarchean gneiss basement is −25 [70], and the εNd(t) of the Early Cretaceous (125–127 Ma) basic magmatic rocks (e.g., the Fangcheng basalt and Yinan gabbro) in Western Shandong, which directly originate from the partial melting of an enriched mantle, ranges from −15.4 to −12.6 [67,69]. Such a condition indicates that the Chishan alkaline rocks could never have derived from the crust. Rather, these rocks must have been associated with an enriched mantle. The εNd(t) value of the perovskite and monazite in the Chishan REE ore veins is −10.1

to −8.2 [32], which agrees with the Nd isotopic composition of the alkaline rocks, confirming that the alkaline rocks are of the same origin as the REE ores.

6.3. Tectonic Setting

Intense tectonic activities, including three major geodynamic events (collisional orogeny, mass lithospheric thinning, and tectonic regime transition within the North China Craton since the Mesozoic), have resulted in extensive magmatic and metallogenic processes [11,67]. During the Triassic, the collision between the North China and Yangtze Plates gave rise to the Dabie-Sulu ultrahigh-pressure metamorphic zone. In the Early Jurassic (180–130 Ma), the lithosphere started thinning. The Yangtze Plate underplated the North China Plate after a deep subduction. Mantle-derived magmas reached the Tanlu fault zone to the crust–mantle boundary. Crustal thickening and remelting triggered a string of magmatic processes, giving rise to high K calc-alkaline to high K alkaline rocks. The lithospheric mantle changed to an EMII-type deposit [67]. In the Early Cretaceous (130–90 Ma), when Western Shandong was in a backarc extensional environment caused by the subduction of the Yangtze Plate toward the North China Plate, the lithosphere thinned enormously, and extensive enriched mantle melting occurred. The tectonic stress regime also changed from extrusion toward extension. Mass uptrusion of asthenospheric mantle materials, along with mass diagenetic processes, caused the formation of late Mesozoic alkaline rocks [11,65,71–74].

From the $\lg(CaO/Na_2O + K_2O) - SiO_2$ discrimination diagram (Figure 10a) [75], the rock projection points of the Chishan alkaline complex mostly fall within the extensional alkaline–alkalilime rock zone, demonstrating that the rocks formed in an extensional tectonic setting. On the $\omega(Rb) - \omega(Y + Nb)$ discrimination diagram (Figure 10b) [76], all the samples fall within the post-collisional granite zone, and the diagenetic background is an active continental margin. These findings also confirm that the Chishan alkaline complex formed in an extensional tectonic setting. The source region of the rocks is closely associated with the tectonic setting of their origin. The source region of post-collisional granite normally results from the mantle contaminated by a part of the crustal materials. The negative Nb, Ta, and Ti anomalies in the rocks are comparative to those of an island arc tectonic setting [77], concurring with the mass lithospheric detachment and thinning background in Eastern China during the Mesozoic [78–81]. This condition suggests that the formation of the Chishan alkaline complex was controlled by the lithospheric thinning of the North China Craton; it must have originated from a backarc extensional setting caused by plate subduction.

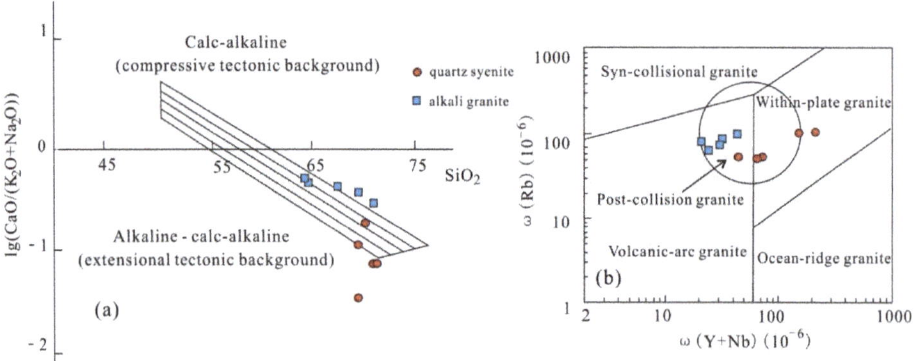

Figure 10. $\lg(CaO/(Na_2O+K_2O)) - SiO_2$ (a) and Rb − (Y + Nb) (b) discrimination diagrams of the Chishan alkaline complex. (a) is from [75], (b) is from [76].

7. Conclusions

(1) The Chishan alkaline complex formed in the Early Cretaceous. The quartz syenite and alkali granite formed at 125.8 ± 1.2 and 127.3 ± 1.0 Ma, respectively, and are closely associated with the mineralization of the Weishan REE deposit.

(2) The Chishan alkaline complex shows the geochemistry of mantle-derived magmas features and originated from partial of the EMI-type lithospheric mantle with assimilation of ancient crustal materials. The complex must be of the same origin as the REE deposit.

(3) The formation of the Chishan alkaline complex developed in an extensional setting that resulted from subduction of the North China and Yangtze Plates, and lithospheric thinning and upwelling in the eastern the North China Craton.

Author Contributions: Conceptualization: P.W.; formal analysis: Y.Z. and Y.S.; investigation: Q.L., L.Y., and N.C.; data curation: Z.L. and K.G.; writing—original draft preparation: P.W.; writing—review and editing: X.Y., D.L., and L.Y.

Funding: This study was financially supported by the National Natural Science Foundation of China, grant nos. 41503038, 41140025, 41672084, and 41372086; Postdoctoral Innovation Project of Shandong Province, grant no. 201703091; Special fund for "Taishan scholars" project in Shandong Province; the Key R and D Program of Shandong Province, Grant No. 2017CXGC1601, 2017CXGC1602, 2017CXGC1603; Key R and D Program of China, grant nos. 2016YFC0600105-04, 2016YFC0600606; and the Shandong Provincial Key Laboratory of Depositional Mineralization and Sedimentary Mineral, grant no. DMSM2018028.

Acknowledgments: The authors are grateful to manager XinZhu Zhang and Dejian Li of the Shandong Weishan Lake REE Co. Ltd. for permission to access the mining project. Tingbao Han and Guangsuo Cheng are thanked for their help during the connection with REE Co. Ltd. Xiaowei Li (China University of Geosciences Beijing) are thanked for his help during the revising process.

Conflicts of Interest: The authors declare no conflict of interest.

References

1. Deng, J.F.; Mo, X.X.; Zhao, H.L.; Luo, Z.H.; Du, Y.S. Lithospheric root/de-rooting and activation. *J. Grad. Sch. China Univ. Geosci.* **1994**, *8*, 350–356, (In Chinese with English abstract).

2. Deng, J.F.; Zhao, H.L.; Mo, X.X.; Wu, Z.X.; Luo, Z.H. *Continental Roots-Plume Tectonics of China—Key to the Continental Dynamics*; Geological Publishing House: Beijing, China, 1996; pp. 1–110. (In Chinese)

3. Wu, F.Y.; Sun, D.Y.; Lin, Q. Petrogenesis of the Phanerozoic granites and crustal growth in the Northeast China. *Acta Pet. Sin.* **1999**, *15*, 181–189, (In Chinese with English abstract).

4. Wu, F.Y.; Sun, D.Y. The Mesozoic magmatism and lithospheric thinning in the eastern China. *J. Chang. Univ. Sci. Tech.* **1999**, *29*, 313–318, (In Chinese with English abstract).

5. Shao, J.A.; Liu, F.T.; Chen, H.; Han, Q.J. Relationship between Mesozoic magmatism and subduction in Da Hinggan-Yanshan Area. *Acta Geol. Sin.* **2001**, *75*, 56–63, (In Chinese with English abstract).

6. Zhang, Q.; Wang, Y.; Yang, J.H.; Wang, Y.L.; Zhao, T.P.; Guo, G.J. The Characteristics and tectonic-metallogenetic significances of the adakites in the Yanshanian Period from eastern China. *Acta Pet. Sin.* **2001**, *17*, 236–244, (In Chinese with English abstract).

7. Zhang, Q.; Wang, Y.; Wang, Y.L. Preliminary study on the components of the lower crust in east China plateau during Yanshanian period: Constraints on Sr and Nd isotopic compositions of adakite-like rocks. *Acta Pet. Sin.* **2001**, *17*, 505–513, (In Chinese with English abstract).

8. Zhai, M.G.; Fan, Q.C. Mesozoic replacement of bottom crust in North China Craton: An orogenic mantle-crust interaction. *Acta Pet. Sin.* **2002**, *18*, 1–8, (In Chinese with English abstract).

9. Hu, S.X.; Zao, Y.Y.; Hu, Z.H.; Guo, J.C.; Xu, B. Evolution and decelopment of tectonics and magmatism at the active continental margin of the East China(E106°) during Mesozoic and Cenozoic. *Acta Pet. Sin.* **1994**, *10*, 370–381.

10. Chen, Y.J.; Guo, Y.J.; Guo, G.J.; Li, X. Metallogenic geodynamic background of Mesozoic gold deposits in the huabei craton granite greenstone terrane. *Sci. China Earth Sci.* **1998**, *28*, 35–40.

11. Mao, J.W.; Zhang, Z.H.; Yu, J.J.; Wang, Y.T.; Niu, B.G. Geodynamics of Mesozoic large-scale mineralization in north China and its adjacent areas: implications from precise dating of metallic deposit. *Sci. China Earth Sci.* **2003**, *4*, 289–299.

12. Li, J.W.; Bi, S.J.; Selby, D.; Chen, L.; Vasconcelos, P.; Thiede, D.; Zhou, M.F.; Zhao, X.F.; Li, Z.K.; Qiu, H.Z. Giant Mesozoic gold procinces related to the destruction of the North China carton. *Earth Planet Sci. Lett.* **2012**, *349–350*, 26–37. [CrossRef]

13. Zhai, M.; Santosh, M. Metallogeny of the North China carton: Link with secular changes in the evolving Earth. *Gondwana Res.* **2013**, *24*, 275–297. [CrossRef]

14. Chen, Y.J.; Franco, P.; Lai, Y.; Li, C. Metallogenic time and tectonic setting of the Jiaodong gold province, eastern China. *Acta Pet. Sin.* **2004**, *20*, 907–922.

15. Mao, J.W.; Pirajno, F.; Xiang, J.F.; Gao, J.J.; Ye, H.S.; Li, Y.F.; Guo, B.J. Mesozoic molybdenum deposits in the east Qinling-Dabie oro-genic belt: Characteristics and tectonic settings. *Ore Geol. Rev.* **2011**, *43*, 264–293. [CrossRef]

16. Goldfarb, R.J.; Santosh, M. The dilemma of the Jiaodong gold deposits: Are they unique? *Geosci. Front.* **2014**, *5*, 139–153. [CrossRef]

17. Yang, J.H.; Zhou, X.H. Rb–Sr, Sm–Nd and Pb isotope systematics of pyrite: implications for the age and genesis of lode gold deposits. *Geology* **2001**, *29*, 711–714. [CrossRef]

18. Chen, Y.J.; Pirajno, F.; Qi, J.P. Origin of gold metallogeny and sources of ore-forming fluids, in the Jiaodong province, eatern China. *Int. Geol. Rev.* **2005**, *47*, 530–549. [CrossRef]

19. Li, J.J.; Luo, Z.K.; Liu, X.Y.; Xu, W.D.; Chen, A.S. The zircon Shrimp U–Pb age of the granite–porphyry veins after mineralization in linglong gold deposit restricts the evolution of the Jiaobei terrains. *Earth Sci. Front.* **2005**, *12*, 317–324.

20. Wu, F.Y.; Lin, J.Q.; Wilde, S.A.; Zhang, X.; Yang, J.H. Nature and significance of the Early Cretaceous giant igneous event in eastern China. *Earth Planet. Sci. Lett.* **2005**, *233*, 103–119. [CrossRef]

21. Hou, M.L.; Jiang, S.Y.; Jiang, Y.H.; Li, H.F. S–Pb isotope geochemistry and Rb–Sr geochronology of the Penglai gold field in the eastem Shangdong province. *Acta Pet. Sin.* **2006**, *22*, 2525–2533, (In Chinese with English abstrcat).

22. Jiang, S.Y.; Dai, B.Z.; Jiang, Y.H.; Zhao, H.X.; Hou, M.L. Jiaodong and Xiaoqinling: Two orogenic gold provinces formed in different tectonic settings. *Acta Pet. Sin.* **2009**, *25*, 2727–2738, (In Chinese with English abstract).

23. Lan, T.G.; Fan, H.R.; Santosh, M.; Hu, F.F.; Yang, K.F.; Yang, Y.H. Early Jurassic high-K calc-alkaline and shoshonitic rocks from the Tongshi intrusive complex, eastern North China Craton: Implication for crust-mantle interaction and post-collisional magmatism. *Lithos* **2012**, *140–141*, 183–199. [CrossRef]

24. Tian, J.X.; Li, X.X.; Song, Z.Z.; Liu, H.D.; Huang, Y.B.; Zhu, D.C. Environment, formation age and material sources of Mesozoic Gold deposits in Western Shandong: A Synthesis. *Acta Geol. Sin.* **2015**, *89*, 1530–1537.

25. Tian, J.X.; Zhang, R.T.; Fan, Y.C.; Li, X.Z.; Xu, H.Y.; Wang, B.Y. Geological characteristics and relation with rare earth elements of alkalic complex in Chishan of Shandong Province. *Shandong Geol.* **2002**, *18*, 21–25.

26. Kong, Q.Y.; Li, J.K.; Yu, X.F. *Shandong Deposit*; Shandong Science & Technology Press: Jinan, China, 2006; pp. 1–902.

27. Yan, G.H.; Cai, J.H.; Ren, K.X.; Mu, B.L.; Chu, Z.Y. Nd, Sr and Pb isotopic geochemistry of late-Mesozoic alkaline-rich intrusions from the Tanlu Fault zone: evidence of the magma soucre. *Acta Pet. Sin.* **2008**, *24*, 1223–1236.

28. Yu, X.F.; Tang, H.S.; Han, Z.Z.; Li, C.Y. Geological characteristics and origin of rare earth elements deposits related with alkaline rock in the Chishan-Longbaoshan area, Shandong Province. *Acta Geol. Sin.* **2010**, *84*, 407–417.

29. Yuan, Z.X.; Li, J.K.; Wang, D.H.; Zheng, G.D.; Lou, D.B.; Chen, Z.H.; Zhao, Z.; Yu, Y. *Metallogenetic Regularity of Rare Earth Deposits in China*; Geological Publishing House: Beijing, China, 2012; pp. 1–116.

30. Wang, J.F.; Sun, M.T.; Du, X.B.; Gan, Y.J.; Zhang, G.Q.; Wang, Z.L. Geological characteristics and prospecting potentiality of Xishan rare earth deposit in Shandong Province. *Shandong Land Res.* **2016**, *32*, 32–40.

31. Li, J.K.; Yuan, Z.X.; Bai, G.; Chen, Y.C.; Wang, D.H.; Ying, L.J.; Zhang, J. Ore-forming fluid evolvement and its controlling to REE (AG) mineralizing in the Weishan deposit, Shandong. *J. Min. Pet.* **2009**, *29*, 60–68.

32. Lan, T.G.; Fan, H.R.; Hu, F.F.; Yang, K.F.; Wang, Y. Genesis of the Weishan REE deposit, Shandong Province: Evidences from Rb–Sr isochron age, LA-MC-ICPMS Nd isotopic compositions and fluid inclusions. *Geochimica* **2011**, *40*, 428–442.

33. Zhou, W.W.; Cai, J.H.; Yan, G.H. The geochemical characteristics and geological significance of alkaline complex in Chishan of Shandong Province. *Northwestern Geol.* **2013**, *46*, 93–105.

34. Yu, X.F.; Tang, H.S.; Li, D.P. Study on the Mineralization of Chishan Rare Earth Element Deposit Related to Alkaline Rock, Shandong Province. *Acta Geol. Sin. (Engl. Ed.)* **2014**, *88*, 480–481. [CrossRef]

35. Liang, Y.W.; Lai, Y.; Hu, H.; Zhang, F. Zircon U–Pb Ages and Geochemical Characteristics Study of Syenite from Weishan REE Deposit, Western Shandong. *Acta Sci. Nat. Univ. Pekin* **2017**, *53*, 652–666.

36. Zhu, G.; Liu, G.S.; Niu, M.L.; Song, C.Z.; Wang, D.X. Transcurrent movement and genesis of the Tan–Lu Fault Zone. *Geol. Bull. China* **2003**, *22*, 200–207, (In Chinese with English abstract).

37. Zhang, X.M.; Zhang, Y.Q.; Ji, W. Fault distribution patterns of the Luxi Block, Shandong, and Mesozoic sedimentary–magmatic–structural evolution sequence. *J. Geomech.* **2007**, *13*, 163–172, (In Chinese with English abstract).

38. Chen, L.; Tao, W.; Zhao, L.; Zhen, T.Y. Distinct lateral variation of lithospheric thickness in the Northeastern North China Craton. *Earth Planet. Sci. Lett.* **2008**, *267*, 56–68. [CrossRef]

39. Zhao, G.; Wilde, S.A.; Guo, J.H.; Cawood, P.A.; Sun, M.; Li, X.P. Single zircon grains record two Paleoproterozoic collisional events in the North China Craton. *Precambrian Res.* **2010**, *177*, 266–276. [CrossRef]

40. Zhao, G.C.; Wilde, S.A.; Cawood, P.A.; Sun, M. Archean blocks and their boundaries in the North China Craton: Lithological, geochemical, structural and P–T pathconstraints and tectonic evolution. *Precambrian Res.* **2001**, *107*, 45–73. [CrossRef]

41. Li, J.; Jin, A.; Hou, G. Timing and implications for the late Mesozoic geodynamic settings of eastern North China Craton: Evidences from K-Ar dating age and sedimentary-structural characteristics records of Lingshan Island, Shandong Province. *J. Earth Syst. Sci.* **2017**, *126*, 117. [CrossRef]

42. Wang, S.J.; Li, X.P.; Schertl, H.P.; Feng, Q.D. Petrogenesis of early cretaceous andesite dykes in the Sulu orogenic belt, eastern China. *Mineral. Petrol.* **2019**, *113*, 77–97. [CrossRef]

43. Li, R.; Albert, N.N.; Yun, M.; Meng, Y.; Du, H. Geological and Geochemical Characteristics of the Archean Basement-Hosted Gold Deposit in Pinglidian, Jiaodong Peninsula, Eastern China: Constraints on Auriferous Quartz-Vein Exploration. *Minerals* **2019**, *9*, 62. [CrossRef]

44. Han, C.; Han, M.; Jiang, Z.X.; Han, Z.Z.; Li, H.; Song, Z.G.; Zhong, W.J.; Liu, K.X.; Wang, C.H. Source analysis of quartz from the Upper Ordovician and Lower Silurian black shale and its effects on shale gas reservoir in the Southern Sichuan Basin and its periphery, China. *Geol. J* **2019**, *54*, 439–449. [CrossRef]

45. Wang, J.; Wang, X.; Liu, J.; Liu, Z.; Zhai, D.; Wang, Y. Geology, geochemistry, and geochronology of gabbro from the Haoyaoerhudong gold deposit, Northern Margin of the North China Craton. *Minerals* **2019**, *9*, 63. [CrossRef]

46. Kong, Q.Y.; Zhang, T.Z.; Yu, X.F.; Xu, J.X.; Pan, Y.L.; Li, X.S. *Deposits in Shandong Province*; Shandong Science and Technology Press: Jinan, China, 2006; pp. 1–902. (In Chinese)

47. Song, B.; Zhang, Y.H.; Wan, Y.S.; Jian, P. Mount Making and procedure of the SHRIMP Dating. *Geol. Rev.* **2002**, *48*, 26–30, (In Chinese with English abstract).

48. Yuan, H.; Gao, S.; Liu, X.; Li, H.; Günther, D.; Wu, F. Accurate U–Pb age and trace element determinations of zircon by laser ablation-inductively coupled plasma-mass spectrometry. *Geostand. Geoanal. Res.* **2010**, *28*, 353–370. [CrossRef]

49. Liu, Y.S.; Hu, Z.C.; Gao, S.; Xu, J.; Gao, C.G.; Chen, H.H. In suit analysis of major and trace elements of anhydrous minerals by LA-ICP-MS without applying an internal standard. *Chem. Geol.* **2008**, *257*, 34–43. [CrossRef]

50. Ludwig, K.R. *User's Manual for Isoplot 3.70, A Geolocro-Nological Toolkit for Microsoft Excel*; Berkeley Geochronology Center Special Publication: Berkeley, CA, USA, 2003; Volume 4, pp. 25–32.

51. Geng, J.Z.; Li, H.K.; Zhang, J.; Zhou, H.Y.; Li, H.M. Zircon Hf isotope analysis by means of LA-MC-ICP-MS. *Geol. Bull. China* **2011**, *30*, 1508–1513, (In Chinese with English abstract).

52. Soderlund, U.; Patchett, P.J.; Vervoort, J.D.; Isachsen, C.E. The ^{176}Lu decay constant determined by Lu-Hf and U-Pb isotope systematics of Precambrian mafic instrusions. *Earth Planet Sci. Lett.* **2004**, *219*, 311–324. [CrossRef]

53. Bichert-Toft, J.; Albarede, F. The Lu–Hf geochemistry of chondrites and the evolution of the mantle-crust system. *Earth Planet. Sci. Lett.* **1997**, *148*, 243–258. [CrossRef]

54. Griffin, W.L.; Pearson, N.J.; Belousova, E.; Jackson, S.E.; Achterbergh, E.V.; O'Reilly, S.Y.; Shee, S.R. The Hf isotope composition of cratonic mantle: LAM-MC-ICPMS analysis of zircon megacrysts in kimberlites. *Geochim. Cosmochim. Acta.* **2000**, *64*, 133–147. [CrossRef]

55. Wu, F.Y.; Li, X.H.; Zheng, Y.F.; Gao, S. Lu–Hf isotopic systematics and their applications in petrology. *Acta Pet. Sin.* **2007**, *23*, 185–220, (In Chinese with English abstract).

56. Middlemost, E.A.K. Naming materials in the magma/igneous rock system. *Earth Sci. Rev.* **1994**, *37*, 215–224. [CrossRef]

57. Wright, J.B. A simple alkalinity ratio and its application to questions of nonorogenic granite genesis. *Geol. Mag.* **1969**, *106*, 370–384. [CrossRef]

58. Peccerillo, A.; Taylor, S.R. Geochemistry of Eocene calc-alkaline volcanic rocks from the Kastamonu, northern Turkey. *Contrib. Mineral. Petrol.* **1976**, *58*, 63–81. [CrossRef]

59. Sun, S.S.; McDonough, W.F. Chemical and isotopic systematics of oceanic basalts: implications for mantle composition and processes. In *Magmatism in the Oceanic Basalts*; Saunders, A.D., Norry, M.J., Eds.; Geological Society Special Publication: London, UK, 1989; pp. 313–345.

60. Vervootr, J.D.; Patchett, P.J.; Gehrels, G.E.; Nutman, A.P. Constraints on Earth differentiation from hafnium and neodymium. *Nature* **1996**, *379*, 624–627.

61. Guo, P. A Study on the Geodynamic Background of Mesozoic Gold Mineralization in Western Shandong. Ph.D. Thesis, China University of Geosciences (Beijing), Beijing, China, 2014.

62. Wilson, M. *Igneous Petrogenesis: Agobal Tectonic Approach*; Unwin Hyman: London, UK, 1989; pp. 1–466.

63. Kinny, P.D. Lu–Hf and Sm–Nd isotope systems in zircon. *Rev. Mineral. Geochem.* **2003**, *53*, 327–341. [CrossRef]

64. Sui, Z.M.; Ge, W.C.; Wu, F.Y.; Xu, X.C.; Zhang, J.H. Hf isotopic characteristics and geological significance of the Chahayan Pluton in northern Daxing´anling Mountains. *J. Jilin Univ. Earth Sci. Ed.* **2009**, *39*, 849–867, (In Chinese with English abstract).

65. Lan, T.G.; Fan, H.R.; Hu, F.F.; Tomkins, A.G.; Yang, K.F.; Liu, Y. Multiple crust-mantle interactions for the destruction of the north china craton: Geochemical and Sr–Nd–Pb–Hf isotopic evidence from the longbaoshan alkaline complex. *Lithos* **2011**, *122*, 87–106. [CrossRef]

66. Xu, Y.G. Thermo-tectonic destruction of the Archean lithospheric keel beneath the Sino-Korean Craton: Evidence, timing and mechanism. *Phys. Chem. Earth Part A Solid Earth Geod.* **2001**, *26*, 747–757. [CrossRef]

67. Xu, Y.G.; Huang, X.L.; Ma, J.L.; Wang, Y.B.; Iizuka, Y.; Xu, J.F. Crust-mantle interaction during the tectono-thermal reactivation of the North China Craton: constraints from SHRIMP zircon U–Pb chronology and geochemistry of Mesozoic plutons from western Shandong. *Contrib. Mineral. Petrol.* **2004**, *147*, 750–767. [CrossRef]

68. Xu, Y.G.; Ma, J.L.; Huang, X.L.; Iizuka, Y.; Chung, S.L.; Wang, Y.B.; Wu, X.Y. Early Cretaceous gabbroic complex from Yinan, Shandong Province: petrogenesis and mantle domains beneath the North China Craton. *Int. J. Earth Sci.* **2004**, *93*, 1025–1041. [CrossRef]

69. Zhang, H.F.; Sun, M.; Zhou, X.H.; Fan, W.M.; Zhai, M.G.; Yin, J.F. Mesozoic lithosphere destruction beneath the North China Craton: Evidence from major-trace-element and Sr–Nd–Pb isotope studies of Fangcheng basalts. *Contrib. Mineral. Petrol.* **2002**, *144*, 241–253. [CrossRef]

70. Xu, G.Z.; Zhou, R.; Wang, Y.F.; She, H.Q.; Li, B.; Du, B.M.; Song, M.C. The intrinsic factors causing the significant differences in Mesozoic mineralization between Jiaodong and Luxi areas. *Geoscience* **2002**, *16*, 9–18, (In Chinese with English abstract).

71. Li, Z.X. Collision between the North and South China Blocks: a Crustal-detachment model for the suturing in the region east of the Tanlu fault. *Geology* **1994**, *22*, 739–742. [CrossRef]

72. Hu, H.B.; Mao, J.W.; Niu, S.Y.; Li, Y.F.; Li, M.W. Geology and geochemistry of telluride-bearing Au deposits in the Pingyi area, Western Shandong, China. *Mineral. Petrol.* **2006**, *87*, 209–240. [CrossRef]

73. Chen, Y.J.; Zhai, M.G.; Jiang, S.Y. Significant achievements and open issues in study of orogenesis and metallogenesis surrounding the North China continent. *Acta Pet. Sin.* **2009**, *25*, 2695–2726.

74. Dong, G.; Santosh, M.; Li, S.; Shen, J.; Mo, X.; Scott, S. Mesozoic magmatism and metallogenesis associated with the destruction of the North China Craton: Evidence from U–Pb geochronology and stable isotope geochemistry of the Mujicun porphyry Cu–Mo deposit. *Ore Geol. Rev.* **2013**, *53*, 434–445. [CrossRef]

75. Brown, G.C. *Calcalkaline Intrusive Rocks: Their Diversity, Evolution and Relation to Volcanic Arcs*; John Wiley & Sons Inc.: Hoboken, NJ, USA, 1982; pp. 437–461.

76. Pearce, J.A. Source and settings of granite genesis. *Geol. Mag.* **1969**, *106*, 370–384.

77. Zhao, Z.H. Use the trace element diagram to discriminate tectonic settings. *Geotectonica Et Metalogenia* **2007**, *31*, 92–103.

78. Wu, F.Y.; Sun, D.Y.; Zhang, G.L.; Ren, X.W. Deep geodynamics of Yanshain Movement. *Geol. J. China Univ.* **2000**, *6*, 379–388.

79. Hong, D.W.; Wang, T.; Tong, Y.; Wang, X.X. Break-up of the North China Craton through Lithospheric Thinning. *Earth Sci. Front.* **2003**, *10*, 232–249.

80. Zhou, X.H.; Zhang, H.F. High chemical heterogeneity of lithospheric mantle and continental lithospheric transition in north China during Mesozoic. *Earth Sci. J. China Univ. Geosci.* **2006**, *31*, 8–13.

81. Ji, S.C.; Wan, Q.; Xu, Z.Q. Break-up of the North China Craton through Lithospheric Thinning. *Acta Geol. Sin.* **2008**, *82*, 175–193.

Article

Petrogenesis of the Early Cretaceous Tiantangshan A-Type Granite, Cathaysia Block, SE China: Implication for the Tin Mineralization

Ru-Ya Jia [1,2], Guo-Chang Wang [3,*], Lin Geng [4], Zhen-Shan Pang [1,2], Hong-Xiang Jia [5], Zhi-Hui Zhang [1,2], Hui Chen [1,2] and Zheng Liu [6]

[1] Development and Research Center, China Geological Survey, Beijing 100037, China; cgsjruya@163.com (R.-Y.J.); pzs927@163.com (Z.-S.P.); zzh1102114@126.com (Z.-H.Z.); chenhui_nju@163.com (H.C.)
[2] Technical Guidance Center for Mineral Resources Exploration, Ministry of Natural Resources of the People's Republic of China, Beijing 100120, China
[3] Yunnan Key Laboratory for Palaeobiology, Yunnan University, Kunming 650091, China
[4] China Geological Survey, Beijing 100037, China; genglin66@126.com
[5] School of Earth Sciences and Resources, China University of Geosciences, Beijing 100083, China; cugbjiahongxiang@126.com
[6] School of Resource Environment and Earth Science, Yunnan University, Kunming 650091, China; liuzheng.0311@163.com
*   Correspondence: gcwang@ynu.edu.cn

Received: 14 March 2019; Accepted: 24 April 2019; Published: 29 April 2019

Abstract: The newly discovered Tiantangshan tin polymetallic deposit is located in the southeast Nanling Range, Cathaysia block, Southeast China. The tin orebodies are mainly hosted in the greisen and the fractured alteration zones of the tufflava and trachydacite. However, the genetic relationship between the hidden alkali-feldspar granite and volcanic rocks and the tin mineralization remains poorly understood. This paper presents SHRIMP zircon U–Pb dating, whole-rock major and trace element analyses, as well as Nd isotopic data of the trachydacite and alkali-feldspar granite. The SHRIMP zircon U–Pb dating of the alkali-feldspar granite and trachydacite yields weight mean $^{206}Pb/^{238}U$ ages of 138.4 ± 1.2, and 136.2 ± 1.2 Ma, respectively. These granitic rocks have high levels of SiO_2 (64.2–75.4 wt%, mostly > 68 wt%), alkalis ($K_2O + Na_2O > 8.3$ wt%), REE (except for Eu), HFSE (Zr + Nb + Ce + Y > 350 ppm) and Ga/Al ratios (10,000 × Ga/Al > 2.6), suggesting that they belong to the A-type granite. According to the high Y/Nb and Yb/Ta ratios, they can be further classified into A_1 subtype. Their ε_{Nd} (T) range from −3.8 to −6.5. They were likely generated by the assimilation-fractional crystallization (AFC) of the coeval oceanic island basalts -like basaltic magma. This study suggests that the A_1 type granite is also a potential candidate for the exploration of tin deposits.

Keywords: A-type granite; tin mineralization; Tiantangshan tin polymetallic deposit; SE China

1. Introduction

The enrichment mechanism of Sn in tin deposits is of great importance for studies on its mineralization and may provide useful guidance for tin prospecting. Previous studies have revealed that most primary tin deposits are closely related to granitic rocks in time and space [1], indicating a possible genetic relationship. The fractional crystallization of granitic rocks plays a vital role in the origin of the tin deposit [2], whereas the geochemical heritage of granitic rocks has been shown to exert a relatively limited influence on tin mineralization [3]. The in situ fractional crystallization of tin granitic rocks is characterized by significant depletion of Ba and Sr, a drop in the TiO_2/Ta ratio [4], and an increase in the level of Rb [2]. As result, the resultant tin granitic rocks display a distinct geochemical specialization in Sn and other incompatible elements (e.g., F, Li, B, Cs) [1]. The extreme differentiation of these tin granitic rocks suggests that they might represent the most evolved phase of mantle-derived magmas [3]. Recent studies showed that many major tin deposit-related granitic rocks can be classified as A-type granite [5–7], which are geochemically characterized by high levels of K_2O, high incompatible trace element contents (including REE, Zr, Nb and Ta), elevated Fe/(Fe + Mg) and K_2O/Na_2O, as well as low concentrations of trace elements compatible in mafic silicate and feldspar [8,9]. A-type granite can be further divided into two subtypes, A_1 and A_2, based on their geochemical differences [10]. The A_1-type granite has been postulated to be generated by the assimilation-fractional crystallization (AFC) of oceanic-island basalt, whereas its A_2-type counterpart is thought to be derived by the partial melting of continental crustal sources [9]. Based on the results of previous investigations, the A_1-type granite may present the highly differentiated mantle-derived magma [9]. However, its genetic relationship with tin mineralization remains poorly understood.

The Nanling Range, located in the middle area of the South China, is well known for its abundant tin-polymetallic ore deposits. This region is characterized by widespread and intensive late Mesozoic magmatism, as well as the associated mineralization. The tin deposits are spatially, temporally and metallogenically associated with the coeval granites [11]. Most tin granites in the Nanling Range are metaluminous to weakly peraluminous biotite (hornblende) granite with an A-type granite affinity [7,12]. They are enriched in Nb, Zr, Ce, Y and HFSEs with relatively high alkali contents [7,12]. Most can be further classified as A_2 type granite [13,14]. They have been interpreted as derived from the partial melting of Precambrian crustal rocks [15]. However, A_1-type tin granite in the Nanling Range is rarely reported in previous studies.

The Tiantangshan tin polymetallic deposit, located in the southeast of the Nanling Range (Figure 1), has a tin reserve of 25 kt with an average grade of 0.5% and a WO_3 reserve of 7 kt with an average grade of 0.3%. $^{40}Ar/^{39}Ar$ dating of the biotite that coexists with the cassiterite in the ore samples yielded an isochron age of 133.5 ± 0.75 Ma [16]. The orebodies are mainly hosted in the greisen and the fractured alteration zones of tufflava and trachydacite (Figure 2). Further geological examination by drilling revealed a hidden alkali-feldspar granite underneath the greisen (Figure 3). Nevertheless, the genetic relationship between tin mineralization and the intrusive-volcanic rocks remains enigmatic. In this contribution, we present detailed SHRIMP zircon U–Pb dating, whole-rock major and trace element analyses, as well as Nd isotope data of trachydacite and alkali-feldspar granite in Tiantangshan deposit to explore their genetic relationships. Our experiment results led us to tentatively conclude that the alkali-feldspar granite and trachydacite can both be categorized as A_1-type generated from the AFC of the coeval OIB-like basaltic magma. Furthermore, the current study suggests that A_1-type Tiantangshan alkali-feldspar granite is genetically associated with tin mineralization.

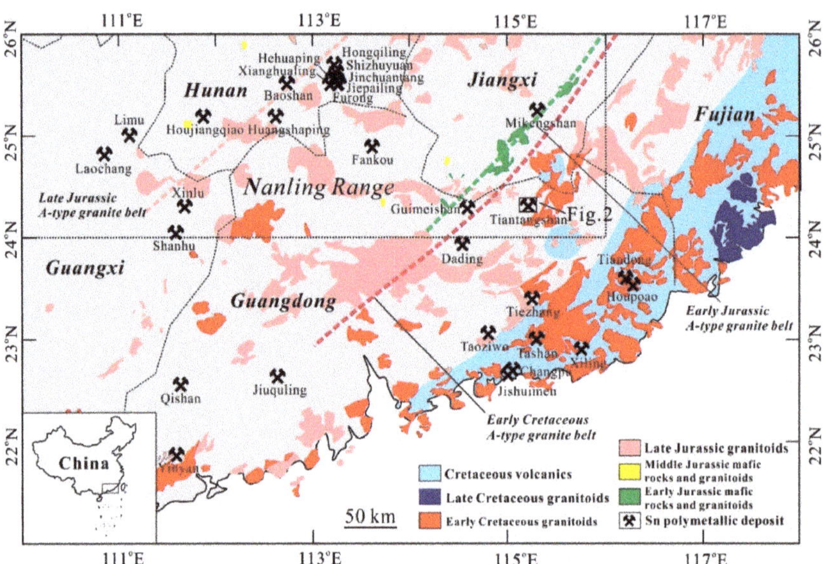

Figure 1. A sketch map of tin polymetallic deposits in the Nanling Range and neighboring area. Also shown are the late Mesozoic volcanic-intrusive complex and the A-type granite belts, modified after [15].

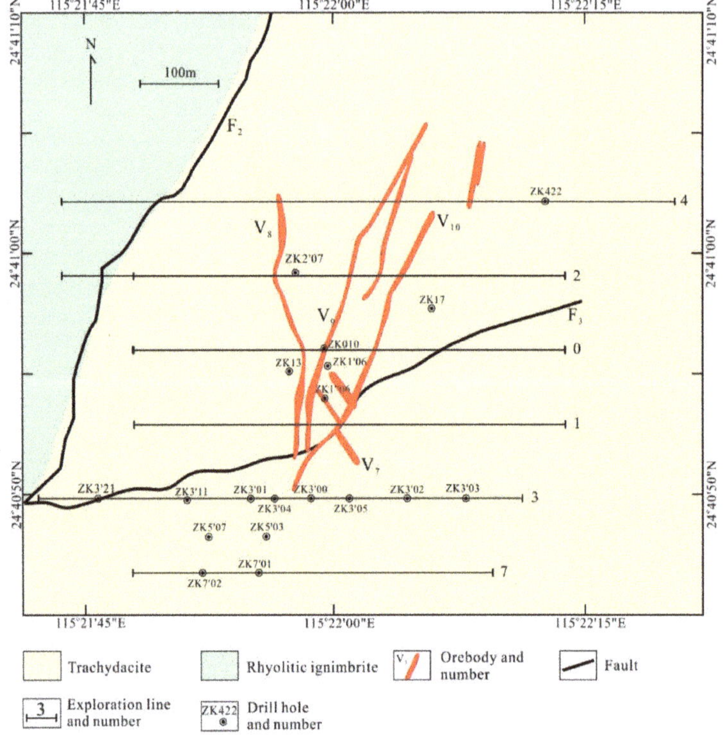

Figure 2. Geological schematic map of the Tiantangshan tin polymetallic deposit.

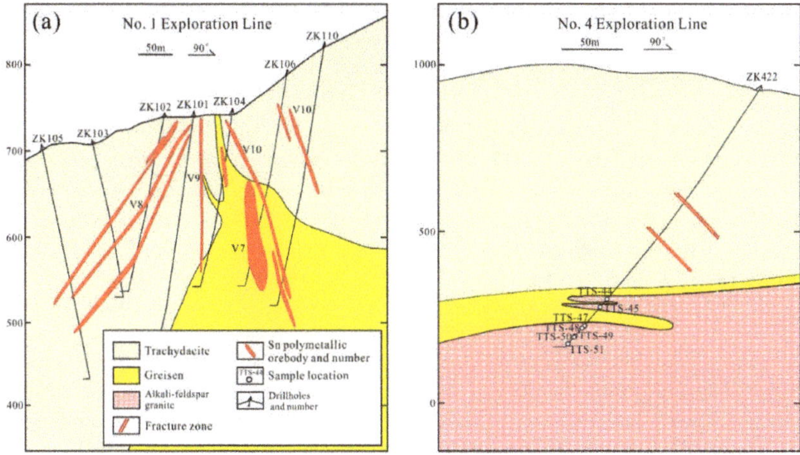

Figure 3. Geological cross-section along the exploration line No.1 (**a**) and No.4 (**b**).

2. Tectonic Settings

2.1. Regional Geology

The South China Block is composed of the Yangtze Block in the north and the Cathaysia Block in the south. The Cathaysia Block consists of Paleoproterozoic to Early Neoproterozoic metamorphic rocks, Late Neoproterozoic to Paleozoic continental to neritic marine sediments, and Mesozoic terrestrial clastics. The South China Block experienced multiple tectono-magmatic events during the Mesozoic, including its collision with the Indochina Block at ca. 240–230 Ma [17,18], its collage with the North China block along the Dabie–Sulu orogenic belt at ca. 220–245 Ma [19], and the westward subduction of the Palaeo–Pacific Plate that likely started in the Late Triassic [15]. These events triggered the intensive and extensive Mesozoic magmatism as well as the related tin mineralization in SE China [20]. Triassic granitoids are mainly exposed over the inland regions of the South China, whereas their Jurassic to Cretaceous equivalents are distributed in the form of 600 km wide volcanic-intrusive complex belt, parallel to the present coastline. Studies have indicated that these granitoids mostly belong to the (high-K) calc-alkaline series, and can be genetically classified as I- and S-types [20,21]. In addition, Mesozoic A-type granite formed over different periods have also been identified. The Late Triassic (229–221 Ma) A-type granite occur as an ENE-trending belt, and are coincident with the Late Triassic mafic magmatism. On the other hand, the Early Jurassic (192–188 Ma), Late Jurassic (163–153 Ma), Early Cretaceous (141–124 Ma) and Late Cretaceous (101–91 Ma) A-type granite belts are all NNE-trending in parallel to the present coastline [15]. All these five belts were produced by repeated slab-advance-retreat of the Palaeo-Pacific plate [15].

The Nanling Range, a major E–W trending W–Sn metallogenic province, is situated within the northwestern margin of the Cathaysia Block (Figure 1), and includes most of the border areas of Hunan, Jiangxi, Fujian, Guangdong and Guangxi Provinces. The region is characterized by the late Mesozoic intensive multistage magmatism and large-scale mineralization of tin, tungsten, and other rare-metals. The tin polymetallic deposits mainly comprise of skarn and wolframite-bearing quartz veins with minor greisen types [11,22]. It has been suggested that the formation of these tin-bearing deposits can be divided into three distinct stages, including the Late Triassic (230–210 Ma), Mid–Late Jurassic (170–150 Ma) and Early–Mid Cretaceous (120–80 Ma) [22]. The Early Cretaceous tin polymetallic deposits in the Nanling Range are mainly located in the volcanic basins of Southeast Jiangxi and have not been vigorously investigated in previous studies possibly due to their comparatively smaller scale. The Tiantangshan tin polymetallic deposit is located in the southeast region of the Nanling Range, but the petrogenesis of the hidden alkali-feldspar granite and trachydacite needs further study, and their possible genetic relationship with the tin deposit remains enigmatic.

2.2. Geology and Sampling of the Tiantangshan Tin Polymetallic Deposit

The newly explored Tiantangshan tin polymetallic deposit is located in the Mabugang town of Northeastern Guangdong (Figure 1). The stratigraphic sequence in the Tiantangshan deposit consists of the Late Jurassic Gaojiping Formation. The Gaojiping Formation is dominated by intermediate felsic lava and pyroclastic rocks, and can be further divided into three members from bottom to top that differ in their eruption cycles and the type of rocks that they contain [16]. The first member is more than 340 m thick and is exposed in the northwest of the mining area (Figure 2), consisting mainly of rhyolitic ignimbrite with minor breccia-bearing ignimbrite. The second member is about 280 to 350 m thick and hosts the tin orebodies (Figure 2), with a mineral composition of trachydacite, rhyolitic tuff, tuffaceous lava and crystal lithic tuff. The third member, over 550 m thick, mainly comprises rhyolitic breccia-bearing lava and rhyolitic lithic crystal tuff. Drilling reveals the concealed alkali-feldspar granite underneath the volcanic rocks (Figure 3b). The main structures in the mining area include NE-striking (F_3) and NNE-striking (F_2) faults. Three types of W-Sn orebodies can be identified in the Tiantangshan deposit, including greisen-, quartz vein- and fracture zones veined type. Typical hydrothermal alterations in the area encompass albitization, silicification, topazization, greisenization, biotitization, chloritization, sericitization, and fluoritization [16]. According to field and microscopic observations, the formation of the tin ores can be divided into four stages that can be distinguished by mineral assemblage: (1) Greisenization stage (stage I), (2) Quartz–cassiterite–wolframite stage (stage II), (3) Quartz–fluorite–cassiterite–sulfides stage (stage III) and (4) Post-ore stage (stage IV) [16].

In this study, the alkali-feldspar granite and trachydacite samples were collected from the core which drilled from the ZK422, ZK010 and −711 m underground tunnels. The alkali-feldspar granite consists of K-feldspar, plagioclase, quartz and biotite with a medium-grained granitic texture (Figure 4a,b). Greisen is made up of quartz, muscovite, topaz, chlorite, cassiterite, and rare pyrite (Figure 4c,d). On the other hand, the trachydacite mainly contains phenocrysts and matrix which consists of alkali-feldspar, plagioclase and quartz (Figure 4e,f).

Figure 4. (**a**,**c**,**e**) Photographs and (**b**,**d**,**b**) Photomicrographs (under crossed polar) of intrusive-volcanic rocks in Tiantangshan tin polymetallic deposits: (**a**) and (**b**) are alkali-feldspar granite; (**c**) and (**d**) are greisen; and (**e**) and (**f**) are trachydacite. Kfs, K-feldspar; Q, quartz; Bt, biotite; Toz, topaz; Mus, muscovite; Pl, plagioclase; Max, matrix.

3. Analytical Methods

3.1. SHRIMP Zircon U–Pb Dating

Zircon in situ U–Th–Pb isotope analyses were conducted using a SHRIMP-II at the Beijing SHRIMP Center, Chinese Academy of Geological Sciences, Beijing, China, following the similar analytical procedure described by [23]. A primary 20–30 μm·O^{2-} ion beam with an intensity of 3–6 nA was used to bombard the zircon surfaces. The raster time was set to 120–200 s. Each analysis consisted of five scans. Reference zircons for elemental abundance calibration included 91,500 (U = 91 ppm), SL13 (U = 238 ppm), and M257 (U = 840 ppm) [23–25]. TEMORA with a $^{206}Pb/^{238}U$ age of 417.0 ± 1.8 Ma was used for calibration [26]. Data were processed by SQUID and Isoplot [27]. Common Pb corrections were based on the measured [204]Pb contents. Uncertainties for individual analyses were quoted at 1σ, whereas the errors for the weighted mean ages were quoted at 2σ (95% confidence).

3.2. Major, Trace Elements and Nd Isotope of Intrusive-Volcanic Rocks

The alkali-feldspar granite and trachydacite samples were crushed and powdered to 200 meshes in an agate mortar. Whole-rock major and trace elements analyses of the alkali-feldspar granite samples were performed at the Key Laboratory of Orogenic Belts and Crustal Evolution, Ministry of Education, School of Earth and Space Sciences, Peking University. Major elements were analyzed with an X-ray fluorescence spectrometer on fused glass beads. The analytical uncertainty of oxide over 10 wt% is <1% while that of oxides below 10 wt% is <10%. Detailed methods for trace elements analyses were presented in [28]. Trace elements were characterized on a PerkinElmer Elan9000/DRCII/DRC-e ICP-MS. In brief, exactly 50 mg of sample powder were dissolved at 190 °C in a closed high-pressure Savillex Teflon beakers containing a mixture of 1 mL HF + 1 mL HNO_3 for 36 h. The resultant solution was then evaporated to dryness, followed by the addition of 1.5 mL of HNO_3, 1.5 mL of HF and 0.5 mL $HClO_4$. The beaker was subsequently capped and placed in an oven to allow the digestion to proceed at 180 °C for at least 48 h. Once the powders were completely digested, the residue was diluted in 50 mL of 1% HNO_3 for further analysis. The international reference samples GSR-1 (granite), GSR-2 (andesite) and GSR-9 (granodiorite) were used as controls. All of trace elements measurement showed an error below 5%.

Whole rock major and trace elements analyses of the trachydacite samples were conducted at the Analytical Laboratory of Beijing Research Institute of Uranium Geology, China National Nuclear Corporation. Whole rock major elements were analyzed by were analyzed on a Philips (Philips PW2404, Amsterdam, The Netherlands) X-ray fluorescence spectrometer (XRF). The test methods were based on GB/T 14506-2010 and had a precision that was better than 1% for the oxides over 10 wt% and 10% for the oxides below 10 wt%. Meanwhile, trace elements were analyzed by inductively coupled plasma mass spectrometry (ICP–MS) on an Agilent 7500a system using the same procedure described above, with a measurement uncertainty of 5%. Nd isotope analyses were performed at the Analytical Laboratory of Beijing Research Institute of Uranium Geology, China National Nuclear Corporation, following a previously established protocol [29,30]. Briefly, 100 mg of the sample powder was dissolved in a Teflon beaker containing a mixture of HF + HNO_3, followed by the separation and purification of Nd with conventional cation-exchange. The mass fractionation corrections of $^{143}Nd/^{144}Nd$ ratios were on the basis of $^{146}Nd/^{144}$ Nd of 0.7219. Total analytical blanks were 5×10^{-11} g for Sm-Nd.

4. Results

4.1. SHRIMP Zircon U–Pb Geochronology

SHRIMP zircon U–Pb dating results of two samples are summarized in Table 1 and illustrated in Figure 5. The zircons exhibit regular oscillatory magmatic zoning with a size distribution mostly between 100 and 200 μm and high Th/U ratios in the range of 0.4 to 0.98 (Table 1). Nine analyses of sample TTS-50 (alkali-feldspar granite) plot in a group on the Concordia curve and yield a weighted mean $^{206}Pb/^{238}U$ age of 138.4 ± 1.2 Ma (MSWD = 0.62) (Figure 5a). Eleven analyses for sample TTS-89 (trachydacite) plot in a group on the Concordia curve and yield a weighted mean $^{206}Pb/^{238}U$ age of 136.2 ± 1.2 Ma (MSWD = 0.49) (Figure 5b).

Table 1. SHRIMP zircon U–Pb dating for the Tiantangshan alkali-feldspar granite and trachydacite.

Spot	U (ppm)	Th (ppm)	Th/U	^{206}Pb* (ppm)	^{207}Pb/^{206}Pb	±%	^{207}Pb*/^{235}U	±%	^{206}Pb*/^{238}U	±%	^{206}Pb/^{238}U (Age/Ma)	±1σ
					Sample TTS-50 (alkali-feldspar granite)							
1.1	1179	513	0.45	22.1	0.0489	4.8	0.1453	5	0.02155	1.3	137.4	1.7
2.1	1859	976	0.54	35.5	0.0497	5.5	0.1487	5.6	0.02168	1.3	138.3	1.7
3.1	1034	399	0.4	19.6	0.0481	4.9	0.1442	5.1	0.02173	1.3	138.6	1.8
6.1	1241	1099	0.91	23.4	0.0488	3	0.1473	3.2	0.02188	1.3	139.5	1.7
7.1	1005	460	0.47	19.1	0.0496	3.4	0.151	3.7	0.02208	1.3	140.8	1.8
9.1	983	933	0.98	18.2	0.0492	3.5	0.1457	3.7	0.02145	1.3	136.8	1.7
12.1	1058	584	0.57	20.4	0.0485	8.4	0.146	8.5	0.0218	1.3	139	1.8
13.1	789	410	0.54	14.5	0.0482	3.4	0.1423	3.7	0.02139	1.3	136.4	1.8
15.1	1143	767	0.69	21.8	0.048	4.8	0.1443	4.9	0.02182	1.3	139.2	1.8
					Sample TTS-89 (trachydacite)							
1.1	357	254	0.74	6.53	0.0489	6.5	0.1434	6.7	0.02125	1.4	135.5	1.9
2.1	401	312	0.8	7.48	0.0487	5	0.1453	5.2	0.02164	1.4	138	1.9
4.1	328	225	0.71	6.07	0.0502	7.8	0.148	7.9	0.02136	1.5	136.3	2.0
5.1	260	167	0.66	4.82	0.0478	6	0.1416	6.1	0.02147	1.5	136.9	2.0
6.1	361	228	0.65	6.58	0.0521	7	0.15	7.1	0.02091	1.5	133.4	1.9
7.1	248	208	0.87	4.62	0.0493	11	0.146	11	0.0214	1.6	136.5	2.1
8.1	226	176	0.8	4.16	0.0478	10	0.14	10	0.02123	1.6	135.4	2.1
9.1	224	167	0.77	4.14	0.0503	13	0.148	13	0.02132	1.7	136	2.3
10.1	487	311	0.66	8.98	0.0473	7.8	0.139	7.9	0.02125	1.4	135.5	1.9
11.1	147	117	0.82	2.7	0.0477	6.9	0.1401	7	0.02133	1.6	136	2.2
12.1	305	266	0.9	5.72	0.0477	4.4	0.1427	4.6	0.0217	1.4	138.4	1.9

All the spots are localized on the around the zircon rims, common lead corrected using ^{204}Pb, ^{206}Pb* is radiogenic lead. The error estimate is 1σ.

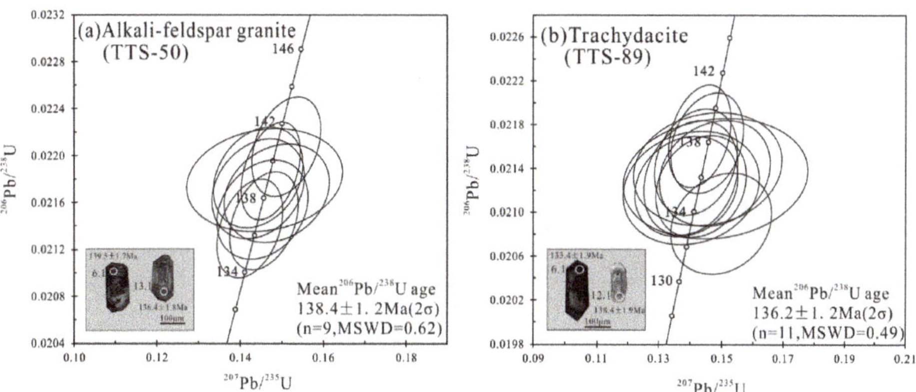

Figure 5. SHRIMP zircon U–Pb concordant curves for the Tiantangshan alkali-feldspar granite (**a**) and trachydacite (**b**).

4.2. Major, Trace Elements of Whole Rocks

The alkali-feldspar granite samples have SiO_2 contents of 73.0–75.4 wt% (Tables 2 and 3), and are characterized by relatively high heavy rare earth elements (HREEs), significant depletion of Eu, Ba and Sr, as well as notable negative Eu, Ba and Sr anomalies (Figure 6a,b). In comparison, the trachydacite contain lower abundances of SiO_2 that range from 68.5 to 70.0 wt.%, and are enriched in light rare earth elements (LREE) but depleted in HREEs with notable negative Eu anomalies (Figure 6c). Furthermore, the trachydacite samples are enriched in large ion lithophile elements (LILE) and depleted in high field strength elements (HFSE), while showing notable negative Ta–Nb and Ti anomalies (Figure 6d).

Table 2. Major (wt%), trace element (ppm), Nd isotope compositions and $T_{DM2}(Nd)$ ages of the Tiantangshan alkali-feldspar granite and trachydacite.

Sample	TTS-44		TTS-45		TTS-47		TTS-48		TTS-49		TTS-50	
Rock Type	Alkali-feldspar Granite											
	wt %	± wt %	wt.%	± wt %	wt %	± wt %	wt %	± wt %	wt %	± wt %	wt %	± wt %
SiO_2	75.2	0.8	73.0	0.7	73.7	0.7	75.4	0.8	74.8	0.7	75.2	0.8
TiO_2	0.10	0.01	0.11	0.01	0.10	0.01	0.11	0.01	0.10	0.01	0.11	0.01
Al_2O_3	12.7	0.1	13.7	0.1	13.2	0.1	12.8	0.1	12.8	0.1	12.9	0.1
FeO	1.11	0.11	2.2	0.2	1.65	0.17	1.65	0.17	1.31	0.13	1.37	0.14
Fe_2O_3	1.58	0.16	2.36	0.24	1.99	0.20	1.72	0.17	1.53	0.15	1.58	0.16
FeO^T	2.54	0.25	4.32	0.43	3.45	0.35	3.20	0.32	2.69	0.27	2.79	0.28
MnO	0.04	0.004	0.10	0.01	0.11	0.01	0.06	0.01	0.05	0.005	0.04	0.004
MgO	0.15	0.02	0.19	0.02	0.08	0.01	0.06	0.01	0.08	0.01	0.05	0.01
CaO	0.65	0.07	0.99	0.10	0.89	0.09	0.48	0.05	0.58	0.06	0.62	0.06
Na_2O	3.31	0.33	2.80	0.28	3.26	0.33	4.10	0.41	4.80	0.48	4.52	0.45
K_2O	5.13	0.51	5.49	0.55	5.55	0.56	4.58	0.46	4.59	0.46	4.23	0.42
P_2O_5	0.02	0	0.02	0	0.01	0	0.02	0	0.02	0	0.02	0
LOI	0.93		1.12		0.90		0.56		0.67		0.54	
	ppm	± ppm	ppm	± ppm	ppm	± ppm	ppm	± ppm	ppm	± ppm	ppm	± ppm
Rb	754	38	851	43	749	37	670	34	806	40	616	31
Ba	68.1	3.4	61.2	3.1	75.6	3.8	50.1	2.5	63.9	3.2	54.3	2.7
Th	59.0	3.0	65.5	3.3	64.8	3.2	64.2	3.2	74.5	3.7	64.3	3.2
U	18.9	0.9	16.9	0.8	19.9	1.0	12.0	0.6	18.9	0.9	18.7	0.9
Nb	117	6	143	7	122	6	136	7	132	7	123	6
Ta	20.3	1.0	23.9	1.2	18.1	0.9	18.8	0.9	16.9	0.8	17.3	0.9
Zr	177	9	204	10	203	10	212	11	169	8	199	10
Hf	10.6	0.5	12.3	0.6	12.5	0.6	12.8	0.6	12.3	0.6	11.3	0.6
Sr	16.5	0.8	26.8	1.3	20.5	1.0	16.1	0.8	20.1	1.0	21.2	1.1
Y	73.0	3.7	77.6	3.9	78.1	3.9	74.4	3.7	77.6	3.9	84.0	4.2
Pb	26.1	1.3	14.2	0.7	51.5	2.6	22.4	1.1	39.3	2.0	26.7	1.3
La	55.7	2.8	56.9	2.8	59.7	3.0	60.1	3.0	55.1	2.8	59.4	3.0
Ce	109	5	111	6	116	6	113	6	109	5	113	6
Pr	12.9	0.6	13.0	0.7	13.4	0.7	13.3	0.7	12.9	0.6	13.7	0.7
Nd	44.8	2.2	40.3	2.0	45.1	2.3	46.7	2.3	45.0	2.3	49.1	2.5
Sm	8.99	0.45	8.27	0.4	9.18	0.46	9.55	0.48	8.87	0.44	10.2	0.5
Eu	0.20	0.01	0.22	0.01	0.23	0.01	0.18	0.01	0.26	0.01	0.21	0.01
Gd	8.20	0.41	8.27	0.41	8.59	0.43	8.59	0.43	8.35	0.42	9.07	0.45
Tb	1.86	0.09	1.84	0.09	1.95	0.10	1.95	0.10	1.99	0.10	2.10	0.11
Dy	11.8	0.6	11.9	0.6	11.7	0.6	11.4	0.6	12.7	0.6	13.1	0.7
Ho	2.50	0.13	2.65	0.13	2.60	0.13	2.57	0.13	2.77	0.14	2.80	0.14
Er	7.10	0.36	7.73	0.39	7.68	0.38	7.25	0.36	8.37	0.42	8.17	0.41
Tm	1.41	0.07	1.55	0.08	1.54	0.08	1.48	0.07	1.77	0.09	1.54	0.08
Yb	8.66	0.43	9.85	0.49	9.81	0.49	9.29	0.46	11.3	0.6	10.2	0.5
Lu	1.13	0.06	1.31	0.07	1.32	0.07	1.27	0.06	1.50	0.1	1.37	0.07
Ga	28.9	1.4	32.1	1.6	31.3	1.6	32.1	1.6	31.7	1.6	31.7	1.6
W	6.22	0.31	8.71	0.44	5.76	0.29	4.83	0.24	3.00	0.15	4.41	0.22
Sn	18.3	0.9	27.8	1.4	31.5	1.6	19.2	1.0	21.3	1.1	17.1	0.9
Cu	2.77	0.14	2.40	0.12	2.87	0.14	2.05	0.10	2.06	0.10	1.84	0.09
Zn	325	16	537	27	250	13	52.4	2.6	87.7	4.4	50.3	2.5
Mo	0.81	0.04	0.97	0.05	0.63	0.03	0.44	0.02	0.43	0.02	0.54	0.03
REE	274		281		290		289		280		292	
LREE	231		236		245		245		231		244	
HREE	42.7		45.1		45.2		43.8		48.8		48.4	
LREE/HREE	5.42		5.24		5.41		5.59		4.74		5.04	
$(La/Yb)_N$	4.61		4.14		4.37		4.64		3.50		4.18	
δEu	0.07		0.07		0.08		0.06		0.09		0.07	
$^{147}Sm/^{144}Nd$	0.1212		0.1240		0.1230		0.1236				0.1255	
$^{143}Nd/^{144}Nd$	0.512352		0.512379		0.512275		0.512354				0.512374	
$\pm 2\sigma$	0.000006		0.000006		0.000012		0.000007				0.000009	
$\varepsilon_{Nd}(T)$	−4.2		−3.8		−5.8		−4.2				−3.9	
$T_{DM2}(Nd)(Ma)$	1277		1238		1402		1277				1249	

± wt %, ± ppm: analytical uncertainty.

Table 3. Major (wt%), trace element (ppm), Nd isotope compositions and $T_{DM2}(Nd)$ ages of the Tiantangshan alkali-feldspar granite and trachydacite.

Sample	TTS-51		TTS-56		TTS-60		TTS-41		TTS-89	
Rock Type	Alkali-feldspar Granite		Trachydacite							
	wt %	± wt %	wt %	± wt %	wt %	± wt %	wt %	± wt %	wt %	± wt %
SiO_2	75.3	0.8	69.9	0.7	64.2	0.6	70.0	0.7	68.5	0.7
TiO_2	0.10	0.01	0.27	0.03	0.64	0.06	0.32	0.03	0.26	0.03
Al_2O_3	12.8	0.1	14.7	0.1	15.1	0.2	15.0	0.1	14.7	1.5
FeO	1.37	0.14	1.65	0.17	2.83	0.28	1.14	0.11	1.98	0.20
Fe_2O_3	1.47	0.15	1.62	0.16	2.74	0.27	1.17	0.12	1.76	0.18
FeO^T	2.69	0.27	3.46	0.35	5.90	0.59	2.44	0.24	3.97	0.40
MnO	0.04	0.004	0.16	0.02	0.12	0.01	0.03	0.003	0.08	0.01
MgO	0.04	0.004	0.38	0.04	0.42	0.04	0.31	0.03	0.22	0.02
CaO	0.63	0.06	0.45	0.04	1.55	0.15	0.65	0.07	1.31	0.13
Na_2O	4.16	0.42	3.42	0.34	5.03	0.50	3.50	0.35	2.36	0.24
K_2O	4.72	0.47	6.32	0.63	4.94	0.49	6.74	0.67	6.57	0.66
P_2O_5	0.01	0	0.04	0	0.07	0.01	0.06	0.01	0.05	0
LOI	0.61		0.74		1.38		0.74		1.37	
	ppm	± ppm	ppm	± ppm	ppm	± ppm	ppm	± ppm	ppm	± ppm
Rb	709	35	757	38	1222	61	863	43	819	41
Ba	49.8	2.5	548	27	396	20	329	16	257	13
Th	68.3	3.4	37.2	1.9	44.1	2.2	34.4	1.7	36.5	1.8
U	21.5	1.1	6.10	0.31	5.90	0.30	6.80	0.34	6.30	0.32
Nb	116	6	52.1	2.6	49.3	2.5	45.3	2.3	43.4	2.2
Ta	16.6	0.8	3.70	0.19	2.50	0.13	2.70	0.14	2.40	0.12
Zr	163	8	248	12	224	11	184	9	155	8
Hf	8.8	0.4	7.60	0.38	6.90	0.35	5.70	0.29	5.30	0.27
Sr	18.6	0.9	120	6	106	5	74.9	3.7	97.2	4.9
Y	83.8	4.2	50.4	2.5	49.0	2.4	27.4	1.4	41.1	2.1
Pb	37.5	1.9	40.8	2.0	149	7	21.6	1.1	23.1	1.2
La	59.8	3.0	83.2	4.2	91.4	4.6	92.3	4.6	97.5	4.9
Ce	115	6	169	8	181	9	186	9	209	10
Pr	13.8	0.7	17.6	0.9	19.4	1.0	19.8	1.0	22.0	1.10
Nd	44.7	2.2	63.0	3.2	68.8	3.4	68.6	3.4	76.4	3.8
Sm	9.41	0.47	10.1	0.5	12.1	0.6	12.0	0.6	12.4	0.6
Eu	0.20	0.01	1.07	0.05	0.91	0.05	0.63	0.03	0.70	0.04
Gd	8.89	0.44	8.70	0.44	9.28	0.46	8.38	0.42	9.37	0.47
Tb	2.10	0.11	1.36	0.07	1.38	0.07	1.06	0.05	1.29	0.06
Dy	12.5	0.6	7.99	0.40	7.85	0.39	4.85	0.24	6.84	0.34
Ho	2.76	0.14	1.62	0.08	1.56	0.08	0.84	0.04	1.30	0.07
Er	8.02	0.40	4.70	0.24	4.57	0.23	2.26	0.11	3.59	0.18
Tm	1.53	0.08	0.72	0.04	0.72	0.04	0.32	0.02	0.52	0.03
Yb	10.0	0.5	4.57	0.23	4.78	0.24	2.14	0.11	3.26	0.16
Lu	1.30	0.07	0.71	0.04	0.72	0.04	0.34	0.02	0.50	0.03
Ga	29.8	1.5	21.3	1.1	23.2	1.2	20.3	1.0	23.2	1.2
W	4.20	0.21	11.4	0.6	29.7	1.5	19.4	1.0	29.7	1.5
Sn	21.0	1.0	6.59	0.33	24.9	1.2	9.11	0.46	45.1	2.3
Cu	2.77	0.14	3.21	0.16	35.1	1.8	4.02	0.20	199	10
Zn	55.6	2.8	46.7	2.3	390	20	35.2	1.8	315	16
Mo	0.26	0.01	3.59	0.18	13.0	0.7	12.2	0.6	3.03	0.15
REE	293		375		405		400		434	
LREE	246		344		374		380		408	
HREE	47.1		30.4		30.9		20.2		26.7	
LREE/HREE	5.23		11.3		12.1		18.8		15.3	
$(La/Yb)_N$	4.30		13.1		13.7		31.0		21.5	
δEu	0.07		0.34		0.25		0.18		0.21	
$^{147}Sm/^{144}Nd$	0.1272								0.0981	
$^{143}Nd/^{144}Nd$	0.512336								0.512216	
$\pm 2\sigma$	0.000009								0.000007	
$\varepsilon_{Nd}(T)$	−4.7								−6.5	
$T_{DM2}(Nd)(Ma)$	1311								1460	

± wt %, ± ppm: analytical uncertainty.

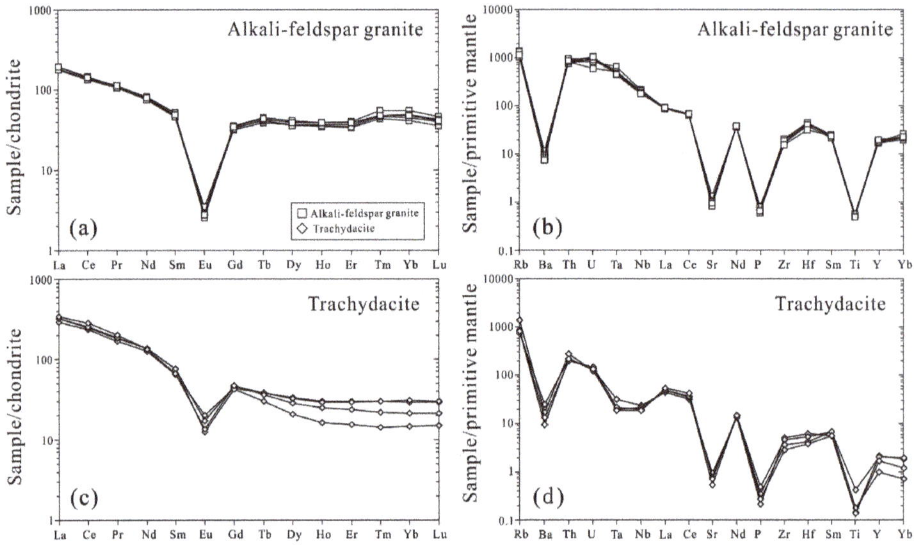

Figure 6. (**a,c**) Chondrite-normalized [31] REE patterns and (**b,d**) primitive mantle-normalized [32] trace element patterns for the Tiantangshan alkali-feldspar granite and trachydacite.

4.3. Nd Isotopes of Whole Rocks

The alkali-feldspar granites exhibit relatively narrow Nd isotopic compositions with ε_{Nd} (T) ranging from −3.8 to −5.8 (Figure 7), whereas the average ε_{Nd} (T) of trachydacite is comparatively lower at −6.5. All the granitic rocks yield Mesoproterozoic T_{DM2} ages between 1.06 and 1.40 Ga.

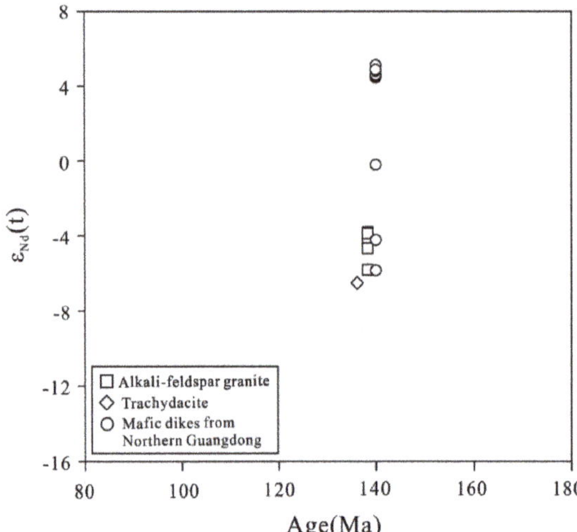

Figure 7. Age vs. ε_{Nd} (T) diagrams for the Tiantangshan alkali-feldspar granite and trachydacite. Also shown are the fields of the Early Cretaceous gabbros in northern Guangdong [33].

5. Discussion

5.1. Duration of the Magmatism and Hydrothermal Activities

Unlike zircon U–Pb age, biotite ^{40}Ar/^{39}Ar ages tend to reflect the cooling history of the minerals rather than their crystallization. In addition, they generally do not record early magmatic and hydrothermal events at >350 °C. The time intervals between zircon U–Pb ages and biotite ^{40}Ar/^{39}Ar ages could indicate a prolonged magmatic-hydrothermal process and cooling history. Previous ^{40}Ar/^{39}Ar dating of biotite coexisting with cassiterite in the Tiantangshan deposit yielded an isochron age of 133.5 ± 0.75 Ma [16]. In our present study, the ^{206}Pb/^{238}U ages of the alkali-feldspar granite and trachydacite in the same deposit are measured to be 138.4 ± 1.2 and 136.2 ± 1.2 Ma, respectively. Evidently, these felsic rocks have slightly older emplacement ages when compared to the biotite ^{40}Ar/^{39}Ar data. An alternative interpretation is that the relatively low closure temperature of the biotite rendered it sensitive and vulnerable to thermal events, which could result in apparent diffusive argon loss. Compared to the zircon U–Pb ages, the younger biotite ^{40}Ar/^{39}Ar ages imply a possible thermal disturbance. However, the biotite ^{40}Ar*/^{39}Ar ratios are relatively consistent in each fractionation across different heating steps and yield similar apparent ages with a flat age spectrum [16]. The plateau ages have a close relationship with the isochron ages [16]. The observation thus precludes the possibility of a subsequent thermal disturbance. The biotite ^{40}Ar/^{39}Ar geochronology is sufficiently reliable to constrain the timing of cooling. The average biotite ^{40}Ar/^{39}Ar age is ~3 million years younger than that of alkali-feldspar granite, which corresponds to the estimated duration from magma emplacement to biotite Ar-Ar closure. We therefore hypothesize that the Tiantangshan tin polymetallic deposit experienced a prolonged magmatic-hydrothermal process that might have lasted for at least 3 million years.

Recent advances in the dating of tin deposits, such as the employment of zircon or cassiterite U-Pb, Molybdenite Re-Os and muscovite or biotite Ar-Ar isotopic ages, have enabled researchers to characterize the magmatic-hydrothermal process and cooling history of tin deposits with excellent precision. The biotite ^{40}Ar-^{39}Ar plateau age (135.1 ± 0.8 Ma) is ~3 million years younger than the zircon U–Pb age of the biotite monzonitic granite porphyry in the Feie'shan W-Sn deposit (139.2 ± 1.7 Ma) [35]. The muscovite ^{40}Ar/^{39}Ar plateau age (140.6±1.0 Ma) is ~4 million years younger than the cassiterite U–Pb age (145.8 ± 0.6 Ma) in the Xiling tin deposit [36]. Two muscovite samples collected from greisen W-Mo-Be ores and W-Mo-bearing muscovite-quartz stockwork ores yield ^{40}Ar/^{39}Ar plateau ages of 146.85 ± 0.8 Ma and 146.31 ± 0.75 Ma, respectively, which are both ~6 million years younger than that of the granite in the Xiatongling W-Mo-Be deposit (154.7 ± 1.7 Ma) [37]. The Sn–Nb–Ta bearing Jinzhuyuan granite in limu mining district show a U–Pb age of 218.3 ± 2.4 Ma with a muscovite ^{40}Ar/^{39}Ar plateau age of 212.4 ± 1.4 Ma and isochron age of 213.2 ± 2.2 Ma [38]. The magmatic crystallization, ore formation, and the related alteration all occurred over a short interval of less than 6 million years in the Taoxikeng tungsten deposit [39].

5.2. Genetic Type, Origin and Tectonic Background of the Tiantangshan A-Type Granite

Both the Tiantangshan alkali-feldspar granite and trachydacite exhibit a strong A-type affinity [10,34,40,41], as evidenced by their enrichment in REEs (except for Eu, Figure 6) and HFSEs (Zr + Nb + Ce + Y > 350 ppm), depletion in Ba and Sr (Figure 6), as well as high Ga/Al ratios (Figure 8a). The relatively low Yb/Ta and Y/Nb ratios further indicate that these granitic rocks should be classified as A_1 subtype (Figure 8b).

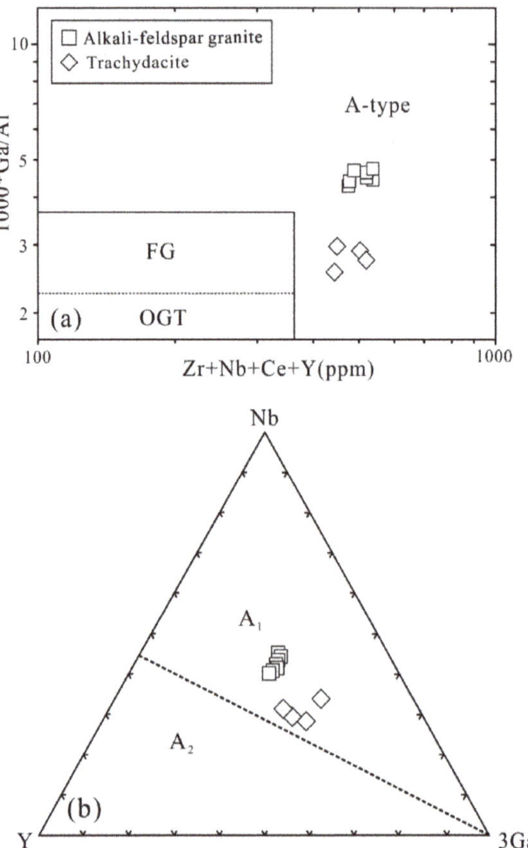

Figure 8. (**a**) Zr + Nb + Ce + Y vs. 10000*Ga/Al [34] and (**b**) representative triangular plots for distinguishing between A_1 and A_2 granitoids [9] for Tiantangshan alkali-feldspar granite.

A-type magmas were suggested to result from the partial melting of specific crustal protoliths [34,40,42] or from the extensive fractional crystallization from mantle-derived basaltic magmas [43]. The A_1 subtype generally share similar Nb/Ta and Y/Nb ratios as oceanic-island basalts, whereas their counterparts bear a geochemical resemblance to average crust and island-arc basalts [9] (Figure 8b). The Tiantangshan alkali-feldspar granite and trachydacite belong to the A_1 subtype and plot in the OIB field, presumably indicating that they were derived from an OIB-like source. Furthermore, the Nd isotopic compositions of these granite are similar to those of coeval (~140 Ma) OIB-like mafic rocks in northern Guangdong (−5.89 to 5.16), suggesting that they might have derived from the AFC of basaltic magmas (Figure 7). The fact that of the alkali-feldspar granite and OIB share similar Nb/Ta and Y/Nb ratios strongly implies that they may represent the mantle differentiates with less continental crust contamination in their origin. The evolved elemental ratios of the trachydacite suggest that they were contaminated with continental crust to a greater extent than the alkali-feldspar granite (Figure 9). This is further supported by the fact that the trachydacite exhibit high $\varepsilon_{Nd}(t)$ and negative Nb-Ta anomalies (Figure 7). Taken together, we conclude that the Tiantangshan A_1-type granite were generated from the AFC of the coeval OIB-like mafic magma. The A-type granite is hypothesized to have formed at extensional tectonic background [9,40]. The Late Mesozoic magmatism and mineralization of SE China are generally considered to be related to the subduction of the Palaeo-Pacific plate, for which a number of models have been proposed [20,44–46]. More recently, a new repeated slab-advance-retreat model of the Palaeo-Pacific plate was developed based on geochronological and geochemical data of Late Triassic to Early Jurassic mafic rocks and Early Jurassic A-type granites in southern Jiangxi and western Fujian Provinces [15]. According to this repeated slab-advance-retreat model, the formation of the Early Cretaceous (141–124 Ma) A-type granite belt (Figure 1) was a result of the regional extension caused by the progressive slab rollback. Our new data presented in this paper lent further support to this model. The origin of the Early Cretaceous Tiantangshan A-type granite suggests the development of a back-arc extension along the Early Cretaceous A-type granite belt since the beginning of Early Cretaceous as a consequence of slab rollback [15]. Such an extension caused lithosphere thinning and the concomitant asthenosphere upwelling. The underplating of basaltic magma could trigger partial melting of the thinned lower-crust rocks, leading to the formation of the Early Cretaceous A_2 type granites. On the other hand, the AFC of the underplated basaltic magma could subsequently give rise to the Tiantangshan A_1 type granite. Therefore, the Tiantangshan tin polymetallic deposit might have generated in an extensional tectonic regime caused by slab rollback of the Paleo–Pacific Plate.

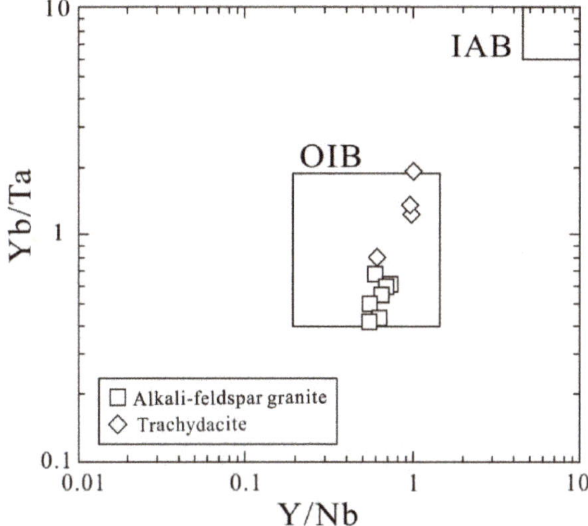

Figure 9. Y/Ta vs. Y/Nb diagrams for A-type granite [9].

5.3. A-Type Granite and Tin Mineralization

Magmatic differentiation plays a crucial role in the formation of tin bearing granite [3]. The in situ fractional crystallization of tin granitic rocks is characterized by a remarkable decrease in the level of Rb and depletion of Ba and Sr [2]. Consistent with this, the Tiantangshan alkali-feldspar granite generally exhibit low abundance of Sr and Ba, as well as increased levels of SiO_2, suggesting that they were derived from fractional crystallization. Continued fractional crystallization in the rest of the melts led to enrichment of Sn, W, Ta, Rb, Cs, F and Li, as well as depletion of Ti, U, Th, Ba, Sr, Zr. Among these elements, Ti and Ta are representative of distinct enrichment or depletion. TiO_2/Ta ratio is, thus, a good indicator of the evolution degree of granitic magma differentiation [4]. Moreover, Rb and Li may be affected by the hydrothermal processes, whereas Ta and Ti are relatively stable. The TiO_2/Ta ratios show a progressive decline from 4900 in the less differentiated granodiorite to <1 in the most differentiated granite from the Marche area [4]. In general, the TiO_2/Ta ratio of the Tiantangshan alkali-feldspar granite is inversely correlated to the level of Sn (Figure 10). In contrast with alkali-feldspar granite, trachydacite has a comparatively lower content of SiO_2, but elevated levels of Sr, Ba (Tables 2 and 3) and TiO_2/Ta ratio (Figure 10). All of these geochemical features imply that trachydacite experienced less differentiation than the alkali-feldspar granite. Furthermore, all except one trachydacite sample (which might have been affected by greisenization at a later stage) that we analyzed contain less Sn than the alkali-feldspar granite (Tables 2 and 3). We, therefore, suggest that Sn enrichment of Tiantangshan tin deposit is more likely related to the Tiantangshan alkali-feldspar granite than the trachydacites. Another possible explanation is that these granitic rocks could have inherited their geochemical specialization from their source rocks during the partial melting. However, TiO_2-Sn and Rb/Sr-Sn binary diagrams of the granitic fractionation series from various tin and non-tin provinces do not seem to support this hypothesis [1]. Previous studies showed that source rocks generally contribute a maximum of 5–10 ppm of Sn. Instead, the observed Sn enrichment is more likely to have resulted from the fractional crystallization of tin bearing granitic rocks. Both Tiantangshan alkali-feldspar granite and the trachydacite are derived from same source rock via AFC. The observation that the less differentiated trachydacite generally have lower Sn contents than the highly differentiated alkali-feldspar granite provides further support to this view.

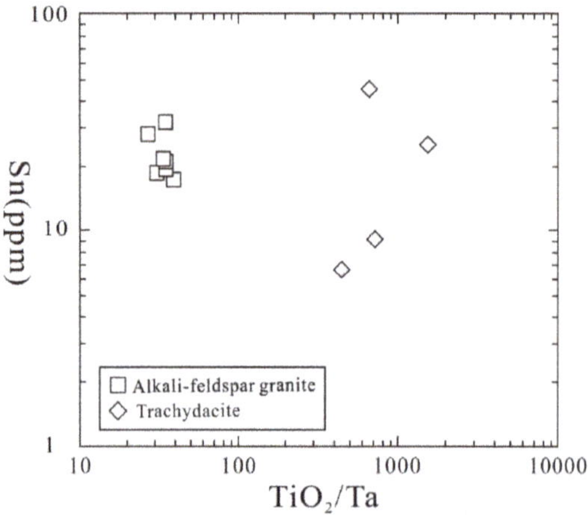

Figure 10. TiO_2/Ta vs. Sn diagrams of the Tiantangshan granitic rocks.

Prolonged fractional crystallization leads to concentration of H_2O and other volatiles in the rest liquid. Free water could react with the granite causing it to be gradually converted into greisen. The beginning of greisenization is characterized by the alteration of K-feldspar into muscovite (Figure 4d), which is illustrated by the following reaction: $3KAlSi_3O_8 + 2H^+ = KAl_3Si_3O_{10}(OH)_2 + 6SiO_2 + 2K^+$. Sn is generally considered to exist as Sn^{2+} or Sn^{4+} in complexation with Cl^- in the fluid [47]. Lowering the HCl activity in the fluid system leads to the precipitation of the cassiterite as manifested by $(SnCl_4[H_2O]_2)^0 = SnO_2 + 4HCl$ and $(SnCl_3)^- + H^+ + 2H_2O = SnO_2 + 3HCl + H_2$. Greisenization is, therefore, a good catalytic reaction for the precipitation of cassiterite. The magmatic-hydrothermal process lasted at least ~3 million years and produced at least one hydrothermal pulse, which resulted in the generation of the hydrothermal biotite at ~133 Ma.

6. Conclusions

1. The SHRIMP zircon U–Pb dating of the alkali-feldspar granite and the trachydacite in the Tiantangshan tin polymetallic deposit yields $^{206}Pb/^{238}U$ ages of 138.4 ± 1.2 and 136.2 ± 1.2 Ma, respectively. The Tiantangshan tin polymetallic deposit experienced a prolonged magmatic-hydrothermal process over a period of ~3 million years from the emplacement of granitic rocks to the origination of the hydrothermal biotite.

2. Both the Tiantangshan alkali-feldspar granite and trachydacite can be classified as A_1-type granite based on geochemical evidence. These granitic rocks were derived from the AFC of the coeval

OIB-like basaltic magma in an extensional setting, which was most likely caused by the rollback of the Paleo-pacific plate.

3. The tin polymetallic mineralization is associated with the Tiantangshan A_1 type alkali-feldspar granite.

Author Contributions: R.-Y.J., L.G., Z.-S.P., H.-X.J., Z.-H.Z., H.C., G.-C.W., Z.L. did the field work; R.-Y.J., G.-C.W., Z.L. analyzed results of all the experiments; R.-Y.J., G.-C.W. wrote the paper; R.-Y.J., G.-C.W., L.G., Z.-S.P., Z.L. revised the manuscript. All authors read and approved the manuscript.

Funding: This study was financially supported by National Key Research and Development Program of China (Grant 2017YFC0601506), Geological Survey Project of China Geological Survey (Grant DD20190570) National Natural Science Foundation of China (Grant 41703022), Fundamental Research Funds for the Central Universities (lzujbky-2018-52), Plateau mountain ecology and earth's environment discipline construction project (Grant C176240107), Joint Foundation Project between Yunnan Science and Technology Department and Yunnan University (Grant C176240210019), Geology Discipline Construction Project of Yunnan University (C176210227),Science (Engineering) Research Project of Yunnan University(2017YDQN08).

Acknowledgments: We thank the senior engineers Hongren Chen and other local geologists from the Bureau of Geology for Nuclear Industry of Guangdong Province for their assistance in the field.

Conflicts of Interest: The authors declare no conflict of interest.

References

1. Lehmann, B. Metallogeny of tin; magmatic differentiation versus geochemical heritage. *Econ. Geol.* **1982**, *77*, 50–59. [CrossRef]

2. Groves, D.I.; McCarthy, T.S. Fractional crystallization and the origin of tin deposits in granitoids. *Miner. Depos.* **1978**, *13*, 11–26. [CrossRef]

3. Lehmann, B. Tin granites, geochemical heritage, magmatic differentiation. *Geol. Rundsch* **1987**, *76*, 177–185. [CrossRef]

4. Boissavy-Vinau, M.; Roger, G. The TiO_2/Ta ratio as an indicator of the degree of differentiation of tin granites. *Miner. Depos.* **1980**, *15*, 231–236. [CrossRef]

5. Liverton, T.; Botelho, N.F. Fractionated alkaline rare-metal granites: Two examples. *J. Asian Earth Sci.* **2001**, *19*, 399–412. [CrossRef]

6. Haapala, I.; Lukkari, S. Petrological and geochemical evolution of the Kymi stock, a topaze granite cupola within the Wiborg rapakivi batholith, Finland. *Lithos* **2005**, *80*, 247–362. [CrossRef]

7. Chen, J.; Wang, R.C.; Zhu, J.C.; Lu, J.J.; Ma, D.S. Multiple-aged granitoids and related tungsten-tin mineralization in the Nanling Range: South China. *Sci. China Earth Sci.* **2013**, *56*, 2045–2055. [CrossRef]

8. Loiselle, M.C.; Wones, D.R. Characteristics and origin of anorogenic granites. *Geol. Soc. Am. Abstr. Programs* **1979**, *11*, 468.

9. Eby, G.N. Chemical subdivision of the A-type granitoids: Petrogenetic and tectonic implications. *Geology* **1992**, *20*, 641–644. [CrossRef]

10. Eby, G.N. The A-type granitoids: A review of their occurrence and chemical characteristics and speculations on their petrogenesis. *Lithos* **1990**, *26*, 115–134. [CrossRef]

11. Mao, J.W.; Cheng, Y.B.; Chen, M.H.; Pirajno, F. Major types and time–space distribution of Mesozoic ore deposits in South China and their geodynamic settings. *Miner. Depos.* **2013**, *48*, 267–294. [CrossRef]

12. Chen, J.; Lu, J.J.; Chen, W.F.; Wang, R.C.; Ma, D.S.; Zhu, J.C.; Zhang, W.L.; Ji, J.F. W-Sn-Nb-Ta-bearing granites in the Nanling Range and their relationship to metallogengesis. *Geol. J. China Univ.* **2008**, *14*, 459–473. (in Chinese with English abstract).

13. Zhou, Y.; Liang, X.Q.; Liang, X.R.; Wu, S.C.; Jiang, Y.; Wen, S.N.; Cai, Y.F. Geochronology and geochemical characteristics of the Xitian tungsen–tin- bearing A-type granite, Hunan Province, China. *Geotecton. Metallog.* **2013**, *37*, 517–535. (in Chinese with English abstract).

14. Zheng, W.; Mao, J.W.; Zhao, H.J.; Zhao, C.S.; Yu, X.F. Two Late Cretaceous A-type granites related to the Yingwuling W–Sn polymetallic mineralization in Guangdong province, South China: Implications for petrogenesis, geodynamic setting, and mineralization. *Lithos* **2017**, *274*, 106–122. [CrossRef]

15. Jiang, Y.H.; Wang, G.C.; Liu, Z.; Ni, C.Y.; Qing, L.; Zhang, Q. Repeated slab-advance retreat of the Palaeo-Pacific plate underneath SE China. *Int. Geol. Rev.* **2015**, *57*, 472–491. [CrossRef]

16. Jia, H.X.; Pang, Z.S.; Chen, R.Y.; Xue, J.L.; Chen, H.; Lin, L.J. Genesis and hydrothermal evolution of the Tiantangshan tin-polymetallic deposit, south-eastern Nanling Range, South China. *Geol. J.* **2018**, 1–22. [CrossRef]
17. Carter, A.; Roques, D.; Bristow, C.; Kinny, P. Understanding Mesozoic accretion in Southeast Asia: Significance of Triassic thermotectonism (Indosinian orogeny) in Vietnam. *Geology* **2001**, *29*, 211–214. [CrossRef]
18. Lepvrier, C.; Maluski, H.; Tich, V.; Leyreloup, A.; Truong Thi, P.; Van Vuong, N. The Early Triassic Indosinian orogeny in Vietnam (Truong Son Belt and Kontum Massif); implications for the geodynamic evolution of Indochina. *Tectonophysics* **2004**, *393*, 87–118. [CrossRef]
19. Ernst, W.G.; Tsujimori, T.; Zhang, R.; Liou, J.G. Permo-Triassic collision, subduction zone metamorphism, and tectonic exhumation along the East Asian continental margin. *Ann. Rev. Earth Planet. Sci.* **2007**, *35*, 73–110. [CrossRef]
20. Zhou, X.M.; Sun, T.; Shen, W.Z.; Shu, L.S.; Niu, Y.L. Petrogenesis of Mesozoic granitoids and volcanic rocks in South China: A response to tectonic evolution. *Episodes* **2006**, *29*, 21–26.
21. Zhou, X.M.; Li, W.X. Origin of late Mesozoic igneous rocks in southeastern China: Implications for lithosphere subduction and underplating of mafic magmas. *Tectonophysics* **2000**, *326*, 269–287. [CrossRef]
22. Mao, J.W.; Xie, G.Q.; Guo, C.L.; Yuan, S.D.; Cheng, Y.B.; Chen, Y.C. Spatial-temporal distribution of Mesozoic ore deposits in South China and their metallogenic settings. *Geol. J. China Univ.* **2008**, *14*, 510–526, (in Chinese with English abstract).
23. Williams, I.S. U-Th-Pb geochronology by ion microprobe. *Rev. Econ. Geol.* **1998**, *7*, 1–35.
24. Ballard, J.R.; Palin, J.M.; Williams, I.S.; Campbell, I.H.; Faunes, A. Two ages of porphyry intrusion resolved for the super-giant Chuquicamata copper deposit of northern Chile by ELA-ICP-MS and SHRIMP. *Geology* **2001**, *29*, 383–386. [CrossRef]
25. Nasdala, L.; Hofmeister, W.; Norberg, N.; Martinson, J.M.; Corfu, F.; Dörr, W.; Kamo, S.L.; Kennedy, A.K.; Kronz, A.; Reiners, P.W.; et al. Zircon M257-a homogeneous natural reference material for the ion microprobe U–Pb analysis of zircon. *Geostand Geoanal Res.* **2008**, *32*, 247–265. [CrossRef]
26. Black, L.P.; Kamo, S.L.; Allen, C.M.; Aleinikoff, J.N.; Davis, D.W.; Korsch, R.J.; Foudoulis, C. TEMORA 1: A new zircon standard for Phanerozoic U–Pb geochronology. *Chem. Geol.* **2003**, *200*, 155–170. [CrossRef]
27. Ludwig, K.R. *Squid 1.02: A User's Manual*; Berkeley Geochronology Centre Special Publication: Berkeley, CA, USA, 2001; pp. 1–19.
28. Gao, J.F.; Lu, J.J.; Lai, M.Y.; Lin, Y.P.; Pu, W. Analysis of trace elements in rock samples using HR-ICPMS (Natural Sciences). *J. Nanjing Univ.* **2003**, *39*, 844–850.
29. Pu, W.; Zhao, K.D.; Ling, H.F.; Jiang, S.Y. High precision Nd isotope measurement by Triton TI Mass Spectrometry. *Acta Geosci. Sin.* **2004**, *25*, 271–274, (in Chinese with English abstract).
30. Pu, W.; Gao, J.F.; Zhao, K.D.; Ling, H.F.; Jiang, S.Y. Separation method of Rb–Sr, Sm–Nd using DCTA and HIBA (Natural Sciences). *J. Nanjing Univ.* **2005**, *41*, 445–450.
31. Boynton, W.V. Cosmochemistry of the rare earth elements: Meteorite studies. In *Rare Earth Element Geochemistry*; Henderson, P., Ed.; Elsevier: Amsterdam, The Netherlands, 1984; pp. 63–114. [CrossRef]
32. McDonough, W.F.; Sun, S.S. Isotopic and geochemical systematics in Tertiary–Recent basalts from southeastern Australia and implication for the sub-continental lithosphere. *Geochim. Cosmochim. Acta* **1985**, *49*, 2051–2067. [CrossRef]
33. Li, X.H.; McCulloch, M.T. Geochemical characteristics of Cretaceous mafic dikes from northern Guangdong, SE China: Age, origin and tectonic significance. Mantle Dynamics and Plate Interactions in East Asia. *AGU Geodyn. Ser.* **1998**, *27*, 405–419.
34. Whalen, J.B.; Currie, K.L.; Chappell, B.W. A-type granites: Geochemical characteristics, discrimination and petrogenesis. *Contrib. Mineral. Petrol.* **1987**, *95*, 407–419. [CrossRef]
35. Liu, P.; Mao, J.W.; Cheng, Y.B.; Yao, W.; Wang, X.Y.; Hao, D. An Early Cretaceous W-Sn deposit and its implications in southeast coastal metallogenic belt: Constraints from U–Pb, Re-Os, Ar-Ar geochronology at the Feie'shan W-Sn deposit, SE China. *Ore Geol. Rev.* **2017**, *81*, 112–122. [CrossRef]
36. Liu, P.; Mao, J.W.; Santosh, M.; Xu, L.G.; Zhang, R.Q.; Jia, L.H. The Xiling Sn deposit, Eastern Guangdong Province, Southeast China: A new genetic model from $^{40}Ar/^{39}Ar$ muscovite and U–Pb cassiterite and zircon geochronology. *Econ. Geol.* **2018**, *113*, 511–530. [CrossRef]

37. Li, X.F.; Yi, X.K.; Huang, C.; Wang, C.Z.; Wei, X.L.; Zhu, Y.T.; Xu, J. Zircon U–Pb, molybdenite Re-Os and muscovite Ar-Ar geochronology of the Yashan W-Mo and Xiatongling W-Mo-Be deposits: Insights for the duration and cooling history of magmatism and mineralization in the Wugongshan district, Jiangxi, South China. *Ore Geol. Rev.* **2018**, *102*, 1–17. [CrossRef]

38. Feng, Z.; Kang, Z.; Yang, F.; Liao, J.; Wang, C. Geochronology of the Limu W–Sn–Nb–Ta-Bearing Granite Pluton in South China. *Resour. Geol.* **2013**, *63*, 320–329. [CrossRef]

39. Guo, C.; Mao, J.; Bierlein, F.; Chen, Z.; Chen, Y.; Li, C.; Zeng, Z. SHRIMP U–Pb (zircon), Ar–Ar (muscovite) and Re–Os (molybdenite) isotopic dating of the Taoxikeng tungsten deposit, South China Block. *Ore Geol. Rev.* **2011**, *43*, 26–39. [CrossRef]

40. Collins, W.J.; Beams, S.D.; White, A.J.R. Nature and origin of A-type granites with particular reference to southeastern Australia. *Contrib. Mineral. Petrol.* **1982**, *80*, 189–200. [CrossRef]

41. Jiang, Y.H.; Jiang, S.Y.; Ling, H.F.; Zhou, X.R.; Rui, X.J.; Yang, W.Z. Petrology and geochemistry of shoshonitic plutons from the western Kunlun orogenic belt, northwestern Xinjiang, China: Implications for granitoid geneses. *Lithos* **2002**, *63*, 165–187. [CrossRef]

42. Creaser, R.A.; Price, R.C.; Wormald, R.J. A-type granites revisited: Assessment of a residual source model. *Geology* **1991**, *19*, 163–166. [CrossRef]

43. Turner, S.P.; Foden, J.D.; Morrison, R.S. Derivation of some A-type magmas by fractionation of basaltic magma: An example from the Padthaway Ridge, South Australia. *Lithos* **1992**, *28*, 151–179. [CrossRef]

44. Lapierre, H.; Jahn, B.M.; Charvet, J. Mesozoic felsic arc magmatism and continental olivine tholeiites in Zhejiang province and their relationship with the tectonic activity in southeastern China. *Tectonophys* **1997**, *274*, 321–338. [CrossRef]

45. Jiang, Y.H.; Jiang, S.Y.; Dai, B.Z.; Liao, S.Y.; Zhao, K.D.; Ling, H.F. Middle to Late Jurassic felsic and mafic magmatism in southern Hunan Province, Southeast China: Implications for a continental arc to rifting. *Lithos* **2009**, *107*, 185–204. [CrossRef]

46. Li, Z.X.; Li, X.H. Formation of the 1300-km-wide intracontinental orogen and postorogenic magmatic province in Mesozoic South China: A flat-slab subduction model. *Geology* **2007**, *35*, 179–182. [CrossRef]

47. Wilson, G.A. Cassiterite solubility and Metal Choride Speciation in Supercritical Solutions. Ph.D. Thesis, John Hopkins University, Baltimore, MD, USA, 1986.

 *minerals*

Article

Petrogenesis of Low Sr and High Yb A-Type Granitoids in the Xianghualing Sn Polymetallic Deposit, South China: Constrains from Geochronology and Sr–Nd–Pb–Hf Isotopes

ChangHao Xiao * , YuKe Shen and ChangShan Wei

Laboratory of Dynamic Diagenesis and Metallogenesis, Institute of Geomechanics,
Chinese Academy of Geological Sciences, Beijing 100081, China; shenyuke@geomech.ac.cn (Y.K.S.);
weichangshan@geomech.an.cn (C.S.W.)
* Correspondence: xiaochanghao1986@126.com or xiaochanghao@geomech.ac.cn; Tel.: +86-010-88815612

Received: 16 January 2019; Accepted: 11 March 2019; Published: 15 March 2019

Abstract: The nature and origin of the early Yanshanian granitoids, widespread in the South China Block, shed light on their geodynamic setting; however, understanding their magmatism processes remains a challenge. In this paper, we present both major and trace elements of bulk rock, Sr–Nd–Pb isotopic geochemistry, and zircon U–Pb–Hf isotopes of the low Sr and high Yb A_2-type granites, which were investigated with the aim to further constrain their petrogenesis and tectonic implications. Zircon U–Pb dating indicates that these granites were emplaced at ca. 153 Ma. The granites are characterized by high SiO_2 (>74 wt.%) and low Al_2O_3 content (11.0 wt.%–12.7 wt.%; <13.9 wt.%). They are enriched in large ion lithophile elements (LILEs) (e.g., Rb, Th, U, and K) and Yb, but depleted in high field-strength elements (HFSEs) (e.g., Nb, Ta, Zr and Hf), Sr, Ba P, Ti and Eu concentrations. They exhibit enriched rare earth elements (REEs) with pronounced negative Eu anomalies. They have $\varepsilon_{Nd}(t)$ values in a range from −6.5 to −9.3, and a corresponding T_{DM} model age of 1.5 to 1.7 Ga. They have a $(^{206}Pb/^{204}Pb)_t$ value ranging from 18.523 to 18.654, a $(^{207}Pb/^{204}Pb)_t$ value varying from 15.762 to 15.797, and a $(^{208}Pb/^{204}Pb)_t$ value ranging from 39.101 to 39.272. The yield $\varepsilon_{Hf}(t)$ ranges from −6.1 to −2.1, with crustal model ages (T_{DMC}) of 1.3 to 1.6 Ga. These features indicate that the low Sr and high Yb weakly peraluminous A_2-type granites were generated by overlying partial melting caused by the upwelling of the asthenosphere in an extensional tectonic setting. The rollback of the Paleo-Pacific Plate is the most plausible combined mechanism for the petrogenesis of A_2-type granites, which contributed to the Sn–W polymetallic mineralization along the Shi-Hang zone in South China.

Keywords: zircon U–Pb geochronology; Sr–Nd–Pb–Hf isotopes; low Sr and high Yb A_2-type granite; Xianghualing; South China

---

1. Introduction

The Mesozoic Age marks an important period in the geologic evolution of mainland South East Asia (Figure 1a) [1–4], during which extensive magmatism took place in the South China Block (SCB) parallel to the present-day coastline (Figure 1b). This magmatic zone was preliminarily named the "Shi-Hang zone" after Gilder et al. [4]. As a granitoid belt, it reveals the Nd-depleted mantle model ages (T_{DM} = 1.4 ± 0.3 Ga) and negative $\varepsilon_{Nd}(t)$ values (−4 to −8) in composition [5]. Such Sn–W polymetallic metallogenic domains related to widespread granitoids in South China represents one of the largest tungsten-tin polymetallic ore provinces on Earth [6–15]. Previous studies demonstrated that these temporally and spatially associated intrusions are different in terms of petrography, and elemental and isotopic compositions, and argued for different petrogenesis and tectonic regimes, ranging from an extension [11], rift-related and convergent [5,16–19], or subduction-related models (Figure 1d) [20–24].

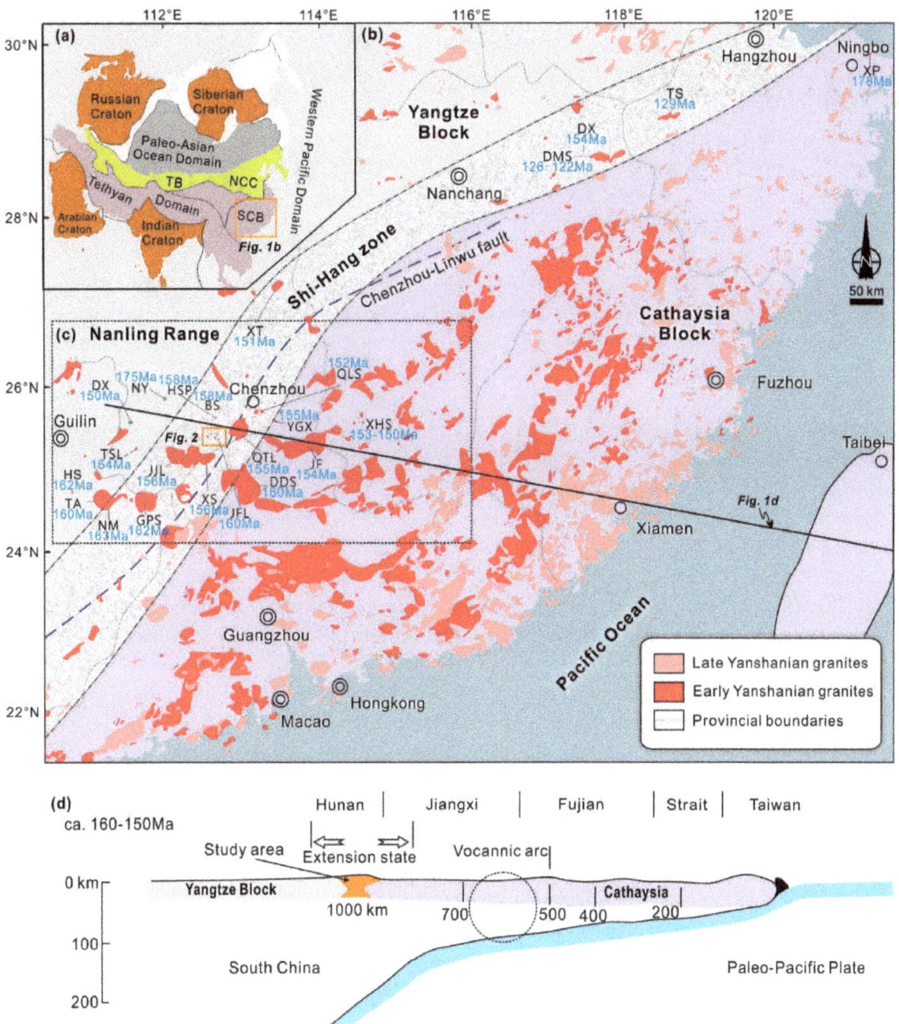

Figure 1. (**a**) Simplified tectonic map of Asia showing the framework of the joint area of the Paleo–Asian, Tethyan, and Pacific domains [4]. TB, Tibet Block; NCC, North China Block; SCB, South China Block. (**b**) Distribution of the Mesozoic intrusive rocks in South China [7,15,21,24,25]. The locations of the Shi-Hang zone and Nangling range are after [19,26,27]. (**c**) Sketch map of the Nanling Range [13]. (**d**) Simplified map showing the Late Jurassic subduction in Southeast China [28]. NM, Niumiao granodiorite [29]; TA, Tong'an monzonite [29]; HS, Huashan granite [30,31]; GPS, Guposhan granite [30,31]; TSL, Tongshanling granodiorite [19,32]; JJL, Jinjiling granite [19,33]; XS, Xishan granite [33]; DDS, Dadongshan granite [34]. JFL, Jianfengling (our no published data); DX, Daoxian basalt [19,35]; NY, Ningyuan basalt [19,35]; QTL, Qitianling granite [36]; HSP, Huangshaping granite [37]; BS, Baoshan granodiorite [32,38]; JF, Jiufeng granite [39,40]; QLS, Qianlishan granite [41]; YGX, Yaogangxian granite [42]; XHS, Xihuashan granite [43,44]; XT, Xitian granite [15]; DMS, Damaoshan granite [45]; DX, Dexing granodiorite [46]; TS, Tongshan granite [45]; XP, Xiepu granite [24].

The Xianghualing Sn polymetallic deposit is situated at the boundary between the Paleo-Pacific and the Tethyan tectonic domains (Figure 1a). The Late Jurassic granites related to this deposit could provide constraints on the petrogenesis of Late Jurassic granite-related W–Sn mineralization and their tectonic settings. In this contribution, we present major and trace element geochemistry, Sr–Nd–Pb isotope data for these granites, and zircon U–Pb dating and Hf isotopic evidence.

2. Geological Background

The Yangtze Block to the northwest and the Cathaysia Block to the southeast were amalgamated to form the South China Block during the Neoproterozoic Era (Figure 1b) [7,11]. Numerous granites are distributed in the Cathaysia Block and the Shi-Hong zone. The study area is located in the central part of Shi-Hang zone (Figure 1b). The NE–SW trending of the Chenzhou-Lingwu fault zone represents a regional fault in the study area (Figure 1b) [25,47]. The zone was originally formed at 970 Ma [1] and reactivated during the Triassic and the Cretaceous Eras [25]. A number of granite plutons and associated Sn polymetallic mineralization were distributed along the Chenzhou-Lingwu fault belt to the west of Shi-Hong zone (Figure 1b). This Sn-metallogenic region, called the Nanling Range, is characterized by multiple and diverse mineral deposits (W, Sn, Cu, Pb–Zn, etc.) and the Jurassic-Cretaceous intrusions [13]. The granites include the Baoshan, Tongshanling, Niumiao, Yuanzhuding, Guposhan, Huashan, Xishan, Jinjiling, Qianlishan, and Qitianling plutons (Figure 1b). Ore deposits associated to this belt are: Shizhuyuan (W–Sn–Mo–Bi–F) [48], Furong (Sn) [6,49,50], Yaogangxian (W) [49], Xianghualing (Sn–W–Pb–Zn) [12,51], Huangshaping (Pb–Zn–W–Sn) [7,52], Baoshan (Cu–Pb–Zn–W) [38], and Yuanzhuding (Cu–Mo) [14]. Basaltic rocks also crop out near the Chenzhou-Lingwu fault, such as the Daoxian and Ningyuan basalts (Figure 1b) [19,35].

The Xianghualing Sn-polymetallic deposit is situated at the Midwestern point of the Nanling Range (Figure 1c). The Xinfeng mine is one of the most important mines in the Xianghualing deposit. The Xianghualing area is a tectono-magmatic dome (Figure 2a). The exposed basement comprises the Cambrian slate and metasandstone, sandstone and shale in the Middle Devonian Tiaomajian Formation, limestone and dolomitic limestone in the Middle Devonian Qiziqiao Formation, dolomitic limestone and sandstone in the Upper Devonian of the Shetianqiao Formation, and the Carboniferous carbonate and clastic rocks. The Qiziqiao Formation is the major ore bed, and the Laiziling and Jianfengling are the two largest granites which intruded into the Cambrian and Devonian Eras, respectively. The Xianghualing Sn polymetallic deposit is a typical skarn deposit related Laiziling Pluton [12]. Skarn-type ore bodies and vein-like bodies formed in the contact zone of the granites.

3. Sampling and Analytical Methods

The granite samples in this study included ZK65-1, ZK65-2, ZK65-3 and ZK57-1 and were collected from drill cores ZK65 and ZK57, one of the deepest holes of Laiziling pluton prior to 2015, in the Xinfeng mine, Xianghualing deposit (Figure 2b). The sequence from the granite to the country rocks is as follows: granite → altered sandstone → skarn → Sn polymetallic ores in the drill cores ZK65, or, alternatively, granite → skarn → tungsten–bearing quartz veins → ore-bearing marble → fracture in drill core ZK57 (Figure 2b). The pluton in the mine connects with the Laiziling pluton that exposes on the surface. Three granite samples collected from the drill core ZK65 are mainly composed of fine to medium grained porphyritic biotite monzonitic granites with a porphyritic texture. They contain 30–35% rounded, variably resorbed quartz phenocrysts of 1–8 mm in diameter, 10–15% euhedral-subhedral K-feldspar phenocrysts of 5–10 mm in length, about 5–10% equant and tabular plagioclase phenocrysts of 0.5–2 mm in length, biotite (3–8%), and minor to trace amounts of hornblende. (Figure 3a,b). Some K-feldspar phenocrysts have partially experienced sericitization. Sample ZK57-1 is two-mica monzonitic granite with small phenocrysts and finer-grained quartz (30–35%), K-feldspar (10–15%), plagioclase (10–15%), biotite (3–8%), and muscovite (3–8%) (Figure 3c,d). Accessory minerals are composed of zircon with minor apatite.

Details of analytical methods, namely whole-rock major and trace element compositions, Sr–Nd–Pb isotopes, and zircon U–Pb and Lu–Hf isotope analyses are presented in the Supplementary material.

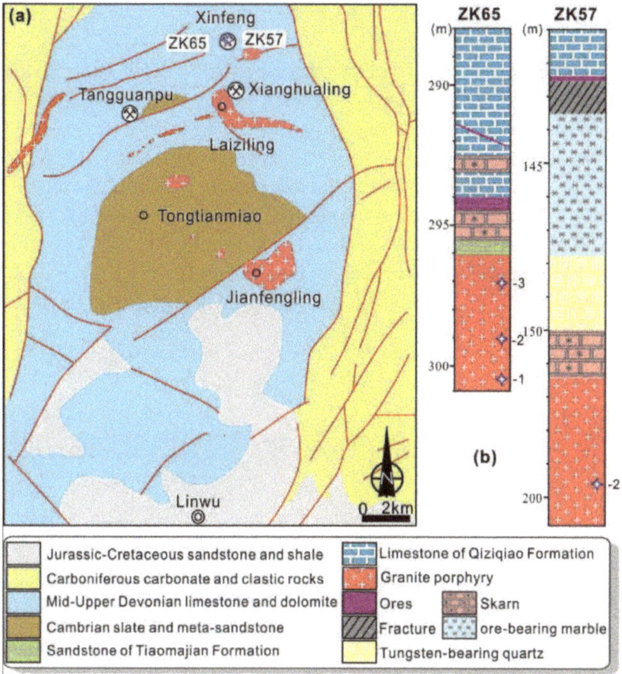

Figure 2. (**a**) The simplified geological map of the Xianghualing Sn-polymetallic deposit, South China [12,53]. (**b**) Drill core profile showing the sample locations.

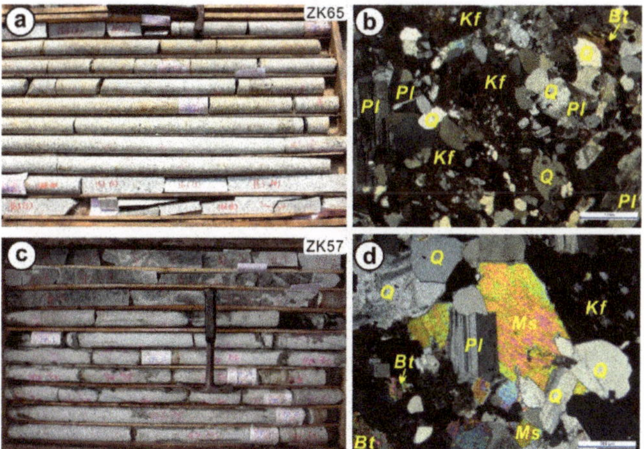

Figure 3. Drill cores and micrographics of ZK65 and ZK57 from the Xinfeng mine, Xianghualing area, South China. (**a**,**b**) Biotite monzonitic granites from ZK65. (**c**,**d**) Two-mica monzonitic granite from ZK57. Q, quartz; Kf, feldspar; Bt, biotite; Ms, muscovite; Pl, plagioclase.

4. Results

4.1. Whole-Rock Geochemistry

4.1.1. Major and Trace Elements

The analytical results of the four granite samples are presented in Table S1. These granites have characteristics of high SiO_2 content (74.1 wt.%–78.0 wt.%) (Figure 4a), similar to that of the Jianfengling granites (73.6 wt.%–75.2 wt.%) [53], and Qitianling granites (65.9 wt.%–75.7 wt.%) in this region [54,55]. They are enriched in alkalis, with the $K_2O + Na_2O$ contents ranging from 7.2 wt.% to 9.2 wt.% (average 8.3 wt.%), and K_2O/Na_2O ratios higher than 1. The samples are plotted in the high K calc-alkaline field in the $Na_2O + K_2O - CaO$ vs. SiO_2 diagram (Figure 4a). They exhibit low Al_2O_3 contents (11.0 wt.% to 12.7 wt.%), and A/CNK ratios (1.0–1.1) lower than 1.1, indicating their weak peraluminous affinity in the A/NK vs. A/CNK diagram (Figure 4b). They have low Fe_2O_3 (0.1 wt.%–0.5 wt.%), MgO (0.1 wt.%), CaO (0.6 wt.%–0.8 wt.%), TiO_2 (0.1 wt.%), and P_2O_5 contents.

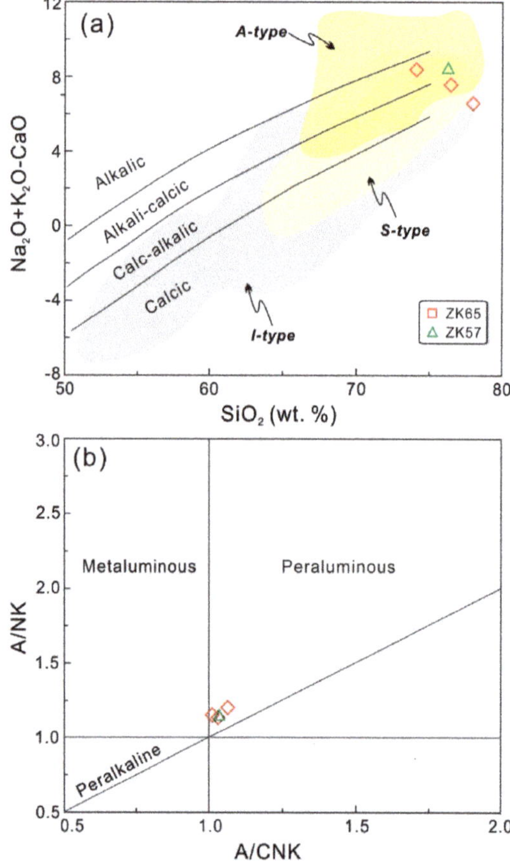

Figure 4. (**a**) Diagram of SiO_2 vs. $Na_2O + K_2O - CaO$ [56]; (**b**) Diagram of Al/(K+Na) vs. Al/(Ca + K + Na) [57].

A primitive mantle-normalized trace element diagram and chondrite-normalized rare earth element (REE) patterns for the four granite samples are illustrated in Figure 5a,b, respectively. All samples have low Sr (3.6–7.3 ppm) and Nb (0.3–2.5 ppm, except ZK65–1) contents and extremely high Rb (1606–2351 ppm) and Yb (7.9–14 ppm) contents. They are enriched in large-ion lithophile elements (LILEs) (e.g., Rb, Th, U, and K) and depleted in high field strength elements (HFSEs) (e.g., Nb, Ta, Zr and Hf), Sr, and Ba (Figure 5a). The total contents of Zr, Nb, Ce, Y of four samples range from 430 to 646 ppm (Figure 6). The granites exhibit enriched REEs (except Eu) with a total REE ranging from 233 to 312 ppm. They are characterized by a slight enrichment of light REE (LREE) $((La/Yb)_N = 1.8–5.0)$ and flat heavy REE (HREE) with pronounced negative Eu anomalies (δEu = 0.02) (Figure 5b). The differentiation indexes (DI) of these samples are 93.7–94.58 (Table S1).

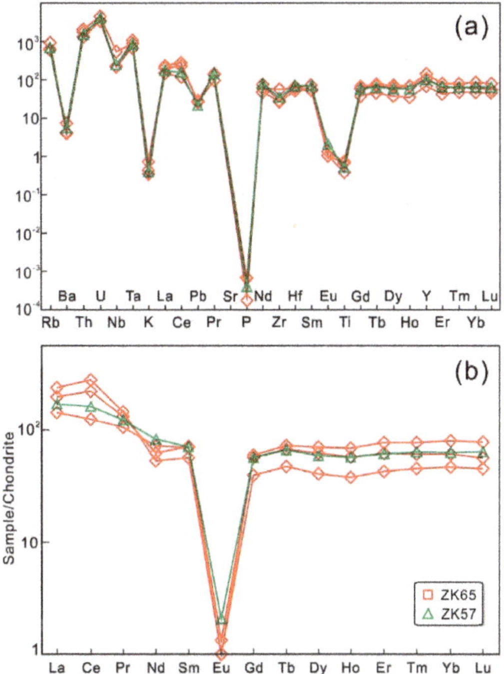

Figure 5. (**a**) Primitive mantle-normalized trace element spider diagram; (**b**) Chondrite-normalized REE patterns for the granites from the Xinfeng mine, Xianghualing area, South China. Normalized values for primitive mantle and chondrite are from Sun and McDonough [58].

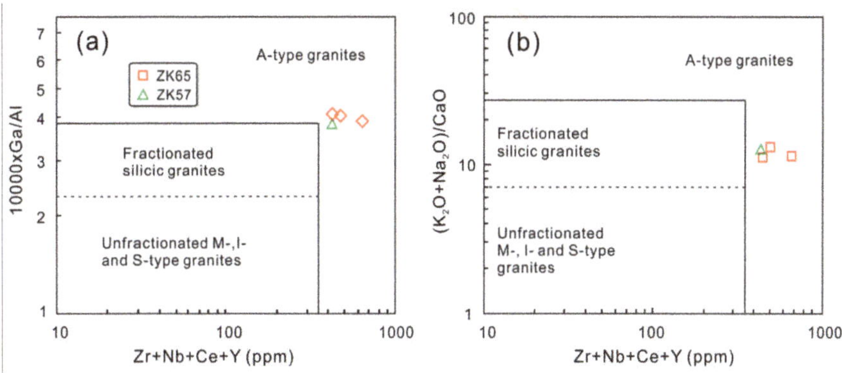

Figure 6. (**a**) 10,000 × Ga/Al and; (**b**) $(K_2O + Na_2O)/CaO$ vs. Zr + Nb + Ce + Y (ppm) diagrams showing that the Jurassic low Sr and high Yb granites are A-type granites [59].

4.1.2. Sr–Nd–Pb Isotopic Compositions

The Sr–Nd–Pb isotopic composition data for the granites are presented in Table S2. Initial values of $(^{206}Pb/^{204}Pb)_t$, $(^{207}Pb/^{204}Pb)_t$, and $(^{208}Pb/^{204}Pb)_t$ are calculated using zircon ages of 153 Ma.

The granites have an extremely high $^{87}Sr/^{86}Sr$ ratio (up to 1.89), which may be due to their extreme Sr (also Eu and Ba) depletion, and thus a very high Rb/Sr ratio. Previous studies pointed that the initial of Sr isotope is unreliable of high Rb granite [60]. We therefore do not discuss the initial ratio of Sr isotopes on the petrogenesis of the Xianghualiang granites. They have $\varepsilon_{Nd}(t)$ values ranging from -9.3 to -6.5 and $(^{143}Nd/^{144}Nd)_i$ from 0.511963 to 0.512107, corresponding to depleted mantle model ages $(T_{DM2}–Nd)$ of 1471–1702 Ma. The plots of $\varepsilon_{Nd}(t)$ vs. $^{206}Pb/^{238}U$ age is exhibited in Figure 7a.

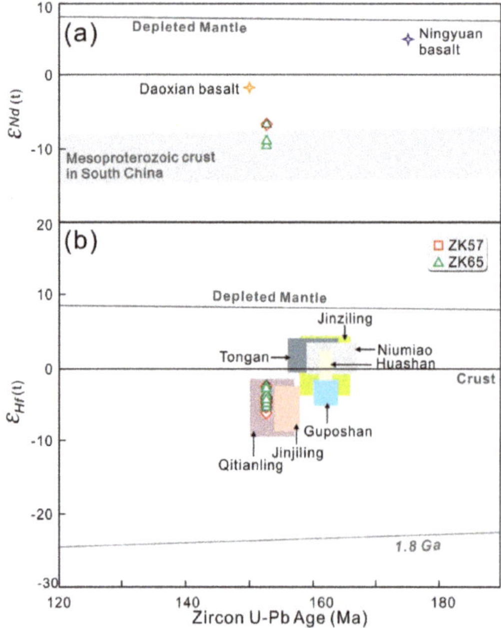

Figure 7. (a) $\varepsilon_{Nd}(t)$-age diagram and (b) $\varepsilon_{Hf}(t)$-age diagram for zircons from Xianghualing area. The data of zircon U–Pb age, $\varepsilon_{Nd}(t)$ and $\varepsilon_{Hf}(t)$ are from [7,15,21,24–45].

Four samples are characterized by high radiogenic Pb isotopic composition, with present-day rock $^{206}Pb/^{204}Pb$, $^{207}Pb/^{204}Pb$, and $^{208}Pb/^{204}Pb$ ratios of 19.624 to 19.791, 15.816 to 15.847, and 39.478 to 39.795, and their corresponding initial ratios are 18.523 to 18.654, 15.762 to 15.797, and 39.101 to 39.272, respectively.

4.2. Zircon U–Pb Ages

Most zircon grains separated from four granite samples are euhedral and prismatic with aspect ratios of 1:1–1:2 and lengths of 50–150 μm. They are transparent and light yellow under an optical microscope. Ubiquitous simple internal oscillatory zoning and little inherited cores are observed by Cathodoluminescence (CL) images (Figure 8). Such characteristics indicate they are magmatic zircon in origin [61]. A few zircon crystals exhibit complex secular zonings (Figure 8).

LA-ICP-MS zircon U–Pb isotopic data for these granitoids are shown in Table S3. The zircon U–Pb concordia and weighted mean diagrams are illustrated in Figure 9. These zircons have varied U (154–6771 ppm) and Th (70–4505 ppm) contents, with Th/U ratios ranging from 0.3 to 0.8 (Table S3), indicating that they are magmatic in origin [62]. Thirteen analyses fall on the Concordia in a single group from the sample ZK57, yielding a Concordia age of 152.8 ± 0.6 Ma (MSWD = 0.2, *n* = 13). The remaining three analyses have a ^{206}Pb/^{238}U age of 148.3 ± 1.5 Ma and 147.1 ± 1.8 Ma (Figure 9a). Fifteen spot analyses of zircons from sample ZK65 yield a single ^{206}Pb/^{238}U age population of 105–156 Ma with Concordia age of 152.7 ± 2.0 Ma (MSWD = 2.9, *n* = 9) (Figure 9b and Table S3). The two Concordia ages may represent the crystallization age of the granites.

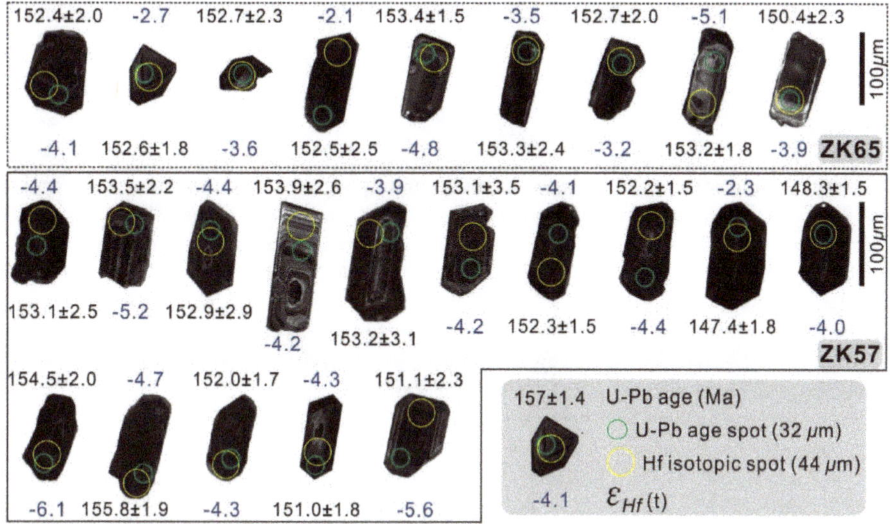

Figure 8. CL electron micrographs of representative zircons separated from the drill cores of ZK65 and ZK57 from the Xinfeng mine, Xianghualing area, South China. Spots are tested by in situ U–Pb age and Hf isotope.

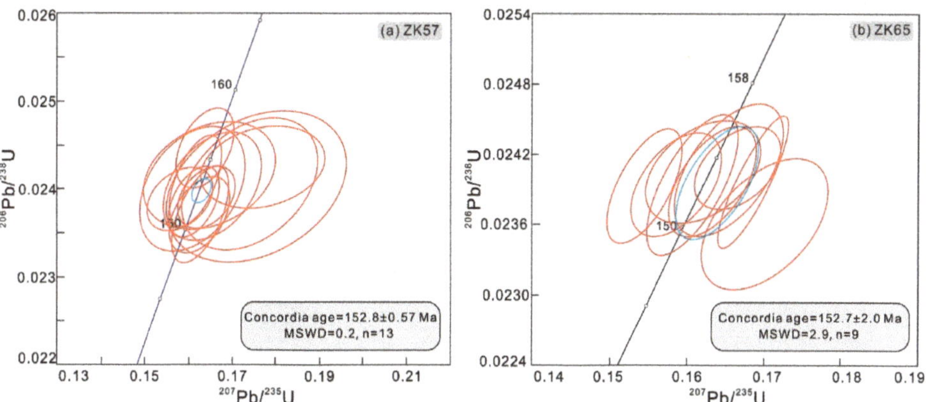

Figure 9. LA–CP–MS zircon U–Pb Concordia diagrams for the drill cores of ZK57 (**a**) and ZK65 (**b**) from the Xinfeng mine, Xianghualing area, South China.

4.3. Zircon Hf Isotopes

The Lu–Hf isotopic compositions of zircon crystals separated from the two granite samples from the Xianghualing area in South China are listed in Table S4. The plots of $\varepsilon_{Hf}(t)$ vs. ^{206}Pb/^{238}U age is illustrated in Figure 7b.

The zircons have $^{176}Lu/^{177}Hf$ ratios from 0.000776 to 0.003445 and $^{176}Hf/^{177}Hf$ ratios from 0.282514 to 0.282628 (Table S4). Most of $^{176}Lu/^{177}Hf$ ratios of zircon from ZK57 are below 0.002, and therefore the accumulation of radiogenic Hf after the formation of zircons can be ignored. $^{176}Lu/^{177}Hf$ ratios of zircon from ZK65 are a little large than 0.002 except for two analyses (ZK65-1, ZK65-5), indicating a minor radiogenic production of ^{176}Hf. The $\varepsilon_{Hf}(t)$ values for analyses of zircon from the samples ZK57 and ZK65 vary from −6.1 to −2.1 (Figure 7b) and yield crustal model ages (T_{DMC}) for the zircon crystals from the samples that have a spectrum from 1336 Ma to 1588 Ma. The $\varepsilon_{Hf}(t)$ values of ZK57 are close and smaller than those of ZK56 (Table S4).

5. Discussion

5.1. Origin of Granitic Rocks: An A_2-Type Affinity

The later Jurassic granitic rocks from the Xianghualing area exhibit features of high silica (74.1–78.0 wt.%) and weak peraluminosity (A/CNK ratios = 1.0–1.1). They have high K_2O contents (K_2O/Na_2O ratios = 1.4–4.3) and low CaO, Ba, and Sr contents. They are enriched in Nb, and Ta, and HFSE contents (Zr + Nb + Ce + Y = 430–646 ppm) (Figure 6b). They have HFSE contents higher than those of A-type granites (350 ppm) [59]. Chondrite-normalized REE plots show relatively flat patterns with large negative Eu anomalies. These features are consistent with those of A-type granites rather than S and I-types [63–66]. They are characterized by extremely low P_2O_5 abundances and limited phosphate minerals, compared with S-type granites.

Strontium contents have been acknowledged as a discriminating parameter to classify granites of A-type granites [63]. The granitic rocks from the Xianghualing area have extremely lower Sr contents (<10 ppm) than typical calc-alkaline I-type granites. Zhang et al. demonstrated that A-type granites in South China are characterized by low Sr and high Yb, using a Sr–Yb diagram with obvious V-type REE patterns [67]. The much lower Sr contents therefore suggest that the granites from the Xianghualing area belong to A-type granites. Biotite and hornblendes commonly exist as interstitial clots or grains (Figure 10a). Moreover, micrographic intergrowths of quartz and myrmekite usually develop in or around the alkali feldspar (Figure 10). These features are consistent with those of typical A-type granites [59,68]. In the 10,000×Ga/Al and (K_2O + Na_2O)/CaO vs. Zr + Nb + Ce + Y (ppm) diagrams, as well as the SiO_2 vs. Na_2O + K_2O − CaO diagrams, they plot within A-type granites field (Figures 4a and 6a,b). All the samples fall into the A_2-type granites field in the plot of Ce/Nb versus Y/Nb and Nb-Y-3Ga discrimination diagrams (Figure 11a,b), and this is also evident in the plot of the R1-R2 diagram [69] (Figure 11c).

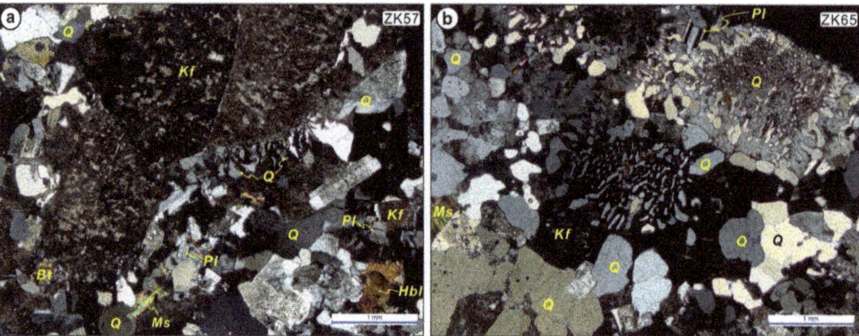

Figure 10. (a) Micrographic intergrowths of quartz and myrmekite develop around the alkali feldspar from the ZK57; (b) The alkali feldspar was replaced by the quartz from the ZK65. Q–quartz, Kf–feldspar, Bt–biotite, Ms–muscovite, Pl–plagioclase; Hbl–hornblende.

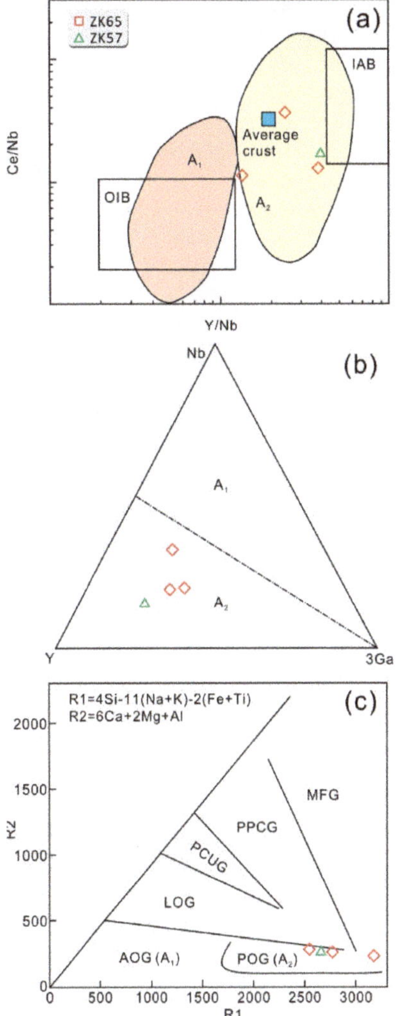

Figure 11. (**a**) Plots of Ce/Nb vs. Y/Nb [70]; (**b**) Nb-Y-3Ga [70]; (**c**) the diagrams of Ta-Yb, "R1-R2" [69] for the Xianghualing granites. OIB, ocean island basalt; IAB, island arc basalt; MFG, mantle-fractionated granite; PPG, pre-plate collision granitoids; PUCG, post-collision uplift granitoids; LOG, late-orogenic granitoids; AOG, anorogenic granitoids; POG, post-collision granitoids.

5.2. Petrogenesis

The petrogenesis of A-type granites are still debated, and several petrogenetic schemes have been proposed [64,69–75]. The Xianghualing granites have low A/CNK values and no aluminous minerals, which is inconsistent with the metasedimentary-melting petrogenetic model [76]. The rocks exhibit flat heavy REE patterns and high Y contents (116–235 ppm) indicating that garnet was absent from the source reservoir during the partial melting process [73]. The rocks from Xianghualing show weakly peraluminous affinities, which are consistent with crustal magmas produced by partial melting and heated by the underplating of mantle-derived magma [77–79]. This is evident by the obviously different Nd–Hf isotopic compositions between the samples and the coeval mafic rocks (Figures 7a and 12). Figures 7a and 12 also indicate that the extensive fractional crystallization from coeval mafic magmas for the origin of the A_2-type granites can be ruled out.

The granites exclusively display negative zircon $\varepsilon_{Hf}(t)$ values, ranging from -6.1 to -2.1, with a corresponding crustal model age of 1.34 to 1.59 Ga (Figure 7b and Table S4). The whole rock Nd model age is 1.47 Ga–1.70 Ga, consistent with the Hf model age. These suggest that the sources of materials had a relatively simple recycling process. The relatively young Nd model ages are a little younger than or similar to that of Mesoproterozoic sediments in South China Block (Figure 7a; T_{DM} = 1.8 Ga) [80]. This indicates a dominated ancient crust material in South China contributed during the magmatic process. The $\varepsilon_{Hf}(t)$ vs. $\varepsilon_{Nd}(t)$ plots display that the range of granites from Xianghualiang area is similar to the Shi-Hang zone [29–31,33,54,81,82], and the range falls into the field of superimposed areas between global lower crust and global sediments (Figure 12). It suggests the lower crust and/or sediments could be involved during the magmatic process. The lead isotopic values of the four samples from the Xianghualing area display limited variations near the upper crust curve in Figure 13, which is similar to the crust materials in the Shi-Hang zone and the western Pacific subducted sediments at trenches, suggesting that the nature source rocks could be dominated by crustal materials [83]. The high radiogenic lead isotopic compositions in the granites from the Xianghualing area support that the magmas for these granites are predominately derived from crustal materials.

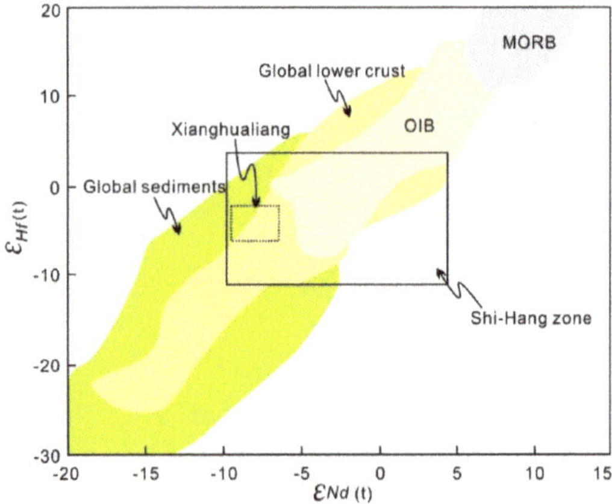

Figure 12. $\varepsilon_{Nd}(t)$ vs. $\varepsilon_{Hf}(t)$ for the granites from Xianghualing area South China. Plots of $\varepsilon_{Nd}(t)$ vs. $\varepsilon_{Hf}(t)$ fall into the quadrangle. The average values concentrate in the dotted quadrangle. The data of $\varepsilon_{Nd}(t)$ and $\varepsilon_{Hf}(t)$ are from [29–31,33,54,81,82]. MORB, mid-ocean ridge basalt.

As shown in Table S1 and Figure 11a, the low Gd/Yb ratios and fairly high Y/Nb ratios indicate that the granites formed in the extensional setting [63,84–86]. This is consistent with the support formative processes of A_2-type granites, which are considered to crystallize at a high temperature [72,74,75]. The presence of micrographic texture in these A_2-type granites is also indicative of a high-level emplacement and provides evidence for an extensional regime (Figure 10) [69,72] and that the magmas formed near the earth surface [68]. A normal geothermal gradient, however, cannot produce high-temperature A-type granites by crustal melting; therefore, an exotic heat source from mantle is a prerequisite. In conclusion, a reasonable explanation for the granites with the signatures of high temperature from the Xianghualing area was generated by partial melting of the crustal materials with minor subducted sediments, and further caused by upwelling of asthenosphere in an extension tectonic setting. The presence of coeval basaltic rocks near the Chenzhou-Lingwu fault supports that a lithospheric extension event could occur during the Early Jurassic Era, such as the Ningyuan alkaline basalts (175 Ma) and the Daoxian basalt (150 Ma) [87–89]. These upwelling mantle materials might provide heat energy for the melting of the crustal materials and subducted sediments.

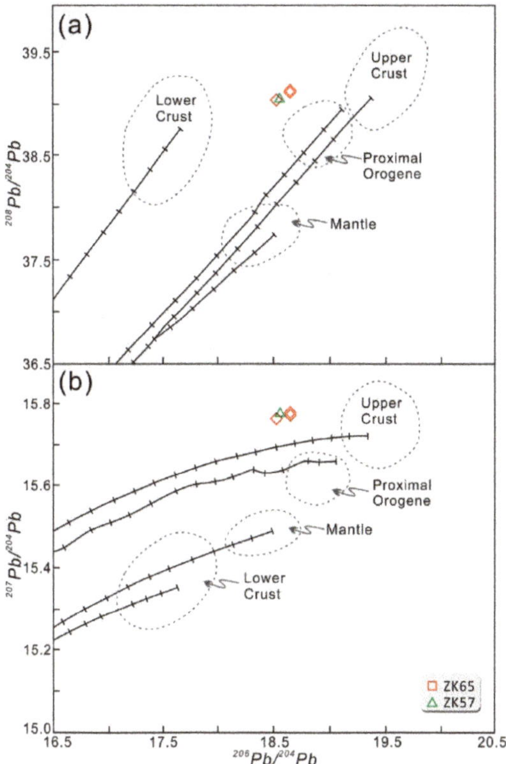

Figure 13. Pb isotopic composition of the Jurassic weak peraluminous granites are A-type granites. (a) ^{206}Pb/^{204}Pb vs. ^{208}Pb/^{204}Pb diagram of samples; (b) ^{206}Pb/^{204}Pb vs. ^{207}Pb/^{204}Pb diagram of samples [83].

5.3. Tectonic Implications

The A$_2$-type granite has been generally considered to be derived from the crust in the subduction zone or the continent–continent collision zone [15]. Various models for the formation of A-type granites in South China are proposed in terms of tectonic setting, and these have been correlated with the subduction of the Pale-Pacific Plate for the petrogenesis of the late Mesozoic magma zone [15,24,28,90–94]. Zircon U-Pb dating on granites in the Xianghualing area demonstrate that the granites were emplaced at ~153 Ma. The age is close to those of A-types granites in the Shi-Hang zone, such as Xihuashan (150–153 Ma, [43,44]), Jiufeng (154 ± 1 Ma, [39,40]), Xitian (152 ± 1 Ma, [15]), Qianlishan (152 ± 2 Ma, [18]), Qitianling (155 ± 1 Ma, [36]), Xishan (156 ± 2 Ma, [33]), Jinjiling (156 ± 2 Ma, [19,33]), Huashan (162 ± 2 Ma, [30,31]), and Guposhan (162 ± 2 Ma, [19,32]). These granites formed a north east trending A-type granite belt in the Shi-Hang zone.

We thus propose that these A$_2$-type granites could be formed in an extensional setting (Figure 1d). The Paleo-Pacific Plate subducted underneath the SE China Block at a very low angle, beginning in the Late Triassic Era and reaching southern Hunan at ca. 174 Ma (Figure 14a) [24,94,95]. The low angle subduction could have caused crustal thickening in the coastal area. Due to the temperature and pressure increases, the ferroan (A-type) granitoids were emplaced in the orogenic thermal relaxation regime [24]. From ca. 174 to 164 Ma, the slab dip angle increased and subsequently caused the subducting slab to dehydrate and release fluids. The slab-released fluids triggered partial melting of mantle wedge and generated basaltic magmas. Around or above the melt zone, the Middle

Jurassic I- and S-type magmas formed in the region, such as in Tongshanling (~164 Ma) in Southern Hunan (Figures 1d and 14b) [19]. The rollback of the Paleo-Pacific plate led to a regional extension during the late Jurassic Era (163–150 Ma) (Figure 14c) [18,19,61,92,93,96]. The extension caused the crust and lithospheric mantle to become thinner, with an accompanying asthenosphere upwelling. The upwelling of the basaltic magmas might have provided heat energy and triggered a partial melting of the thinned crust rocks and subducted sediments to form the ca. 163–150 Ma A$_2$-type granites along the Shi-Hang zone.

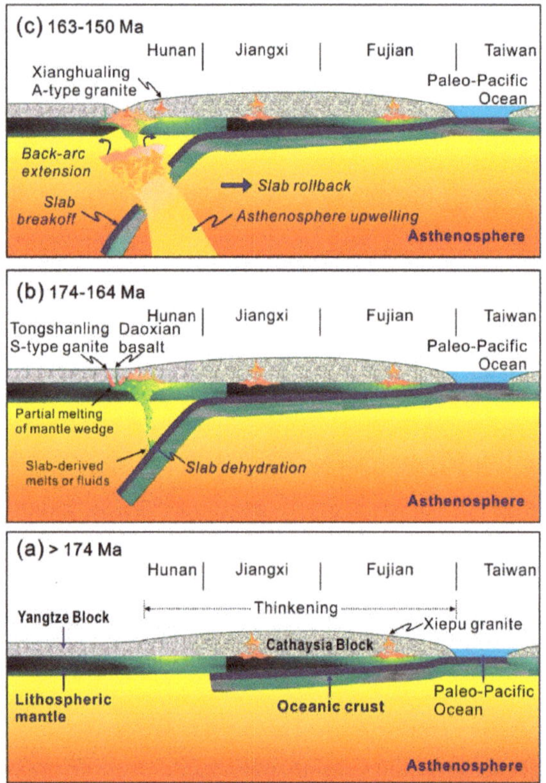

Figure 14. Schematic illustrations showing the generation of the late Jurassic A-type granites in South China. (**a**) Low angle subduction before ~174 Ma, forming the middle Jurassic S-type granites such as the Xiepu granite; (**b**) High-angle subduction during 174 to 164 Ma, forming S-type granites and basalt such as the Tongshanling S-type granite and the Daoxian basalt; (**c**) During the late Jurassic Era, slab rollback and slab breakoff formed the A-type granite such as the Xianghualing granite.

5.4. Implication for the Generation of Tin Mineralization

Yuan et al. [12] gave a cassiterite U–Pb weighted average age of 156 ± 4 Ma. The ore-forming age coincidesdwith the other W–Sn–Pb–Zn deposits in the Shi-Hang zone, such as the Shizhuyuan W deposit (Re–Os, 151 ± 4 Ma, [97]; Sm–Nd, 161 ± 2 Ma, [98]; Sm–Nd, 149 ± 2 Ma, [41]), the Furong tin ore field (^{40}Ar–^{39}Ar, 160–150 Ma, [50]), the Yaoganxian tungsten deposit (Re–Os, 150 ± 3 Ma, [99]), the Huangshaping Pb–Zn–W–Sn–Ag deposit (Re–Os, 155 ± 2 Ma, [38]), the Yuanzhuding Cu–Mo deposit (Re–Os, 157 ± 4 Ma, [47]), and the Xitian W–Sn deposit (^{40}Ar–^{39}Ar, 157–150 Ma and Cassiterite U–Pb, 154–157 Ma [100–102]). These ages indicate that a globally significant W-Sn polymetallic mineralization has a close relationship with the A-type granites in the Shi-Hang zone. It is further supported by the fact that the Sn, W, Pb, and Zn contents of the Mesozoic granites are more than ten times their Clark values, respectively. For examples, the Sn, W, Pb, and Zn contents of the A-type granites in the study area are 8.7–39.7 ppm, 34.4–66.7 ppm, 48.2–66.1 ppm, and 15.0–44.9 ppm, respectively (Table S1). It suggests that they could provide ore-forming materials for mineralization. Moreover, the later Jurassic A-type granites are characterized by high Si (74.1 wt.% to 78.0 wt.%) and Rb (1606 ppm to 2351 ppm) contents, low Ti, P, Sr, and Ba contents, extremely high Rb/Sr ration and flat heavy REE (HREE) with strong negative Eu anomalies. These features imply that an intense fluid–magma interaction is favorable to the formation of the Sn–W-polymetallic mineralization [103, 104]. In the Xianghualing deposit, skarn rocks containing fluorite markedly developed around the granite porphyry, indicating the involvement of an F-rich fluid. The dramatic negative Eu anomaly may indicate that the ore-forming fluids were developed in oxidizing conditions [12]. All of these indicate that there is huge W–Sn prospecting potential related to the granites in the Xianghualing deposit.

6. Conclusions

(1) The low Sr and high Yb A_2-type granites from the Xianghualing Sn polymetallic deposit were emplaced in the later Jurassic Era (~153 Ma) with a typical enrichment in SiO_2, REEs (except Eu), HFSE, and Yb, and a depletion in Al_2O_3, CaO, MgO, and Sr.

(2) The geochemical data and isotopic composition (Sr–Nd–Pb–Hf) suggest that the granites from the Xianghualing deposit were derived from predominantly crustal materials (Mesoproterozoic basement rocks in South China Block and subducted sediments) and some minor subducted sediments.

(3) Crustal partial melting in the extensional tectonic setting was induced by subduction of the Paleo-Pacific Plate, accompanied by the decompression melting in the localized mantle wedge and a rollback of the Paleo-Pacific Plate. These would be the likely accepted combined mechanisms for the petrogenesis of A_2-type granites.

Supplementary Materials: The following are available online at http://www.mdpi.com/2075-163X/9/3/182/s1, Table S1: Whole-rock major and trace element of the low-Sr and high-Yb granites in the Xianghualing, SW China; Table S2: Sr, Nd and Pb isotope data for selected granites from Xiahualing, South China. Table S3: LA–ICP–MS zircon U–Pb age data of the late Jurassic low-Sr and high-Yb A-type granites from Xianghualing, South China; Table S4: In situ zircon Hf isotope data for the later Jurassic low-Sr and high-Yb A-type granites in the Xianghualing Sn-polymetallic deposit, South China.

Author Contributions: Conceptualization, C.H.X. and C.S.W.; Data curation, C.H.X. and Y.K.S.; Investigation, C.H.X. and C.S.W.; Project administration, Y.K.S.; Writing—original draft, C.H.X.; Writing—review & editing, C.H.X.

Funding: This research was jointly funded by the National Key R&D Program of China (No. 2016YFC0600107) and the Basic Research Fund of the Chinese Academy of Geological Sciences (No. JYYWF20180602, JYYWF20183702).

Acknowledgments: We would like to thank Qing Zhang, Kun-Feng Qiu, and two anonymous reviewers for the constructive comments and suggestions on this paper.

References

1. Li, X.H.; McCulloch, M.T. Nd isotopic evolution of sediments from the southern margin of the Yangtze Block and its tectonic significance. *Acta Petrol. Sin.* **1996**, *12*, 359–369. (In Chinese with English abstract)

2. Metcalfe, I. Palaeozoic and Mesozoic tectonic evolution and palaeogeography of East Asian crustal fragments: The Korean Peninsula in context. *Gondwana Res.* **2006**, *9*, 24–46. [CrossRef]

3. Metcalfe, I. Tectonic framework and Phanerozoic evolution of Sundaland. *Gondwana Res.* **2011**, *19*, 3–21. [CrossRef]

4. Zheng, Y.F.; Xiao, W.J.; Zhao, G.C. Introduction to tectonics of China. *Gondwana Res.* **2013**, *23*, 1189–1206. [CrossRef]

5. Gilder, S.A.; Gill, J.; Coe, R.S.; Zhao, X.; Liu, Z.; Wang, G.; Yuan, K.; Liu, W.; Kuang, G.; Wu, H. Isotopic and paleomagnetic constraints on the Mesozoic tectonic evolution of South China. *J. Geophys. Res.* **1996**, *101*, 16137–16154. [CrossRef]

6. Mao, J.W.; Li, X.F.; Chen, W.; Lan, X.M.; Wei, S.L. Geological characteristics of the Furong tin orefield, Hunan, ^{40}Ar–^{39}Ar dating of tin ores and related granite and its geodynamic significance for rock and ore formation. *Acta Geol. Sin. (Engl. Ed.)* **2004**, *78*, 481–491.

7. Ding, T.; Ma, D.S.; Lu, J.J.; Zhang, R.Q.; Zhang, S.T. S, Pb and Sr isotope geochemistry and genesis of Pb–Zn mineralization in the Huangshaping polymetallic ore deposit of southern Hunan Province, China. *Ore Geol. Rev.* **2016**, *77*, 117–132. [CrossRef]

8. Hu, R.Z.; Mao, J.W.; Fan, W.M.; Hua, R.M.; Bi, X.W.; Zhong, H.; Song, X.Y.; Tao, Y. Some scientific questions on the intra–continental metallogeny in the South China continent. *Earth Sci. Front.* **2010**, *17*, 13–26. (In Chinese)

9. Hua, R.M.; Chen, P.R.; Zhang, W.L.; Liu, X.D.; Lu, J.J.; Lin, J.F.; Yao, J.M.; Qi, H.W.; Zhang, Z.S.; Gu, S.Y. Metallogenic systems related to Mesozoic and Cenozoic granitoids in South China. *Sci. China D* **2003**, *46*, 816–829. [CrossRef]

10. Pirajno, F.; Bagas, L. Gold and silver metallogeny of the South China Fold Belt: A consequence of multiple mineralizing events? *Ore Geol. Rev.* **2002**, *20*, 109–126. [CrossRef]

11. Shu, X.J.; Wang, X.L.; Sun, T.; Xu, X.S.; Dai, M.N. Trace elements, U–Pb ages and Hf isotopes of zircons from Mesozoic granites in the western Nanling range, South China: Implications for petrogenesis and W–Sn mineralization. *Lithos* **2011**, *127*, 468–482. [CrossRef]

12. Yuan, S.D.; Peng, J.T.; Hu, R.Z.; Li, H.M.; Shen, N.P.; Zhang, D.L. A precise U–Pb age on cassiterite from the Xianghualing tin–polymetallic deposit (Hunan, South China). *Miner. Depos.* **2008**, *43*, 375–382. [CrossRef]

13. Zaw, K.; Peters, S.G.; Cromie, P.; Burrett, C.; Hou, Z. Nature, diversity of deposit types and metallogenic relations of South China. *Ore Geol. Rev.* **2007**, *31*, 3–47. [CrossRef]

14. Zhong, L.F.; Li, J.; Peng, T.P.; Xia, B.; Liu, L.W. Zircon U–Pb geochronology and Sr–Nd–Hf isotopic compositions of the Yuanzhuding granitoid porphyry within the Shi–Hang Zone, South China: Petrogenesis and implications for Cu–Mo mineralization. *Lithos* **2013**, *177*, 402–415. [CrossRef]

15. Zhou, Y.; Liang, X.Q.; Wu, S.C.; Cai, Y.F.; Liang, X.R.; Shao, T.B.; Wang, C.; Fu, J.G.; Jiang, Y. Isotopic geochemistry, zircon U–Pb ages and Hf isotopes of A–type granites from the Xitian W–Sn deposit, SE China: Constraints on petrogenesis and tectonic significance. *J. Asian Earth Sci.* **2015**, *105*, 122–139. [CrossRef]

16. Gilder, S.A.; Keller, G.R.; Luo, M.; Goodell, P.C. Timing and spatial distribution of rifting in China. *Tectonophysics* **1991**, *197*, 225–243. [CrossRef]

17. Li, X.H. Cretaceous magmatism and lithospheric extension in Southeast China. *J. Asian Earth Sci.* **2000**, *18*, 293–305. [CrossRef]

18. Jiang, Y.H.; Ling, H.F.; Jiang, S.Y.; Fan, H.H.; Shen, W.Z.; Ni, P. Petrogenesis of a Late Jurassic peraluminous volcanic complex and its high–Mg, potassic, quenched enclaves at Xiangshan, Southeast China. *J. Petrol.* **2005**, *46*, 1121–1154. [CrossRef]

19. Jiang, Y.H.; Jiang, S.Y.; Dai, B.Z.; Liao, S.Y.; Zhao, K.D.; Ling, H.F. Middle to late Jurassic felsic and mafic magmatism in southern Hunan province, southeast China: Implications for a continental arc to rifting. *Lithos* **2009**, *107*, 185–204. [CrossRef]

20. Chen, J.F.; Jahn, B.M. Crustal evolution of southeastern China: Nd and Sr isotopic evidence. *Tectonophysics* **1998**, *284*, 101–133. [CrossRef]

21. Zhou, X.M.; Sun, T.; Shen, W.Z.; Shu, L.S.; Niu, Y.L. Petrogenesis of Mesozoic granitoids and volcanic rocks in South China: A response to tectonic evolution. *Episodes* **2006**, *29*, 26–33.

22. Zhu, K.Y.; Li, Z.X.; Xu, X.S.; Wilde, S.A. Late Triassic melting of a thickened crust in southeastern China: Evidence for flat–slab subduction of the Paleo–Pacific plate. *J. Asian Earth Sci.* **2013**, *74*, 265–279. [CrossRef]

23. Zhu, K.Y.; Li, Z.X.; Xu, X.S.; Wilde, S.A. A Mesozoic Andean–type orogenic cycle in southeastern China as recorded by granitoid evolution. *Am. J. Sci.* **2014**, *314*, 187–234. [CrossRef]

24. Zhu, K.Y.; Li, Z.X.; Xu, X.S.; Wilde, S.A.; Chen, H.L. Early Mesozoic ferroan (A–type) and magnesian granitoids in eastern South China: Tracing the influence of flat–slab subduction at the western Pacific margin. *Lithos* **2016**, *240*, 371–381. [CrossRef]

25. Wang, Y.J.; Fan, W.M.; Cawood, P.A.; Li, S.Z. Sr–Nd–Pb isotopic constraints on multiple mantle domains for Mesozoic mafic rocks beneath the South China Block hinterland. *Lithos* **2008**, *106*, 297–308. [CrossRef]

26. Li, X.H.; Li, Z.X.; Li, W.X.; Wang, X.C.; Gao, Y. Revisiting the "C–type adakites" of the Lower Yangtze River Belt, central eastern China: In–situ zircon Hf–O isotope and geochemical constraints. *Chem. Geol.* **2013**, *345*, 1–15. [CrossRef]

27. Mao, J.W.; Cheng, Y.B.; Chen, M.H.; Pirajno, F. Major types and time–space distribution of Mesozoic ore deposits in South China and their geodynamic settings. *Lithos* **2013**, *48*, 267–294.

28. Zhou, X.M.; Li, W.X. Origin of late Mesozoic igneous rocks in southeastern China: Implications for lithosphere subduction and underplating of mafic magmas. *Tectonophysics* **2000**, *326*, 269–287. [CrossRef]

29. Zhu, J.C.; Xie, C.F.; Zhang, P.H.; Yang, C.; Gu, C.Y. Niemiao and Tong'an intrusive bodies of NE Guangxi: Petrology, zircon SHRIMP U–Pb geochronology and geochemistry. *Acta Petrol. Sin.* **2005**, *21*, 665–676. (In Chinese)

30. Zhu, J.C.; Zhang, P.H.; Xie, C.F.; Zhang, H.; Yang, C. The Huashan–Guposhan A type granitoid belt in the western part of the Nanling Mountains: Petrology, geochemistry and genetic interpretations. *Acta Geol. Sin.* **2006**, *80*, 529–542. (In Chinese)

31. Zhu, J.C.; Zhang, P.H.; Xie, C.F.; Zhang, H.; Yang, C. Magma mixing origin of the mafic enclaves in Lisong Granite, NE Guangxi, western Nanling Mountains. *Geochimica* **2006**, *35*, 506–516. (In Chinese)

32. Wang, Y.J.; Fan, W.M.; Guo, F. Geochemistry of early Mesozoic potassium–rich diorites–granodiorites in southeastern Hunan Province, South China: Petrogenesis and tectonic implications. *Geochem. J.* **2003**, *37*, 427–448. [CrossRef]

33. Fu, J.M.; Ma, C.Q.; Xie, C.F.; Zhang, Y.M.; Peng, S.B. Shrimp U–Pb zircon dating of the Jiuyishan composite granite in Hunan and its geological significance. *Geotecton. Metallog.* **2004**, *28*, 370–378. (In Chinese)

34. Ma, T.Q.; Bai, D.Y.; Kuang, J.; Peng, X.J.; Wang, X.H. ^{40}Ar–^{39}Ar dating and geochemical characteristics of the granites in north Dadongshan pluton, Nangling Mountains. *Geochimica* **2006**, *35*, 333–345.

35. Li, X.H.; Chung, S.L.; Zhou, H.W.; Lo, C.H.; Liu, Y.; Chen, C.H. Jurassic intraplate magmatism in southern Hunan–eastern Guangxi: 40Ar/39Ar dating geochemistry, Sr–Nd isotopes and implications for the tectonic evolution of SE China. *Geol. Soc. Lond. Spec. Publ.* **2004**, *226*, 193–215. [CrossRef]

36. Zhao, G.D.; Jiang, S.Y.; Liu, D.Y. SHRIMP U–Pb dating of the Furong unit of Qitangling granite from southeast Hunan province and their geological implications. *Acta Petrol. Sin.* **2006**, *22*, 2611–2616.

37. Li, H.; Koichiro, W.; Kotaro, Y. Geochemistry of A–type granites in the Huangshaping polymetallic deposit (South Hunan, China): Implications for granite evolution and associated mineralization. *J. Asian Earth Sci.* **2014**, *88*, 149–167. [CrossRef]

38. Lu, Y.F.; Ma, L.Y.; Qu, W.J.; Mei, Y.P.; Chen, X.Q. U–Pb and Re–Os isotope geochronology of Baoshan Cu–Mo polymetallic ore deposit in Hunan Province. *Acta Petrol. Sin.* **2006**, *22*, 2483–2492. (In Chinese)

39. Li, X.H. Geochronology of Wanyangshan–Zhuguangshan granitoid batholith: Implication for the crust development. *Sci. China (Ser. B)* **1991**, *34*, 620–629.

40. Yin, Z.P.; Ling, H.F.; Huang, G.R.; Shen, W.Z.; Wang, L. Research on geochemical characteristics and genesis of Jiufeng rock mass in Northern Guangdong Province. *J. East China Inst. Technol.* **2010**, *33*, 15–21. (In Chinese)

41. Li, X.H.; Liu, D.Y.; Sun, M.; Li, W.X.; Liang, X.R.; Liu, Y. Precise Sm–Nd and U–Pb isotopic dating of the supergiant Shizhuyuan polymetallic deposit and its host granite, SE China. *Geol. Mag.* **2004**, *141*, 225–231. [CrossRef]

42. Li, C.F.; Li, X.H.; Guo, J.F.; Li, X.F.; Li, H.K.; Zhou, H.Y.; Li, Z.G. Single–step separation of Rb–Sr and Pb from minor rock samples and high precision determination using thermal ionization mass spectrometry. *Geochimica* **2011**, *40*, 399–406. (In Chinese)

43. Lv, K.; Wang, Y.; Xiao, J. Geochemistry characteristics of Xihuashan granite and structural environment discussion. *J. East China Inst. Technol.* **2011**, *34*, 117–128. (In Chinese)

44. Shen, W.Z.; Xu, S.J.; Wang, Y.X.; Yang, J.D. Research of the Nd–Sr isotopic composition of Xihuashan granite. *Chin. Sci. Bulltin* **1994**, *39*, 154–156. (In Chinese)

45. Jiang, Y.H.; Zhao, P.; Zhou, Q.; Liao, S.Y.; Jin, G.D. Petrogenesis and tectonic implications of Early Cretaceous S– and A–type granites in the northwest of the Gan–Hang rift, SE China. *Lithos* **2011**, *121*, 57–73. [CrossRef]

46. Zhou, Q.; Jiang, Y.H.; Liao, S.Y.; Zhao, P.; Jin, G.D.; Jia, R.Y.; Liu, Z.; Xu, S.M. SHRIMP zircon U–Pb dating and Hf isotope studies of the diorite porphyrite from the Dexing copper deposit. *Acta Geol. Sin.* **2012**, *86*, 1726–1734.

47. Zhong, L.F.; Liu, L.W.; Xia, B.; Li, J.; Lin, X.G.; Xu, L.F.; Lin, L.Z. Re–Os geochronology of molybdenite from Yuanzhuding porphyry Cu–Mo deposit in South China. *Resour. Geol.* **2010**, *60*, 389–396. [CrossRef]

48. Lu, H.Z.; Liu, Y.M.; Wang, C.L.; Xu, Y.Z.; Li, H.Q. Mineralization and fluid inclusion study of the Shizhuyuan W–Sn–Bi–Mo–F skarn deposit, Hunan Province, China. *Econ. Geol.* **2003**, *98*, 955–974. [CrossRef]

49. Li, Z.L.; Hu, R.Z.; Peng, J.T.; Bi, X.W.; Li, X.M. Helium isotopic geochemistry of ore–forming fluids tin deposit in Hunan Province, China. *Resour. Geol.* **2006**, *56*, 9–16. [CrossRef]

50. Peng, J.T.; Hu, R.Z.; Bi, X.W.; Dai, T.M.; Li, Z.L.; Li, X.M.; Shuang, Y.; Yuan, S.D.; Liu, S.R. 40Ar/39Ar isotopic dating of tin mineralization in Furong deposit, Hunan and its geological significance. *Miner. Depos.* **2007**, *26*, 237–248. (In Chinese)

51. Yuan, S.D.; Peng, J.T.; Shen, N.P.; Hu, R.Z.; Dai, T.M. 40Ar–39Ar isotopic dating of the Xianghualing, Hunan, Sn–polymetallic orefield and its geological implications. *Acta Geol. Sin. (Engl. Ed.)* **2007**, *81*, 278–286.

52. Yao, J.M.; Hua, R.M.; Du, A.D.; Qi, H.W. Re–Os isotopic dating of the molybdenite from Huangshaping Pb–Zn deposit in Southern Hunan and its implications. *Sci. China (Ser. D)* **2007**, *37*, 471–477. (In Chinese)

53. Xuan, Y.S.; Yuan, S.D.; Yuan, Y.B.; Mi, J.R. Zircon U–Pb age, geochemistry and Petrogenesis of Jianfengling pluton in southern Hunan Province. *Miner. Depos.* **2014**, *33*, 1379–1390.

54. Bai, D.Y.; Chen, J.C.; Ma, T.Q.; Wang, X.H. Geochemical characteristics and tectonic setting of Qitianling A–type granitic Pluton in southeast Hunan. *Acta Petrol. Et Mineral.* **2005**, *24*, 255–272. (In Chinese with English abstract)

55. Li, Z.L.; Hu, R.Z.; Yang, J.S.; Peng, J.T.; Li, X.M.; Bi, X.W. He, Pb and S isotopic constraints on the relationship between the A–type Qitianling granite and the Furong tin deposit, Hunan Province, China. *Lithos* **2007**, *97*, 161–173. [CrossRef]

56. Frost, B.R.; Brans, C.G.; Collins, W.J.; Arculus, R.J.; Ellis, D.J.; Frost, C.D. Geochemical classification for granitic rocks. *J. Petrol.* **2001**, *42*, 2033–2048. [CrossRef]

57. Maniar, P.D.; Piccoli, P.M. Tectonic discrimination of granitoids. *Geol. Soc. Am. Bull.* **1989**, *101*, 635–643. [CrossRef]

58. Sun, S.S.; McDonough, W. Chemical and isotopic systematics of oceanic basalts: Implications for mantle composition and processes. *Geol. Soc. Lond. Spec. Publ.* **1989**, *42*, 313–345. [CrossRef]

59. Whalen, J.B.; Currie, K.L.; Chappell, B.W. A–type granites: Geochemical characteristics, discrimination and petrogenesis. *Contrib. Mineral. Petrol.* **1987**, *95*, 407–419. [CrossRef]

60. Wu, S.P.; Wu, C.L.; Chen, Q.L.; Zhang, C.Q.; Li, D.S. Error estimation of initial ratios of Sr–Nd isotope and Nd model age. *Acta Geol. Sin.* **2009**, *83*, 1031–1038. (In Chinese)

61. Wong, J.; Sun, M.; Xing, G.F.; Li, X.H.; Zhao, G.C.; Wong, K.; Yuan, C.; Xia, X.P.; Li, L.M.; Wu, F.Y. Geochemical and zircon U–Pb and Hf isotopic study of the Baijuhuajian metaluminous A–type granite: Extension at 125–100 Ma and its tectonic significance for South China. *Lithos* **2009**, *112*, 289–305. [CrossRef]

62. Hoskin, P.W.O.; Schaltegger, U. The composition of zircon and igneous and metamorphic petrogenesis. *Rev. Mineral. Geochem.* **2003**, *53*, 27–55. [CrossRef]

63. Douce, A.E.P. Generation of metaluminous A–type granites by low–pressure melting of calc–alkaline granitoids. *Geology* **1997**, *25*, 743–746. [CrossRef]

64. Eby, G.N.; Kochhar, N. Geochemistry and petrogenesis of the Malani igneous suite, North Peninsular India. *J. Geol. Soc. India* **1990**, *36*, 109–130.

65. Loiselle, M.C.; Wones, D.R. Characteristics and origin of anorogenic granites. *Geol. Soc. Am. Abstr. Programs* **1979**, *11*, 468.

66. Shen, X.M.; Zhang, H.X.; Wang, Q.; Wymanc, D.A.; Yang, Y.H. Late Devonian–Early Permian A–type granites in the southern Altay Range, Northwest China: Petrogenesis and implications for tectonic setting of "A2–type" granites. *J. Asian Earth Sci.* **2011**, *42*, 986–1007. [CrossRef]

67. Zhang, Q.; Li, C.D.; Wang, Y.; Wang, Y.L.; Jin, W.J.; Jia, X.Q.; Han, S. Mesozoic high–Sr and Low–Yb granitoids and low–Sr and high–Yb granitoids in eastern China: Comparison and geological implications. *Acta Petrol. Sin.* **2005**, *21*, 1527–1537.

68. Williams, I.S. Some observations on the use of zircon U–Pb geochronology in the study of granitic–rocks. *Earth Environ. Sci. Trans. R. Soc. Edinb.* **1987**, *83*, 447–458. [CrossRef]

69. de la Roche, H.; Leterrier, J.; Grand Claude, P. and Marchal, M. A classification of volcanic and plutonic rocks using R1–R2 diagrams and major element analyses—Its relationships with current nomenclature. *Chem. Geol.* **1980**, *29*, 183–210. [CrossRef]

70. Eby, G.N. Chemical subdivision of the A–type granitoids: Petrogenetic and tectonic implication. *Geology* **1992**, *20*, 641–644. [CrossRef]

71. Bonin, B. A–type granites and related rocks: Evolution of a concept, problems and prospects. *Lithos* **2007**, *97*, 1–29. [CrossRef]

72. King, P.L.; White, A.J.R.; Chappell, B.W.; Allen, C.M. Characterization and origin of aluminous A–type granites from the Lachlan Fold Belt, Southeastern Australia. *J. Petrol.* **1997**, *38*, 371–391. [CrossRef]

73. Zhang, Q.; Wang, Y.; Li, C.D.; Wang, Y.L.; Jin, W.J.; Jia, X.Q. Granite classification on the basis of Sr and Yb contents and its implications. *Acta Petrol. Sin.* **2006**, *22*, 2249–2269. (In Chinese)

74. Clemens, J.D.; Holloway, J.R.; White, A.J.R. White Origin of an A–type granite; experimental constraints. *Am. Mineral.* **1986**, *71*, 317–324.

75. Collins, W.J.; Beams, S.D.; White, A.J.R.; Chappell, B.W. Nature and origin of A–type granites with particular reference to southeastern Australia. *Contrib. Mineral. Petrol.* **1982**, *80*, 189–200. [CrossRef]

76. Chappell, B.W. Granitoids from the Moonbi district, New England Batholith, Eastern Australia. *J. Geol. Soc. Aust.* **1987**, *25*, 267–283. [CrossRef]

77. Creaser, R.A.; Price, R.C.; Wormald, R.J. A–type granites revisited: Assessment of a residual–source model. *Geology* **1991**, *19*, 163–166. [CrossRef]

78. Gaudemer, Y.; Jaupart, C.; Tapponnier, P. Thermal control on post–orogenic extension in collision belts. *Earth Planet. Sci. Lett.* **1988**, *89*, 48–62. [CrossRef]

79. Patiño Dounce, A.E.; Beard, J.S. Dehydration–melting of biotite gneiss and quartze amphibolite from 3 to 15 kbar. *J. Petrol.* **1995**, *36*, 707–738. [CrossRef]

80. Li, H.Y.; Mao, J.W.; Sun, Y.L.; Du, A.D. Re–Os isotopic chronology of molybdenites in the Shizhuyuan polymetallic tungsten deposit. *Geol. Rev.* **1996**, *42*, 261–267. (In Chinese with English abstract)

81. Mao, J.W.; Li, H.Y.; Pei, R. Nd–Sr isotopic and petrogenetic studies of the Qianlishan granite stock. Hunan Province. *Miner. Depos.* **1995**, *14*, 235–242. (In Chinese with English abstract)

82. Zhu, J.C.; Li, X.D.; Shen, W.Z.; Wang, Y.X.; Yang, J.D. Sr, Nd and O isotope studies on the genesis of the Huashan granite complex. *Acta Geol. Sin.* **1989**, *3*, 225–235. (In Chinese with English abstract)

83. Zartman, R.E.; Haines, S.M. The plumbotectonic model for Pb isotopic systematics among major terrestrial reservoirs—A case for bi–directional transport. *Geochim. Cosmochim. Acta* **1988**, *52*, 1327–1339. [CrossRef]

84. Dall'Agnol, R.; Scaillet, B.; Pichavant, M. An experimental study of a Lower Proterozoic A-type granite from the Eastern Amazonian Craton, Brazil. *J. Petrol.* **1999**, *40*, 1673–1698. [CrossRef]

85. Rajesh, H.M. Characterization and origin of a compositionally zoned aluminous A-type granite from South India. *Geol. Mag.* **2000**, *137*, 291–318. [CrossRef]

86. Wu, F.Y.; Sun, D.Y.; Li, H.; Jahn, B.M.; Wilde, S. A-type granites in northeastern China: Age and geochemical constraints on their petrogenesis. *Chem. Geol.* **2002**, *187*, 143–173. [CrossRef]

87. Dai, B.Z.; Jiang, S.Y.; Jiang, Y.H.; Zhao, K.D.; Liu, D.Y. Geochronology, geochemistry and Hf–Sr–Nd isotopic compositions of Huziyan mafic xenoliths, southern Hunan Province, South China: Petrogenesis and implications for lower crust evolution. *Lithos* **2008**, *102*, 65–87. [CrossRef]

88. Zhao, Z.H.; Bao, Z.W.; Zhang, B.Y.; Xiong, X.L. Crust–mantle interaction and its contribution to the Shizhuyuan tungsten–polymetallic mineralization. *Sci. China (Ser. D) Earth Sci.* **2001**, *44*, 266–276. [CrossRef]

89. Ye, H.M.; Mao, J.R.; Zhao, X.L.; Liu, K.; Chen, D.D. Revisiting Early–Middle Jurassic igneous activity in the Nanling Mountains, South China: Geochemistry and implications for regional geodynamics. *J. Asian Earth Sci.* **2013**, *72*, 108–117. [CrossRef]

90. Lapierre, H.; Jahn, B.M.; Charvet, J.; Yu, Y.W. Mesozoic felsic arc magmatism and continental olivine tholeiites in Zhejiang province and their relationship with the tectonic activity in southeastern China. *Tectonophysics* **1997**, *274*, 321–338. [CrossRef]

91. Li, Z.X.; Li, X.H. Formation of the 1300–km–wide intracontinental orogen and postorogenic magmatic province in Mesozoic South China: A flat–slab subduction model. *Geology* **2007**, *35*, 179–182. [CrossRef]

92. Jiang, S.Y.; Zhao, K.D.; Jiang, Y.H.; Dai, B.Z. Characteristics and genesis of Mesozoic A–type granites and associated mineral deposits in the southern Hunan and northern Guangxi provinces along the Shi–Hang belt, South China. *Geol. J. China Univ.* **2008**, *14*, 496–509. (In Chinese with English abstract)

93. Jiang, Y.H.; Wang, G.C.; Liu, Z.; Ni, C.Y.; Qing, L.; Zhang, Q. Repeated slab–advanceretreat of the Palaeo–Pacific plate underneath SE China. *Int. Geol. Rev.* **2015**, *57*, 472–491. [CrossRef]

94. Wang, G.C.; Jiang, Y.H.; Liu, Z.; Ni, C.Y.; Qing, L.; Zhang, Q.; Zhu, S.Q. Multiple origins for the Middle Jurassic to Early Cretaceous high–K calc–alkaline I–type granites in northwestern Fujian province, SE, China and tectonic implications. *Lithos* **2016**, *246–247*, 197–211. [CrossRef]

95. Tatsumi, Y.; Eggins, S. *Subduction Zone Magmatism*; Blackwell Science: New York, NY, USA, 1995; pp. 1–211.
96. Sun, F.; Xu, X.; Zou, H.; Xia, Y. Petrogenesis and magmatic evolution of ~130 Ma A–type granites in Southeast China. *J. Asian Earth Sci.* **2015**, *98*, 209–224. [CrossRef]
97. Li, X.H.; McCulloch, M.T. Secular variation in the Nd isotopic composition of Neoproterozoic sediments from the southern margin of the Yangtze Block: Evidence for a Proterozoic continental collision in southeast China. *Precambrian Res.* **1996**, *76*, 67–76. [CrossRef]
98. Liu, Y.M.; Dai, T.M.; Lu, H.Z.; Xu, Y.Z.; Wang, C.L.; Kang, W.Q. Isotopic data of 40Ar/39Ar and Sm–Nd for diagenesis–metallogenesis of the Qianlishan granite. *Sci. China (Ser. D)* **1997**, *27*, 425–430. (In Chinese)
99. Peng, J.T.; Zhou, M.F.; Hu, R.Z.; Shen, N.P.; Yuan, S.D.; Bi, X.W.; Du, A.D.; Qu, W.J. Precise molybdenite Re–Os and mica Ar–Ar dating of the Mesozoic Yaogangxian tungsten deposit, central Nanling district, South China. *Miner. Depos.* **2006**, *41*, 661–669. [CrossRef]
100. Fu, J.M.; Cheng, S.B.; Lu, Y.Y.; Wu, S.C.; Ma, L.Y.; Chen, X.Q. Geochronology of the greisen–quartz–vein type tungsten–tin deposit and its host granite in Xitian, Hunan Province. *Geol. Explor.* **2012**, *48*, 0313–0320. (In Chinese with English abstract)
101. Liu, Y.S.; Hu, Z.C.; Gao, S.; Günther, D.; Xu, J.; Gao, C.G.; Chen, H.H. In situ analysis of major and trace elements of anhydrous minerals by LA–ICP–MS without applying an internal standard. *Chem. Geol.* **2008**, *257*, 34–43. [CrossRef]
102. Jiang, W.C.; Li, H.; Wu, J.H.; Zhou, Z.K.; Kong, H.; Cao, J.Y. A newly Found Biotite Syenogranite in the Huangshaping Polymetallic Deposit, South China: Insights into Cu Mineralization. *J. Earth Sci.* **2018**, *29*, 537–555. [CrossRef]
103. Ma, L.Y.; Fu, J.M.; Wu, S.C.; Xu, D.M.; Yang, X.J. 40Ar/39Ar isotopic dating of the Longshang tin–polymetallic deposit, Xitian orefield, eastern Hunan. *Geol. China* **2008**, *35*, 706–713. (In Chinese with English abstract)
104. Bastos Neto, A.C.; Pereira, V.P.; Ronchi, L.H.; DeLima, E.F.; Frantz, J.C. The world–class Sn, Nb, Ta, F (Y, REE, Li) deposit and the massive cryolite associated with the albite–enriched facies of the Madeira A–type granite, Pitinga mining district, Amazonas state, Brazil. *Can. Mineral.* **2009**, *47*, 1329–1357. [CrossRef]

Article

Metallogenic Epoch and Tectonic Setting of Saima Niobium Deposit in Fengcheng, Liaoning Province, NE China

Nan Ju [1,2,*], Yun-Sheng Ren [1], Sen Zhang [2], Zhong-Wei Bi [2], Lei Shi [2], Di Zhang [2], Qing-Qing Shang [1], Qun Yang [1], Zhi-Gao Wang [1], Yu-Chao Gu [2], Qiu-Shi Sun [2] and Tong Wu [2]

[1] College of Earth Sciences, Jilin University, Changchun 130061, China; renys@jlu.edu.cn (Y.-S.R.); shangqq15@163.com (Q.-Q.S.); yangq2009jlu@163.com (Q.Y.); wzg_jldx@163.com (Z.-G.W.)
[2] Shenyang Geological Survey, CGS, Shenyang 110034, China; zhangsen556@163.com (S.Z.); bzw_cgs@163.com (Z.-W.B.); shih1101@163.com (L.S.); zhangdi9521@163.comm (D.Z.); guyi1224@126.com (Y.-C.G.); sunqiushi2007@163.com (Q.-S.S.); jesnet1981@163.com (T.W.)
* Correspondence: junan_cgs@163.com; Tel.: +86-024-81847053

Received: 17 December 2018; Accepted: 24 January 2019; Published: 29 January 2019

Abstract: The Saima deposit is a newly discovered niobium deposit which is located in the eastern of Liaoning Province, NE China. Its mineralization age and geochemical characteristics are firstly reported in this study. The Nb orebodies are hosted by the grey–brown to grass-green aegirine nepheline syenite. Detailed petrographical studies show that the syenite consists of orthoclase (~50%), nepheline (~30%), biotite (~15%) and minor arfvedsonite (~3%) and aegirine (~2%), with weak hydrothermal alteration dominated by silicification. In situ LA-ICP-MS zircon U-Pb dating indicates that the aegirine nepheline syenite was emplaced in the Late Triassic (229.5 $\pm$ 2.2 Ma), which is spatially, temporally and genetically related to Nb mineralization. These aegirine nepheline syenites have SiO_2 contents in the range of 55.86–63.80 wt. %, low TiO_2 contents of 0.36–0.64 wt. %, P_2O_5 contents of 0.04–0.11 wt. % and Al_2O_3 contents of more than 15 wt. %. They are characterized by relatively high ($K_2O + Na_2O$) values of 9.72–15.51 wt. %, K_2O/Na_2O ratios of 2.42–3.64 wt. % and Rittmann indexes ($\sigma = [\omega(K_2O + Na_2O)]^2/[\omega(SiO_2 - 43)]$) of 6.84–17.10, belonging to the high-K peralkaline, metaluminous type. These syenites are enriched in large ion lithophile elements (LILEs, e.g., Cs, Rb and Ba) and light rare earth elements (LREEs) and relatively depleted in high field strength elements (HFSEs, e.g., Nb, Zr and Ti) and heavy rare earth elements (HREEs), with transitional elements showing an obvious W-shaped distribution pattern. Based on these geochronological and geochemical features, we propose that the ore-forming intrusion associated with the Nb mineralization was formed under post-collision continental-rift setting, which is consistent with the tectonic regime of post-collision between the North China Craton and Paleo-Asian oceanic plate during the age in Ma for Indosinian (257–205 Ma). Intensive magmatic and metallogenic events resulted from partial melting of lithospheric mantle occurred during the post-collisional rifting, resulting in the development of large-scale Cu–Mo mineralization and rare earth deposits in the eastern part of Liaoning Province.

Keywords: syenite; post-collisional; niobium mineralization; Saima deposit; Liaoning Province

---

1. Introduction

China's reserves of rare earth metals account for up to 87% of the world total, while the reserves of niobium metals are relatively scarce, which only account for 18.19% of the world [1]. Geological survey in recent years have shown that most of the niobium (Nb) and tantalum (Ta) deposits are distributed in south China, such as Hunan, Guangdong, Shanxi and particularly Jiangxi Province in which host the most abundant Nb and Ta resources. Minor similar deposits have been sporadically discovered in the Panzhihua-Xichang district, northern Xinjiang and Inner Mongolia [2]. In contrast, few Nb and Ta deposits had been discovered in NE China, which was attributed to the restricted prospecting and research at present. However, numerous rare earth deposits associated with alkali rocks represented by Ba'erzhe (inner Mongolia province) have been recently discovered in NE China [3] and rare metals such as Nb and Ta occurred in alkaline intrusive rocks have received extensive attention.

The Saima deposit, an niobium deposit with medium-scale prospecting potentials, is located about 2 km to the south of Saima town, Fengcheng city, Liaoning province. Nevertheless, systematic studies on the metallogenic of the newly discovered Saima niobium deposit remain inadequate. In this study, we present new in situ LA-ICP-MS zircon U-Pb geochronological data and whole-rock geochemical data for the ore-forming alkaline intrusive rocks to constrain the timing and origin of the niobium mineralization and in particular, to shed light on the tectonic environment of regional deposition by the views of geological characteristics of the Saima deposit. Moreover, this study provides a scientific basis for further mineral prospecting and future prospecting of new rare earth deposits in NE China.

2. Geological Setting and Ore Deposit Geology

The Saima deposit is located in the intersection of the Taizihe–Hunjiang fault depression and Yingkou–Kuandian uplift in the Liaodong Massif, northeast of the North China Craton. Stratigraphic units are dominated by the Neoproterozoic, Cambrian–Ordovician, Jurassic and Quaternary alluvium. Massive Late Archean to Mesozoic granitic intrusions are exposed in the region, with particularly intensive magmatism during Late Triassic and Jurassic. Alkaline rocks closely related to the rare earth deposits are distributed along the intersection between the E–W-trending fault and NNE–SSW-trending fault. This alkaline rock belt extends about 40 km from east to west, with the width of 15 km, covering an area of about 200 km^2. These rocks occurred as small stocks and apophysis and intruded into the Pre-Ordovician formations, showing angular unconformity contact with the overlying Jurassic strata. Four alkaline intrusions and abundant alkaline volcanic-subvolcanic rocks constitute the alkaline rock belt, with their alkalis gradually increasing from east to west and rock types changing from syenite–miascite (biotite–nepheline) to aegirine nepheline syenite. Regional faults are dominated by the E–W-trending faults, accompanied by the NE–SW-trending, NW–SE-trending and NNE–SSW-trending faults. The Saima deposit is located in the intersection of EW-trending fault and NE–SW-trending fault (Figure 1a), as these faults acted as the conduits for the migration and enrichment of ore-forming elements and provided favorable conditions for the formation of an alkaline rock type niobium deposit.

The strata in the Saima district consist of the Lower Proterozoic Gaixian Formation and Dashiqiao Formation, the Upper Proterozoic Xihe Group and Cambrian, Ordovician and Jurassic formations. Intense magmatism developed in the Saima district, including the Proterozoic magmatic rocks exposed in the eastern portion of the district and the Mesozoic magmatic rocks exposed in the western part. The Mesozoic magmatic rocks composed of Fengcheng alkaline intrusion as well as Nanda and Beida granitic intrusions, are controlled by the regional EW-trending faults. The ore-forming alkaline rocks occurred as dikes and stocks along the EW-trending structural zone and are characterized by medium- to fine-grained and pegmatitic textures, which had been interpreted to be the intrusive and flooding products of the Indosinian movement. These rocks are dominated by the aegirine nepheline syenite, accompanied by biotite-bearing aegirine nepheline syenite, black aegirine nepheline syenite, black grenadine syenite, miascite and green or grass-green aegirine nepheline syenite. They belong to the higher K$_2$O, lower Na$_2$O, alkaline to peralkaline series. The rare earth mineralization mainly occurred in the green, grass-green and grey–brown aegirine nepheline syenite. It develops EW-, NS-, NE–SW- and NW–SE-trending faults in the Saima district, with the NE–SW-trending faults as the ore-controlling faults [4,5] (Figure 1b). Extensive wall-rock alteration including microclinization, nephelinization, Na-zeolitization and carbonatization are well developed in the Saima deposit. Various kinds of primary nepheline syenite have been extensively altered to syenite, forming the main part of the Saima alkaline intrusion. The Nb and Ta mineralization was closely related to the nephelinization and Ca-nephelinization.

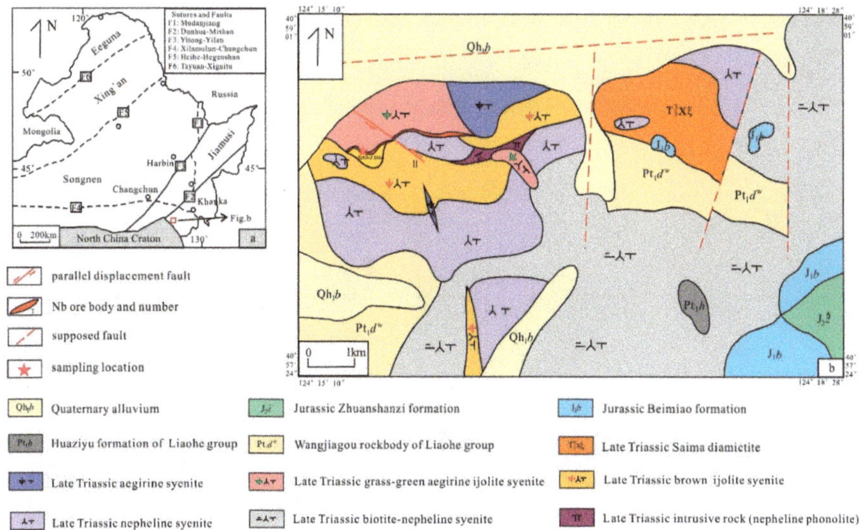

Figure 1. Geotectonic location map (**a**) and Geological sketch map (**b**) of Saima deposit in Fengcheng, Liaoning Province (modified from [5]).

Currently, eight orebodies have been found in the Saima district, which mainly occurred in grey–brown to grass-green aegirine nepheline syenite. The No. I and II orebodies are about 500 m and 160 m long, with the width of 6 m and 18 m, respectively. Most of the orebodies are stratiform or lenticular, inclined to 45–115° and dip at 25–75°, with the maximum thickness up to 40 m (Figure 2). Fergusonite ($YNbO_4$) and brocenite are the dominant ore minerals with high Nb grades of 0.03–0.06%, Ce grades of about 0.1%. Gangue minerals include nepheline, aegirine, rutile and pyrite. There is no sharp boundary between the orebodies and wall-rocks but rather a gradual transitional relationship. The rare earth ores show lamellated, cataclastic and subhedral pegmatitic textures and veined-disseminated structures.

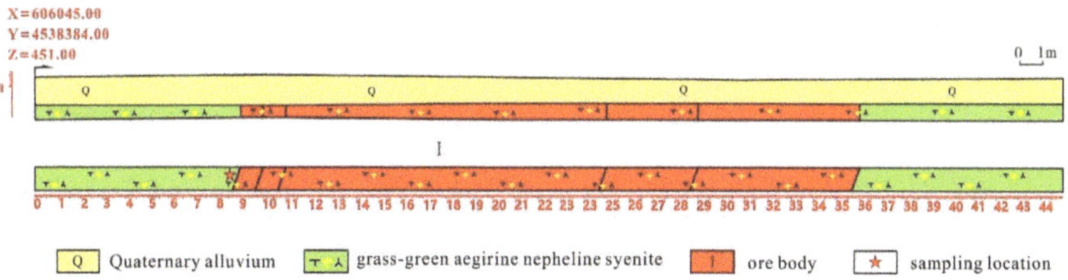

Figure 2. TCG10-3 trench exploration map of the Saima deposit in Fengcheng, Liaoning Province.

3. Petrography and Analytical Methods

3.1. Petrographic Features

Samples analyzed in this study were collected from the No. II orebody in the Saima district (Figure 1). No (ore) mineralization and alteration had been distinguished in the sample SMB1 and it was considered as the original ore-forming rock. Thin sections were made from these samples and subsequently observed by using an OLYMPUS microscope at the Laboratory of Rock-mineral

Determination, Jilin University. Detailed petrographic studies indicate that the ore-forming rocks are mainly medium-coarse grained aegirine nepheline syenite composed of subhedral-anhedral orthoclase (~50%) with carlsbad twinning, nepheline (~30%) with tabular or granular textures, biotite (~10%) partly replaced by nepheline, aegirine (~7%) and arfvedsonite (~3%; Figure 3). The aegirine nepheline syenites are in conformable and direct contact with the orebodies as shown in sketch map of TCG10-3 exploratory trench (Figure 2). Combined with the distribution characteristics of orebodies in the deposit scale (Figure 1), it is suggested that the aegirine nepheline syenite is spatially, temporally and genetically related to Nb mineralization.

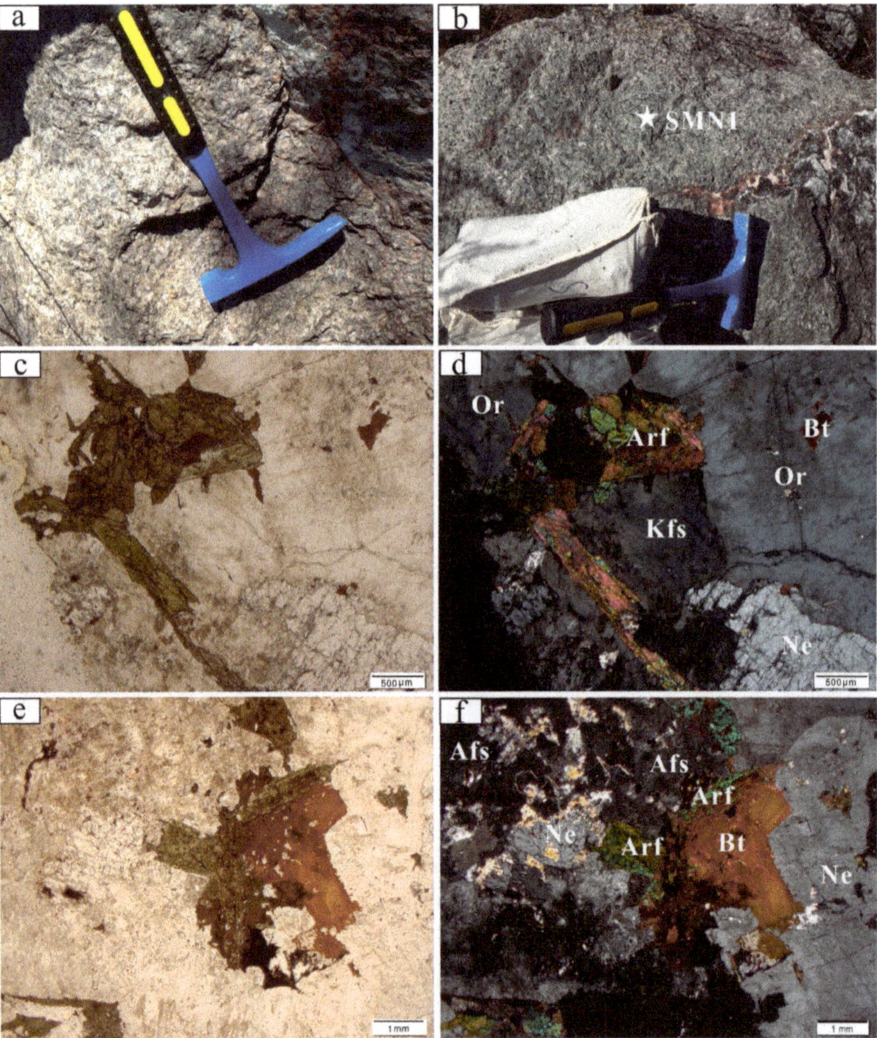

Figure 3. Representative outcrop-pictures and photomicrographs of the ore-forming rocks in Saima niobium deposit. (**a**) Brown aegirine nepheline syenite; (**b**) Grass-green aegirine–nepheline syenite(SMN1 is the sample number.); (**c**) Medium-coarse grained nepheline syenite (PPL); (**d**) Medium-coarse grained nepheline syenite (CPL); (**e**) Medium-coarse grained aegirine nepheline syenite (PPL); (**f**) Medium-coarse grained aegirine nepheline syenite (CPL). Mineral abbreviations: Afs, alkalifeldspar; Arf, arfvedsonite; Bt, biotite; Kfs, K-feldspar; Ne, nepheline; Or, orthoclase.

3.2. Analytical Methods

3.2.1. Zircon LA-ICP-MS U-Pb Dating

Sample SMN1 collected from the aegirine nepheline syenite in the Saima district was selected for zircon U–Pb dating. Zircons were extracted using conventional heavy liquid and magnetic separation techniques and then handpicked under a binocular microscope at the Langfang Regional Geological Survey, Hebei Province, China. Their transmitted and reflected light images were obtained using a standard polarizing microscope and cathodoluminescence (CL) images were obtained using a JEOL scanning electron microscope (JEOL, Tokyo, Japan). LA-ICP-MS zircon U-Pb analyses were performed using an Agilent 7900 ICP-MS equipped with a 193 nm laser, housed at the Key Laboratory of Mineral Resources Evaluation in Northeast Asia, Ministry of Land and Resources of China, Changchun, China. The zircon 91,500 was used as an external standard for age calibration and the NIST SRM 610 silicate glass was applied for instrument optimization. The beam diameter was 32 μm during the analyses. Other instrument parameters and detailed procedures were described by Yuan et al. [6]. Recommended values of U-Th-Pb isotope ratios for zircon standard 91,500 were referred to Wiedenbeck et al. [7] and common Pb was corrected by applying the methods introduced by Andersen [8]. Weighted averages and intercept ages were calculated using the program Isoplot (Version 4.0, United States Geological Survey, Washington, DC, USA). Errors on individual analyses by LA-ICP-MS are quoted at the 2σ level, while errors on the average weighted ages are 2σ level.

3.2.2. Whole-Rock Geochemical Analyses

Five samples (ore-forming rocks) were selected for major and trace element analyses. After the removal of altered surfaces, fresh whole-rock samples were crushed to 200 meshes in an agate mill. Major oxides were determined by using an inductively coupled plasma–optical emission spectroscopy (ICP-OES) system with high-dispersion echelle optics at the China University of Geosciences, Beijing. Loss on ignition was determined by placing 100 mg of sample in a furnace at 980 °C for several hours, cooling the sample in a desiccator and then reweighing the sample. Trace element analyses were performed using an Agilent-7500a inductively coupled plasma-mass spectrometer (ICP-MS) at the China University of Geosciences, Beijing. The rock reference materials AGV-2 (USGS) and GSR-3 (National Geological Standard Reference Materials of China) were used to monitor analytic accuracy and precision. Precision and accuracy are better than 5% for major elements and 10% for trace elements shown from repeated analyses.

4. Analytical Results

4.1. Zircon U-Pb Geochronology

Most of the zircons from the aegirine nepheline syenite (SMN1) are transparent, euhedral and colorless, with length/width ratios of more than 2:1. Typical oscillatory zoning of magmatic crystallization is common in most of the zircon grains under CL (Figure 4). In addition to their relatively high Th/U ratios (0.5–1.0), a magmatic origin is indicated [9,10]. Thirteen zircon U-Pb analyses are summarized in Table 1 and have yielded concordant $^{206}Pb/^{238}U$ ages of 233 to 222 Ma, with a weighted mean $^{206}Pb/^{238}U$ age of 229.5 ± 2.2 Ma (MSWD = 0.54; Figure 4).

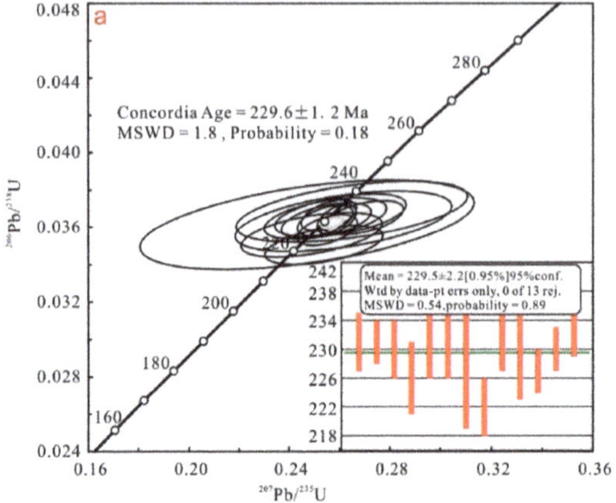

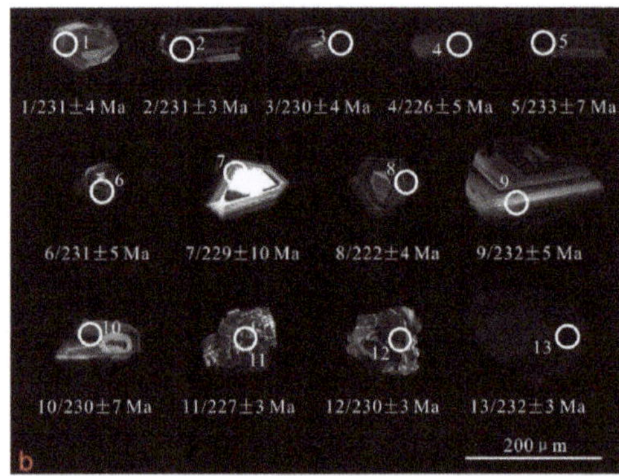

Figure 4. LA-ICP-MS zircon U-Pb concordia diagrams (**a**) and CL images (**b**) for the syenite in the Saima deposit.

4.2. Major and Trace Element Geochemistry

The whole-rock major and trace element data from the Saima syenites are presented in Tables 2 and 3. The aegirine nepheline syenites have variable SiO_2 contents of 55.86–80.94 wt. % (concentrating in the range of 55.86–63.80 wt. %) and contain lower TiO_2 of 0.36–0.64 wt. %, P_2O_5 of 0.04–0.11 wt. % and Al_2O_3 of more than 15 wt. % in most of the samples. They are characterized by relatively high total $(K_2O + Na_2O)$ values of 9.72–15.51 wt. %, K_2O/Na_2O ratios of 2.42–3.64 wt. % and Rittmann indexes $(\sigma = [\omega(K_2O + Na_2O)]^2/[\omega(SiO_2 - 43)])$ of 6.84–17.10 wt. %, belonging to the high-K peralkaline, metaluminous syenitoid. These rock samples plot into fields of syenite or nepheline syenite in the TAS and R1-R2 diagrams (Figure 5a,b), into field of calc-alkaline series in the FMA diagram (Figure 5c) and into field of metaluminous rocks in the $(Fe + Mg + Ti) - [Al - (Na + K + 2Ca)]$ and A/CNK-A/NK diagrams (Figure 5d,e).

Table 1. LA-ICP-MS zircon U-Pb analysis results for the syenite in the Saima deposit.

Sample No.	Isotopic Ratios									Age (Ma)						
	$^{206}Pb/^{238}U$	1σ	$^{207}Pb/^{235}U$	1σ	$^{207}Pb/^{206}Pb$	1σ	$^{208}Pb/^{232}Th$	1σ	$^{232}Th/^{238}U$	$^{206}Pb/^{238}U$	1σ	$^{207}Pb/^{235}U$	1σ	$^{208}Pb/^{232}Th$	1σ
1	0.0365	0.0006	0.2695	0.0106	0.0535	0.0023	0.0118	0.0002	0.8691	231	4	242	8	238	5
2	0.0364	0.0005	0.2589	0.0051	0.0515	0.0013	0.0122	0.0002	0.3079	231	3	234	4	246	3
3	0.0364	0.0006	0.2548	0.0117	0.0508	0.0025	0.0114	0.0002	0.8899	230	4	230	9	229	5
4	0.0356	0.0008	0.2489	0.0185	0.0507	0.0039	0.0131	0.0003	1.2228	226	5	226	15	263	5
5	0.0369	0.0011	0.2648	0.0303	0.0521	0.0061	0.0108	0.0004	0.9987	233	7	239	24	218	9
6	0.0365	0.0007	0.2544	0.0160	0.0506	0.0033	0.0112	0.0004	0.7179	231	5	230	13	226	9
7	0.0361	0.0016	0.2470	0.0438	0.0496	0.0090	0.0130	0.0012	0.5357	229	10	224	36	262	23
8	0.0351	0.0007	0.2538	0.0154	0.0525	0.0033	0.0109	0.0004	0.5554	222	4	230	12	219	7
9	0.0367	0.0008	0.2606	0.0173	0.0516	0.0036	0.0116	0.0003	1.7506	232	5	235	14	232	5
10	0.0363	0.0011	0.2615	0.0284	0.0522	0.0059	0.0097	0.0005	0.6265	230	7	236	23	195	10
11	0.0359	0.0006	0.2551	0.0078	0.0516	0.0018	0.0112	0.0002	1.1001	227	3	231	6	226	3
12	0.0364	0.0005	0.2543	0.0072	0.0507	0.0016	0.0115	0.0001	1.5393	230	3	230	6	232	3
13	0.0366	0.0006	0.2558	0.0072	0.0506	0.0016	0.0124	0.0001	4.8276	232	3	231	6	249	2

Table 2. Whole-rock major oxides (wt. %) of syenites in the Saima deposit.

Sample No.	SiO_2	Al_2O_3	TFe_2O_3	MgO	CaO	Na_2O	K_2O	MnO	TiO_2	P_2O_5	LOI	Total
SMH1-1	63.80	15.25	3.50	0.51	1.59	2.57	9.36	0.09	0.36	0.10	2.89	100.02
SMH1-2	80.94	1.16	2.64	0.03	1.26	2.53	7.18	0.07	0.46	0.04	3.69	100.00
SMH1-3	57.16	19.11	3.72	0.45	1.57	4.53	10.98	0.08	0.41	0.08	1.90	99.99
SMH1-4	57.38	18.11	3.68	0.53	1.95	4.07	10.25	0.09	0.64	0.11	3.18	99.98
SMH1-5	55.86	18.81	3.96	0.70	2.11	4.25	10.58	0.10	0.62	0.11	2.91	100.00

Table 3. Whole-rock trace and rare-earth elements (ppm) of syenites in the Saima deposit.

Sample No.	Li	Ti	Sc	V	Cr	Co	Ni	Cu	Zn	Ga	Rb	Sr	Y	Mn	Zr	Nb	Hf	Cs	Ba
SMH1-1	69.94	2608.20	2.30	64.87	15.26	7.07	0.92	30.94	85.41	29.11	312.20	2846.00	73.54	804.80	410.80	37.89	9.06	2.90	3366.00
SMH1-2	61.38	4044.60	1.40	71.03	12.84	4.79	0.12	28.83	73.26	24.17	214.20	1171.20	77.84	706.20	888.20	106.59	20.30	2.93	2528.00
SMH1-3	40.37	2552.40	1.88	61.56	14.45	4.46	0.91	25.42	98.48	30.28	290.20	2750.00	49.60	676.40	380.40	34.79	7.84	2.50	2708.00
SMH1-4	55.55	4221.00	2.45	60.91	15.32	5.14	2.45	29.01	98.51	30.49	287.60	2900.00	112.72	766.80	629.80	69.39	14.05	2.95	2912.00
SMH1-5	51.15	3943.80	2.47	59.11	13.10	5.953	0.51	26.30	103.14	32.06	292.80	2742.00	154.16	774.80	532.60	74.74	11.89	2.93	2396.00

Sample No.	La	Ce	Pr	Nd	Sm	Eu	Gd	Tb	Dy	Ho	Er	Tm	Yb	Lu	Ta	Pb	Th	U
SMH1-1	227.60	446.00	48.36	159.80	26.62	7.38	19.04	2.56	13.27	2.48	6.65	0.83	4.85	0.56	1.41	60.88	127.10	18.02
SMH1-2	91.50	186.02	26.80	95.36	19.87	6.02	15.95	2.73	16.60	3.41	9.95	1.37	8.16	0.94	2.66	48.80	124.42	38.53
SMH1-3	176.90	320.60	33.80	110.90	19.39	5.53	14.37	1.95	9.95	1.80	4.61	0.55	3.01	0.34	1.42	69.53	89.85	13.03
SMH1-4	340.00	667.60	77.48	233.40	38.34	10.59	27.82	3.81	19.91	3.70	9.78	1.22	6.98	0.81	2.60	79.18	189.32	32.25
SMH1-5	466.00	946.00	108.26	331.60	55.38	15.14	40.24	5.56	29.08	5.35	13.87	1.67	9.08	0.99	2.35	56.45	275.50	41.75

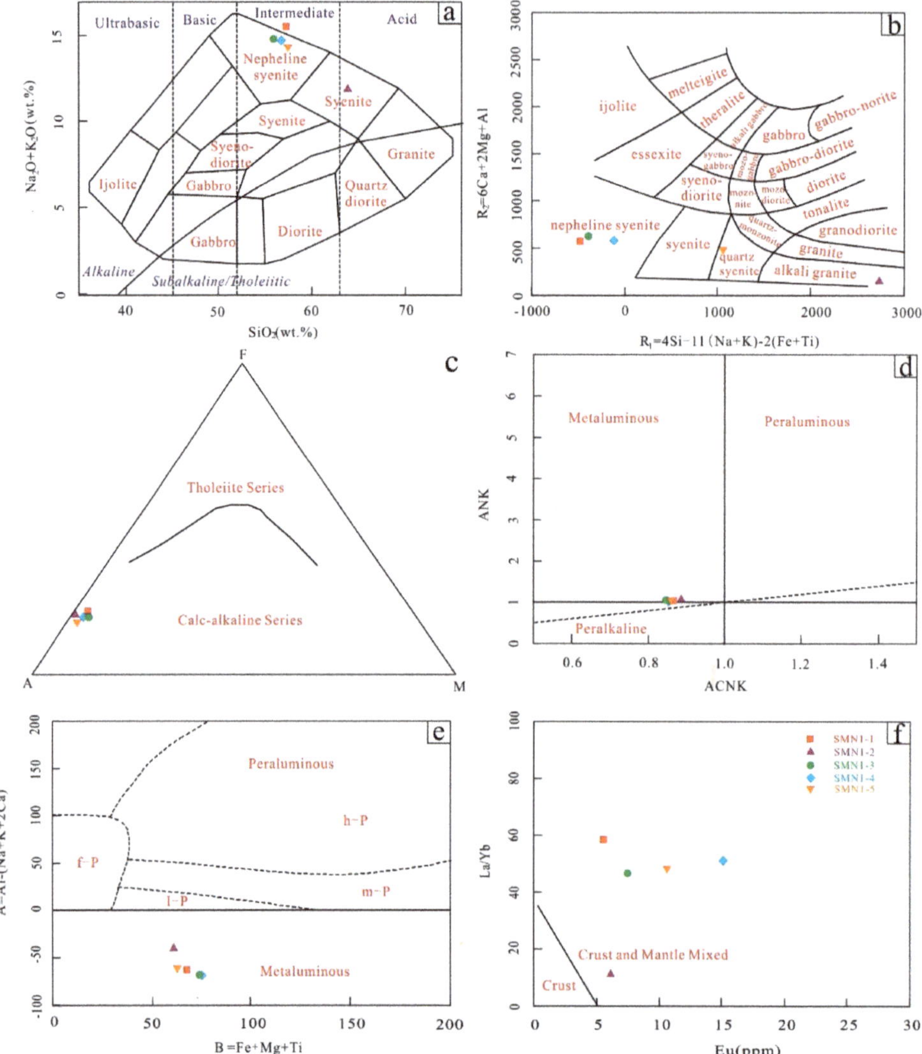

Figure 5. Geochemical classification diagrams for the syenites in the Saima deposit. (**a**) TAS diagram. (**b**) R2 vs. R1 diagram. (**c**) FAM diagram. (**d**) A/NK vs. A/CNK diagram. (**e**) [Al − (Na + K + 2Ca)] vs. (Fe + Mg + Ti) diagram. (**f**) La/Yb vs. Eu diagram (after Le Maître 1989; after Middlemost, 1994; after Irvine, 1971) [11–13].

In the primary mantle (PM)-normalized trace element diagram (Figure 6a), the aegirine nepheline syenites are enriched in LILEs (e.g., Cs, Rb and Ba) and depleted in HFSEs (e.g., Nb, Zr, Ti), suggesting that the origin of associated parent magmas is related to the enriched mantle. In addition, these samples have relatively high Nb contents of 34.79–106.59 ppm and are potentially able to provide ore-forming materials for the Nb mineralization. Moreover, the transitional elements show an obvious W-shaped distribution pattern, which are consistent with those from the mantle-derived granite and mafic intrusions formed under continental-rift setting [14].

The aegirine nepheline syenites have variable REE contents varying from 0.34 to 946.00 ppm, with high LREE/HREE ratios of 7.20–18.48 (Table 3). In the chondrite-normalized REE diagram (Figure 6b), these syenites are enriched in LREEs and depleted in HREEs, with high $(La/Yb)_N$ ratios of 7.56–39.58, Nb/Ta ratios of 24.56–40.04 and low $(Rb/La)_N$ ratios of 0.31–1.14. All of the samples have insignificant negative Sr and Eu anomalies (Eu/Eu* = 0.94–1.00). Considering that the depletions of Sr and Eu were caused by the crystallization of plagioclase or residue in the source [5,15–18], it is suggested that the plagioclase is absent in the magmatic source of aegirine nepheline syenites. Additionally, the relatively enriched LREEs indicate the presence of garnet as residual mineral in the source [19–24].

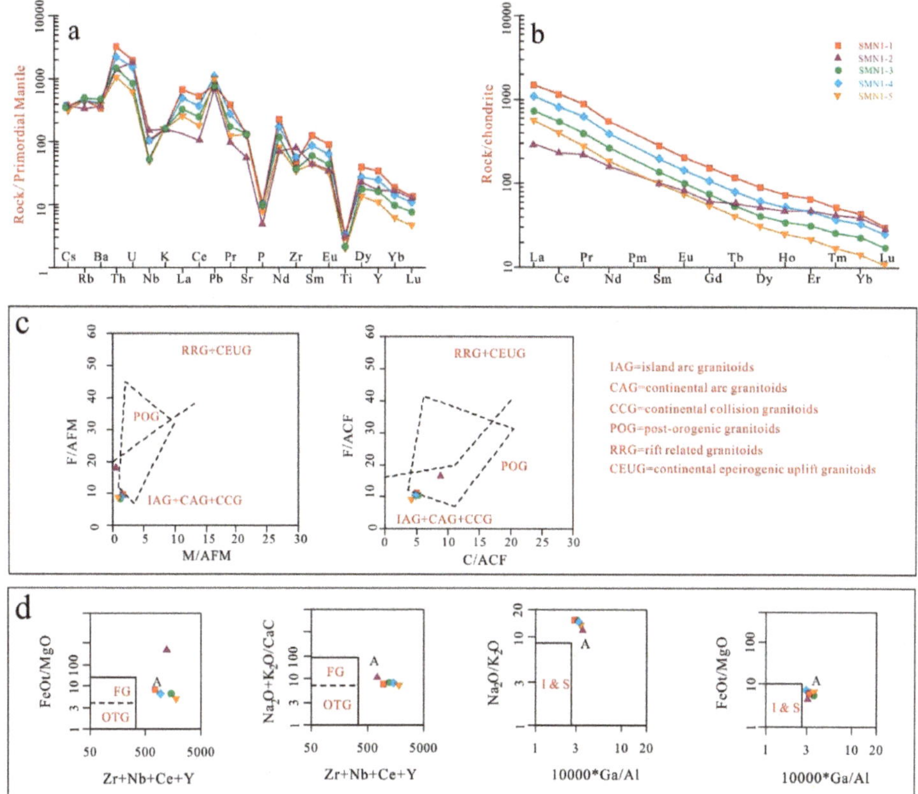

Figure 6. Geochemical classification diagrams for the syenites in the Saima deposit. (**a**) Primitive mantle-normalized trace-element diagram (normalized data from Sun and Mcdonough,1989); (**b**) Chondrite-normalized rare-earth element (REE) patterns (normalized data from Boynton, 1984); (**c**,**d**) Structural discriminant diagram (after Defant and Drummond, 1990; modified from Pearce 1982) [25–28].

5. Discussion

5.1. Timing of Nb Mineralization

Based on the Mesozoic diagenetic and metallogenic ages obtained from previous studies in the NE China, three large-scale mineralization events have been identified [29–33]. Chen and Liu et al. pointed out that the Jia–Meng Continental Block collided with the North China Craton along the Xra Moron–Changchun–Yanji Suture Zone, resulted in the intracontinental crustal shortening and thickening and the formation of crustal-derived granitoids and related Cu–Mo–REE mineralization during 255–190 Ma. During 200–130 Ma, a series of EW-, NE-trending regional faults and their secondary faults were developed and reactivated due to the subduction of the Palaeo-Pacific Plate beneath the Eurasian Continent. These faults have provided channels for the ascent of magmas and ore-forming materials, leading to the most intensive mineralization in NE China. North China and Siberian plates collided during 150–120 Ma, resulting in the crustal thickening and the generation of intermediate to felsic magmas and ore-forming fluids. At the same time, the subduction of Pacific oceanic plate entered the peak period at about 140 Ma. The regional tectonic framework in eastern China was then evolved into the coastal western Pacific tectonic domain, leading to the development of the coastal Pacific metallogenic system. These above three large-scale magmatic-mineralized events occurred in NE China, corresponding to the initial subduction and subduction peak of the Palaeo-Pacific Plate, as well as the closure of the Paleo-Asian Ocean Plate, respectively. This reflects that the Mesozoic mineralization events in NE China are associated with the conjunction of multiple plates.

In terms of the Late Triassic intrusive rocks in the eastern Liaoning Province, they were formed under an intracontinental orogenic setting with thickened continental crust. During the earliest Late Triassic, the Liaodong Peninsula was in the retention stage of uplifted mantle, causing the emplacement of massive ultramafic-mafic complex and diorite-quartz diorite-tonalite association along the large-scale lateral ductile shear zone. With the development of intracontinental orogenesis, abundant calc-alkaline granitic rocks originated from the remelting of thickened continental crust emplaced in the Liaodong Peninsula. While, post-orogenic collapse occurred after intracontinental orogeny in the latest Triassic [34–36], crustal thinning and the upwelling of mantle-derived magmas resulted in the formation of A-type granites and alkaline complex represented by Saima syenites, which marked the end of the Indosinian magmatism and orogenic cycle.

Sixteen zircon grains with typical magmatic origin from Saima syenite have yielded a weighted mean ^{206}Pb/^{238}U age of 229.5 ± 2.2 Ma (Figure 4), indicating that the ore-related aegirine nepheline syenite emplaced in the Late Triassic. In other words, the metallogenic age of Saima Nb deposit is the Late Triassic. The Nb mineralization at Saima was the product of latest Triassic large-scale mineralization in NE China [31].

5.2. Implications for Tectonic Setting

Recently, it is widely acknowledged that the distribution of REE deposits and their ore-forming rocks were controlled by regional faults in the margins of plates and tectonic conjunctions. This type of REE deposit was genetically related to the magmatism, which shows that the distribution of Nb and Ta metallogenic belt was spatially consistent with the occurrence of magmatic rocks in Liaoning Province [1,37–39]. With the deepening of the research on the alkaline rocks, a series of accurate data have been acquired, indicating that the alkali-rich igneous rocks are generally derived from the deep source. These alkaline rocks involve the messages of deep geodynamic processes in the shallow crust, which could reflect the compositions and geodynamic settings of deep earth, the superstructure of the crust and physico-chemical conditions, as well as the concentrations of rare metals in melts. Alkalic complex was known as the product of stable continental rift magmatism and deep fault movements and this property could bring important information about the composition, evolution, tectonic setting and physico-chemistry of deep earth [40,41].

Previous studies have shown that the abnormal geochemical behaviors of Nb and Ta and the ranges of Nb/Ta values during magmatic-hydrothermal processes can reflect the genesis and origin of magmas. The Nb/Ta values from the C1 type chondrite, primitive mantle (PM), depleted mantle (DM) and continental crust are 17.3–17.6, 17.5, 15.5 and 10–14, respectively. Igneous rocks distributed along the deep fault or rift zone show strong enrichments of Nb and Ta, with remarkable high Nb/Ta values than those of primitive mantle [42–46]. The ore-forming syenites in the Saima deposit have obviously higher Nb/Ta values of 24.56–40.04 than those of primitive mantle, which further demonstrated that the Saima syenites were derived from the deep mantle in an extensional environment. Moreover, the syenite samples plot in the field of crust-mantle mixed source in the La/Yb-Eu diagram (Figure 5f). In addition, the Saima syenites show enrichments in Sr and Nd and depletion in Hf, revealing that the asthenospheric mantle had contributed to the generation of ore-related alkaline magmas with

Nb and Ta enrichments and high Nb/Ta values. All of the rock samples obtained from the Saima deposit plot in the fields of within-plate, post-collision and rift- granitoids in the tectonic environment discriminant diagrams (Figure 6c,d). The above geochemical characteristics are consistent with those of ore-forming rocks from the same type of deposits and we thus infer that the aegirine nepheline syenite is the ore-forming rock of Saima Nb deposit formed under post-collisional rift setting. The basic characteristics of the deposit are basically consistent with those of typical alkaline-type rare earth deposits, for example, Strange Lake Nb deposit (Canada), Ghurayyah Nb deposit (Saudi Arabia), Khaldzan Buregte (Mongolia) and so on [47,48].

As discussed above, the formation of ore-related aegirine nepheline syenite in the Saima Nb deposit is mainly attributed to crust-mantle mixing. The mineralized processes are summarized as follows. The crust relaxed and thinned after the termination of intracontinental orogeny in the Late Indosinian, causing the upwelling of mantle-derived magmas along the weak zone of a fault system and rising abundant ore-forming materials from deep mantle. The low-degree melting of mantle had resulted in the enrichment of incompatible elements (e.g., LILEs and REEs) and volatile components for the Saima deposit. The aegirine nepheline syenites are enriched in high field-strength elements (particularly for Nb, Ta, Zr and Hf) and LREEs that were higher than those of background values of crust [49–52], which provide preferable conditions for the formation of Nb mineralization.

6. Conclusions

(1) The Nb orebodies mainly occurred in grey-brown to grass-green aegirine nepheline syenite in the Saima deposit, with fergusonite and brocenite as the dominant ore minerals, which are consistent with the typical alkaline-type REE deposit in China.

(2) In situ LA-ICP-MS zircon U-Pb dating indicates that the aegirine nepheline syenite was emplaced in Late Triassic (229.5 ± 2.2 Ma), which are spatially, temporally and genetically related to Nb mineralization.

(3) The medium-coarse grained aegirine nepheline syenites are high-K peralkaline, metaluminous syenitoid, with relatively high Nb/Ta ratios.

(4) The Saima Nb deposit was formed under the post-collision continental-rift setting, which is consistent with the tectonic regime of post-collision between the North China Craton and Paleo-Asian oceanic plate during the Indosinian. Intensive magmatic and metallogenic events caused by partial melting of lithospheric mantle occurred under the post-collisional rifting, resulting in the development of large-scale Cu–Mo mineralization and rare earth deposits in the eastern of Liaoning Province.

Author Contributions: Conceptualization, Z.-W.B.; Data curation, S.Z., Q.-Q.S., Q.Y. and Z.-G.W.; Methodology, Y.-C.G. and Q.-S.S.; Software, L.S., D.Z. and T.W.; Writing—original draft, Y.-S.R.; Writing—review & editing, N.J.

Funding: This research was funded by the China Geological Survey Project, grant No. DD20160131.

Acknowledgments: We thank the geologists from the Shenyang Geological Survey of CGS with their assistance in the field work for the Saima Nb deposit.

Conflicts of Interest: The authors declare no conflict of interest.

References

1. Chen, G.; Zhao, J.; Zhai, Y. Shaolahazi alkaline rock Nb-Ta mineralization characteristics and prospecting direction of alkaline rocks in the central part of Jilin Province. *Jilin Geol.* **2011**, *30*, 36–39. (In Chinese with English abstract).

2. Cai, X.; Song, Y. Distribution of niobium and tantalum minerals and metallogenic conditions in northeast China. *Ore Depos. Geol.* **2014**, *33*, 1155–1156. (In Chinese with English abstract).

3. Mao, C.X.; Zheng, C.Q.; Bi, Z.W.; Yang, Y.J.; Cai, L.; Zhang, C.P. Preliminary Study on the Prospectng Potentlal of Niobium-Tantalum Deposite in Daxinganling Region. *Geol. Resour.* **2016**, *25*, 269–274. (In Chinese with English abstract).

4. Sun, L.J.; Zhang, Y.H.; Yu, J.H. Discussion Geological Features and Age on the Alkaline-Complex of Fengcheng of the Eastern Liaoning. *J. Liaoning Commun. Coll.* **2008**, *10*, 41–44. (In Chinese with English abstract).

5. Ju, N.; Zhang, S.; Bi, Z.W.; Shi, L.; Zhang, D. Study on ore-forming conditions of fengcheng horse niobium deposit in Liaoning Province. *J. Miner. Sci.* **2017**, supplement, 189. (In Chinese with English abstract).

6. Yuan, H.L.; Gao, S.; Liu, X.M.; Li, H.M.; Gunther, D.; Wu, F.Y. Accurate U-Pb age and trace element determinations of zircon by laser ablation-inductively coupled plasma-mass spectrometry. *Geostand. Geoanal. Res.* **2004**, *28*, 353–370. [CrossRef]

7. Wiedenbeck, M.; Alle, P.; Corfu, F.; Griffin, W.L.; Meier, M.; Oberli, F.; Spiegel, W. Three natural zircon standards for U–Th–Pb, Lu–Hf, trace element and REE analyses. *Geostand. Geoanal. Res.* **1995**, *19*, 1–23. [CrossRef]

8. Andersen, T. Correction of common lead in U-Pb analyses that do not report 204 Pb. *Chem. Geol.* **2002**, *192*, 59–79. [CrossRef]

9. Belousova, E.A.; Griffin, W.L.; O'Reilly, S.Y. Igneous zircon: Trace element composition as an indicator of source rock type. *Contrib. Mineral. Petrol.* **2002**, *143*, 602–622. [CrossRef]

10. Hoskin, P.W.O.; Schaltegger, U. The composition of zircon and igneous and metamorphic petrogenesis. *Rev. Mineral. Geochem.* **2003**, *53*, 27–62. [CrossRef]

11. Middlemost, E.A.K. Naming materials in the magma/igneous rock system. *Earth Sci. Rev.* **1994**, *37*, 215–224. [CrossRef]

12. Le Maitre, R.W.; Bateman, P.; Dudek, A.; Keller, J.; Lameyre, J.; Le Bas, M.J.; Sabine, P.A.; Schmid, R.; Sorensen, H.; Streckeisen, A.; et al. *A Classification of Igneous Rocks and Glossary of Terms*; Blackwell: Oxford, UK, 1989.

13. Irvine, T.N.; Baragar, W.R.A. A guide to the chemical classification of the common volcanic rocks. *Can. J. Earth Sci.* **1971**, *8*, 523–548. [CrossRef]

14. Niu, H.C.; Shan, Q.; Chen, X.M.; Zhang, H.Y. The relation between light rare earth deposits and slow earth processes in the pangxi rift zone. *Sci. China* **2002**, *32*, 33–40. (In Chinese with English abstract).

15. Bakker, R.J.; Elburg, M.A. A magmatic-hydrothermal transition in Arkaroola (northern Flinders Ranges, South Australia): From diopside-titanitepegmatites to hematite-quartz growth. *Contrib. Mineral. Petrol.* **2006**, *152*, 541–569. [CrossRef]

16. Qiu, K.F.; Yu, H.C.; Gou, Z.Y.; Liang, Z.L.; Zhang, J.L.; Zhu, R. Nature and origin of Triassic igneous activity in the Western Qinling Orogen: The Wenquan composite pluton example. *Int. Geol. Rev.* **2018**, *60*, 242–266. [CrossRef]

17. Bonin, B. Alkaline rocks and geodynamics. *Turk. J. Earth Sci.* **1998**, *7*, 105–118.

18. Crockettand, R.N.; Sutphin, D.M. *International Strategic Minerals Inventory Summary Report—Niobium and Tantalum*; Circular 930-M; U.S. Geological Survey: Reston, VA, USA, 1993; 36p.

19. Cai, J.H.; Yan, G.H.; Xu, B.L.; Wang, G.Y.; Mou, B.L.; Zhao, Y.C. The Late Mesozoic Alkaline Intrusive Rocks at the East Foot of the Taihang-Da Hinggan Mountains: Lithogeochemical Characteristics and Their Implications. *J. Earth* **2006**, *27*, 447–459. (In Chinese with English abstract).

20. Deng, J.; Wang, Q.F.; Li, G.J.; Santosh, M. Cenozoic tectono-magmatic and metallogenic processes in the Sanjiang region, southwestern China. *Earth-Sci. Rev.* **2014**, *138*, 268–299. [CrossRef]

21. Liu, X.Y.; Cai, J.H.; Yan, G.H. Petrogeochemical characteristics and geological significance of ancient middle proterozoic alkali rocks in the southern margin of north China craton. *Ore Depos. Geol.* **2010**, *29*, 1109–1110. (In Chinese with English abstract).

22. Geng, J.Z.; Qiu, K.F.; Gou, Z.Y.; Yu, H.C. Tectonic regime switchover of Triassic Western Qinling Orogen: Constraints from LA-ICP-MS zircon U–Pb geochronology and Lu–Hf isotope of Dangchuan intrusive complex in Gansu, China. *Chem. Erde Geochem.* **2017**, *77*, 637–651. [CrossRef]

23. Wang, F.L.; Zhao, T.P.; Chen, W. Advances in study of Nb-Ta ore deposits in Panxi area and tentative discussion on genesis of these ore deposits. *Ore Depos. Geol.* **2012**, *31*, 293–308. (In Chinese with English abstract).

24. Wang, P.X.; Bao, M.W. General Situation and Prospecting Revelation of Tantalum-Niobium Rare Metal Deposits in China. *Met. Mines* **2015**, *468*, 92–97. (In Chinese with English abstract).

25. Sun, S.S.; McDonough, W.F. Chemical and isotopic systematics of oceanic basalts: Implications for mantle composition and processes. In *Magmatism in Oceanic Basins*; Saunders, A.D., Norry, M.J., Eds.; Geological society London Special Publications: London, UK, 1989; Volume 42, pp. 313–345.

26. Boynton, W.V. Cosmochemistry of the rare earth elements: Meteorite studies. In *Rare Earth Element Geochemistry: Development in Geochemistry*; Henderson, P., Ed.; Elsevier: Amsterdam, The Netherlands, 1984; pp. 63–107.

27. Defant, M.J.; Drummond, M.S. Derivation of some modern arc magmas by melting young subducted lithosphere. *Nature* **1990**, *347*, 662–665. [CrossRef]

28. Pearce, J.A. Trace element characteristics of lavas from destructive plate boundaries. In *Andesits. Chochester*; Thorpe, R.S., Ed.; Wiley: Hoboken, NJ, USA, 1982; pp. 525–548.

29. Wu, F.Y.; Jahn, B.M.; Wilde, S.; Sun, D.Y. Phanerozoic continental crustal growth: U-Pb and Sr-Nd isotopic evidence from the gran-ites in northeastern China. *Tectonophysics* **2000**, *328*, 89–113. [CrossRef]

30. Zhang, C.H.; Kang, Z.; Zhang, G.Y.; Liu, Y.C. Sequential characteristic of regional mineralization and magmatism of NE, China in Mesozoic. *Geol. Resour.* **2009**, *18*, 87–99. (In Chinese with English abstract).

31. Chen, Y.J.; Zhang, C.; Li, N.; Yang, Y.F.; Deng, K. Geology of the Mo Deposits in Northeast China. *J. Jilin Univ. (Earth Sci. Ed.)* **2012**, *42*, 1223–1268. (In Chinese with English abstract).

32. Zhang, Y. Research on Characteristics of Geology, Geochemistry and Metallogenic Mechanism of the Jurassic Molybdenum Deposits in the Mid-East Area of Jilin. Ph.D. Thesis, Jilin University, Changchun, China, 2013. (In Chinese with English abstract).

33. Cao, Z.Q.; Hou, G.J. The Late Mesozoic Alkaline Intrusive Rocks at the North of the Da Hinggan Mountains: Lithogeochemical Characteristics and Their Implications. *Miner. Rock Geochem. Bull.* **2009**, *28*, 209–216. (In Chinese with English abstract).

34. Deng, J.; Wang, Q.F. Gold mineralization in China: Metallogenic provinces, deposit types and tectonic framework. *Gondwana Res.* **2016**, *36*, 219–274. [CrossRef]

35. Deng, J.; Wang, C.M.; Bagas, L.; Carranza, E.J.M.; Lu, Y.J. Cretaceous-Cenozoic tectonic history of the Jiaojia Fault and gold mineralization in the Jiaodong Peninsula, China: Constraints from zircon U-Pb, illite K-Ar and apatite fission track thermochronometry. *Miner. Depos.* **2015**, *50*, 987–1006.

36. Cheng, X.H.; Xu, J.H.; Zhang, H. The inclusions in quartz veins in the alkaline rock area of berchon in eastern Liaoning Province. *Ore Depos. Geol.* **2014**, *33*, 503–504. (In Chinese with English abstract).

37. He, J.L. Ore-forming Geological Conditions and Prospecting Potential for Nb-Ta Mineral Deposits in Panzhihua-Xichang Region, Sichuan. *Geol. J. Sichuan* **2004**, *24*, 206–211. (In Chinese with English abstract).

38. Li, S.P.; Zhan, S.Z.; Jin, T.T.; Chen, J.; Ren, H.; Qiu, W. REE Geochemical Characteristics and Provenance Analysis of the Shaliuquan Niobium Tantalum Pegmatite Ore, Qinghai Province. *Rare Earths* **2016**, *37*, 39–46. (In Chinese with English abstract).

39. Yuan, Z.X.; Bai, G. Temporal and Spatial Distribution of Endogenic Rare and Rare Earth Mineral Deposits of China. *Ore Depos. Geol.* **2001**, *20*, 347–354. (In Chinese with English abstract).

40. Yang, L.Q.; Deng, J.; Dilek, Y.; Qiu, K.F.; Ji, X.Z.; Li, N.; Taylor, R.D.; Yu, J.Y. Structure, geochronology and petrogenesis of the Late Triassic Puziba granitoid dikes in the Mianlue suture zone, Qinling Orogen, China. *Geol. Soc. Am. Bull.* **2015**, *127*, 1831–1854. [CrossRef]

41. Yang, L.Q.; Deng, J.; Qiu, K.F.; Ji, X.Z.; Santosh, M.; Song, K.R.; Song, Y.H.; Geng, J.Z.; Zhang, C.; Hua, B. Magma mixing and crust–mantle interaction in the Triassic monzogranites of Bikou Terrane, central China: Constraints from petrology, geochemistry and zircon U–Pb–Hf isotopic systematics. *J. Asian Earth Sci.* **2015**, *98*, 320–341. [CrossRef]

42. Zhang, P.S.; Yang, Z.M.; Tao, K.J.; Song, R.K. Niobian-Tantalian and Rare Earth Mineralogy in China and Their Industrial Utilization. *Rare Met.* **2005**, *29*, 206–210. (In Chinese with English abstract).

43. Nan, J.; Yunsheng, R.; Sen, Z.; Linlin, K.; Di, Z.; Yuchao, G.; Qun, Y.; Hui, W.; Lei, S.; Qiushi, S.; et al. The Early Jurassic Chang'anbu porphyry Cu–Mo deposit in Northeastern China: Constraints from zircon U-Pb geochronology and H-O-S-Pb stable isotopes. *Geol. J.* **2018**, *53*, 2437–2448.

44. Ran, Q.C.; Liu, X.Z. Significance of Contrasting Between Fengcheng Alkali Complex and Related Diamond-Bearing Rocks. *J. Chang. Inst. Geol.* **1993**, *23*, 279–285. (In Chinese with English abstract).

45. Ren, J.; Lu, S.C. Survey of world niobium resources and their characteristics. *Nonferr. Metall.* **1997**, *5*, 1–3. (In Chinese with English abstract).

46. Qiu, K.F.; Marsh, E.; Yu, H.C.; Pfaff, K.; Gulbransen, C.; Gou, Z.Y.; Li, N. Fluid and metal sources of the Wenquan porphyry molybdenum deposit, Western Qinling, NW China. *Ore Geol. Rev.* **2017**, *86*, 459–473. [CrossRef]

47. Zhu, J.C.; Li, R.K.; Li, F.C.; Xiong, X.L.; Zhou, F.Y.; Huang, X.L. Topaz-albite granites and rare-metal mineralization in the Limu District, Guangxi Province, Southeast China. *Miner. Desposita* **2001**, *36*, 393–405. [CrossRef]

48. Kempe, U.; Gotze, J.; Dandar, S.; Habermann, D. Magmatic and metasomatic processes during formation of the Nb-Zr-REE deposits Khaldzan Buregte and Tsakhir (Mongolian Altai): Indications from a combined CL-SEM study. *Mineral. Mag.* **1999**, *63*, 165–177. [CrossRef]

49. Pal, D.C.; Mishra, B.; Bernhardt, H.J. Mineralogy and geochemistry of pegmatite-hosted Sn-Ta-Nb-and Zr-Hf-bearing minerals from the southeastern part of the Bastar-Malkangiri Pegmatite Belt, Central India. *Ore Geol. Rev.* **2007**, *30*, 30–55. [CrossRef]

50. Qiu, K.F.; Yang, L.Q. Genetic feature of monazite and its U–Th–Pb dating: Critical considerations on the tectonic evolution of Sanjiang Tethys. *Acta Petrol. Sin.* **2011**, *27*, 2721–2732. (In Chinese with English abstract).

51. Sørensen, H. *Alkaline Rocks*; Wiley: Hoboken, NJ, USA, 1974; 622p.

52. Shand, S.J. The problem of the alkaline rocks. *Proc. Geol. Soc. S. Afr.* **1922**, *25*, 19–33.

Article

Petrogenesis of Ore-Hosting Diorite in the Zaorendao Gold Deposit at the Tongren-Xiahe-Hezuo Polymetallic District, West Qinling, China

Zong-Yang Gou [1,2], Hao-Cheng Yu [1,2], Kun-Feng Qiu [1,2,3,*], Jian-Zhen Geng [1], Ming-Qian Wu [1,4], Yong-Gang Wang [5], Ming-Hai Yu [6] and Jun Li [1]

[1] State Key Laboratory of Geological Processes and Mineral Resources, and School of Earth Science and Resources, China University of Geosciences, Beijing 100083, China; zongygou@cugb.edu.cn (Z.-Y.G.); haochengyu@cugb.edu.cn (H.-C.Y.); mumu1270@163.com (J.-Z.G.); aria.wu1990@gmail.com (M.-Q.W.); 2001170100@cugb.edu.cn (J.L.)

[2] Key Laboratory of Mineral Resources, Institute of Geology and Geophysics, Chinese Academy of Sciences, Beijing 100029, China

[3] Department of Geology and Geological Engineering, Colorado School of Mines, Golden, CO 80401, USA

[4] Department of Earth and Environmental Sciences, University of Windsor, Windsor, On N9B 3P4, Canada

[5] Guangxi Geological Exploration Institute of CCGMB, Nanning 530001, China; Yonggwang@163.com

[6] Zhaojin Ming Industry co., LTD, Zhaoyuan 265400, China; minghyu@126.com

* Correspondence: kunfengqiu@cugb.edu.cn; Tel.: +86-10-82321383

Received: 15 December 2018; Accepted: 24 January 2019; Published: 28 January 2019

Abstract: The newly discovered Zaorendao gold deposit is in the Tongren-Xiahe-Hezuo polymetallic district in the westernmost West Qinling orogenic belt. The estimated pre-mining resource is approximately 13.6 t of Au at an average grade of 3.02 g/t. Mineralization is predominantly controlled by NW-trending and EW-trending faults within diorite intrusions and surrounding sedimentary rocks. In the present study, in situ zircon U–Pb geochronology and Lu–Hf isotopic analyses of the ore-hosting diorite at Zaorendao were measured using LA-ICP-MS. The data suggest that the diorite was emplaced at ca. 246.5 ± 1.9 Ma. The large variation of zircon Hf isotopic composition ($\varepsilon_{Hf}(t)$ values ranging from −12.0 to −1.8) indicates a two-stage model age (T_{DM2}) that ranges from 1.4 Ga to 2.0 Ga. Such Lu–Hf isotopic compositions indicate that the diorite was dominantly derived from a Paleo- to Meso-Proterozoic continental crust. The wide range of $\varepsilon_{Hf}(t)$ and the presence of inherited zircon can be interpreted to suggest the mixing of Paleo- to Meso-Proterozoic continental crust with a mantle component. Combining such characteristics with the geochemistry of coeval rocks that are associated with the diorite, we therefore proposed that the gold-hosting Triassic diorite in the Zaorendao gold deposit formed in an active continental margin that was associated with the northward subduction of the paleo-Tethyan ocean.

Keywords: geochronology; Hf isotopes; petrogenesis; paleo-Tethyan ocean closure; Zaorendao gold deposit; West Qinling

---

1. Introduction

The WNW-trending West Qinling orogen, the western half of the Qinling orogen, is connected to the Kunlun Orogen and the Qaidam terrane in the west [1,2]. The West Qinling orogen evolved over a protracted period from the Proterozoic up to the late Mesozoic. The tectonic history evolved in response to subduction and closure of the paleo-Tethys ocean and subsequent continental collision between the South China block and North China block during Late Triassic (Figure 1) [1–6]. This region contains abundant Cu-Au, Cu-Mo, Au, Au-Sb, and Pb-Zn deposits, and is known to be one of the most important polymetallic belts in China [3,7–14]. Most of the gold deposits are hosted in Cambrian to Early

Triassic marine sedimentary rocks and these deposits have been suggested to be either orogenic [3,15,16] or Carlin-like [9,17]. Widespread magmatic activity in the West Qinling orogen is temporally and/or spatially associated with several types of gold mineralization, which include porphyry Cu-Au deposits, skarn-type Cu-Au (-Fe) deposits, and lode/disseminated Au deposits [12,18–20]. Thus, information on the widespread Triassic granitoids closely associated with the deposits in the West Qinling is important to constrain the tectono-magmatic and mineralization processes.

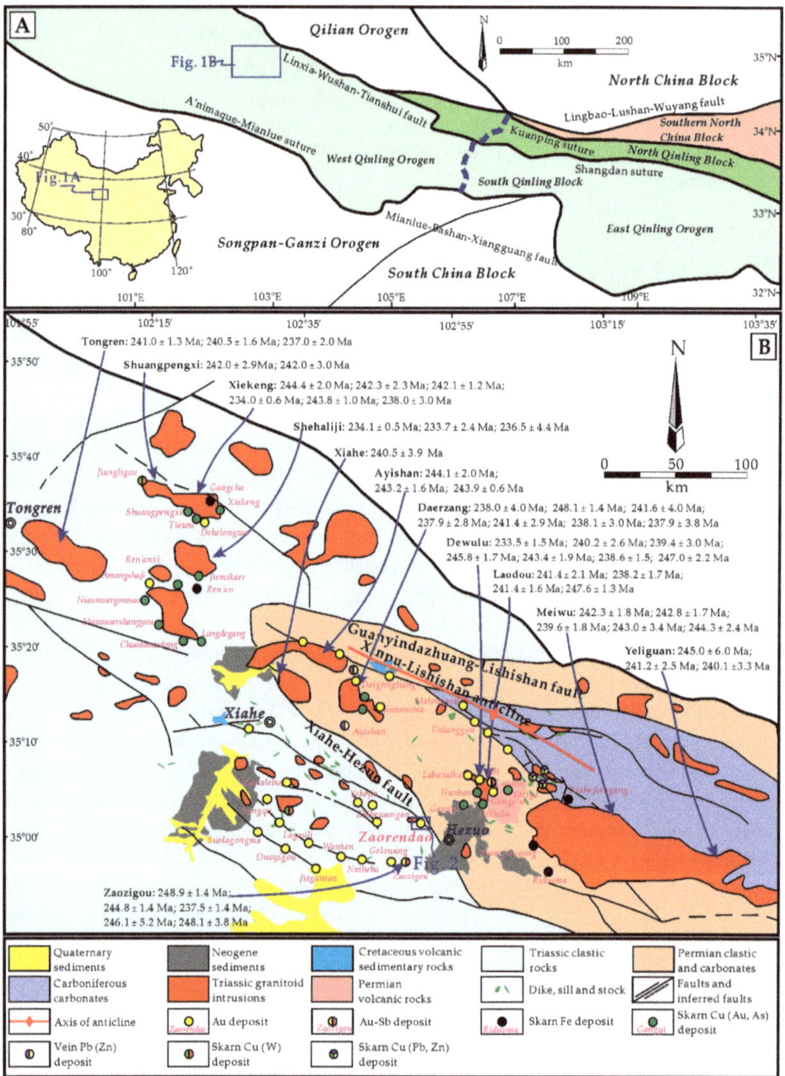

Figure 1. (**A**) Major tectonic domains of China and the location of the Qinling Orogenic Belt. Inset shows the location of the study area (modified after [12]). (**B**) Simplified geological map of the Tongren-Xiahe-Hezuo district of the West Qinling orogen, showing distributions and ages of Early Triassic granitoids and ore deposits (modified after [12] and [21]).

The Tongren-Xiahe-Hezuo district, located to the northwestern section of the West Qinling, contains numerous sediment- and/or dike-hosted disseminated and lode gold deposits, as well as many Cu-Au-W-Fe skarn deposits, and has been one of the most popular districts for exploring metal resources over the last decade (Figure 1) [12,19,22]. The relationship between the disseminated gold deposits to the south of Xiahe-Hezuo fault and magmatism is controversial and poorly understood, due to the rare constraints on the emplacement age and petrogenesis of stocks and dikes.

The newly discovered Zaorendao gold deposit is one of largest in the Tongren-Xiahe-Hezuo polymetallic district. The diorite stock and surrounding sedimentary rocks that are cut by NW-trending and nearly EW-trending fracture belts host the dominant gold mineralization. However, no petrography, geochronology, or geochemistry research on the diorite has been conducted until now. Based on systematic field investigations and petrographic observations, we present zircon LA-ICP-MS U–Pb geochronology, and Lu–Hf isotopic composition of the ore-hosting diorite in the Zaorendao deposit, to: a) establish the time of its emplacement, b) understand the petrogenesis of the diorite.

2. Geological Background

2.1. Regional Geology

The E-W-trending Qinling orogen in central China, which extends for more than 1500 km and is ~200-250 km wide, is tectonically situated between the North China and the South China blocks (Figure 1A) [1,23]. The Lingbao-Lushan-Wuyang fault separates the West Qinling orogen from the Qilian Orogen and North China block to the north. The West Qinling orogen is separated from the Songpan-Ganzi Orogen and the South China block in the south by the Mianlue-Bashan-Xiangguang fault [2,23,24]. The Qinling orogen is further divided into Southern North China block, North Qinling and South Qinling blocks by the Kuanping and Shangdan sutures [1,25–27].

The West Qinling orogen, the western part of the Qinling orogen, is traditionally separated from the East Qinling orogen by the Huicheng basin or Foping dome, which can be observed as roughly marked by the Baoji-Chengdu railway line (shown as blue dash line in Figure 1A) [25,28]. The West Qinling orogen is bounded by the Linxia-Wushan-Tianshui fault (Shangdan suture) on the north and the A'nimaque–Mianlue suture on the south (Figure 1A) [24,29,30]. The West Qinling had a protracted and complex evolution history from the Proterozoic up to the late Mesozoic, with the final collision between the South China block and the South Qinling block along the A'nimaque–Mianlue suture zone [1,3,23,30–32]. The West Qinling orogen is dominantly covered by Paleozoic to early Mesozoic marine sedimentary rocks that were deformed in the early Mesozoic [2,33], and intruded by numerous Mesozoic granitoid intrusions. The widespread Mesozoic igneous rocks are grouped into subduction-related granitoids with ages ranging from ca. 250–235 Ma, syn-collision granitoids with ages ranging from ca. 228–215 Ma and post-collisional granitoids with ages ranging from ca. 210–190 Ma [4,11,13,34–36]. This is interpreted to be due to the northward subduction of paleo-Tethys ocean and subsequent syn-collision to post-collision between the North China and South China blocks.

The Tongren-Xiahe-Hezuo district is located to the northwestern section of West Qinling (Figure 1B) [12,19,21,37]. The exposed strata are dominated by weakly metamorphosed to un-metamorphosed marine sedimentary rocks of Carboniferous to Triassic age (Figure 1B) [12,19,21,37]. The Carboniferous rocks consist mainly of sandstone, siltstones, thin- to thick-bedded limestones and clastic limestone, whereas the Permian slope to deep-marine basinal facies rocks are mainly composed of carbonaceous/argillaceous siltstone, sandstone, conglomerate, limestone and calcirudite. The Lower to Middle Triassic strata comprise interbedded sandstones, siltstones, calcareous siltstones, and mudstones, with minor amounts of siliceous nodules hosted by siltstone layers [12,19,21]. The Carboniferous carbonates and siltstones, and Permian marine clastic and carbonates are exposed mainly in the core and both flanks of the Xinpu-Lishishan anticline, respectively, and cover the central to southeast part of the district. The Triassic strata are separated from Carboniferous and Permian rocks by the Guanyindazhuang-Lishishan fault to the north and Xiahe-Hezuo fault to the south [12], and are exposed in the west of the district. Several dioritic to granitic plutons, represented by the Tongren, Jiangligou, Shuangpengxi, Xiekeng, Shehaliji, Xiahe, Ayishan, Daerzang, Dewulu, Laodou,

Meiwu and Yeliguan plutons/batholiths, intrude the strata. These intrusions were emplaced between 250 and 234 Ma (Table 1) [12,37–41], and form a discontinuous NW-trending magmatic belt between the axial trace of the Xinpu-Lishishan anticline and the Xiahe-Hezuo fault. The magmatic belt extends northwest to the Tongren district [19,39]. Numerous stocks and dikes occur mainly to the south of the Xiahe-Hezuo fault, and these stocks and dikes are similar in composition to the plutons in the north. The emplacement age and petrogenesis of stocks and dikes is poorly constrained, but recent studies on zircon U–Pb age of dikes in the Zaozigou gold deposit indicate that they were emplaced in the 238–249 Ma interval, coeval with the plutons in the north [20,41]. A few tens of hydrothermal ore deposits and occurrences have been identified in the Tongren-Xiahe-Hezuo district, and these deposits are temporally and/or spatially associated with the Triassic granitoids (Figure 1B). Mineralization to the south of the Xiahe-Hezuo fault is mainly hosted by Triassic marine sedimentary rocks and, less significantly in the intermediate to felsic stocks and/or dikes. These deposits are characterized by disseminated gold deposits [42–44]. In the north part of the district, dozens of lode gold deposits and Cu-Au-(W)-(Fe) skarn deposits have been discovered hosted within or close proximity of intrusions [12,18,19].

Table 1. Compilation of previously published zircon U–Pb ages of the Triassic magmatism in the Tongren-Xiahe-Hezuo polymetallic district.

Pluton	Location	Rock	Age (Ma)	Method	Reference
Tongren	Tongren, Qinhai	Granodiorite	241.0 ± 1.3	LA-ICP-MS	Li et al., 2015 [37]
Tongren	Tongren, Qinhai	Granodiorite	240.5 ± 1.6	LA-ICP-MS	Li et al., 2015 [37]
Tongren	Tongren, Qinhai	Granodiorite	237.0 ± 2.0	LA-ICP-MS	Luo, 2013 [45]
Shuangpengxi	Tongren, Qinhai	Granodiorite	242.0 ± 2.9	LA-ICP-MS	Luo et al., 2012 [39]
Shuangpengxi	Tongren, Qinhai	Granodiorite	242.0 ± 3.0	SHRIMP	Zhang et al., 2014 [46]
Xiekeng	Tongren, Qinhai	Granodiorite	244.4 ± 2.0	LA-ICP-MS	Luo et al., 2012 [39]
Xiekeng	Tongren, Qinhai	Granodiorite	242.3 ± 2.3	LA-ICP-MS	Luo et al., 2012 [39]
Xiekeng	Tongren, Qinhai	Andesite	242.1 ± 1.2	LA-ICP-MS	Guo et al., 2012 [18]
Xiekeng	Tongren, Qinhai	Diorite	234.0 ± 0.6	LA-ICP-MS	Guo et al., 2012 [18]
Xiekeng	Tongren, Qinhai	Gabbro diorite	243.8 ± 1.0	LA-ICP-MS	Guo et al., 2012 [18]
Xiekeng	Tongren, Qinhai	Gabbro diorite	238.0 ± 3.0	SHRIMP	Zhang et al., 2014 [46]
Shehaliji	Tongren, Qinhai	Quartz monzonite	234.1 ± 0.5	LA-ICP-MS	Huang et al., 2014 [47]
Shehaliji	Tongren, Qinhai	Granodiorite porphyry	233.7 ± 2.4	LA-ICP-MS	Luo, 2013 [45]
Shehaliji	Tongren, Qinhai	Mafic enclave	236.5 ± 4.4	LA-ICP-MS	Luo, 2013 [45]
Ayishan	Xiahe, Gansu	Diorite	244.1 ± 2.0	LA-ICP-MS	Luo, 2013 [45]
Ayishan	Xiahe, Gansu	Granodiorite	243.2 ± 1.6	LA-ICP-MS	Luo, 2013 [45]
Ayishan	Xiahe, Gansu	Granodiorite	243.9 ± 0.6	LA-ICP-MS	Wei, 2013 [48]
Xiahe	Xiahe, Gansu	Granodiorite	240.5 ± 3.9	LA-ICP-MS	Luo, 2013 [45]
Daerzang	Xiahe, Gansu	Quartz diorite porphyry	238.0 ± 4.0	SHRIMP	Jin et al., 2005 [38]
Daerzang	Xiahe, Gansu	Granodiorite	248.1 ± 1.4	LA-ICP-MS	Wei, 2013 [48]
Daerzang	Xiahe, Gansu	Granodiorite	241.6 ± 4.0	LA-ICP-MS	Xu et al., 2014 [40]
Daerzang	Xiahe, Gansu	Granodiorite	237.9 ± 2.8	LA-ICP-MS	Luo, 2013 [45]
Daerzang	Xiahe, Gansu	Quartz diorite	241.4 ± 2.9	LA-ICP-MS	Luo, 2013 [45]
Daerzang	Xiahe, Gansu	Biotite granite	238.1 ± 3.0	LA-ICP-MS	Luo, 2013 [45]
Daerzang	Xiahe, Gansu	Mafic enclave	237.9 ± 3.8	LA-ICP-MS	Luo, 2013 [45]
Dewulu	Hezuo, Gansu	Granodiorite	233.5 ± 1.5	LA-ICP-MS	Xu et al., 2014 [40]
Dewulu	Hezuo, Gansu	Granodiorite	240.2 ± 2.6	LA-ICP-MS	Luo, 2013 [45]
Dewulu	Hezuo, Gansu	Granodiorite	239.4 ± 3.0	LA-ICP-MS	Luo, 2013 [45]
Dewulu	Hezuo, Gansu	Quartz diorite	245.8 ± 1.7	LA-ICP-MS	Zhang et al., 2015 [49]
Dewulu	Hezuo, Gansu	Quartz diorite	243.4 ± 1.9	LA-ICP-MS	Zhang et al., 2015 [49]
Dewulu	Hezuo, Gansu	Quartz diorite	238.6 ± 1.5	LA-ICP-MS	Jin et al., 2013 [50]
Dewulu	Hezuo, Gansu	Dioritic Mafic enclave	247.0 ± 2.2	LA-ICP-MS	Qiu and Deng, 2017 [12]
Laodou	Hezuo, Gansu	Granodiorite	241.4 ± 2.1	LA-ICP-MS	Zhang et al., 2015 [49]
Laodou	Hezuo, Gansu	Granodiorite	238.2 ± 1.7	LA-ICP-MS	Zhang et al., 2015 [49]
Laodou	Hezuo, Gansu	Granodiorite	241.4 ± 1.6	LA-ICP-MS	Zhang et al., 2015 [49]
Laodou	Hezuo, Gansu	Quartz diorite porphyry	247.6 ± 1.3	LA-ICP-MS	Jin et al., 2013 [50]
Yeliguan	Hezuo, Gansu	Quartz diorite	245.0 ± 6.0	SHRIMP	Jin et al., 2005 [38]
Yeliguan	Hezuo, Gansu	Quartz diorite	241.2 ± 2.5	LA-ICP-MS	Luo, 2013 [45]
Yeliguan	Hezuo, Gansu	Quartz diorite	240.1 ± 3.3	LA-ICP-MS	Luo, 2013 [45]
Meiwu	Hezuo, Gansu	Granodiorite	242.3 ± 1.8	LA-ICP-MS	Luo, 2013 [45]
Meiwu	Hezuo, Gansu	Granodiorite	242.8 ± 1.7	LA-ICP-MS	Luo, 2013 [45]
Meiwu	Hezuo, Gansu	Mafic enclave	239.6 ± 1.8	LA-ICP-MS	Luo, 2013 [45]
Meiwu	Hezuo, Gansu	Mafic enclave	243.0 ± 3.4	LA-ICP-MS	Luo, 2013 [45]
Meiwu	Hezuo, Gansu	Biotite granite	244.3 ± 2.4	LA-ICP-MS	Luo, 2013 [45]
Zaozigou	Hezuo, Gansu	Granodiorite	248.9 ± 1.4	LA-ICP-MS	Sui et al., 2018 [20]
Zaozigou	Hezuo, Gansu	Quartz diorite porphyry	244.8 ± 1.4	LA-ICP-MS	Sui et al., 2018 [20]
Zaozigou	Hezuo, Gansu	Diorite porphyry	237.5 ± 1.4	LA-ICP-MS	Sui et al., 2018 [20]
Zaozigou	Hezuo, Gansu	Porphyritic dacite	246.1 ± 5.2	LA-ICP-MS	Yu et al., 2019 [41]
Zaozigou	Hezuo, Gansu	Porphyritic dacite	248.1 ± 3.8	LA-ICP-MS	Yu et al., 2019 [41]
Zaorendao	Hezuo, Gansu	Diorite	246.6 ± 1.9	LA-ICP-MS	This study

2.2. Geology of the Zaorendao Gold Deposit

The Zaorendao gold deposit (102°50′59″E, 35°01′57″N, also referred to as Jiademu deposit), is located about 7 km northwester of Hezuo in Gansu Province (Figure 1B). The estimated pre-mining resource of the Zaorendao deposit is approximately 13.6 t Au at an average grade of 3.02 g/t [51].

The stratigraphic sequence in the Zaorendao area is dominantly composed of slate, sandy slate and arkose of late Triassic age. The sedimentary sequence is up to 3047m in thickness and represents a shore to shallow sea facies. These sedimentary rocks are intruded by two diorite stocks and numerous intermediate to felsic dikes (Figure 2). The dioritic stocks, named the West rock and the East rock, are exposed in the central and northwestern part of the mining area. The East rock outcrops have a rectangle-shape with approximately area of 0.19 km², while the square-shaped West rock has an exposed area of 0.12 km². Another barren, small diorite stock is also exposed to the northeast of the East rock. The dikes mainly consist of porphyritic diorite, porphyritic granodiorite, and porphyritic quartz diorite. These dikes are 20~250 m long, 0.3~16 m wide, and cut across bedding of the Early Triassic slate.

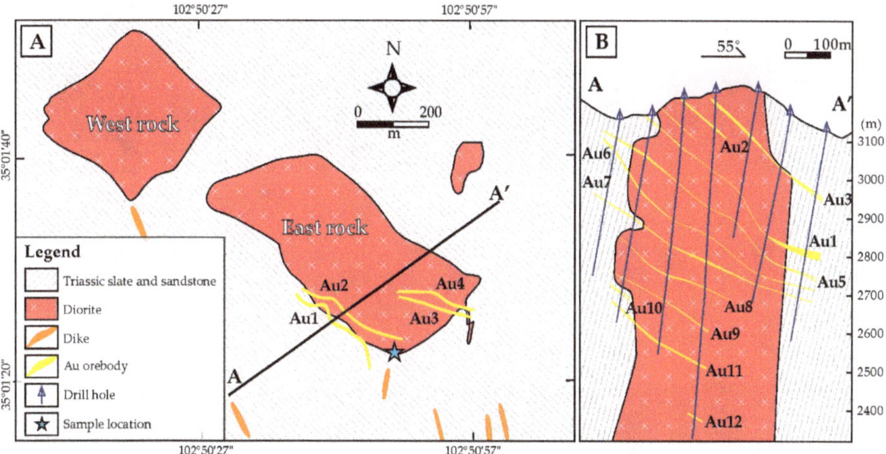

Figure 2. (**A**) Sketch geologic map of the Zaorendao gold deposit and (**B**) selected exploration cross section showing the occurrence of gold ores (after [51]).

One hundred and ninety-nine orebodies have been delineated in the Zaorendao mine, among which are six major orebodies, including Au1, Au2, Au5, Au6, Au7 and Au8 orebodies. These orebodies contain approximately 85% of the total gold reserves. Mineralization is controlled predominantly by the NW-trending faults cutting through the diorite and surrounding sedimentary rocks. These orebodies, striking NW 310° to 330° and dipping NE, and are a few meters thick and hundreds of meters long. The downward extent of the orebodies dip from 3227 m to 2366 m above sea level.

Hydrothermal alteration mineral assemblages in the mining area includes hydrothermal pyrite, arsenopyrite, stibnite, sericite, silicification, hematite and carbonate. Ore minerals are dominated by pyrite, arsenopyrite, and stibnite, with lesser sphalerite, chalcocite, galena, and hematite. The gangue minerals include sericite, quartz, epidote, biotite, plagioclase and calcite (Figure 3).

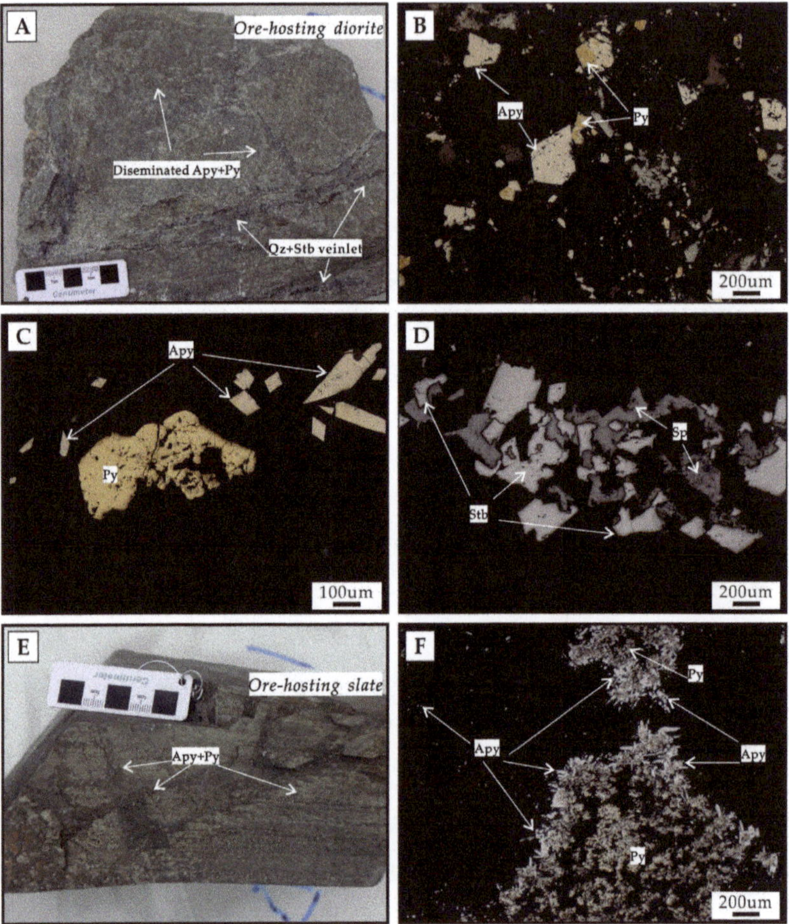

Figure 3. Photographs (**A**,**E**) and photomicrographs under reflected light (**B–D**,**F**) showing mineralization styles and mineral assemblages. Disseminations of pyrite and arsenopyrite in diorite (**A–C**). Stibnite veinlet in diorite (**A**,**D**). Disseminations of pyrite and arsenopyrite in slate (**E**,**F**). Mineral abbreviations: Apy = arsenopyrite, Py = pyrite, Qz = quartz, Sp = sphalerite, Stb = stibnite.

3. Sampling Strategy and Analytical Procedures

3.1. Sampling Strategy

One representative, minimally altered and relatively pristine sample of diorite (17ZRD01) was collected from surface on the Zaorendao gold mine premises (Figure 4A,B). This samples were used for zircon separation and laser ablation-inductively coupled plasma-mass spectrometry (LA-ICP-MS) U–Pb dating and Lu–Hf isotope analyses. The sampling locations are shown in Figure 2. The grey to sage-green diorite has medium-grained texture, and is mainly composed of plagioclase (~65 vol. %, 0.5~1.5 mm), biotite (~20 vol. %), and quartz (5–10 vol. %) with minor hornblende and pyroxene (Figure 4C,D). Accessory minerals are titanite, apatite and zircon. Plagioclase shows polysynthetic twinning and oscillatory zoning, and is locally altered to sericite (Figure 4C).

Figure 4. Photographs and photomicrographs of the diorite in the Zaorendao deposit showing its geologic relationship with Triassic slate (**A**) and texture and mineral association (**B–D**). Blue box lines in Figure (**B**) show locations of doubly polished thin sections. Mineral abbreviations: Bt = biotite, Hbl = hornblende, Pl = plagioclase, Qz = quartz, Ser = sericite.

3.2. Analytical Procedures

3.2.1. Zircon LA-ICP-MS U–Pb Dating and Trace-Element Analyses

Zircon crystals were separated from crushed diorite sample through standard magnetic and density separation techniques. The crystals were then carefully handpicked under a binocular microscope, mounted in epoxy, and polished down to approximately half sections to expose internal structures. To identify the internal structure and texture of the zircon grains and to select potential locations for U–Pb analysis, the mount was photographed in transmitted and reflected light, and cathodoluminescence (CL). The CL images were taken on a JXA-880 electron microscope (JEOL Ltd., Akishima, Tokyo, Japan) and an image analysis software was used under operating conditions of 20 kV and 20 nA at the Institute of Mineral Resources, Chinese Academy of Geological Sciences, Beijing, China.

Zircons were analyzed in situ for U, Th, and Pb isotope composition by a LA-ICP-MS system at the Isotopic Laboratory, Tianjin Institute of Geology and Mineral Resources of China Geological Survey, using a Neptune multiple-collector inductively coupled plasma-mass spectrometer (Thermo Fisher Ltd., Waltham, MA, USA) to a NEW WAVE 193 nm-FX ArF Excimer laser-ablation system (Elemental Scientific, Inc., Omaha, NE, USA). The analytical procedures are described in Reference [4]. The analyses were carried out with a pulse rate of 8 Hz, beam energy of 11 J/cm^2, and a spot diameter of 30 μm. Helium was used as a carrier gas. The NIST SRM 610 glass standard was used as an external standard to normalize U, Th, and Pb concentrations of the unknowns. Plešovice zircon, a new natural reference material for U–Pb isotopic microanalysis, and 91500 zircon were used as internal standards for U–Pb dating, and were both analyzed twice every 8 unknown zircons analyses, to normalize isotopic fractionation during isotope analysis. Common-Pb corrections were made using the method of Anderson [52]. $^{207}Pb/^{206}Pb$, $^{206}Pb/^{238}U$, $^{207}Pb/^{235}U$ and $^{208}Pb/^{232}Th$ ratios were calculated from measured ion intensities using ICPMSDataCal 8.4 [53]. Concordia diagrams and weighted mean U–Pb ages were processed using ISOPLOT 3 [54]. Age data and concordia plots were reported at 1σ error, whereas the uncertainties for weighted mean ages are given at 95% confidence level.

3.2.2. In situ Zircon Lu–Hf Isotope Analyses

The same domains or the same zircon zones that had been analyzed for U-Pb were analyzed for their Lu–Hf isotopic composition at the Isotopic Laboratory, Tianjin Institute of Geology and Mineral Resources of China Geological Survey, following the method described in Reference [4] and using a Thermo Finnigan Neptune MC-ICP-MS system coupled to a New Wave UP193 nm laser-ablation system. The analyses were carried out with a laser repetition rate of 11 Hz at 100 mJ, and a spot diameter of 50 μm. Helium was used as a carrier gas. During the analyses, the zircon GJ-1 was analyzed as quality control every 8 unknown samples, and yielded ^{176}Hf/^{177}Hf ratios of 0.282023 ± 30 (2σ, n = 12). Results excellently coincide with the recommended ^{176}Hf/^{177}Hf ratios of 0.282015 ± 19 [55] for GJ-1.

For calculation of initial epsilon Hf ($\varepsilon_{Hf}(t)$) values, the decay constant of 1.865×10^{-11} [56] and the values for the chondritic uniform reservoir (CHUR, ^{176}Lu/^{177}Hf = 0.0332, ^{176}Hf/^{177}Hf = 0.282772 [57]) were used. The initial ^{176}Hf/^{177}Hf ratios and $\varepsilon_{Hf}(t)$ for all analyzed zircon domains were calculated by the corresponding ^{206}Pb/^{238}U age. The single stage depleted mantle model ages (T_{DM}) were determined for each sample by calculating the intersection of the zircon/parent-rock growth trajectory with the depleted mantle evolution curve [58] calculated with a present-day ^{176}Hf/^{177}Hf ratio of 0.28235 (similar to mid-ocean ridge basalts values [59] and ^{176}Lu/^{177}Hf of 0.0384 [60] for the depleted mantle (DM). The two stages DM Hf model age (T_{DM2}) was calculated for the source rock of the magma from the initial ^{176}Hf/^{177}Hf of each zircon at the time of crystallization (in terms of the apparent ^{206}Pb/^{238}U age) by using ^{176}Lu/^{177}Hf= 0.015 for the average crust [61].

4. Analytical Results

4.1. Zircon U-Pb Morphology and Geochronology

Cathodoluminescence images of representative analyzed zircons for U–Pb and Lu–Hf analyses are illustrated in Figure 5. The LA-ICP-MS zircon U–Pb analytical data for diorite sample 17ZRD01 are presented in Table 2 and graphically illustrated in the concordia diagrams (Figure 6).

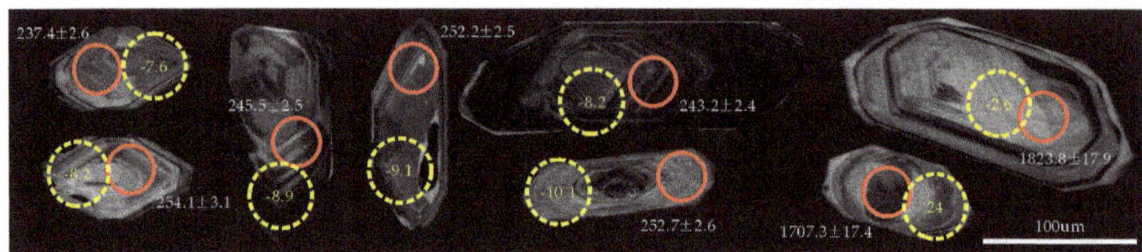

Figure 5. Representative CL images of zircons separated from diorite sample 17ZRD01 with identified analytical spot, U–Pb age (Ma) and $\varepsilon_{Hf}(t)$ value. Red circle: U–Pb beam. Yellow dash circle: Lu–Hf beam.

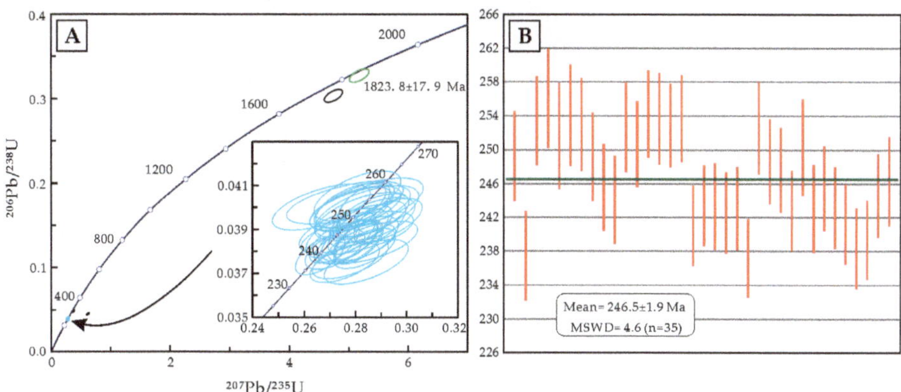

Figure 6. Zircon LA-ICP-MS U–Pb concordia diagrams for the ore-hosting diorite sample 17ZRD01.

Zircons separated from sample 17ZRD01 are mostly colorless to pale yellow, transparent, and commonly euhedral prismatic to elongated prismatic. The length of zircons varies from 90 to 250 μm with aspect ratios of 5:1 to 2:1. Most zircon grains show oscillatory zoning in CL images (Figure 5), consistent to that of igneous zircons [62]. Some grains display a core-rim structure consisting of an inherited core with distinct oscillatory and patchy zoning (Figure 5).

Thirty-nine analyses were carried out for U–Pb dating on sample 17ZRD01, yielding U-Pb ages ranging from 237.2 ± 2.3 Ma to 1823.8 ± 17.9 Ma. Thirty-seven analyses were obtained from the zircons with oscillatory zoning, among which, two analyses (spot 17ZRD01.10, 17ZRD01.17) give $^{206}Pb/^{238}U$ dates of 285.0 ± 3.0 Ma and 306.4 ± 3.2 Ma, and are excluded because they are statistical outliers. The remaining 35 analyses are concordant, with $^{206}Pb/^{238}U$ dates ranging from 237.2 ± 2.3 Ma to 256.1 ± 2.9 Ma with a weighted mean of 246.5 ± 1.9 (MSWD = 4.6, Figure 6). The Th/U ratios of the zircons are between 0.15 and 0.55, with an average of 0.34. The high Th/U values for the zircons suggest that they are magmatic in origin. The Th content ranges from 25.17 to 324.02 ppm, with an average of 104.43 ppm. The U content ranges from 97.60 to 1404.46 ppm, with an average of 344.91 ppm. The U-Pb age is therefore interpreted as the crystallization age of the diorite. One analysis (17ZRD01.27) on the zircon core yield $^{206}Pb/^{238}U$ age of 1707.3 ± 17.4 Ma is excluded because it is a statistical outlier. The remaining one analysis (spot 17ZRD01.36), gives a $^{206}Pb/^{238}U$ dates of 1823.8 ± 17.9 Ma with Th/U ratio of 0.17 (Table 2, Figures 5 and 6A). This is concordant and interpreted to be inherited during its emplacement.

4.2. Zircon Lu–Hf Isotopic Composition

In situ Hf isotope analyses have been carried out on zircons at the same spots used for zircon U–Pb dating. Zircon Lu–Hf isotopic data are presented in Table 3, and shown in Figure 7. The Hf isotopic data, expressed as values of epsilon Hf ($\varepsilon_{Hf}(t)$) at the time of emplacement, from the 17ZRD01 are plotted versus time in Figure 7.

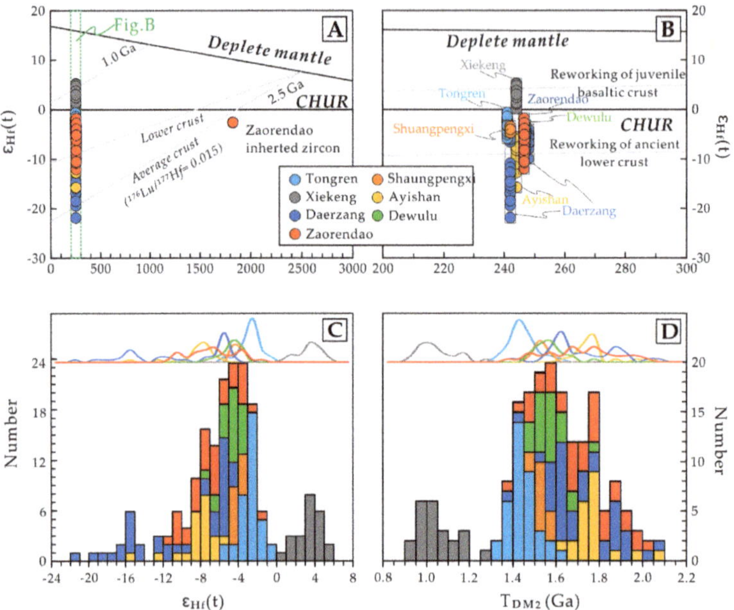

Figure 7. Plots of $\varepsilon_{Hf}(t)$ values vs. U-Pb ages for zircons from the diorite in the Zaorendao deposit (**A**, **B**). (**C**) Histogram, and frequency curves of $\varepsilon_{Hf}(t)$ values. (**D**) Histogram, and frequency curves of corresponding two-stage model age (T_{DM2}). Date from the igneous rocks of Tongren-Xiahe-Hezuo district are shown (Date are from Reference [12] and references therein.

The 35 analyses for an estimated age at ca. 247 Ma yield ^{176}Lu/^{177}Hf and ^{176}Hf/^{177}Hf ratios of 0.000093 to 0.001847 and 0.282278 to 0.282575. They display $\varepsilon_{Hf}(t)$ values varying from -12.0 to -1.8, with an average of -6.7, and the two-stage model age (T_{DM2}) range from 1.4 Ga to 2.0 Ga with an average of 1.7 Ga. It may indicate the contributions of both juvenile and ancient crustal materials. The inherited core gives ^{176}Lu/^{177}Hf and ^{176}Hf/^{177}Hf ratios of 0.001655 and 0.281608, with $\varepsilon_{Hf}(t)$ values of T_{DM2} of 2.6 Ga.

Table 2. Zircon LA-ICP-MS U–Pb data of the ore-hosting diorite in the Zaorendao deposit.

No. Spot	Element			Isotopic Ratios						Age (Ma)			
	Th (ppm)	U (ppm)	Th/U	$^{207}Pb/^{206}Pb$	1σ	$^{207}Pb/^{235}U$	1σ	$^{206}Pb/^{238}U$	1σ	$^{207}Pb/^{235}U$	1σ	$^{206}Pb/^{238}U$	1σ
17ZRD01.1	37.81	144.43	0.26	0.0528	0.0007	0.2867	0.0068	0.0394	0.0004	256.0	6.1	249.2	2.6
17ZRD01.2	43.84	160.35	0.27	0.0550	0.0013	0.2844	0.0067	0.0375	0.0004	254.1	6.0	237.4	2.6
17ZRD01.3	48.76	135.67	0.36	0.0503	0.0011	0.2783	0.0056	0.0401	0.0004	249.3	5.0	253.4	2.6
17ZRD01.4	41.56	155.62	0.27	0.0503	0.0008	0.2809	0.0069	0.0405	0.0005	251.4	6.2	256.1	2.9
17ZRD01.5	128.56	255.67	0.50	0.0519	0.0008	0.2846	0.0083	0.0398	0.0005	254.3	7.4	251.6	3.1
17ZRD01.6	36.38	110.39	0.33	0.0515	0.0008	0.2856	0.0086	0.0402	0.0005	255.1	7.7	254.1	3.0
17ZRD01.7	42.30	137.12	0.31	0.0484	0.0008	0.2669	0.0078	0.0400	0.0004	240.2	7.0	252.9	2.7
17ZRD01.8	72.33	160.50	0.45	0.0536	0.0009	0.2915	0.0060	0.0394	0.0004	259.7	5.4	249.1	2.6
17ZRD01.9	51.89	185.91	0.28	0.0525	0.0013	0.2809	0.0090	0.0388	0.0004	251.4	8.0	245.5	2.6
17ZRD01.10 *	443.87	1804.48	0.25	0.0983	0.0013	0.6127	0.0098	0.0452	0.0005	485.2	7.8	285.0	3.0
17ZRD01.11	147.61	312.24	0.47	0.0534	0.0007	0.2843	0.0051	0.0386	0.0004	254.0	4.6	244.1	2.6
17ZRD01.12	41.05	131.09	0.31	0.0508	0.0007	0.2802	0.0073	0.0400	0.0004	250.8	6.5	252.7	2.6
17ZRD01.13	324.02	1363.07	0.24	0.0525	0.0007	0.2868	0.0058	0.0396	0.0004	256.0	5.2	250.7	2.5
17ZRD01.14	40.50	155.23	0.26	0.0498	0.0009	0.2761	0.0093	0.0402	0.0004	247.6	8.4	254.2	2.5
17ZRD01.15	174.51	318.25	0.55	0.0514	0.0007	0.2842	0.0045	0.0401	0.0004	253.9	4.1	253.6	2.7
17ZRD01.16	79.95	152.98	0.52	0.0506	0.0013	0.2790	0.0062	0.0400	0.0004	249.9	5.5	252.9	2.5
17ZRD01.17 *	282.42	506.15	0.56	0.0555	0.0008	0.3723	0.0050	0.0487	0.0005	321.4	4.3	306.4	3.2
17ZRD01.18	164.32	385.89	0.43	0.0512	0.0007	0.2835	0.0041	0.0401	0.0004	253.4	3.7	253.7	2.6
17ZRD01.19	74.11	244.10	0.30	0.0518	0.0007	0.2721	0.0048	0.0381	0.0004	244.4	4.3	241.0	2.4
17ZRD01.20	126.05	304.49	0.41	0.0523	0.0007	0.2776	0.0045	0.0385	0.0004	248.8	4.0	243.4	2.4
17ZRD01.21	64.70	195.65	0.33	0.0522	0.0007	0.2771	0.0049	0.0385	0.0004	248.3	4.4	243.3	2.5
17ZRD01.22	38.30	144.45	0.27	0.0530	0.0008	0.2800	0.0065	0.0383	0.0004	250.6	5.8	242.6	2.4
17ZRD01.23	36.20	157.47	0.23	0.0542	0.0010	0.2870	0.0071	0.0384	0.0004	256.2	6.4	243.0	2.5
17ZRD01.24	77.77	191.50	0.41	0.0534	0.0022	0.2759	0.0056	0.0375	0.0004	247.4	5.0	237.2	2.3
17ZRD01.25	25.79	97.60	0.26	0.0512	0.0017	0.2820	0.0080	0.0400	0.0004	252.2	7.2	252.6	2.7
17ZRD01.26	307.45	1404.46	0.22	0.0534	0.0008	0.2896	0.0042	0.0393	0.0004	258.3	3.7	248.6	2.5
17ZRD01.27 *	125.41	340.95	0.37	0.1134	0.0015	4.7409	0.0627	0.3032	0.0031	1774.5	23.5	1707.3	17.4
17ZRD01.28	198.35	418.84	0.47	0.0523	0.0015	0.2821	0.0051	0.0392	0.0004	252.3	4.5	247.6	2.5
17ZRD01.29	149.27	784.68	0.19	0.0514	0.0008	0.2719	0.0040	0.0384	0.0004	244.2	3.6	242.8	2.4
17ZRD01.30	25.17	107.51	0.23	0.0518	0.0009	0.2827	0.0075	0.0396	0.0004	252.8	6.7	250.3	2.8
17ZRD01.31	185.97	498.31	0.37	0.0531	0.0008	0.2813	0.0040	0.0384	0.0004	251.7	3.5	243.0	2.6
17ZRD01.32	155.40	608.82	0.26	0.0521	0.0021	0.2787	0.0038	0.0388	0.0004	249.6	3.4	245.5	2.5
17ZRD01.33	199.16	1334.77	0.15	0.0531	0.0009	0.2813	0.0037	0.0384	0.0004	251.7	3.3	243.2	2.4
17ZRD01.34	138.04	272.51	0.51	0.0540	0.0011	0.2837	0.0049	0.0381	0.0004	253.6	4.4	241.2	2.4
17ZRD01.35	115.54	242.56	0.48	0.0536	0.0013	0.2781	0.0061	0.0377	0.0004	249.2	5.4	238.3	2.4
17ZRD01.36	44.76	257.28	0.17	0.1149	0.0020	5.1806	0.0673	0.3270	0.0032	1849.4	24.0	1823.8	17.9
17ZRD01.37	157.94	351.17	0.45	0.0511	0.0010	0.2664	0.0039	0.0378	0.0004	239.8	3.6	239.4	2.3
17ZRD01.38	54.58	318.64	0.17	0.0509	0.0010	0.2711	0.0042	0.0387	0.0004	243.6	3.8	244.6	2.5
17ZRD01.39	49.94	129.80	0.38	0.0513	0.0014	0.2756	0.0071	0.0389	0.0004	247.1	6.4	246.3	2.6

Note: Data marked with * are excluded while being calculated due to their discordance.

Table 3. Lu–Hf isotopic composition of zircons from the ore-hosting diorite in the Zaorendao deposit.

Sample No.	Age (Ma)	$^{176}Yb/^{177}Hf$	$^{176}Lu/^{177}Hf$	$^{176}Hf/^{177}Hf$	2σ	$(^{176}Hf/^{177}Hf)_i$	$\varepsilon_{Hf}(0)$	$\varepsilon_{Hf}(t)$	T_{DM} (Ma)	T_{DM2} (Ma)	$f_{Lu/Hf}$
17ZRD01.1	249	0.028163	0.000842	0.282440	0.000044	0.282436	−11.7	−6.4	1144	1686	−0.97
17ZRD01.2	237	0.025482	0.000826	0.282413	0.000118	0.282409	−12.7	−7.6	1182	1754	−0.98
17ZRD01.3	253	0.036162	0.001121	0.282404	0.000045	0.282399	−13.0	−7.6	1203	1766	−0.97
17ZRD01.4	256	0.024469	0.000753	0.282394	0.000040	0.282390	−13.4	−7.9	1206	1784	−0.98
17ZRD01.5	252	0.031765	0.000994	0.282320	0.000052	0.282316	−16.0	−10.6	1317	1954	−0.97
17ZRD01.6	254	0.025813	0.000886	0.282386	0.000045	0.282382	−13.7	−8.2	1221	1804	−0.97
17ZRD01.7	253	0.011117	0.000376	0.282405	0.000040	0.282403	−13.0	−7.5	1179	1757	−0.99
17ZRD01.8	249	0.032364	0.000968	0.282443	0.000042	0.282439	−11.6	−6.3	1144	1680	−0.97
17ZRD01.9	246	0.027803	0.000895	0.282440	0.000047	0.282436	−11.7	−6.5	1145	1687	−0.97
17ZRD01.10 *	285	0.042466	0.001449	0.282504	0.000045	0.282496	−9.5	−3.5	1072	1528	−0.96
17ZRD01.11	244	0.019020	0.000679	0.282429	0.000046	0.282426	−12.1	−6.9	1154	1711	−0.98
17ZRD01.12	253	0.017053	0.000537	0.282332	0.000047	0.282329	−15.6	−10.1	1285	1923	−0.98
17ZRD01.13	251	0.040060	0.001254	0.282533	0.000045	0.282527	−8.4	−3.2	1025	1479	−0.96
17ZRD01.14	254	0.033660	0.001277	0.282363	0.000041	0.282357	−14.5	−9.1	1267	1860	−0.96
17ZRD01.15	254	0.027274	0.000823	0.282423	0.000047	0.282419	−12.3	−6.9	1167	1721	−0.98
17ZRD01.16	253	0.013977	0.000388	0.282278	0.000052	0.282276	−17.5	−12.0	1354	2042	−0.99
17ZRD01.17 *	306	0.059984	0.002180	0.282262	0.000046	0.282250	−18.0	−11.8	1443	2065	−0.93
17ZRD01.18	254	0.032326	0.001056	0.282524	0.000043	0.282519	−8.8	−3.4	1033	1497	−0.97
17ZRD01.19	241	0.020187	0.000663	0.282418	0.000041	0.282415	−12.5	−7.3	1169	1737	−0.98
17ZRD01.20	243	0.038346	0.001120	0.282536	0.000045	0.282531	−8.3	−3.2	1017	1476	−0.97
17ZRD01.21	243	0.020875	0.000660	0.282499	0.000044	0.282496	−9.7	−4.4	1057	1556	−0.98
17ZRD01.22	243	0.025903	0.000942	0.282485	0.000035	0.282481	−10.1	−5.0	1084	1589	−0.97
17ZRD01.23	243	0.019343	0.000649	0.282456	0.000044	0.282453	−11.2	−5.9	1116	1651	−0.98
17ZRD01.24	237	0.023885	0.000747	0.282499	0.000049	0.282496	−9.7	−4.6	1059	1560	−0.98
17ZRD01.25	253	0.026457	0.000947	0.282381	0.000049	0.282377	−13.8	−8.4	1230	1816	−0.97
17ZRD01.26	249	0.024925	0.000797	0.282515	0.000050	0.282511	−9.1	−3.8	1039	1518	−0.98
17ZRD01.27 *	1707	0.010608	0.000320	0.282385	0.000046	0.282374	−13.7	24.0	1205	890	−0.99
17ZRD01.28	248	0.059339	0.001847	0.282363	0.000059	0.282354	−14.5	−9.3	1286	1870	−0.94
17ZRD01.29	243	0.021977	0.000730	0.282575	0.000038	0.282572	−7.0	−1.8	952	1385	−0.98
17ZRD01.30	250	0.026909	0.000810	0.282318	0.000040	0.282314	−16.1	−10.7	1313	1958	−0.98
17ZRD01.31	243	0.036243	0.001157	0.282474	0.000049	0.282469	−10.5	−5.4	1105	1616	−0.97
17ZRD01.32	245	0.004126	0.000112	0.282369	0.000047	0.282368	−14.3	−8.9	1220	1841	−1.00
17ZRD01.33	243	0.005910	0.000165	0.282390	0.000041	0.282390	−13.5	−8.2	1192	1794	−1.00
17ZRD01.34	241	0.039751	0.001183	0.282550	0.000045	0.282545	−7.8	−2.7	999	1446	−0.96
17ZRD01.35	238	0.038658	0.001139	0.282519	0.000048	0.282514	−9.0	−3.9	1042	1518	−0.97
17ZRD01.36	1824	0.056523	0.001655	0.281608	0.000051	0.281550	−41.2	−2.6	2345	2642	−0.95
17ZRD01.37	239	0.052997	0.001538	0.282334	0.000050	0.282327	−15.5	−10.5	1317	1936	−0.95
17ZRD01.38	245	0.003130	0.000093	0.282448	0.000047	0.282448	−11.5	−6.1	1111	1663	−1.00
17ZRD01.39	246	0.015306	0.000500	0.282477	0.000046	0.282475	−10.4	−5.1	1083	1601	−0.98

Note: Data marked with * are excluded while being calculated due to their discordance.

5. Discussion

5.1. Early to Middle Triassic Magmatism in Tongren-Xiahe-Hezuo Area

Zircon U–Pb dating of diorite in the Zaorendao suggests that the emplacement of diorite occurred at ca. 246.5 Ma, coeval with the widespread Triassic magmatism in the Tongren-Xiahe-Hezuo district (Table 1). Similar ages were obtained for the Tongren granodiorite (241 Ma [37]; 237 Ma [45]), Shuangpengxi granodiorite (242 Ma [39,46]), Xiekeng diorite–granodiorite (242–244 Ma [39]), Shehaliji quartz monzonite, Granodiorite porphyry and mafic microgranular enclaves (MME) (234 Ma [47]; 234–237 Ma [45]), Ayishan granodiorite (244 Ma [48]; 243–244 Ma [45]), Xiahe granodiorites (241 Ma [45]), Daerzang granodiorites (238 Ma [38]; 238–241 Ma [45]; 242 Ma [40]; 248 Ma [48]), Dewulu quartz diorite, quartz diorite porphyry and dioritic MME (238–247 Ma [12,49,50]), Laodou Granodiorite and Quartz diorite porphyry (238-241 Ma [49]; 248 Ma [50]), Meiwu granodiorite, biotite granite, and dioritic MME (240–244 Ma [45]), Yeliguan quartz diorite (240–241 Ma [45]; 245 Ma [38]). In addition, a recent study on the zircon U-Pb geochronology of granodiorite, quartz diorite porphyry diorite porphyry, and porphyritic dacites dikes in Zaozigou Au-Sb deposit to the northeast of the Zaorendao deposit indicate that the corresponding magmas were emplaced during 238–249 Ma [20,41]. We thus propose that the diorite of the Zaorendao deposit was emplaced during the Early Triassic, and was triggered by the same geodynamic processes as the widespread magmatism throughout the Tongren-Xiahe-Hezuo district.

5.2. Petrogenesis of Ore-Hosting Diorite

The diorite from the Zaorendao deposit yields relatively low $^{176}Lu/^{177}Hf$ and $^{176}Hf/^{177}Hf$ ratios (0.000093 to 0.001847 and 0.282278 to 0.282575), and has negative zircon $\varepsilon_{Hf}(t)$ values varying from -12.0 to -1.8 with an average of -6.7 (Figure 7). The Lu–Hf isotopic composition is a sensitive tracer to detect the evolutionary history of the crust and mantle [58], by virtue of the fact that Hf partitions more strongly into the melt than Lu during mantle melting [63]. During the production of granitoid magmas, according to Belousova et al. [63], high values of $^{176}Hf/^{177}Hf$ ($\varepsilon_{Hf} \gg 0$) indicate a 'juvenile' mantle input, either directly via mantle-derived mafic melts, or by remelting of a young mantle-derived mafic lower crust. The low values of $^{176}Hf/^{177}Hf$ (negative $\varepsilon_{Hf}(t)$ values) provide evidence for crustal reworking. Mixing of crustal-derived and mantle-derived magmas during granite production can also be detected by inhomogeneity (including zoning) or wide variations in the Hf isotope composition and trace-element abundances in zircon populations [61,63]. The Hf isotopic composition therefore indicates that the source material of the diorite in the Zaorendao deposit is mainly the ancient crust-derived material.

Zircon Hf model ages represent a time when isotopic composition of zircon was the same as that of the parental magma that precipitated zircon. The two-stage Hf isotope model age (T_{DM2}), from which we assume that the sample was derived via melting of an average continental crust ($^{176}Lu/^{177}Hf$ = 0.015) following derivation from the DM [63], range from 1.4 Ga to 2.0 Ga, with an average of 1.7 Ga. This indicates that the source material is mainly the Paleo- to Meso-Proterozoic crustal component. This is generally consistent with the analyses on the core (spot 17ZRD01.36) that yields a $^{206}Pb/^{238}U$ age of 1823.8 ± 17.9 Ma, and has been interpreted to be inherited. This indicates that this zircon grain could be inherited from a Paleoproterozoic basement during magma crystallization at ca. 246.5 Ma.

Another important feature of the Hf isotope data is the wide range of $\varepsilon_{Hf}(t)$ values over 10 units. Such variations within a single sample can only be reconciled by the operation of open-system processes, such as magma mixing and/or assimilation [64–66]. The Hf isotope composition of the diorite from the Zaorendao deposit is very similar to the widespread coeval igneous rocks in the Tongren-Xiahe-Hezuo district (Figure 7). The Tongren granodiorite exhibits $\varepsilon_{Hf}(t)$ values of -5.8 to -0.6, with a corresponding T_{DM2} age of 1.31 to 1.64 Ga [37]. The Shuangpengxi granodiorite shows $\varepsilon_{Hf}(t)$ values of -4.7 to -3.4 and T_{DM2} values of 1.49 to 1.57 Ga [39]. The Ayishan granodiorites have $\varepsilon_{Hf}(t)$ values that range from -15.8 to -5.6, with a corresponding T_{DM2} of 1.63 to 2.27 Ga [48].

The Da'erzang granodiorite exhibits $\varepsilon_{Hf}(t)$ values ranging from -22.0 to -4.0, with a corresponding T_{DM2} of 1.38 to 2.18 Ga [40,48]. The Dewulu granites have $\varepsilon_{Hf}(t)$ values of -8.2 to -4.2 [67], while the dioritic MME have negative $\varepsilon_{Hf}(t)$ values in the range of -8.0 to -3.3 with two-stage model ages that range from 1.48 to 1.78 Ga [12]. These granitoids turned out to be generated by melting of a lower crustal component with additional input of a mafic component derived from the mantle, indicating a regional magma mixing process [12]. Furthermore, the high Mg# values (58.25 and 60.00 [51]) of the diorite from the Zaorendao deposit are significantly higher than the melts produced by melting of metabasalts and eclogites (usually, Mg# < 45 [68]), indicating that mixing of a basic magma is required for the origin. Thus, taking into account the regional occurrence of coeval igneous rocks (250–235 Ma), we propose that the Zaorendao diorite was derived mainly from the partial melting of ancient Paleo- to Meso-Proterozoic crustal material and has been mixed with a lesser number of mantle-derived components.

5.3. Geodynamic Implications

The widespread early to middle Triassic magmatism provide a window to better understand the geodynamic evolution of the Tongren-Xiahe-Hezuo polymetallic district. Li et al. [37] proposed that the Tongren granodiorite shows geochemical features of arc-related granitoids, and formed in a subduction-related regime in response to slab roll-back of the northward-subducting A'nimaque–Mianlue oceanic lithosphere. Combining zircon U–Pb dating with geochemical study of Shuangpengxi granodiorite pluton and the Xiekeng diorite–granodiorite pluton, Luo et al. [39] proposed that such magmatism resulted from break-off of the subducted oceanic slab after collision in the early stages of Indosinian orogeny. Guo et al. [18] argued that they are formed as a continental margin arc in northeastern Tibet due to northward subduction during consumption of the paleo-Tethys ocean. Huang et al. [47] reported zircon U–Pb ages of 234.1 Ma from the Shehaliji quartz monzonite, and suggested the Early Indosinian granitoids in the West Qinling were formed in an active continental margin. Wei [48] reported the Xiahe granitoids were emplaced between 244 to 248 Ma, and proposed that the Xiahe granitoids were formed in a continental margin arc setting. The Yeliguan quartz diorite and Xiahe quartz diorite porphyry show adakitic geochemical features, and have been interpreted to have formed in the active plate margin [38]. The petrology, geochronology and geochemistry of quartz diorite, quartz diorite porphyry and dioritic MME indicate that the granitoids in the Dewulu skarn copper deposit were products of arc magmatism in an active continental margin setting [12,50]. We therefore propose that the ore-hosting diorite from the Zaorendao deposit in this study and the coeval magmatism throughout the Tongren-Xiahe-Hezuo polymetallic district, formed along an active continental margin, and define a regionally similar geodynamic process associated with the northward subduction of the paleo-Tethyan ocean (Figure 8).

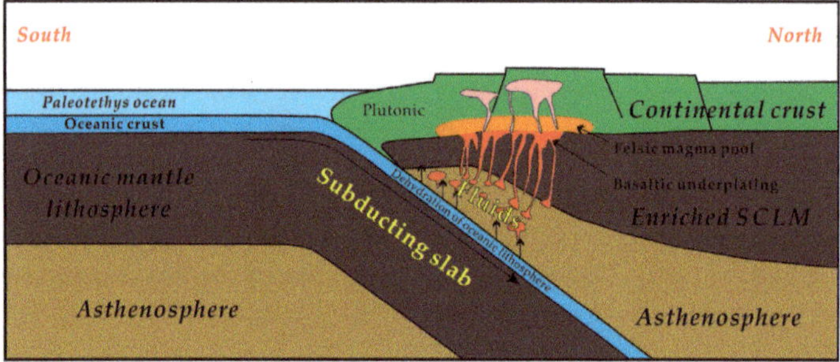

Figure 8. Schematic cartoon illustrating the petrogenesis of the ore-hosting diorite at Zaorendao and regional Triassic magmatism in the Tongren-Xiahe-Hezuo polymetallic district (modified after [12]). SCLM = sub-continental lithospheric mantle.

6. Conclusions

(1) The gold-hosting diorite in the Zaorendao deposit yielded zircon LA-ICP-MS U-Pb age at 246.5 ± 1.9 Ma, coeval with the widespread Early Triassic magmatism in the Tongren-Xiahe-Hezuo district.

(2) The ca. 247 Ma ore-hosting diorite is interpreted to have originated from the Paleo to Meso-Proterozoic continental crust, with lesser number of mantle-derived components.

(3) The widespread Early Triassic magmatism in the Tongren-Xiahe-Hezuo polymetallic district probably formed along an active continental margin, corresponding to the northward subduction of the paleo-Tethyan ocean.

Author Contributions: Conceptualization, K.-F.Q.; Data curation, Z.-Y.G.; Investigation, Z.-Y.G., H.-C.Y., K.-F.Q. and Y.-G.W.; Methodology, J.-Z.G. and J.L.; Project administration, K.-F.Q.; Resources, Y.-G.W. and M.-H.Y.; Writing—original draft, Z.-Y.G. and K.-F.Q.; Writing—review and editing, H.-C.Y. and M.-Q.W.

Funding: This research was funded by the National Natural Science Foundation of China (41702069), China Postdoctoral Science Foundation (2016M591221, 20170108), CAS Key Laboratory of Mineral Resources (KLMR2017-03), State Key Laboratory of Geological Processes and Mineral Resources (MSFGPMR201804), and the Fundamental Research Funds for the Central Universities China (292018125, 292018141).

Acknowledgments: The authors thank Zhi-Lu Liang, Ding-Guo Jin, and Hong-Shun Ma from the Bureau of Geology and Mineral Exploration and Development of Gansu Province for their assistance in the field. Xue-Feng Li at the China University of Geosciences in Beijing is acknowledged for his assistance with petrography. The authors would also like to thank Duncan McIntire at Colorado School of Mines for assistance in editing this manuscript, and for scientific feedback.

Conflicts of Interest: The authors declare no conflict of interest.

References

1. Meng, Q.R.; Zhang, G.W. Geologic framework and tectonic evolution of the Qinling orogen, central China. *Tectonophysics* **2000**, *323*, 183–196. [CrossRef]

2. Dong, Y.P.; Zhang, G.W.; Neubauer, F.; Liu, X.M.; Genser, J.; Hauzenberger, C. Tectonic evolution of the Qinling orogen, China: Review and synthesis. *J. Asian Earth Sci.* **2011**, *41*, 213–237. [CrossRef]

3. Chen, Y.J.; Santosh, M. Triassic tectonics and mineral systems in the Qinling Orogen, China. *Geol. J.* **2014**, *49*, 338–358. [CrossRef]

4. Geng, J.Z.; Qiu, K.F.; Gou, Z.Y.; Yu, H.C. Tectonic regime switchover of Triassic Western Qinling Orogen: Constraints from LA-ICP-MS zircon U–Pb geochronology and Lu–Hf isotope of Dangchuan intrusive complex in Gansu, China. *Chem. Erde-Geochem.* **2017**, *77*, 637–651. [CrossRef]

5. Deng, J.; Wang, Q.F.; Li, G.J.; Santosh, M. Cenozoic tectono-magmatic and metallogenic processes in the Sanjiang region southwestern China. *Earth Sci. Rev.* **2014**, *138*, 268–299. [CrossRef]

6. Liu, Y.; Chakhmouradian, A.R.; Hou, Z.Q.; Song, W.L.; Kynický, J. Development of REE mineralization in the giant Maoniuping deposit (Sichuan, China): Insights from mineralogy, fluid inclusions, and trace-element geochemistry. *Miner. Deposita* **2018**. [CrossRef]

7. Deng, J.; Wang, Q.F.; Li, G.J. Tectonic evolution, superimposed orogeny, and composite metallogenic system in China. *Gondwana Res.* **2017**, *50*, 216–266. [CrossRef]

8. Liu, J.J.; Liu, C.H.; Carranza, E.J.M.; Li, Y.J.; Mao, Z.H.; Wang, J.P.; Wang, Y.H.; Zhang, J.; Zhai, D.G.; Zhang, H.F.; et al. Geological characteristics and ore-forming process of the gold deposits in the western Qinling region, China. *J. Asian Earth Sci.* **2014**, *103*, 40–69. [CrossRef]

9. Liu, J.J.; Dai, H.Z.; Zhai, D.G.; Wang, J.P.; Wang, Y.H.; Yang, L.B.; Mao, G.J.; Liu, X.H.; Liao, Y.F.; Yu, C.; et al. Geological and geochemical characteristics and formation mechanisms of the Zhaishang Carlin-like type gold deposit, western Qinling Mountains, China. *Ore Geol. Rev.* **2015**, *64*, 273–298. [CrossRef]

10. Deng, J.; Wang, Q.F. Gold mineralization in China: Metallogenic provinces, deposit types and tectonic framework. *Gondwana Res.* **2016**, *36*, 219–274. [CrossRef]

11. Qiu, K.F.; Taylor, R.D.; Song, Y.H.; Yu, H.C.; Song, K.R.; Li, N. Geologic and geochemical insights into the formation of the Taiyangshan porphyry copper-molybdenum deposit, Western Qinling Orogenic Belt, China. *Gondwana Res.* **2016**, *35*, 40–58. [CrossRef]

12. Qiu, K.F.; Deng, J. Petrogenesis of granitoids in the Dewulu skarn copper deposit: Implications for the evolution of the Paleotethys ocean and mineralization in Western Qinling, China. *Ore Geol. Rev.* **2017**, *90*, 1078–1098. [CrossRef]

13. Qiu, K.F.; Marsh, E.; Yu, H.C.; Pfaff, K.; Gulbransen, C.; Gou, Z.Y.; Li, N. Fluid and metal sources for the Wenquan porphyry molybdenite deposit, Western Qinling, NW China. *Ore Geol. Rev.* **2017**, *86*, 459–473. [CrossRef]

14. Deng, J.; Gong, Q.J.; Wang, C.M.; Carranza, E.J.M.; Santosh, M. Sequence of late Jurassic-early cretaceous magmatic-hydrothermal events in the Xiong'ershan region, central china: an overview with new zircon U-Pb geochronology data on quartz porphyries. *J. Asian Earth Sci.* **2014**, *79*, 161–172. [CrossRef]

15. Mao, J.W.; Qiu, Y.M.; Goldfarb, R.J.; Zhang, Z.C.; Garwin, S.; Ren, F.S. Geology, distribution, and classification of gold deposits in the western Qinling belt, central China. *Miner. Deposita* **2002**, *37*, 352–377. [CrossRef]

16. Yang, L.Q.; Deng, J.; Li, N.; Zhang, C.; Ji, X.Z.; Yu, J.Y. Isotopic characteristics of gold deposits in the yangshan gold belt, west Qinling, central china: Implications for fluid and metal sources and ore genesis. *J. Geochem. Explor.* **2016**, *168*, 103–118. [CrossRef]

17. Liu, D.S.; Tan, Y.J.; Wang, J.Y.; Wei, L.M. Carlin-type gold deposits in China. In *Chinese Carlin-Type Gold Deposits*; Liu, D.S., Tan, Y.J., Wang, J.Y., Eds.; Nanjing University Press: Nanjing, China, 1994; pp. 1–38. (In Chinese)

18. Guo, X.Q.; Yan, Z.; Wang, Z.; Wang, T.; Hou, K.; Fu, C.; Li, J. Middle Triassic arc magmatism along the northeastern margin of the Tibet: U–Pb and Lu–Hf zircon characterization of the Gangcha complex in the West Qinling terrane, Central China. *J. Geol. Soc.* **2012**, *169*, 327–336. [CrossRef]

19. Jin, X.Y.; Li, J.W.; Hofstra, A.H.; Sui, J.X. Magmatic-hydrothermal origin of the early Triassic Laodou lode gold deposit in the Xiahe-Hezuo district, West Qinling orogen, China: Implications for gold metallogeny. *Miner. Deposita* **2017**, *52*, 883–902. [CrossRef]

20. Sui, J.X.; Li, J.W.; Jin, X.Y.; Vasconcelos, P.; Zhu, R. ^{40}Ar/^{39}Ar and U-Pb constraints on the age of the Zaozigou disseminated gold deposit, Xiahe-Hezuo district, West Qinling orogen, China: Implications for the early Triassic reduced intrusion-related gold metallogeny. *Ore Geol. Rev.* **2018**, *101*, 885–899. [CrossRef]

21. Sui, J.X.; Li, J.W.; Wen, G.; Jin, X.Y. The Dewulu reduced Au-Cu skarn deposit in the Xiahe-Hezuo district, West Qinling orogen, China: Implications for an intrusion-related gold system. *Ore Geol. Rev.* **2017**, *80*, 1230–1244. [CrossRef]

22. Li, J.W.; Sui, J.X.; Jin, X.Y.; Wen, G.; Chang, J. A magmatic related gold system in the xiahe-hezuo district, western qinling orogen, china. *Acta Geol. Sin. Engl.* **2014**, *88*, 751–752. [CrossRef]

23. Yang, L.Q.; Ji, X.Z.; Santosh, M.; Li, N.; Zhang, Z.C.; Yu, J.Y. Detrital zircon U–Pb ages, Hf isotope, and geochemistry of Devonian chert from the Mianlue suture: Implications for tectonic evolution of the Qinling orogen. *J. Asian Earth Sci.* **2015**, *113*, 589–609. [CrossRef]

24. Ratschbacher, L.; Hacker, B.R.; Calvert, A.; Webb, L.E.; Grimmer, J.C.; McWilliams, M.O.; Ireland, T.; Dong, S.W.; Hu, J.M. Tectonics of the Qinling (central China): Tectonostratigraphy, geochronology, and deformation history. *Tectonophysics* **2003**, *366*, 1–53. [CrossRef]

25. Zhang, G.W.; Meng, Q.R.; Lai, S.C. Tectonics and structure of the QinlingOrogenic belt. *Sci. China Ser. B* **1995**, *38*, 1379–1394. (In Chinese)

26. Cao, H.H.; Li, S.Z.; Zhao, S.J.; Yu, S.; Li, X.Y.; Somerville, I.D. Detrital zircon geochronology of Neoproterozoic to early Paleozoic sedimentary rocks in the North Qinling Orogenic Belt: Implications for the tectonic evolution of the Kuanping Ocean. *Precambrian Res.* **2016**, *279*, 1–16. [CrossRef]

27. Li, S.Z.; Zhao, S.J.; Liu, X.; Cao, H.H.; Yu, S.; Li, X.Y.; Somerville, I.; Yu, S.Y.; Suo, Y.H. Closure of the Proto-Tethys Ocean and Early Paleozoic amalgamation of microcontinental blocks in East Asia. *Earth-Sci. Rev.* **2017**. [CrossRef]

28. Feng, Y.M.; Cao, X.D.; Zhang, E.P.; Hu, Y.X.; Pan, X.P.; Yang, J.L.; Jia, Q.Z.; Li, W.M. Tectonic evolution framework and nature of the west Qinling orogenic belt. *Northwest. Geol.* **2003**, *36*, 1–10, (In Chinese with English Abstract).

29. Deng, J.; Wang, Q.F.; Li, G.J.; Li, C.S.; Wang, C.M. Tethys tectonic evolution and its bearing on the distribution of important mineral deposits in the Sanjiang regionn, SW China. *Gondwana Res.* **2014**, *26*, 419–437. [CrossRef]

30. Dong, Y.P.; Yang, Z.; Liu, X.M.; Sun, S.S.; Li, W.; Cheng, B.; Zhang, F.F.; Zhang, X.N.; He, D.F.; Zhang, G.W. Mesozoic intracontinental orogeny in the Qinling Mountains, central China. *Gondwana Res.* **2016**, *30*, 144–158. [CrossRef]

31. Zhang, G.W.; Dong, Y.P.; Lai, S.C.; Guo, A.L.; Meng, Q.R.; Liu, S.F.; Cheng, S.Y.; Yao, A.P.; Zhang, Z.Q.; Pei, X.Z.; et al. Mianlue tectonic zone and Mianlue suture zone on southern margin of Qinling-Dabie orogenic belt. *Sci. China Ser. D* **2004**, *47*, 300–316. [CrossRef]

32. Li, S.Z.; Kusky, T.M.; Wang, L.; Zhang, G.W.; Lai, S.C.; Liu, X.C.; Dong, S.W.; Zhao, G.C. Collision leading to multiple-stage large-scale extrusion in the Qinling orogen: Insights from the Mianlue suture. *Gondwana Res.* **2007**, *12*, 121–143. [CrossRef]

33. Yang, L.Q.; Deng, J.; Dilek, Y.; Qiu, K.F.; Ji, X.Z.; Li, N.; Taylor, R.D.; Yu, J.Y. Structure, geochronology, and petrogenesis of the Late Triassic Puziba granitoid dikes in the Mianlue suture zone, Qinling Orogen, China. *Geol. Soc. Am. Bull.* **2015**, *11/12*, 1831–1854. [CrossRef]

34. Wang, X.X.; Wang, T.; Zhang, C.L. Neoproterozoic, Paleozoic, and Mesozoic granitoid magmatism in the Qinling Orogen, China: Constraints on orogenic process. *J. Asian Earth Sci.* **2013**, *72*, 129–151. [CrossRef]

35. Li, N.; Chen, Y.J.; Santosh, M.; Pirajno, F. Compositional polarity of Triassic granitoids in the Qinling orogen, China: Implication for termination of thenorthernmost paleo-Tethys. *Gondwana Res.* **2015**, *27*, 244–257. [CrossRef]

36. Qiu, K.F.; Yu, H.C.; Gou, Z.Y.; Liang, Z.L.; Zhang, J.L.; Zhu, R. Nature and origin of Triassic igneous activity in the Western Qinling Orogen: The Wenquan composite pluton example. *Int. Geol. Rev.* **2018**, *60*, 242–266. [CrossRef]

37. Li, X.W.; Mo, X.X.; Huang, X.F.; Dong, G.C.; Yu, X.H.; Luo, M.F.; Liu, Y.B. U–Pb zircon geochronology, geochemical and Sr-Nd-Hf isotopic compositions of the Early Indosinian Tongren Pluton in West Qinling: Petrogenesis and geodynamic implications. *J. Asian Earth Sci.* **2015**, *97*, 38–50. [CrossRef]

38. Jin, W.J.; Zhang, Q.; He, D.F.; Jia, X.Q. SHRIMP dating of adakites in western Qinling and their implications. *Acta Petrol. Sin.* **2005**, *21*, 959–966, (In Chinese with English Abstract).

39. Luo, B.J.; Zhang, H.F.; Lü, X.B. U-Pb zircon dating, geochemical and Sr-Nd-Hf isotopic compositions of Early Indosinian intrusive rocks in West Qinling, central China: Petrogenesis and tectonic implications. *Contrib. Mineral. Petr.* **2012**, *164*, 551–569. [CrossRef]

40. Xu, X.Y.; Chen, J.; Gao, T.; Li, P.; Li, T. Granitoid magmatism and tectonic evolution in northern edge of the western Qinling terrane, NW China. *Acta Petrol. Sin.* **2014**, *30*, 371–389, (In Chinese with English Abstract).

41. Yu, H.C.; Guo, C.A.; Qiu, K.F.; McIntire, D.; Jiang, G.P.; Gou, Z.Y.; Geng, J.Z.; Pang, Y.; Zhu, R.; Li, N.B. Geochronological and Geochemical Constraints on the Formation of the Giant Zaozigou Au-Sb Deposit, West Qinling, China. *Minerals* **2019**, *9*, 37. [CrossRef]

42. Liu, C.X.; Li, L.; Sui, J.X. Mineralization Characteristics and Ore Genesis of the Zaozigou Gold Deposit, Gansu Province. *Geol. Sci. Tech. Inf.* **2011**, *30*, 66–74, (In Chinese with English Abstract).

43. Cao, X.F.; Sanogo, M.L.S.; Lv, X.B.; He, M.C.; Chen, C.; Zhu, J.; Tang, R.K.; Liu, Z.; Zhang, B. Ore-forming process of the Zaozigou gold deposit: Constraints from geological characteristics, gold occurrence and stable isotope compositions. *J. Jilin Univ. (Earth Sci. Ed.)* **2012**, *42*, 1039–1054, (In Chinese with English Abstract).

44. Zhang, X.H.; Ren, F.S.; Yu, C.; Liu, J.H.; Zhao, Y.Q.; Zhang, F.R.; Jia, Z.L. Breakthrough in geological prospecting based on study of metallogenic regularity. *Miner. Depos.* **2015**, *34*, 1130–1142, (In Chinese with English Abstract).

45. Luo, B.J. Petrogenesis and Geodynamic Processes of the Indosinian Magmatism in the West Qinling Orogenic Belt, Central China. Ph.D. Thesis, China University of Geosciences, Wuhan, China, 2013. (In Chinese)

46. Zhang, T.; Zhang, D.H.; Yang, B. SHRIMP zircon U–Pb dating of Gangcha intrusions in Qinghai and its geological significance. *Acta Petrol. Sin.* **2014**, *30*, 2739–2748, (In Chinese with English Abstract).

47. Huang, X.F.; Mo, X.X.; Yu, X.H.; Li, X.W.; Yang, M.C.; Luo, M.F.; He, W.Y.; Yu, J.C. Origin and geodynamic settings of the Indosinian Sr/Y granitoids in the West Qinling: An example from the Shehaliji pluton in Tonren area. *Acta Petrol. Sin.* **2014**, *30*, 3255–3270, (In Chinese with English Abstract).

48. Wei, P. Petrogenesis and Tectonic Setting of the Indosinian Granites from Xiahe Area, West Qinling. Master's Thesis, China University of Geosciences, Beijing, China, 2013. (In Chinese)

49. Zhang, D.X.; Cao, H.; Shu, Z.X.; Lu, A.H. Indosinian magmatism and tectonic setting of Xiahe-Hezuo area, western Qinling Mountains—Implications from the Dewulu quartz diorite and Laodou quartz dioritic porphyry. *Geol. China* **2015**, *42*, 1257–1273, (In Chinese with English Abstract).

50. Jin, X.Y.; Li, J.W.; Sui, J.X.; Wen, G.; Zhang, J.Y. Geochronological and geochemical constraints on the genesis and tectonic setting of Dewulu intrusive complex in Xiahe-Hezuo District of Western Qinling. *J. Earth Sci. Environ.* **2013**, *35*, 20–38, (In Chinese with English Abstract).

51. Wang, Y.G.; Chi, J.; Wei, X.J.; Bi, S.T.; Han, X.Q.; Dong, C.R.; Wang, L.R. *Geological Survey Report of the Jiademu Gold-Polymetallic Deposit, Hezuo Country, Gansu Province*; Investigation Report; Guangxi Geological Exploration Institute of China Chemical Geology and Mine Bureau: Nanning, China, 2017. (In Chinese)

52. Anderson, T. Correction of common lead in U–Pb analyses that do not report ^{204}Pb. *Chem. Geol.* **2002**, *192*, 59–79. [CrossRef]

53. Liu, Y.S.; Gao, S.; Hu, Z.C.; Gao, C.G.; Zong, K.Q.; Wang, D.B. Continental and oceanic crust recycling-induced melt–peridotite interactions in the Trans-North China Orogen: U–Pb dating, Hf isotopes and trace elements in zircons from mantle xenoliths. *J. Petrol.* **2010**, *51*, 537–571. [CrossRef]

54. Ludwig, K.R. Isoplot v. 3.0: A geochronological toolkit for Microsoft Excel. In *Berkeley Geochronology Center*; Special Publication: Berkeley, CA, USA, 2003; pp. 1–70.

55. Elhlou, S.; Belousova, E.; Griffin, W.L.; Pearson, N.J.; O'Reilly, S.Y. Trace element and isotopic composition of GJ–red zircon standard by laser ablation. *Geochim. Cosmochim. Acta* **2006**, *70*, A158-A158. [CrossRef]

56. Scherer, E.; Münker, C.; Mezger, K. Calibration of the lutetium-hafniumclock. *Science* **2001**, *293*, 683–687. [CrossRef] [PubMed]

57. Blichert-Toft, J.; Albarède, F. The Lu-Hf isotope geochemistry of chondrites and the evolution of the mantle–crust system. Earth Planet. *Sci. Lett.* **1997**, *148*, 243–258.

58. Vervoort, J.D.; Blichert-Toft, J. Evolution of the depleted mantle: Hf isotopeevidence from juvenile rocks through time. *Geochim. Cosmochim. Acta* **1999**, *63*, 533–556. [CrossRef]

59. Nowell, G.M.; Kempton, P.D.; Noble, S.R.; Fitton, J.G.; Saunders, A.D.; Mahoney, J.J.; Taylor, R.N. High precision Hf isotope measurements of MORB and OIB by thermal ionisation mass spectrometry: Insights into the depleted mantle. *Chem. Geol.* **1998**, *149*, 211–233. [CrossRef]

60. Griffin, W.L.; Pearson, N.J.; Belousova, E.; Jackson, S.E.; Achterbegrh, E.V.; O'Reilly, S.Y.; Shee, S.R. The Hf isotope composition of cratonic mantle: LAM-MC-ICPMS analysis of zircon megacrysts in kimberlites. *Geochim.Cosmochim. Acta* **2000**, *64*, 133–147. [CrossRef]

61. Griffin, W.L.; Wang, X.; Jackson, S.E.; Pearson, N.J.; O'Reilly, S.Y.; Xu, X.; Zhou, X. Zircon chemistry and magma mixing, SE China: In-situ analysis of Hf isotopes, Tonglu and Pingtan igneous complexes. *Lithos* **2002**, *61*, 237–269. [CrossRef]

62. Corfu, F.; Hanchar, J.M.; Hoskin, P.W.O.; Kinny, P. Atlas of zircon textures. In *Mineralogical Society of America; Reviews in Mineralogy and Geochemistry*; Hanchar, J.M., Hoskin, P.W.O., Eds.; Zircon: Washington, DC, USA, 2003; pp. 469–500.

63. Belousova, E.A.; Griffin, W.L.; O'Reilly, S.Y. Zircon crystal morphology, trace element signatures and Hf isotope composition as a tool for petrogenetic modelling: Examples from Eastern Australian granitoids. *J. Petrol.* **2006**, *47*, 329–353. [CrossRef]

64. Kemp, A.I.S.; Hawkesworth, C.J.; Foster, G.L.; Paterson, B.A.; Woodhead, J.D.; Hergt, J.M.; Gray, C.M.; Whitehouse, M.J. Magmatic and crustal differentiation history of granitic rocks from Hf–O isotopes in zircon. *Science* **2007**, *315*, 980–983. [CrossRef]

65. Zhu, D.C.; Mo, X.X.; Wang, L.Q.; Zhao, Z.D.; Niu, Y.L.; Zhou, C.Y.; Yang, Y.H. Petrogenesis of highly fractionated I-type granites in the Chayu area of eastern Gangdese, Tibet: Constraints from zircon U-Pb geochronology, geochemistry and Sr-Nd-Hf isotopes. *Sci. China Ser. D* **2009**, *52*, 1223–1239, (In Chinese with English Abstract). [CrossRef]

66. Yang, L.Q.; Deng, J.; Qiu, K.F.; Ji, X.Z.; Santosh, M.; Song, K.R.; Song, Y.H.; Geng, J.Z.; Zhang, C.; Hua, B. Magma mixing and crust–mantle interaction in the Triassic monzogranites of Bikou Terrane, central China: Constraints from petrology, geochemistry, and zircon U–Pb–Hf isotopic systematics. *J. Asian Earth Sci.* **2015**, *98*, 320–341. [CrossRef]

67. Huang, X.; Mo, X.X.; Yu, X. Middle Triassic magma mixing in an active continental margin: Evidence from mafic enclaves and host granites from the Dewulu pluton in West Qinling, central China. In Proceedings of the American Geophysical Union Fall Meeting, San Francisco, CA, USA, 14–18 December 2015.

68. Rapp, R.P.; Watson, E.B. Dehydration melting of metabasalt at 8–32 kbar: Implications for continental growth and crust-mantle recycling. *J. Petrol.* **1995**, *36*, 891–931. [CrossRef]

Article

Geology, Geochemistry, and Geochronology of Gabbro from the Haoyaoerhudong Gold Deposit, Northern Margin of the North China Craton

Jianping Wang *, Xiu Wang, Jiajun Liu, Zhenjiang Liu, Degao Zhai and Yinhong Wang

School of Earth Sciences and Resources, China University of Geosciences, Beijing 100083, China; wangxiucugb@163.com (X.W.); liujiajun@cugb.edu.cn (J.L.); lzj@cugb.edu.cn (Z.L.); dgzhai@cugb.edu.cn (D.Z.); wyh@cugb.edu.cn (Y.W.)
* Correspondence: jpwang@cugb.edu.cn; Tel.: +86-10-8232-2264

Received: 5 December 2018; Accepted: 17 January 2019; Published: 21 January 2019

Abstract: The Haoyaoerhudong gabbro is a mafic intrusion located in the Haoyaoerhudong gold deposit, which is a giant gold deposit (148 t Au) hosted in Proterozoic strata on the northern margin of the North China Craton. In this paper, we present integrated SHRIMP U–Pb, geochemical and Sr–Nd isotopic data from gabbro of the Haoyaoerhudong gold deposit to reveal the magmatic processes behind its origin. SHRIMP zircon U–Pb dating constrains the timing of crystallization of the Haoyaoerhudong gabbro to 278.8 ± 0.81 Ma. Whole-rock geochemical results indicate that the Haoyaoerhudong gabbro has calc-alkaline features with enrichments of large-ion lithophile elements (LILE) and light rare-earth elements (REE) as well as depletions of high-field strength elements (HFSE). The relatively high $(^{87}Sr/^{86}Sr)_i$ (0.7053 to 0.7078) and low $\varepsilon Nd(t)$ (−4.6 to −15.1) values of the gabbro indicate the involvement of crustal materials. Low Ce/Pb ratios (1.35 to 7.38), together with nearly constant La/Sm and Th/Yb ratios and variable Ba/Th and Sr/Nd ratios, suggest that the ancient mantle was modified by slab dehydration fluids. Based on new geochemical data and regional geological investigations, we propose that both the Haoyaoerhudong gold deposit and the Haoyaoerhudong gabbro formed in a post-orogenic extensional setting.

Keywords: geochemistry; U–Pb zircon age; Sr–Nd isotopes; Haoyaoerhudong gabbro; Inner Mongolia

---

1. Introduction

The North China Craton (NCC) is a well-known Precambrian block in China [1–3]. The amalgamation between the eastern and western blocks ca. 1.85 Ga marked the cratonization process of the NCC [4]. Its northern margin underwent a strong rifting process, forming the Langshan–Zhaertaishan Rift and the Bayan Obo Rift during the Proterozoic era. Affected by the evolution of the Central Asian Orogenic Belt (CAOB), a giant magmatic rock belt formed along the northern margin of the NCC during the Paleozoic era (Figure 1) [5–8]. Several large orogenic gold deposits, including the Zhulazhaga and Haoyaoerhudong gold deposits, formed during the Paleozoic orogenic process.

The Haoyaoerhudong gold deposit is a super large (148 t Au) low-grade deposit in the western part of the NCC. Gold is hosted by pyrite and pyrrhotite, which were mainly deposited on bedding and schistosity planes of the Proterozoic black shales [9]. Hercynian intrusions are common in the Haoyaoerhudong gold district. A small gabbro intrusion is located west of the Haoyaoerhudong gold deposit (called the Haoyaoerhudong gabbro below). Most previous research focused on the gold deposit itself and the granitoids around the deposit [10–14], but little is known about the Haoyaoerhudong gabbro.

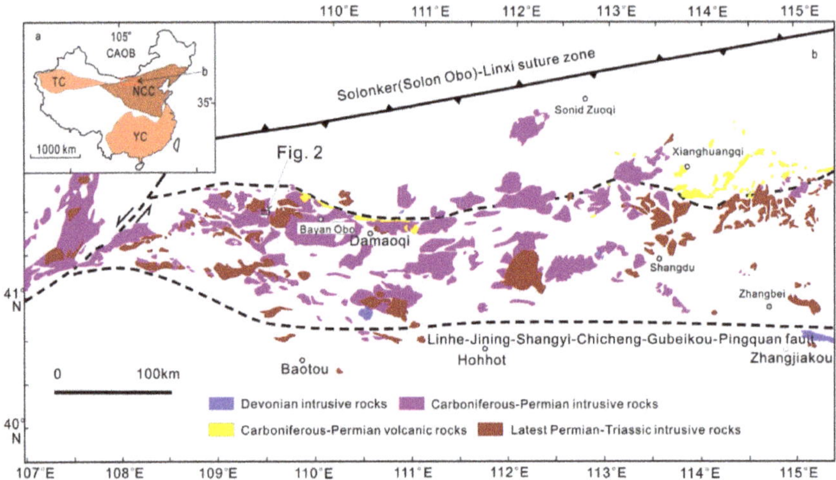

Figure 1. (**a**) Tectonic framework of China, showing the location of (**b**) Distribution of Paleozoic magmatic rocks along the northern margin of the North China Craton (NCC) (modified after [6]). TC, Tarim Craton; YC, Yangtze Craton.

Mafic rocks, although in a small volume on the earth's surface, are very important study subjects to explore deep mantle heterogeneity, crust-mantle interactions [15], and the tectonic properties of different blocks in ancient orogens [16]. The presence of the Haoyaoerhudong gabbro gives us an opportunity to investigate the composition of the ancient mantle, crust-mantle interaction, and crustal differentiation [17] of the northern margin of the NCC. We report new data of whole-rock geochemistry, zircon U–Pb dating, and Sr–Nd isotopes of the Haoyaoerhudong gabbro with the goal of constraining the petrogenesis and timing of the gabbro as well as the geodynamic setting of the Permian magmatism and related gold mineralization.

2. Geological Setting

A "double rift" system developed on the northern margin of the NCC in the Proterozoic era [18]. The Langshan–Zhaertaishan Rift (inside) and the Bayan Obo Rift (outside), host gigantic thick Proterozoic rift sediments called the Zhaertaishan and Bayan Obo Groups, respectively. The supracrustal succession of the Zhaertaishan and Bayan Obo Groups are thought to be contemporaneous with different types of magmatism [19–21]. Carbonates characterized the Bayan Obo Rift [22–24], whereas double-peak volcanism occurred in the Langshan–Zhaertaishan Rift [25]. The giant Bayan Obo REE–Nb–Fe deposit is located in the Bayan Obo Group, and the superlarge Langshan–Zhaertaishan SEDEX (sedimentary exhalative) Pb–Zn–Cu–S ore belt (including the Dongshengmiao, Huogeqi, Tanyaokou, and Jiashengpan ore deposits) is hosted in the Zhaertaishan Group. Carbonate dikes and bimodal magmatism indicate that the northern margin of the NCC experienced rifting in the Middle Mesoproterozoic, which is considered to be the result of the final breakup of the Columbia Supercontinent [21], and induced the formation of the Bayan Obo REE–Nb–Fe deposit and the giant Langshan–Zhaertaishan SEDEX ore belt. Along with the evolution of the CAOB, the northern margin of the NCC experienced an orogenic process with strong tectono-magmatism during the Paleozoic. Large volumes of magmatic rocks currently crop out along the northern margin of the NCC (Figure 1). Zhang et al. [6] summarized three important magmatic events: Devonian (from 400 to 360 Ma), late Early Carboniferous to Middle Permian (from 330 to 265 Ma), and latest Permian to Triassic (from 250 to 200 Ma). The most striking regional structure is the Solonker suture zone. This zone, which is composed of ophiolites and olistostromes, is thought to be the remnants of the Paleo-Asian Ocean. Additional information about the regional tectonic evolution is given in references [2,26,27].

The geology of the Haoyaoerhudong gold deposit is similar to the northern margin of the NCC, which is characterized by Proterozoic strata divided by Paleozoic magmatic rocks. The Bayan Obo Group exposed in the gold district include the Jianshan, Halahuogete, and Bilute formations [9,14]. The Bilute Formation, which is mainly composed of black shales (rich in organic material, with an average TOC of 3.4%), is the major gold-hosting rock (Figure 2). Most intrusions are Permian in age, and the main lithologies include granitic porphyry, monzogranitic porphyry, granodiorite, and biotite granite. The granite porphyry and monzonitic granite porphyry have zircon U–Pb ages of 290.9 ± 2.8 Ma and 287.5 ± 1.9 Ma, respectively. The biotite granite samples yield zircon U–Pb ages of 267.9 ± 1.2 Ma and 274.0 ± 2.3 Ma [11,12]. Gold-related alteration ages are reported as 270.1 ± 2.5 Ma (biotite Ar–Ar [9]) and 250.9 ± 1.5 Ma (muscovite Ar–Ar [13]). Most researchers have inferred a genetic relation between the gold mineralization and the Permian tectono-magmatism [9,13,14,28].

The Haoyaoerhudong gabbro is located in the Haoyaoerhudong gold ore district, approximately 2 km southwest of the west open pit (Figure 2). It is an oval intrusion with an area of about 2.5 km^2. The gabbro mainly intrudes into the Halahuogete Formation, nearly parallel to the border between the Halahuogete Formation and the Bilute Formation to the north. The gabbro truncates the gray-white tonalite in the southwest area (Figure 2). Several NE-SW-trending gabbro dikes are present in the Haoyaoerhudong gold district (Figure 3A).

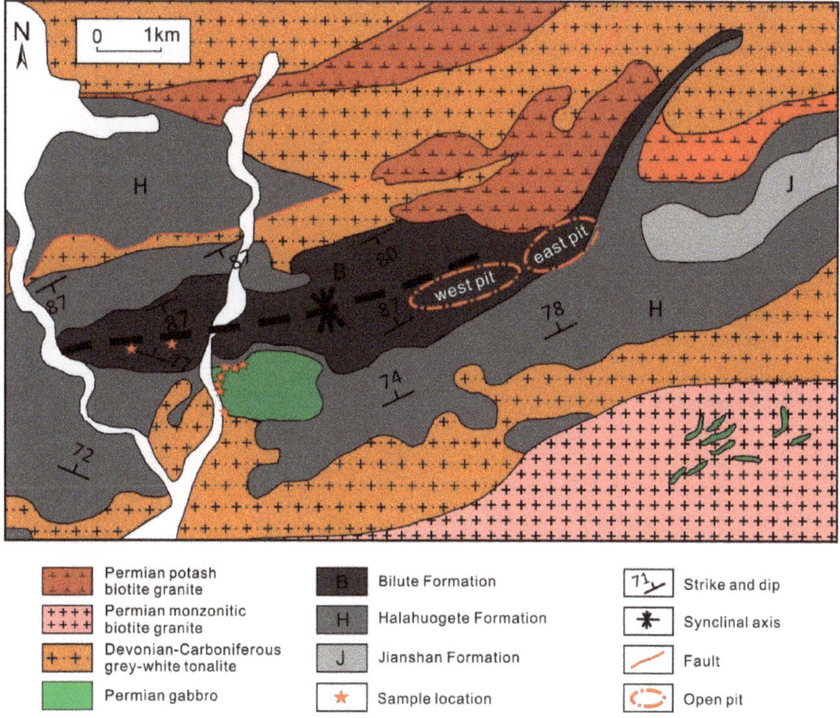

Permian potash biotite granite

Permian monzonitic biotite granite

Devonian-Carboniferous grey-white tonalite

Permian gabbro

B — Bilute Formation

H — Halahuogete Formation

J — Jianshan Formation

★ Sample location

71 Strike and dip

✶ Synclinal axis

Fault

Open pit

Figure 2. Geological map of the Haoyaoerhudong gold deposit showing the gabbro intrusion (modified after [14]).

In outcrops, the gabbro is characteristically dark-gray to black due to weathering (Figure 3B). Fresh surfaces of the rocks are gray or whitish gray with different contents of mafic minerals. No deformation has been observed in the oval intrusion or in the dikes, but some small felsic veins cut the gabbro intrusions (Figure 3C). Thin section studies of the rocks show a medium-grained gabbroic assemblage, mainly including plagioclase, clinopyroxene, amphibole, and biotite (Figure 3D,E). Small amounts of pyrite and chalcopyrite were observed for the first time (Figure 3F) in this study. The mineral crystallization sequence of the Haoyaoerhudong gabbro appears to be clinopyroxene first, then plagioclase, and finally amphibole and biotite.

Figure 3. Representative field photographs and photomicrographs (cross-polarized light, **D** and **E**; single-polarized light, **F**) of the Haoyaoerhudong gabbro. (**A**) EW-trending gabbro dike; (**B**) Outcrop of the Haoyaoerhudong gabbro; (**C**) Thin felsic vein cut through the gabbro; (**D,E**) Mineral assemblage of the Haoyaoerhudong gabbro; (**F**) Metal minerals of the gabbro. Mineral abbreviations: Cpx, clinopyroxene; Bi, biotite; Hb, hornblende; Pl, plagioclase; Cpy; chalcopyrite; Py, pyrite.

3. Samples and Analytical Methods

In this study, a total of 10 representative samples were collected (8 from the Haoyaoerhudong gabbro and 2 from gabbro dikes 1 km west of the Haoyaoerhudong gabbro) for whole-rock major and trace elements, zircon dating, and bulk Sr–Nd isotopes. The samples were carefully examined to have the least alteration, cracks, and inclusions. The rock is massive, medium-fine grained, and equigranular. No cumulate textures were observed.

3.1. Whole-Rock Major and Trace Elements

Whole-rock major and trace elements were determined at the Langfang Institute of Geophysical and Geochemical Exploration, Hebei Province, China. Major elements were analyzed by X-ray fluorescence spectrometry. REEs and trace elements were tested by inductively coupled plasma-mass spectrometry. The general precision was better than 3% and 5% for major oxides and trace elements, respectively.

3.2. Zircon U–Pb Dating

Zircons were selected from sample RB25 using traditional techniques (density and magnetic separation) at the Langfang Institute of Regional Geology Survey, Hebei Province, China. All of the zircons were studied with micrographs and cathodoluminescence (CL) images to illustrate their microstructures. The zircon U–Th–Pb compositions were determined using a sensitive high-resolution ion microprobe at the Beijing SHRIMP Center. The analytical spot sizes were 30 μm, but each spot was rastered over 120 μm for three minutes to remove common Pb on the zircon surfaces. Five consecutive scans were performed for each zircon spot. The detailed analytical procedures are given in Jian et al. [29].

3.3. Sr–Nd Isotope

Sr–Nd isotopes were tested using an Isoprobe-T thermal ionization mass spectrometer (TIMS) at the Beijing Research Institute of Uranium Geology (BRIUG). Powder samples were mixed for isotope dilution and dissolved using HF + HNO_3 + $HClO_4$ in sealed Teflon capsules on a hot plate for 24 h. After separation of the Rb, Sr, and light REEs in a cation-exchange column, the Sm and Nd were further purified using a cation-exchange column, conditioned and eluted with dilute HCl. The TIMS was operated in static mode according to the standard procedures of GB/T17672-1999. Total chemical blanks were lower than 200 pg for strontium and lower than 50 pg for neodymium. The normalization standards are $^{146}Nd/^{144}Nd$ of 0.7219 and $^{88}Sr/^{86}Sr$ of 0.1194. Additional analytical details are given in Zhang et al. [30].

4. Results

4.1. SHRIMP U–Pb Age

The zircons from sample RB25 are euhedral to subhedral with length/width ratios of 1:1 to 2:1 (Figure 4). CL images show that some of the zircons have clear oscillatory zoning, indicating their magmatic origin [31]. The zircon SHRIMP U–Pb analytical data are presented in Table A1. The measured Th and U contents are 137–658 ppm and 153–477 ppm, respectively, and the Th/U ratios vary between 0.65 and 1.43. Sixteen analyses show clustered $^{206}Pb/^{238}U$ ages with a concordia age of 278.8 ± 0.81 Ma (MSWD = 0.117) (Figure 4), which is considered the crystallization age of the Haoyaoerhudong gabbro.

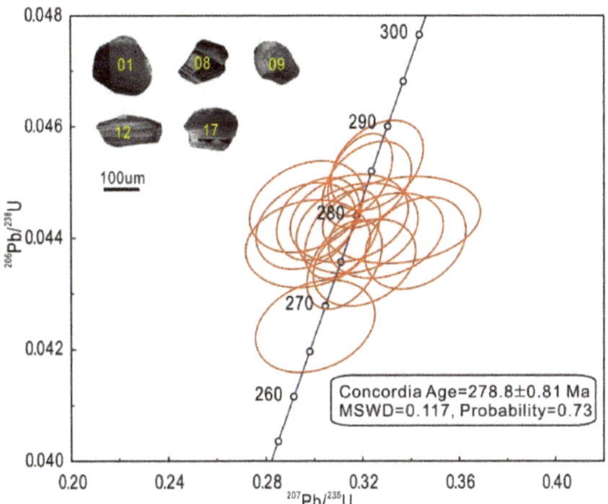

Figure 4. Representative cathodoluminescence (CL) images and U–Pb concordia diagram of zircons from sample RB25.

4.2. Major and Trace Element Data

As discussed above, the freshest available rock samples were selected for the chemical analysis. However, the sulfur minerals found in some thin sections suggest that the Haoyaoerhudong gabbro may suffer from hydrothermal alteration. Thus, it is important to evaluate the influence of postmagmatic alteration. In this study, all of the samples have LOI values much lower than 6 wt.% (ranging from 1.29 to 2.83 wt.%) and Ce/Ce* values concentrated in a narrow range from 0.9 to 1.0 (except samples RB04 and RB05, which have similar values of 0.87 and 1.03, respectively), indicating that these samples still have their primary chemical signatures without strong modification [32].

The major and trace element contents of the Haoyaoerhudong gabbro are reported in Table A2. The results show moderate variations in SiO_2 (44.76–51.0 wt.%) and relatively large variations of other oxides. The samples are characterized by relatively high contents of Al_2O_3 (average 13 wt.%, range 5.36–18.34 wt.%), CaO (average 11.12 wt.%, range 5.47–13.96 wt.%), and MgO (average 12.52 wt.%, range 4.98–24.78 wt.%) and low contents of Na_2O (average 1.25 wt.%, range 0.66–2.39 wt.%) and K_2O (average 0.43 wt.%, range 0.16–0.90 wt.%). Their Mg numbers ($100 \times Mg/[Mg + Fe^{2+}]$) range from 61 to 74 (excluding RB22, which has a very low value of 36). All of the points fall into the gabbro field in the Total Alkali vs Silica (TAS) diagram (Figure 5a). Most of the data plot in the calc-alkaline field on the Co–Th classification diagram [33], indicating calc-alkaline affinity (Figure 5b). On the MgO vs. major element oxides diagrams, CaO, K_2O, and Na_2O decrease and Fe_2O_3 (total iron) increases with increasing MgO (Figure 6), indicating fractionation processes of clinopyroxene and plagioclase with some Fe–Ti oxides.

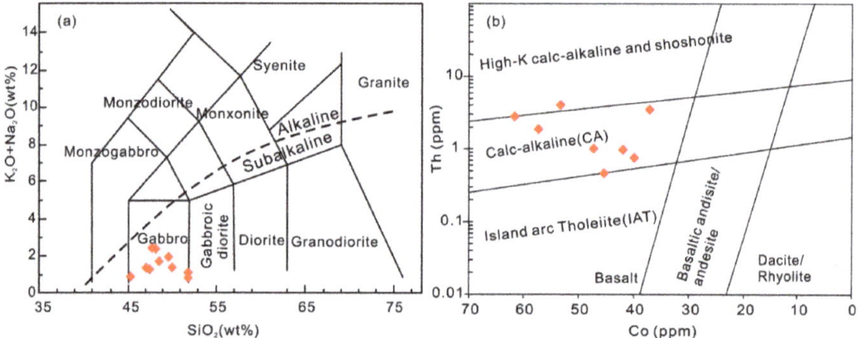

Figure 5. (**a**) TAS classification diagram and (**b**) Th–Co classification diagram (after [33]) for the Haoyaoerhudong gabbro.

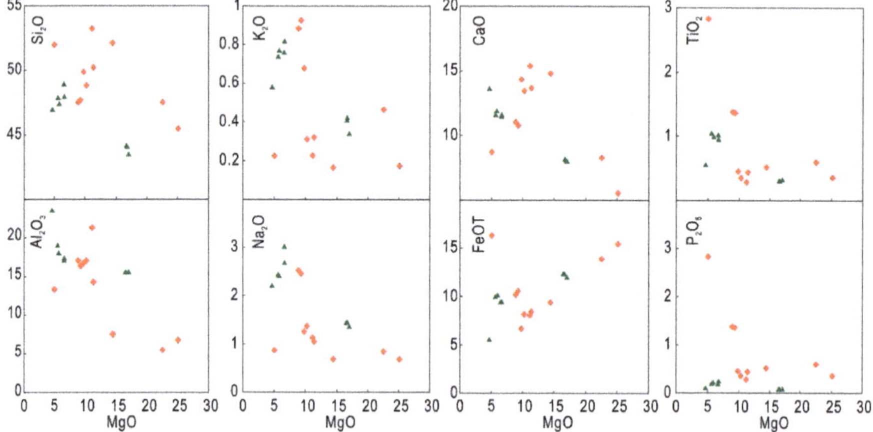

Figure 6. Harker diagrams for the Haoyaoerhudong gabbro. The data for the green triangles are from [34].

For compatible trace elements, the Cr contents vary in the range of 114–986 ppm (average 80 ppm), and Ni averages 161 ppm with a range of 30–530 ppm. For the high field strength elements, Nb averages 5.03 ppm with a large range of 0.51–16.79 ppm, and Zr averages 65 ppm with a range of 23.9–88.9 ppm. The large ion lithophile elements are relatively high. Ba ranges from 28.5 to 462.8 ppm (average of 251 ppm), and Sr ranges from 178 to 905 ppm (average of 560 ppm). On the primitive mantle-normalized diagram (Figure 7a), all of the samples show similar features: enrichment of LILEs (Rb, Sr, and Ba) and depletion of HFSEs (Nb, Ta and P), indicating geochemical affinity to arc-type igneous rocks [35,36].

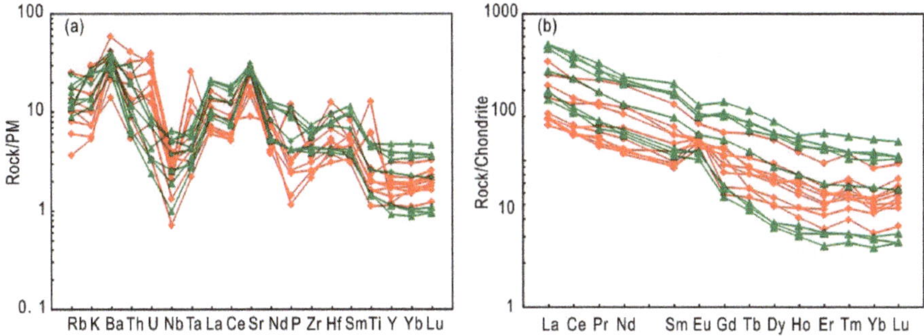

Figure 7. (a) PM-normalized incompatible element abundances and (b) chondrite-normalized REE patterns for the Haoyaoerhudong gabbro. The data for the green triangles are from [34]. The chondrite and PM values are from [37].

The REE distribution patterns show moderate fractionation between LREEs and HREEs with positive Eu anomalies (Figure 7b). The total REEs of the samples vary between 27.13 and 65.85 ppm with $(La/Yb)_N$ ratios of 3.13 to 10.44. The Eu/Eu* ratios have a range of 0.90–1.83 and an average of 1.25.

4.3. Sr–Nd Isotopic Data

Five samples (RB04, RB05, RB20, RB22, RB23) from the Haoyaoerhudong gabbro were tested for Sr–Nd isotopes (Table A3). The initial isotopic compositions were recalculated to 279 Ma based on the zircon U–Pb age from this study. The $(^{87}Sr/^{86}Sr)_i$ ratios of the five tested samples vary between 0.7053 and 0.7078. The Nd isotopic compositions are relatively homogeneous, and the $^{143}Nd/^{144}Nd$ ratios have a small range of 0.5118–0.5123, corresponding to εNd(t) varying between −4.6 and −15.1.

5. Discussion

5.1. Timing of a Hercynian Mafic Magmatic Event

Our new SHRIMP zircon data provide a precise age for the Haoyaoerhudong gabbro. The morphologies and internal structures, as well as the high Th/U values of the zircons from the Haoyaoerhudong gabbro, are suggestive of their magmatic origin. The 278.8 ± 0.81 Ma age (concordia age) indicates that the Haoyaoerhudong gabbro was emplaced in the Early Permian.

Previous studies have obtained age data for gabbros in the areas adjacent to the Haoyaoerhudong gold deposit. A 275.5 ± 1.8 Ma zircon U–Pb age (LA-ICP-MS) was obtained by Xiao [38] for the gabbro in the Nuoergong area (sample location N 39°56′42″, E 104°47′20″), which is thought to be a Permian continental magmatic arc. Zircon U–Pb dating (LA-ICP-MS) of two samples selected from the Taohaotuoxiquan gabbro yielded a consistent age of 276–275 Ma, suggesting emplacement in the Early Permian [39]. Zhao et al. [35] identified olive gabbro and amphibole gabbro in the Beiqigetao intrusion in the Wengeng area. They obtained a concordant SHRIMP U–Pb age of 269 ± 8 Ma for the olive gabbro (sample location N 41°31′44″, E 108°05′24″). All of these data demonstrate a wide range of Hercynian mafic magmatism in the studied area.

5.2. Magma Source and Petrogenesis

Magmatic rocks sourced from primary magma generally have high MgO, Mg#, Cr, and Ni contents (>15 wt.%, >65, >2000 ppm, and >500 ppm, respectively) [40]. The samples from the Haoyaoerhudong gabbro have variable MgO contents (4.98–24.78, average of 13 wt.%), a wide Mg# range (35.7–74.4), Cr <1000 ppm, and Ni <500 ppm (Table A2). These data indicate that the Haoyaoerhudong gabbro underwent a differentiation process before its emplacement.

Several immobile elements, including Zr, Nb, and Yb, are widely used to study the attributes of the magma source, and element ratios (e.g., Nb/Yb) are considered to be indexes of mantle fertility of the rocks [41]. The Nb/Yb values of the studied gabbro cover a relatively wide range of 0.6–5.49 with an average of 2.95, which is similar to that of enriched mid-ocean ridge basalts (E-MORB, 3.5 [42]), indicating that the melt may be sourced from the mantle. The strong positive Sr and Eu anomalies of the Haoyaoerhudong gabbro are also similar to gabbros sourced from mantle-derived basaltic melt [43] (Figure 7a,b). Most of the samples plot in areas far from that of ocean island basalts (OIB) in the La/Ba vs. La/Nb diagram (Figure 8a), indicating lithospheric affinity rather than an asthenospheric mantle origin [44]. The Dy/Yb ratios of the tested samples are concentrated in a small range of 1.92–2.35, and the La/Yb ratios have a range of 5.11–14.36, indicating partial melting of spinel peridotite facies (Figure 8b) from a relatively shallow source [44,45].

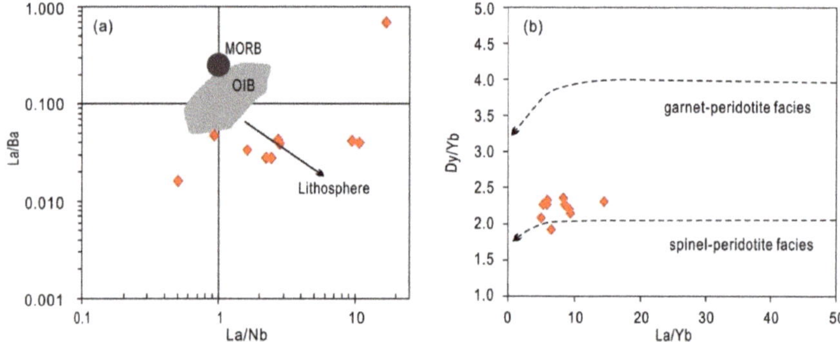

Figure 8. (**a**) La/Ba vs. La/Nb diagram and (**b**) Dy/Yb vs. La/Yb diagram showing the petrogenetic features of the Haoyaoerhudong gabbro. The fields in (**a**) are from [46]. The trends of the spinel-peridotite and garnet-peridotite facies in (**b**) are from [47].

It is widely accepted that crustal contamination should be considered in any intraplate setting [48]. Crustal contamination may lead to an increase in $(^{87}Sr/^{86}Sr)_i$ and a decrease in $\varepsilon Nd(t)$ in the magma [49] due to high $^{87}Sr/^{86}Sr$ and SiO_2 values and low $\varepsilon Nd(t)$ and MgO values in crustal materials [50]. Thus, high $(^{87}Sr/^{86}Sr)_i$ and low $\varepsilon Nd(t)$ values may suggest the involvement of crustal materials. As illustrated by the $(^{87}Sr/^{86}Sr)_i$ vs. $\varepsilon Nd(t)$ plot, most of the points straddle the area from OIB to continental volcanics (Figure 9), indicating the influence of crustal materials in the formation of the Haoyaoerhudong gabbro.

Elemental ratios, such as Y/Nb, Nb/Zr, and La/Zr, are thought to be indexes of crustal contamination [51,52]. In the present case, the nearly constant or small ranges of these ratios (i.e., 1.77–5.52 for Y/Nb, except for one value of 15.18; 0.02–0.12 for Nb/Zr; and 0.10–0.22 for La/Zr) indicate that crustal contamination may have played a crucial role in the generation of the Haoyaoerhudong gabbro. It has been suggested that Lu–Yb ratios will not change dramatically during partial melting or fractional crystallization [53]. Mantle-derived magma and the continental crust have relatively low (0.14–0.15) and high (0.16–0.18) Lu/Yb ratios, respectively [42,54]. The Lu/Yb ratios in this study (0.14–0.19, average of 0.17) indicate crustal addition during the formation of the Haoyaoerhudong gabbro. The high Th/Ta ratios (1.83–19.22, average of 8.23) and low Nb* ($Nb/Nb^* = Nb_{PM}/\sqrt{(Th_{Pm} \times La_{PM})}$) values of the gabbro (0.09–0.84, average of 0.36) also suggest the involvement of crustal materials [55].

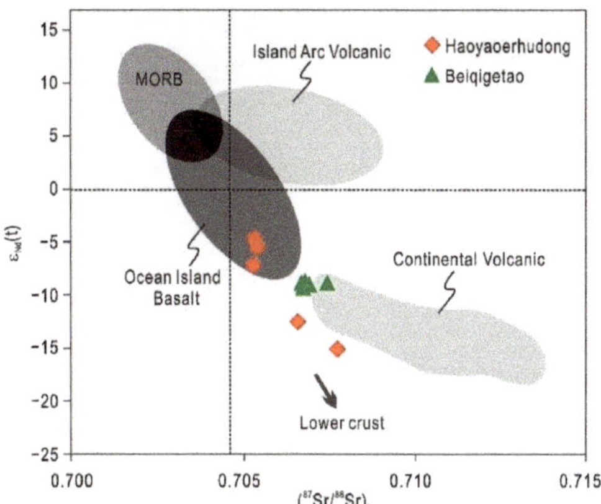

Figure 9. εNd(t) vs. $(^{87}Sr/^{86}Sr)_i$ diagram for the Haoyaoerhudong gabbro. The contents of MORB, OIB, island arc volcanics and continental volcanics are from [15,56,57]. The trend to the lower crust is from [58]. The data for the green triangles are from [34].

The decoupling between Nb and Ta and between Zr and Hf are widely used as indexes for subduction processes [59]. The scattered Nb/Ta ratios (5.67–26.12) and Zr/Hf ratios (15.45–30.21) of the studied samples differ from that of the primitive mantle (Nb/Ta of 17.8 and Zr/Hf of 37) [37] and crust (Nb/Ta of 11 and Zr/Hf of 33) [60], indicating that they might be related to subduction processes. Generally, subducted slab-released fluids lead to high Rb, Sr, Ba, U, and Pb contents, whereas subducted oceanic sediment-derived melts lead to high concentrations of LREEs and Th in the magma. Experiments show that the Ce/Pb values of slab-dehydrated fluids are usually lower than 0.1 [61], and Ce/Pb ratios less than 20 may indicate the involvement of subduction-dehydrated fluids [62]. Therefore, the Ce/Pb ratios of the studied samples (1.35–7.38, average of 3.84) which are lower than that of mantle-sourced magmas (e.g., oceanic basalts have Ce/Pb ratios of approximately 25) might be caused by the addition of Pb via subduction fluids [61]. Mantle metasomatism of the Haoyaoerhudong gabbro was further indicated by the Ba/Th versus La/Sm plot [63] and the Th/Yb versus Sr/Nd plot [64]. As shown in Figure 10a,b, the points of the Haoyaoerhudong samples are consistent with the slab dehydration trend and distinctly different from the sediment melting trend, indicating that the mantle was modified by fluids from slab dehydration. In addition, the presence of hornblende and biotite in most thin sections showed water-rich features of the parent magma, suggesting the effect of slab hydration during magma genesis.

Both the major and trace element contents of the Haoyaoerhudong gabbro vary over large ranges, indicating fractional crystallization processes. Ni and Cr are positively related to MgO in the Haoyaoerhudong gabbro, indicating fractional crystallization of clinopyroxene. The negative correlation of MgO and Al_2O_3 as well as the positive Eu anomaly of the Haoyaoerhudong gabbro demonstrates plagioclase fractional crystallization. In summary, we suggest that the Haoyaoerhudong gabbro originated from partial melting of a spinel peridotite facies of the mantle, which was metasomatized by fluids that dehydrated from a subducted slab. Fractionation crystallization is the most likely petrogenesis mechanism for the Haoyaoerhudong gabbro.

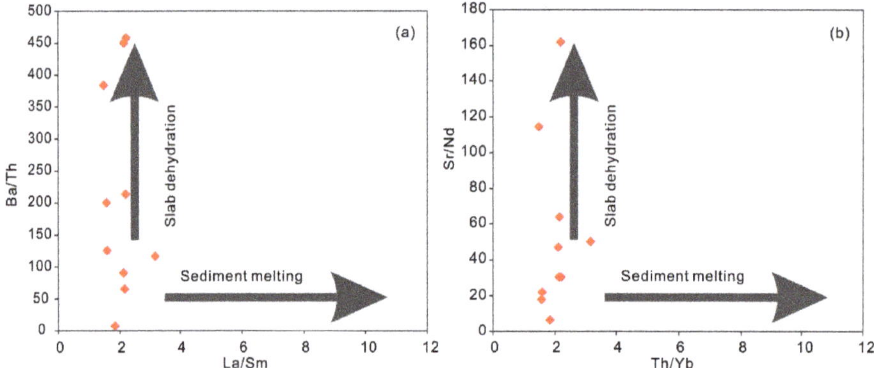

Figure 10. (**a**) Plot of Ba/Th vs. La/Sm for the Haoyaoerhudong gabbro (after [63]). (**b**) Plot of Th/Yb vs. Sr/Nb for the Haoyaoerhudong gabbro (after [64]).

5.3. Implications for the Tectonic Setting

Generally, Th and Nb are geochemically similar. However, Th becomes mobile in subduction magmatic processes due to the addition of hydrous fluids from the subducted slab, whereas Nb is stable during this process [65]. Thus, Th and Nb are widely used to identify mafic rocks of different tectonic units [55]. To understand the tectonic setting of the Haoyaoerhudong gabbro, we applied the Th_N–Nb_N discrimination plot [66]. Three samples spread along the MORB-OIB array, two points plot in the forearc and intra-arc field, and five samples fall into the continental margin volcanic arc field (Figure 11). This plot also indicates that the Haoyaoerhudong magma was affected by fluids that dehydrated from the subducting slab.

Jenner et al. reported that arc gabbros and non-arc gabbros have different Nb/Th values (<7.5 and >8.5, respectively) [67]. Most of the Nb/Th values of the Haoyaoerhudong gabbro are <7.5 (eight of the 10 tested samples range from 0.65 to 4.18, and two samples are slightly higher than 7.5), which indicates their arc affinities. In the Hf–Nb–Th diagram (Figure 12), two samples fall into the E-MORB field, and the other eight points fall into the CAB (calc-alkaline basalts) field, indicating that they likely formed in a continental arc environment. Other geochemical characteristics of the Haoyaoerhudong gabbro that are consistent with rocks from a continental arc include the enrichment of LILEs and LREEs, depletion of HSFEs, high La/Nb ratios (1.2–5.0 with an average of 3.5, which is higher than that of N-MORB, which has La/Nb ratios lower than 1.4), and negative $\varepsilon Nd(t)$ values of −4.6 to −15.1 [41].

The time that the Paleo-Asian Ocean (PAO) closed in the Xingmeng Orogenic Belt (XOB) has long been a controversial topic. A diachronous closure of the PAO, from west (Tarim) to east (Changchun), was suggested by Xiao et al. [26] and Wilde [68]. Luo et al. obtained a U–Pb zircon age of 277 ± 3 Ma for the Wuliangsitai intrusion, which was identified as an A_2-type (postcollisional) granitoid, indicating a post-orogenic extensional setting [69]. Recently, Liu et al. noticed a sharp change of the $\varepsilon Hf(t)$ (in zircon) and $\varepsilon Nd(t)$ (whole-rock) values of the 280–265 Ma gabbros from the Alxa Terrane and inferred a tectonic change from subduction to post-collision [70]. The Haoyaoerhudong gabbro has similar REE and trace element distributions and Sr–Nd isotopic features (Figures 7 and 9) as the Beiqigetao gabbro [34], which was interpreted as a Permian postcollisional intrusion. In addition, no obvious deformation has been observed in the Haoyaoerhudong gabbro. The gabbroic dikes are steep and sharp. All of this information suggests that the Haoyaoerhudong gabbro is the product of post-collisional mafic magmatism. Thus, it can be inferred that the PAO closed before 278.8 ± 0.81 Ma in the western section of the XOB.

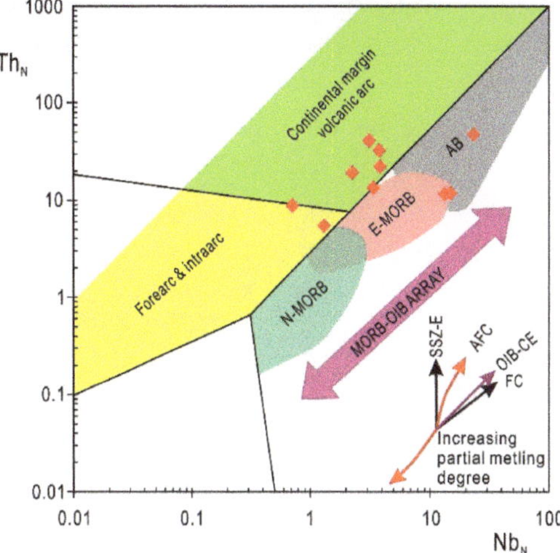

Figure 11. Th_N vs. Nb_N plot for the Haoyaoerhudong gabbro (after [66]). The N-MORB composition is from [42].

The post-collisional setting is also supported by the gold mineralization features. In the Haoyaoerhudong gold deposit, most of the thin gold-bearing sulfide veins (films) are located on E-W-trending open bedding planes and schistosity planes, which suggests that the gold mineralization, along with the late stage magmatic activities, formed in a post-collisional extensional setting.

6. Conclusions

1. The SHRIMP U–Pb zircon age shows that the Haoyaoerhudong gabbro in the Haoyaoerhudong gold deposits intruded into Proterozoic strata at 279 ± 2 Ma, indicating a Permian mafic magmatic event in the western section of the northern margin of the NCC.

2. The Haoyaoerhudong gabbro is subalkaline with calc-alkaline affinity, and its source magma was probably the result of the partial melting of the spinel peridotite zone of the lithospheric mantle, which was modified by subducting fluids.

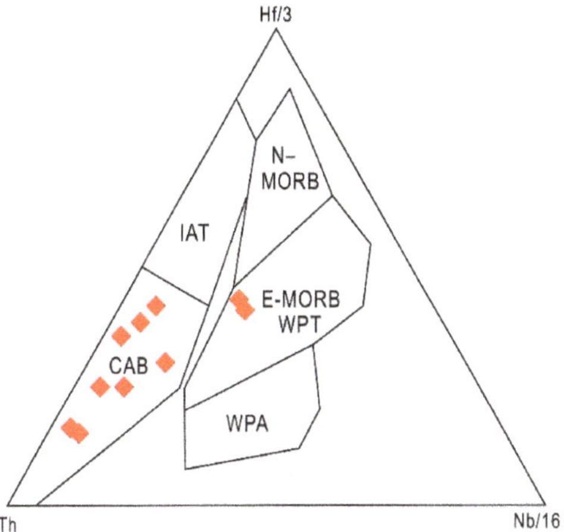

Figure 12. Th–Hf–Nb triangular discrimination diagram (after [71]) for the Haoyaoerhudong gabbro.

3. Both the Haoyaoerhudong gold deposit and the Haoyaoerhudong gabbro were generated in a post-collisional geodynamic setting, most likely induced by subduction and rollback of the paleo-oceanic plate beneath the NCC.

Author Contributions: Conceptualization, J.W.; Data curation, X.W.; Formal analysis, Z.L.; Investigation, J.W. and X.W.; Methodology, J.L.; Supervision, J.L.; Writing—original draft, J.W.; Writing—review and editing, Z.L., D.Z. and Y.W.

Funding: This study was funded by the National Natural Science Foundation of China (No. 41730426, 41272106, 41030423).

Acknowledgments: The authors are grateful to manager Xiangdong Jiang and Qiang Li of the Inner Mongolia Pacific Mining Co. Ltd. for permission to access the mining project. Engineer Hailong Gao and Zhihui Dou are thanked for their help during the fieldwork.

Conflicts of Interest: The authors declare no conflict of interest.

Appendix A

Table A1. SHRIMP zircon U–Pb isotope data for the Haoyaoerhudong gabbro.

Spot	U	Th	U/Th	Pb*	Pbc	$^{206}Pb/^{238}U$	±%	$^{207}Pb/^{235}U$	±%	$^{207}Pb/^{206}Pb$	±%	$^{206}Pb/^{238}U$age (Ma)	±%
RB25-1	188	203	1.11	7.03	–	0.04353	1.3	0.313	3.3	0.0585	2.6	274.7	3.4
RB25-2	257	293	1.17	9.84	0.61	0.04421	1.2	0.297	4.7	0.0538	3.8	278.9	3.2
RB25-3	214	200	0.97	8.04	–	0.04390	1.2	0.342	4.0	0.0548	3.7	277.0	3.3
RB25-4	405	521	1.33	15.5	0.49	0.04438	1.1	0.315	5.2	0.0574	2.8	279.9	3
RB25-5	306	331	1.12	11.6	0.16	0.04403	1.2	0.330	4.6	0.0571	2.3	277.7	3.2
RB25-6	186	212	1.18	6.80	0.29	0.04242	1.3	0.300	5.4	0.0532	3.0	267.8	3.4
RB25-9	378	471	1.29	14.6	–	0.04514	1.1	0.3193	2.6	0.0553	2.0	284.6	3
RB25-10	285	368	1.33	10.9	0.54	0.04432	1.2	0.337	6.4	0.0631	3.8	279.6	3.2
RB25-11	328	349	1.10	12.8	0.19	0.04532	1.1	0.326	3.9	0.0556	2.1	285.8	3.1
RB25-13	259	275	1.10	9.69	0.15	0.04355	1.2	0.321	4.4	0.0541	2.6	274.8	3.2
RB25-14	343	350	1.05	13.1	0.56	0.04416	1.1	0.318	5.3	0.0597	2.0	278.5	3.1
RB25-15	324	388	1.24	12.5	0.05	0.04495	1.1	0.322	3.9	0.0572	2.6	283.4	3.1
RB25-17	380	436	1.18	14.5	0.29	0.04432	1.1	0.302	3.8	0.0542	2.0	279.6	3
RB25-18	191	215	1.16	7.23	–	0.04409	1.2	0.312	3.3	0.0506	3.0	278.1	3.4
RB25-19	207	239	1.20	7.95	0.59	0.04456	1.2	0.298	5.1	0.0581	2.6	281.0	3.4
RB25-20	272	290	1.10	10.3	0.32	0.04391	1.2	0.298	5.8	0.0512	2.5	277.0	3.2

Errors are 1σ. Pbc and Pb* are common Pb and radiogenic Pb, respectively. ^{204}Pb correction used for the zircon ages.

Table A2. Major (wt.%) and trace (ppm) element compositions for the Haoyaoerhudong gabbro.

Sample	RB4	RB5	RB7	RB8	RB9	RB10	RB20	RB22	RB23	RB25
SiO_2	46.44	46.30	44.76	48.80	50.64	46.19	51.00	47.33	45.93	48.51
Al_2O_3	15.95	16.61	6.70	13.87	7.36	5.36	13.07	16.50	18.34	16.24
TiO_2	1.32	1.34	0.35	0.42	0.50	0.58	2.77	0.34	0.24	0.44
Fe_2O_3	1.97	1.74	1.53	0.44	0.71	1.49	2.43	1.40	1.26	0.66
FeO	8.50	8.36	13.79	7.78	8.46	12.15	13.80	6.65	5.80	5.93
CaO	10.49	10.75	5.47	13.27	14.36	8.03	8.56	13.02	13.28	13.96
MgO	9.10	8.71	24.78	11.11	14.11	21.99	4.98	9.97	10.91	9.56
K_2O	0.90	0.86	0.17	0.31	0.16	0.45	0.22	0.30	0.22	0.66
Na_2O	2.39	2.46	0.67	1.02	0.66	0.82	0.85	1.33	1.09	1.22
MnO	0.15	0.14	0.21	0.11	0.17	0.17	0.22	0.14	0.11	0.10
P_2O_5	0.26	0.24	0.05	0.07	0.07	0.06	0.24	0.03	0.03	0.07
LOI	2.31	2.28	1.29	2.55	2.59	2.44	1.72	2.83	2.63	2.46
Total	99.78	99.79	99.77	99.75	99.79	99.71	99.87	99.83	99.84	99.82
Be	0.73	0.60	0.20	0.37	0.41	0.23	0.82	0.20	0.22	0.38
Co	47.10	41.85	115.80	61.50	57.10	92.90	53.10	39.70	45.20	37.00
Ni	86.00	70.85	530.10	233.80	150.00	365.70	71.00	32.20	37.70	29.70
Cr	227.30	181.80	868.80	985.60	526.60	117.00	114.20	196.10	289.70	294.40
V	279.50	264.40	106.90	233.60	244.30	172.70	487.60	212.60	140.60	159.00
Sc	38.34	36.24	21.05	46.40	66.63	33.84	37.90	37.71	30.54	34.67

Table A2. *Cont.*

Sample	RB4	RB5	RB7	RB8	RB9	RB10	RB20	RB22	RB23	RB25
Li	10.83	9.44	5.95	3.98	4.10	4.96	22.73	5.22	6.91	9.56
Cs	0.50	0.42	0.19	0.99	0.26	0.45	0.11	0.56	0.40	1.10
Ga	21.33	19.45	7.41	14.21	9.91	6.73	25.27	15.83	16.01	12.11
In	0.07	0.07	0.03	0.04	0.06	0.04	0.11	0.04	0.03	0.03
Tl	0.11	0.09	0.05	0.05	0.17	0.09	0.10	0.06	0.04	0.13
Cu	41.20	37.49	51.10	149.80	88.00	81.50	20.00	14.00	15.70	37.00
Pb	6.10	6.00	3.10	6.80	3.80	4.10	12.00	5.80	5.20	16.10
Zn	106.70	96.11	134.80	69.80	75.80	112.60	183.40	60.10	47.40	59.40
Rb	13.30	10.90	3.80	8.50	2.30	11.20	6.90	6.60	5.10	16.00
Sr	825.50	798.90	306.60	615.60	334.70	190.60	177.90	887.00	904.50	554.80
Zr	88.90	88.80	28.70	59.40	60.10	34.40	190.10	23.90	27.80	52.00
Nb	10.71	9.51	1.63	2.74	2.81	2.45	16.79	0.51	0.94	2.25
Ta	0.41	0.53	0.17	0.17	0.23	0.13	1.06	0.09	0.12	0.18
Ba	462.80	436.40	148.20	180.80	235.40	227.80	28.50	288.10	98.20	404.50
Hf	3.64	3.58	0.95	2.95	3.89	1.30	6.53	1.32	1.14	2.07
W	0.17	0.19	0.14	0.14	0.43	0.19	0.53	0.17	0.25	0.22
Bi	0.05	0.04	0.03	0.22	0.09	0.04	0.21	0.08	0.06	0.08
Th	1.01	0.97	1.64	2.77	1.87	1.14	4.02	0.75	0.46	3.46
U	0.41	0.31	0.26	0.72	0.53	0.32	0.82	0.07	0.16	0.66
La	18.54	18.10	4.92	7.67	9.08	6.31	19.48	4.59	4.58	11.21
Ce	45.00	43.41	10.57	16.21	22.05	14.89	44.39	10.20	9.37	21.70
Pr	6.38	6.18	1.48	2.22	3.35	2.33	6.24	1.62	1.33	2.73
Nd	27.14	26.36	6.53	9.64	15.19	10.65	27.52	7.75	5.59	11.11
Sm	5.41	5.46	1.50	2.29	3.67	2.58	6.76	1.98	1.35	2.28
Eu	1.77	1.77	0.47	0.82	1.03	0.72	2.26	0.82	0.71	0.79
Gd	4.53	4.43	1.35	1.92	3.18	2.23	6.47	1.70	1.20	1.92
Tb	0.73	0.74	0.24	0.33	0.57	0.39	1.24	0.30	0.21	0.30
Dy	4.18	4.27	1.42	2.01	3.49	2.40	7.93	1.92	1.27	1.77
Ho	0.80	0.80	0.27	0.38	0.64	0.45	1.47	0.35	0.23	0.33
Er	2.02	1.98	0.70	0.95	1.58	1.10	3.77	0.85	0.56	0.83
Tm	0.34	0.35	0.12	0.16	0.28	0.19	0.68	0.14	0.10	0.15
Yb	1.95	1.94	0.74	0.89	1.50	1.06	3.81	0.85	0.54	0.77
Lu	0.37	0.36	0.13	0.15	0.24	0.19	0.52	0.13	0.09	0.14
Y	18.93	18.35	6.31	8.55	14.21	10.06	34.39	7.74	5.19	7.77
ΣREE	29.58	29.47	30.84	45.64	65.85	45.49	28.94	33.20	27.13	56.03
LREE	22.16	22.28	25.87	38.85	54.37	37.48	23.15	26.96	22.93	49.82
HREE	7.42	7.19	4.97	6.79	11.48	8.01	5.79	6.24	4.20	6.21
LREE/HREE	2.99	3.10	5.21	5.72	4.74	4.68	4.00	4.32	5.46	8.02
LaN/YbN	3.43	3.13	4.77	6.18	4.34	4.27	3.97	3.87	6.08	10.44
δEu	1.11	1.24	1.83	1.16	0.90	0.90	1.19	1.33	1.67	1.12
δCe	0.87	1.03	0.95	0.95	0.98	0.95	0.96	0.92	0.92	0.93

Table A3. Rb–Sr and Sm–Nd isotopes for the Haoyaoerhudong gabbro.

Sample	Age (Ma)	SiO_2 (wt.%)	Rb (µg/g)	Sr (µg/g)	$^{87}Rb/^{86}Sr$	$^{87}Sr/^{86}Sr$	Error	$(^{87}Sr/^{86}Sr)_i$	Sm (µg/g)	Nd (µg/g)	$^{147}Sm/^{144}Nd$	$^{143}Nd/^{144}Nd$	Error	$(^{143}Nd/^{144}Nd)_i$	$\varepsilon Nd(0)$	$\varepsilon Nd(t)$	$f_{Sm/Nd}$	T_{DM}
RB04	279	46.44	6.3	525.5	0.033844	0.70789	0.00002	0.70776	1.41	5.14	0.173697	0.51182	0.00001	0.511503	−16	−15.1	−0.12	5017
RB05	279	46.3	6.9	598.9	0.032524	0.70673	0.00002	0.7066	1.46	5.36	0.172474	0.51195	0.00001	0.511635	−13.4	−12.6	−0.12	4401
RB20	279	51	6.9	377.9	0.051544	0.70551	0.00003	0.70531	1.76	5.52	0.201888	0.51228	0.00004	0.511911	−7	−7.2	−0.03	10,968
RB22	279	47.33	6.6	587	0.031741	0.70554	0.00002	0.70541	1.98	7.75	0.161771	0.51229	0.00002	0.511995	−6.8	−5.5	−0.18	2519
RB23	279	45.93	5.1	504.5	0.028538	0.70544	0.00002	0.70533	1.35	5.59	0.152918	0.51232	0.00002	0.512041	−6.2	−4.6	−0.22	2080

References

1. Liu, D.Y.; Nutman, A.P.W.; Compston, W.; Wu, J.S.; Shen, Q.H. Remnants of ≥3800 Ma crust in the Chinese part of the Sino-Korean Craton. *Geology* **1992**, *20*, 339–342. [CrossRef]

2. Zhai, M.G.; Santosh, M. The early Precambrian odyssey of the North China Craton: A synoptic overview. *Gondwana Res.* **2011**, *20*, 6–25. [CrossRef]

3. Zhai, M.G.; Santosh, M. Metallogeny of the North China Craton: Link with secular changes in the evolving Earth. *Gondwana Res.* **2013**, *24*, 275–297. [CrossRef]

4. Zhao, G.C.; Sun, M.; Wilde, S.A.; Li, S.Z. Late Archean to Paleoproterozoic evolution of the North China Craton: Key issues revisited. *Precambrian Res.* **2005**, *136*, 177–202. [CrossRef]

5. Li, J.Y.; Zhang, J.; Yang, T.N.; Li, Y.P.; Sun, G.H.; Zhu, Z.X.; Wang, L.J. Crustal tectonic division and evolution of the southern part of the North Asian Orogenic Region and its adjacent areas. *J. Jilin Univ.* **2009**, *39*, 584–605.

6. Zhang, S.H.; Zhao, Y.; Liu, J.M.; Hu, J.M.; Song, B.; Liu, J.; Wu, H. Geochronology, geochemistry and tectonic setting of the Late Paleozoic–Early Mesozoic magmatism in the northern margin of the North China Block: A preliminary review. *Acta Petrol. Mineral.* **2010**, *29*, 824–842. (In Chinese with English abstract)

7. Zhang, J.J.; Wang, T.; Zhang, Z.C.; Tong, Y.; Zhang, L.; Shi, X.J.; Zeng, T. Magma mixing origin of Yamatu granite in Nuoergong–Langshan area, western part of the Northern Margin of North China Craton: Petrological and geochemical evidences. *Geol. Rev.* **2012**, *58*, 53–66. (In Chinese with English abstract)

8. Shi, Y.R.; Liu, C.; Deng, J.F.; Jian, P. Geochronological frame of granitoids from Central Inner Mongolia and its tectonomagmatic evolution. *Acta Petrol. Sin.* **2014**, *30*, 3155–3171.

9. Wang, J.P.; Liu, J.J.; Peng, R.M.; Liu, Z.J.; Zhao, B.S.; Li, Z.; Wang, Y.F.; Liu, C.H. Gold mineralization in Proterozoic black shales: Example from the Haoyaoerhudong gold deposit, northern margin of the North China Craton. *Ore Geol. Rev.* **2014**, *63*, 150–159. [CrossRef]

10. Wang, Y.F.; Wang, J.P.; Li, Y.M.; Jiang, X.D.; Wang, B.; Jiang, S.M. Organic geochemical characteristics and metallogenetic significance of the ore-hosting strata of the Haoyaoerhudong gold deposit, Inner Mongolia. *Acta Miner. Sin.* **2011**, *31*, 405–407. (In Chinese)

11. Wang, Y.F.; Wang, J.P.; Wang, H.; Sun, H.; Jiang, X.D.; Wang, B.; Jiang, S.M. Geochemical characteristics of pluton in the Haoyaoerhudong gold deposit of Inner Mongolia and its ore-forming significance. *Geoscience* **2013**, *27*, 56–66, (In Chinese with English abstract).

12. Xiao, W.; Nie, F.J.; Liu, Y.F.; Liu, Y. Isotope geochronology study of the granitoid intrusions in the Changshanhao gold deposit and its geological implications. *Acta Petrol. Sin.* **2012**, *28*, 535–543.

13. Cao, Y.; Nie, F.J.; Xiao, W.; Liu, Y.F.; Zhang, W.B.; Wang, F.X. Metallogenic age of the Changshanhao gold deposit in Inner Mongolia, China. *Acta Petrol. Sin.* **2014**, *30*, 2092–2100. [CrossRef]

14. Liu, Y.F.; Nie, F.J.; Jiang, S.H.; Bagas, L.; Xiao, W.; Cao, Y. Geology, geochronology and sulphur isotope geochemistry of the black schist-hosted Haoyaoerhudong gold deposit of Inner Mongolia, China: Implications for ore genesis. *Ore Geol. Rev.* **2016**, *73*, 253–269. [CrossRef]

15. Hart, S.R.; Gerlach, D.C.; White, W.M. A possible new Sr–Nd–Pb mantle array and consequences for mantle mixing. *Geochim. Cosmochim. Acta* **1986**, *50*, 1551–1557. [CrossRef]

16. Robinson, P.T.; Zhou, M.F. The origin and tectonic setting of ophiolites in China. *J. Asian Earth Sci.* **2008**, *32*, 301–307. [CrossRef]

17. Halama, R.; Wenzel, T.; Upton, B.G.J.; Siebel, W.; Markl, G. A geochemical and Sr–Nd–O isotopic study of the Proterozoic Eriksfjord Basalts, Gardar Province, South Greenland: Reconstruction of an OIB signature in crustally contaminated rift-related basalts. *Miner. Mag.* **2003**, *67*, 831–853. [CrossRef]

18. Wang, J.; Li, S.Q.; Wang, B.L.; Li, J.J. *The Langshan-Baiyunebo Rift System*; Peking University Press: Beijing, China, 1992; pp. 1–129.

19. Gong, W.B.; Hu, J.M.; Li, Z.H.; Dong, X.P.; Liu, Y.; Liu, S.C. Detrital zircon U–Pb dating of Zhaertai Group in the north margin rift zone of North China Craton and its implications. *Acta Petrol. Sin.* **2016**, *32*, 2151–2165.

20. Meng, Q.R.; Wei, H.H.; Qu, Y.Q.; Ma, S.X. Stratigraphic and sedimentary records of the rift to drift evolution of the northern North China craton at the Paleo- to Mesoproterozoic transition. *Gondwana Res.* **2011**, *20*, 205–218. [CrossRef]

21. Liu, C.H.; Liu, F.L. The Mesoproterozoic rifting in the North China Craton: A case study for magmatism and sedimentation of the Zhaertai-Bayan Obo-Huade rift zone. *Acta Petrol. Sin.* **2015**, *31*, 3107–3128.

22. Bai, G.; Yuan, Z.X.; Wu, C.Y.; Zhang, Z.Q.; Zheng, L.X. *Demonstration on the Geological Features and Genesis of the Bayan Obo ore Deposit*; Geological Publishing House: Beijing, China, 1996; pp. 1–92.

23. Yang, X.M.; Le, B.M.J. Chemical compositions of carbonate minerals from Bayan Obo, Inner Mongolia, China: Implications for petrogenesis. *Lithos* **2004**, *72*, 97–116. [CrossRef]

24. Smith, M.P.; Campbell, L.S.; Kynicky, J. A review of the genesis of the world class Bayan Obo Fe-REE-Nb deposits, Inner Mongolia, China: Multistage processes and outstanding questions. *Ore Geol. Rev.* **2015**, *64*, 459–476. [CrossRef]

25. Peng, R.M.; Zhai, Y.S. The confirmation of the metamophic double-peaking volcanic rocks in langshan Group of the Dongshengmiao ore district, Inner Monglia, and its significance. *Earth Sci. J. China Univ. Geosci.* **1997**, *22*, 589.

26. Xiao, W.J.; Windley, B.F.; Hao, J.; Zhai, M.G. Accretion leading to collision and the Permian Solonker suture, Inner Mongolia, China: Termination of the central Asian orogenic belt. *Tectonics* **2003**, *22*, 1069. [CrossRef]

27. Xiao, W.J.; Kroner, A.; Windley, B. Geodynamic evolution of Central Asia in the Paleozoic and Mesozoic. *Int. J. Earth Sci.* **2009**, *98*, 1185–1188. [CrossRef]

28. Nie, F.J.; Jiang, S.H.; Huo, W.R.; Liu, Y.F.; Xiao, W. Geological features and genesis of gold deposits hosted by low-grade metamorphic rocks in central-western Inner Mongolia. *Miner. Depos.* **2010**, *29*, 58–70.

29. Jian, P.; Kröner, A.; Zhou, G.Z. SHRIMP zircon U–Pb ages and REE partition for high-grade metamorphic rocks in the north Dabie complex: Insight into crustal evolution with respect to Triassic UHP metamorphism in east-central China. *Chem. Geol.* **2012**, *328*, 49–69. [CrossRef]

30. Zhang, L.H.; Guo, Z.F.; Zhang, M.L.; Cheng, Z.H.; Sun, Y.T. Post-collisional potassic magmatism in the eastern Lhasa terrane, South Tibet: Products of partial melting of melanges in a continental subduction channel. *Gondwana Res.* **2017**, *41*, 9–28. [CrossRef]

31. Hoskin, P.W.O.; Schaltegger, U. The composition of zircon and igneous and metamorphic petrogenesis. *Rev. Miner. Geochem.* **2003**, *53*, 27–62. [CrossRef]

32. Polat, A.; Hofmann, A.W. Alteration and geochemical patterns in the 3.7–3.8 Ga Isua greenstone belt, West Greenland. *Precambrian Res.* **2003**, *126*, 197–218. [CrossRef]

33. Hastie, A.R.; Kerr, A.C.; Pearce, J.A.; Mitchell, S.F. Classification of altered volcanic island arc rocks using immobile trace elements: Development of Th–Co discrimination diagrams. *J. Petrol.* **2007**, *48*, 2341–2357. [CrossRef]

34. Zhao, L.; Wu, T.R.; Luo, H.L. SHRIMP U–Pb dating, geochemistry and tectonic implications of the Beiqigetaogabbros in UradZhongqi area, Inner Mongolia. *Acta Petrol. Sin.* **2011**, *27*, 3071–3082. (In Chinese with English abstract)

35. Rudnick, R.L. Making continental crust. *Nature* **1995**, *378*, 571–578. [CrossRef]

36. Taylor, S.R.; McLennan, S.M. The geochemical evolution of the continental crust. *Rev. Geophys.* **1995**, *33*, 241–265. [CrossRef]

37. McDonough, W.F.; Sun, S.S. The composition of the earth. *Chem. Geol.* **1995**, *120*, 223–253. [CrossRef]

38. Xiao, J.; Sun, P.; Xu, L. LA-ICP-MS zircon U–Pb age, geochemistry and genesis of the early Permian gabbros in the Nuoergong area, northern Alex. *West. Resour.* **2011**, *4*, 55–60. (In Chinese)

39. Zhang, L.; Shi, X.J.; Zhang, J.J.; Yang, Q.D.; Tong, Y.; Wang, T. LA-ICP-MS zircon U–Pb age and geochemical characteristics of the Taohaotuoxiquan gabbro in northern Alxa, Inner Mongolia. *Geol. Bull. China* **2013**, *32*, 1536–1547. (In Chinese)

40. Winter, J. *Introduction to Igneous and Metamorphic Petrology*; Prentice Hall: Upper Saddle River, NJ, USA, 2001; p. 796.

41. Pearce, J.A.; Stern, R.J. The origin of back-arc basin magmas: Trace element and isotope perspectives. In *Back-Arc Spreading Systems: Geological, Biological, Chemical and Physical Interactions*; Christie, C.R., Lee, S.-M., Givens, S., Eds.; American Geophysical Union Geophysical Monograph; John Wiley & Sons, Inc.: Hoboken, NJ, USA, 2006; Volume 166, pp. 63–86.

42. Sun, S.S.; McDonough, W.F. Chemical and isotopic systematics of oceanic basalts: Implications for mantle composition and processes. In *Magmatism in the Ocean Basin*; Geological Society of London Special Publication: London, UK, 1989; Volume 42, pp. 313–345.

43. Marchesi, C.; Garrido, C.J.; Godard, M.; Proenza, J.A.; Gervilla, F.; Blanco-Moreno, J. Petrogenesis of highly depleted peridotites and gabbroic rocks from the Mayarí-Baracoa ophiolitic belt (eastern Cuba). *Contrib. Miner. Petrol.* **2006**, *151*, 717–736. [CrossRef]

44. Thirlwall, M.F.; Upton, B.G.J.; Jenkins, C. Interaction between continental litho- sphere and the Iceland plume: Sr–Nd–Pb isotope chemistry of Tertiary basalts, NE Greenland. *J. Petrol.* **1994**, *35*, 839–897. [CrossRef]
45. Bogaard, P.F.J.; Wörner, G. Petrogenesis of basanitic to tholeiitic volcanic rocks from the Miocene Vogelsberg, Central Germany. *J. Petrol.* **2003**, *44*, 569–602. [CrossRef]
46. Saunders, A.D.; Storey, M.; Kent, R.W.; Norry, M.J. Consequences of plume-lithosphere interactions. In *Magmatism and the Causes of Continental Break-up*; Alabaster, T., Storey, B.C., Pankhurst, R.J., Eds.; Geological Society: London, UK, 1992; Volume 68, pp. 41–60.
47. Jung, C.; Jung, S.; Hoffer, E.; Berndt, J. Petrogenesis of tertiary mafic alkaline magmas in the Hocheifel, Germany. *J. Petrol.* **2006**, *47*, 1637–1671. [CrossRef]
48. Wilson, M. *Igneous Petrogenesis*; Chapman and Hall: London, UK, 1989; p. 466.
49. Rogers, N.; Macdonald, R.; Fitton, J.G.; George, R.; Smith, M.; Barreiro, B. Two mantle plumes beneath the east African rift system: Sr, Nd and Pb isotope evidence from Kenya Rift basalts. *Earth Planet. Sci. Lett.* **2000**, *176*, 387–400. [CrossRef]
50. Rudnick, R.L.; Fountain, D.M. Nature and composition of the continental crust—A lower crustal perspective. *Rev. Geophys.* **1995**, *33*, 267–309. [CrossRef]
51. Weaver, S.D.; Sceal, J.C.; Gibson, I.L. Trace element data relevant to the origin of trachytic and pantenlleritic lavas in the East African Rift System. *Contrib. Mineral. Petrol.* **1971**, *36*, 181–194. [CrossRef]
52. Lippard, S.J. The petrology of phonolites from Kenyan rift. *Lithos* **1973**, *6*, 217–234. [CrossRef]
53. Dai, J.G.; Wang, C.S.; Hebert, R.; Li, Y.L.; Zhong, H.T.; Guillaume, R.; Bezard, R.; Wei, Y.S. Late Devonian OIB alkaline gabbro in the YarlungZangbo suture zone: Remnants of the paleo-Tethys. *Gondwana Res.* **2011**, *19*, 232–243. [CrossRef]
54. Rudnick, R.; Gao, S. Composition of the continental crust. In *Treatise on Geochemistry*; Rudnick, R., Ed.; Elsevier-Pergamon: Oxford, UK, 2003; Volume 3, pp. 1–64.
55. Pearce, J.A. Geochemical fingerprinting of oceanic basalts with applications to ophiolite classification and the search for Archean oceanic crust. *Lithos* **2008**, *100*, 14–48. [CrossRef]
56. Zindler, A.; Hart, S. Chemical geodynamics. *Annu. Rev. Earth Planet. Sci.* **1986**, *14*, 493–571. [CrossRef]
57. Hart, S.R.; Hauri, E.H.; Oschmann, L.A.; Whitehead, J. Mantle plumes and entrainment: Isotopic evidence. *Science* **1992**, *256*, 517–520. [CrossRef]
58. Jahn, B.M.; Wu, F.Y.; Lo, C.H.; Tsai, C.H. Crust-mantle interaction induced by deep subduction of the continental crust: Geochemical and Sr–Nd isotopic evidence from post-collisional mafic-ultramafic intrusions of the northern Dabie complex, central China. *Chem. Geol.* **1999**, *157*, 119–146. [CrossRef]
59. Stolz, A.J.; Jochum, K.P.; Hofmann, A.W. Fluid- and melt-related enrichment in the subarc mantle: Evidence from Nb/Ta variations in island-arc basalts. *Geology* **1996**, *24*, 587–590. [CrossRef]
60. Taylor, S.R.; McLennan, S.M. *The Continental Crust: Its Composition and Evolution*; Blackwell Scientific Publications: London, UK, 1985; p. 312.
61. Chung, S.L.; Wang, K.L.; Crawford, A.J.; Kamenetsky, V.S.; Chen, C.H.; Lan, C.Y.; Chen, C.H. High-Mg potassic rocks from Taiwan: Implications for the genesis of orogenic potassic lavas. *Lithos* **2001**, *59*, 153–170. [CrossRef]
62. Seghedi, I.; Downes, H.; Szakacs, A.; Mason, P.; Thirlwall, M.; Rosu, E.; Pecskay, Z.; Marton, E.; Panaiotu, C. Neogene–Quaternary magmatism and geodynamics in the Carpathian–Pannonian region: A synthesis. *Lithos* **2004**, *72*, 117–146. [CrossRef]
63. Labanieh, S.; Chauvel, C.; Germa, A.; Quidelleur, X. Martinique: A clear case for sediment melting and slab dehydration as a function of distance to the trench. *J. Petrol.* **2012**, *53*, 2441–2464. [CrossRef]
64. Woodhead, J.D.; Eggins, S.M.; Johnson, R.W. Magma genesis in the New Britain Island Arc: Further insights into melting and mass transfer processes. *J. Petrol.* **1998**, *39*, 1641–1668. [CrossRef]
65. Leat, P.T.; Pearce, J.A.; Barker, P.F.; Millar, I.L.; Barry, T.L.; Larter, R.D. Magmagenesis and mantle flow at a subducting slab edge: The South Sandwicharc-basin system. *Earth Planet. Sci. Lett.* **2004**, *227*, 17–35. [CrossRef]
66. Saccani, E. A new method of discriminating different types of post-Archean ophiolitic basalts and their tectonic significance using Th–Nb and Ce–Dy–Yb systematics. *Geosci. Front.* **2015**, *6*, 481–501. [CrossRef]
67. Jenner, G.A.; Dunning, G.R.; Malpas, J.; Brown, M.; Brace, T. Bay of Islands and Little Port complexes, revisited: Age, geochemical and isotopic evidence confirm supra-subduction zone origin. *Can. J. Earth Sci.* **1991**, *28*, 135–162. [CrossRef]

68. Wilde, S.A. Final amalgamation of the Central Asian Orogenic Belt in NE China: Paleo-Asian Ocean closure versus Paleo-Pacific plate subduction—A review of the evidence. *Tectonophysics* **2015**, *662*, 345–362. [CrossRef]
69. Luo, H.L.; Wu, T.R.; Zhao, L. Zicron SHRIMP U–Pb dating of Wuliangsitai A-type granite on the northern margin of the North China Plate and tectonic significance. *Acta Petrol. Sin.* **2009**, *25*, 515–526.
70. Liu, Q.; Zhao, G.C.; Han, Y.G.; Eizenhöfer, P.R.; Zhu, Y.L.; Hou, W.Z.; Zhang, X.R. Timing of the final closure of the Paleo-Asian Ocean in the Alxa Terrane: Constraints from geochronology and geochemistry of Late Carboniferous to Permian gabbros and diorites. *Lithos* **2017**, *274*, 19–30. [CrossRef]
71. Wood, D.A. The application of a Th–Hf–Ta diagram to problems of tectonomagmatic classification and to establishing the nature of crustal contamination of basaltic lavas of the British Tertiary volcanic province. *Earth Planet. Sci. Lett.* **1980**, *50*, 11–30. [CrossRef]

Article

Importance of Magmatic Water Content and Oxidation State for Porphyry-Style Au Mineralization: An Example from the Giant Beiya Au Deposit, SW China

Xinshang Bao, Liqiang Yang *, Wenyan He and Xue Gao

Affiliation State Key Laboratory of Geological Processes and Mineral Resources, China University of Geosciences, Beijing 100083, China; baoxinshang@163.com (X.B.); wyhe@cugb.edu.cn (W.H.); xuegao@cugb.edu.cn (X.G.)
* Correspondence: lqyang@cugb.edu.cn; Tel.: +86-(010)-82321937

Received: 26 July 2018; Accepted: 6 October 2018; Published: 10 October 2018

Abstract: The Beiya Au deposit is the largest Cenozoic Au deposit in the Jinshajiang-Ailaoshan porphyry metallogenic belt. Numerous studies document that high water content and fO_2 are vital factors for the generation of Au mineralization. In this belt, only the Wandongshan and Hongnitang districts are considered to be of economic importance, while the other districts, such as Bailiancun, are barren. So in order to reveal the importance of water content and oxidation state for Beiya porphyry-style Au mineralization, the amphiboles and zircons compositions are used to evaluate the physicochemical conditions (e.g., pressure, temperature, fO_2, and water content) of the Wandongshan ore-fertile porphyries and Bailiancun ore-barren porphyries observed in the Beiya Au deposit. The results show that the water content of the Wandongshan parent magma ($\leq 4.11 \pm 0.4$ wt %) are slightly higher than those of the parent magma at Bailiancun ($\leq 3.91 \pm 0.4$ wt %), while the emplacement pressure of the Wandongshan parent magma (31.5–68.6 MPa) is much lower than that of the parent magma at Bailiancun (142.3–192.8 MPa), indicating that the Wandongshan magma reached water saturation earlier. In addition, the Wandongshan porphyries crystallized from more oxidized magma (average of ΔFMQ = +3.5) with an average temperature of 778 °C compared to the Bailiancun porphyries (average of ΔFMQ = +1.5) with a mean magmatic temperature of 770 °C. The Ce^{4+}/Ce^{3+} ratio of zircon in the Wandongshan ore-related intrusions (average Ce^{4+}/Ce^{3+} of 62.00) is much higher than that of the Bailiancun barren porphyries (average Ce^{4+}/Ce^{3+} of 23.15), which further confirmed Wandongshan ore-related magma is more oxidized than the Bailiancun barren magma. Therefore, melts that are more enriched in water and with a high oxidation state will be more fertile to form an economic porphyry-style Au system.

Keywords: oxidation state; hydrous melts; constraints on mineralization; Beiya Au deposit; SW China

1. Introduction

Porphyry systems supply three-quarters of the world's Cu and one-fifth of the world's Au resources [1]. It's widely accepted that high water content (>4 wt %) [2,3] and a high oxidation state (>FMQ + 1.3–2) [4,5] of the magma are critical factors for the generation of Cu-Au mineralization. The solubility of Cu and Au in hydrous silicate melts was proven to be affected by the amount and proportion of dissolved sulfate and sulfide in S-bearing magma, whereas fO_2 controls the speciation of sulfur [5–7]. Under high fO_2 conditions, the majority of sulfur exists as SO_2 or SO_4^{2-} with very low concentrations of S^{2-} in a magma [6], which prevents saturation of immiscible Cu- and Au-sulfides during magmatic fractionation [4,6]. High magmatic water content of the magma (H_2O > 4 wt %) is also an essential factor for the generation of porphyry Cu-Au deposits [8–11]. The high water content of the magma not only promotes magmatic water reaching saturation earlier [10], but also increases the partitioning of Au from melt to aqueous volatile phases [12,13]. However, the impacts of water content and magma oxidation state on Au mineralization are still debated.

The Beiya Au deposit is the largest Cenozoic Au deposit in the Jinshajiang-Ailaoshan porphyry metallogenic belt [14,15], with proven Au reserves of more than 377 t, and by-products of Cu, Fe, Ag, Pb, and Zn [16]. Numerous studies have documented the petrography, chronology, geochemistry, and fluid inclusions of the porphyry intrusion from the Beiya Au deposit [17–19]. These studies have revealed that the monzogranite porphyry is directly related to skarn formation and mineralization [17,18]. The ore-forming monzogranite porphyry at Wandongshan is water-rich [13,19] and has a high oxidation state [20,21]. However, systematic studies on the relationship of water content and oxidation state on Au mineralization at Beiya are lacking. Comparing the water content and oxidation state of the mineralized and barren porphyries in the Beiya Au deposit is important for understanding the importance of water content and oxidation state for Beiya porphyry-style Au mineralization.

Both the monzogranite porphyries at Wandongshan and Bailiancun contain amphibole phenocrysts [22], even though the Wandongshan is considered to be of economic importance and the Bailiancun is barren [22–24]. Thus, we choose the Wandongshan and Bailiancun as the focus of our study. We investigated the mineralogical and geochemical characteristics of the amphiboles and zircons from the Wandongshan and Bailiancun monzogranite porphyries. In order to determine the relationship of fO_2 and water content on Au mineralization, we compared the water content and oxidation state of the mineralized Wandongshan ore-bearing and Bailiancun barren monzogranite porphyries.

2. Geological Setting

The Sanjiang (Nujiang-Lancangjiang-Jinshajiang) region is an important part of the Tethys tectonic domain, located in Southwestern China [25,26]. The Jinshajiang-Ailaoshan suture zone is located in the eastern Himalaya-Tibetan plateau of the Tethyan domain and is bounded by the western Yangtze Craton to the southeast (Figure 1). The regional tectonic evolution includes subduction of the Paleo-Tethyan oceanic slab, and magmatic arc formation followed by arc-continent collision during the Paleozoic-Mesozoic era, and then Indo-Asian continental collision during the Cenozoic [27,28]. The collision between the Indo and Asian continents have caused more than 300 km of strike-slip movement along the Ailaoshan-Red River shear zone since the Paleocene [29,30]. As a consequence of this strike-slip, lithospheric-scale extension has occurred, causing a set of secondary NNW-trending strike-slip faults and folds, and the emplacement of numerous alkali-rich intrusions along the Jinshajiang suture zone [31–33]. These intrusions form a 2000-km-long and 50- to 80-km-wide alkaline magmatic belt, and are associated with several important porphyry and skarn polymetallic (Cu, Au, Mo, Zn, Pb, and Ag) deposits [34,35] in the Jinshajiang-Ailaoshan metallogenic belt (Figure 1b) [7], from north to south: the Yulong Cu, the Beiya Au, the Machangqing Cu-Mo-Au, the Yao'an Au, the Habo Cu-Mo-Au, and the Tongchang Cu-Mo deposits.

The Beiya Au deposit is located in the center of the Jinshajiang-Ailaoshan magmatic belt (Figure 1a), which covers an area of ~800 km^2 and is distributed along the limbs of the N–S trending Beiya syncline [16]. The Beiya Au deposit consists of two zones [22,23], with the Bijiashan (BJS), Weiganpo (WGP), and Guogaishan (GGS) ore districts to the east, the Wandongshan (WDS), Hongnitang (HNT), and Jingouba (JGB) ore zones to the west, and the Bailiancun (BLC) occur around the periphery of the Beiya Au deposit (Figure 2). The formation hosting the Wandongshan district are dominated by Permian basalt, Triassic sandstone and limestone, with Quaternary alluvial cover, while the strata hosting the Bailiancun district are mainly Permian basalt and Quaternary alluvial cover. There are abundant Tertiary alkaline porphyry intrusions distributed in the Wandongshan and Bailiancun districts. The Tertiary alkaline porphyry intrusions include monzogranite porphyries, biotite monzogranite porphyries, and lamprophyres (Figure 2a). The monzogranite porphyries represent the volumetrically most important igneous rock and are genetically related to the mineralization both at Wandongshan and Bailiancun districts [15,25]. Though the monzogranite porphyries at Wandongshan and Bailiancun districts have similar zircon U-Pb ages (36.2 ± 0.6 Ma and 36.1 ± 0.15 Ma, respectively) [15,17] and the same magma source [22], their mineralization types are different. The Wandongshan monzogranite porphyries form economical porphyry-skarn Fe-Cu-Au mineralization [22–24], while the Bailiancun monzogranite porphyries form hydrothermal Pb-Zn veins [22].

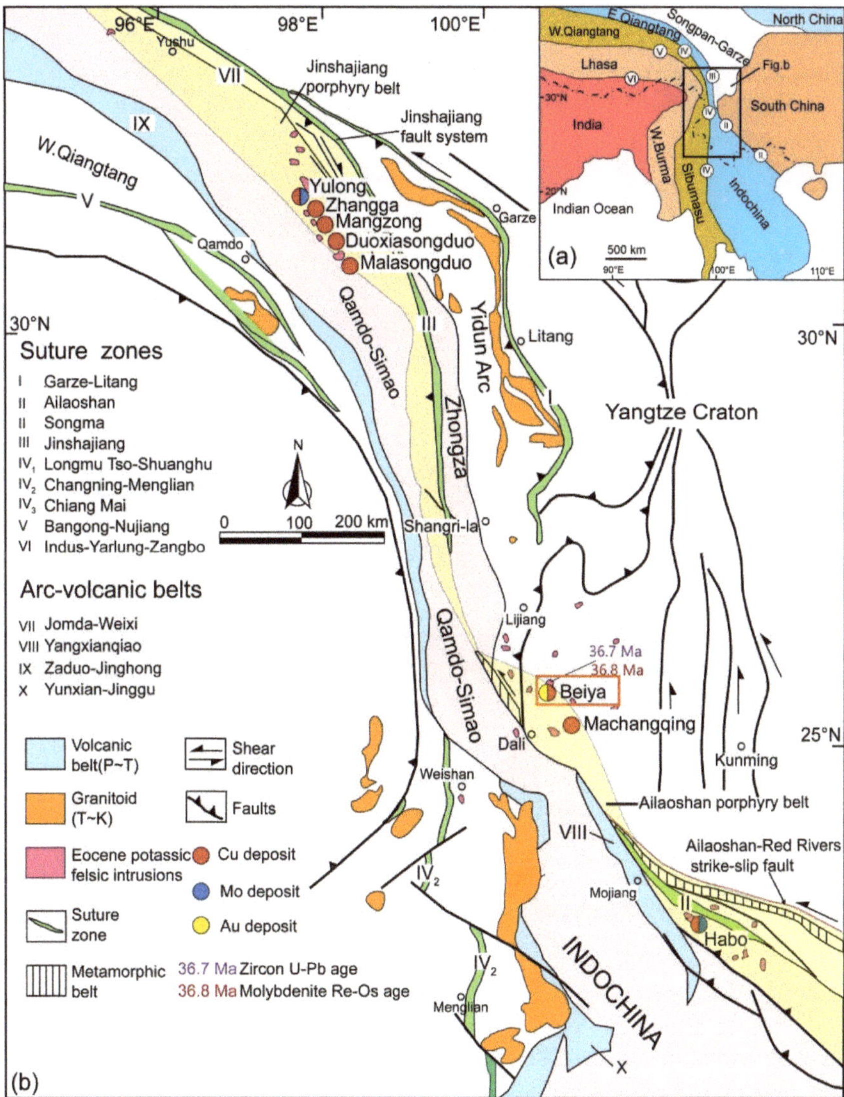

Figure 1. Regional tectonics of the Jinshajiang-Ailaoshan alkaline porphyry belt showing the distribution of Cenozoic deposits [17].

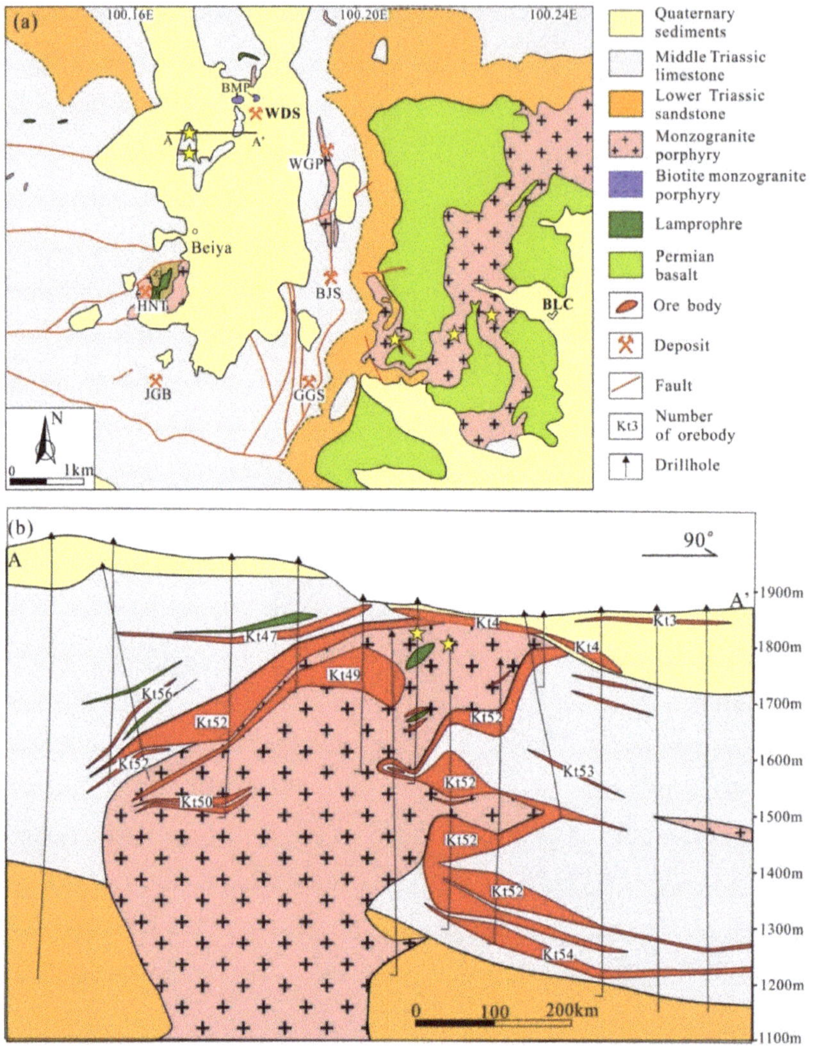

Figure 2. Geology of the Beiya Au deposit (**a**) and representative crosssection of Wandongshan district showing the location of the skarn ore body (**b**) [23]. WDS: Wandongshan district; HNT: Hongnitang district; JGB: Jin'gouba district; WGP: Weiganpo district; GGS: Guogaishan district; BJS: Bijiashan district; BLC: Bailiancun district.

3. Petrography of Alkaline Intrusions

3.1. The Wandongshan Monzogranite Porphyry

The Wandongshan monzogranite porphyry intrusion has a gray-white color and porphyritic textures. Phenocrysts form about 50 vol % of the rock, and include K-feldspar, plagioclase, quartz, biotite, and minor amphiboles (Table 1). The K-feldspar phenocrysts make up about 5–15 vol % and usually have euhedral to subhedral shapes that range in size from 0.2 to 0.5 mm, with occasional Carlsbad twinning (Figure 3a). The plagioclase phenocrysts (1–5 mm) are mostly euhedral in shape, showing tabular biotite inclusions (Figure 3c). The quartz phenocrysts are irregular, with a circular shape. The amphibole phenocrysts are mostly euhedral, usually measuring 0.4 mm in size (Figure 3b). The groundmass comprises fine-grained feldspar and quartz. The accessory minerals are apatite, magnetite, titanite, and zircon. The mineralization is dominated by pyrite and chalcopyrite within quartz veins, as well as disseminated in the monzogranite.

Table 1. Petrography of the monzogranite porphyry from the Wandongshan and Bailiancun ore districts.

Ore Deposits	Rock Type	Phenocrysts	Groundmass	Accessory
Wandongshan	Monzogranite porphyry	Kfs + Qtz + Pl + Bt + Amp	Pl + Qtz	Ap + Zrc + Ttn + Mag
Bailiancun	Monzogranite porphyry	Qtz + Kfs + Pl + Amp	Kfs + Qtz + Amp	Ap + Zrc

Qtz: quartz; Kfs: K-feldspar; Pl: plagioclase; Bt: biotite; Amp: amphibole; Ap: apatite; Zrc: zircon; Ttn: titanite; Mag: magnetite.

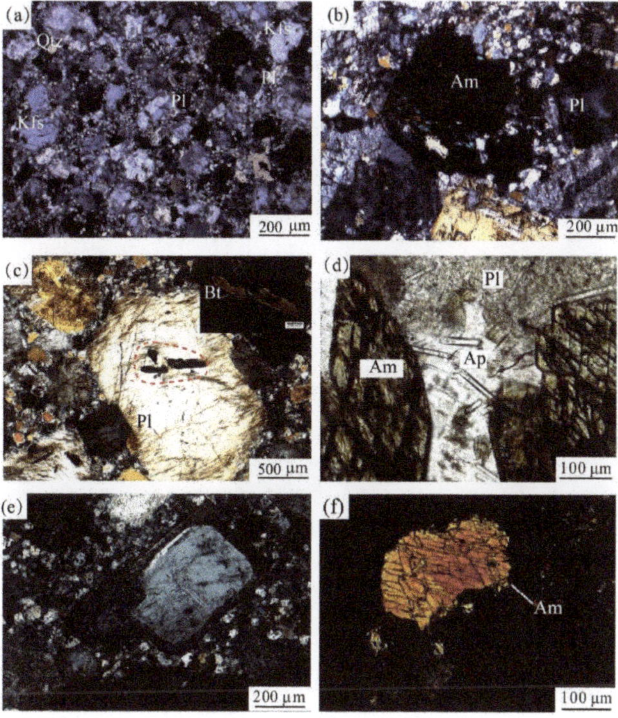

Figure 3. Photomicrographs of the monzogranite porphyries at the Wandongshan and Bailiancun districtdistricts. (**a**) The monzogranite porphyries at the Wandongshan deposit; cross-polarized light; (**b**) Plagioclase and amphibole phenocrysts of the Wandongshan monzogranite porphyries; cross-polarized light; (**c**) Biotite is inside the plagioclase; cross-polarized light; (**d**) Dioritic enclaves from Wangdongshan; cross-polarized light [23]; (**e,f**) are photographs of monzogranite porphyries at the Bailiancun district; cross-polarized light. Qtz: quartz; Kfs: K-feldspar; Pl: plagioclase; Bt: biotite; Am: amphibole; Ap: apatite.

3.2. The Bailiancun Monzogranite Porphyry

The monzogranite porphyry intrusion at the Bailiancun deposit is characterized by porphyritic texture. Phenocrysts consist of K-feldspar, quartz, plagioclase, and amphibole. The groundmass includes K-feldspar, quartz, and amphiboles (Table 1). The accessory minerals are dominated by apatite and zircon. The K-feldspar phenocrysts make up about 30 vol % with a euhedral to subhedral habitus, and measuring 0.4 mm in size (Figure 3e). The quartz phenocrysts have subrounded circular shapes. The plagioclase phenocrysts account for 25 vol % and are mostly euhedral. The amphibole phenocrysts are small (0.3 mm), with euhedral shapes (Figure 3f).

4. Analytical Methods and Results

4.1. Analytical Methods

Electron microprobe analysis was conducted on amphiboles using a JEOL JXA-8800 Superprobe at the Institute of Mineral Resources, Chinese Academy of Geological Sciences. The microprobe was operated at an accelerating voltage of 15 kV, a beam current of 20 nA, and a beam size of 5 μm. Matrix corrections were performed using the ZAF correction program supplied by the instrument manufacturer. Chemical compositions of amphiboles can be used to evaluate the physicochemical conditions (e.g., pressure, temperature, fO_2, and water content) of the parental magma from which the amphibole crystallized [25,36,37].

Zircon trace elements were determined by laser ablation inductively coupled plasma mass spectrometry (LA-ICP-MS) at the Key Laboratory of Crust-Mantle Materials and Environments, Chinese Academy of Sciences (CAS), Hefei. The LA-ICP-MS analyses used a Finnigan Neptune MC-IPS-MS with a high-performance interface coupled with a New-Wave UP-213nm Nd: YAG UV laser. The diameter of the laser ablation pit was 45 μm and the mean power output was 4 W. Signal and background measuring durations were about 70 and 40 s, respectively. Further analytical details are given in Liu et al. [38]. Zircon is a widespread accessory mineral that retains its primary chemical and isotope compositions from the time of crystallization [39], which means that the magma oxygen state during crystallization can be recorded by the zircon composition. Zircon solubility is mainly controlled by the temperature and major oxide concentrations of the magma [40]. Titanium exists as Zr^{4+} and Si^{4+} isotopes, so the Ti content are controlled by the temperature of the zircon crystallization [41–43]. This study uses the Ti content of the zircon crystals in zircon-saturated melts to estimate the temperature of the melt ($\log(Ti) = 6.01 \pm 0.03 - (5080 \pm 30)/T(K)$) [42].

4.2. Mineral Chemistry

4.2.1. Amphibole Compositions

Major element compositions and chemistry of amphibole from the monzogranite porphyry at Wandongshan and Bailiancun districts are reported in Supplementary Materials, Table S1.

The amphiboles from the Wandongshan ore deposit are characterized by high Mg (MgO = 12.46–14.58 wt %), high Ca (CaO = 10.50–12.31 wt %), high Fe (FeO = 13.61–16.29 wt %), but low K (K_2O = 0.33–0.94 wt %) and Na (Na_2O = 0.55–2.18 wt %) content (Table S1). Compared to the Wandongshan amphiboles, the Bailiancun amphiboles are characterized by lower content of Mg (MgO = 9.57–9.90 wt %) and Ca (CaO = 9.54–10.85 wt %), higher content of Fe (FeO = 20.20–21.16 wt %) and K (K_2O = 1.22–1.44 wt %). The amphiboles from the Wandongshan porphyry are all edenites (Figure 4a) with high $Mg/(Mg + Fe^{2+})$ ratios (0.67–0.79) based on the classification of Leake et al. [41], whereas the amphiboles from the Bailiancun porphyry are all silicic edenites with low $Mg/(Mg + Fe^{2+})$ ratios (0.50–0.52; Table S1).

The emplacement pressure of the parental magma at Wandongshan and Bailiancun can be estimated using the Al-in-amphibole geobarometer [25,36]. These results show that the emplacement pressures of Wandongshan porphyries range from 31.5 MPa to 68.6 MPa, which corresponds to a paleodepth of 1.2–2.4 km (average 1.8 km) under lithostatic pressure (Table 2) [36]. In contrast, the emplacement pressures of Bailiancun porphyries are significantly higher (142.3–192.8 MPa), which corresponds to a deeper paleodepth of 5.4–7.3 km (average 6.7 km). Furthermore, water content are directly measure by the component of Wandongshan amphiboles (H_2O_{melt} = 5.215[6] Al* + 12.28) [25]. These results show that the water content of the Wandongshan parent magma exhibit a wider range of 3.05 ± 0.4 wt % to 4.11 ± 0.4 wt %, whereas those of the Bailiancun parent magma have a tight range of 3.46 ± 0.52 wt % to 3.91 ± 0.4 wt % (Table 2, Figure 4b). Moreover, the amphibole composition is affected by the oxidation state of its parental melt, thus they allow the direct analysis of the oxygen state of their parental magmas [25]. The oxidation state deduced from amphiboles of the Wandongshan porphyry range from ΔNNO + 1.00 to ΔNNO + 1.89 (NNO is the nickel-nickel oxide oxygen buffer), slightly higher than that of the Bailiancun porphyries (ΔNNO + 0.39 to ΔNNO + 0.67, Table S1).

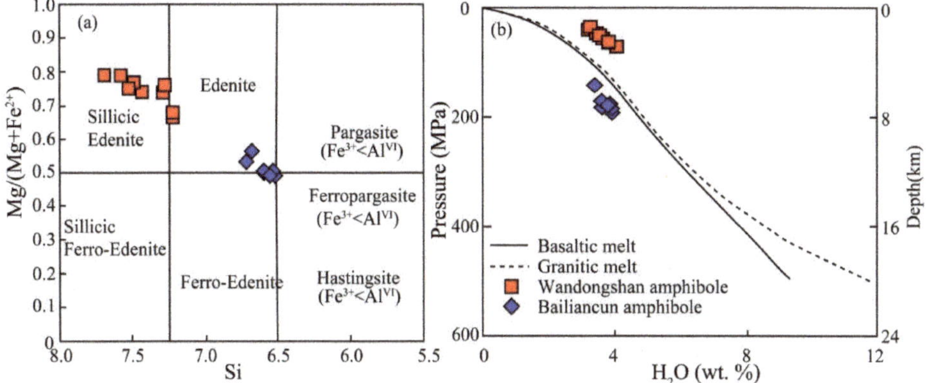

Figure 4. Composition of amphiboles from the Wandongshan and Bailiancun monzogranite porphyries, using the nomenclature of Leake et al. [42] (**a**), and experimentally determined solubilities of H_2O in silicate melts as a function of granitic pegmatite pressure (**b**) [44]. WDS: Wandongshan; BLC: Bailiancun.

Table 2. The water content and oxidation state of the parental magmas from the Wandongshan and Bailiancun districts.

| | Amphibole Phenocryst | | Zircon | | | |
| | | | Oxidation State Index | | | |
Magmatic Geochemistry	Pressure (MPa)	Water Content (wt %)	ΔNNO	ΔFMQ	Ce^{4+}/Ce^{3+}	δCe
Wandongshan	31.5~68.6	3.05~4.11	+1.62	+3.5	12.1~144.6	9.25~203.83
Bailiancun	142.3~192.8	3.46~3.91	+0.5	+1.5	11.5~39.4	11.42~53.49

4.2.2. Zircon Trace Element Composition

Zircons from the Wandongshan monzogranite porphyries have ΣREE ranging from 149.0 to 1096.6 ppm, LREE/HREE (Low Rare Earth Elements/Heavy Rare Earth Elements) ratios varying from 0.01–0.15 (averaging 0.06), Ti content of 2.83–48.3 ppm, Ce anomalies varying from 9.25 to 203.83 (averaging 76.11), and Eu anomalies varying from 0.24 to 0.87 (averaging 0.6). However, zircons from the Bailiancun monzogranite have a lower average LREE/HREE ratio (0.04), average Ce anomaly (30.98), and higher average Eu anomaly (0.7) (Figure 5d, Table S2).

The zircon Ce^{4+}/Ce^{3+} ratios (12.1–144.6, averaging 62.00) of the Wandongshan monzogranite porphyry intrusions are much higher than those of the Bailiancun monzogranite porphyry (11.5–39.4, averaging 23.15; Figure 5d). Moreover, the average ΔFMQ value of Wandongshan monzogranite porphyry is ΔFMQ + 3.5, slightly higher than that of the Bailiancun monzogranite porphyry (ΔFMQ + 1.5; Figure 5a). In addition, the Ti-in zircon temperatures of the Wandongshan monzogranite porphyry intrusions yield a wider range of 670.84–988.42 °C, averaging 778 °C, whereas those of the Beiya porphyry system exhibit a tight range of 680.17–818.53 °C, averaging 770 °C (Table 2).

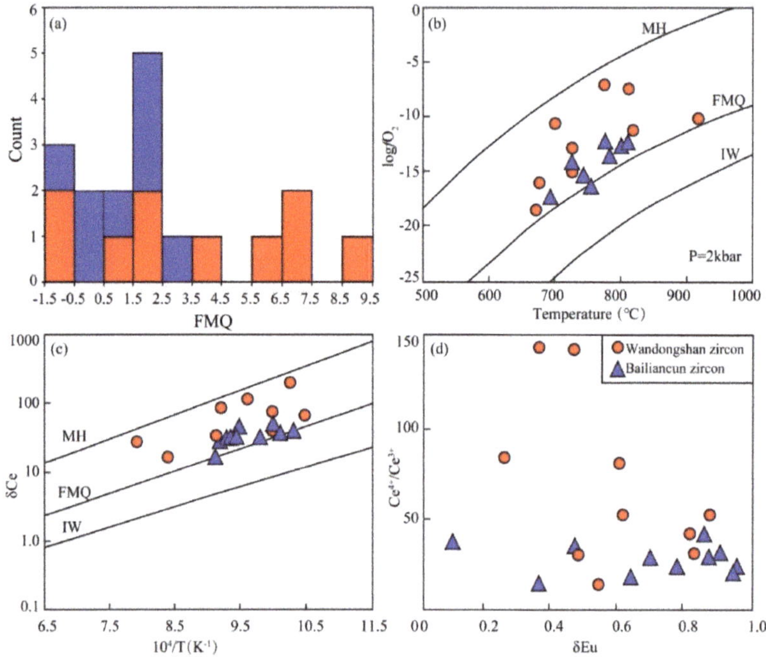

Figure 5. Magmatic oxidation state of intrusions from the Wandongshan and Bailiancun district, estimated from zircon. (**a**) Magma oxidation state of the Wandongshan and Bailiancun porphyries [41]; (**b**) Zircon logfO_2 vs. temperature; (**c**) linear relations for zircon Ce anomaly vs. 10^4/T; (**d**) zircon Ce^{4+}/Ce^{3+} vs. δEu [44]. MH: magnetite-hematite buffer curve; NNO: nickel-nickel oxide buffer curve; FMQ: fayalite-magnetite-quartz buffer curve; IW: iron-wustite.

5. Discussion

As mentioned above, the Wandongshan Au-bearing porphyries and Bailiancun barren porphyries have different physicochemical conditions (e.g., pressure, temperature, fO_2, and water content). It is widely accepted that high oxidation state (high fO_2) and high H_2O content are two key factors responsible for Au mineralization [4,12,13].

5.1. The Water Content of the Parental Magmas

A high magmatic fluid flux ensures the exsolution of an aqueous volatile phase, which is considered as the sine qua non for magmatic hydrothermal ore-forming systems [2]. Both the Wandongshan and Bailiancun monzogranite porphyries contain amphibole phenocrysts. The amphibole phenocrysts at Wandongshan and Bailiancun are mostly euhedral in shape, and are characterized by embayed textures indicative of resorption (Figure 3b,f), suggesting that these amphiboles crystallized early. The presence of amphibole during the early stages of crystallization demonstrates that the parental magma at Wandongshan and Bailiancun districts are hydrous melts (>4 wt % H_2O) at elevated pressures [45,46]. Furthermore, the water content of the Wandongshan parent magma are characterized by a wider range of 3.05 ± 0.4 wt % to 4.11 ± 0.4 wt %, whereas those of the Bailiancun parent magma have a tight range of 3.46 ± 0.52 wt % to 3.91 ± 0.4 wt % (Table 2), which further confirmed the parental magma at Wandongshan and Bailiancun districts are relatively water-rich.

The emplacement pressure of the parental magma at Wandongshan and Bailiancun can be estimated using the Al-in-amphibole geobarometer [25,36]. These results show that the emplacement pressures of Wandongshan porphyries range from 31.5 MPa to 68.6 MPa, but those of Bailiancun porphyries are significantly higher, ranging from 142.3 MPa to 192.8 MPa (Table 2). The solubility of H_2O in silicate magmas is mainly controlled by pressure and to a lesser extent by temperature [45]. The Wandongshan parent magma could reach water saturation at 31.5–70.9 MPa if they contain ~2.5% H_2O (Figure 4b). By contrast, the Bailiancun parental magma would need more water when they reach water saturation at 142.3–192.8 MPa. Therefore, it is expected that the Wandongshan parental magma will achieve water saturation easier, and the fluid exsolution would take place at 1.2–2.4 km, which consistent with the estimated trapping depth for Stage I S-type inclusions (~2 km depth under lithostatic conditions) [17].

5.2. Magmatic Oxidation State

The $\log(fO_2)$ vs. T and Ce anomaly vs. $1/T$ diagrams are divided into several oxygen fugacity fields by curves, which can be used to estimate the oxidation state of the magma. The data points of the Wandongshan monzogranite porphyry are mainly plotted within the field between magnetite-hematite buffer curve (MH) and fayalite-magnetite-quartz buffer curve (FMQ), while the fO_2 of the intrusions from the Bailiancun area are lower than those of the Wandongshan porphyries (Figure 5b,c). This result confirms that the porphyries from the Wandongshan district are more oxidized than those the porphyries from the Bailiancun district.

Furthermore, direct measurement of fO_2 also implies that the Wandongshan porphyries are more oxidized, the fO_2 deduced from amphibole chemistry of the Wandongshan porphyry range from $\Delta NNO + 1.00$ to $\Delta NNO + 1.89$ (NNO is the nickel-nickel oxide oxygen buffer), slightly higher than that of the Bailiancun porphyries ($\Delta NNO + 0.39$ to $\Delta NNO + 0.67$, Figure 6a). This result is consistent with the values calculated by zircon chemistry. The estimated fO_2 using zircon chemistry for the Wandongshan porphyry is $\Delta FMQ + 3.5$, slightly higher than that of the Bailiancun monzogranite porphyry ($\Delta FMQ + 1.5$; Figure 5a). Moreover, the Wandongshan monzogranite porphyries have an average zircon Ce^{4+}/Ce^{3+} ratio of 62.00 (Table 2), and an average zircon Ce anomaly of 76.11 (Figure 5c), much higher than those of the Bailiancun monzogranite porphyry (23.15, 30.98, respectively). Those differences above indicate the formation of Au-bearing porphyries requires elevated fO_2.

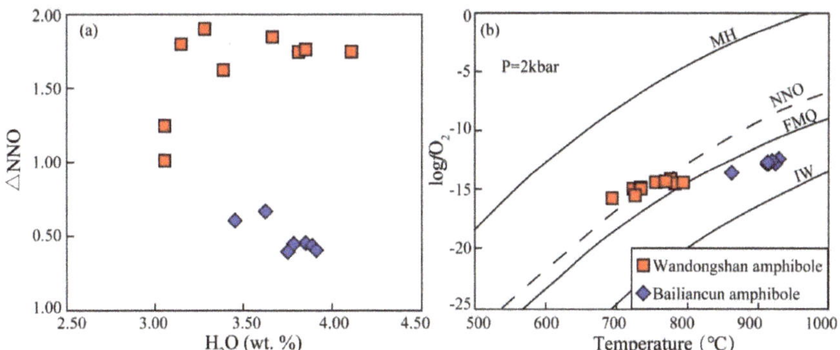

Figure 6. Magmatic oxidation state of intrusions in the Wandongshan and Bailiancun districts, estimated from amphibole compositions. (**a**) Magma oxidation state (ΔNNO) of Wandongshan and Bailiancun porphyries; (**b**) magmatic fO_2 ($\log fO_2$) vs. magmatic temperature. MH: magnetite-hematite buffer curve; NNO: nickel-nickel oxide buffer curve; FMQ: fayalite-magnetite-quartz buffer curve; IW: iron-wustite.

5.3. Implications for Au Mineralization

Beiya is one of the largest Au deposits in China, with by-products Cu, Fe, Ag, Pb, and Zn [23–25]. Au and other ore metals were derived from the alkaline melt as documented by Yang et al. [47], and the mineralized fluids at Beiya were mainly derived from magmatic water [16,17]. Those indicate that the Beiya alkaline melt provides ore-forming fluids and metallic elements for the formation of porphyry Au mineralization [16].

In contrast to Bailiancun porphyry, the Wandongshan porphyries are characterized by higher water content and higher fO_2, but lower emplacement pressure (Table 2, Figure 7). Au solubility was investigated as a function of fO_2 in sulfur-saturated silicate melts [12], indicating Au solubility in sulfur-free silicate melt increases with increasing of fO_2 [48]. In addition, high fO_2 favors sulfur as sulfate dissolved in the evolved magmas and suppresses magmatic sulfide precipitation [4,7], which can cause sequestration of metals before they partition into the aqueous phase [10]. On the other hand, a high H_2O content and a low pressure results in the Wandongshan parental magma reaching water saturation easier, because water solubility increases with increasing pressure [44], hence ensuring exsolution of aqueous volatile phases and the partitioning of Au between sulfides and silicate melt efficiently [12,49,50]. Therefore, melts that are more enriched in water and with a high oxidation state will be more fertile to generate Au mineralization.

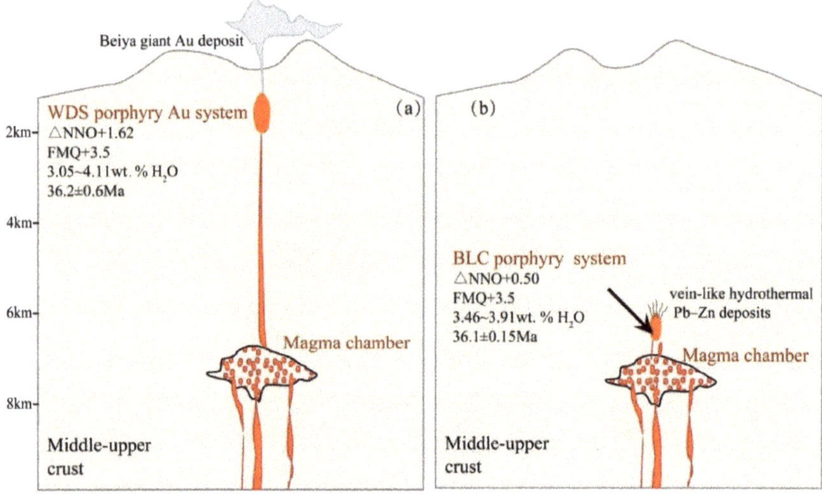

Figure 7. Cartoons illustrating the different water content, magmatic oxidation state, and emplacement depths of the parental magmas at Wandongshan (**a**) and Bailiancun (**b**), respectively. Previous studies show that the parent magma at Wandongshan (**a**) and Bailiancun (**b**) districts are derived from the same magma source [22]. This study reveals that the magma of Wandongshan experienced fluid saturation at about 1.19–2.68 km, while the magma of Wandongshan experienced fluid saturation at about 6.48–7.28 km. The water content and magmatic oxidation state of Wandongshan parental magmas are both higher than those of the Bailiancun parental magmas.

6. Conclusions

Chemical compositions of zircons and amphiboles are used to evaluate the physicochemical conditions (e.g., pressure, temperature, fO_2, and water content) of the parental magma. We found the water content of the Wandongshan parent magma ($\leq 4.11 \pm 0.4$ wt %) are slightly higher than those of the parent magma at Bailiancun ($\leq 3.91 \pm 0.4$ wt %). Furthermore, the fO_2 deduced from amphibole chemistry of the Wandongshan porphyry range from $\Delta NNO + 1.00$ to $\Delta NNO + 1.89$, slightly higher than that of the Bailiancun porphyries ($\Delta NNO + 0.39$ to $\Delta NNO + 0.67$). This result is consistent with the values calculated by zircon chemistry. The Wandongshan porphyry has higher zircon Ce^{4+}/Ce^{3+} ratio (average of 62) and high fO_2 value (average of $\Delta FMQ = +3.5$) than those of Bailiancun porphyry (average of $Ce^{4+}/Ce^{3+} = 23.15$; $\Delta FMQ = +1.5$). Obviously, the Wandongshan ore-related porphyry has higher water content and more oxidized than the Bailiancun barren porphyry. Therefore melts that are more enriched in water and with a high oxidation state will be more fertile to form an economic hydrothermal system especially when emplaced at shallow levels.

Supplementary Materials: The following are available online at http://www.mdpi.com/2075-163X/8/10/441/s1, Table S1: The compositions of amphiboles from the monzogranite porphyries at Wandongshan and Bailiancun, respectively, Table S2: The Zircon trace elements in the monzogranite porphyry from Wandongshan and Bailiancun districts.

Author Contributions: X.B., L.Y. and X.G. conceived and designed the ideas; X.B. performed the experiments; X.B. and W.H. analyzed the data; X.B. prepared the original draft; X.B., L.Y., W.H. and X.G. reviewed and edited the draft.

Funding: This study was financially supported by the National Basic Research Program of China (Grant No. 2015CB452605 and 2009CB421008), the Natural Science Foundation of China (41602089), the Geological investigation work project of China Geological Survey (Grant No. 12120114013501), 111 Project of the Ministry of Education, China (Grant No. B07011) and China Postdoctoral Science Foundation (2015M581143).

Acknowledgments: We would like to thank Zhonghua He, Rui Yang and Congming Wang at Yunnan Au Corporation for their help during field work, as well as Ruigang Zhang and Mengmeng Li for their assistance with the data collection. We are grateful for constructive comments by three reviewers.

Conflicts of Interest: The authors declare no conflict of interest.

References

1. Sillitoe, R.H. Porphyry copper systems. *Econ. Geol.* **2010**, *105*, 3–41. [CrossRef]
2. Richards, J.P. High Sr/Y arc magmas and porphyry Cu-Mo-Au deposits: Just add water. *Econ. Geol.* **2011**, *106*, 1075–1081. [CrossRef]
3. Robb, L. *Introduction to Ore-Forming Processes*; Blackwell Publishing: Hoboken, NJ, USA, 2005; pp. 1–386.
4. Sun, W.D.; Arculus, R.J.; Kamenetsky, V.S.; Binns, R.A. Release of gold-bearing fluids in convergent margin magmas prompted by magnetite crystallization. *Nature* **2004**, *431*, 975–978. [CrossRef] [PubMed]
5. Richards, J.P. The oxidation state, and sulfur and Cu contents of arc magmas: Implications for metallogeny. *Lithos* **2015**, *233*, 27–45. [CrossRef]
6. Zhou, T.C.; Zeng, Q.D.; Chu, S.X.; Zhou, L.L.; Yang, Y.H. Magmatic oxygen fugacities of porphyry Mo deposits in the East Xing'an-Mongolian Orogenic Belt (NE China) with metallogenic implications. *J. Asian Earth Sci.* **2018**, in press. [CrossRef]
7. Yang, Z.; Yang, L.Q.; He, W.Y.; Gao, X.; Liu, X.D.; Bao, X.S.; Lu, Y.G. Control of magmatic oxidation state in intracontinental porphyry mineralization: A case study from Cu (Mo-Au) deposits in the Jinshajiang–Red River metallogenic belt, SW China. *Ore Geol. Rev.* **2017**, *90*, 827–846. [CrossRef]
8. Kelley, K.A.; Cottrell, E. Water and the oxidation state of subduction zone magmas. *Science* **2009**, *325*, 605–607. [CrossRef] [PubMed]
9. Yang, Z.M.; Lu, Y.J.; Hou, Z.Q.; Chang, Z.S. High-Mg diorite from Qulong in southern Tibet: Implications for the genesis of adakite-like intrusions and associated porphyry Cu deposits in collisional orogens. *J. Petrol.* **2015**, *56*, 227–254. [CrossRef]
10. Richards, J.P.; Spell, T.; Rameh, E.; Razique, A.; Fletcher, T. High Sr/Y magmas reflect arc maturity, high magmatic water content, and porphyry Cu-Mo-Au potential: Examples from the Tethyan arcs of Central and Eastern Iran and Western Pakistan. *Econ. Geol.* **2012**, *107*, 295–332. [CrossRef]
11. Yang, L.Q.; Deng, J.; Guo, L.N.; Wang, Z.L.; Li, X.Z.; Li, J.L. Origin and evolution of ore fluid, and Au deposition processes at the giant Taishang Au deposit, Jiaodong Peninsula, eastern China. *Ore Geol. Rev.* **2016**, *72*, 582–602. [CrossRef]
12. Li, Y.; Audetat, A. Gold solubility and partitioning between sulfide liquid, monosulfide solid solution and hydrous mantle melts: Implications for the formation of Au-rich magmas and crust-mantle differentiation. *Cosmochim. Acta* **2013**, *118*, 247–262. [CrossRef]
13. Zajacz, Z.; Candela, P.A.; Piccoli, P.M.; Walle, M.; Sanchez-Valle, C. Gold and copper in volatile saturated mafic to intermediate magmas: Solubilities, partitioning, and implications for ore deposit formation. *Geochem. Cosmochim. Acta* **2012**, *91*, 140–159. [CrossRef]

14. Deng, J.; Wang, Q.F. Gold mineralization in China: Metallogenic provinces, deposit types and tectonic framework. *Gondwana Res.* **2016**, *36*, 219–274. [CrossRef]

15. He, W.Y.; Yu, X.H.; Mo, X.X.; He, Z.H.; Li, Y.; Huang, X.K.; Su, G.S. Genetic types and the relationship between alkali-rich intrusion and mineralization of Beiya Au-polymetallic ore field, western Yunnan province, China. *Acta Petrol. Sin.* **2012**, *5*, 1401–1412. (In Chinese)

16. He, W.Y. The Beiya Giant Au-Polymetallic Deposit: Magmatism and Metallogenic Model. Ph.D. Thesis, China University of Geosciences (Beijing), Beijing, China, 2014. (In Chinese)

17. He, W.Y.; Yang, L.Q.; Brugger, J.; McCuaig, T.C.; Lu, Y.J.; Bao, X.S.; Gao, X.Q.; Lu, Y.G.; Xing, Y.L. Hydrothermal evolution and ore genesis of the Beiya giant Au polymetallic deposit, western Yunnan, China: Evidence from fluid inclusions and H-O-S-Pb isotopes. *Ore Geol. Rev.* **2017**, *90*, 847–862. [CrossRef]

18. Xu, X.W.; Cai, X.P.; Xiao, Q.B. Porphyry Cu-Au and associated polymetallic Fe-Cu-Au deposits in the Beiya area, western Yunnan province, South China. *Ore Geol. Rev.* **2007**, *31*, 224–246. [CrossRef]

19. Bao, X.S.; He, W.Y.; Gao, X. The Beiya Gold deposit: Constrainition of water-rich magmas to mineralization. *Acta Petrol. Sin.* **2017**, *33*, 2175–2188. (In Chinese)

20. Gao, X.Q.; He, W.Y.; Gao, X.; Bao, X.S.; Yang, Z. Constraints of magmatic oxidation state on mineralization in the Beiya alkali-rich porphyry Au deposit, western Yunnan, China. *Solid Earth Sci.* **2017**, *2*, 65–78. [CrossRef]

21. Meng, X.Y.; Mao, J.W.; Zhang, C.Q.; Zhang, D.Y.; Liu, H. Melt recharge, fO_2-T conditions, and metal fertility of felsic magmas: Zircon trace element chemistry of Cu-Au porphyries in the Sanjiang orogenic belt, southwest China. *Miner. Depos.* **2018**, *53*, 649–663. [CrossRef]

22. Liu, B.; Liu, H.; Zhang, C.Q.; Mao, Z.H.; Zhou, Y.M.; Huang, H.; He, Z.H.; Su, G.S. Geochemistry and geochronology of porphyries from the Beiya Au-polymetallic orefield, western Yunnan, China. *Ore Geol. Rev.* **2015**, *69*, 360–379. [CrossRef]

23. He, W.Y.; Mo, X.X.; Yang, L.Q.; Xing, Y.L.; Dong, G.C.; Yang, Z.; Gao, X.; Bao, X.S. Origin of the Eocene porphyries and mafic microgranular enclaves from the Beiya porphyry Au polymetallic deposit, western Yunnan, China: Implications for magma mixing/mingling and mineralization. *Gondwana Res.* **2016**, *40*, 230–248. [CrossRef]

24. He, W.Y.; Mo, X.X.; He, Z.H.; White, N.C.; Chen, J.B.; Yang, K.H.; Wang, R.; Yu, X.H.; Dong, G.C.; Huang, X.F. The Geology and mineralogy of the Beiya Skarn Au Deposit in Yunnan, Southwest China. *Econ. Geol.* **2015**, *110*, 1625–1641. [CrossRef]

25. Ridolfi, F.; Renzulli, A.; Puerini, M. Stability and chemical equilibrium of amphibole in calc-alkaline magmas: An overview, new thermobarometric formulations and application to subduction-related volcanoes. *Contrib. Mineral. Petrol.* **2010**, *160*, 45–66. [CrossRef]

26. Yang, L.Q.; He, W.Y.; Gao, X.; Xie, S.X.; Yang, Z. Mesozoic multiple magmatism and porphyry–skarn Cu-polymetallic systems of the Yidun Terrane, Eastern Tethys: Implications for subduction-and transtension-related metallogeny. *Gondwana Res.* **2018**, *62*, 144–162. [CrossRef]

27. Deng, J.; Wang, Q.F.; Li, G.J.; Santosh, M. Cenozoic tectono-magmatic and metallogenic processes in the Sanjiang region, southwestern China. *Earth Sci. Rev.* **2014**, *138*, 268–299. [CrossRef]

28. Deng, J.; Liu, X.F.; Wang, Q.F.; Dilek, K.; Yang, L.Q. Isotopic characterization and petrogenetic modeling of early cretaceous mafic diking-lithospheric extension in the North China craton, eastern Asia. *GSA Bull.* **2017**, *129*, 1379–1407. [CrossRef]

29. Deng, J.; Wang, Q.F.; Li, G.J. Tectonic evolution, superimposed orogeny, and composite metallogenic system in China. *Gondwana Res.* **2017**, *50*, 216–266. [CrossRef]

30. Yang, L.Q.; Deng, J.; Qiu, K.F.; Ji, X.Z.; Santosh, M.; Song, K.R.; Song, Y.H.; Geng, J.Z.; Zhang, C.; Hua, B. Magma mixing and crust-mantle interaction in the Triassic monzogranites of Bikou Terrane, central China: Constraints from petrology, geochemistry, and zircon U-Pb-Hf isotopic systematics. *J. Asian Earth Sci.* **2015**, *98*, 320–341. [CrossRef]

31. Tapponnier, P.; Lacassin, R.; Leloup, P.H.; Shharer, U.; Zhong, D.L.; Wu, H.X.; Liu, X.H.; Ji, S.C.; Zhang, L.S.; Zhong, J.Y. The Ailaoshan-Red River metamorphic belt: Tertiary left-lateral shear between Indochina and South China. *Nature* **1990**, *343*, 431–437. [CrossRef]

32. Deng, J.; Yang, L.Q.; Li, R.H.; Groves, D.I.; Santosh, M.; Wang, Z.L.; Sai, S.X.; Wang, S.R. Regional structural control on the distribution of world-class gold deposits: An overview from the Giant Jiaodong Gold Province, China. *Geol. J.* **2018**, *3*. [CrossRef]

33. Yang, L.Q.; Deng, J.; Wang, Z.L.; Zhang, L.; Aufarb, R.J.; Yuan, W.M.; Weinberg, R.F.; Zhang, R.Z. Thermochronologic constraints on evolution of the Linglong Metamorphic Core Complex and implications for Au mineralization: A case study from the Xiadian Au deposit, Jiaodong Peninsula, eastern China. *Ore Geol. Rev.* **2016**, *72*, 165–178. [CrossRef]

34. Deng, J.; Wang, Q.F.; Li, G.J.; Li, C.S.; Wang, C.M. Tethys tectonic evolution and its bearing on the distribution of important mineral deposits in the Sanjiang region, SW China. *Gondwana Res.* **2014**, *26*, 419–437. [CrossRef]

35. Deng, J.; Wang, C.M.; Li, G.J. Style and process of the superimposed mineralization in the Sanjiang Tethys. *Acta Petrol. Sin.* **2012**, *28*, 1349–1361. (In Chinese)

36. Schmidt, M.W. Amphibole composition in tonalite as a function of pressure; an experimental calibration of the Al-in-hornblende barometer. *Contrib. Mineral. Petrol.* **1992**, *110*, 304–310. [CrossRef]

37. Liu, Y.S.; Hu, Z.C.; Gao, S.; Günther, D.; Xu, J.; Gao, C.G.; Chen, H.H. In situ analysis of major and trace elements of anhydrous minerals by LA-ICP-MS without applying an internal standard. *Chem. Geol.* **2008**, *257*, 34–43. [CrossRef]

38. Cherniak, D.J.; Watson, E.B.; Hanchar, J.M. Rare-earth diffusion in zircon. *Chem. Geol.* **1997**, *134*, 289–301. [CrossRef]

39. Loader, M.A.; Wilkinson, J.J.; Armstrong, R.N. The effect of titanite crystallisation on Eu and Ce anomalies in zircon and its implications for the assessment of porphyry Cu deposit fertility. *Earth Planet. Sci. Lett.* **2017**, *472*, 107–119. [CrossRef]

40. Watson, E.B.; Wark, D.A.; Thomas, J.B. Crystallization thermome-ters for zircon and rutile. *Contrib. Mineral. Petrol.* **2006**, *151*, 413–433. [CrossRef]

41. Leake, B.E.; Woolley, A.R.; Arps, C.E.S.; Birch, W.D.; Gilbert, M.C.; Grice, J.D.; Hawthorne, F.C.; Kato, A.; Kisch, H.J.; Krivovichev, V.G.; et al. Nomenclature of amphiboles: Report of the subcommittee on amphiboles of the International Mineralogical Association Commission on New Minerals and Mineral Names. *Can. Mineral.* **1997**, *35*, 219–246.

42. Ballard, J.R.; Palin, J.M.; Campbell, I.H. Relative oxidation states of magmas inferred from Ce(IV)/Ce(III) in zircon: Application to porphyry copper deposits of northern Chile. *Contrib. Mineral. Petrol.* **2002**, *144*, 347–364. [CrossRef]

43. Qiu, J.T.; Yu, X.Q.; Santosh, M.; Zhang, D.H.; Chen, S.Q.; Li, P.J. Geochronology and magmatic fO_2 of the Tongcun molybdenum deposit, northwest Zhejiang, SE China. *Miner. Depos.* **2013**, *48*, 545–556. [CrossRef]

44. Burnham, C.W. Magmas and hydrothermal fluids. In *Geochemistry of Hydrothermal Ore Deposits*, 2nd ed.; Wiley: Hoboken, NJ, USA, 1979; pp. 71–136.

45. Rohrlach, B.D.; Loucks, R.R.; Porter, T.M. Multi-million-year cyclic ramp-up of volatiles in a lower crustal magma reservoir trapped below the tampakan Cu-Au deposit by mio-pliocene crustal compression in the southern philippines. *Super Porphyry Copp. Gold Depos.* **2005**, *2*, 369–407.

46. Whitney, J.A.; Naldrett, A.J. Ore deposition associated with magmas. *Econ. Geol.* **1989**, *4*, 250.

47. Yang, L.Q.; Deng, J.; Dilek, Y.; Meng, J.Y.; Gao, X.; Santosh, M.; Wang, Da.; Yan, H. Melt source and evolution of I-type granitoids in the SE Tibetan Plateau: Late Cretaceous magmatism and mineralization driven by collision-induced transtensional tectonics. *Lithos* **2016**, *245*, 258–273. [CrossRef]

48. Bell, A.S.; Simon, A.; Guillong, M. Gold solubility in oxidized and reduced, water-saturated mafic melt. *Geochim. Cosmochim. Acta* **2011**, *75*, 1718–1732. [CrossRef]

49. Yang, L.Q.; Deng, J.; Wang, Z.L.; Guo, L.N.; Li, R.H.; Groves, D.I.; Danyushevsky, L.V.; Zhang, C.; Zheng, X.L.; Zhao, K. Relationships between Au and pyrite at the Xincheng Au deposit, Jiaodong Peninsula, China: Implications for Au source and deposition in a brittle epizonal environment. *Econ. Geol.* **2016**, *111*, 105–126. [CrossRef]

50. Humphreys, M.C.; Brooker, R.A.; Fraser, D.G.; Burgisser, A.; Mangan, M.T.; McCammon, C. Coupled interactions between volatile activity and Fe oxidation state during arc crustal processes. *J. Petrol.* **2015**, *56*, 795–814. [CrossRef]

Article

Metal Sources of World-Class Polymetallic W–Sn Skarns in the Nanling Range, South China: Granites versus Sedimentary Rocks?

Weicheng Jiang [1], Huan Li [2,*], Noreen J. Evans [3], Jinghua Wu [1] and Jingya Cao [4]

1   Department of Resources Science and Engineering, Faculty of Earth Resources, China University of Geosciences, Wuhan 430074, China; wcjiang@cug.edu.cn (W.J.); jhwu@cug.edu.cn (J.W.)
2   Key Laboratory of Metallogenic Prediction of Nonferrous Metals and Geological Environment Monitoring, Ministry of Education, School of Geosciences and Info-Physics, Central South University, Changsha 410083, China
3   School of Earth and Planetary Science, John de Laeter Centre, TIGeR, Curtin University, Bentley, WA 6845, Australia; noreen.evans@curtin.edu.au
4   CAS Key Laboratory of Crust–Mantle Materials and Environments, School of Earth and Space Sciences, University of Science and Technology of China, Hefei 230026, China; jingyacao@126.com
*   Correspondence: lihuan@csu.edu.cn

Received: 25 May 2018; Accepted: 20 June 2018; Published: 24 June 2018

Abstract: Widespread, large-scale polymetallic W–Sn mineralization occurs throughout the Nanling Range (South China) dated 160–150 Ma, and related to widely developed coeval granitic magmatism. Although intense research has been carried out on these deposits, the relative contribution of ore-forming elements either from granites or from surrounding strata is still debated. In addition, the factors controlling the primary metallogenic element in any given skarn deposit (e.g., W-dominated or Sn-dominated) are still unclear. Here, we select three of the most significant skarn-deposits (i.e., Huangshaping W–Mo–Sn, Shizhuyuan W–Sn–Mo–Bi and Xianghualing Sn), and compare their whole-rock geochemistry with the composition of associated granites and strata. The contents of Si, Al and most trace elements in skarns are controlled by the parent granite, whereas their Fe, Ca, Mg, Mn, Ti, Sr and REE patterns are strongly influenced by the wall rock. Samples from the Huangshaping skarn vary substantially in elemental composition, probably indicating their varied protoliths. Strata at the Shizhuyuan deposit exerted a strong control during metasomatism, whereas this occurred to a lesser degree at Huangshaping and Xianghualing. This correlates with increasing magma differentiation and increasing reduction state of granitic magmas, which along with the degree of stratigraphic fluid circulation, exert the primary control on dominant metallogenic species. We propose that wall rock sediments played an important role in the formation of W–Sn polymetallic mineralization in South China.

Keywords: W–Sn skarn; whole-rock geochemistry; Huangshaping; Shizhuyuan; Xianghualing

---

1. Introduction

Most skarn deposits are formed by the interaction between shallow magmatic systems and carbonate rocks [1–4]. Since the 1990s, the trace element (and particularly rare earth element; REE) characteristics of skarn-associated plutons and their constituent minerals have been the focus of much research given that these signatures reveal much about deposit genesis [5–10]. Major and accessory mineral phases in skarns (such as garnet and scheelite) inherit their REE patterns from magmatic fluids and they control the REE signatures of the host skarn [11]. However, Giuliani et al. [12] propose that strong REE variations in skarns, characterized by a pronounced negative Eu anomaly and low

La/Yb ratios, result from the interaction between scheelite and the surrounding metamorphic rocks. In addition, REE signatures are affected by the distance of the skarn from the related intrusive body, with REE concentrating in skarns during prograde alteration and migrating outward of the skarns during retrograde alteration [13]. Furthermore, Alirezaei et al. [8] note that skarns vary in texture from massive to banded and that textural contrast is reflected in their whole rock composition: Fe, Si and S are significantly enriched, and Na, large ion lithophile elements (LILE) and light REE (LREE) are strongly depleted in the massive skarns while Na, K, Si, and S are slightly enriched in the banded skarns. Siesgesmund et al. [14] recently note that LILE (K, Rb, Ba, Sr) and REE are moderately depleted and high field strength elements (HFSE) are slightly enriched in Cu–Au skarns, relative to the intrusive rocks. Correlations between major and trace elements in skarns are weak or absent, but skarn REE patterns can be used as a prospectivity indicator in a metallogenic system [15]. However, it is still confused whether the chemical composition of ore-related skarns is correlated with the size of relevant deposit and the primary metallogenic elements enriched in skarns. Determining whether the composition of skarns can be used to indicate the metal source of skarn-type deposits requires further research.

The Nanling Range (Figure 1a), located in the central part of South China, represents the largest W–Sn ore province in the world. More than 30 Mesozoic granite-related skarn-type polymetallic deposits are distributed across this region, represented by the large Shizhuyuan, Xianghualing, Huangshaping, Xintianling, Furong, Baoshan and Yaogangxian deposits (Figure 1b) [16–20]. These deposits mainly comprise nonferrous, rare and base metal elements, such as W, Sn, Mo, Bi, Li, Be, Nb, Ta, Cu, Au, Ag, Pb, Zn, U and REE [20–27]. Given the economic significance of these deposits, a great deal of research has focused on skarn minerals, including the study of mineralogy, geochemistry, geochronology, formation temperature and pressure, and fluid and ore associations [28–37]. It is traditionally believed that extensive Mesozoic magmatism is one of the primary factors responsible for the formation of coeval polymetallic deposits [9,16,34,38–40], and that ore-forming elements are mainly derived from parent granitic plutons with minor input from the surrounding strata [11,28,33,34,41–43]. However, the strata contribution to the ore element budget may previously have been underestimated. Chen et al. [44] propose that ore-forming metals such as W and Sn are extracted from country rocks by high temperature circulating water and metamorphic fluids released from deep crustal rocks. In addition, systematic and comprehensive studies comparing the behavior of major and trace elements in different types of skarn-related bodies (granites and host strata) at the time of ore formation are lacking. Understanding the behavior of elements during metasomatism, in both the pluton and wallrock, would elucidate metal sources and enhance current exploration models.

In this paper, the whole rock major and trace element geochemistry of three typical Nanling Range skarn-type deposits with different primary metal enhancements are studied (Huangshaping W–Mo–Sn, Shizhuyuan W–Sn–Mo–Bi and Xianghualing Sn deposits). By comparing the chemical composition of these skarns, the related granites and the sedimentary host rocks, this study aims to elucidate the controls on the nature of the metals and deposit size and providing new insights into the genesis of large-scale polymetallic skarn-type deposits in the Nanling Range.

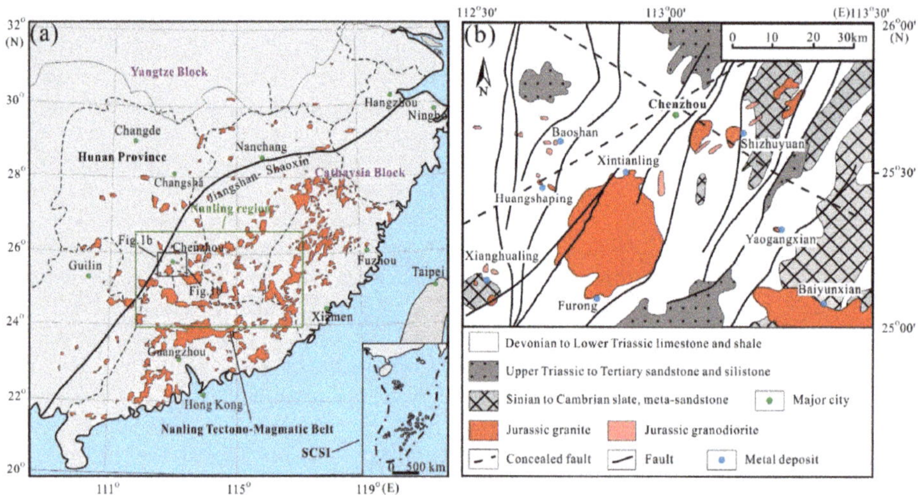

Figure 1. Regional geological map. (**a**) Simplified geological sketch map showing the distribution of Jurassic granitoids in South China (after [24]); (**b**) Geological sketch map of W–Sn polymetallic deposits in the central Nanling Range, southern Hunan (modified from [22]).

2. Geological Setting

2.1. Regional Geology

The South China Block comprises by two major terranes: the Yangtze block in the northwest and the Cathaysia block in the southeast (Figure 1a) [45], amalgamated along the Jiangnan Orogen during the Neoproterozoic [46–51].

The Nanling Range is in the center of the South China Block (Figure 1a). Geographically, it contains most of the bordering areas of the Hunan, Jiangxi, Fujian, Guangdong and Guangxi provinces. Mesozoic (especially Jurassic) granitoids, characterized by calc-alkaline A-type granites, are distributed throughout the area [39,52,53]. The sedimentary rocks within the Nanling Range can be divided into three groups: (1) Late Proterozoic basement composed of thick-bedded (metamorphic) detrital sedimentary rocks that are overlain by (2) Devonian–Triassic carbonate rocks and marls intercalated with clastic rock and then by (3) Jurassic–Cretaceous detrital and volcanic rocks deposited in rifted basins [17,54]. The Nanling Range has been affected by multiple episodes of tectonic activity since the Phanerozoic [55,56]. The structural framework of the Nanling Range was primarily controlled by both the Tethyan tectonic domain and the Indosinian orogenesis pre-Jurassic, but later overprinted by Pacific plate tectonism and intracontinental deep structures [57,58]. A three-stage Mesozoic tectonic evolution is proposed by Shu et al. [59] based on sedimentary basin analysis: (1) Late Triassic–Early Jurassic post-orogenic stage; (2) Middle Jurassic tectonic rifting stage and (3) Cretaceous intracontinental extension and faulted-depression stage.

2.2. Deposit Geology

2.2.1. The Huangshaping W–Mo–Sn Polymetallic Skarn Deposit

The Huangshaping deposit, located southwest of Chenzhou City (Figure 1b), is one of the most important W–Mo–Sn polymetallic skarn deposits in the Hunan province. The outcropping strata in this area are mainly clastic and carbonate rocks, consisting of the Upper Devonian Shetianqiao and Xikuangshan Formations and Lower Carboniferous Ceshui, Menggong'ao, Shidengzi and Zimenqiao Formations, respectively (Figure 2a). Among them, the Shidengzi limestones are the richest host rocks, followed by the Ceshui sandstones and shales and the Zimenqiao dolomite. There are several episodes of magmatic activity in the Huangshaping area between 180 and 150 Ma. Among them, the granite porphyry, the latest and most evolved intrusive phase, is regarded as the most important metal source and host rock of the skarns [60–62].

At Huangshaping, the dominant mineralization type hosted by the granite porphyry is W–Mo–Sn skarn-type ore (WO_3 = 0.15 Mt and its grade = 0.2%; Mo = 0.043 Mt and its grade = 0.06%) [24,61,62]. The ore is mainly located between granite porphyry and carbonate rocks and along faults (Figures 3a and 4a,b). The skarn-forming metasomatism can be further subdivided into two stages in terms of the mineral assemblages: the early stage prograde skarn contains garnet, diopside and fluorite while the late stage retrograde skarn is dominated by tremolite and magnetite [35,63].

2.2.2. The Shizhuyuan W–Sn–Mo–Bi Polymetallic Skarn Deposit

The Shizhuyuan deposit is located to the southeast of Chenzhou City (Figure 1b), and is often referred to as the "nonferrous metal museum" due to the variety of mineralizing elements. The Middle-Upper Devonian Qiziqiao and Shetianqiao limestones and dolomitic limestones are the most favorable hosts of metal mineralization at Shizhuyuan (Figure 2b). The formation of the large Shizhuyuan deposit is related to the emplacement of the Qianlishan granite complex, with significant metal reserves of W (0.71 Mt with grade of 0.31%), Sn (0.48 Mt with grade of 0.14%), Mo (0.12 Mt with grade of 0.054%) and Bi (0.27 Mt with grade of 0.11%) (unpublished data from [64]). The Qianlishan complex is characterized by a small outcrop area (9.7 km^2) and multistage magmatic-hydrothermal activity from 158 to 151 Ma [44,65–68]. Three main episodes of magmatism can be distinguished in chronological order: (1) porphyritic biotite granite; (2) equigranular biotite granite and (3) granite porphyry [39,67–69]. Among them, the equigranular biotite granite is believed to have the closest relationship to the skarn-forming metasomatism [28].

The skarn mineralization at Shizhuyuan can be divided into massive-type and vein-type based on their textures, and vesuvianite-wollastonite skarn (ore-free) and garnet-diopside skarn (ore-bearing) by mineral assemblages, respectively [25,70,71]. Skarn mineralization can also be divided into two stages: (1) early stage massive W–Sn–Mo–Bi (Figures 3b and 4c,d); and (2) late stage vein-type W–Sn–Mo–Bi–Be–Pb–Zn–Ag [70].

2.2.3. The Xianghualing Sn Polymetallic Skarn Deposit

The Xianghualing deposit is located about 60 km from Chenzhou City. The exposed strata in the Xianghualing area comprise Cambrian to Quarternary sequences, with no Ordovician and Silurian strata present. The main ore-hosting layers are Cambrian slates and meta-sandstones, as well as Devonian carbonate rocks of the Qiziqiao and Shetianqiao Formations (Figure 2c). The Sn mineralization of the Xianghualing deposit is hosted by the Laiziling pluton, with albite granite as its major rock type [72]. Based on geochronological studies, the crystallization age of the Laiziling granite is about 155–150 Ma [23,33,39,73,74].

Mainly mineralized by Sn, the skarn orebodies in the Xianghualing deposit are mainly stratiform, pipe or vein-like, distributed in the contact zone between albite granite and sedimentary rocks along faults (Figure 3c), or within the formation unconformity. The total metal reserve of Sn is estimated at 0.76 Mt (unpublished mining data from the local geological survey) and skarn-type cassiterite-arsenopyrite-pyrrhotite ore is the most economic mineralization (Figure 4e,f).

2.3. Petrology and Mineralogy of Skarns

2.3.1. The Huangshaping Skarn

The Huangshaping skarn is dark green with bronze or brown irregular mineral lumps on the surface, blastic texture and massive structure, dominated by W–Mo–Sn mineralization (Figure 5a). The skarn can be divided into garnet skarn and pyroxene skarn, which formed in the proximal and distal portions of the granite porphyry, respectively. The garnet is mainly grossularite (Figure 5b) and andradite, whereas the pyroxene is dominated by hedenbergite. The garnet skarn can be classified into two sub-types: (1) granite-related type with rufous color, inhomogeneous granularity and dense massive structure; and (2) exoskarn with coarse-grained texture and dark brown or green color. The latter is associated with abundant hydrous skarn minerals (e.g., actinolite, vesuvianite and hornblende) [35].

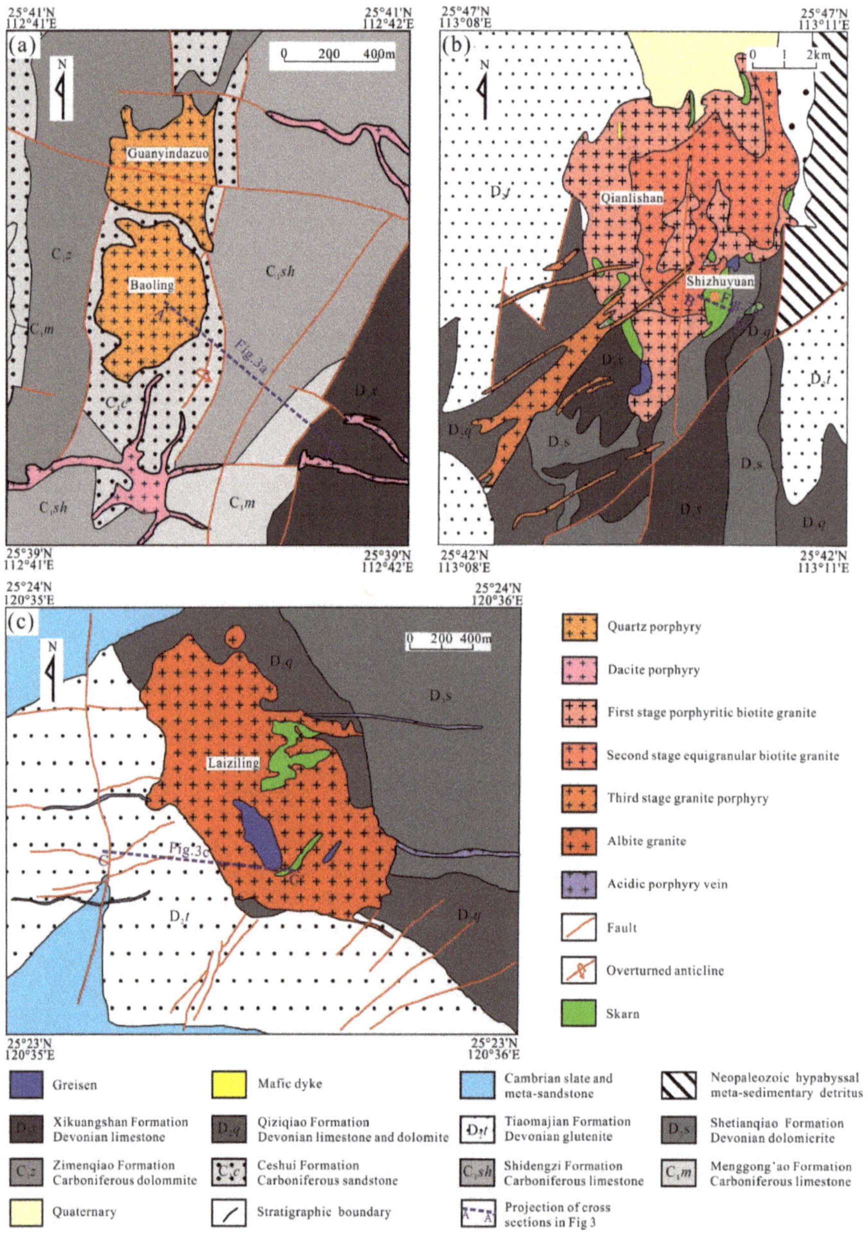

Figure 2. Geological maps of (**a**) the Huangshaping deposit (modified from [24]); (**b**) the Shizhuyuan deposit (modified from [68]); and (**c**) the Xianghualing deposit.

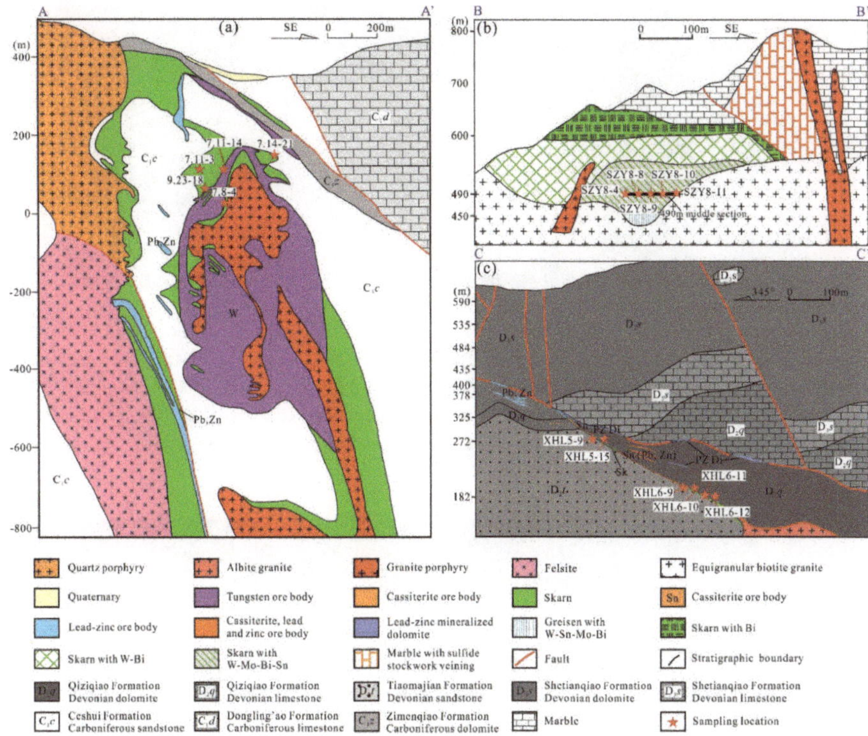

Figure 3. Cross-sections of (**a**) the Huangshaping deposit, (**b**) the Shizhuyuan deposit (modified from [11]); and (**c**) the Xianghualing deposit, respectively.

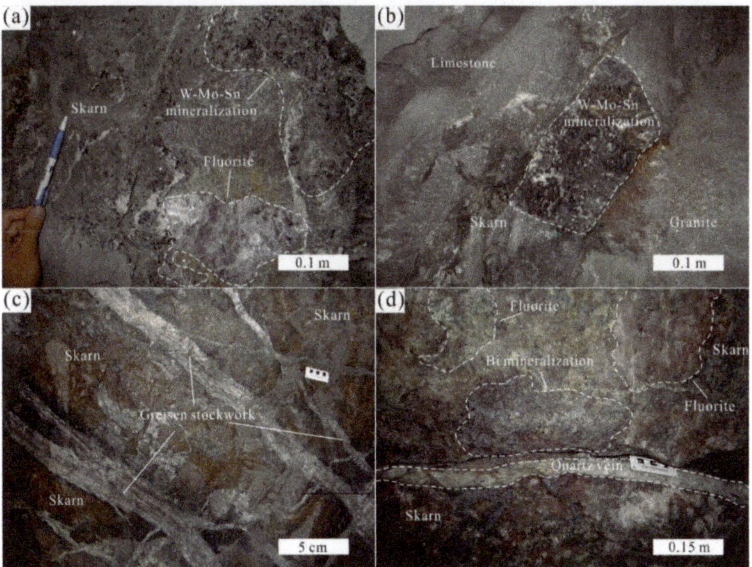

Figure 4. *Cont.*

Figure 4. Field occurrences of ore-related skarns. The Huangshaping deposit: (**a**) fluorite and W–Mo–Sn mineralization developed in massive skarn; (**b**) W–Mo–Sn mineralization located at the contact zone between limestone and granite. The Shizhuyuan deposit: (**c**) greisen stockwork interweave within massive skarn; (**d**) fluoritization, Bi mineralization and quartz veins in massive skarn. The Xianghualing deposit: (**e**) contacting relationship between carbonate rock and skarn; (**f**) massive skarn mineralized by galena, sphalerite, pyrite, chalcopyrite and cassiterite.

Figure 5. Photos of hand specimen and thin sections of skarn samples. The Huangshaping deposit: (**a**) garnet-fluorite skarn, (**b**) grossularite and fluorite in garnet skarn, (**c**) scheelite mineralization in garnet skarn; The Shizhuyuan deposit: (**d**) Bi-mineralized skarn, (**e**) grossularite, chloritoid, quartz and feldspar in ore-bearing garnet skarn, (**f**) grossularite, andradite and fluorite in ore-bearing garnet skarn; The Xianghualing deposit: (**g**) Sn-mineralized skarn ore, (**h**) zoisite and sericite in skarn, (**i**) zoisite and tremolite in skarn. Adr: andradite; Cld: Chloritoid; Fl: fluorite; Gn: galena; Grs: grossularite; Kfs: K-feldspar; Sh: scheelite; Sp: sphalerite; Srt: sericite; Py: pyrite; Q: quartz; Tr: tremolite; Zo: zoisite.

Besides garnet and pyroxene, other minerals in the Huangshaping skarn include chlorite, calcite, fluorite, feldspar and quartz. In addition, metal-bearing minerals include scheelite (Figure 5c), molybdenite, cassiterite, arsenopyrite, pyrite, chalcopyrite, pyrrhotite, magnetite, galena and sphalerite.

2.3.2. The Shizhuyuan Skarn

The Shizhuyuan skarn is brownish-red to gray-green in color, with granoblastic, granular and crystalloblastic textures and massive, comb and banded structures (Figure 5d).

The dominant minerals of the Shizhuyuan skarns are garnet, followed by pyroxene, wollastonite and vesuvianite. Compositionally, pyroxene mainly comprises hedenbergite and diopside, while the garnet belongs to the grossularite-andradite series. According to the mineral paragenesis, garnets from the Shizhuyuan skarn can be classified into early and late stages. The early garnet is characterized by euhedral granular texture whereas the late-stage garnet is a veinlet-type with bright red color and larger single crystal [75].

Additional skarn minerals are chlorite, epidote, actinolite and hornblende as well as fluorite. Most metallic minerals are hosted by massive garnet-pyroxene skarns (Figure 5e,f), including scheelite, wolframite, molybdenite, cassiterite, bismuthinite, pyrite, pyrrhotite, chalcopyrite, magnetite, galena and sphalerite.

2.3.3. The Xianghualing Skarn

The gray to dark green Xianghualing skarn has a granular to crystalloblastic texture and a massive structure (Figure 5g). The skarn has a variety of mineral assemblages: (1) skarn developed near the metasomatic contact zone is dominanted by diopside, wollastonite, vesuvianite and zoisite (Figure 5h) with minor garnet; (2) skarn controlled by fractures or structures is mainly composed of diopside, actinolite, tremolite (Figure 5i) and vesuvianite; and (3) skarn located at the unconformity between Devonian and Cambrian strata mainly consists of actinolite, associated with silicification [33]. Ore minerals in the skarn are dominated by cassiterite, followed by pyrite, pyrrhotite, arsenopyrite, stannite, galena, sphalerite, magnetite and chalcopyrite.

3. Sampling and Analytical Methods

All skarn samples were collected from underground adits or drill cores (Figure 3), and only the freshest material was analyzed (supplementary materials). From the Huangshaping skarn, samples 7.11–3 and 7.11–14 were collected at Level 96 m, from the west side of the granite porphyry pluton; sample 7.8–4 was collected from Level 20 m; sample 7.14–21 was collected from Level 136 m; and sample 9.23–18 was collected at Level 56 m. All the Shizhuyuan skarn samples (SZY8-4, SZY8-8, SZY8-9, SZY8-10 and SZY8-11) were collected at Level 490 m, close to the equigranular biotite granite. From the Xianghualing skarn, samples XHL5-9, XHL5-15, XHL6-9 and XHL6-10 were collected at Level 272 m near the fault F1, whereas samples XHL6-11 and XHL6-12 were collected at Level 182 m, closely related to the albite granite.

After sampling, each sample was cut for thin sections and the remainder was crushed to >200 mesh size using an iron pestle and mortar. From the Huangshaping skarn, major element analysis was conducted at Kyushu University (Fukuoka, Japan), using X-ray fluorescence spectrometry (XRF, Rigaku RIX 3100, Tokyo, Japan), and the trace elements were determined by inductively coupled plasma mass spectrometry (ICP-MS, Perkin Elmer Elan 9000, Perkin, Waltham, MA, USA) at the ALS Laboratory in Vancouver, Canada. The XRF (Panalytical, Almelo, The Netherlands) and ICP-MS (Perkin Elmer Elan 9000, Perkin, Waltham, MA, USA) analysis of the whole-rock compositions for the Xianghualing and Shizhuyuan skarns was carried out at the ALS Laboratory in Guangzhou, China. For analysis of major elements, samples were decomposed by lithium borate fusion, and the major elements were analyzed by XRF with a determination of loss-on-ignition at 1000 °C. The "total" is a combination of all data. The analytical process for trace element analysis is as follows: A prepared sample was added to a lithium metaborate/lithium tetraborate flux, mixed well and fused at 1025 °C. The melt was cooled and dissolved in a nitric, hydrochloric and hydrofluoric acid mixture and diluted for ICP-MS to determine trace element content. Standard sample JG-2 was used to detect the reliability of analytical results with errors for major elements and most trace elements of about 1–5% and 5–10%, respectively.

4. Results

Major element compositions of the skarns from the Huangshaping, Shizhuyuan and Xianghualing deposits are presented in Table S1, and trace element compositions are given in Tables S2–S4. Previously published geochemical data on these skarns are discussed. The loss-on-ignition (LOI) for most samples was <10 wt %. The skarns from the three deposits are characterized by variable chemical compositions, notably for SiO_2, Al_2O_3, total Fe, MgO and CaO. The oxides vary between deposits too, with the Huangshaping skarn displaying the widest ranges of SiO_2 (28.77–87.18%), total Fe (0.98–46.25%), MgO (0.09–20.07%), CaO (1.33–37.78%), Na_2O (0.03–3.95%) and K_2O (0.12–6.88%), the Shizhuyuan skarn showing the most variable MnO contents (0.12–2.46%), the Xianghualing skarn yielding the widest range of Al_2O_3 (2.89–21.06%), TiO_2 (<0.01–1.11%) and P_2O_5 (0.01–0.20%), respectively (Figure 6a–i). When comparing the compositional range of skarns with those of skarn-related granites and host sedimentary strata, the Huangshaping skarn is characterized by a relatively wide elemental distribution, with only a few samples plotting around or within the granite range, while other samples plot close to the range defined by the host strata. By contrast, the Shizhuyuan skarn has a relatively narrow range of major elements that corresponds most closely to sedimentary strata. In addition, the SiO_2 contents of skarn samples from different deposits show distinct linear trends with other oxides, such as Al_2O_3, total Fe, MnO and CaO (Figure 6).

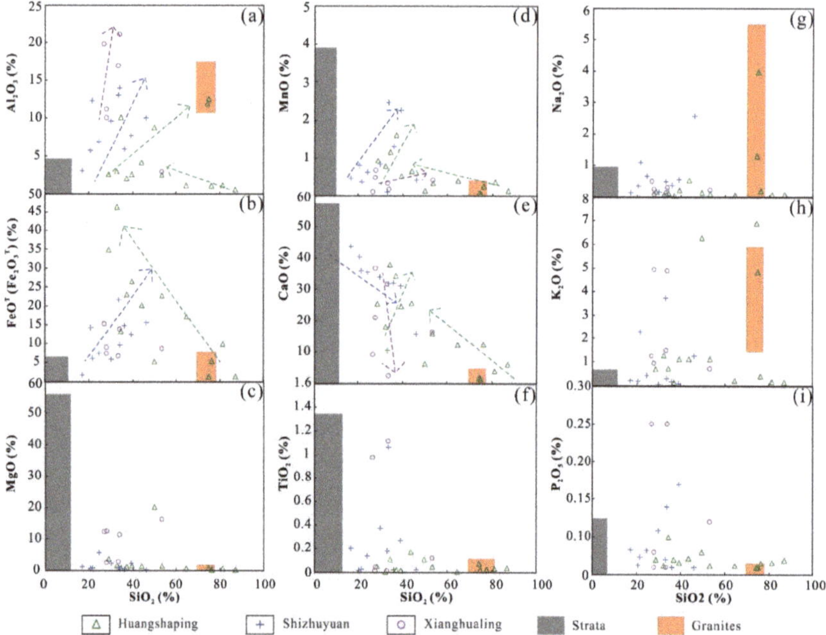

Figure 6. Correlation plots of skarn oxide compositions, compared to those of granites and strata of the Huangshaping, Shizhuyuan and Xianghualing deposits. (**a**) SiO_2 vs. Al_2O_3 diagram; (**b**) SiO_2 vs. FeO^T ($Fe_2O_3{}^T$) diagram; (**c**) SiO_2 vs. MgO diagram; (**d**) SiO_2 vs. MnO diagram; (**e**) SiO_2 vs. CaO diagram; (**f**) SiO_2 vs. TiO_2 diagram; (**g**) SiO_2 vs. Na_2O diagram; (**h**) SiO_2 vs. K_2O diagram and (**i**) SiO_2 vs. P_2O_5 diagram. Colored arrows show correlation trends for skarns (color coded to match data points). The orange box shows the range of composition of granites from all three deposits. The gray box shows the composition of the Carboniferous limestone at Huangshaping and Devonian limestone at other deposits in the Nanling range. No strata analysis was available at the Shizhuyuan and Xianghualing deposits. Data for the granites are reference from [24,44,68,76]. Data for the strata are from [31,77].

Likewise, the trace element compositions of all skarn samples are also highly variable (Figure 7a–i). The Xianghualing skarn has the highest enrichments of LILE such as Ba (10.1–1045 ppm), Rb (139.5–2440 ppm) and Cs (14.8 to 610 ppm), followed by the Huangshaping and Shizhuyuan skarns. In addition, Sr is extremely enriched in the Shizhuyuan skarn (20.4–680 ppm), followed by the Xianghualing (20.1–154 ppm) and Huangshaping (6.8–99.6 ppm) skarns. Regarding the HFSE, the Huangshaping skarn contains the most variable contents of Nb (0.30–88.2 ppm), Hf (0.2–28.3 ppm) and Th (0.82–48 ppm), respectively. The Xianghualing skarn has the highest Zr (up to 534 ppm) and U (up to 90.3 ppm) concentrations, followed by the Shizhuyuan (up to 365 and 36.7 ppm, respectively) and Huangshaping skarns (up to 130 and 32 ppm, respectively). On primitive mantle-normalized spider diagrams (Figure 8a–c), the Huangshaping skarn samples are strongly depleted in Ti, but enriched in Hf, while the Shizhuyuan skarn samples show moderate values for most elements. The weakest Ba anomaly and most significant U anomalies are recorded in the Xianghualing skarn. By comparing skarn compositions with the associated granite and sedimentary strata, it can be concluded that the Huangshaping skarn is most closely related to the granite, whereas the Shizhuyuan skarn seems to be more similar to regional strata in terms of its trace element composition.

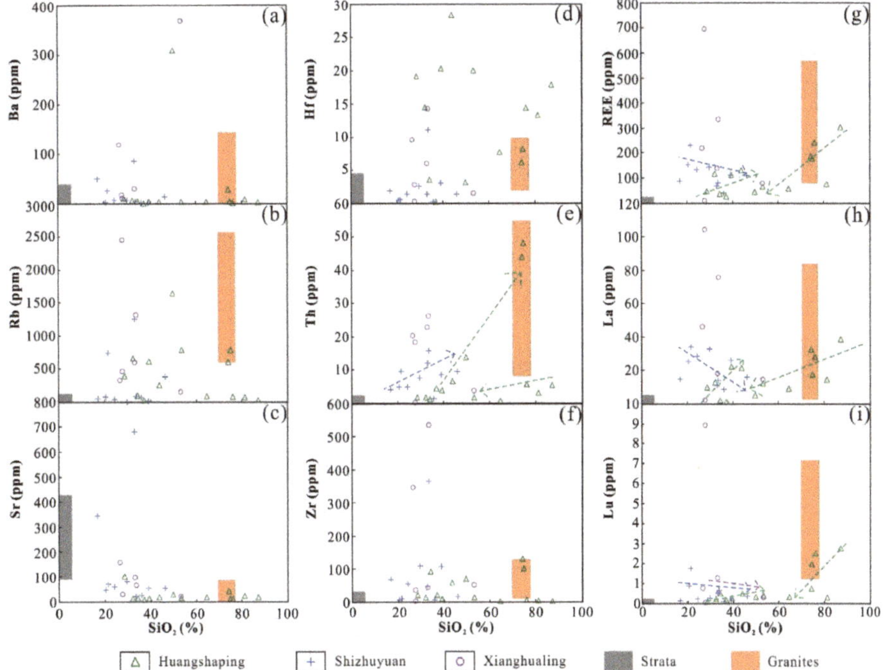

Figure 7. Correlation plots of trace elements vs. SiO$_2$ in skarns. Colored arrows show correlation trends for skarns (color coded to match data points). (**a**) SiO$_2$ vs. Ba diagram; (**b**) SiO$_2$ vs.Rb diagram; (**c**) SiO$_2$ vs. Sr diagram; (**d**) SiO$_2$ vs. Hf diagram; (**e**) SiO$_2$ vs. Th diagram; (**f**) SiO$_2$ vs. Zr diagram; (**g**) SiO$_2$ vs. REE diagram; (**h**) SiO$_2$ vs. La diagram and (**i**) SiO$_2$ vs. Lu diagram. The orange box shows the range of composition of granites from all three deposits. The box shows the composition of the Carboniferous limestone at Huangshaping and Devonian limestone at other deposits in the Nanling range. No strata analysis was available at the Shizhuyuan and Xianghualing deposits. Data sources for the granites and strata are the same as in Figure 6.

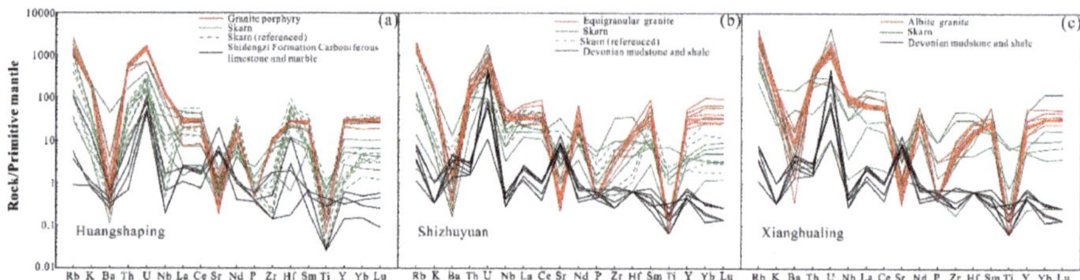

Figure 8. Primitive mantle-normalized trace element diagrams for granites, strata and skarns from the (**a**) Huangshaping, (**b**) Shizhuyuan and (**c**) Xianghualing deposit. Skarn data are from this study, and are reference from [25,31]. Strata composition of the Shidengzi Formation Carboniferous limestone are reference from [31], Devonian mudstone and shale composition are reference from [77]. Granite composition of the Huangshaping granite porphyry are reference from [24], Shizhuyuan equigranular biotite granite are reference from [44,68], Xianghualing albite granite are reference from [76]. Normalized values for primitive mantle are from [78].

The rare earth elements (REE) of three skarn samples are characterized by variable total REE contents and fractionation between LREE and HREE (Figure 9a–c). The Xianghualing skarn samples have the most variable and highest total REE contents (up to 693.5 ppm), with high La (1.8–104 ppm), Ce (3.5–245 ppm) (Figure 7g–i) and Yb (0.13–60.2 ppm). The average total REE content of the Huangshaping and Shizhuyuan skarns is lower. In addition, the fractionation between LREE and HREE in the Shizhuyuan skarn (highest average $(La/Yb)_N$ = 7.7) is stronger than that at the Huangshaping and Xianghualing skarns (Figure 9). The negative Eu anomaly in the Xianghualing skarn samples is the weakest and shows a similar REE pattern to its hosted granite. By contrast, the Huangshaping skarn samples have the largest Eu anomalies, which are also similar to its parent granite (Figure 9a).

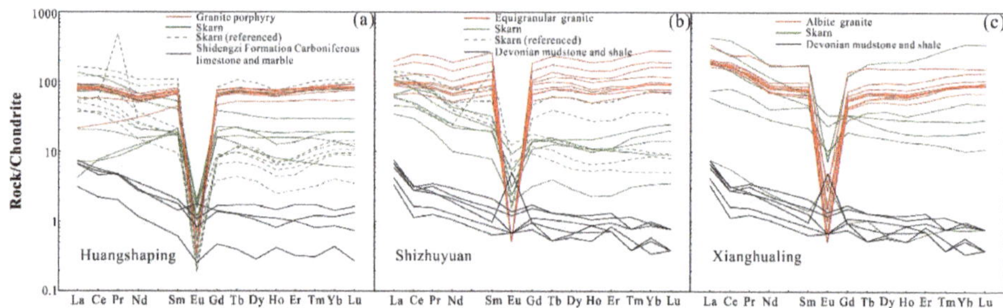

Figure 9. Chondrite-normalized REE patterns for granites, strata and skarns of the (**a**) Huangshaping, (**b**) Shizhuyuan and (**c**) Xianghualing deposits. Data sources are the same as in Figure 8. Normalized values for chondrite are from [78].

5. Discussion

5.1. Magma Differentiation and Metasomatism

The skarns at the Huangshaping, Shizhuyuan and Xianghualing deposits are located at the contact zones between granitic intrusions and strata (mainly carbonate rocks) (Figures 2–4). This reflects the undisputed role of granites in the formation of skarns. It is important to examine the differences in magma composition of the skarn-related granites at each deposit and to assess the possible impact on the respective metal contents.

The skarn-related granites in the three deposits are highly differentiated and can be classified as reduced, high-K-calc-alkaline, A_2-subtype granitoids, formed in an extensional tectonic setting [24,33,34,44,63,68].

In addition, the granites are believed to have contributed a large proportion of the volatile components (especially fluorine, [24,25,36,42,44,62,68,72]) and ore-forming elements [16,40,47,52]. Although the parental magmas of the three granitic plutons are highly fractionated, the degree of differentiation, which can be inferred from whole-rock element compositions, could vary between the plutons. In general, the Rb content will increase with increasing degree of differentiation, whereas Ba and Sr are usually enriched in the early stage during magmatic evolution without an obvious variation of K. This means that K/Rb and Rb/Sr ratios can be used to evaluate the degree of fractionation of the magmas [24,79,80]. The Xianghualing granite has the lowest K/Rb ratio (mean = 22), but the highest Rb/Sr ratio (mean = 146.9), indicating that it is the most differentiated (Table S5, [81]). Moreover, the Laiziling pluton (i.e., the Xianghualing granite) can be divided into several magmatic-hydrothermal zones from the base to the top [73], also suggesting strong crystal fractionation. On the other hand, previous studies reveal that the Huangshaping granite porphyry is coarse-grained and porphyritic in texture, and crystallized over an extended period (~10 Myr; [82]) and at a high crystallization temperatures (~915 °C; [34]). This contrasts with the relatively low crystallization temperatures (680–725 °C) for both the Shizhuyuan and Xianghualing granite which have a hypidiomorphic granular texture [33,68]. In summary, the Xianghualing granite appears to exhibit the strongest degree of magma differentiation, followed by the Huangshaping granite and finally, the Shizhuyuan granite.

By comparing the compositions of the skarns with the composition of the associated granites and sedimentary strata, it is evident that the Huangshaping skarn, which can be classified as a siliceous skarn, has the most variable elemental composition, and one that shows affinities to both the granite and strata compositions (Figures 6 and 7), probably indicating varied protoliths of the skarn components. The average content of Si in the Huangshaping skarn is much higher than those in other skarns. This may imply that Si was sourced from the highly evolved granite in this deposit. Due to the very low abundances of both Si and Al in the host strata (Table S6), the migration or exchange of these two elements between two end members (granites and strata) is probably mainly controlled by fluids derived from the granites. Both the Shizhuyuan and Xianghualing skarns exhibit positive correlations between Al, Mn and Ca oxides and SiO_2, whereas the Huangshaping skarns show two opposite trends (see arrows in Figure 6a,d,e), indicating contributions from both the granites and strata. The negative correlation between SiO_2 and CaO in the Shizhuyuan and Xianghualing skarns can be attributed to their protolith properties (carbonate rocks). Highly differentiated magma-related fluids may provide strong exchange capacity and thus mobilize these major elements.

The average Al_2O_3 content in the skarns show a decreasing trend from Xianghualing (average = 13.62%) to Shizhuyuan (average = 8.80%) and then to Huangshaping (average = 4.54%), which is inconsistent with other oxides. This may indicate that the Xianghualing skarn inherited most of its metal elements from the granite. By contrast, this effect is less pronounced at the Huangshaping and Shizhuyuan skarns, respectively. It is also reasonable to speculate that the activity of Al was suppressed by the gradual enrichment of Fe during metasomatism (Table S5). This is evidenced by the gradation of garnet types from grossular to almandine with increasing metasomatism, and the dominant pyroxene composition being hedenbergite [35].

Trace element contents may also reflect variations in the granite composition of the studied skarns. Some trace elements, such as Ba, Rb, Cr, Cs, V, Ga and most REEs, are enriched in skarns and in the associated granites, especially in the Xianghualing samples (Table S6). This implies the dominant control of granites on these elements during metasomatism. Although the average contents of other trace elements (such as Nb, Y, Zr, Th and U) in skarns are lower than those of granites, their maximum values are comparable to as those in granites. This further suggests that granites influenced the enrichment of those trace elements in the skarns. Moreover, consistent linear trends for these skarns can be observed on the Zr vs. Hf, Ba vs. Sr, Zr vs. Ti and U vs. Th diagrams (Figure 10), probably indicating that these elements may actived in pairs during metasomatism, aided by the intrusion of granites. In addition, the two distinct and contrasting trends of the Huangshaping skarn on the Zr vs. Ti diagram (green arrows, Figure 10c) indicates at least two types of skarns in this area.

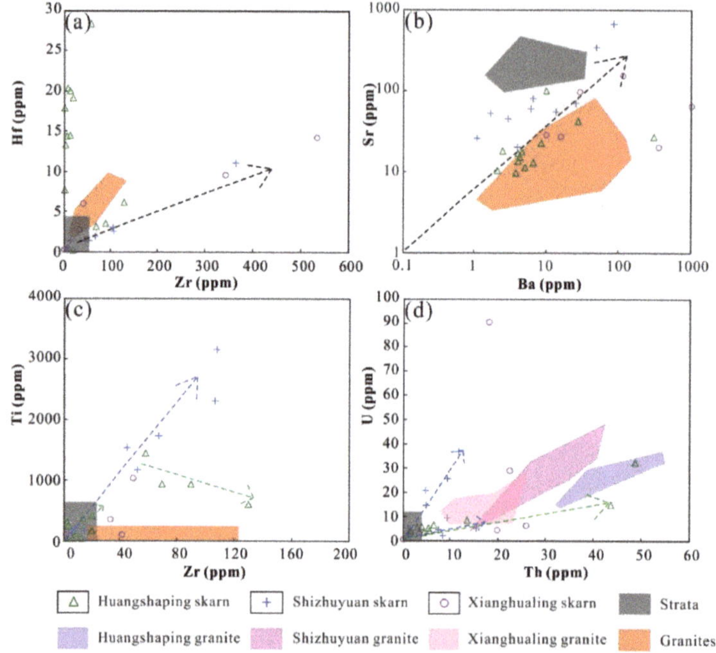

Figure 10. Correlation of trace elements in skarns, illustrating the geochemical behavior of typical trace elements during metasomatism. (**a**) Zr vs. Hf diagram; (**b**) Ba vs. Sr diagram; (**c**) Zr vs. Ti diagram and (**d**) Th vs. U diagram. The colored arrows are provided to highlight correlation trends (black arrow shows overall trend for all skarns, with the green arrow for Huangshaping and blue arrow for Shizhuyuan). The orange box shows the range of composition of all granites from the three deposits and the gray box shows the range of composition of the Carboniferous limestone at Huangshaping and Devonian limestone at other deposits in the Nanling range. Data sources are the same as in Figure 8.

In summary, the Xianghualing skarn composition is most strongly influenced by the parent granite composition, by contrast, this effect is less pronounced at the Shizhuyuan and Huangshaping skarns, respectively. Major elements such as Si and Al, and most trace elements such as Ba, Rb, Cs, Cr, V, Ga, Nb, Zr, U and Th in the skarns are dominantly controlled by the composition of the related granites.

5.2. Strata Contribution to Skarns

The sedimentary strata also has an impact on the composition of skarns. The mean total FeO content of the skarns is 16.26%, 10.86% and 10.02% at Huangshaping, Shizhuyuan and Xianghualing, respectively (Table S5). These Fe contents can be can be primarily ascribed to stratigraphic fluids released from host sediments or regional strata, due to the very low contents of Fe in the three related granites. High contents of CaO (43.62%) and MgO (20.07%) in the skarns may stem from the carbonate rocks at each deposit, and the similarity of MnO and TiO_2 contents in skarns and strata indicate that such components are dominantly inherited from the local strata (Figure 6d,f).

The Sr contents are relatively high in both the sedimentary strata and skarns (average = 143.3 ppm) at Shizhuyuan (Table S6), indicating the dominant role of strata in Sr enrichment. The $\sum$REE content varies widely within individual skarns, especially at Xianghualing where the $\sum$REE content ranges from 8.0 to 693.5 ppm, also suggesting the combined control of both granite and strata on its composition at this site. The ratios of $(LREE/HREE)_T$ (6.1 to 8.4) and $(La/Yb)_N$ (3.8 to 7.7) in skarns are also in accordance with the average ratios of the strata. Moreover, the steepness of the REE pattern from

LREE to HREE at both Shizhuyuan and Xianghualing is more similar to the parent strata than the corresponding granite on chondrite-normalized REE diagrams (Figure 9), suggesting that the host strata influenced the REE signatures of the skarns. By contrast, the REE pattern of the Huangshaping skarn can be classified into two types: (1) one shows a steep REE pattern and significant negative Eu anomaly, similar to that of the parent granite; and (2) the other shows a shallower REE pattern and a moderate Eu anomaly, reflecting control by both granite and strata. The REE contents of most skarns moderately decrease along with increasing SiO_2, except for some Huangshaping samples which show the opposite trend for La and Lu (Figure 7h,i). All of this could imply varied protoliths of the Huangshaping skarn.

In summary, the REE signatures of the Shizhuyuan and Xianghualing skarns are more strongly influenced by the sedimentary strata, while the REE signatures at the Huangshaping skarn suggest influence of both strata and granite. Major elements such as Fe, Mn, Mg and Ca in the skarns are dominated by the host sedimentary strata, and Sr and REE patterns are mainly inherited from the surrounding sedimentary rocks.

5.3. Source of Metals and Scales of Skarn Mineralization

As described below, numerous previous studies have focused on the temperature, pressure, crystallization depth and ore-forming relationships in the skarns at the Huangshaping, Shizhuyuan and Xianghualing deposits. For the Huangshaping skarn, two crystallization stages (>500 °C and 250–476 °C, respectively) are distinguished [32]. Homogenization temperatures of fluid inclusions in the Shizhuyuan skarn minerals range from 350 °C to 535 °C [28]. Based on the analysis of hessonite inclusions, the formation temperature of the Xianghualing skarn is estimated to be 430–570 °C [29]. Thus, the diagenetic pressures are estimated to be 0.2–0.85 kbar at Huangshaping [32,83], 1.8–3.2 kbar at Shizhuyuan [30] and 0.2–1.0 kbar at Xianghualing [29], corresponding to the formation depth of skarns and related mineralization at 0.7–3.0 km for Huangshaping, 5.4–12 km for Shizhuyuan, and 0.7–3.5 km for Xianghualing. In summary, the Shizhuyuan skarns formed under the highest temperatures and pressures and the deepest formation depth, consistent with most W skarns (formation depth of 5–20 km; [84]. Consequently, it can be inferred that the precipitation of multiple metals (especially W) in the Shizhuyuan area occurred in a deep, high-pressure setting, in which magma differentiation was largely suppressed. By contrast, local strata composition strongly influenced skarn formation and mineralization.

Skarn mineralization in the Nanling Range can be divided into three types in terms of dominant mineralization element: (1) the Huangshaping W–Mo–Sn; (2) Shizhuyuan W–Sn–Mo–Bi; and (3) the Xianghualing Sn. The W skarns are closely related to calc-alkaline plutons and can be further divided into reduced and oxidized types [4]. The reduced type W skarns are composed of hedenbergite, grandite garnet and low-molybdenum scheelite in the early stages. During the later stages, garnet is subcalcic, and dominated by spessartine and almandine. Moreover, enrichment of W is likely to be favored by reduced environments and under more reduced conditions, there is strong affinity between W and Sn [85]. The Sn skarn is closely related to high-silica magma under reducing conditions [4]. For the W–Mo–Cu assemblage, a reducing environment with high fluorine is favored. Therefore, it can be concluded that reducing environments were dominated in the skarn mineralization in the Nanling Range, with an increasing reduction states from the Xianghualing Sn to the Huangshaping W–Mo–Sn and then to the Shizhuyuan W–Sn–Mo–Bi deposit. The abundant Bi contents in the Shizhuyuan skarn may be attributed to the strong influence of the sedimentary wall rocks at Shizhuyuan.

As discussed above, the Xianghualing and Huangshaping skarns have a greater affinity to their strongly differentiated parent granites, compared to the Shizhuyuan skarn which has strong affinities to the host sediments. The Xianghualing granite is the most evolved, enriching in most trace elements (e.g., Ba, Rb, Cr, Cs, V, Ga and most REEs), and its related skarn is also the richest. However, Sn is the only ore-forming element at Xianghualing and metal reserves are relatively low compared to the W–Mo–Sn skarn at Huangshaping. Compared to the Xianghualing and Huangshaping skarns, the Shizhuyuan skarn is characterized by the largest ore-forming scale, most varied mineral species and smallest pluton size (~10 km^2). It is interesting that such a small pluton contributed to the formation of a giant polymetallic deposit. There are many similarities between these three deposits, such as well-developed structures and multiphase magmatic-hydrothermal activity. We suggest that strongly evolved granites (e.g., the Xianghualing granite) may be less favorable for the development of mineralization, hence, resulting in a relatively small size of deposits. By contrast, less differentiated, smaller plutons formed at greater depths (e.g., the Shizhuyuan granite) may favor the formation of mineralized skarns where substantial amount of ore-forming metals stems from host sedimentary strata. The sedimentary rocks at Shizhuyuan have high concentrations of metals (average W contents of 5–6 ppm and average Sn contents of 2–8 ppm) [86], and the hot environment resulting from deep magmatic activity probably creating suitable conditions for widespread fluid convection in the sedimentary strata as well as strong water-rock interaction [70]. Although the abundance of ore metals in the strata could be lower than in the granite, fluid circulation through a large volume of strata may concentrate these elements more effectively than a limited volume of granite. The Shizhuyuan deposit was less influenced by granite-derived fluids due to the relatively low degree of differentiation of the magma; however, continuous magmatic activity and an extended hot environment likely accounted for the prolonged circulation of stratigraphic fluids. These fluids carried ore-forming elements through fractures widely developed in the deposit area, constantly assimilating ore-forming material from surrounding strata, and eventually forming the giant Shizhuyuan skarn deposits (Figure 11). By contrast, the Xianghualing granite experienced the strongest degree of magmatic differentiation, resulting in relatively low abundance and a single element (Sn) skarn mineralization with a minor contribution from the strata. The metasomatism and mineralization at Xianghualing has more affinities with local magmatism. At Huangshaping deposit, both granite and strata may equally have contributed in forming the W–Mo–Sn mineralization (Figure 11).

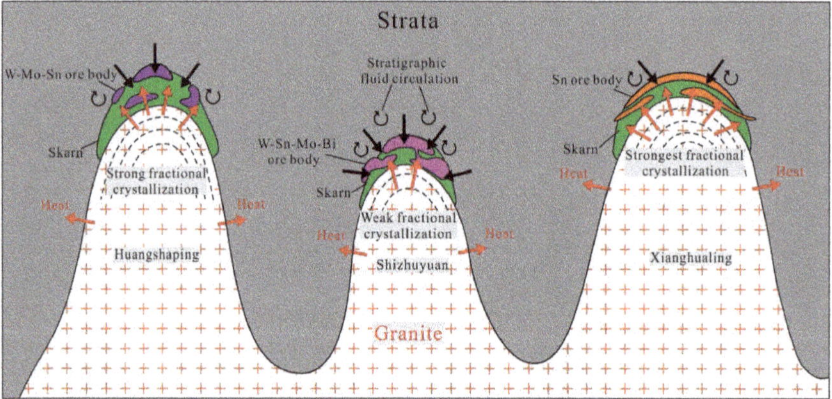

Figure 11. A genetic model for W–Sn polymetallic skarn deposits in the Nanling Range, South China, showing relative material contributions from granite and strata, respectively.

6. Conclusions

(1) Concentrations of Si, Al and most trace elements (e.g., Ba, Rb, Cs, Cr, V, Ga, Nb, Zr, U and Th) in ore-related skarns controlled by related granites but FeO, CaO, MgO, MnO, TiO$_2$, Sr and REE signatures in most skarns are significantly controlled by the composition of the host sediments. The composition of strata plays a crucial role in the enrichment of polymetallic elements, scale and metallogenic ore species of the W–Sn skarn mineralization in South China.

(2) The formation of the Shizhuyuan W–Sn–Mo–Bi polymetallic skarn deposit is strongly controlled by strata during metasomatism. The Xianghualing Sn skarn is controlled by the granite composition. Granite and strata may have contributed almost equally to the Huangshaping W–Mo–Sn deposit, however.

Supplementary Materials: The following are available online at http://www.mdpi.com/2075-163X/8/7/265/s1, Table S1: Major element compositions of skarns from the Huangshaping, Shizhuyuan and Xianghualing deposit (%), Table S2: Trace element compositions of skarns in the Huangshaping deposit (ppm), Table S3: Trace element compositions of skarns in the Shizhuyuan deposit (ppm), Table S4: Trace element compositions of skarns in the Xianghualing deposit (ppm), Table S5: Minimum, maximum and average values of trace element compositions in skarns, granites and strata (ppm), Table S6: Minimum, maximum and average values of major element compositions in skarns, granites and strata (%).

Author Contributions: H.L., W.J. and J.W. conceived and designed the experiments; H.L., W.J., J.W. and J.C. took part in the field investigation; W.J., H.L. and N.J.E. interpreted the data and took part in the discussion; W.J. wrote the original draft; H.L. and N.J.E. reviewed and edited the paper.

Funding: This work was funded by the National Natural Science Foundation of China (Grant Nos. 41502067).

Acknowledgments: The authors are thankful to Hua Kong, Qilin Peng and Yong Wang for their kind cooperation during the fieldwork. The authors are also grateful to the editorial board members and other two reviewers for their constructive comments that greatly improved this manuscript.

Conflicts of Interest: The authors declare no conflict of interest.

References

1. Einaudi, M.T.; Meinert, L.D.; Newberry, R.J. Skarn deposits. *Econ. Geol.* **1981**, *964*, 317–391.
2. Einaudi, M.T.; Burt, D.M. Introduction-terminology, classification, and composition of skarn deposits. *Econ. Geol.* **1982**, *77*, 745–754. [CrossRef]
3. Mueller, A.G. The Savage Lode magnesian skarn in the Marvel Loch gold–silver mine, Southern Cross greenstone belt, Western Australia. Part 1: Structural setting, petrography, and geochemistry. *Can. J. Earth Sci.* **1991**, *28*, 659–685. [CrossRef]
4. Meinert, L.D. Skarns and skarn deposits. *Geosci. Can.* **1992**, *19*, 145–162.
5. Jamtveit, B.; Wogelius, R.A.; Fraser, D.G. Zonation patterns of skarn garnets: Records of hydrothermal system evolution. *Geology* **1993**, *21*, 113–116. [CrossRef]
6. Gaspar, M.; Knaack, C.; Meinert, L.D.; Moretti, R. REE in skarn systems: A LA-ICP-MS study of garnets from the Crown Jewel gold deposit. *Geochim. Cosmochim. Acta* **2008**, *72*, 185–205. [CrossRef]
7. Ismail, R.; Ciobanu, C.L.; Cook, N.J.; Teale, G.S.; Giles, D.; Mumm, A.S.; Wade, B. Rare earths and other trace elements in minerals from skarn assemblages, Hillside iron oxide–copper–gold deposit, Yorke Peninsula, South Australia. *Lithos* **2014**, *184–187*, 456–477. [CrossRef]
8. Alirezaei, S.; Einali, M.; Jones, P.; Hassanpour, S.; Arjmandzadeh, R. Mineralogy, geochemistry, and evolution of the Mivehrood skarn and the associated pluton, northwest Iran. *Int. J. Earth Sci.* **2015**, *105*, 849–868. [CrossRef]
9. Hu, X.L.; Gong, Y.J.; Pi, D.H.; Zhang, Z.J.; Zeng, G.P.; Xiong, S.F.; Yao, S.Z. Jurassic magmatism related Pb–Zn–W–Mo polymetallic mineralization in the central Nanling Range, South China: Geochronologic, geochemical, and isotopic evidence from the Huangshaping deposit. *Ore Geol. Rev.* **2017**, *91*, 877–895. [CrossRef]
10. Yin, S.; Ma, C.Q.; Xu, J.N. Geochronology, geochemical and Sr–Nd–Hf–Pb isotopic compositions of the granitoids in the Yemaquan orefield, East Kunlun orogenic belt, northern Qinghai-Tibet Plateau: Implications for magmatic fractional crystallization and sub-solidus hydrothermal alteration. *Lithos* **2017**, *294–295*, 339–355. [CrossRef]
11. Chen, J.; Halls, C.; Stanley, C.J. Rare earth element contents and patterns in major skarn minerals from Shizhuyuan W, Sn, Bi and Mo deposit, South China. *Geochem. J.* **1992**, *26*, 147–158. [CrossRef]
12. Giuliani, G.; Cheilletz, A.; Mechiche, M. Behaviour of REE during thermal metamorphism and hydrothermal infiltration associated with skarn and vein-type tungsten ore bodies in central Morocco. *Chem. Geol.* **1987**, *64*, 279–294. [CrossRef]

13. Ma, W.; Liu, Y.C.; Yang, Z.S.; Li, Z.Q.; Zhao, X.Y.; Fei, F. Alteration, mineralization, and genesis of the Lietinggang–Leqingla Pb–Zn–Fe–Cu–Mo skarn deposit, Tibet, China. *Ore Geol. Rev.* **2017**, *90*, 897–912. [CrossRef]

14. Siesgesmund, S.; López-Doncel, R.; Sieck, P.; Wilke, H.; Wemmer, K.; Frei, D.; Oriolo, S. Geochronological and geochemical constraints on the genesis of Cu-Au skarn deposits of the Santa María de la Paz district (Sierra del Fraile, Mexico). *Ore Geol. Rev.* **2018**, *94*, 310–325. [CrossRef]

15. Baker, J.H.; Hellingwerf, R.H. Rare earth element geochemistry of W-Mo-(Au) skarns and granites from Western Bergslagen, central Sweden. *Miner. Petrol.* **1988**, *39*, 231–244. [CrossRef]

16. Hua, R.M.; Chen, P.R.; Zhang, W.L.; Junming, Y.; Lin, J.F.; Zhang, Z.S.; Gu, S.Y.; Liu, X.D.; Qi, H.W. Metallogenesis related to Mesozoic granitoids in the Nanling Range, South China and their geodynamic settings. *Acta Geol. Sin-Engl. Ed.* **2005**, *79*, 810–820.

17. Mao, J.W.; Xie, G.Q.; Guo, C.L.; Chen, Y.C. Large scale tungsten-tin mineralization in the Nanling region, South China: Metallogenic ages and corresponding geodynamic processes. *Acta Petrol. Sin.* **2007**, *23*, 2329–2338. (In Chinese)

18. Cao, J.Y.; Wu, Q.H.; Yang, X.Y.; Kong, H.; Li, H.; Xi, X.S.; Huang, Q.F.; Liu, B. Geochronology and Genesis of the Xitian W-Sn Polymetallic Deposit in Eastern Hunan Province, South China: Evidence from Zircon U-Pb and Muscovite Ar-Ar Dating, Petrochemistry, and Wolframite Sr-Nd-Pb Isotopes. *Minerals* **2018**, *8*, 111. [CrossRef]

19. Jiang, W.C.; Li, H.; Wu, J.H.; Zhou, Z.K.; Kong, H.; Cao, J.Y. A newly found biotite syenogranite in the Huangshaping polymetallic deposit, South China: Insights into Cu mineralization. *J. Earth Sci.* **2018**. [CrossRef]

20. Yang, J.H.; Kang, L.F.; Peng, J.T.; Zhong, H.; Gao, J.F.; Liu, L. In-situ elemental and isotopic compositions of apatite and zircon from the Shuikoushan and Xihuashan granitic plutons: Implication for Jurassic granitoid-related Cu-Pb-Zn and W mineralization in the Nanling Range, South China. *Ore Geol. Rev.* **2018**, *93*, 382–403. [CrossRef]

21. Lu, Y.F.; Ma, L.Y.; Qu, W.J.; Mei, Y.P.; Chen, X.Q. U-Pb and Re-Os isotope geochronology of Baoshan Cu–Mo polymetallic ore deposit in Hunan Province. *Acta Petrol. Sin.* **2006**, *22*, 2483–2492. (In Chinese)

22. Peng, J.T.; Zhou, M.-F.; Hu, R.Z.; Shen, N.P.; Yuan, S.D.; Bi, X.W.; Du, A.D.; Qu, W.J. Precise molybdenite Re–Os and mica Ar–Ar dating of the Mesozoic Yaogangxian tungsten deposit, central Nanling district, South China. *Miner. Depos.* **2006**, *41*, 661–669. [CrossRef]

23. Yuan, S.D.; Peng, J.T.; Hu, R.Z.; Li, H.M.; Shen, N.P.; Zhang, D.L. A precise U–Pb age on cassiterite from the Xianghualing tin-polymetallic deposit (Hunan, South China). *Miner. Depos.* **2007**, *43*, 375–382. [CrossRef]

24. Li, H.; Watanabe, K.; Yonezu, K. Geochemistry of A-type granites in the Huangshaping polymetallic deposit (South Hunan, China): Implications for granite evolution and associated mineralization. *J. Asian Earth Sci.* **2014**, *88*, 149–167. [CrossRef]

25. Cheng, Y.S. Petrogenesis of skarn in Shizhuyuan W-polymetallic deposit, southern Hunan, China: Constraints from petrology, mineralogy and geochemistry. *Trans. Nonferr. Met. Soc. China* **2016**, *26*, 1676–1687. [CrossRef]

26. Ye, Z.H.; Wang, P.; Xiang, X.K.; Wang, A.J.; Yan, Q.; Li, Y.K. Re-Os dating and H-O isotope geochemistry of the Shimensi deposit, northern Jiangxi, China: Implication for ore genesis. *Geochem. J.* **2016**, *50*, 139–152. [CrossRef]

27. Liu, J.P. Indium Mineralization in a Sn-Poor Skarn Deposit: A Case Study of the Qibaoshan Deposit, South China. *Minerals* **2017**, *7*, 76. [CrossRef]

28. Lu, H.Z.; Liu, Y.M.; Wang, C.L.; Xu, Y.Z.; Li, H.Q. Mineralization and fluid inclusion study of the Shizhuyuan W-Sn–Bi–Mo–F skarn deposit, Hunan Province, China. *Econ. Geol.* **2003**, *98*, 955–974. [CrossRef]

29. Zhou, T. *Research on the Geochemistry and Thermodynamics Characteristics in Xianghualing Mining Area in Hunan Province*; Central South University: Changsha, China, 2009. (In Chinese)

30. Cheng, X.Y. *Research on the Skarns Formation Mechanism of the Shizhuyuan W–Sn Polymetallic Deposit, Hunan Province*; Kunming University of Science and Technology: Kunming, China, 2012. (In Chinese)

31. Qi, F.Y.; Zhang, Z.; Zhu, X.Y.; Li, Y.S.; Zhen, S.M.; Gong, F.Y.; Gong, X.D.; He, P. Skarn geochemistry of the Huangshaping W–Mo polymetallic deposit in Hunan and its geological significance. *Geol. China* **2012**, *39*, 338–348. (In Chinese)

32. Huang, C.; Li, X.F.; W, L.F.; Liu, F.P. Fluid inclusion study of the Huangshaping polymetallic deposit, Hunan Province, South China. *Acta Petrol. Sin.* **2013**, *29*, 4232–4244. (In Chinese)

33. Lai, S.H. *Research on Mineralization of the Xianghualing Tin Polymetallic Deposit, Hunan Province, China*; China University of Geoscience: Beijing, China, 2014. (In Chinese)

34. Yuan, Y.B. *The Genetic Difference between Two Metallogenic Granites and the Ore-Forming Material Source of the Huangshaping Deposit in Southern Hunan*; China University of Geoscience: Beijing, China, 2015. (In Chinese)

35. Zhao, F.; Yin, J.W.; Wang, M.Y.; Zhang, Z.H.; Sun, Y.D.; Zhang, P.; Gao, Y.W.; Wang, L.F.; Zong, Z.H. Skarn mineral characteristics and their geological significance of the Huangshaping deposit in Hunan Province. *Geoscience* **2016**, *30*, 1038–1050. (In Chinese)

36. Liu, J.P.; Rong, Y.N.; Zhang, S.G.; Liu, Z.F.; Chen, W.K. Indium Mineralization in the Xianghualing Sn-Polymetallic Orefield in Southern Hunan, Southern China. *Minerals* **2017**, *7*, 173. [CrossRef]

37. Liu, J.P.; Rong, Y.N.; Gu, X.P.; Shao, Y.J.; Lai, J.Q.; Chen, W.K. Indium Mineralization in the Yejiwei Sn-Polymetallic Deposit of the Shizhuyuan Orefield, Southern Hunan, China. *Resour. Geol.* **2018**, *68*, 22–36. [CrossRef]

38. Wang, Y.J.; Fan, W.M.; Guo, F. Geochemistry of early Mesozoic potassium-rich diorites-granodiorites in southeastern Hunan Province, South China: Petrogenesis and tectonic implications. *Geochem. J.* **2003**, *37*, 427–448. [CrossRef]

39. Shu, X.J.; Wang, X.L.; Sun, T.; Xu, X.S.; Dai, M.N. Trace elements, U–Pb ages and Hf isotopes of zircons from Mesozoic granites in the western Nanling Range, South China: Implications for petrogenesis and W–Sn mineralization. *Lithos* **2011**, *127*, 468–482. [CrossRef]

40. Mao, J.W.; Cheng, Y.B.; Chen, M.H.; Pirajno, F. Major types and time–space distribution of Mesozoic ore deposits in South China and their geodynamic settings. *Miner. Depos.* **2013**, *48*, 267–294. [CrossRef]

41. Zhu, X.Y.; Wang, J.B.; Wang, Y.L.; Cheng, X.Y.; Fu, Q.B. Sulfur and lead isotope constraints on ore formation of the Huangshaping W–Mo–Bi–Pb–Zn polymetallic ore deposit, Hunan Province, South China. *Acta Petrol. Sin.* **2012**, *28*, 3809–3822. (In Chinese)

42. Ding, T.; Ma, D.S.; Lu, J.J.; Zhang, R.Q.; Zhang, S.T. S, Pb, and Sr isotope geochemistry and genesis of Pb–Zn mineralization in the Huangshaping polymetallic ore deposit of southern Hunan Province, China. *Ore Geol. Rev.* **2016**, *77*, 117–132. [CrossRef]

43. Wang, Q.; Mo, N.; Mao, Y.D. S-isotope geochemical study of Shizhuyuan deposit. *Land Resour. Her.* **2017**, *14*, 64–68. (In Chinese)

44. Chen, B.; Ma, X.H.; Wang, Z.Q. Origin of the fluorine-rich highly differentiated granites from the Qianlishan composite plutons (South China) and implications for polymetallic mineralization. *J. Asian Earth Sci.* **2014**, *93*, 301–314. [CrossRef]

45. Zheng, Y.F.; Xiao, W.J.; Zhao, G.C. Introduction to tectonics of China. *Gondwana Res.* **2013**, *23*, 1189–1206. [CrossRef]

46. Li, X.H.; Li, W.X.; Li, Z.X.; Lo, C.H.; Wang, J.; Ye, M.F.; Yang, Y.H. Amalgamation between the Yangtze and Cathaysia Blocks in South China: Constraints from SHRIMP U–Pb zircon ages, geochemistry and Nd–Hf isotopes of the Shuangxiwu volcanic rocks. *Precambrian Res.* **2009**, *174*, 117–128. [CrossRef]

47. Wang, X.L.; Shu, L.S.; Xing, G.F.; Zhou, J.C.; Tang, M.; Shu, X.J.; Qi, L.; Hu, Y.H. Post-orogenic extension in the eastern part of the Jiangnan orogen: Evidence from ca 800–760Ma volcanic rocks. *Precambrian Res.* **2012**, *222–223*, 404–423. [CrossRef]

48. Zhang, S.B.; Wu, R.X.; Zheng, Y.F. Neoproterozoic continental accretion in South China: Geochemical evidence from the Fuchuan ophiolite in the Jiangnan orogen. *Precambrian Res.* **2012**, *220–221*, 45–64. [CrossRef]

49. Zhang, S.B.; Zheng, Y.F. Formation and evolution of Precambrian continental lithosphere in South China. *Gondwana Res.* **2013**, *23*, 1241–1260. [CrossRef]

50. Zhao, G. Jiangnan Orogen in South China: Developing from divergent double subduction. *Gondwana Res.* **2015**, *27*, 1173–1180. [CrossRef]

51. Li, H.; Wu, Q.H.; Evans, N.J.; Zhou, Z.K.; Kong, H.; Xi, X.S.; Lin, Z.W. Geochemistry and geochronology of the Banxi Sb deposit: Implications for fluid origin and the evolution of Sb mineralization in central-western Hunan, South China. *Gondwana Res.* **2018**, *55*, 112–134. [CrossRef]

52. Zhou, X.M.; Sun, T.; Shen, W.Z.; Shu, L.S.; Niu, Y.L. Petrogenesis of Mesozoic granitoids and volcanic rocks in South China: A response to tectonic evolution. *Episodes* **2006**, *29*, 26–33.

53. Li, H.; Palinkaš, L.A.; Watanabe, K.; Xi, X.S. Petrogenesis of Jurassic A-type granites associated with Cu–Mo and W–Sn deposits in the central Nanling region, South China: Relation to mantle upwelling and intra-continental extension. *Ore Geol. Rev.* **2018**, *92*, 449–462. [CrossRef]

54. Xu, X.; O'Reilly, S.Y.; Griffin, W.L.; Deng, P.; Pearson, N.J. Relict Proterozoic basement in the Nanling Mountains (SE China) and its tectonothermal overprinting. *Tectonics* **2005**, *24*. [CrossRef]

55. Wang, Y.J.; Fan, W.M.; Sun, M.; Liang, X.Q.; Zhang, Y.H.; Peng, T.P. Geochronological, geochemical and geothermal constraints on petrogenesis of the Indosinian peraluminous granites in the South China Block: A case study in the Hunan Province. *Lithos* **2007**, *96*, 475–502. [CrossRef]

56. Li, J.H.; Dong, S.W.; Zhang, Y.Q.; Zhao, G.C.; Johnston, S.T.; Cui, J.J.; Xin, Y.J. New insights into Phanerozoic tectonics of south China: Part 1, polyphase deformation in the Jiuling and Lianyunshan domains of the central Jiangnan Orogen. *J. Geophys. Res. Solid Earth* **2016**, *121*, 3048–3080. [CrossRef]

57. Shu, L.S.; Zhou, X.M.; Deng, P.; Yu, X.Q.; Wang, B.; Zu, F.P. Geological features and tectonic evolution of Meso-Cenozoic basins in southeastern China. *Geol. Bull. China* **2004**, *23*, 876–884. (In Chinese)

58. Shu, L.S.; Wang, D.Z. A comparison study of basin and range tectonics in the Western North America and southeastern China. *Geol. J. China Univ.* **2006**, *12*, 1–13. (In Chinese)

59. Shu, L.S.; Zhou, X.M.; Deng, P.; Wang, B.; Jiang, S.Y.; Yu, J.H.; Zhao, X.X. Mesozoic tectonic evolution of the Southeast China Block: New insights from basin analysis. *J. Asian Earth Sci.* **2009**, *34*, 376–391. [CrossRef]

60. Yao, J.M.; Hua, R.M.; Lin, J.F. Zircon LA-ICP-MS U-Pb dating and geochemical characteristics of Huangshaping granite in southeast Hunan Province, China. *Acta Petrol. Sin.* **2005**, *21*, 688–696. (In Chinese)

61. Li, H.; Watanabe, K.; Yonezu, K. Zircon morphology, geochronology and trace element geochemistry of the granites from the Huangshaping polymetallic deposit, South China: Implications for the magmatic evolution and mineralization processes. *Ore Geol. Rev.* **2014**, *60*, 14–35. [CrossRef]

62. Ding, T.; Ma, D.S.; Lu, J.J.; Zhang, R.Q.; Zhang, S.T.; Gao, S.Y. Petrogenesis of Late Jurassic granitoids and relationship to polymetallic deposits in southern China: The Huangshaping example. *Int. Geol. Rev.* **2016**, *58*, 1646–1672. [CrossRef]

63. Ding, T.; Ma, D.S.; Lu, J.J.; Zhang, R.Q.; Zhang, S.T. Mineral geochemistry of granite porphyry in Huangshaping pollymetallic deposit, southern Hunan Province, and its implications for metallogensis of skarn scheelite mineralization. *Acta Petrol. Sin.* **2017**, *33*, 716–728. (In Chinese)

64. *Verification Report on the Reserves of the Shizhuyuan Deposit*; Southern Hunan Geological Survey Institute: Chenzhou, China, Unpublished work; 2015. (In Chinese)

65. Mao, J.W.; Li, H.Y.; Guy, B.; Raimbault, L. Geology and metallogeny of the Shizhuyuan skarn-greisen W–Sn–Mo–Bi deposit, Hunan Province. *Miner. Depos.* **1996**, *15*, 1–15. (In Chinese)

66. Jiang, Y.H.; Jiang, S.Y.; Zhao, K.D.; Ling, H.F. Petrogenesis of Late Jurassic Qianlishan granites and mafic dykes, Southeast China: Implications for a back-arc extension setting. *Geol. Mag.* **2006**, *143*, 457–474. [CrossRef]

67. Guo, C.L.; Wang, R.C.; Yuan, S.D.; Wu, S.H.; Yin, B. Geochronological and geochemical constraints on the petrogenesis and geodynamic setting of the Qianlishan granitic pluton, Southeast China. *Miner. Petrol.* **2014**, *109*, 253–282. [CrossRef]

68. Chen, Y.X.; Li, H.; Sun, W.D.; Ireland, T.; Tian, X.F.; Hu, Y.B.; Yang, W.B.; Chen, C.; Xu, D.R. Generation of Late Mesozoic Qianlishan A2-type granite in Nanling Range, South China: Implications for Shizhuyuan W–Sn mineralization and tectonic evolution. *Lithos* **2016**, *266–267*, 435–452. [CrossRef]

69. Li, X.H.; Liu, D.Y.; Sun, M.; Li, W.X.; Liang, X.R.; Liu, Y. Precise Sm–Nd and U–Pb isotopic dating of the supergiant Shizhuyuan polymetallic deposit and its host granite, SE China. *Geol. Mag.* **2004**, *141*, 225–231. [CrossRef]

70. Mao, J.W. Metallogenic speciality of super giant polymetallic tungsten deposit: Taking the Shizhuyuan deposit as an example. *Sci. Geol. Sin.* **1997**, *32*, 351–363. (In Chinese)

71. Zhu, X.Y.; Wang, J.B.; Wang, Y.L.; Chen, X.Y. The role of magma-hydrothermal transition fluid in the skarn-type tungsten mineralization process: A case study from the Shizhuyuan tungsten and tin polymetallic ore deposit. *Acta Petrol. Sin.* **2015**, *31*, 891–905. (In Chinese)

72. Xiong, X.L.; Rao, B.; Chen, F.R.; Zhu, J.C.; Zhao, Z.H. Crystallization and melting experiments of a fluorine-rich leucogranite from the Xianghualing Pluton, South China, at 150 MPa and H_2O-saturated conditions. *J. Asian Earth Sci.* **2002**, *21*, 175–188. [CrossRef]

73. Zhu, J.C.; Wang, R.C.; Lu, J.J.; Zhang, H.; Zhang, W.L.; Xie, L.; Zhang, R.Q. Fractionation, evolution, petrogenesis and mineralization of Laiziling granite pluton, southern Hunan Province. *Geol. J. China Univ.* **2011**, *17*, 381–392. (In Chinese)

74. Yang, L.Z.; Wu, X.B.; Cao, J.Y.; Hu, B.; Zhang, X.W.; Gong, Y.S.; Liu, W.D. Geochronology, Petrology, and Genesis of Two Granitic Plutons of the Xianghualing Ore Field in South Hunan Province: Constraints from Zircon U–Pb Dating, Geochemistry, and Lu–Hf Isotopic Compositions. *Minerals* **2018**, *8*, 213. [CrossRef]

75. Yin, J.W.; Lee, H.K.; Chio, K.K.; Kim, S.J. Characteristics of garnet in Shizhuyuan skarn deposit, Hunan province. *Earth Sci.-J. China Univ. Geosci.* **2000**, *25*, 163–171. (In Chinese)

76. Li, H.; Wu, J.H.; Evans, J.E.; Jiang, W.C.; Zhou, Z.K. Zircon geochronology and geochemistry of the Xianghualing A-typegranitic rocks: Insights into multi-stage Sn-polymetallic mineralization in South China. *Lithos* **2018**, *312–313*, 1–20. [CrossRef]

77. Cheng, Y.S.; Huang, H.M. Geochemical characteristics and mineralization indication of Devonian strata in Dachang ore field, Guangxi. *Chin. J. Nonferr. Met.* **2012**, *23*, 2649–2658. (In Chinese)

78. Sun, S.S.; McDonough, W.F. Chemical and isotopic systematics of oceanic basalts: Implications for mantle compositions and processes. In *Magmatism in the Ocean Basins*; Saunders, A.D., Norry, M.J., Eds.; Geological Society, London Special Publication: London, UK, 1989; Volume 32, pp. 313–345.

79. Li, H.; Myint, A.Z.; Yonezu, K.; Watanabe, K.; Algeo, T.J.; Wu, J.H. Geochemistry and U–Pb geochronology of the Wagone and Hermyingyi A-type granites, southern Myanmar: Implications for tectonic setting, magma evolution and Sn–W mineralization. *Ore Geol. Rev.* **2018**, *95*, 575–592. [CrossRef]

80. Halliday, A.N.; Davidson, J.P.; Hildreth, W.; Holden, P. Modelling the petrogenesis of high Rb/Sr silicic magmas. *Chem. Geol.* **1991**, *92*, 107–114. [CrossRef]

81. Li, F.L.; Chen, D.X.; Zhang, B.R. Geochemical characteristics and metallization of the strata in South Hunan. *Earth Sci.-J. China Univ. Geosci.* **1996**, *21*, 535–540. (In Chinese)

82. Ai, H. Zircon U–Pb geochronology and Hf isotopic compositions of ore-related granites from Huangshaping polymetallic deposit of Hunan Province. *Miner. Depos.* **2013**, *32*, 545–563. (In Chinese)

83. Wang, H. *Huangshaping Lead-Zinc Deposit Ore-Forming Fluid Geochemistry and Depth Estimation*; China University of Geoscience: Beijing, China, 2013. (In Chinese)

84. Meinert, L.D. Application of skarn deposit zonation models to mineral exploration. *Explor. Min. Geol.* **1997**, *6*, 185–208.

85. Ishihara, S.; Lee, D.S.; Kim, S.Y. Comparative study of Mesozoic granitoids and related W-Mo mineralization in Southern Korea and Southwestern Japan. *Min. Geol.* **1981**, *31*, 311–320.

86. Liu, Y.M.; Lu, H.Z.; Wang, C.L.; Xu, Y.Z.; Kang, W.Q.; Zeng, T. On the ore-forming conditions and ore-forming model of the superlarge multimetal deposit in Shizhuyuan. *Sci. China Ser. D* **1998**, *41*, 502–512. [CrossRef]

MDPI

St. Alban-Anlage 66

4052 Basel

Switzerland

Tel. +41 61 683 77 34

Fax +41 61 302 89 18

www.mdpi.com

Minerals Editorial Office

E-mail: minerals@mdpi.com

www.mdpi.com/journal/minerals

9 783039 212934